BASIC BUSINESS STATISTICS

CONCEPTS AND APPLICATIONS

MARK L. BERENSON
DAVID M. LEVINE

Department of Statistics
Baruch College
City University of New York

PRENTICE-HALL, INC.
Englewood Cliffs, N.J. 07632

Library of Congress Cataloging in Publication Data

Berenson, Mark L.
Basic business statistics.

Includes bibliographies and index.
1. Statistics. 2. Commercial statistics.
I. Levine, David M. (date) Joint author.
II. Title
HA29.B395 519.5 78-25734
ISBN 0-13-057596-8

Cover design by: Wanda Lubelska, based on a theme suggested by the authors
Manufacturing buyer: Trudy Pisciotti

Printed in the United States of America

10 9 8 7 6 5

Prentice-Hall International, Inc., *London*

Prentice-Hall of Australia Pty. Limited, *Sydney*

Prentice-Hall of Canada, Ltd., *Toronto*

Prentice-Hall of India Private Limited, *New Delhi*

Prentice-Hall of Japan, Inc., *Tokyo*

Prentice-Hall of Southeast Asia Pte. Ltd., *Singapore*

Whitehall Books Limited, *Wellington, New Zealand*

To our wives, Rhoda B. and Marilyn L.
and to our children,
Kathy B., Lori B., and Sharyn L.

CONTENTS

PREFACE

Whenever a new textbook is being planned the authors must resolve the issues of how the new text will differ from those already available and what contribution it would make to a field of study. There are several ways in which these issues are resolved in this textbook. First of all, it contains a set of data based upon the results of a survey that was developed and completed by a population of 2,202 students. The survey data, used for examples and problems throughout the text, serves as a means of integrating material from all parts of the course—especially between descriptive statistics, probability, and statistical inference.[1] In addition to using the survey as an integrating theme in the text, each student could select his or her own random sample from this population data base for analysis throughout the course, or the instructor could use the questionnaire included in Chapter 2 to compare the results of his or her class to those presented in the textbook. The use of an actual survey, examined from beginning to end, will aid the student in conducting basic re-

[1] The results of the survey are examined and analyzed in fourteen chapters of this textbook. Only in three of the seventeen chapters (Bayesian Decision Making, Index Numbers, and Time-Series Analysis) was it not pedagogically feasible to evaluate the survey.

search in future endeavors such as other courses, theses, or occupational situations. Moreover, the use of an actual survey throughout the various chapters provides continuity, a great amount of student interest in the material, and hopefully, a better understanding of concepts and applications.

In addition to the integrated survey throughout the chapters, this textbook includes other important and desirable features. One such desirable feature is the case study method which is used in examining the concepts of statistical inference for pedagogical purposes. This approach enables the student to better appreciate the overall value of estimation and hypothesis testing procedures.

Another desirable feature of the text is that in most chapters numerous realistic problems are presented. That is, many of these problems apply to realistic situations (using real data whenever possible) in various fields of endeavor including accounting, economics, management, and marketing, while others are keyed to the results of the survey. All end-of-chapter problems pertaining to the data base are indicated by the symbol □ . In addition, as an aid to both the instructor and the student, the solutions to selected odd-numbered problems (indicated by *) appear at the end of the book.

Moreover, another important feature of this book is the inclusion of a discussion of computer packages in order to convey the usefulness of high-speed data processing equipment for the solution of statistical problems. The applicability of these packages is demonstrated particularly for descriptive statistics, inferential techniques, and regression and correlation.

An additional helpful feature is that the textbook contains several useful appendixes. Appendix A describes the population data base and provides the responses to the 26 questions in the survey for all 2,202 students in the population. Appendix B consists of a brief review of rules for arithmetic and algebraic operations, as well as a discussion of summation notation, while Appendix C provides commonly used statistical symbols and the Greek alphabet. Appendix D presents a useful listing of conversions to the metric system of measurement, while Appendix E provides an extensive set of statistical tables.

A final important feature of this book is its flexibility. The textbook is written for students in either a one-semester or two-semester basic statistics course. However, there are numerous ways in which the instructor could adapt material to meet specific needs. For example, an introductory, one-semester (or two-quarter) course might consist of Chapters 1, 2, 3 (except Section 3.5), 4 (except Section 4.6), 5 (Sections 5.1 to 5.3), 6 (Sections 6.1 to 6.5 and 6.8 to 6.10), 7, 8, 9 (except Section 9.10), 10 (Sections 10.1, 10.2 and 10.4 to 10.7) and 14 (Sections 14.1 to 14.7). However, the material is organized so that instructors who do not wish to devote time to the development of the questionnaire could skim or omit the data collection phase of the survey (Chapter 2) and begin the course with Chapter 3 merely by using the results of the survey. If such a course were to emphasize inferential statistics, it could skim Chapters 1 and 2 and really begin with sections of Chapters 3, 4, and 5 before covering Chapters 6 through 10 and 12 through 14. However, if the course were primarily to emphasize descriptive statistics and probability, then Chapters 1 through 6, 11, 14 (Sections 14.1 to 14.7), 15 (Sections 15.1 to 15.3 and 15.10), 16, and 17 would be included. Moreover, for a two-semester course, the entire book could be covered: descriptive statistics, probability and probability distributions, decision making, index numbers and time series (Chapters 1 through 6, 11, 16 and 17) could be stressed in the first half and inferential statistics, correlation, regression and multiple regression could be studied in the second half (Chapters 7 through 10 and 12 through 15). Regardless of which topics are stressed by each instructor, the *primary emphasis is* on the *concepts and applications of basic statistical methods* to business subjects such as accounting, economics, management, and marketing.

It is our hope and anticipation that the unique approaches taken in this textbook will make the study of basic statistics more meaningful, rewarding, and comprehensible for all students who use it.

We wish to express our thanks to our colleagues at Baruch College, not only for their encouragement but also for their assistance in obtaining the population data base. We are particularly grateful to Ms. Macy Pon and also to Mr. Lawrence Latour and Ms. Miriam Frost-Gordon of the Educational Computer Center at Baruch College for their assistance in compiling the data base and to Ms. Vini Cigliano and Ms. Carol Adamson for their careful typing of the manuscript. Moreover, we wish to express our thanks to the editorial staff at Prentice-Hall for their continued encouragement—to Ms. Judy Rothman, Mr. Ron Ledwith, Mr. Ken Cashman. We also wish to thank Professors Jay Strum (New York University), Barbara Price (Wayne State University), Richard H. Haase (Drexel University), Charles H. Kriebel (Carnegie-Mellon University), Charles M. Schaninger (University of Massachusetts), and Charles T. Clark (University of Texas at Austin) for their ideas and constructive comments during the development of this textbook. Finally, we wish to express our thanks to our wives and children for their patience, understanding, love, and assistance in making this book a reality. It is to them that we dedicate this book.

Mark L. Berenson
David M. Levine

CHAPTER 1

INTRODUCTION

1.1 WHAT IS STATISTICS?

Everyday we are "bombarded" with a vast assortment of numerical information. As examples, we may hear the radio or television commentator telling us:

That the prices we pay for goods and services are 6.6% higher than they were last year

That the Dow-Jones Industrial Average fell 10 points yesterday

That 63.2% of the residents of a large city favor the construction of additional mass-transit facilities

That by the year 1990 the number of school-aged children will have decreased while, in contrast, the number of Americans receiving Social Security payments will have increased.

How have such findings as these been reached and, more importantly, what implications might these kinds of information have on our lives?

These few examples are only a small selection of the statistical information that is being continually collected, processed, analyzed, and utilized in making rational decisions in all fields of application ranging from anthropology to business, political science, and psychology through zoology. **The science of statistics can, therefore, be viewed as the application of the scientific method in the analysis of numerical data for the purpose of making rational decisions.**

In the past 30 years there has been a virtual "explosion" in the use of statistical methods. This has been particularly true in recent years with the advent and accessibility of high-speed digital computers which have the capacity to process large amounts of information. Many problems that were unapproachable because of their computational magnitude, as well as those that were computationally unwieldy, are now routinely solved by using electronic data-processing equipment. The widespread usage of statistical methods can be better understood by listing several applications. In politics, we have all seen how polls are used to predict the outcome of elections. In market research, statistical methodology plays a key role in determining market strategy and the selection of desirable product features. In the pharmaceutical industry, systematic experimentation helps determine the effectiveness of new drugs. In agriculture, statistical techniques are used to improve crop yield. In accounting, statistical methods enable the auditor to determine the rate of errors in a set of records. In zoology, statistics can be used to estimate the true population size of a particular animal species.

1.2 POPULATIONS AND SAMPLES

In order to understand how statistical methods can be applied, we must distinguish between a population and a sample. **A population (or universe) is the totality of items or things under consideration. A sample is the portion of the population that has been selected for analysis.**

A summary measure that is computed to describe a characteristic of an entire population is called a **parameter.** A summary measure that is computed to describe a characteristic from only a sample of the population is called a **statistic.**

One of the fundamental purposes of statistical methods is to use sample statistics to estimate population parameters. The necessity for sampling, and thereby using statistics to estimate parameters, is evidenced by the fact that it is usually too costly, too time consuming, or too cumbersome to deal with an entire population. This can be seen in the case of the political poll. Clearly, if the political scientist wishes to estimate the percentage of the vote that a candidate will receive in a presidential election, he or she will not interview each of the millions of possible voters. Instead, a sample of these voters would be selected. Based upon the outcome of the sample, conclusions will be drawn concerning the entire population of voters in the country. In a similar manner, the production manager of a company manufacturing automobile tires will certainly not want to "waste" all the tires being produced in order to study their performance quality. Hence only a sample of tires will be selected, and based upon the results, conclusions will be drawn concerning the quality of the entire population of tires produced.

This process of using sample statistics to draw conclusions about the true population parameters is called statistical inference. The inferential process is a form of inductive reasoning. We may remember that inductive reasoning attempts to generalize specific conclusions, while deductive reasoning applies what is true in general to specific situations. Although deductive reasoning is often used to apply general rules to specific applications, it is inductive reasoning through the process of statistical inference that provides the foundation of modern statistics. It is the process of inferential reasoning that enables the quality control manager to decide whether the quality of a product is acceptable, the political pollster to predict the outcome of an election, the market researcher to determine the superiority of two competing product alternatives, and the economist to make predictions of future economic conditions.

1.3 DESCRIPTIVE AND INFERENTIAL STATISTICS

From a historical point of view, modern statistics grew out of two separate phenomena, the mathematics of probability theory and the need to collect data on a national basis. The collection of national statistics goes as far back as recorded history (References 2 and 6). Many types of information were recorded by the Egyptians, the Greeks, and the Romans, primarily for the purposes of taxation and military conscription. In the Middle Ages various records concerning land, births, deaths, and marriages were maintained principally by church institutions. In the middle of the seventeenth century the work of John Graunt ["Observations on the London Bills of Mortality," 1662 (Reference 2)] led to the development of the theory of annuities and the concept of insurance. In America, although various records were kept back to colonial times (Reference 7), it was the United States Federal Constitution (1787) that called for the taking of a decennial census with the first occurring in 1790. These (and other) national statistics were closely intertwined with the development of the area of statistical methodology called descriptive statistics. **Descriptive statistics can be defined as those methods involving the collection, presentation, and characterization of a set of data in order to properly describe the various features of that set of data.**

Although descriptive statistics are important in characterizing and presenting information (see Chapters 2 to 4), it has been the development of statistical inference that has led to the great expansion in the application of statistical methods.

The initial impetus for the development of probability theory, the cornerstone of statistical inference, came from the investigation of games of chance during the Renaissance. As early as the middle of the seventeenth century, correspondence between the mathematician Pascal and the gambler Chevalier de Méré led to the foundations of probability theory (Reference 2). These and other developments in probability by such mathematicians as Bernoulli, De Moivre, and Gauss were the forerunners of statistical inference. **Inferential statistics can be defined as those methods that make possible the estimation of a characteristic of a population or the making of a decision concerning a population based only upon sample results.**

However, it has only been since the turn of this century that statisticians such as Pearson, Fisher, Student, Neyman, and Wald have pioneered in the development of the methods of statistical inference that are so widely applied in so many fields of endeavor today.

1.4 THE ROLE OF STATISTICS IN THE FUTURE

Many years ago H. G. Wells commented that "statistical thinking will one day be as necessary for efficient citizenship as the ability to read and write." With continued technological progress in communications and electronics, the use of statistical methods as an aid to decision making in the face of uncertainty or imperfect sample information has grown dynamically and certainly will continue to grow. Usage of computers and computer packages for solving and/or refining problems of large magnitude in all fields of endeavor will play an ever-increasing role. Thus quantitatively oriented students who seek challenges and who are interested in the application of statistical techniques toward the solution of business problems will find rewarding careers and experiences in the field of statistics. On the other hand, for those students who would rather pursue other interests than careers in statistics, the concepts and methods described in this text will provide the framework for understanding both the role of the statistician in an organizational structure and the quantitative tools necessary to solve basic types of business problems. Thus the book is not intended to make all of us statisticians—much more training would be necessary. It is, however, intended to provide us with an awareness and feeling for research as an aid to problem solving. To that end, a survey is designed by a researcher and a dean of a business school in Chapter 2, and various hypotheses are presented, examined, analyzed, and tested throughout the text as an integrating theme. This integration of descriptive and inferential statistics by using the survey (and its ensuing data base) throughout the text is intended to show how research and/or experimentation is conceived, perceived, developed, and conducted from its inception to its completion. This should prove most valuable for us, regardless of our major field of interest, because we will find that we are learning and applying a most useful body of knowledge—one that will remain with us throughout our careers—as an aid to making decisions in the light of imperfect information.

PROBLEMS

1.1 Describe three statistical applications to medicine or medical research.

1.2 Describe three statistical applications to marketing research.

1.3 Describe three statistical applications to advertising.

1.4 Describe three statistical applications to sports.

1.5 Describe three statistical applications to political science or public administration.

1.6 Describe three statistical applications to economics or finance.

REFERENCES

1. KENDALL, M. G., AND R. L. PLACKETT, eds., *Studies in the History of Statistics and Probability, Volume II* (London: Charles W. Griffin & Co., 1977).
2. KIRK, R. E., ed., *Statistical Issues; A Reader for the Behavioral Sciences* (Monterey: Brooks/Cole Publ., 1972).
3. PEARSON, E. S., AND M. G. KENDALL, eds., *Studies in the History of Statistics and Probability* (Darien, Conn.: Hafner, 1970).
4. TANUR, J., F. MOSTELLER, W. H. KRUSKAL, R. F. LINK, R. S. PIETERS, AND G. R. RISING, eds., *Statistics: A Guide to the Unknown* (San Francisco: Holden Day, 1972).
5. TUKEY, J., *Exploratory Data Analysis* (Reading, Mass.: Addison-Wesley, 1977).
6. WALKER, H. M., *Studies in the History of the Statistical Method* (Baltimore: Williams and Wilkins, 1929).
7. WATTENBERG, B., ed., *Statistical History of the United States: From Colonial Times to the Present* (New York: Basic Books, 1976).

CHAPTER 2

DATA COLLECTION

2.1 INTRODUCTION

An interesting comparison may be made between the work of a sculptor and that of a statistical researcher. Both require some basic resource with which to work. For the sculptor, that resource is some type of medium or material. For the researcher, the basic resource necessary for any statistical endeavor is data. Whereas the sculptor uses various implements to create a design out of a chosen material, the researcher, like the artist, must also choose appropriate tools to mold his "material"—the data—into a finished product. Here, however, the analogy ends. While the sculptor's finished product is to be appreciated for its own intrinsic value, as an end in itself rather than as a means to an end, the researcher's finished product must be appreciated extrinsically—as to how it will be of aid toward the solution of various managerial, marketing, and other business problems. However, for both the sculptor and the researcher the quality of the final product depends upon the quality of the raw material used. Thus proper data collection is extremely important to the researcher. To underscore this, researchers have adopted the term **GIGO, garbage-in, garbage-out.** If the data lack substance because of biases, ambiguities, or other types of errors, all the fancy, sophisticated tools selected by the researcher to mold that data may not be very useful toward the ultimate solution of a problem.

2.2 GOALS

The following then are the six goals of this chapter:

1. To develop an understanding for formulating a research problem and conducting research
2. To examine various sources of data for research
3. To obtain an understanding of different types of data
4. To obtain an appreciation for the art of questionnaire design and its problems
5. To obtain an appreciation for and the knowledge of conducting a random probability sample
6. To obtain an appreciation for the problems involved in data collection with respect to coding, editing, and keypunching and with respect to response and nonresponse

2.3 FORMULATING THE RESEARCH PROBLEM: THE NEED FOR RESEARCH

Throughout history the inquisitiveness of the human spirit has lead to research—whether formal or informal—in order to effectuate the decision-making process. The maturation of a person may be measured by his or her ability to learn from previous experiences by drawing proper analogies from past events and making correct decisions. If the "economic person," acting rationally, strives to maximize his own benefits, that individual will continually search for more efficient and effective methods of goal attainment. The decision-making process, involving the solution of problems and the attainment of goals, is a common denominator to all fields of endeavor. The football, baseball, or basketball coach needs to select the correct play and players for a given set of conditions; the personnel director wants to hire the right person for the right job; the airline executive seeks the most effective time schedule for particular routes; the surgeon wishes to compare different surgical procedures based on their postoperative results; the pharmaceutical manufacturer needs to determine whether or not a new drug is more effective than those on the market; the market researcher looks for the characteristics that distinguish a product from its competitors; the advertising director wishes to know if an advertisement is both attracting attention and plausible so that sales may be increased; consumer protection groups want to know if consumers are getting their money's worth based on manufacturing claims; potential investors want to determine which firms within which industries are likely to have accelerated growth in a period of economic recovery; public officials want information pertaining to the attitudes and desires of their constituents for legislative purposes (and for being retained in office); and a college administrator wishes to know information pertaining to the attributes, attitudes, and interests of the student body in order to implement decisions regarding the quality of campus life.

In every case a specific situation leads to the formulation of a particular problem and its ramifications. Defining one's goals leads to various courses of action to consider. The rational decision maker seeks to evaluate information (that is, data) in order to maximize objectives.

2.4 SOURCES OF DATA FOR RESEARCH

There are three basic methods by which the researcher may obtain needed data. First, the researcher may seek data already published by governmental, industrial, or individual sources. Second, the researcher may design an experiment to obtain the necessary data. Third, the researcher may conduct a survey.

2.4.1 Using Published Sources of Data

The federal government is a major collector of data for both public and private purposes. The Bureau of Labor Statistics is responsible for collecting data on employment as well as establishing the well-known monthly Consumer Price and Wholesale Price Indexes. In addition to its constitutional requirement for conducting a decennial census,[1] the Bureau of the Census is also concerned with a variety of ongoing surveys regarding population, housing, and manufacturing, and from time to time undertakes special governmental studies such as those on crime, travel, and health care. Much of its findings are published as special reports or can be located in such publications as the monthly *Survey of Current Business* or in the annual *Statistical Abstract of the United States.*

In addition to the federal government, various trade publications often present data pertaining to specific industrial groups, while individual annual reports often provide illuminating information on company-level activities. Various investment services such as Moody's, Standard and Poor's, and Value Line also accumulate a wealth of data on a company basis, and, of course, the daily newspapers are filled with information regarding stock prices, weather conditions, and sports statistics. Furthermore, many companies subscribe to syndicated services such as A. C. Nielsen to obtain information on their products and their competitors. Throughout this text various applications and problems will utilize data obtained from such published sources.

2.4.2 Designing an Experiment

A second method of obtaining needed data is through experimentation. Proper experimental designs are usually the subject matter of more advanced

[1] Because of the overwhelming need for up-to-date data, a law was passed in October 1976 requiring a census to be taken every five years beginning in 1985.

texts since they often involve extremely sophisticated statistical procedures. However, in order to develop a feeling for testing and experimentation, the fundamental experimental design concepts will be considered in Chapters 9, 10, 12, and 13.

2.4.3 Conducting a Survey

A third method of obtaining data is to conduct a survey. Suppose a newly appointed dean of a business school, faced with the task of proposing policy which may have a drastic effect on the quality of campus life, wishes to conduct a survey in order to obtain information on the composition of the student body as well as to ascertain information pertaining to the students' physical attributes (characteristics), attitudes, and interests. If the survey is to be useful, the questionnaire instrument must be valid, that is, the "right" questions must be asked in a manner that will elicit meaningful response (remember GIGO). Hence to aid in the construction of the questionnaire the dean might call upon the services of the survey statistician or researcher. In order to design the questionnaire, an understanding of different types of data is needed.

2.5 TYPES OF DATA

The survey statistician will most likely want to develop an instrument that asks several questions and deals with a variety of phenomena. Hence the subject matter with which the statistician deals is **random variables** or phenomena of interest whose observed outcomes (data) may differ from response to response. As outlined in Figure 2.1, there are basically two types of random variables yielding two types of data: **qualitative** and **quantitative.** The difference between them is that qualitative random variables yield **categorical** responses while quantitative random variables yield **numerical** responses. For example, the response to the question "Do you currently own United States Government (Series E) Savings Bonds?" is categorical. The choices are clearly "yes" or "no." On the other hand, responses to questions such as "How many magazines do you currently subscribe to?" or "How tall are you?" are clearly numerical.

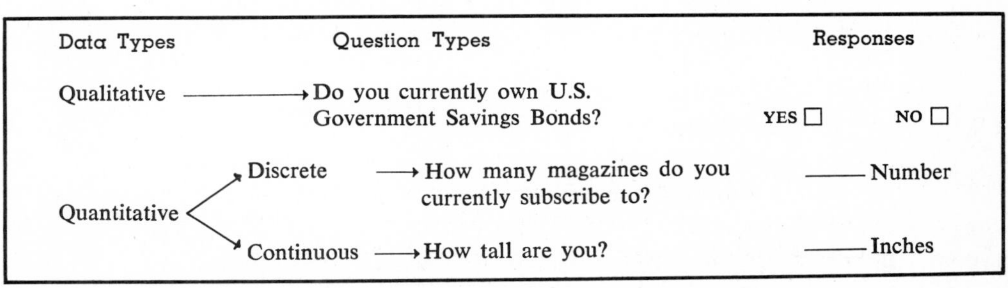

Figure 2.1 Types of data.

It should be pointed out that quantitative data may be considered as discrete or continuous. **Discrete quantitative data are numerical responses which arise from a counting process while continuous quantitative data are numerical responses which arise from a measuring process.** "The number of magazines subscribed to" is an example of a discrete quantitative variable since the response takes on one of a (finite) number of countable values. The individual either currently subscribes to no magazine, one magazine, two magazines, etc. On the other hand, "the height of an individual" is an example of a continuous quantitative variable since the response can take on any value within a continuum or interval, depending on the precision of the measuring instrument. Using "eyeball judgment" an observer may guess at the height of an individual. However, the use of a ruler, an architectural scale, or a micrometer would produce a more precise measurement of a person's height. Thus depending on which of the sophisticated measuring devices mentioned is employed, a person whose height is reported as 67 inches may be more precisely measured as 67 1/4 inches, 67 7/32 inches, or 67 58/250 inches respectively. Therefore we can see that height is a continuous phenomenon which can take on any value within an interval. It is interesting to note that theoretically no two persons could have exactly the same height, since the finer the measuring device used, the greater the likelihood of detecting differences between them. However, measuring devices used by researchers may not be sophisticated enough to detect small differences and hence *tied observations* are often found in experimental or survey data. With discrete phenomena ties are common since only certain values can occur.

2.6 DESIGNING THE QUESTIONNAIRE INSTRUMENT

Questionnaire development is an art that improves with experience. In constructing the questionnaire, consideration must be given to both the length of the instrument and the mode of obtaining the responses. Remember that the goal is to obtain meaningful responses that will be of aid in the decision-making process. In our particular case the dean of the business school wants to learn something about the makeup of the student body. After planning out various groupings of questions pertaining to socioeconomic and demographic characteristics, physical attributes, attitudes, and interests, a first effort is made at developing a series of questions within each grouping.

2.6.1 Length of Questionnaire

Before very long a large number of questions will have been created. Unfortunately, however, there is an inverse relationship between the length of a questionnaire and the rate of response to the survey. That is, the longer the questionnaire, the lower will be the rate of response; the shorter the questionnaire, the higher will be the rate of response. It is, therefore, imperative that the researcher and the dean determine the merits of each question:

Is the question clearly presented?

Does the question have fewer than 20 words?

For qualitative questions, are the response categories nonoverlapping and complete?

For quantitative questions, what kinds of responses are likely to be obtained?

Will the responses to a question be meaningful and helpful toward the decision-making process (satisfying the study's objectives)?

Is the question really essential to the survey?

Hard answers to such questions as these will both improve the instrument and help reduce its length.

2.6.2 Mode of Response

The particular questionnaire format to be selected and the specific question wording are affected by the intended mode of response. There are essentially three modes through which survey work is accomplished: personal interview, telephone interview, and mail.

The personal interview is the most costly method. It involves a large training effort in order to ensure that the interviewer (enumerator) may skillfully ascertain desired information. The chief advantages of the personal interview are that the response rate is usually highest, and any unforeseen ambiguities are more easily rectified by the interviewer. Like the personal interview, the telephone interview is also a costly method because it requires detailed interviewer training. Unfortunately, rates of response are not usually as high since personal contact between interviewer and interviewee is not as readily established. On the other hand, while the mail survey is least costly, it usually suffers from the lowest rate of response. Sometimes individuals who eventually respond procrastinate in returning the forms, causing delays in project completion. In addition, it is imperative to provide both a clear set of instructions and a clear set of categorical choice and/or fill-in type questions with mail surveys since respondents who find problems and/or ambiguities are not readily able to rectify the situation. For example, *open-ended questions* such as:

"Why did you choose to attend this college over other colleges?" ______________

____

which require the subject to respond freely in essay form are usually undesirable for mail surveys because the subject may perceive such questions as too time consuming. In addition, since such questions are time consuming to answer, they are also time consuming to evaluate. On the other hand, when surveys are conducted by personal or telephone interview, the open-ended question is more appropriate since a well-trained interviewer would be recording the pertinent information which will facilitate the editing and coding processes.

After careful examination of the various modes of response, the researcher and the dean determined that the mail survey, conducted early in the semester, would be the most practical means of obtaining the desired information.

2.6.3 Wording the Questions

Question development is not exempted from Murphy's law; if something can go wrong it will! The researcher, then, must continuously play the role of the "devil's advocate" and search for all possible ambiguities in each question. As examples, what could be simpler than such questions as:

1. Do you smoke? YES ☐ NO ☐
2. How old are you? ____________ years

For question 1, which pertains to smoking habits, there are several possible ambiguities. First of all, does the desired response pertain to cigarettes, to cigars, to pipes, or to combinations thereof? Second, quantity is an important factor in the respondent's perception of smoking. An individual who smokes only a few cigarettes a day may simply respond "no" to the above question since it may be perceived that relative to heavy smokers or perhaps to the amount the individual used to smoke that he or she does no longer really smoke! If the researcher were only interested in current cigarette consumption, perhaps it would be better to ask "About how many cigarettes do you currently smoke each day?"______The response might then be categorized during the editing and coding stage of the survey into such groups as "does not currently smoke," "light smoker," and "heavy smoker."

For question 2, which pertains to age, the respondent may be confused as to whether to base the answer on the last birthday or the nearest birthday—especially if the birthday is in the next few days. In addition, it is interesting to note that previous studies have indicated age to be a sensitive issue and that the question of age has yielded a disproportionate number of responses ending in the digits 0 or 5. These problems may be avoided if the respondent is merely asked to

State your date of birth ________ ________ ________
month day year

and the editors and coders make the desired age recording.

2.6.4 Asking "Very Personal" Questions: The Randomized Response Technique

It is often important to ascertain information on very personal or sensitive issues—those which might embarrass the subject and therefore cause a great deal of anxiety or discomfort in response. Questions pertaining to sexual habits are of this nature, as are questions regarding other personal habits, attitudes, or beliefs. Problems arise with such kinds of questions because the subject may be "turned off" and thus may fail to respond, or the subject may willfully give false answers, both of which then distort (bias) the results of the survey. To alleviate these problems, Warner devised the **randomized response technique** (Reference 11). A later modification of this technique permits the subject to respond at random to one of two questions, a very personal question or a control question, so that only the subject is aware of which question was actually answered, and his or her privacy is therefore maintained (Reference 5). For example, suppose that in a medical study pertaining to women it is desirable to obtain an estimate of the true percentage of women who have had one or more abortions during their lifetime. The subject might be instructed as follows:

> For the following set of questions flip a coin; if the coin lands "heads," only answer question A; if the coin lands "tails," only answer question B. Do not divulge the result of the coin flip or the question you are answering.
> Place answer here: YES ☐ NO ☐
>
> A. Have you ever had an abortion?
> B. Was your father born in the month of January?

Therefore, when employing this technique, anonymity is preserved and the subject is more likely to respond honestly. Using basic rules of probability theory the statistician is then able to make predictions regarding the responses of an entire population to sensitive questions without ever knowing how any particular individual responded. This topic will be referred to again in Chapter 5 (Basic Probability) and Chapter 8 (Estimation).

2.6.5 Testing the Questionnaire Instrument

Once the researcher and the dean have discussed the pros and cons of each question as described in Section 2.6.3, the instrument is properly organized and made ready for *pilot testing* so that the document may be examined for its clarity and its length. Such pilot testing on a small group of subjects is an essential phase in conducting a survey. Not only will this group of individuals be providing an estimate of the time needed for responding to the survey, but they also will be asked to comment on any perceived ambiguities in each question and to recommend other additional questions. Once the researcher and the dean have digested these results, changes are made, and, if time and budget permit, a second pilot study can be undertaken on a fresh sampling of subjects to further improve the final document. Since a questionnaire is a subjective instrument which can never really be perfect, the pilot-testing procedure enables the researcher to be more certain that the final instrument will yield the kinds of information desired to aid in the decision-making process.

Figure 2.2 depicts the questionnaire devised by the researcher and the dean in its final form.

Codes	STUDENT QUESTIONNAIRE
1 2 3 4	
5	1. What is your sex? Male [1] Female [2]
6	2. What is your current registered class designation? Freshman [1] Sophomore [2] Junior [3] Senior [4]
7	3. At the present time do you plan to attend graduate school? Yes [1] No [2] Not sure [3]
8	4. What is your major area of study? Undecided [1] Accounting [2] Economics or Finance [3] Management [4] Marketing [5] Statistics or Computer Systems [6] Other ____7____ (Specify)
9 10 11 12	5. What is your college grade-point index? ________
13 14	6. What was your high-school average? ________ (*To nearest integer value*)
15 16 17 18 19	7. What would you expect your starting annual salary to be if you were to seek employment immediately after obtaining your baccalaureate? ________
20 21 22 23	8. What is the maximum yearly tuition you would be willing to pay at this college before you would seriously consider transferring or leaving college? ________
24	9. What is your current employment status? Full-time [1] Part-time [2] Unemployed [3]
25	10. Would you favor or oppose legislation *prohibiting* Sunday shopping? Favor [1] Oppose [2]
26	11. If you were given a $5,000 gift, how would you *primarily* consider using it this year? Place in savings bank [1] Spend for goods and services [2] Invest in bonds [3] Donate to charity [4] Invest in stocks [5] Other ____6____ (Specify)
27	12. If you were to purchase an automobile this year, what size car would you *primarily* be interested in buying? Station wagon or van [1] Standard or intermediate [2] Sports car [3] Compact or subcompact [4] Other ____5____ (Specify)
28 29	13. How many pairs of jeans do you own? ________
30	14. Do you own a calculator? Yes [1] No [2]
31 32	15. What is your date of birth? ________ Month ________ Day ________ Year
33 34	16. What is your height? ________ (*To nearest inch*)

Figure 2.2 Questionnaire.

35 36 37 17. What is your weight? ________________ (*To nearest pound*)

38 18. Has a doctor ever told you that you have high blood pressure?
Yes [1] No [2]

39 19. About how many cigarettes are you currently smoking each day?

None [1] Less than one pack [2] At least one pack but less than two [3] Two or more packs [4]

40 20. Should smoking be permitted in our classrooms?
Yes [1] No [2]

41 21. Which *one* of the following minimum standards of admission should be adopted by our college?

No minimum standards; open to all students with high school diploma [1]
Eighth-grade mathematics and English levels [2]
Tenth-grade mathematics and English levels [3]
Twelfth-grade mathematics and English levels [4]

42 22. Should a more stringent retention standard be placed upon our students in order to show adequate progress toward a degree?
Yes [1] No [2]

43 44 23. How many clubs, groups, organizations, or teams are you currently affiliated with at the college? ________________

45 24. How satisfied are you with the Student Personnel Services at the college? (circle choice)

Extremely unsatisfied	1	2	3	4 Neutral	5	6	7	Extremely satisfied

46 25. How satisfied are you with the Library Services at the college? (circle choice)

Extremely unsatisfied	1	2	3	4 Neutral	5	6	7	Extremely satisfied

The following set of questions contains a personal question and a control question. To safeguard the privacy of each student, the following instructions are to be followed:

1. Flip a coin. Answer 26A if the coin toss is "heads" and answer 26B if the coin toss is "tails."
2. Record the appropriate answer next to the question number, but do not divulge either the result of the coin toss or the specific question answered.

47 26. Yes [1] No [2]

26A. Are you receiving any scholarship, stipend, or financial assistance to attend college?

26B. Was your mother born in a year ending in an odd number (1, 3, 5, 7, 9)?

2.7 CHOOSING THE SAMPLE SIZE FOR THE SURVEY

Since the researcher and the dean had already determined that a mail survey, to be conducted early in the semester, was the most economical way of obtaining their desired information, it was then necessary for them to determine the appropriate sample size for the study. Rather than taking a complete census, statistical sampling procedures have become the preferred tool in most survey situations. There are three main reasons for drawing a sample. First of all, it is usually too time consuming to perform a complete census. Second, it is too costly to do a complete census. Third, it is just too cumbersome and inefficient to obtain a complete count of the target population. Thus the goal of the researcher and the dean is to make inferences about the entire population of business students at the college based on the results obtained from the sample. After we determine the most essential qualitative and quantitative questions, the sample size needed will be based on satisfying the question with the most stringent requirements. In our case, the dean and the researcher have determined that questions 8 and 22 are the most essential quantitative and qualitative questions, respectively. However, calculation of the sample size required for a given survey is a matter that will be more appropriately examined in Chapter 8. As will be described in Chapter 8, the required sample size is 94 students out of a business student body (population) of 2,202 full-time, matriculated, day-session students currently enrolled.

2.8 TYPES OF SAMPLES

As depicted in Figure 2.3 there are basically two kinds of samples: the **probability sample** and the **nonprobability sample.** The latter, which is usually much simpler and cheaper to obtain, comprises a grouping of procedures such as **judgment samples, quota samples,** and the **chunk.**

2.8.1 Nonprobability Samples

A major drawback of both the judgment sample and the quota sample is that the interviewer is given too much discretion in the process of subject selection. While this may be both efficient and economical as compared to probability sampling methods, there is no probabilistic way of estimating how representative such selected samples are. Hence it is incorrect to use such samples to make inferences to the entire population—which is the objective of sampling! In the judgment sample the interviewer selects any subjects whom he or she desires, while in the quota sample this type of selecting is constrained by various pre-established quotas regarding sex, race, age, etc., which try to simulate known population characteristics. In either case it is unlikely that the interviewer will choose to operate in poorer neighborhoods, depressed areas, or places not readily accessible. The results then are automatically biased since entire groups would be omitted from the selection process. At the other extreme, however, is the chunk in which the interviewer plays no role in the selection process. The chunk is composed by a self-selecting

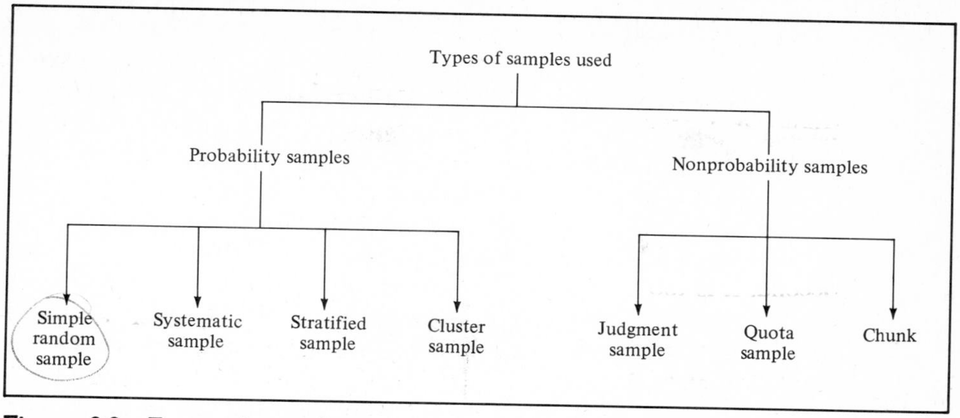

Figure 2.3 Types of samples.

process, that is, a chunk is merely a "convenience sample"—a collection of subjects easily grouped such as the members of a particular class, attendees at a particular theater, persons responding to a coupon or advertisement, or individuals who visit a display in a shopping center. Again, the main disadvantage of such **nonprobability sample** procedures is that there is no probabilistic way of interpreting how representative the particular sample is of the overall population. Since such procedures are less time consuming and costly to obtain, all too often it is assumed, however erroneously, that the data are quite representative of the overall population. But that is a big assumption! The only correct way in which the researcher could make statistical inferences from a sample to a population and interpret the results probabilistically is through the use of a **probability sample.**

2.8.2 Probability Samples

The subject matter of probability sampling techniques is of such breadth that entire courses are offered in sampling theory and methods. Imagine trying to forecast the results of a national election, or attempting to predict automobile sales for the coming year, or estimating the number of consumers who would purchase a new product. It will be sufficient here, however, to develop an understanding of the basic concepts of probability sampling and, in particular, the four most commonly used types: the **simple random sample,** the **systematic sample,** the **stratified sample,** and the **cluster sample.**

A probability sample is one in which the subjects of the sample are chosen based on known probabilities. In particular, the simple random sample is one in which every subject has the same chance of selection as every other subject at each successive stage of the selection process. Moreover, a simple random sample may also be construed as one in which each possible sample that is drawn had the same chance of selection as any other sample.

Stratified sampling procedures, cluster sampling procedures, and systematic sampling procedures, however, are more sophisticated versions of probability sampling. For example, in a stratified sample homogeneous groupings or strata are first developed across the entire population, and then a simple random sample is taken within each stratum to ensure complete coverage of all types of subjects in the population—something not necessarily achieved by the simple random sample. In our example, if the researcher and the dean wish to ensure a proportionate response from the freshman, sophomore, junior, and senior students, a stratified sample could be achieved by drawing simple random samples within each of the four groups and combining the results. While more costly and sophisticated, the stratified probability sample usually yields more efficient results than simple random sampling. However, a more detailed discussion of stratified sampling, cluster sampling, or systematic sampling procedures is beyond the scope of this textbook. For further information on these topics, see References 1 through 4, 6, and 7.

2.9 DRAWING THE SIMPLE RANDOM SAMPLE

In this section we will be concerned with the process of selecting a simple random sample which, while not the most efficient of the probability sampling procedures, provides the base from which the more sophisticated sampling procedures have evolved. The key to proper sample selection is the adequacy of the listing of all the subjects from which the sample will be drawn. If no such listing exists, it must be constructed; if a listing is out of date, it must be revised. Recall again the concept of GIGO. This population listing will serve as the target population so that if many different probability samples were to be drawn from such a list, hopefully each sample would be a miniature representation of the population and yield reasonable estimates of its characteristics. If the listing is inadequate because certain groups of subjects in the population were not properly included, the random probability samples will only be providing estimates of the characteristics of the target population—not the actual population—and biases in the results will occur.

Suppose, for our survey, that the researcher and the dean decided (see Section 2.7) to draw a simple random sample of 94 students out of the population of 2,202 full-time, matriculated, day-session business students who are currently enrolled at the college. Access to the listing of these students would be provided by the college registrar. Any such file or listing would most likely include such information as the name, Social Security number, current residence and telephone number, permanent residence, major category of study (for example, business), and current class designation (for example, freshman, sophomore, etc.).

Let N represent the population size and n the sample size. To draw a simple random sample of size $n = 94$ one could conceivably record the names of each of the $N = 2{,}202$ subjects in the population on a separate index card, place the N index cards in a large fish bowl, thoroughly mix the cards, and then randomly select the n sample subjects from the fish bowl. Naturally, in order to ensure that each subject (as well as each sample) had the same chance of being selected, all the index cards used must be of the same dimensions.

2.9.1 Sampling with or without Replacement from Finite Populations

It should be noted here that there are two basic methods which could be used for selecting the sample: the sample could be obtained **with replacement** or **without replacement** from the finite population. The method employed must be clearly stated by the researcher since various formulas subsequently used for purposes of statistical inference (to be discussed in Chapter 8) are dependent upon the selection method.[2] When sampling *with* replacement the chance that any particular subject in the population, say Mary Doe, is selected on the first draw from the fish bowl is $1/N$ (or 1/2,202 for our example). Regardless of whoever is actually selected on the first draw, pertinent information is recorded on a master file and then the particular index card is *replaced* in the bowl (sampling *with* replacement). The N cards in the bowl are then well shuffled and the second card is drawn. Since the first card had been replaced, the chance for selection of any particular subject, including Mary Doe, on the second draw—regardless of whether or not that individual had already been previously selected—is still $1/N$. Again, the pertinent information is recorded on a master file and the index card is replaced in order to prepare for the third draw. Such a process is repeated until the desired sample size is obtained. It is thus noted that when sampling with replacement, every subject on every draw will always have the same 1 out of N chance of being selected.

But should the researcher and the dean want to have the same individual subject possibly selected more than once? When sampling from human populations, it is generally felt more appropriate to have a sample of different subjects than to permit repeated measurements of the same subject. Thus the method of selection that the researcher and the dean would employ is the method of sampling *without* replacement, whereby once a subject is drawn, the same subject cannot be selected again. As before, when sampling without replacement, the chance that any particular subject in the population, say Mary Doe, is selected on the first draw from the fish bowl is $1/N$. Whoever is selected, the pertinent information is recorded on a master file and then the particular index card is set aside rather than replaced in the bowl (sampling *without* replacement). The remaining $N - 1$ cards in the bowl are then well shuffled, and the second card is drawn. The chance that any individual not previously selected will be selected on the second draw is now 1 out of $N - 1$. This process of selecting a card, recording the information on a master file, shuffling the remaining cards, and then drawing again continues until the desired sample of size n is obtained.

[2] It is interesting to note that whether sampling **with** replacement from **finite** populations or sampling **without** replacement from **infinite** populations (such as some continuous, on-going production process), the formulas used are the same.

"Fish bowl" methods of sampling, while easily understandable, are not very efficient. Less cumbersome and more scientific methods of selection are desirable. One such method utilizes a table of random numbers (see Appendix E, Table E.1) for obtaining the sample.

2.9.2 Using a Table of Random Numbers

Such a table consists of a series of digits randomly generated (by electronic impulses) and listed in the sequence in which the digits were generated. Since our numeric system uses 10 digits (0, 1, 2, . . . , 9) the chance of randomly generating any particular digit is equal to the probability of generating any other digit. This probability is 1 out of 10. Hence if a sequence of 500 digits were generated, we would expect about 50 of them to be the digit 0, 50 of them to be the digit 1, and so on. In fact, researchers who use tables of random numbers usually test out such generated digits for randomness before employing them. Table E.1 in Appendix E has met all such criteria for randomness. Since every digit or sequence thereof in the table is random, we may use the table by reading either horizontally or vertically. The margins of the table designate row numbers and column numbers. The digits themselves are grouped into sequences of five for the sole purpose of facilitating the viewing of the table.

To use such a table in lieu of a fish bowl for selecting the sample, it is first necessary to assign code numbers to the individual members of the population. Suppose that the registrar's file contained a listing, in alphabetical sequence, of all $N = 2{,}202$ currently enrolled, fully matriculated, day-session business students at the college. Since the population size (2,202) is a four-digit number, each assigned code number must also be four digits so that every subject has an equal chance for selection. Thus a code of 0001 is given to the first subject in the population listing, a code of 0002 is given to the second subject in the population listing, . . . , a code of 0875 is given to the 875th subject in the population listing, and so on until a code of 2202 is given to the Nth subject in the listing. In order to select the random sample, a random starting point for the table of random numbers (Appendix E, Table E.1) must be established. One such method is to close one's eyes and strike the table of random numbers with a pencil. Suppose that the researcher and the dean utilized such a procedure and thereby selected row 29, column 01 as the starting point. Reading from left to right in sequences of four digits without skipping, groupings are obtained as shown in Table 2.1.

Since $N = 2{,}202$ is the largest possible coded value, all four-digit code sequences greater than N (2203 through 9999 and 0000) are discarded. Hence the subject with code number 0877 is the first person in the sample (row 29 and columns 25 through 28) since the six previous four-digit code sequence in row 29 are all discarded. Subject 1793 (row 30 and columns 01 through 04) is the second individual selected. Subject 0007 is third; subjects with code numbers 0261, 1246, 1495, 0023, 0397, 0172, and 0703 are selected fourth through tenth, respectively. The selecting process continues in

Table 2.1
USE OF A TABLE OF RANDOM NUMBERS

	Column							
Row	00000 12345	00001 67890	11111 12345	11112 67890	22222 12345	22223 67890	33333 12345	33334 67890
.								
29	41613	42375	00403	03656	77580	87772	86877	57085
30	17930	00794	53836	53692	67135	98102	61912	11246
31	24649	31845	25736	75231	83808	98917	93829	99430
32	79899	34061	54308	59358	56462	58166	97302	86828
33	76801	49594	81002	30397	52728	15101	72070	33706
.	.	.	.	.	.	.	.	.
.	.	.	.	.	.	.	.	.
.	.	.	.	.	.	.	.	.
73	11100	02340	12860	74697	96644	89439	28707	25815
.								
.								

SOURCE: Data are partially extracted from Appendix E, Table E.1.

a similar manner until the desired sample of $n = 94$ subjects is obtained. If during the selection process any four-digit coded sequence repeats, the subject corresponding to that coded sequence would be included again as part of the sample if sampling *with* replacement; however, the repeating coded sequence is merely discarded if sampling *without* replacement. This was noted to first happen when the coded sequence 1793 appeared in row 30, columns 01 through 04 and then again in row 31, columns 29 through 32. Since the researcher and the dean are sampling without replacement, the repeating sequences were discarded and a sample of 94 unique individuals was obtained.

For the problem at hand the sampling process described is both time consuming and wasteful of data since 45 rows of the random numbers table (Appendix E, Table E.1) had to be used to draw the desired sample of 94 subjects. This has occurred here because the population size of $N = 2{,}202$ is the largest acceptable four-digit coded sequence, and, consequently, approximately four out of every five four-digit coded sequences (numbers 2203 through 9999 and 0000) are excluded. As can be verified in Problem 2.12, had the population size been only 902, almost 19 out of every 20 three-digit coded sequences would have been acceptable, and only nine rows of the random numbers table would have been utilized. On the other hand, with a population size of 2,202 many statisticians would advise a *two-stage* use of the table of random numbers in order to draw the sample. As can be verified in Problem 2.13, one such approach would require only 16 rows to draw a sample that required 45 rows above and was so time consuming.

2.10 COLLECTING THE DATA

Now that the sample of 94 individuals has been selected, their responses must be obtained. For a mail survey sufficient time must be permitted for initial response. The questionnaire and any set of instructions should be mailed with a covering letter and with a self-addressed stamped envelope for the convenience of the respondent.

2.10.1 The Covering Letter

The covering letter should be brief and to the point. It should state the goal or purpose of the survey, how it will be used, and why it is important that selected subjects promptly respond. Moreover, it should give any necessary assurance of respondent anonymity and offer a note of appreciation for the respondent's time.

2.10.2 Rate of Response

The rate of response to mail surveys, while usually lower than with personal or telephone interviews, is a function not only of questionnaire length but also of subject interest. The greater the subject's perception of the questionnaire's importance, the greater (and more prompt) will be the rate of response. This is why the covering letter is deemed so important.

2.10.3 Survey Errors: Response and Nonresponse

There are two kinds of survey errors: **errors of response** and **errors of nonresponse.** Errors of response may be honest or willful. The subject may make an honest mistake in recalling an event, expressing a belief, or in recording an answer. On the other hand, willful errors of response may occur when the subject finds a question too personal (see Section 2.6.4) or when the subject wants to either please or deceive rather than give an honest answer. With personal or telephone interviews the interviewer may make an honest recording error or, if improperly trained, willful recording errors as well. Errors of nonresponse may either be willful or the subject may be unavailable due to death, illness, change of address, or travel. Since it cannot be generally assumed that persons who do not respond to surveys are similar to those that do, it is extremely important to follow up on the nonresponses after a specified period of time. Several attempts should be made, either by mail or by telephone, to locate the unavailable subject or to convince the willful nonrespondent to change his or her mind. Based on these results, an effort is made to tie together the estimates obtained from the initial respondents with those obtained from the follow-ups so that we can be reasonably sure that the inferences made from the survey are valid (Reference 1).

For our study we recall that the researcher and the dean had determined that 94 responses were needed to satisfy the requirements of the most stringent question in the survey. Hence depending on the rate of response, it may become necessary for the researcher and the dean to take a larger sample to ensure that the requirements are satisfied.

2.11 DATA PREPARATION: EDITING, CODING, AND TABULATING

Once the set of data is collected it must be prepared for tabular and chart presentation, analysis, and interpretation. The editing and coding processes are extremely important. Responses are scrutinized for completeness and for errors and, if necessary, response validation is obtained by recontacting subjects whose answers appear inconsistent or unusual. In addition, responses to open-ended questions are properly classified or scored, while responses to both quantitative and qualitative questions are coded for tabulation either by keypunch or by optical scanning equipment. The advantage of the latter over keypunching is the elimination of the risk of keypunching error. The importance of careful editing, coding, and keypunching cannot be over-

Question Number	Type of Question	Card Columns for Keypunching	Jamie Bonds' Responses	Coded Response
—	File code number	1, 2, 3, 4	—	0007
1	Sex	5	Male	1
2	Class designation	6	Sophomore	2
3	Graduate school intent	7	Not sure	3
4	Major	8	Accounting	2
5	Grade-point index	9, 10, 11, 12	3.17	3.17
6	High-school average	13, 14	87.2	87
7	Anticipated starting salary (dollars)	15, 16, 17, 18, 19	$11,000	11000
8	Tuition charge (dollars)	20, 21, 22, 23	$700	0700
9	Employment status	24	Part-time	2
10	Prohibition of Sunday shopping	25	Oppose	2
11	Investment	26	Savings bank	1
12	Auto preference	27	Standard or Intermediate	2
13	Number of pairs of jeans owned	28, 29	9	09
14	Ownership of calculator	30	Yes	1
15	Age (as of Jan. 1, 1979)	31, 32	9/26/59	19
16	Height	33, 34	6 feet 1¼ inches	73
17	Weight	35, 36, 37	175 pounds	175
18	High-blood-pressure history	38	No	2
19	Smoking habit	39	Don't smoke	1
20	Attitude toward public smoking	40	Should permit	1
21	Minimum admission standards	41	10th year math and English	3
22	Stringent retention standards	42	Yes	1
23	Number of groups affiliated	43, 44	None	00
24	Satisfaction with SPS	45	Unsatisfied	2
25	Satisfaction with library	46	Neutral	4
26	Personal question	47	Yes	1

Figure 2.4 Coding the questionnaire responses of Jamie Bonds, #0007.

stated. Recall once more GIGO. What good is it to collect properly recorded data and then, through faulty editing, coding, or keypunching, improperly transcribe the initial desired information? Thus, in our example, each process required validation by the researcher and the dean to ensure that the collected data not only were properly recorded initially, but also properly prepared. Figure 2.4 represents the responses of Jamie Bonds, coded sequence 0007, who was the third subject selected in the sample.

Notice how the questionnaire responses are coded for keypunching. Each qualitative question requires a one-digit code such as is observed with question 3, "At the present time do you plan to attend graduate school?" Jamie Bonds is "not sure" and this response is given a code of 3. For quantitative questions, however, the number of spaces to allocate for a response must be based on the most extreme answer possible. Thus, for example, in question 13 two spaces are required because a subject can own more than nine pairs of jeans. Since Jamie Bonds owns nine pairs, a coded value of 09 is recorded. Similarly, for question 17, involving weight to the nearest pound, three spaces are required (although some subjects weigh less than 100 pounds) for coding. The coded responses then for each subject require 47 spaces. Since a punch card has 80 columns, a keypunched data card is obtained for each subject, and it is observed from Figure 2.5 that the coded data are punched in the first 47 columns.

Figure 2.6 is a computer printout of the data corresponding to the responses collected from the sample of 94 students obtained by the researcher and the dean. (The **data base** for the entire population of $N = 2{,}202$ students, from which the sample of 94 was drawn, is given in Appendix A). In the following two chapters, methods of tabular and chart presentation will be demonstrated, and various descriptive measures useful for data analysis and interpretation will be developed.

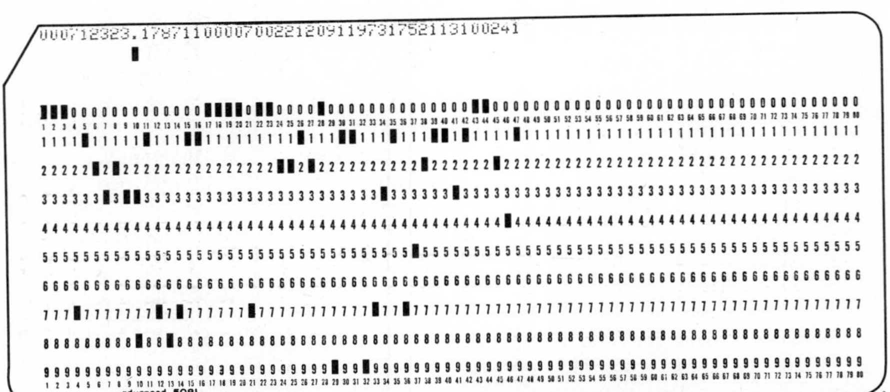

Figure 2.5 Data card for Jamie Bonds, #0007.

Sample number	File code number	1 Sex	2 Class	3 Grad. school	4 Major	5 Grade-point index	6 H. S. average	7 Est. starting salary $	8 Tuition $	9 Employment	10 Sunday shop	11 Investment	12 Car pref.	13 No. jeans owned	14 Calculator	15 Age	16 Height	17 Weight	18 Blood pressure	19 Smoking status	20 Smoking belief	21 Entrance standards	22 Retention standards	23 No. clubs and groups	24 SPS rating	25 Library rating	26 Personal question
1	0877	1	2	2	5	2.00	84	12000	600	3	2	2	2	3	2	22	76	195	2	2	1	4	1	0	4	4	2
2	1793	2	3	1	2	2.80	85	12000	2000	3	2	2	4	4	1	25	60	115	2	2	1	3	1	0	4	6	2
3	0007	1	2	3	2	3.17	87	11000	700	2	2	1	2	9	1	19	73	175	2	1	1	3	1	0	2	4	1
4	0261	1	2	1	4	3.43	72	12000	2000	2	2	4	4	1	2	24	68	165	2	1	2	3	1	2	2	4	2
5	1246	1	2	3	2	2.70	70	12000	500	3	2	2	2	2	1	17	67	148	2	1	2	3	1	2	4	6	2
6	1495	2	3	2	4	3.51	88	15000	3000	2	2	1	4	15	2	20	62	98	2	1	2	4	1	2	4	4	1
7	0023	2	4	2	5	2.30	87	15000	800	3	2	2	2	5	1	20	61	115	1	1	2	1	2	1	5	6	1
8	0397	2	3	2	4	3.32	86	15000	800	2	2	5	4	5	1	22	69	125	2	2	2	4	2	1	3	4	2
9	0172	1	4	1	4	2.80	88	11000	1000	2	2	2	2	3	2	22	69	145	2	2	1	4	1	0	4	4	1
10	0703	1	3	3	2	2.00	80	12000	1000	2	2	2	4	3	1	23	69	155	1	2	2	1	1	0	4	4	1
11	1406	2	2	3	1	2.40	82	18000	700	3	1	2	2	9	1	18	67	131	2	1	1	3	2	2	2	4	2
12	0739	1	2	2	2	2.90	88	8000	600	2	2	2	2	3	1	18	70	170	2	2	2	3	2	0	3	6	2
13	0432	2	3	2	5	2.36	76	10000	700	3	1	2	4	4	1	19	63	162	2	2	1	2	2	0	3	6	1
14	0861	1	3	3	4	3.82	75	18000	800	2	1	2	3	1	1	20	69	180	2	1	2	3	1	0	4	4	2
15	1515	2	3	2	3	2.82	85	10000	500	2	2	1	3	3	1	23	69	140	2	1	2	3	1	1	4	4	1
16	0810	1	3	1	3	3.00	81	11000	800	1	2	2	4	5	1	27	70	160	2	1	2	3	2	0	4	5	2
17	1514	1	1	1	2	3.20	93	11000	1200	2	2	1	4	5	2	18	68	155	2	1	2	3	1	0	4	6	1
18	1166	2	2	1	2	2.56	84	10000	2000	2	2	1	3	5	1	20	65	103	2	4	1	3	1	0	4	4	1
19	1658	2	4	2	5	2.90	80	10000	600	2	2	5	3	7	1	23	61	95	2	2	2	4	1	1	2	4	1
20	1173	1	2	3	7	2.64	83	15000	200	2	2	2	4	4	1	20	69	162	2	1	2	3	2	1	4	4	2
21	1788	1	3	3	4	2.45	89	11500	800	3	2	5	4	7	1	19	72	185	2	1	1	2	2	4	4	5	1
22	0452	2	4	1	4	3.00	83	15000	700	2	2	2	4	14	1	18	60	125	2	2	2	4	2	0	4	4	2
23	1170	2	2	2	2	2.28	75	15000	1000	3	2	2	4	6	1	17	68	132	1	2	2	3	2	1	4	4	1
24	1239	1	3	3	2	2.92	88	12000	1000	2	2	5	4	3	1	24	71	175	2	1	1	3	1	0	4	5	1
25	1576	1	2	3	2	3.00	87	12000	1000	1	2	5	4	6	1	28	71	155	2	1	2	3	1	0	4	4	1
26	1789	1	2	1	5	2.45	80	15000	500	1	2	2	4	5	1	19	72	160	2	1	2	3	2	0	2	5	2
27	0282	1	4	1	7	2.82	80	10000	1000	3	1	1	4	6	1	21	69	165	2	2	2	3	2	3	4	4	1
28	0103	1	2	1	2	3.86	87	15000	750	2	2	5	4	7	2	18	69	160	2	1	2	3	1	0	4	2	2
29	0847	2	2	3	2	2.00	84	12000	1000	1	2	2	4	10	1	19	65	120	2	3	1	4	2	0	4	4	2
30	1684	1	2	3	2	2.51	83	12000	1000	2	2	1	4	4	2	18	71	140	2	1	2	3	2	0	2	5	1
31	0969	1	3	2	2	2.62	83	12000	1000	2	1	2	2	2	2	23	64	175	2	1	2	3	2	1	2	2	2
32	1475	1	3	1	2	2.76	88	12000	1800	2	2	2	4	12	1	19	67	140	2	1	2	4	1	0	4	4	2
33	1131	2	3	2	2	2.91	85	32000	1000	2	2	2	4	4	1	20	61	145	2	1	1	4	2	0	4	4	1
34	1150	2	4	1	4	3.86	98	9000	600	2	2	1	4	6	1	21	65	127	2	1	2	3	2	2	4	6	2
35	1437	2	2	3	3	2.75	85	15000	600	1	2	2	2	2	1	27	66	114	2	1	2	4	1	0	4	4	1
36	1852	1	2	1	2	3.20	91	18000	1200	1	2	2	4	5	1	30	70	167	1	1	2	4	2	0	4	4	1
37	1392	1	4	3	7	2.60	78	18000	800	1	2	1	3	5	1	28	68	178	2	1	1	4	2	0	4	4	1
38	0559	2	3	2	7	2.50	82	13000	600	3	2	2	4	5	1	20	60	115	2	2	2	3	2	3	2	4	1
39	1480	2	4	1	5	3.34	85	8000	400	2	2	2	2	4	2	22	61	110	2	1	2	4	2	2	4	4	1
40	0292	2	2	1	5	2.30	80	10000	1000	3	1	1	3	3	1	21	64	115	2	3	1	3	2	2	4	3	2
41	0244	2	2	3	4	2.80	83	12000	1000	2	2	2	2	3	1	20	63	160	2	1	2	2	2	0	4	2	2
42	0467	2	4	3	5	2.50	80	10000	800	3	2	1	4	5	2	21	63	125	2	1	2	4	1	2	1	4	2
43	0216	2	4	3	5	3.00	92	10000	1000	2	2	2	2	1	1	22	61	110	2	1	2	1	2	1	4	4	1
44	1498	1	4	1	5	3.20	76	10000	500	2	1	2	2	4	1	24	65	140	2	1	2	4	2	0	4	4	2
45	0242	1	4	3	7	3.20	81	10000	1000	2	2	5	4	3	1	22	72	176	2	2	1	3	1	0	2	3	1
46	1058	2	2	1	4	3.75	75	15000	500	3	2	1	4	2	2	20	67	135	2	1	1	1	1	0	4	6	2
47	1786	2	2	1	5	3.00	86	12000	440	2	2	2	2	4	2	19	61	105	2	1	2	4	1	1	4	5	1
48	0560	1	3	3	4	2.69	84	16000	400	3	2	5	4	2	2	20	69	150	2	2	1	2	2	4	4	6	2
49	0593	1	4	1	3	2.64	82	13000	1000	2	2	2	4	3	1	21	69	165	2	3	2	3	2	2	4	5	1
50	0025	1	4	3	3	3.21	89	12000	1000	2	2	2	2	3	1	22	68	165	2	3	1	3	2	1	4	5	1
51	0033	2	3	1	7	2.94	85	15000	2000	3	2	2	4	5	2	21	61	105	2	1	2	3	1	1	4	5	2
52	1389	1	2	1	5	2.86	88	18000	600	3	2	2	4	11	1	18	72	152	2	1	2	2	2	3	4	6	1
53	0889	2	2	1	1	3.39	85	17000	1000	2	1	2	4	3	1	20	69	118	2	1	2	3	2	0	4	4	2
54	0283	2	3	1	5	2.85	89	15000	800	3	2	2	2	4	2	19	60	105	2	1	2	3	1	1	4	4	1
55	0416	1	2	3	2	2.91	83	15000	1000	2	2	2	2	7	1	19	74	200	2	3	2	3	2	0	4	4	2
56	0788	1	2	3	3	3.01	87	14000	800	3	2	2	4	4	1	20	67	180	1	2	2	4	2	0	4	5	1
57	1230	2	3	2	1	2.25	85	9500	800	2	2	2	2	4	2	21	65	140	2	1	2	4	1	0	2	3	2
58	0585	1	2	1	5	2.80	85	12333	1600	3	2	5	2	3	2	20	69	140	2	1	2	4	1	0	1	3	2
59	1708	1	3	1	5	2.89	92	16000	600	3	2	6	2	0	2	21	65	163	1	1	2	3	2	2	5	6	1
60	0959	2	2	3	1	2.51	80	15000	2000	2	2	2	4	3	1	22	70	145	2	1	2	3	2	0	4	4	1
61	0871	2	2	3	3	3.07	92	15000	1000	3	2	5	4	6	1	19	65	123	2	2	2	4	2	0	2	3	1
62	0072	1	1	3	2	3.60	86	13000	1000	3	1	6	2	5	1	18	71	182	2	1	2	3	2	1	4	4	2
63	0646	1	4	1	5	2.66	78	14000	1000	2	2	2	2	3	1	22	69	152	2	2	2	2	2	0	3	6	1
64	0461	2	3	3	5	2.81	89	12000	2000	2	2	2	2	9	2	19	66	121	2	3	1	4	2	0	4	4	2
65	0900	1	2	3	4	2.83	88	13000	500	2	2	2	2	5	2	21	64	140	2	1	1	3	1	2	6	4	1
66	1812	2	3	1	2	3.75	81	15000	2000	3	2	1	4	3	1	20	62	120	2	1	2	4	2	3	4	7	2
67	0908	2	2	1	5	3.89	90	25000	3000	2	1	1	4	7	1	22	64	105	2	2	2	3	2	1	5	6	1
68	1405	1	1	3	6	3.08	85	15000	500	3	2	1	4	7	1	18	72	160	2	1	2	4	1	2	4	4	2
69	0785	1	3	3	2	2.55	72	11000	400	3	2	1	4	3	2	19	69	175	1	1	2	2	2	2	2	6	2
70	0167	2	3	3	2	2.73	87	12000	1000	2	2	2	2	5	1	20	60	110	2	3	1	3	2	1	5	3	1
71	0353	1	2	3	5	1.83	80	14500	1200	2	2	2	2	5	1	20	67	155	2	1	2	2	2	0	4	6	1
72	0228	1	3	2	2	3.12	87	12000	1500	3	2	1	2	11	1	20	67	145	2	1	2	3	1	1	4	5	1
73	1309	1	2	2	2	2.48	83	15000	1000	2	2	2	4	7	1	22	70	170	2	1	2	3	2	0	4	4	2
74	0392	2	2	1	6	3.53	89	16000	3500	2	2	2	4	3	2	19	63	117	2	1	2	4	1	0	3	3	1
75	0692	1	2	1	2	3.09	80	15000	1000	2	2	2	2	6	1	26	68	142	2	1	2	4	1	1	4	5	2
76	0840	2	2	1	2	3.54	90	12000	800	3	2	1	2	4	1	19	71	135	2	1	2	4	1	1	4	7	2
77	1262	2	2	3	2	3.18	90	10000	500	2	2	1	2	11	2	18	64	108	2	1	2	4	1	1	4	4	1
78	0245	1	2	3	2	2.64	87	14000	800	3	2	2	4	14	1	20	66	155	2	1	2	3	2	2	4	4	1
79	1186	2	2	1	5	2.75	87	12000	1000	3	2	2	4	11	2	20	63	115	2	1	2	4	1	0	4	4	2
80	1042	1	1	1	5	3.25	72	16000	2000	2	2	5	4	1	2	18	69	150	2	3	1	1	1	1	4	5	1
81	1718	2	2	2	2	3.62	87	12000	2000	3	2	1	4	7	1	18	62	115	2	2	2	3	2	2	4	4	1
82	0654	1	3	1	5	3.61	82	25000	1500	2	2	1	4	6	1	21	68	149	2	2	2	4	1	0	3	5	2
83	1308	1	2	3	5	2.76	85	10000	1000	3	2	2	2	5	1	19	67	140	2	1	2	4	1	3	4	6	2
84	0989	1	2	3	3	2.85	78	10000	1000	1	2	2	2	5	1	28	68	165	2	1	2	3	1	0	4	4	2
85	0514	1	2	3	2	3.20	80	10000	1500	3	2	2	2	7	1	19	76	197	2	1	2	2	1	0	4	6	2
86	1628	1	3	3	2	3.00	90	12000	500	2	1	2	3	6	2	29	69	160	2	1	2	4	2	0	4	4	2
87	0899	1	2	2	2	2.10	77	12000	1000	3	2	2	4	6	1	18	64	145	2	1	1	3	2	0	4	1	1
88	0892	2	2	1	4	2.84	84	12500	450	3	2	2	2	7	1	18	67	144	2	2	1	1	2	3	4	5	1
89	0048	2	2	3	1	2.90	84	13500	1500	3	2	5	2	12	1	19	65	142	2	1	2	3	1	4	5	5	1
90	1403	2	2	1	2	3.00	89	10000	800	2	2	2	2	26	1	19	60	93	2	2	2	3	1	1	5	6	1
91	0823	2	2	1	1	2.00	79	15000	500	2	2	1	4	6	1	27	63	133	2	1	2	3	2	2	1	4	1
92	1048	2	2	1	2	2.00	81	18000	1600	3	2	1	4	2	1	30	62	145	2	4	1	2	2	5	4	5	2
93	0408	1	2	3	2	2.93	85	12500	1200	2	2	2	4	4	1	21	68	135	2	1	2	4	1	1	4	6	2
94	1110	1	3	3	7	2.50	75	15000	1000	3	2	1	3	7	2	19	71	130	2	1	1	1	2	3	4	4	1

Figure 2.6 Computer listing of responses to questionnaire from a sample of 94 students.

PROBLEMS

2.1 Explain the differences between qualitative and quantitative random variables and give three examples of each.

2.2 Explain the differences between discrete and continuous random variables and give three examples of each.

* 2.3 Determine whether each of the following random variables is qualitative or quantitative. If quantitative, determine whether the phenomenon of interest is discrete or continuous.
   (a) Net weight (in grams) of packaged dry cereal.
   (b) Useful lifetime of a 100-watt light bulb.
   (c) Political party affiliation of civil service workers.
   (d) Number of bankrupt corporations per month.
   (e) 1977 dividend payments of various utility companies.
   (f) Number of on-time arrivals per hour at a large airport.

2.4 If two students both score a 90 on the same examination, what arguments could be used to show that the underlying random variable (phenomenon of interest)—test score—is continuous?

2.5 Develop a set of five quantitative and five qualitative questions that were not included in the questionnaire (Figure 2.2) and which might also have been of interest to the dean.

2.6 What arguments could be given both for and against the use of the randomized response technique for maintaining one's anonymity and privacy?

2.7 How might a dishonest or untrained interviewer bias a judgment or quota sample? What "checks and balances" might management attempt to impose to avoid such biases?

2.8 Why would the collection of data from the students in your statistics class be described as a **chunk**?

2.9 Suppose that you want to estimate the average size fish in a lake and you are determined to draw a sample of 100 fish and measure each. Explain how you might do this by sampling **with** versus **without** replacement.

2.10 Given a population of $N = 73$, draw a sample of size $n = 13$ **without** replacement by starting in row 04 of the table of random numbers (Appendix E, Table E.1). Reading across the row, list the 13 coded sequences obtained.

* 2.11 Do Problem 2.10 by sampling **with** replacement.

2.12 For a population of 902 students, verify that by starting in row 29 of the table of random numbers (Appendix E, Table E.1) only nine rows are needed to draw a sample of size $n = 94$ **without** replacement.

Note: Asterisk indicates problems included in the answer section.

2.13 For a population of 2,202 students, a **two-stage** usage of the table of random numbers (Appendix E, Table E.1) may be recommended to avoid wasting time and effort. To obtain the sample by a two-stage approach, list the four-digit coded sequences after adjusting the first digit in each sequence as follows.

If the first digit is a 0, 3, or 6, change the digit to 0; if the first digit is a 1, 4, or 7, change the digit to 1; if the first digit is a 2, 5, or 8, change the digit to 2; if the first digit is a 9, discard the sequence. Thus starting in row 29 of the random numbers table (Appendix E, Table E.1), the sequence 4161 becomes 1161, while the sequence 3423 becomes 0423, etc. Verify that only 16 rows are needed to draw a sample of size $n = 94$ without replacement.

2.14 For a population of 1,202 students, what adjustment can you suggest for a two-stage usage of the random numbers table so that each of the 1,202 students have the same chance of selection?

* 2.15 The following computerized output is extracted from a set of data similar to those collected for the population data base (Appendix A). However, each of the five lines, representing the respective responses of five particular subjects, has one error in it. Use the coded statements shown in Figures 2.2 and 2.4 to determine the particular keypunching error in each of the five responses.

3000	1	2	4	2	2.73	80	15000	1000	3	2	2	2	07	1	18	73	210	2	1	1	3	1	02	5	5	1
3001	1	4	1	1	2.92	42	14000	1100	2	1	5	4	05	1	21	70	173	2	2	2	4	1	01	5	6	2
3002	1	6	1	2	2.81	85	12000	1400	3	2	2	4	08	1	18	68	155	2	1	1	4	1	01	4	4	2
3003	2	3	2	2	2.45	78	13000	0900	2	2	1	3	05	1	20	98	126	2	2	2	3	2	00	2	4	1
3004	1	2	3	2	2.70	77	14000	0850	2	2	2	4	06	1	19	67	151	2	2	2	1	3	01	4	6	1

REFERENCES

1. Cochran, W. G., *Sampling Techniques,* 3rd ed. (New York: Wiley, 1977).
2. Deming, W. E., *Sample Design in Business Research* (New York: Wiley, 1960).
3. Frankel, M. R., and L. R. Frankel, "Some Recent Developments in Sample Survey Design," *J. Marketing Res.,* 14 (August 1977), 280–93.
4. Hansen, M. H., W. N. Hurwitz, and W. G. Madow, *Sample Survey Methods and Theory,* vols. I and II (New York: Wiley, 1953).
5. Horvitz, D. G., B. V. Shah, and W. R. Simmons, "The Unrelated Question Randomized Response Model," *Proc. American Statistical Association, Social Statistics Section,* 1967, pp. 65–72.
6. Kish, L., *Survey Sampling* (New York: Wiley, 1965).
7. Raj, D., *The Design of Sample Surveys* (New York: McGraw-Hill, 1972).
8. Rand Corporation, *A Million Random Digits with 100,000 Normal Deviates* (New York: Free Press, 1955).
9. *Statistical Abstract of the United States,* U.S. Department of Commerce, 1977.
10. *Survey of Current Business,* U.S. Department of Commerce, 1977.
11. Warner, S. L., "Randomized Response—A Survey Technique for Eliminating Evasive Answer Bias," *J. Amer. Statistical Assoc.,* 60 (1965), 63–69.

CHAPTER 3

DATA PRESENTATION: TABLES AND CHARTS

3.1 INTRODUCTION

In the previous chapter the researcher and the dean designed a questionnaire in order to ascertain information regarding the socioeconomic conditions, attitudes, interests, and objectives of a business-school student body. A random sample of 94 students was obtained, and their responses to the questionnaire were listed in Figure 2.6. It is the goal of this chapter to show how the researcher and the dean would prepare and organize the set of data so that it may be presented in tabular and chart form.

Suppose that the researcher and the dean were interested in comparing the grade-point indexes of accounting majors versus non-accounting majors in the business school and, in addition, wanted to compare the two groups in terms of their attitudes toward entrance standards and retention standards. The researcher and the dean would scan the data in Figure 2.6 and sort out the responses from the accounting majors versus those from the non-accounting majors as shown in Figure 3.1.[1]

[1] Specifically, all students who did not claim to be majoring in accounting in question 4 of the survey (Figure 2.2) are categorized as non-accounting majors. These include students majoring in economics or finance, management, marketing, statistics or computer systems, other business areas, and students who have not yet decided on a particular major.

(Admission standards) (Retention standards)

Accounting Majors			Non-accounting Majors		
Grade-Point Index	Admission	Retention	Grade-Point Index	Admission	Retention
2.80	3	1	2.00	4	1
3.17	3	1	3.43	3	1
2.70	3	1	3.51	4	1
2.00	1	1	2.30	1	2
2.90	3	2	3.32	4	2
3.20	3	1	2.80	4	1
2.56	3	1	2.40	3	2
2.28	3	2	2.36	2	2
2.92	3	1	3.82	3	1
3.00	3	1	2.82	3	1
3.86	3	1	3.00	3	2
2.00	4	2	2.90	4	1
2.51	3	2	2.64	3	2
2.62	3	2	2.45	2	2
2.76	4	1	3.00	4	2
2.91	4	2	2.45	3	2
3.20	4	2	2.82	3	2
2.91	3	2	3.86	3	2
3.60	3	2	2.75	4	1
3.75	4	2	2.60	4	2
2.55	2	2	2.50	3	2
2.73	3	2	3.34	4	2
3.12	3	1	2.30	3	2
2.48	3	2	2.80	2	2
3.09	4	1	2.50	4	1
3.54	4	1	3.00	1	2
3.18	4	1	3.20	4	2
2.64	3	2	3.20	3	1
3.62	3	2	3.75	1	1
3.20	2	1	3.00	4	1
3.00	4	2	2.69	2	2
2.10	3	2	2.64	3	2
3.00	3	1	3.21	3	2
2.00	2	2	2.94	3	1
2.93	4	1	2.86	2	2
			3.39	3	2
			2.85	3	1
			3.01	4	2
			2.25	4	1
			2.80	4	1
			2.89	3	2
			2.51	3	2
			3.07	4	2
			2.66	2	2
			2.81	4	2
			2.83	3	1
			3.89	3	2
			3.08	4	1
			1.83	2	2
			3.53	4	1
			2.75	4	1
			3.25	1	1
			3.61	4	1
			2.76	4	1
			2.85	3	1
			2.84	1	2
			2.90	3	1
			2.00	3	2
			2.50	1	2

Figure 3.1 Responses to questions pertaining to grade-point indexes, admission standards, and retention standards by major.

SOURCE: Data extracted from Figure 2.6 and from questionnaire of Figure 2.2.

Notes: For admission standards: code 1 = no minimum standard; code 2 = eighth-grade mathematics and English; code 3 = tenth-grade mathematics and English; code 4 = twelfth-grade mathematics and English.

For retention standards: code 1 = more stringent standards imposed
code 2 = maintain status quo

3.2 QUALITATIVE DATA

The responses to categorical questions may be tallied and presented as either **bar charts** or **pie diagrams.** For example, from Figure 3.1 it is seen that of the 35 accounting majors only one student favored "no minimum standard for admission beyond a high school diploma," three students favored an "eighth-grade mathematics and English level," 21 students favored a "tenth-grade mathematics and English level," and 10 students favored a "twelfth-grade mathematics and English level." On the other hand, for the 59 non-accounting majors the tallies were 6, 7, 24, and 22, respectively. Since the sample sizes differ for these two groups, percentage comparisons are necessary. To obtain the percentages, each of the aforementioned tallies are divided by their respective totals, and the result is multiplied by 100. Thus for accounting majors,

$$1/35 \times 100 = 2.9\% \qquad 3/35 \times 100 = 8.6\%$$
$$21/35 \times 100 = 60.0\% \qquad 10/35 \times 100 = 28.6\%$$

and for non-accounting majors,

$$6/59 \times 100 = 10.2\% \qquad 7/59 \times 100 = 11.9\%$$
$$24/59 \times 100 = 40.7\% \qquad 22/59 \times 100 = 37.3\%$$

To facilitate a comparison between the two groups, the **percentage component bar chart** shown in Figure 3.2 is constructed. It is seen that in general the accounting students favor higher admission standards to college than do the non-accounting majors. However, while the overwhelming majority of accounting students who favor higher admission standards support a tenth-grade mathematics and English level, the non-accounting students who favor higher entrance standards are more evenly divided in their support for tenth-

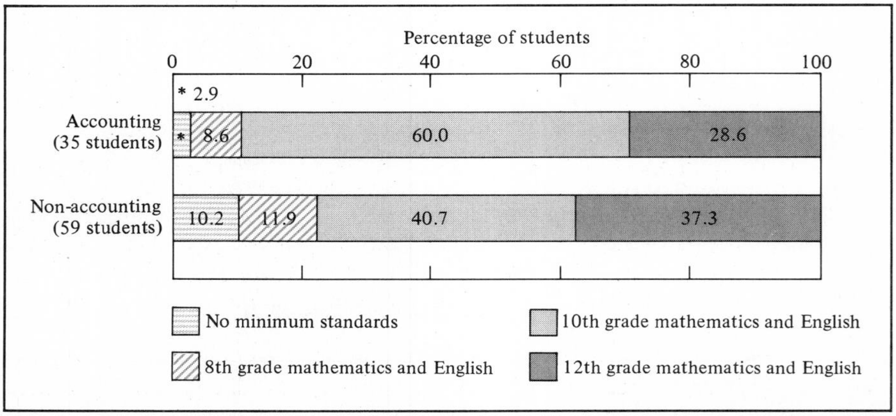

Figure 3.2 Percentage component bar chart of attitudes toward admission standards to college by accounting versus non-accounting students.

grade and twelfth-grade mathematics and English levels. This can also be seen from the **percentage bar chart** in Figure 3.3 or the **pie diagram** in Figure 3.4.

Figures 3.2, 3.3, and 3.4 all attempt to convey the same message with regard to attitudes toward admission standards. Thus the selection of a particular type of bar chart or pie diagram is subjective. The researcher should choose that method of presentation which he or she believes most clearly highlights the pertinent aspects of a set of data. Some experts (Reference 1) believe that component bar charts are usually preferable to pie diagrams. However, for such categorical questions as "investment intent" or "major area of study," for which the response categories cannot in any way be ordered, the component bar chart would likely be misleading since combining or accumulating adjacent categories is not warranted. In these situations the choice should rest between the pie diagram and the bar chart. These are presented in Figure 3.5(a) and (b), respectively, for all 94 responses to the question "What is your major area of study?"

3.2.1 Construction of Bar Charts and Pie Diagrams

To construct bar charts the following suggestions are made.

1. For categorical responses that are qualitative the bars should be constructed horizontally as in Figure 3.5(b); for categorical responses that are numerical the bars should be constructed vertically.

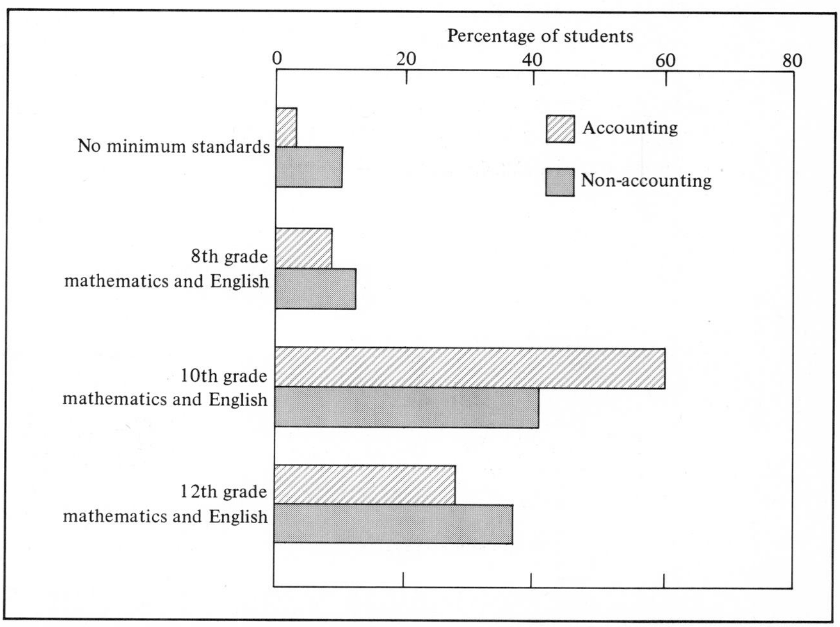

Figure 3.3 Percentage bar chart of attitudes toward admission standards to college by accounting and non-accounting students.

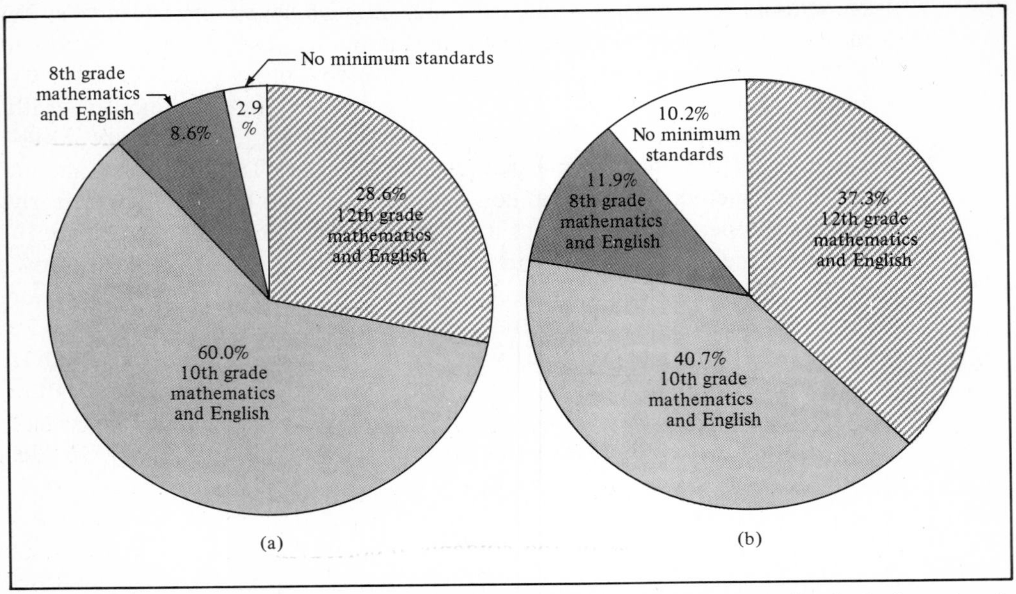

Figure 3.4 Percentage pie diagrams depicting attitudes toward admission standards to college by (a) accounting students and (b) non-accounting students.

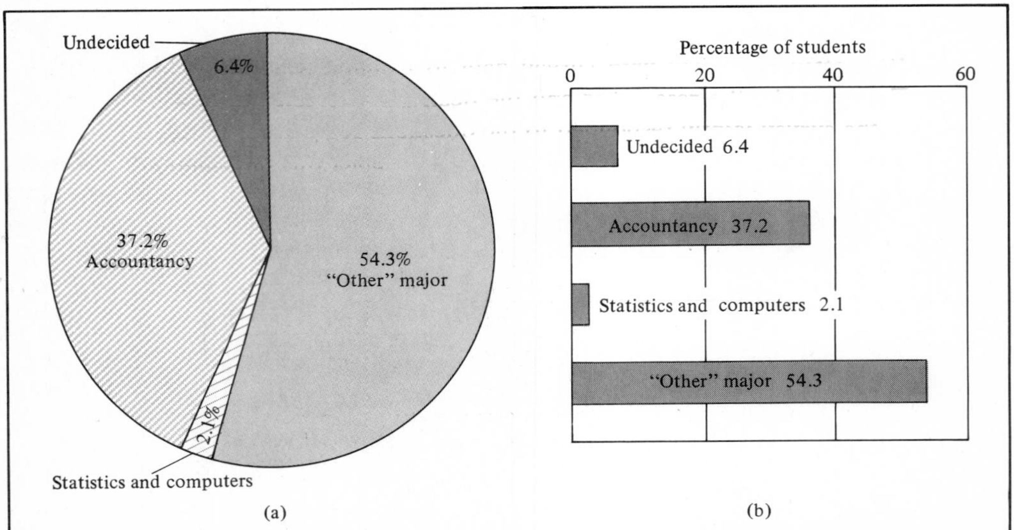

Figure 3.5 Depicting major areas of study of 94 business students by means of (a) a percentage pie diagram and (b) a percentage bar chart.

2. All bars should have the same width [as in Figures 3.3 and 3.5(b)] so as not to mislead the reader. Only the lengths may differ.

3. Spaces between bars should range from one-half the width of a bar to the width of a bar.

4. Scales and guidelines are useful aids in reading a chart and should be included.

5. The axes of the chart should be clearly labeled.

6. Any "keys" to interpreting the chart may be included within the body of the chart as in Figure 3.3 or below the body of the chart as in Figure 3.2.

7. The title of the chart appears below the body.

8. Footnotes or source notes, when appropriate, are given below the title of the chart.

To construct pie diagrams the researcher may use both the compass and the protractor—the former to draw the circle, the latter to measure off the appropriate pie sectors. Since the circle has 360°, the protractor may be used to divide up the pie based on the percentage "slices" desired. As an example, in Figure 3.5(a), 37.2% of the students reported majoring in accounting. Thus the researcher would multiply .372 by 360, mark off the resulting 133.9° with the protractor, and then connect the appropriate points to the center of the pie, forming a slice comprising 37.2% of the area. In such a manner the pie diagram can be constructed. As in Figure 3.4, whenever there are two or more pies to be compared, and each is composed of different sample sizes, such comparisons may be facilitated by expressing all relations on a percentage basis in order to avoid misleading interpretations.

3.2.2 Cross-Classification Tables

It is often necessary with survey data to examine variables simultaneously. For example, the researcher and the dean are interested in how each student in the survey jointly responded to the questions regarding admission standards and retention standards. Tables 3.1 and 3.2 depict this information separately for both accounting majors and non-accounting majors. Such two-way tables

Table 3.1
CROSS-CLASSIFICATION TABLE OF ACCOUNTING MAJORS' ATTITUDES TOWARD ADMISSION AND RETENTION STANDARDS

	Admission Standards Imposed				
Retention Standards Imposed	*No Minimum Standards*	*8th Grade Math and English*	*10th Grade Math and English*	*12th Grade Math and English*	Totals
More stringent than now	1	1	10	5	17
Same as now	0	2	11	5	18
Totals	1	3	21	10	35

Table 3.2
CROSS-CLASSIFICATION TABLE OF NON-ACCOUNTING MAJORS' ATTITUDES TOWARD ADMISSION AND RETENTION STANDARDS

Retention Standards Imposed	Admission Standards Imposed: No Minimum Standards	8th Grade Math and English	10th Grade Math and English	12th Grade Math and English	Totals
More stringent than now	2	0	9	14	25
Same as now	4	7	15	8	34
Totals	6	7	24	22	59

of cross classification are also known as **contingency tables** and will be studied once again in Chapters 5 and 10.

To construct Table 3.1, for example, the joint responses of each of the 35 accounting majors to the admission and retention questions are tallied into one of the eight possible "cells" of the table. Thus from Figure 3.1, the first accounting student responded 3,1 to the admission and retention standards questions, respectively. Since code 3 means an admission standard of tenth-grade mathematics and English and code 1 means more stringent retention standards, this student's responses were tallied into the cell composed of the first row and third column. The second and third students also responded 3,1 and were similarly tallied. The fourth accounting major, however, responded 1,1 and, as can be seen from Table 3.1, is the only accounting student who favors no minimum entrance standards but more stringent retention standards. The remaining 31 joint responses were recorded in a similar manner.

Since the sample sizes for the two groups differ, it would be extremely cumbersome to compare the findings from Table 3.1 (accounting majors) to those of Table 3.2 (non-accounting majors) unless the data are standardized on a percentage basis. One way that this may be achieved is to compute the percentages based on the overall sample group totals—35 for the accounting majors and 59 for the non-accounting majors—as illustrated in Table 3.3. For example, it is seen from Table 3.1 that only 1 of the 35 accountng majors (2.9%) favors no minimum admission standards but more stringent retention standards whereas from Table 3.2 it is observed that 2 of the 59 non-accounting majors (3.4%) hold similar beliefs. Such comparisons can be made on a percentage basis for every cell in Table 3.3 as well as for the row and column margins. Hence in total, 17 out of the 35 accounting majors (48.6%) favor more stringent retention standards as compared to 25 out of the 59 non-accounting majors (42.4%). In an effort to compare and contrast the beliefs of accounting versus non-accounting students in terms of admission and retention standards Table 3.3 is examined for large percentage discrepancies. Thus it is noted that 28.6% of accounting students believe more stringent retention standards should be coupled with a tenth-grade mathematics and English admission criterion, while only 15.3% of the non-

Table 3.3
CROSS-CLASSIFICATION TABLE OF ATTITUDES TOWARD ADMISSION AND RETENTION STANDARDS BY MAJOR
(Percentages Based on Sample Group Totals)

	Accounting Majors (35)					Non-accounting Majors (59)				
	Admission Standards Imposed					Admission Standards Imposed				
Retention Standards Imposed	No Minimum Standards (%)	8th Grade Math and English (%)	10th Grade Math and English (%)	12th Grade Math and English (%)	Totals (%)	No Minimum Standards (%)	8th Grade Math and English (%)	10th Grade Math and English (%)	12 Grade Math and English (%)	Totals (%)
More stringent than now	2.9	2.9	28.6	14.3	48.6*	3.4	0.0	15.3	23.7	42.4
Same as now	0.0	5.7	31.4	14.3	51.4	6.8	11.9	25.4	13.6	57.6*
Totals	2.9	8.6	60.0	28.6	100.0*	10.2	11.9	40.7	37.3	100.0*

* Errors due to rounding.

accounting students hold similar views. On the other hand, only 14.3% of accounting students believe more stringent retention standards should be coupled with twelfth-grade mathematics and English entrance standards as opposed to 23.7% of the non-accounting students who hold such a view. These results highlight previous findings. As was seen from Figures 3.2, 3.3, and 3.4, 88.6% of the accounting students favor high admission standards (tenth- or twelfth-grade mathematics and English) as compared to 78.0% of the non-accounting majors. However, these accounting students overwhelmingly favor tenth-grade mathematics and English while these non-accounting majors are about evenly divided in their views toward tenth- versus twelfth-year mathematics and English.

3.3 QUANTITATIVE DATA

When numerical data are listed in the random order in which they have been collected, the data are said to be in **raw form.** The grade-point indexes of accounting majors and non-accounting majors, listed in Figure 3.1, are examples of data in raw form. As seen from Figure 3.1, as the number of observations gets large, it becomes more difficult to focus on the major characteristics of the overall data. As a rule of thumb, whenever the set of data contains 25 or more observations, the best way to examine such "mass data" is to present the data in summary form by constructing appropriate tables and charts from which the researcher can then extract the important features. Prior to presenting the data in tabular or chart form, however, it is often useful (though not essential) to place the raw data in rank order, from the smallest to the largest observation. Such an ordered sequence of data is called an **ordered array.** Figure 3.6 depicts the ordered arrays for the accounting and non-accounting majors.

When the data are all organized into an ordered array, the researcher's evaluation is facilitated. It becomes easy to pick out extremes, typical values, and concentrations of values. In addition, from the ordered array all de-

Accounting Majors (35 Students)		Non-accounting Majors (59 Students)		
2.00	3.00	1.83	2.75	3.00
2.00	3.00	2.00	2.76	3.01
2.00	3.00	2.00	2.80	3.07
2.10	3.09	2.25	2.80	3.08
2.28	3.12	2.30	2.80	3.20
2.48	3.17	2.30	2.81	3.20
2.51	3.18	2.36	2.82	3.21
2.55	3.20	2.40	2.82	3.25
2.56	3.20	2.45	2.83	3.32
2.62	3.20	2.45	2.84	3.34
2.64	3.54	2.50	2.85	3.39
2.70	3.60	2.50	2.85	3.43
2.73	3.62	2.50	2.86	3.51
2.76	3.75	2.51	2.89	3.53
2.80	3.86	2.60	2.90	3.61
2.90		2.64	2.90	3.75
2.91		2.64	2.94	3.82
2.91		2.66	3.00	3.86
2.92		2.69	3.00	3.89
2.93		2.75	3.00	

Figure 3.6 Ordered arrays of grade-point indexes attained by accounting vs. non-accounting majors.

SOURCE: Data are taken from Figure 3.1.

scriptive measures used in data analysis (Chapter 4) can be calculated. On the negative side of the ledger, however, as the number of observations gets large, the process of organizing the raw data into an ordered array becomes more cumbersome. In such situations it becomes particularly useful to organize the data into **stem-and-leaf displays** (Reference 5) in order to study the aforementioned data characteristics. Figure 3.7 depicts the stem-and-leaf display for the grade-point indexes of all 94 students in the sample, regardless of major. Note that the "stems" are listed in a column to the left of the vertical line, while in each row the "leaves" branch out to the right of the vertical line. Since grade-point indexes are recorded in hundredths, it is seen that the stems consist of all grade-point indexes in tenths, while the leaves give the particular values in hundredths.

Using the data presented in Figure 2.6 the stem-and-leaf display of Figure 3.7 may be constructed. It is noted that the first subject in the sample had an index of 2.00. Therefore, the hundredths digit of 0 in his grade-point index is listed as the first leaf value next to the stem value of 2.0. Moreover, the second subject had an index of 2.80. Here the hundredths digit of 0 is listed as the first leaf value next to the stem value of 2.8. Continuing, the third subject had an index of 3.17 so that the hundredths digit of 7 in his index is listed as the first leaf value next to the stem value of 3.1. At this point in its construction the stem-and-leaf display would have appeared as follows:

1.8	
1.9	
2.0	0
2.1	
2.2	
2.3	
2.4	
2.5	
2.6	
2.7	
2.8	0
2.9	
3.0	
3.1	7
3.2	
3.3	
3.4	
3.5	
3.6	
3.7	
3.8	
3.9	

1.8	3
1.9	
2.0	00000
2.1	0
2.2	85
2.3	060
2.4	8055
2.5	6150010
2.6	2440946
2.7	063556
2.8	0022065091354
2.9	02113040
3.0	09000000178
3.1	728
3.2	0000015
3.3	249
3.4	3
3.5	413
3.6	021
3.7	55
3.8	6269
3.9	
	$n = 94$

Figure 3.7 Stem-and-leaf display of grade-point indexes of 94 students.
SOURCE: Data are taken from Figure 2.6.

1.8 and **1.9**	3
2.0 and **2.1**	000000
2.2 and **2.3**	80605
2.4 and **2.5**	61580550010
2.6 and **2.7**	0263445094656
2.8 and **2.9**	002113020204650913540
3.0 and **3.1**	70298000000178
3.2 and **3.3**	0002400195
3.4 and **3.5**	4313
3.6 and **3.7**	05251
3.8 and **3.9**	6269

Figure 3.8 Stem-and-leaf display of grade-point indexes of 94 students using fewer stems.

SOURCE: Data are taken from Figure 2.6.

As more and more subjects are included, those possessing the same stems and, perhaps, even the same leaves within stems (that is, the same grade-point indexes) will be observed. Such leaf values will be recorded adjacent to the previously recorded leaves, opposite the appropriate stem.

If, however, the researcher desires to alter the size of the stem-and-leaf display, the schema is flexible enough for such an adjustment. Suppose, for example, it is desired to reduce the number of stems so as to permit a heavier concentration of leaves on the remaining stems. This is accomplished in the stem-and-leaf display depicted in Figure 3.8. Note that two adjacent stems from Figure 3.7 are now grouped together as one stem. If boldface is used, as here, it is still easy to pick out extremes, typical values, and concentrations of values corresponding to a particular stem. If boldface cannot readily be used, differentiations may be made by underscoring. These two methods of clarifying the stem-and-leaf display are demonstrated in Figures 3.9 and 3.10, respectively. The two figures indicate how comparative sets of data may be organized and displayed. Thus the grade-point indexes of the 35 accounting majors are compared against those achieved by the 59 non-accounting majors. Figure 3.9 is a *side-by-side* stem-and-leaf display depicting the grade-point indexes of these two groups of students. Boldface is used here to differentiate between the stem groupings. On the other hand, Figure 3.10 is a *back-to-back* stem-and-leaf display comparing the grade-point indexes of accounting versus non-accounting students. Here, however, underscoring rather than boldface is used to distinguish between the stem groupings. The selection between the two types of stem-and-leaf displays presented in Figures 3.9 and 3.10 for organizing comparative data (side-by-side versus back-to-back) and the method of presentation used for distinguishing between the stem groupings (boldface versus underscoring) are both matters of personal choice.

Regardless of whether an ordered array or a stem-and-leaf display is selected for *organizing* the data, as the number of observations gets large it

Accounting Majors		Non-Accounting Majors	
1.8 and 1.9		1.8 and 1.9	3
2.0 and 2.1	0000	2.0 and 2.1	00
2.2 and 2.3	8	2.2 and 2.3	0605
2.4 and 2.5	6158	2.4 and 2.5	0550010
2.6 and 2.7	02634	2.6 and 2.7	45094656
2.8 and 2.9	002113	2.8 and 2.9	020204650913540
3.0 and 3.1	7029800	3.0 and 3.1	0000178
3.2 and 3.3	000	3.2 and 3.3	2400195
3.4 and 3.5	4	3.4 and 3.5	313
3.6 and 3.7	052	3.6 and 3.7	51
3.8 and 3.9	6	3.8 and 3.9	269
	$n = 35$		$n = 59$

Figure 3.9 Side-by-side stem-and-leaf display of grade-point indexes of 35 accounting students and 59 non-accounting students.

SOURCE: Data are taken from Figure 3.1.

becomes necessary to further condense or summarize the data into appropriate summary tables. Thus the researcher may wish to group or arrange the data into **classes** or **categories** according to conveniently established divisions of the range of the observations. Such an arrangement of data in tabular form is called a frequency distribution. **A frequency distribution is a summary**

Accounting Majors		Non-Accounting Majors
	1.8 and 1.9	3
0000	2.0 and 2.1	00
8	2.2 and 2.3	0605
8516	2.4 and 2.5	0550010
43620	2.6 and 2.7	45094656
311200	2.8 and 2.9	020204650913540
0089207	3.0 and 3.1	0000178
000	3.2 and 3.3	2400195
4	3.4 and 3.5	313
250	3.6 and 3.7	51
6	3.8 and 3.9	269
$n = 35$		$n = 59$

Figure 3.10 Back-to-back stem-and-leaf display of grade-point indexes of 35 accounting students and 59 non-accounting students.

SOURCE: Data are taken from Figure 3.1.

table in which the data are grouped or arranged into conveniently established numerically ordered classes or categories.

When the data are grouped or condensed into frequency distribution tables, the process of data analysis and interpretation is made much more manageable and meaningful. In such summary form the salient data characteristics are very easily approximated, thus compensating for the fact that when data are so grouped, the initial information pertaining to individual observations that was previously available is lost through the grouping or condensing process.

3.3.1 Constructing the Frequency Distribution

In constructing the frequency distribution table, attention must be given to:

1. Selecting the appropriate number of class groupings for the table
2. Obtaining a suitable **class interval** or "width" of each class grouping
3. Establishing the **limits** and the **boundaries** of each class grouping to avoid overlapping

SELECTING THE NUMBER OF CLASSES

The number of class groupings to be used is primarily dependent upon the number of observations in the data, that is, the larger the number of observations, the larger the number of class groups and vice versa. In general, however, the frequency distribution should have at least five class groupings, but no more than 15. If there are not enough class groupings there is too much concentration of data; if there are too many groupings there is too little concentration of data. In either case little information would be obtained. As an example, a frequency distribution having but one class grouping could be formed for the grade-point indexes of the accounting students as follows:

Grade-Point Index	Number of Students
2.00–4.00	35
Total	35

This, of course, is equivalent to a stem-and-leaf display having but one stem and all 35 leaves branching out on it. However, no additional information is obtained from such a summary table or stem-and-leaf display that was not already known from either scanning the raw data or the ordered array. Such a table or display, with too much data concentration, is not meaningful. The same would be true at the other extreme—if a table had too many class groupings or if a display had too many stems there would be an underconcentration of data and very little would be learned.

OBTAINING THE CLASS INTERVALS

When developing the frequency distribution table it is desirable to have each class grouping of equal width. To determine the width of each class, the

range of the data (the difference between the largest and the smallest observations) is divided by the number of class groupings desired:

$$\text{Width of Interval} = \frac{\text{Range}}{\text{Number of Class Groupings}}$$

For the accounting majors, five class groupings are desired since there are only 35 observations. From the ordered array in Figure 3.6, the largest observation is 3.86 while the smallest is 2.00. Hence the range is computed as

$$\begin{aligned}\text{Range} &= \text{Largest Observation} - \text{Smallest Observation}\\ &= 3.86 - 2.00 = 1.86\end{aligned}$$

and the class interval is approximated by

$$\text{Width of Interval} = \frac{1.86}{5} = .372$$

For convenience and ease of reading the selected interval or width of each class grouping is rounded to .40.

ESTABLISHING THE LIMITS AND BOUNDARIES OF THE CLASSES

To construct the frequency distribution table, it is necessary to establish clearly defined class limits and boundaries for each class grouping so that the observations either in the raw form (Figure 3.1), in an ordered array (Figure 3.6), or in a stem-and-leaf display (Figure 3.9) can be tallied into the proper class grouping.[2] Overlapping of classes must be avoided.

Since the width of each class interval for the accounting students' grade-point indexes has been set at .40, the "limits and boundaries" of the various class groupings must be established so as to include the entire range of observations. Whenever possible these limits and boundaries should be chosen to facilitate the reading and interpreting of data. Thus the first class interval is established from 2.00 to under 2.40, the second is from 2.40 to under 2.80, etc. The data in their raw form (Figure 3.1) or from the ordered array (Figure 3.6) are then tallied into each class as shown:

Grade-Point Index	Frequency Tallies	
2.00 but less than 2.40	𝍸	5
2.40 but less than 2.80	𝍸 ||||	9
2.80 but less than 3.20	𝍸 𝍸 |||	13
3.20 but less than 3.60	||||	4
3.60 but less than 4.00	||||	4
Totals		35

By establishing the limits and boundaries of each class as above, all 35 observations have been tallied into five classes, each having an interval width

[2] Depending on the manner in which the data are recorded, there may or may not be a distinction between **class limits** and **class boundaries**. For further discussion pertaining to these two concepts see Section 3.5.5.

Table 3.4
FREQUENCY DISTRIBUTION OF GRADE-POINT INDEXES FOR 35 ACCOUNTING MAJORS

Grade-Point Index	Number of Students
2.00 but less than 2.40	5
2.40 but less than 2.80	9
2.80 but less than 3.20	13
3.20 but less than 3.60	4
3.60 but less than 4.00	4
Totals	35

SOURCE: Data are taken from Figure 3.1.

of .40 without overlapping. From this "worksheet" the frequency distribution is presented in Table 3.4.

The major disadvantage of such a summary table is that it is not possible to know how the individual values are distributed within a particular class interval without access to the original data. Thus for the four students whose grade-point indexes are between 3.60 and 4.00, it is not clear from Table 3.4 as to whether the values are distributed throughout the interval, are all close to 3.60, or are all close to 4.00. The class mark or midpoint of the class interval, however, is the value used to represent all the data summarized into a particular interval. **The class mark is the point halfway between the boundaries of each class and is representative of the data within that class.** The class mark for the interval "3.60 but less than 4.00" is 3.80. (The other class marks are, respectively, 2.20, 2.60, 3.00, and 3.40.)

On the other hand, the major advantage of using such a summary table is that the salient data characteristics become immediately clear to the reader. Hence from Table 3.4 it is seen that the approximate range of the 35 grade-point indexes of accounting students is from 2.00 to 4.00 with most indexes tending to cluster between 2.40 and 3.20—especially from 2.80 to under 3.20. Thus the frequency distribution is truly a summary table in which the original data are condensed or grouped to facilitate data analysis. To further facilitate the analysis, however, it is desirable to form either the **relative frequency distribution** or the **percentage distribution,** depending on whether the researcher prefers proportions or percentages. These two equivalent distributions are shown as Tables 3.5 and 3.6, respectively.

The relative frequency distribution depicted in Table 3.5 is formed by dividing the frequencies in each class of the frequency distribution (Table 3.4) by the total number of observations. A percentage distribution (Table 3.6) may then be formed by multiplying each relative frequency or proportion by 100.0. Thus from Table 3.5 it is clear that the proportion of accounting students with grade-point indexes of 2.00 to under 2.40 is .143, while from Table 3.6 it is seen that 14.3% of the accounting students had such grade-point indexes. Working with a base of 1 for proportions or 100.0 for percentages is usually more meaningful than using the frequencies themselves.

Table 3.5
RELATIVE FREQUENCY DISTRIBUTION OF GRADE-POINT INDEXES FOR 35 ACCOUNTING MAJORS

Grade-Point Index	Proportion of Students
2.00 but less than 2.40	.143
2.40 but less than 2.80	.257
2.80 but less than 3.20	.371
3.20 but less than 3.60	.114
3.60 but less than 4.00	.114
Totals	.999*

SOURCE: Data are taken from Table 3.4.
* Error due to rounding.

Indeed, the use of the relative frequency distribution or percentage distribution becomes essential whenever one set of data is being compared to other sets of data, especially if the numbers of observations in each set differ. Hence to compare the 35 accounting students' grade-point indexes with those of the 59 non-accounting students it becomes necessary to establish either the relative frequency distribution or the percentage distribution for the latter group. Table 3.7 then depicts both the frequency distribution and the percentage distribution of the grade-point indexes of 59 non-accounting students. It is both permissible and desirable to construct one such table in lieu of two separate tables to save space.

Note that the class groupings selected for the nonaccounting majors in Table 3.7 match, where possible, those selected for the accounting majors in Tables 3.4 through 3.6. While one additional class was needed here, the limits and boundaries of the classes should match as above or be multiples of each other in order to facilitate comparisons of two or more data sets. Using the percentage distributions of Tables 3.6 and 3.7, it is now meaningful to state

Table 3.6
PERCENTAGE DISTRIBUTION OF GRADE-POINT INDEXES FOR 35 ACCOUNTING MAJORS

Grade-Point Index	Percentage of Students
2.00 but less than 2.40	14.3
2.40 but less than 2.80	25.7
2.80 but less than 3.20	37.1
3.20 but less than 3.60	11.4
3.60 but less than 4.00	11.4
Totals	99.9*

SOURCE: Data are taken from Table 3.4.
* Error due to rounding.

Table 3.7
FREQUENCY DISTRIBUTION AND PERCENTAGE DISTRIBUTION FOR 59 NON-ACCOUNTING MAJORS

Grade-Point Index	Number of Students	Percentage of Students
1.6 but less than 2.0	1	1.7
2.0 but less than 2.4	6	10.2
2.4 but less than 2.8	15	25.4
2.8 but less than 3.2	22	37.3
3.2 but less than 3.6	10	16.9
3.6 but less than 4.0	5	8.5
Totals	59	100.0

SOURCE: Data are taken from Figure 3.1.

that the grade-point indexes "2.80 but less than 3.20" are most typical of both accounting students and non-accounting students; the former class grouping contains 37.1% while the latter contains 37.3%. Since the numbers of observations differ (35 versus 59) it would obviously not be appropriate to compare the 13 accounting majors in this class grouping against the 22 non-accounting students in the same grouping without adjusting for such differences in the totals.

What other kinds of comparisons can be made from Table 3.6 and 3.7? The **range** of grade-point indexes of the non-accounting majors is approximately 2.4 (the difference between the upper boundary of the last class and the lower boundary of the first class = 4.0 − 1.6 = 2.4) while the range of grade-point indexes of the accounting students is approximately 2.0. Moreover, 14.3% of the accounting majors have indexes under 2.4 while 11.9% of the non-accounting students have such indexes. Furthermore, 22.8% of the accounting majors have indexes of at least 3.2 which meet the minimum requirements for the dean's list, while 25.4% of the non-accounting students have such indexes. Although the results for the two groups appear fairly similar, the non-accounting group seems to have slightly higher percentages of students with high grade-point indexes and slightly lower percentages of students with poor grade-point indexes.

3.3.2 Graphing the Data: Histograms and Polygons

There is an old saying that "one picture is worth a thousand words." Indeed, statisticians have employed graphic techniques to more vividly describe sets of data. Bar charts and pie diagrams were presented in Figures 3.2 through 3.5 to describe qualitative data. With quantitative data summarized into frequency, relative frequency, or percentage distributions, however, **histograms** and **polygons** are used to describe the data.

HISTOGRAMS

When plotting histograms, the random variable or phenomenon of interest is plotted along the horizontal axis while the vertical axis represents the number, proportion, or percentage of observations per class interval—depending

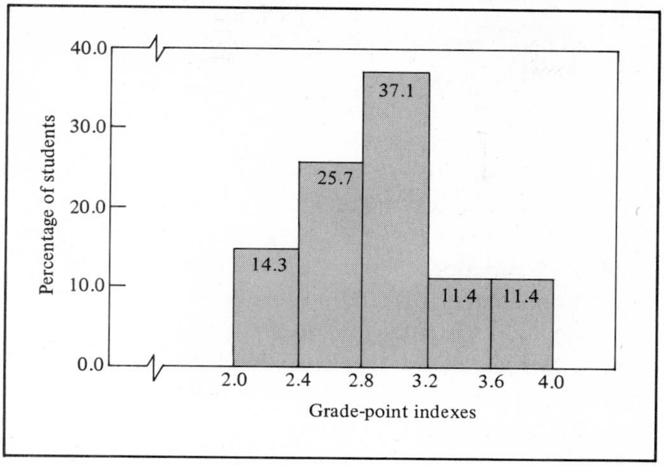

Figure 3.11 Percentage histogram of grade-point indexes of 35 accounting students.

SOURCE: Data are taken from Table 3.6.

Note - Verticle format because data is numerical.

on whether or not the particular histogram is, respectively, a frequency histogram, a relative frequency histogram, or a percentage histogram. Histograms are essentially vertical bar charts in which the rectangular bars are constructed at the boundaries of each class.[3] A percentage histogram is depicted in Figure 3.11 for the 35 accounting students' grade-point indexes.

When comparing two or more sets of data, however, the various histograms cannot be constructed on the same graph because superimposing the vertical bars of one on another would cause difficulty in interpretation. For such cases it is necessary to construct relative frequency or percentage polygons.

POLYGONS

As with histograms, when plotting polygons the phenomenon of interest is plotted along the horizontal axis while the vertical axis represents the number, proportion, or percentage of observations per class interval—depending on whether the particular polygon is, respectively, a frequency polygon, a relative frequency polygon, or a percentage polygon. The percentage polygon, for example, is formed by letting the class mark or midpoint of each class represent the data in that class and then connecting together the sequence of

[3] Interestingly, for quantitative data organized into a stem-and-leaf display, the display itself may be used to pictorially describe the data if it is rotated 90° counterclockwise so that (1) the vertical line becomes the horizontal axis; (2) the stems become the class groupings; and (3) the leaves or accumulations of digits for each stem become the vertical bar representations.

midpoints at their respective class percentages. Figure 3.12 shows the percentage polygon for the 35 accounting students' grade-point indexes and Figure 3.13 compares the percentage polygons of the 35 accounting students versus the 59 non-accounting students in terms of their grade-point indexes. The similarities and slight differences in the indexes between these two groups of students, previously discussed when comparing Tables 3.6 and 3.7, are clearly indicated here.

It should be noted that the polygon is a representation of the shape of the particular distribution. Since the area under the entire curve or percentage distribution must be 100.0%, it is necessary to connect the first and last midpoints with the horizontal axis so as to enclose the area of the observed distribution. In Figure 3.12 this is accomplished by connecting the first observed midpoint with the midpoint of a "fictitious preceding" class (1.8) having 0.0% observations and by connecting the last observed midpoint with the midpoint of a "fictitious succeeding" class (4.2) having 0.0% observations.

It is further noted that when polygons (Figure 3.12) or histograms (Figure 3.11) are constructed the vertical axis must show the true zero (origin) so as not to distort or otherwise misrepresent the character of the data. The horizontal axis, however, does not need to specify the zero point for the phenomenon of interest. As in Figures 3.11 and 3.12 the range of the random variable should constitute the major portion of the chart, and when zero is not included, "breaks" (-∿-) in the axis are appropriate.

3.3.3 Cumulative Distributions and Cumulative Polygons

Other useful methods of presentation to facilitate data analysis and interpretation are the construction of cumulative distribution tables and the plotting

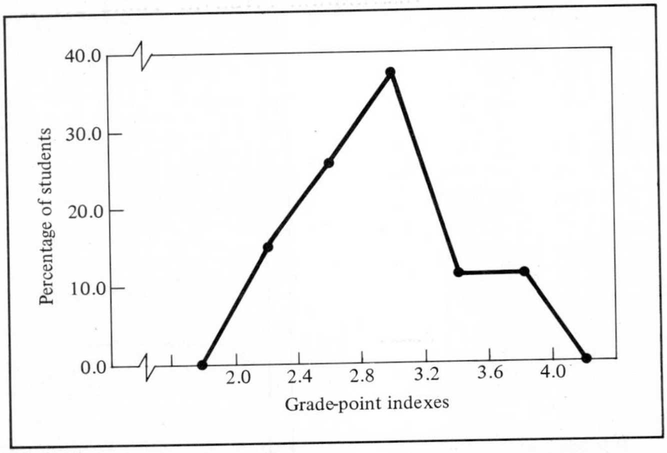

Figure 3.12 Percentage polygon of grade-point indexes of 35 accounting students.

SOURCE: Data are taken from Table 3.6.

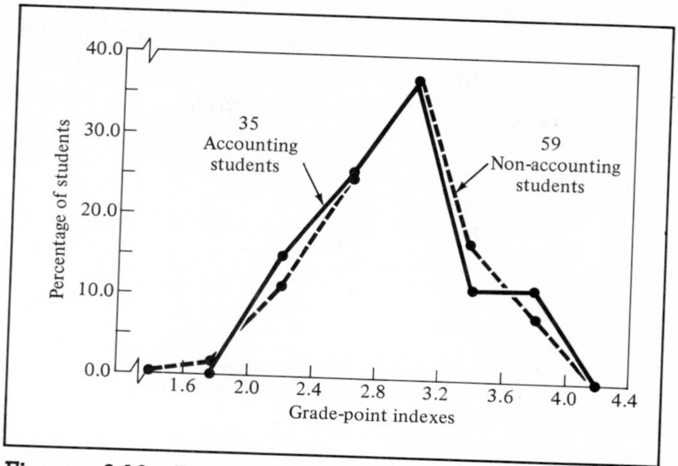

Figure 3.13 Percentage polygons of grade-point indexes of 35 accounting students and 59 non-accounting students.

SOURCE: Data are taken from Tables 3.6 and 3.7.

of cumulative polygons. Both may be developed from the frequency distribution table (Table 3.4), the relative frequency distribution table (Table 3.5), or the percentage distribution table (Table 3.6). When comparing two or more data sets with different numbers of observations either the relative frequency distribution or the percentage distribution is pertinent, again depending upon the researcher's preference for proportions or percentages. The cumulative percentage distribution given in Table 3.8 is based on the percentage distribution of the grade-point indexes of the 35 accounting majors

Table 3.8
CUMULATIVE PERCENTAGE DISTRIBUTION OF GRADE-POINT INDEXES OF 35 ACCOUNTING MAJORS

	Percentage of Students	
Grade-Point Index	*"Less Than" Indicated Value*	*"Equal to or More Than" Indicated Value*
2.0	0.0	100.0
2.4	14.3	85.7
2.8	40.0	60.0
3.2	77.1	22.9
3.6	88.6	11.4
4.0	100.0	0.0

SOURCE: Data are taken from Table 3.6.

Table 3.9
CUMULATIVE PERCENTAGE DISTRIBUTION OF GRADE-POINT INDEXES OF 59 NON-ACCOUNTING MAJORS

	Percentage of Students	
Grade-Point Index	*"Less Than" Indicated Value*	*"Equal to or More Than" Indicated Value*
1.6	0.0	100.0
2.0	1.7	98.3
2.4	11.9	88.1
2.8	37.3	62.7
3.2	74.6	25.4
3.6	91.5	8.5
4.0	100.0	0.0

SOURCE: Data are taken from Table 3.7.

presented in Table 3.6. Similarly, the cumulative percentage distribution shown in Table 3.9 is constructed from the percentage distribution of the grade-point indexes of the 59 non-accounting students shown in Table 3.7.

To construct a cumulative percentage distribution table, record the lower boundaries of each class in the percentage distribution table and "add in" an extra boundary at the end. To compute the cumulative percentages in the "less than" column, examine a particular lower boundary and determine the percentage of observations less than that boundary. Thus using Table 3.6, 0.0% of the observations are less than 2.0; 14.3% of the observations are less than 2.4; 40.0% of the observations are less than 2.8, and so on until all 100.0% of the observations are less than 4.0. The "equal to or more than" column is constructed for each lower boundary value by determining the percentage of observations equal to or greater than that particular value. Hence all 100.0% of the observations are greater than or equal to 2.0; 85.7% of the observations are greater than or equal to 2.4; 60.0% of the observations are greater than or equal to 2.8, and so on until 0.0% of the observations are greater than or equal to 4.0. As a check, for each lower boundary value, the percentage of observations "less than" the value and the percentage of obser-

Table 3.10
FORMING THE CUMULATIVE PERCENTAGE DISTRIBUTION

From Table 3.6		From Table 3.8		
			Percentage of Students	
Grade-Point Index	Percentage of Students	Grade-Point Index	*"Less Than" Indicated Value*	*"Equal to or More Than" Indicated Value*
2.0 but under 2.4	**14.3**	2.0	0.0	100.0
2.4 but under 2.8	25.7	**2.4**	**14.3**	**85.7**

vations "equal to or more than" the value must add up to 100.0%. This is logical; if, for example, as shown in Table 3.6, no observations are less than 2.0 and if 14.3% of the observations are between 2.0 and under 2.4, then (see Table 3.8) 14.3% of the observations are less than 2.4, and the remainder, 85.7% of the observations, must be greater than or equal to 2.4 (Table 3.10).

To construct a cumulative percentage polygon or *ogive*, the phenomenon of interest—the grade-point indexes—is again plotted on the horizontal axis while the cumulative percentages (from the "less than" column and from the "equal to or more than" column) are plotted on the vertical axis. The values are plotted at the lower boundaries as listed in the cumulative percentage distribution table (see Table 3.8). Figure 3.14 presents the cumulative percentage polygon of grade-point indexes of 35 accounting students.

It is noted from Figure 3.14 that the two curves are comprised of a sequence of straight-line segments connecting the plotted points and intersecting at the 50.0% value. It is further noted that if a mirror were held horizontally along this 50.0% axis, the reflection above would match the reflection below. The "less than" curve is always rising (or horizontal) while the "equal to or more than" curve is always declining (or horizontal).

The major advantage of the ogive is the ease with which we can interpolate between the plotted points. The researcher, for example, might wish to approximate the percentage of students with grade-point indexes less than a specified value, say 2.5. A vertical line is projected upward at 2.5 until it intersects the "less than" curve, and the desired percentage is then approxi-

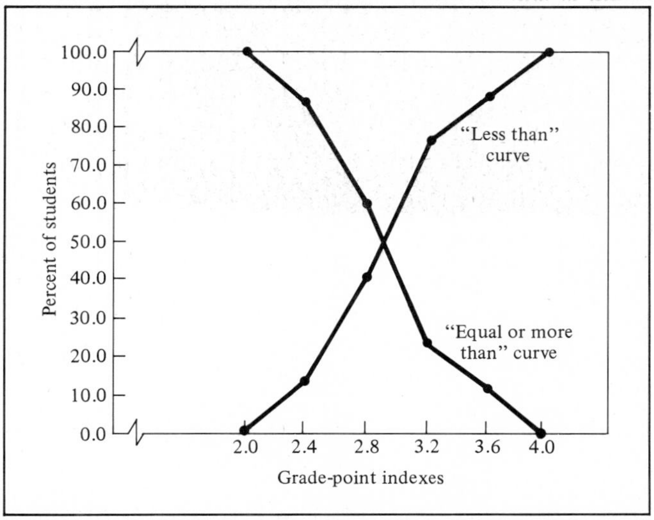

Figure 3.14 Cumulative percentage polygon (ogive) of grade-point indexes of 35 accounting students.

SOURCE: Data are taken from Table 3.8.

mated by reading horizontally to the percentage axis. In this case approximately 20.0% of the accounting students have indexes less than 2.5. (This implies that 80.0% of the accounting students have indexes of 2.5 or more. This is obtained by projecting the vertical line upward at 2.5 until it crosses the "equal to or more than" curve and then reading horizontally to the percentage axis.) As an additional example, the researcher may wish to approximate the grade-point index for which various selected percentages of students have indexes less than that amount. Some commonly considered percentage points (see Chapter 4) are the 25.0% value, the 50.0% value, and the 75.0% value. Starting with a desired percentage point, say 25.0%, a horizontal projection is made until it intersects the "less than" curve. Then the desired grade-point index is approximated by dropping a perpendicular (a vertical line) at the point of intersection, and the result is read along the horizontal axis. In this case 25.0% of the accounting students have, approximately, grade-point indexes less than 2.57. At 50.0% the approximate grade-point-index value is 2.91, while at 75.0% the approximate value is 3.18.

Such approximations as the above are extremely helpful when comparing two or more sets of data. Figure 3.15 presents the cumulative percentage polygons for grade-point indexes of 35 accounting students versus 59 non-accounting students.

From Figure 3.15 it is seen that 25.0% of the non-accounting majors have, approximately, grade-point indexes less than 2.60, 50.0% have, approxi-

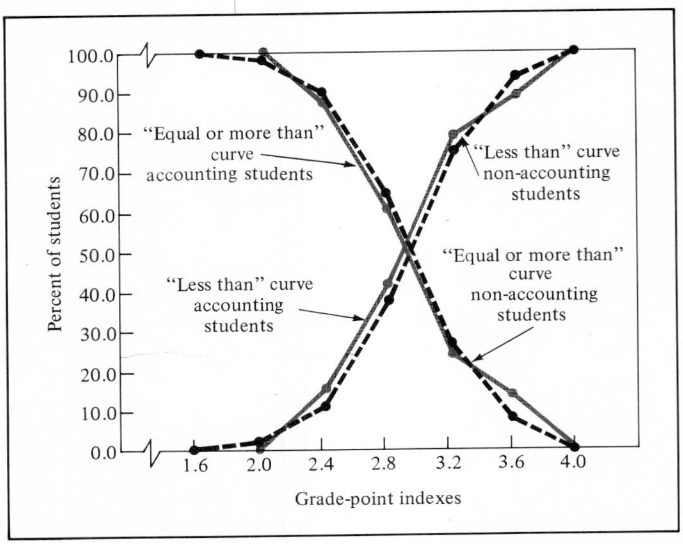

Figure 3.15 Ogives of grade-point indexes of 35 accounting students versus 59 non-accounting students.

SOURCE: Data are taken from Tables 3.8 and 3.9.

mately, indexes less than 2.94, while 75.0% have, approximately, indexes less than 3.22. For a specific index, say 2.5, the researcher may project upward to the "less than" curves and then approximate the desired percentages by reading horizontally to the percentage axis. While approximately 20.0% of accounting majors have indexes less than 2.5, only 16.0% of the non-accounting students have indexes less than 2.5. All these comparisons indicate the similarities in grade-point indexes between the two groups of students with very slight advantages going to the non-accounting majors.

3.4 PRESENTATION OF DATA: A SUMMARY AND OVERVIEW

In this chapter both qualitative and quantitative types of data have been appropriately summarized and presented in the form of tables and charts in order to make the data more manageable and meaningful. In the next chapter a variety of descriptive measures which are helpful for analyzing and interpreting data will be discussed. However, it would be inappropriate, especially for an introductory course in statistics, not to mention some of the problems encountered by the researcher in constructing tables and charts. In fact, it would be misleading if the reader were left with the impression that data presentation was always so clear-cut as described in this chapter. The following section, which may be used primarily for reference, is given here as optional reading.

3.5 OPTIONAL TOPIC—PROBLEMS WITH DATA PRESENTATION

Among the many problems faced by the statistician when presenting data into tables and charts are:

1. The subjectivity in the selection of classes or class limits and boundaries
2. The need for comparing two sets of data on a relative (that is, proportions or percentages) basis
3. The construction of charts from frequency distribution tables having unequal class-interval widths
4. The treatment of frequency distribution tables having open-ended classes
5. The examination of differences between class limits and class boundaries based on the method used for recording data

3.5.1 Subjectivity in Selecting Classes

A major purpose of data presentation is to be informative, not to mislead. Unfortunately, however, the selection of classes or class limits and boundaries for tables and charts is very subjective. Hence for data sets which do not contain very large numbers of observations, the choice of a particular set of classes or a particular set of class limits and boundaries over another might yield an entirely different picture to the reader. For example, for the data on grade-point indexes, using a class interval width of .3 instead of .4 (as was used in Tables 3.6 and 3.7) may cause shifts in data concentration—especially if the number of observations is not very large. However, such shifts in data concentration do not only occur because the width of the class interval is altered. The researcher may keep the interval width at .4 but choose different lower and upper class limits and boundaries. Such manipulation may also cause shifts in data concentration—especially if the number of observations is not very large.

Now if the purpose of data presentation in terms of tables and charts is to provide summary information in a meaningful and manageable manner, how can the researcher determine, from among different possible tables and charts for the same set of data, which is the most appropriate presentation? Since this is subjective, it can only be suggested that the researcher, using his or her experience or intuition, select that set which is believed to most adequately represent the original (raw) data. Fortunately, as the number of observations increases, alterations in the size of the interval, the number of classes, or in the particular selection of limits and boundaries affect the concentration of data less and less.

3.5.2 Data Comparisons on a Relative Basis

Throughout this chapter comparisons between two groups of data with differing numbers of observations have been made on a relative basis. Using frequencies rather than proportions or percentages would be misleading. To show this, Figures 3.16 and 3.17 are the frequency polygons and the cumula-

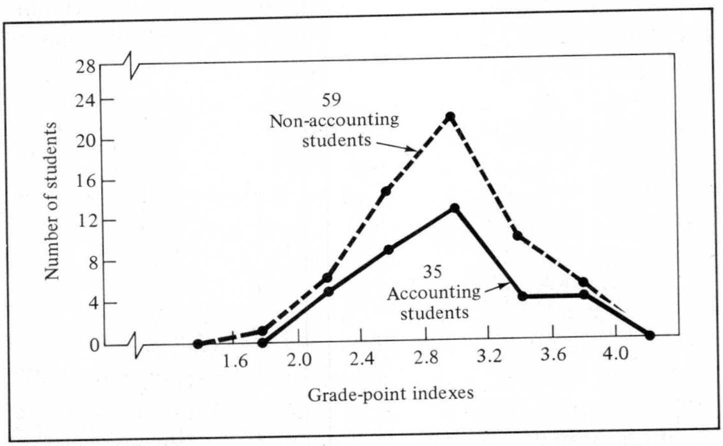

Figure 3.16 "Improper" frequency polygons of grade-point indexes of 35 accounting students and 59 non-accounting students.

SOURCE: Data are taken from Tables 3.4 and 3.7.

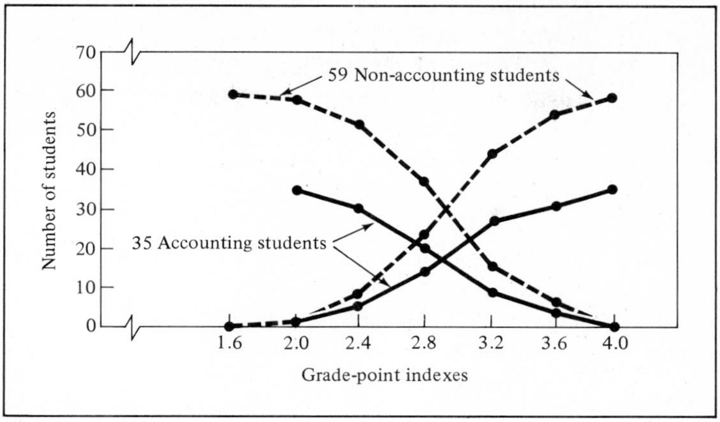

Figure 3.17 "Improper" cumulative frequency polygons of grade-point indexes of 35 accounting students and 59 non-accounting students.

SOURCE: Data are taken from Tables 3.4 and 3.7.

tive frequency polygons of the grade-point indexes of 35 accounting majors versus 59 non-accounting students. The proper charts were presented as Figures 3.13 and 3.15.

As can be seen from Figures 3.16 and 3.17, the polygons (frequency and cumulative frequency) for the 35 accounting students are "overwhelmed" by those for the 59 non-accounting students. Therefore no meaningful comparisons can be made from such distorted charts.[4] Imagine how grossly exaggerated such charts would be if even larger differences were to exist between the numbers of observations in the data sets? The necessity for making such comparisons on a relative basis using proportions or percentages is clearly demonstrated.

3.5.3 Distributions Having Unequal Class-Interval Widths

When constructing the frequency distribution it sometimes happens that the data are grouped into classes with varying class-interval sizes. This usually occurs when the phenomenon of interest, say reported annual income, is likely to have some observations that are extremely spread out at the high end. Rather than having either a few very wide class intervals of equal size (in order to cover the entire range of the data) or too many equal-sized class intervals that are much narrower in width, such tables are often constructed with varying class-interval widths. As an example, Table 3.11 depicts the expected starting annual salaries of the 35 accounting students if they were to seek employment immediately after obtaining their baccalaureate.

[4] Moreover, it is also clear from Figures 3.9 and 3.10 that neither the side-by-side stem-and-leaf display nor the back-to-back stem-and-leaf display overcome this problem of data comparison when the sample sizes differ.

Table 3.11
EXPECTED STARTING SALARIES OF
35 ACCOUNTING STUDENTS

Salary ($)	Number of Students
6,000 but less than 9,000	1
9,000 but less than 12,000	7
12,000 but less than 15,000	18
15,000 but less than 21,000	8
21,000 but less than 33,000	1
Total	35

SOURCE: Data are taken from Figure 2.6.

Caution must be observed, however, when constructing charts or when making descriptive calculations (Chapter 4) from such a distribution. If, for example, the researcher were merely to plot the frequency histogram from class boundary to boundary as in Figure 3.18, a distortion would result because the areas under the rectangular bars for classes with wider interval widths would be overexaggerated.

The proper way to compensate for varying class-interval widths is to ensure that the height of each rectangular bar is represented on a **frequency per standard interval** basis or a **percentage per standard interval** basis, depending on whether the constructed chart is to be a frequency histogram or a percentage histogram. The same holds true when constructing the polygons. This adjustment is seen from Table 3.12.

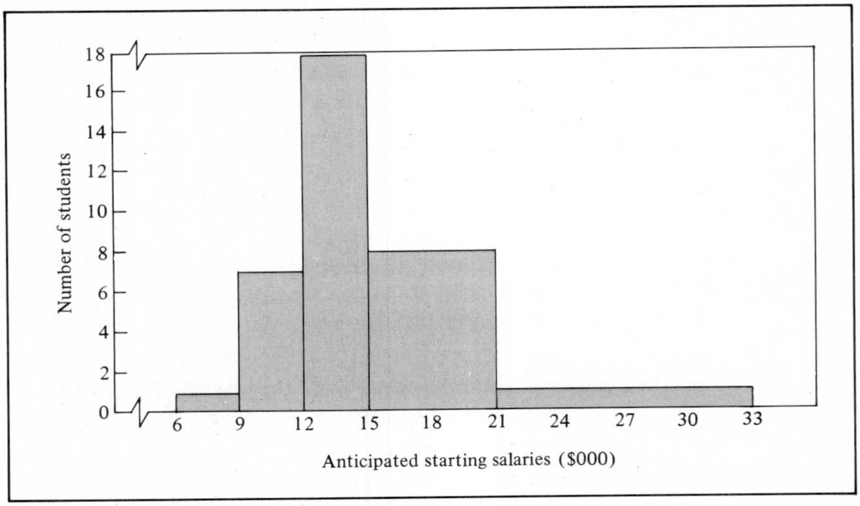

Figure 3.18 "Improper" frequency histogram of anticipated starting salaries of 35 accounting majors.

SOURCE: Data are taken from Table 3.11.

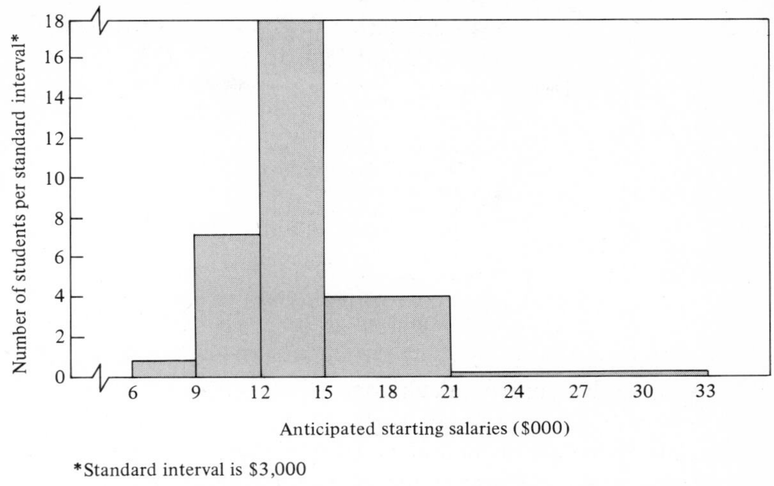

Figure 3.19 Frequency histogram of anticipated starting salaries of 35 accounting majors.

SOURCE: Data are taken from Table 3.11.

Table 3.12
OBTAINING FREQUENCIES PER STANDARDIZED INTERVAL

(1) Class Boundaries	(2) Midpoints	(3) Width of Interval	(4) Number of Standardized Intervals	(5) Frequencies per Interval	(6) Frequencies per Standardized Interval
$6,000 but less than $9,000	$ 7,500	$ 3,000*	1	1	1
9,000 but less than 12,000	10,500	3,000*	1	7	7
12,000 but less than 15,000	13,500	3,000*	1	18	18
15,000 but less than 21,000	18,000	6,000	2	8	4
21,000 but less than 33,000	27,000	12,000	4	1	.25

* The most common interval width, $3,000, is chosen as the standard interval.

Hence rather than plotting frequencies per interval (column 5) as in Figure 3.18, a standard interval width of $3,000 is selected (column 3) as most typical and used to adjust the frequencies on a standard interval basis (column 6). Figure 3.19 then presents the data from Table 3.11 wherein these adjustments have been properly accounted for. Of course, if all classes were of equal-sized interval widths, the frequencies per standardized interval (column 6) would be the same.

3.5.4 Distributions with Open-Ended Class Intervals

Sometimes when constructing frequency distribution tables, some observations are so extreme that rather than choosing classes of varying interval sizes, it

Table 3.13
EXPECTED STARTING SALARIES OF 59 NON-ACCOUNTING STUDENTS

Salary ($)	Number of Students
6,000 but less than 9,000	1
9,000 but less than 12,000	16
12,000 but less than 15,000	16
15,000 but less than 18,000	20
18,000 but less than 21,000	4
21,000 or more	2
Total	59

SOURCE: Data are taken from Figure 2.6.

seems more appropriate to express the last class grouping with an "open end." Constructing such open-ended classes, however, presents problems not only in charting, but also in computing various descriptive summary measures which are useful in analyzing the data (Chapter 4).

As an example, Table 3.13 presents the expected starting annual salaries of the 59 non-accounting students if they were to seek employment immediately following their baccalaureate.

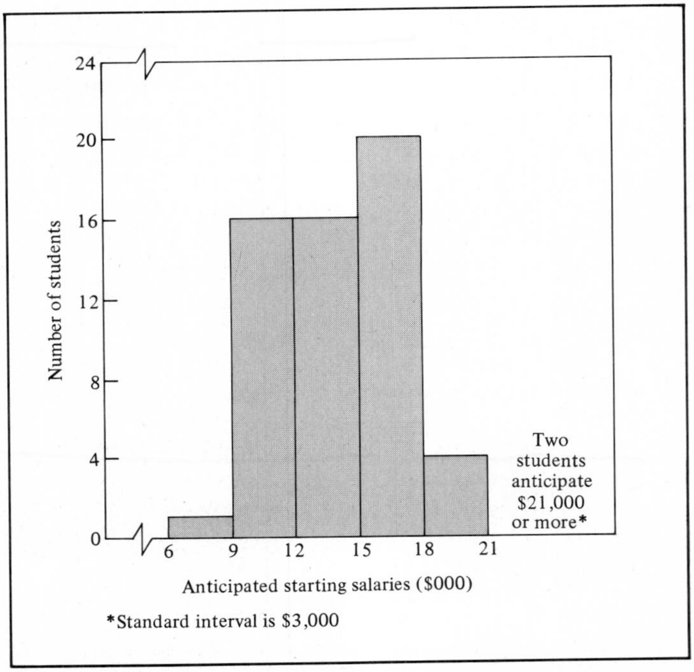

Figure 3.20 Frequency histogram of anticipated starting salaries of 59 non-accounting majors.

SOURCE: Data are taken from Table 3.13.

It is shown from Figure 3.20 that when constructing the histogram for these data, the open-ended class is referenced but not plotted.

3.5.5 Recording Data: Differences Between Class Limits and Class Boundaries

As long as the problem of overlapping classes (Section 3.3) is avoided when constructing frequency distribution tables, for convenience no distinction has been drawn between the class limits and the class boundaries. In general, however, whether or not there is a difference between the limits and the boundaries of the classes is based upon the method used for recording data. For example, if data are collected on the weights of a group of children, and if the weights are recorded to the *last full pound,* children reported as 60 pounds would have weights varying between exactly 60 pounds and 61 pounds and, as a group, they would average 60.5 pounds. Thus when constructing the frequency distribution, if it were deemed desirable to have class-interval widths of 10 pounds (and the weights are reported to the last full pound), it would be correct to write:

Class Boundaries	Class Limits	Class Midpoints
60.0–69.99999 . . .	60 but less than 70	65.0
70.0–79.99999 . . .	70 but less than 80	75.0
80.0–89.99999 . . .	80 but less than 90	85.0

In such cases the class boundaries and class limits are the same.

On the other hand, if the data are recorded to the *nearest pound,* children reported as 60 pounds would have their weights vary between 59.5 pounds and 60.5 pounds, and as a group, they would average 60.0 pounds. Hence when constructing the frequency distribution, if it were deemed desirable to have class-interval widths of 10 pounds (and the weights are reported to the nearest pound) it would be correct to write:

Class Boundaries	Class Limits	Class Midpoints
59.5–69.49999 . . .	60–69	64.5
69.5–79.49999 . . .	70–79	74.5
79.5–89.49999 . . .	80–89	84.5

In such cases the class boundaries and class limits differ and the class midpoints (class marks) are midway between the various class boundaries.

However, if measurements are quite precise and recorded to the *nearest tenth of a pound,* children reported as 60 pounds would have their weights vary between 59.95 pounds and 60.05 pounds, and, as a group, they would average 60.00 pounds so that:

Class Boundaries	Class Limits	Class Midpoints
59.95–69.94999 . . .	60.0–69.9	64.95
69.95–79.94999 . . .	70.0–79.9	74.95
79.95–89.94999 . . .	80.0–89.9	84.95

Thus it is seen that while the class boundaries and the class limits are different, such differences become negligible as more precise measurements are recorded.

This section has been concerned with special kinds of problems facing the statistician when presenting data into tables and charts. The following chapter deals with the understanding and computation of several kinds of descriptive summary measures which will be helpful in characterizing or analyzing either mass data cast into tables and charts or data merely collected in their raw form.

PROBLEMS

3.1 (a) Explain the differences between raw data and an ordered array.
(b) Why is it advantageous to use a stem-and-leaf display instead of an ordered array?

3.2 Explain the differences between frequency distributions, relative frequency distributions, and percentage distributions.

3.3 When comparing two or more sets of data with different sample sizes, why is it necessary to compare their respective relative frequency or percentage distributions?

3.4 Explain the differences between histograms, polygons, and ogives (cumulative polygons).

3.5 Explain the differences between bar charts and histograms.

3.6 Explain the differences between bar charts and pie diagrams.

* 3.7 A well-known newspaper conducted a telephone poll of New Yorkers' attitudes toward New York City. A total of 419 persons were selected in a simple random sample. The following data reflect the responses to a question regarding the adequacy of police and fire protection:

Is the police and fire protection in your neighborhood adequate?	
Yes	293
No	80
Don't know or refused to answer	46
Total	419

(a) Convert the data to percentages and construct:

(1) A percentage component bar chart.
(2) A bar chart.
(3) A pie diagram.

(b) Which of these charts do you prefer to use here? Why?

3.8 The board of directors of a large housing cooperative wish to investigate the possibility of hiring a supervisor for an outdoor playground. All 616 households in the cooperative were polled, with each household having but one vote, regardless of its size. The following data were collected:

Should the co-op hire a supervisor?	
Yes	146
No	91
Not sure	58
No response	321
Total	616

(a) Convert the data to percentages and construct:

(1) A percentage component bar chart.
(2) A bar chart.
(3) A pie diagram.

(b) Which of these charts do you prefer to use here? Why?
(c) Eliminating the "no response" group convert the 295 responses to percentages and construct:

(1) A percentage component bar chart.
(2) A bar chart.
(3) A pie diagram.

(d) Based on your findings in (a) and (c), what would you recommend that the board of directors do?

3.9 For the 94 students in the sample taken by the researcher and the dean:
(a) Use the data for question 2 (Figure 2.6) to determine the breakdown by class designation—freshman, sophomore, junior, and senior.
(b) Convert the data to percentages and construct:

(1) A percentage component bar chart.
(2) A bar chart.
(3) A pie diagram.

3.10 For the 94 students in the sample taken by the researcher and the dean:
(a) Use the data for question 3 (Figure 2.6) to determine the breakdown by graduate school intention—yes, no, not sure.
(b) Convert the data to percentages and construct:

(1) A percentage component bar chart.
(2) A bar chart.
(3) A pie diagram.

3.11 For the 94 students in the sample taken by the researcher and the dean, select from among the other questions with qualitative responses (questions 1, 9,

10, 11, 12, 14, 18, 19, 20, 21, 22) in the questionnaire (Figure 2.2) and use the appropriate codes in the columns shown in Figures 2.4 and 2.6 to:

(a) Determine the breakdowns and convert to percentages.

(b) Construct either a bar chart or a pie diagram.

3.12 Starting in the row of the table of random numbers (Appendix E, Table E.1) corresponding to the day of the month in which you were born (for example, March 8th is row 08, December 29th is row 29), and using four-digit coded sequences, select your own simple random sample, without replacement, of size $n = 50$ from the population data base of $N = 2{,}202$ students presented in Appendix A.

(a) Use your data to do Problem 3.9.

(b) Use your data to do Problem 3.10.

(c) Use your data to do Problem 3.11.

3.13 (**Class Project**) Let each student in your class fill out the questionnaire (Figure 2.2), code the responses as in Figure 2.4, keypunch the data as in Figure 2.5, and then have your teacher collect the cards to obtain listings for your class as in Figure 2.6.

(a) For your own class do Problem 3.9.

(b) For your own class do Problem 3.10.

(c) For your own class do Problem 3.11.

3.14 For the 94 students in the sample taken by the researcher and the dean:

(a) Use the data for questions 2 and 3 in Figure 2.6 *jointly* and form the cross-classification table for class designation and graduate school intention.

(b) Analyze the findings.

3.15 Use your data from Problem 3.12 to do Problem 3.14.

3.16 Use the data from your own class from Problem 3.13 to do Problem 3.14.

3.17 If the annual salaries of city employees varied from \$7,400 to \$40,200:

(a) Indicate the class limits or boundaries of 10 classes into which these values can be grouped.

(b) What class-interval width did you choose?

(c) What are the 10 class midpoints (class marks)?

3.18 Given the sets of data in Table 3.14.

(a) Form the frequency distributions and percentage distributions for each group.

(b) Plot the frequency histograms for each group.

(c) On one graph, plot the percentage polygons for each group.

(d) Form the cumulative percentage distributions for each group.

(e) On one graph, plot the ogives (cumulative percentage polygons) for each group.

(f) Write a brief report comparing and contrasting the two groups.

3.19 Given the sets of data in Table 3.15.

(a) For each group place the raw data in ordered arrays.

Table 3.14
ORDERED ARRAYS OF BATTING AVERAGES*
(Samples of 40 Players from Each League)

American League		National League	
.184	.255	.201	.263
.204	.268	.205	.263
.207	.271	.214	.265
.209	.272	.220	.274
.210	.278	.227	.275
.219	.279	.233	.279
.220	.280	.234	.287
.220	.282	.242	.290
.225	.291	.242	.300
.227	.292	.245	.306
.227	.294	.247	.308
.235	.298	.248	.310
.240	.299	.250	.319
.241	.301	.253	.320
.247	.302	.253	.320
.248	.313	.253	.338
.252	.316	.253	.342
.253	.321	.261	.356
.255	.327	.261	.359
.255	.348	.261	.375

* For individuals with at least 75 "at bats."

Table 3.15
RAW DATA FOR RANDOM SAMPLES OF GAS MILEAGE PER GALLON ATTAINED BY 50 COMPACT CARS OF TYPE A AND TYPE B IN CITY DRIVING

Compact Car Type A		Compact Car Type B	
29.9	23.4	23.0	20.4
22.5	18.7	12.6	20.4
18.5	19.6	24.4	17.2
25.3	28.2	14.2	20.6
25.7	12.2	28.8	15.5
17.5	15.8	15.2	17.9
23.2	19.8	20.8	13.7
20.1	26.8	15.4	22.8
28.1	17.5	15.3	20.9
17.4	27.3	20.6	21.9
24.0	11.3	16.6	18.9
24.7	24.8	22.2	21.3
22.1	25.3	20.8	21.7
20.7	25.9	21.2	14.5
15.5	26.5	19.1	16.8
25.5	24.2	22.0	13.9
26.1	28.9	26.2	16.7
24.8	18.6	17.8	17.6
12.5	19.7	10.8	16.8
24.7	15.2	18.7	20.5
15.6	22.1	15.2	15.2
29.5	29.9	17.6	21.2
16.3	12.3	21.2	19.5
19.3	12.3	17.4	16.4
11.8	11.4	15.8	15.2

(b) Place the raw data in stem-and-leaf displays.

Hint: For the displays, let the leaves be the tenths digits.

(c) Form the frequency distributions and percentage distributions for each group.

(d) Plot the frequency histograms for each group.

(e) On one graph, plot the percentage polygons for each group.

(f) Form the cumulative percentage distributions for each group.

(g) On one graph, plot the ogives (cumulative percentage polygons) for each group.

(h) Write a brief report comparing and contrasting the two groups.

3.20 Given the sets of data in Table 3.16.

(a) Using interval widths of $10 form the frequency distributions and percentage distributions for each group.

(b) Plot the frequency histograms for each group.

(c) On one graph, plot the percentage polygons for each group.

(d) Form the cumulative percentage distributions for each group.

Table 3.16
RAW DATA FOR RANDOM SAMPLES OF CLOSING PRICES OF ISSUES TRADED ON THE AMERICAN AND NEW YORK STOCK EXCHANGES

American Exchange (25 Issues)	New York Exchange (50 Issues)	
$6.88	$36.50	$26.00
.75	23.50	19.00
3.88	8.25	46.00
4.12	57.50	23.50
11.88	27.12	22.62
15.88	3.75	12.88
16.50	25.00	5.50
8.75	15.50	37.50
9.25	36.12	9.88
7.50	6.00	59.12
5.38	9.12	35.25
14.38	33.38	20.62
2.50	22.50	24.00
4.88	8.75	80.50
6.38	8.62	29.38
33.62	5.75	3.75
4.88	21.88	64.75
9.00	6.12	14.25
2.00	25.00	46.38
20.00	15.88	4.75
14.25	24.00	25.00
4.00	10.88	35.00
15.25	18.75	9.00
2.38	53.88	12.38
49.50	20.38	31.00

(e) On one graph, plot the ogives (cumulative percentage polygons) for each group.

(f) Write a brief report comparing and contrasting the two groups.

* 3.21 For the 94 students in the sample taken by the researcher and the dean:

(a) Use the data for question 6 (Figure 2.6) to list the high-school averages of the students.

(b) Construct the ordered array of high-school averages and the stem-and-leaf display. **Hint:** For the display, let the leaves be the units digits of the high-school averages.

(c) Form the frequency distribution and percentage distribution.

(d) Plot the frequency histogram.

(e) Plot the percentage polygon.

(f) Form the cumulative percentage distribution.

(g) Plot the ogive (cumulative percentage polygon).

(h) What is the "most typical" high-school average?

(i) What is the range of high-school averages?

(j) What percentage of students had high-school averages less than 80?

(k) 85% of the students had high-school averages less than what value?

3.22 Use your data from Problem 3.12 to do Problem 3.21. Compare and contrast the results with those from Problem 3.21.

3.23 (**Class Project**) Use the data from your own class from Problem 3.13 to do Problem 3.21. Compare and contrast the results with those from Problem 3.21.

3.24 From the data in Figure 3.1 construct the frequency distributions and percentage distributions for the grade-point indexes of the 35 accounting students and the 59 non-accounting students using class-interval widths of .3 and starting at 1.6 but less than 1.9, 1.9 but less than 2.2, etc. Compare your results to Tables 3.4, 3.6, and 3.7. Verify shifts in data concentration as discussed Section 3.5.

3.25 From the data in Figure 3.1 construct the frequency distributions and percentage distributions for the grade-point indexes of the 35 accounting students and the 59 non-accounting students using class-interval widths of .4 and starting at 1.5 but less than 1.9, 1.9 but less than 2.3, etc. Compare your results to Tables 3.4, 3.6, and 3.7. Verify shifts in data concentration as discussed in Section 3.5.

3.26 Figure 3.21 contains the cumulative relative frequency polygons of family incomes for two random samples (A and B) of 200 families each—drawn from two communities. Based on these data, answer each of the following questions:

(a) How many families in sample A have incomes of $80,000 or more?

(b) What is the percentage of families in sample A with incomes less than $60,000?

(c) Which sample has a larger range of incomes?

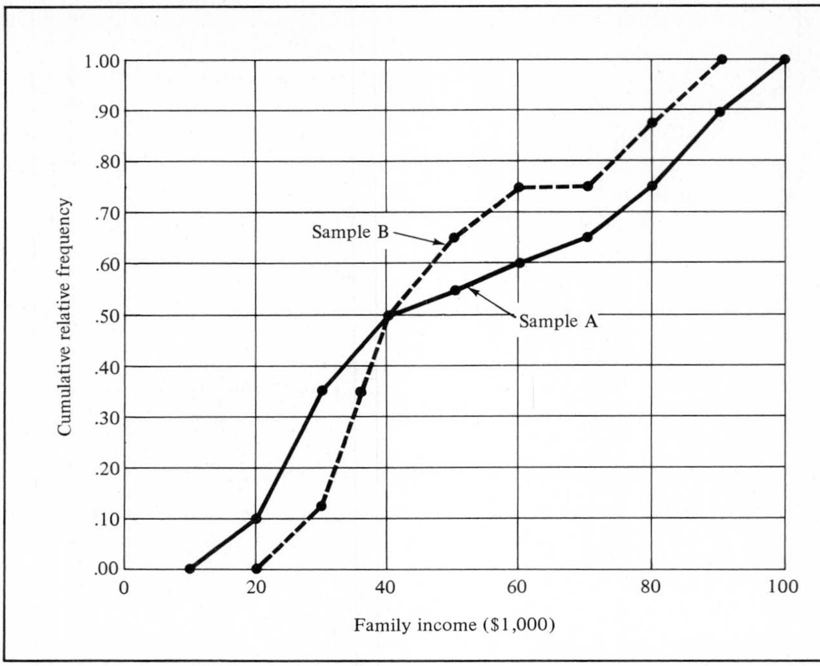

Figure 3.21 Cumulative relative frequency distribution of family incomes for two random samples.

(d) How many families in sample B have an income between $60,000 and $70,000?

(e) Does sample A or sample B have more family incomes of $40,000 or above?

(f) What percentage of sample A families earn less than $40,000?

(g) What percentage of sample A families earn $40,000 or more?

(h) Which sample has more incomes below $80,000?

3.27 A wholesale appliance distributing firm wished to study its accounts receivable for two successive months. Two independent samples of 50 accounts were selected for each of the two months. The results have been summarized in Table 3.17.

(a) Plot the frequency histograms for each month.

(b) On one graph, plot the percentage polygons for each month.

(c) Form the cumulative percentage distributions for each month.

(d) On one graph, plot the ogives (cumulative percentage polygons) for each month.

(e) Write a brief report comparing and contrasting the accounts receivable of the two months.

Table 3.17
FREQUENCY DISTRIBUTIONS FOR ACCOUNTS RECEIVABLE

Amount	Month 1 Frequency	Month 2 Frequency
\$0–under \$2,000	6	10
\$2,000–under \$4,000	13	14
\$4,000–under \$6,000	17	13
\$6,000–under \$8,000	10	10
\$8,000–under \$10,000	4	0
\$10,000–under \$12,000	0	3
Totals	50	50

3.28 A home-heating-oil delivery firm wished to compare how fast the oil bills were paid in two different suburbs. A random sample of 50 vouchers from suburb A and 100 vouchers from suburb B were selected, and the number of days between delivery and payment were recorded as shown in Table 3.18.

Table 3.18
PAYMENT RECORDS OF HOME-HEATING-OIL BILLS

Number of Days	Suburb A Frequency	Suburb B Frequency
0–4	4	6
5–9	14	21
10–14	16	24
15–19	10	30
20–24	5	7
25–29	1	6
30–59	0	6
Totals	50	100

(a) Plot the percentage histograms for each suburb.
(b) On one graph, plot the percentage polygons for each suburb.
(c) Form the cumulative percentage distributions for each suburb.
(d) On one graph, plot the ogives (cumulative percentage polygons) for each suburb.
(e) Write a brief report comparing and contrasting the number of days between delivery and payment in the two suburbs.

REFERENCES

1. ARKIN, H., AND R. COLTON, *Statistical Methods,* 5th ed. (New York: Barnes & Noble College Outline Ser., 1970).
2. CROXTON, F., D. COWDEN, AND S. KLEIN, *Applied General Statistics,* 3rd ed. (Englewood Cliffs, N.J.: Prentice-Hall, 1967).

3. GRIFFIN, J. I., *Statistics: Methods and Applications* (New York: Holt, Rinehart and Winston, 1962).
4. NETER, J., W. WASSERMAN, AND G. WHITMORE, *Fundamental Statistics for Business and Economics,* 4th ed. (Boston: Allyn and Bacon, 1973).
5. TUKEY, J., *Exploratory Data Analysis* (Reading, Mass.: Addison-Wesley, 1977).
6. YAMANE, T., *Statistics, An Introductory Analysis* (New York: Harper and Row, 1964).

CHAPTER 4

DATA CHARACTERISTICS: DESCRIPTIVE SUMMARY MEASURES

4.1 INTRODUCTION

In the previous chapters data were collected and appropriately summarized into tables and charts. In this chapter a variety of descriptive summary measures will be developed. These descriptive measures are useful for analyzing and interpreting quantitative data, whether collected in raw form (**ungrouped data**) or summarized into frequency distributions (**grouped data**). The chapter will be developed in two parts. First, a conceptual understanding of several often used descriptive summary measures will be developed for numerical data collected in their raw form. Second, for data grouped into frequency distributions (as in the problem of interest to the researcher and the dean) analogous formulas will be developed for obtaining these various descriptive summary measures, and, in addition, a graphical approach using the charts constructed in Chapter 3 will also be demonstrated where appropriate.

4.2 PROPERTIES OF DATA

In descending order of importance, the three major properties or characteristics which describe a set of data pertaining to some numerical random variable or phenomenon of interest are:

1. Location
2. Dispersion
3. Shape

In any analysis and/or interpretation of numerical data, a variety of descriptive measures representing the properties of location, dispersion, and shape may be used to extract and summarize the salient features of the data set. If these descriptive summary measures are computed from a sample of data they are called **statistics;** if these descriptive measures are computed from an entire population of data they are called **parameters.** Since statisticians usually take samples rather than use entire populations, our primary emphasis deals with statistics rather than parameters.

4.3 MEASURES OF LOCATION

The most important characteristic that describes or summarizes a set of data is its **location.** Most sets of data show a distinct tendency to group or cluster about a certain point. Thus for any particular set of data, it usually becomes possible to select a typical value to describe or summarize the entire set of data. Such a descriptive typical value is called an *average*. It is a measure of central tendency or location. In the hypothetical diagram depicted in Figure 4.1 the two polygons are shown to be identical except for location, that is, the points of central tendency are located at different positions on the scale.

The three primary measures of location or central tendency are the **arithmetic mean,** the **median,** and the **mode.**

4.3.1 The Arithmetic Mean

The arithmetic mean is the best known, most commonly used average or measure of central tendency. The mean is easy to calculate from data either collected in raw form or placed in an ordered array. The statistic, $\bar{X}$, the arithmetic mean, is found by summing all the values in the sample and then dividing this total by the number of observations in the sample. Thus for a set of n values $X_1, X_2, X_3, \ldots, X_n$ in the sample

$$\bar{X} = \frac{X_1 + X_2 + \cdots + X_n}{n}$$

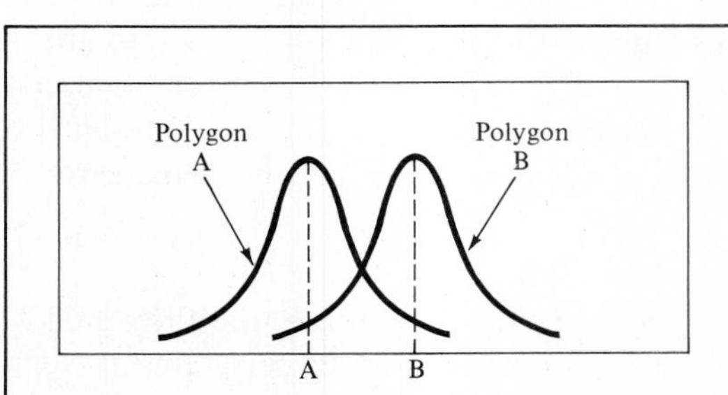

Figure 4.1 Hypothetical polygons differing only in location.

To simplify such an expression, the term $\sum_{i=1}^{n} X_i$ (meaning the "summation of") is conventionally used whenever we wish to add together a series of values.[1] That is,

$$\sum_{i=1}^{n} X_i = X_1 + X_2 + \cdots + X_n$$

Therefore, using this summation notation, the arithmetic mean can be more simply expressed as

$$\bar{X} = \frac{\sum_{i=1}^{n} X_i}{n} \qquad (4.1)$$

where $\bar{X}$ = sample arithmetic mean
n = sample size
X_i = ith observation of the random variable X
$\sum_{i=1}^{n}$ = Greek symbol meaning "summation of" all values from 1 to n (see Appendix B, Section B.3)

As an example, suppose the following data represent the hourly wage rates of six executive secretaries selected in a random sample from a very large company in New York City.

X_1 = \$ 9.50 = hourly wage rate of the *first* secretary in the sample
X_2 = \$ 3.00 = hourly wage rate of the *second* secretary in the sample
X_3 = \$10.00 = hourly wage rate of the *third* secretary in the sample
X_4 = \$ 9.50 = hourly wage rate of the *fourth* secretary in the sample
X_5 = \$ 8.50 = hourly wage rate of the *fifth* secretary in the sample
X_6 = \$ 7.50 = hourly wage rate of the *sixth* secretary in the sample

The mean for this sample is calculated as

$$\bar{X} = \frac{\sum_{i=1}^{n} X_i}{n} = \frac{X_1 + X_2 + X_3 + \cdots + X_n}{n}$$

$$= \frac{\$9.50 + \$3.00 + \$10.00 + \cdots + \$7.50}{6}$$

$$\bar{X} = \frac{\$48.00}{6} = \$8.00$$

[1] See Appendix B, Section B.3, for a discussion of rules pertaining to summation notation.

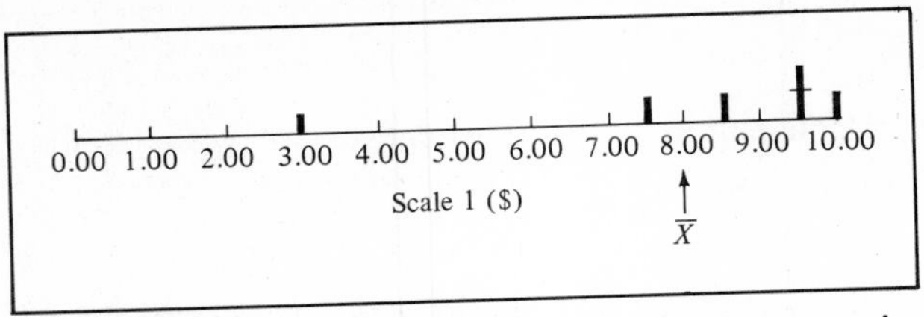

Figure 4.2 Salary scale for six executive secretaries.

Note that the mean is computed as \$8.00 even though not one individual in the sample actually obtained that wage rate. In fact, it is seen from the scale of Figure 4.2 that for this set of data four observations are larger than the mean and two are smaller. The mean acts as a *balancing point* so that observations which are larger balance out those which are smaller.

Suppose that in the same large firm the hourly wage rate of a sample of $n = 6$ senior bookkeepers yields the following results:

$$\begin{aligned} X_1 &= \$\ 9.50 \\ X_2 &= \$\ 6.50 \\ X_3 &= \$10.00 \\ X_4 &= \$\ 6.00 \\ X_5 &= \$\ 8.50 \\ X_6 &= \$\ 7.50 \end{aligned}$$

Again, the mean is \$8.00 and no individual in the sample obtained that wage rate. However, as seen from the scale of Figure 4.3, the data here are less scattered or variable than the first sample.

Suppose further that in this firm the hourly wage rates of a sample of six section managers is taken:

$$\begin{aligned} X_1 &= \$8.00 \\ X_2 &= \$8.00 \\ X_3 &= \$8.00 \\ X_4 &= \$8.00 \\ X_5 &= \$8.50 \\ X_6 &= \$7.50 \end{aligned}$$

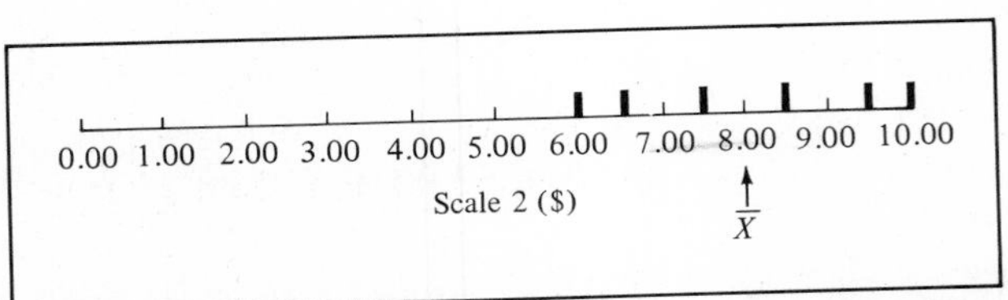

Figure 4.3 Salary scale for six senior bookkeepers.

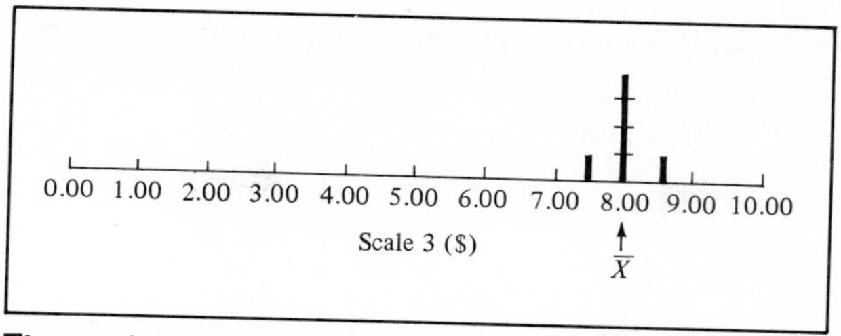

Figure 4.4 Salary scale for six section managers.

Here the mean is \$8.00—as are four of the six hourly wage rates—and this set of data is seen to be the least spread out or variable of the three sets (Figure 4.4).

Thus in each of the above three cases the statistic $\overline{X}$ has been used as a measure of central tendency—to summarize or characterize the three respective samples. In each case the mean is \$8.00, although the three samples differ in dispersion.

The arithmetic mean has four interesting properties. First of all, the sum of the deviations about the mean is zero,[2] that is,

$$\sum_{i=1}^{n} (X_i - \overline{X}) = 0$$

Recall that the mean acts as a balancing point for observations larger and smaller than it. To demonstrate this property, the sample of hourly wage rates of executive secretaries is used (Table 4.1).

Table 4.1
HOURLY WAGES OF SECRETARIES

X_i	$\overline{X}$	$(X_i - \overline{X})$
\$9.50	\$8.00	+\$1.50
3.00	8.00	− 5.00
10.00	8.00	+ 2.00
9.50	8.00	+ 1.50
8.50	8.00	+ .50
7.50	8.00	− .50
		$\sum_{i=1}^{n} (X_i - \overline{X}) = 0$

The reader may verify the results for the second and third samples (see Problem 4.1).

The second property of the mean is that the sum of the squares of the deviations about the mean is a minimum. That is, the summation of squared

[2] See Appendix B, Section B.3 for a proof that $\sum_{i=1}^{n} (X_i - \overline{X}) = 0$.

differences between each observation and its mean must be less than the summation of squared differences between each observation and any other chosen value. Thus,

$$\sum_{i=1}^{n} (X_i - \overline{X})^2 = \text{Minimum Total}$$

To demonstrate this property we return again to the (sample of $n = 6$) executive secretaries' hourly wages. Note that

$$\sum_{i=1}^{n} (X_i - \overline{X})^2$$

$$
\begin{aligned}
(9.50 - 8.00)^2 &= (+1.50)^2 = 2.25\\
(3.00 - 8.00)^2 &= (-5.00)^2 = 25.00\\
(10.00 - 8.00)^2 &= (+2.00)^2 = 4.00\\
(9.50 - 8.00)^2 &= (+1.50)^2 = 2.25\\
(8.50 - 8.00)^2 &= (+.50)^2 = .25\\
(7.50 - 8.00)^2 &= (-.50)^2 = .25
\end{aligned}
$$

$$\text{Minimum total} = \sum_{i=1}^{n} (X_i - \overline{X})^2 = 34.00$$

Now suppose that the summation of squared differences between each of the above observations and some arbitrary value, say \$8.10, is obtained.

$$\sum_{i=1}^{n} (X_i - 8.10)^2$$

$$
\begin{aligned}
(9.50 - 8.10)^2 &= (+1.40)^2 = 1.96\\
(3.00 - 8.10)^2 &= (-5.10)^2 = 26.01\\
(10.00 - 8.10)^2 &= (+1.90)^2 = 3.61\\
(9.50 - 8.10)^2 &= (+1.40)^2 = 1.96\\
(8.50 - 8.10)^2 &= (+.40)^2 = .16\\
(7.50 - 8.10)^2 &= (-.60)^2 = .36\\
& \qquad\qquad\quad 34.66
\end{aligned}
$$

The total, 34.66, is larger than that of $\sum_{i=1}^{n} (X_i - \overline{X})^2$ which is 34.00. If the squared differences are taken about some arbitrarily chosen value smaller than $\overline{X}$ the resulting summation would still be larger than $\sum_{i=1}^{n} (X_i - \overline{X})^2$. Suppose we chose \$7.90 as the arbitrary value smaller than $\overline{X}$. Thus,

$\sum_{i=1}^{n} (X_i - 7.90)^2$
$(9.50 - 7.90)^2 = (+1.60)^2 = 2.56$
$(3.00 - 7.90)^2 = (-4.90)^2 = 24.01$
$(10.00 - 7.90)^2 = (+2.10)^2 = 4.41$
$(9.50 - 7.90)^2 = (+1.60)^2 = 2.56$
$(8.50 - 7.90)^2 = (+.60)^2 = .36$
$(7.50 - 7.90)^2 = (+.40)^2 = .16$
34.06

The third property of the mean is that it may be used to estimate a *total amount* in a population when appropriate. For instance, the mean hourly wage rate for a sample of $n = 6$ executive secretaries is \$8.00. If, in this particular company, there are 200 executive secretaries, the total hourly cost of executive secretarial labor is estimated from

$$\text{Total} = N\overline{X} \tag{4.2}$$

where N = population size
$\overline{X}$ = sample arithmetic mean

Thus,

$$\text{Total} = N\overline{X} = (200)(\$8.00) = \$1{,}600.00$$

The total hourly cost of executive secretarial labor is \$1,600.00 and, for a 40-hour work week, it is estimated that the weekly cost to the firm is \$64,000.

The fourth property of the mean is that its computation is based on every observation. Therefore, as can be observed from the sample of six executive secretary salaries, the mean is greatly affected by any extreme value or values. In this case the mean is only \$8.00 because of one "extreme" hourly wage rate—the \$3.00 obtained by the second secretary in the sample. Perhaps that individual is less experienced at the job and/or receives other fringe benefits. The mean of the other five observations, however, is \$9.00. Hence only one extremely small value results in a large reduction in $\overline{X}$. Of course the reverse would be true if the extreme value were much larger than the other observations.

4.3.2 The Median

Since any extreme value (or values) in a set of data so greatly distorts the arithmetic mean, it is not a good measure of central tendency in such circumstances. Thus whenever any extreme value is present, it is more appropriate to use the median as a measure of central tendency. The median is unaffected by any extreme values in a set of data. **The median is a measure of central tendency which appears in the "middle" of an ordered sequence of values.** That is, half of the observations in a set of data are lower than it and half of the observations are greater than it.

To calculate the median from a set of data collected in its raw form, we must first arrange the data in an ordered array. If the number of observations in the sample is an *odd* number, the median is represented by the numerical value of the (n + 1)/2 ordered observation. On the other hand, if the number of observations in the sample is an *even* number, the median is represented by the mean or average of the two middle values in the ordered array. Thus to obtain the median hourly wage rate for the six executive secretaries, the raw data are cast into an ordered array as follows:

Raw Data	Ordered Array
X_1 = $ 9.50	$X_{(1)}$ = $ 3.00
X_2 = $ 3.00	$X_{(2)}$ = $ 7.50
X_3 = $10.00	$X_{(3)}$ = $ 8.50
	$X_{(3.5)}$ = **Median** = $9.00
X_4 = $ 9.50	$X_{(4)}$ = $ 9.50
X_5 = $ 8.50	$X_{(5)}$ = $ 9.50
X_6 = $ 7.50	$X_{(6)}$ = $10.00

The notation X_i stands for the *i*th observation in the sample (group) while $X_{(i)}$ corresponds to the *i*th ordered observation in the sequence. For example, X_2, the second observation selected in the sample, is $X_{(1)}$, the *first ordered* or smallest observation. The median value, however, appears midway through the ordered array. Using $(n + 1)/2$, the median value is represented by the $(6 + 1)/2 = 3.5$th ordered observation—the average between the third and fourth ordered observations. Therefore the median is estimated to be ($8.50 + $9.50)/2 = $9.00. As seen from above, the median is unaffected by extreme observations. Regardless of whether $X_{(1)}$, the smallest observation, is $0, $3.00, or $7.50, the median would still be $9.00.

Had the sample size been an odd number, the median would merely be represented by the numerical value given to the $[(n + 1)/2]$th observation in the ordered array. Thus in the following ordered array for $n = 5$ students' midterm examination results, the median is the value of $X_{(3)}$ or 88.0:

$X_{(1)} = 64$
$X_{(2)} = 79$
$X_{(3)} = 88$ = **Median**
$X_{(4)} = 90$
$X_{(5)} = 94$

that is,

$$\begin{aligned}\text{Median} &= \frac{n+1}{2} \text{ Ordered Observation}\\ &= \frac{5+1}{2} = \text{Third Ordered Observation}\\ &= X_{(3)} = 88.0\end{aligned}$$

Again we may observe from this set of data that the median would be unaffected by any alteration of the extreme values; $X_{(1)}$ could be 0, 64, or 78—the median would still be 88.0.

Thus to summarize, the median has three interesting characteristics. First, the calculation of the median value is affected by the number of observations, not by the magnitude of any extreme(s). Second, any observation selected at random is just as likely to exceed the median as it is to be exceeded by it. Third (although the proof is beyond the scope of the book), the summation of absolute differences about the median is a minimum, that is,

$$\sum_{i=1}^{n} |X_i - \text{Median}| = \text{Minimum Total}$$

where $|X_i - \text{Median}|$ refers to the absolute difference between the ith observation and the median—the difference without regard to the sign. To demonstrate this property we again use the sample of six secretaries' hourly wage rates in which the mean $\bar{X}$ was computed to be \$8.00 and the median was found to be \$9.00. Comparing the absolutes differences between each observation and its mean versus each observation and its median, we note that the latter total is smaller:

$\sum_{i=1}^{n} \lvert X_i - \bar{X} \rvert$	$\sum_{i=1}^{n} \lvert X_i - \text{Median} \rvert$
9.50 − 8.00 = 1.50	9.50 − 9.00 = .50
3.00 − 8.00 = 5.00	3.00 − 9.00 = 6.00
10.00 − 8.00 = 2.00	10.00 − 9.00 = 1.00
9.50 − 8.00 = 1.50	9.50 − 9.00 = .50
8.50 − 8.00 = .50	8.50 − 9.00 = .50
7.50 − 8.00 = .50	7.50 − 9.00 = 1.50
$\sum_{i=1}^{n} \lvert X_i - \bar{X} \rvert = 11.00$	$\sum_{i=1}^{n} \lvert X_i - \text{Median} \rvert = 10.00$

4.3.3 The Mode

Sometimes, when describing a set of data, the researcher uses the mode as a measure of location or central tendency. **The mode is the most typical or commonly observed value in a set of data.** The mode is not affected by the occurrence of any extreme values. It is easily obtained from an ordered array.

As an example, the ordered array for the hourly wage rates of six senior bookkeepers has no mode. None of the values were "most typical." Now *no mode* is a different result than a mode of zero. Suppose the following ordered array reflects the recorded noon-time temperatures (Fahrenheit) in Duluth, Minnesota, during the first week in January:

$$\begin{aligned} X_{(1)} &= -4^\circ \\ X_{(2)} &= -2^\circ \\ X_{(3)} &= -1^\circ \\ X_{(4)} &= -1^\circ \\ X_{(5)} &= 0^\circ \\ X_{(6)} &= 0^\circ \\ X_{(7)} &= 0^\circ \end{aligned}$$

Here the most typical value or mode is 0°. When no observation is most common, there is no mode. However, if $X_{(2)}$, the second ordered observation, had been -1° instead of -2°, this set of data would have had two most typical values—-1° and 0°—and the data would be described as *bimodal.* Of course, as discussed in Chapter 2, one may theoretically argue that with *continuous data* no two measurements can ever be the same, and so there can never be a most typical value. Therefore, a mode may exist only because of gross measurements in which the resulting temperatures are merely recorded to the nearest whole degree. With *discrete data,* however, modal values can be observed.

The mode has one distinguishing characteristic, it is the only measure of central tendency that can be used with qualitative data. As an example, we recall from Chapter 3 that the researcher and the dean had been examining attitudes toward admission standards to college by accounting students and by non-accounting students. From Figures 3.2, 3.3, and 3.4 it is seen that both groups of students most typically desired tenth-grade mathematics and English levels to be the minimum standard for admission. With quantitative data, however, the mean and the median are the more frequently used measures of central tendency.

4.3.4 Quantiles

Aside from the above measures of central tendency, there exist some useful measures of "noncentral" location which are often employed when summarizing or describing a set of data. These measures are called **quantiles.** The most familiar quantiles are the **quartiles,** the **deciles,** and the **percentiles.** The latter two, however, are more appropriate measures of noncentral location in such fields as educational research or industrial psychology where several hundreds of subjects are given standardized tests and it is deemed desirable to compare the relative position of the subjects' performance on the test. Hence, our concern here will be with the quartiles.

Whereas the median is a value that splits the ordered array in half (50.0%

of the observations are larger and 50.0% of the observations are smaller), the quartiles are descriptive measures that split the ordered data into four quarters. The first quartile, Q_1, is the value such that 25.0% of the observations are smaller and 75.0% of the observations are larger. The second quartile, Q_2, is the median—50.0% of the observations are smaller and 50.0% are larger. The third quartile, Q_3, is the value such that 75.0% of the observations are smaller and 25.0% of the observations are larger. To approximate the quartiles, the following positioning point formulas are used:

$$Q_1 = \text{value corresponding to the } \frac{n+1}{4} \text{ ordered observation}$$

$$Q_2 = \text{median, the value corresponding to the } \frac{2(n+1)}{4} = \frac{n+1}{2} \text{ ordered observation}$$

$$Q_3 = \text{value corresponding to the } \frac{3(n+1)}{4} \text{ ordered observation}$$

If the resulting positioning point value is an integer, the particular numerical observation corresponding to that positioning point is chosen for the quantile. If the resulting positioning point is halfway between two positioning points, the mean of their corresponding values is selected. Finally, if the resulting positioning point is neither an integer nor a value halfway between two other positioning points, a simple rule of thumb used to approximate the particular quantile is to merely round off to the nearest integer positioning point and select the numerical value of the corresponding observation. Thus for example, for the ordered array of seven noon-time temperatures in Duluth, Minnesota, during the first week in January:

$$\begin{aligned} Q_1 &= \frac{n+1}{4} \text{ Ordered Observation} \\ &= \frac{7+1}{4} = \text{Second Ordered Observation} \\ &= X_{(2)} = -2° \end{aligned}$$

$$\begin{aligned} Q_2 &= \frac{2(n+1)}{4} = \frac{n+1}{2} \text{ Ordered Observation} \\ &= \frac{7+1}{2} = \text{Fourth Ordered Observation} \\ &= \text{Median} = X_{(4)} = -1° \end{aligned}$$

$$\begin{aligned} Q_3 &= \frac{3(n+1)}{4} \text{ Ordered Observation} \\ &= \frac{3(7+1)}{4} = \text{Sixth Ordered Observation} \\ &= X_{(6)} = 0° \end{aligned}$$

On the other hand, for the hourly wage rates of six executive secretaries:

$$Q_1 = \frac{n+1}{4} = \frac{6+1}{4} = 1.75\text{th Ordered Observation}$$

$$\cong X_{(2)} = \$7.50$$

$$Q_2 = \frac{2(n+1)}{4} = \frac{n+1}{2} = \frac{7}{2} = 3.5\text{th Ordered Observation}$$

$$= \text{Median} = X_{(3.5)} = \frac{X_{(3)} + X_{(4)}}{2} = \frac{\$8.50 + \$9.50}{2} = \$9.00$$

$$Q_3 = \frac{3(n+1)}{4} = \frac{3(6+1)}{4} = 5.25\text{th Ordered Observation}$$

$$\cong X_{(5)} = \$9.50$$

4.4 MEASURES OF DISPERSION

The second most important characteristic which describes a set of data is dispersion. **Dispersion is the amount of variation, scatter, or spread in the data.** Two sets of data may differ in both central tendency and dispersion or, as shown in the hypothetical diagram of Figure 4.5, two sets of data may have the same measures of central tendency but differ greatly in terms of dispersion. The data set depicted by polygon C is much more concentrated about the measure of central tendency and therefore is less variable than that depicted by polygon A. This property was also noted in the previous section when comparing the wage rates of executive secretaries, senior bookkeepers, and section managers.

In dealing with numerical data then, it is insufficient to summarize that data by merely presenting some descriptive measures of central tendency. We must also characterize the data in terms of their dispersion or variability. Five such measures are the **range,** the **interquartile range,** the **variance,** the **standard deviation,** and the **coefficient of variation.**

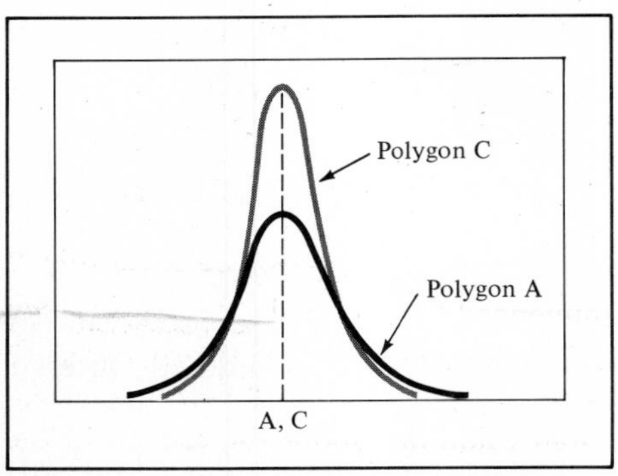

Figure 4.5 Hypothetical polygons differing in dispersion.

4.4.1 The Range

When dealing with ungrouped data, the range can readily be determined from an ordered array. For a sample of size n the range is the difference between the largest and the smallest observations, that is,

$$\text{Range} = X_{(n)} - X_{(1)} \tag{4.3}$$

Although the range is a simple, easily understood, and easily calculated measure of dispersion, its distinct weakness is that it fails to take into account how the data are distributed between the smallest and largest values. It would be improper to use the range as a measure of dispersion if either one or both of its components are extreme observations. Consider, as an example, the sample of six secretaries' hourly wage rates. The range is found to be

$$\text{Range} = X_{(6)} - X_{(1)} = \$10.00 - \$3.00 = \$7.00$$

In this example the range is not a good measure of variability since $X_{(1)}$, \$3.00, is an extreme value compared to the other five observations (where the hourly wage rates are distributed within \$2.50 of each other).

4.4.2 The Interquartile Range

A second measure of dispersion, the interquartile range, avoids the problem of extreme values in the data. This simple measure considers the spread in the middle 50% of the data and thus is in no way influenced by possibly occurring extreme values, that is,

$$\text{Interquartile Range} = Q_3 - Q_1 \tag{4.4}$$

In the above example,

$$\text{Interquartile Range} = \$9.50 - \$7.50 = \$2.00$$

While its major advantages are the facts that it is easy to compute and that it is not influenced by any possibly occurring extreme values, the interquartile range nevertheless possesses two major disadvantages as a measure of dispersion. First, it only measures the spread in the center of the data and therefore states nothing about the spread in the total data. Second, the interquartile range merely measures the "distance" between Q_1 and Q_3 and not how the data are distributed or spread out between Q_1 and Q_3.

4.4.3 The Variance and the Standard Deviation

Two measures of dispersion which take into account how all the observations in the data are distributed are the variance and its square root, the standard deviation.

DEFINITIONAL FORMULAS

The variance of a sample, given by the symbol S^2, measures the average of the squared differences between each observation and its mean.[3] Thus,

$$S^2 = \frac{\sum_{i=1}^{n} (X_i - \bar{X})^2}{n - 1} \quad (4.5)$$

The standard deviation of a sample, given by the symbol S, is merely the square root of the variance. Therefore **the standard deviation measures the square root of the average of squared differences around the mean,** that is,

$$S = \sqrt{\frac{\sum_{i=1}^{n} (X_i - \bar{X})^2}{n - 1}} \quad (4.6)$$

While the variance possesses certain mathematical properties, its computation is measured in squared units—squared dollars, squared inches, squared kilograms, squared percentage points, squared seconds, etc. Thus for practical work our primary measure of dispersion will be the standard deviation, whose measurement is in the original units of the data—dollars, inches, kilograms, percentage points, seconds, etc.

Now what do the variance and standard deviation actually measure? The variance and standard deviation are "sort of" measuring the average scatter around the mean, that is, how larger observations fluctuate above it and how smaller observations distribute below it. The formulas for variance and stan-

[3] Had the denominator been n instead of $n - 1$, the average of the squared differences around the mean would have been obtained. However, $n - 1$ is used here because of the property of **degrees of freedom.** Essentially, in the summation $\sum_{i=1}^{n} (X_i - \bar{X})^2$ only $n - 1$ of the terms are independent since the calculation of the statistic S^2 as defined by Equation (4.5) assumes a prior knowledge of the statistic $\bar{X}$, that is, if we know $\bar{X}$ we lose 1 degree of freedom since we only need to know $n - 1$ of the n terms to determine the remaining observation through subtraction. It is further noted that the sample variance S^2 possesses good mathematical estimation properties which will be considered in Chapter 7.

dard deviation could not merely use $\sum_{i=1}^{n} (X_i - \bar{X})$ as a numerator because we may recall from Section 4.3.1 that this summation is always zero. Hence the numerator of the variance and standard deviation formulas makes use of the **least-squares property** of the mean:

$$\sum_{i=1}^{n} (X_i - \bar{X})^2 = \text{Minimum Total}$$

so that as measures of average squared differences about the mean, the variance and standard deviation must be smaller than any other measure of average squared differences about any other indicator of central tendency.

From the three scales of Figure 4.6 which, respectively, represent the three separate samples of hourly wage rates obtained by six executive secretaries, six senior bookkeepers, and six section managers (presented in Section 4.3.1) it is clear that all four measures of dispersion—range, interquartile range, variance, and standard deviation—similarly reflect on the spread of the data, that is, the wider the spread, the larger the range, interquartile range, variance, and standard deviation; the narrower the spread, the smaller are these measures of dispersion. As seen from the scales in Figure 4.6, the wider the spread, the greater the distance between the observations and the mean. Thus $\sum_{i=1}^{n} (X_i - \bar{X})^2$ is larger than would be the case if the observed data were not as widely dispersed.

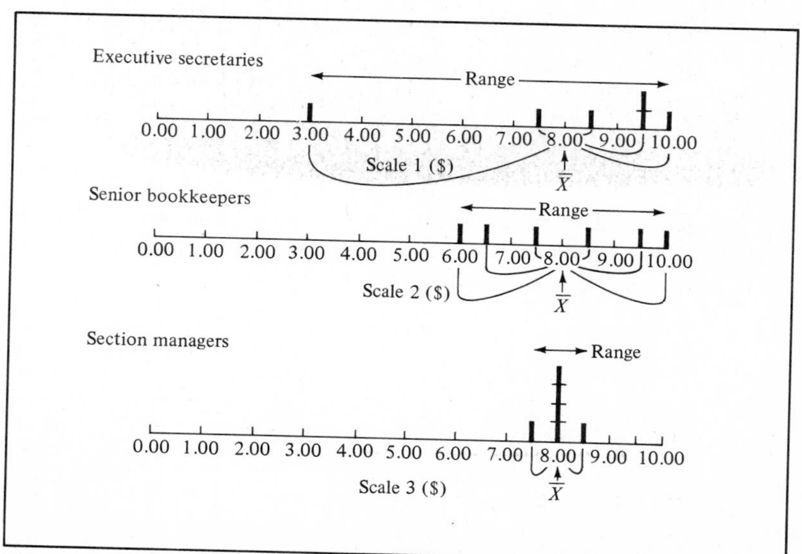

Figure 4.6 A comparison of salary ranges.

To compute the variance we merely take the difference between each observation and the mean; square it; add each squared result together; and then divide this summation by $n - 1$. To obtain the standard deviation we merely take the square root of the variance. The results for each of the three samples are shown below.

Sample 1: Secretaries' Hourly Wages:

$$S^2 = \frac{\sum_{i=1}^{n} (X_i - \bar{X})^2}{n - 1}$$

$$= \frac{(9.50 - 8.00)^2 + (3.00 - 8.00)^2 + (10.00 - 8.00)^2 + (9.50 - 8.00)^2 + (8.50 - 8.00)^2 + (7.50 - 8.00)^2}{6 - 1}$$

$$= \frac{34.00}{5} = 6.80$$

$$S = \sqrt{6.80} \cong \$2.61$$

Sample 2: Bookkeepers' Hourly Wages:

$$S^2 = \frac{\sum_{i=1}^{n} (X_i - \bar{X})^2}{n - 1}$$

$$= \frac{(9.50 - 8.00)^2 + (6.50 - 8.00)^2 + (10.00 - 8.00)^2 + (6.00 - 8.00)^2 + (8.50 - 8.00)^2 + (7.50 - 8.00)^2}{6 - 1}$$

$$= \frac{13.00}{5} = 2.60$$

$$S = \sqrt{2.60} \cong \$1.61$$

Sample 3: Section Managers' Hourly Wages:

$$S^2 = \frac{\sum_{i=1}^{n} (X_i - \bar{X})^2}{n - 1}$$

$$= \frac{(8.00 - 8.00)^2 + (8.00 - 8.00)^2 + (8.00 - 8.00)^2 + (8.00 - 8.00)^2 + (8.50 - 8.00)^2 + (7.50 - 8.00)^2}{6 - 1}$$

$$= \frac{.50}{5} = .10$$

$$S = \sqrt{.10} \cong \$.32$$

Since, in the preceding computations we are squaring the differences, neither the variance nor the standard deviation can ever be negative. The only time S^2 and S can be zero would be if there were no spread at all in the data—if each observation in the sample were exactly the same as shown below for a hypothetical set of six hourly wages:

$$X_1 = \$8.00,\ X_2 = \$8.00,\ X_3 = \$8.00,\ X_4 = \$8.00,\ X_5 = \$8.00,\ X_6 = \$8.00$$

$$\bar{X} = \frac{\sum_{i=1}^{n} X_i}{n} = \frac{\$48.00}{6} = \$8.00$$

$$S^2 = \frac{\sum_{i=1}^{n} (X_i - \bar{X})^2}{n-1}$$

$$= \frac{(8.00 - 8.00)^2 + (8.00 - 8.00)^2 + (8.00 - 8.00)^2 + (8.00 - 8.00)^2 + (8.00 - 8.00)^2 + (8.00 - 8.00)^2}{6 - 1}$$

$$= \frac{0}{5} = 0$$

$$S = \sqrt{0} = \$.00$$

In such an unusual case the range and the interquartile range would also be zero. But data are inherently variable—not constant. Any random phenomena of interest that we could think of usually takes on a variety of values. For example, people have differing IQ's, incomes, weights, ages, pulse rates, etc. It is because data inherently vary that it becomes so important to study not only measures (of central tendency) which summarize the data but also measures of dispersion (such as standard deviation) which reflect how the data are varying.

CALCULATING S^2 AND S: COMPUTATIONAL FORMULAS

The formulas for variance and standard deviation, Equations (4.5) and (4.6), are *definitional formulas,* but they are not often practical to use—even with an electronic calculator. For each of our sets of data regarding hourly wage rates the mean is an integer—$8.00. For more realistic situations where the observations and the mean are unlikely to be integers the following *computational formulas* for the variance and the standard deviation are given for practical use:

$$S^2 = \frac{\sum_{i=1}^{n} X_i^2 - \frac{\left(\sum_{i=1}^{n} X_i\right)^2}{n}}{n-1} \qquad (4.7)$$

$$S = \sqrt{\frac{\sum_{i=1}^{n} X_i^2 - \frac{\left(\sum_{i=1}^{n} X_i\right)^2}{n}}{n-1}} \tag{4.8}$$

where $\sum_{i=1}^{n} X_i^2$ = summation of the squares of each observation

$\left(\sum_{i=1}^{n} X_i\right)^2$ = square of the total summation

The computational formulas, Equations (4.7) and (4.8), are identical to the definitional formulas, Equations (4.5) and (4.6). Since the denominators are the same, it is easy to show through expansion and the use of summation rules (see Appendix B, Section B.3) that

$$\sum_{i=1}^{n} (X_i - \bar{X})^2 = \sum_{i=1}^{n} X_i^2 - \frac{\left(\sum_{i=1}^{n} X_i\right)^2}{n}$$

Moreover, since S^2 (and S) can never be negative, the summation of squares, $\sum_{i=1}^{n} X_i^2$, must always equal or exceed, $\left(\sum_{i=1}^{n} X_i\right)^2 \Big/ n$, the square of the total summation divided by n.

For the sample of six executive secretaries' hourly wage rates, the standard deviation is recomputed below using the computational formula (4.8). The reader may recompute S for the second and third samples (see Problem 4.4) to verify the identity between Equations (4.6) and (4.8).

$$S^2 = \frac{\sum_{i=1}^{n} X_i^2 - \frac{\left(\sum_{i=1}^{n} X_i\right)^2}{n}}{n-1}$$

$$= \frac{(9.50^2 + 3.00^2 + 10.00^2 + 9.50^2 + 8.50^2 + 7.50^2) - \frac{(9.50 + 3.00 + 10.00 + 9.50 + 8.50 + 7.50)^2}{6}}{6-1}$$

$$= \frac{(90.25 + 9.00 + 100.00 + 90.25 + 72.25 + 56.25) - \frac{(48.00)^2}{6}}{5}$$

$$S^2 = \frac{418.00 - \frac{2{,}304.00}{6}}{5} = \frac{418.00 - 384.00}{5}$$

$$= \frac{34.00}{5} = 6.80$$

$$S = \sqrt{6.80} \cong \$2.61$$

It should be noted that the computational formulas for S^2 and S can also be written by multiplying numerator and denominator by n, as follows:

$$S^2 = \frac{n \sum_{i=1}^{n} X_i^2 - \left(\sum_{i=1}^{n} X_i\right)^2}{n(n-1)} \tag{4.7a}$$

$$S = \sqrt{\frac{n \sum_{i=1}^{n} X_i^2 - \left(\sum_{i=1}^{n} X_i\right)^2}{n(n-1)}} \tag{4.8a}$$

Finally, it is of interest to note that for situations in which it is appropriate to use the sample mean to estimate a total amount [see Equation (4.2)], the standard deviation for the total can be obtained by

$$S_{\text{total}} = \sqrt{N}\, S \tag{4.8b}$$

where N = population size

S = sample standard deviation

As an example, if the standard deviation for the sample of six executive secretaries' hourly wage rates is \$2.61 and the firm employs a total of 200 executive secretaries (from which the sample was randomly selected),

$$S_{\text{total}} = \sqrt{N}\, S = \sqrt{(200)}\cdot(\$2.61) = \$36.91$$

Thus the standard deviation of the total amount is \$36.91 on an hourly basis and is \$1,476.40 on a weekly basis (for a 40-hour work week).

USING THE STANDARD DEVIATION: THE BIENAYMÉ–CHEBYSHEV RULE

More than a century ago, the mathematicians Bienaymé and Chebyshev (Reference 4) independently examined the property of data variability around the mean. They found that regardless of how a set of data are distributed, the percentage of observations that are contained within distances of $\pm\ k$ standard deviations around the mean must be at least

$$\left(1 - \frac{1}{k^2}\right)100\%$$

Therefore for data whose polygons take any shape whatsoever, at least $[1 - (1/2^2)]\ 100\% = 75.0\%$ of the observations must be contained within distances of ± 2 standard deviations around the mean; at least $[1 - (1/3^2)]\ 100\% = 88.89\%$ of the observations must be contained within distances of ± 3 standard deviations around the mean; and at least $[1 - (1/4^2)]\ 100\% = 93.75\%$ of all the observations must be included within distances of $\pm$ 4 standard deviations about the mean.[4]

While the Bienaymé–Chebyshev rule is general in nature and applies to any kind of distribution of data, we will see in Chapter 6 that specifically if, as in Figures 4.1 and 4.5, the data were symmetrical and bell-shaped (that is, the normal distribution), 68.26% of all the observations would be contained within distances of $\pm$ 1 standard deviation around the mean, while 95.44%, 99.73%, and 99.99% of the observations would be included, respectively, within distances of $\pm$ 2, $\pm$ 3, and $\pm$ 4 standard deviations around the mean. These results are summarized in Table 4.2.

Table 4.2
HOW DATA VARY AROUND THE MEAN

Number of Standard Deviation Units k	Percentage of Observations Contained Between the Mean and k Standard Deviations	
	Bienaymé–Chebyshev	*Normal Distribution*
1	Not calculable	Exactly 68.26%
2	At least 75.00%	Exactly 95.44%
3	At least 88.89%	Exactly 99.73%
4	At least 93.75%	Exactly 99.99%

Specifically, if we knew that a particular random phenomenon followed the pattern of the normal distribution, we would then know (as will be shown in Chapter 6) *exactly* how likely it is that any particular observation were close to or far from its mean. Generally, however, for any kind of distribution, the Bienaymé–Chebyshev rule tells us *at least* how likely it must be that any particular value falls within a given distance about the mean. Such information is quite valuable in data analysis.

[4] The Bienaymé–Chebyshev rule can only apply to distances beyond $\pm$ 1 standard deviation about the mean.

4.4.4 The Coefficient of Variation

The coefficient of variation is a fifth measure of dispersion. Unlike the previous four measures, the **coefficient of variation is a relative measure.** It is expressed as a percentage rather than in terms of the units of the particular data. As a relative measure the coefficient of variation is particularly useful when comparing the variability of two or more sets of data (distributions) that are expressed in different units of measurement. That is, suppose, as an example from Chapters 2 and 3, that the researcher and the dean were interested in determining whether students' college grade-point indexes are less variable than their high school averages. Since high school averages are usually computed on a percentage basis (0.0%–100.0%) while college grade-point indexes may be computed on a four-point system ($A = 4.0$, $B = 3.0$, $C = 2.0$, $D = 1.0$, and $F = 0.0$), it is impossible to directly compare their respective standard deviations. However, the coefficient of variation, given by the symbol CV, measures scatter relative to the mean and may be computed by

$$\text{CV} = \left(\frac{S}{\overline{X}}\right) 100\% \qquad (4.9)$$

where S = standard deviation of a set of data
$\overline{X}$ = mean of a set of data

Thus after obtaining the mean and standard deviation for the college grade-point indexes and for the high school averages, the researcher and the dean may compute the two coefficients of variation and then determine for students who attend this business school whether or not high school averages are relatively more variable than are college grade-point indexes.

In addition, the coefficient of variation is also very useful when comparing two or more sets of data (distributions) which are measured in the same units but differ to such an extent that a direct comparison of the respective standard deviations is not very helpful. As an example, suppose a potential investor was considering purchasing shares of stock in one of two companies, A or B, which are listed on the American Stock Exchange. If neither company offered dividends to its stockholders and if both companies were rated equally high (by various investment services) in terms of potential growth, the potential investor might want to consider the volatility of the two stocks to aid in the investment decision. Now suppose each share of stock in Company A has averaged \$50 over the past months with a standard deviation of \$10. In addition, suppose that in this same time period the price per share for Company B stock averaged \$12 with a standard deviation of \$4. In terms of the actual standard deviations the price of Company A shares seems to be more volatile than that of Company B. However, since the average prices per share for the two stocks are so different, it would be more appropriate for the potential investor to consider the variability in price relative to the average price in order to examine the volatility/stability of the two stocks. For Company A the coefficient of variation is $CV_A = (\$10/\$50)100\% = 20.0\%$; for Company B the coefficient of variation is $CV_B = (\$4/\$12)100\% = 33.3\%$. Thus relative to the mean, the price of Stock B is much more variable than the price of Stock A.

4.5 SHAPE

A third important characteristic of a set of data is its shape. In describing a set of numerical data it is not only necessary to summarize the data by presenting appropriate measures of central tendency and variability; it is also necessary to consider the shape of the data—the manner in which the data are distributed. Either the distribution of data is symmetrical or it is not. If the distribution of data is not symmetrical, it is called asymmetrical or **skewed.** Thus, skewness refers to the lack of symmetry in a distribution of data. While several measures of shape have been developed (Reference 7), it is sufficient here (for descriptive purposes) to discuss this important property of data in terms of symmetry or lack thereof without employing any particular measure. To describe the property of shape we need only compare the mean and the median. If these two measures are equal, we may consider the data to be symmetrical (or **zero-skewed**). On the other hand, if the mean exceeds the median the data may be described as **positive or right-skewed** while if the mean is exceeded by the median that data can be called **negative or left-skewed,** that is,

$\bar{X} >$ Median	Positive or Right-Skewness
$\bar{X} =$ Median	Symmetry or Zero-Skewness
$\bar{X} <$ Median	Negative or Left-Skewness

Positive skewness occurs when the mean is increased by some unusually high values, while negative skewness arises when the mean is decreased by some extremely low values. Data are symmetrical when there are no really extreme

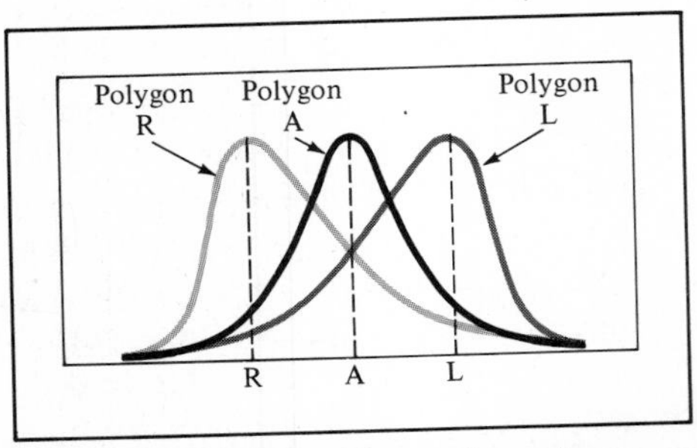

Figure 4.7 Hypothetical polygons differing in shape.

Table 4.3
USING DESCRIPTIVE MEASURES ON THREE DATA SETS

	Hourly Wage Rates		
Descriptive Measures	*Executive Secretaries* n = 6	*Senior Bookkeepers* n = 6	*Section Managers* n = 6
$\bar{X}$	\$8.00	\$8.00	\$8.00
Median	\$9.00	\$8.00	\$8.00
Mode	\$9.50	No mode	\$8.00
Q_1	\$7.50	\$6.50	\$8.00
Q_3	\$9.50	\$9.50	\$8.00
Range	\$7.00	\$4.00	\$1.00
Interquartile range	\$2.00	\$3.00	\$ .00
S^2	6.800	2.600	.100
S	\$2.61	\$1.61	\$ .32
CV	32.6%	20.1%	4.0%
Shape	Negative or left-skew	Symmetrical	Symmetrical

values in a particular direction so that low values and high values balance each other out.

In Figure 4.7 three hypothetical polygons are depicted: polygon L is negative or left-skewed (since the distortion to the left is caused by extremely small values); polygon R is positive or right-skewed (since the distortion to the right is caused by extremely large values); and polygon A is symmetrical (since the mean and the median are equal so that the shape of the curve to their left is the mirror image of the shape of the curve to their right).

As seen from Figure 4.7, the property of shape can best be depicted from the polygon of a particular distribution of data. Hence more will be said about symmetry and skewness in the following section which examines the properties of data and their appropriate descriptive summary measures when the numbers of observations are sufficiently large to warrant their grouping into a frequency distribution.

To conclude this section, however, Table 4.3 summarizes the results of employing the various descriptive measures which we have investigated on the three sets of samples—each containing the hourly wage rates of six individuals.

4.6 OBTAINING DESCRIPTIVE SUMMARY MEASURES FROM GROUPED DATA

It is often necessary to obtain descriptive summary measures from data grouped into frequency distributions. In many cases researchers obtain data from reports published in magazines, newspapers, journals, etc. In these situations the original (raw) data are simply not available. In other cases where a computer is not readily accessible, as the number of observations increases it becomes more and more laborious to extract the salient features of the data unless the "mass data" are first grouped into tables and charts. While the descriptive summary measures computed from **ungrouped** data—data in their raw form or in an ordered array—provide actual results, approximations for these descriptive measures can be obtained from **grouped** data. Therefore in this section the grade-point indexes of the 35 accounting majors and 59 non-accounting majors (for which appropriate tables and charts are constructed in Chapter 3) will be reexamined so that the researcher and the dean will be able to more appropriately compare and analyze the results.

It is recalled that Table 3.4 presents the frequency distribution of grade-point indexes for the 35 accounting majors. From this frequency distribution table approximation results for all the considered descriptive measures will be obtained.

4.6.1 Approximating the Mean: Definitional Method

In the frequency distribution presented in Table 4.4, depicting the 35 accounting students' grade-point indexes, the midpoint or class mark of each class is used to represent all the observations that have been tallied into the class. Thus in the first class, whose boundaries range from 2.00 to under 2.40, the midpoint 2.20 represents the five observations tallied into that class. In the second class the midpoint 2.60 represents the nine students' grade-point indexes which were tallied into that class, and so on. To approximate the mean, each midpoint must first be multiplied by the number of observations it represents; the summation of these resulting products must then be obtained; and lastly, this total must be divided by the number of observations, that is,

$$\bar{X} \cong \frac{\sum_{j=1}^{g} m_j f_j}{n} \tag{4.10}$$

where $\bar{X}$ = mean of the sample
n = number of observations in the sample
g = number of groups or classes in the frequency distribution
m_j = midpoint or class mark of the jth class
f_j = number of observations tallied into the jth class

Table 4.4
CALCULATING THE MEAN BY THE DEFINITIONAL METHOD

Grade-Point Index	Class Marks m_j	Number of Students f_j	$m_j f_j$
2.00 < 2.40	2.20	5	11.00
2.40 < 2.80	2.60	9	23.40
2.80 < 3.20	3.00	13	39.00
3.20 < 3.60	3.40	4	13.60
3.60 < 4.00	3.80	4	15.20
		35	102.20

Hence it is seen from Table 4.4 that

$$\bar{X} \cong \frac{\sum_{j=1}^{g} m_j f_j}{n} = \frac{102.20}{35} = 2.920$$

Now Equation (4.10) is analogous to Equation (4.1) and may be considered as a *definitional formula.* It can be employed for distributions having either equal or unequal class-interval widths.[5] However, for distributions having equal class-interval widths a more practical *coded-computational formula* can be used.

4.6.2 Approximating the Mean: Coded-Computational Method

Provided that the distribution has class intervals of equal width, the computation of $\bar{X}$ can be facilitated by first selecting one of the class midpoints as the arbitrary origin m_a and assigning it a coded value of zero and then replacing all the other class midpoints, m_j with a sequence of consecutive integer values, U_j around the arbitrary origin. Thus, for example, if the first class mark, $m_1 = 2.20$, were designated the arbitrary origin m_a and assigned a code of zero, the five coded values replacing the class marks would be 0, 1, 2, 3, and 4. On the other hand, if the fourth class mark, $m_4 = 3.40$, were designated as the arbitrary origin, the five coded values U_j would be -3, -2, -1, 0, and 1. While the computation of $\bar{X}$ must be identical no matter which class midpoint is selected as the arbitrary origin, some calculation effort can be saved if the arbitrary origin were to be selected near the middle of the distribution. Hence, Table 4.5 demonstrates the computation of $\bar{X}$ using the coded-computational formula

$$\bar{X} \cong m_a + \frac{\left(\sum_{j=1}^{g} U_j f_j\right)}{n} C \tag{4.11}$$

where $\bar{X}$ = mean of the sample
m_a = midpoint of the class interval selected as the arbitrary origin
C = width of each class interval
U_j = integer code assigned to the jth class midpoint
f_j = number of observations tallied into the jth class
n = number of observations in the sample
g = number of groups or classes in the frequency distribution

Thus, as shown in Table 4.5, with the third class mark selected as the arbitrary origin,

[5] With open-ended classes the mean cannot be calculated (unless the particular observations in the open-ended groups are known).

Table 4.5
CALCULATING THE MEAN BY THE CODED-COMPUTATIONAL METHOD

Grade-Point Index	Class Marks m_j	Coded Values U_j	Number of Students f_j	U_jf_j
2.00 < 2.40	2.20	−2	5	−10
2.40 < 2.80	2.60	−1	9	− 9
2.80 < 3.20	3.00 = m_a	0	13	0
3.20 < 3.60	3.40	1	4	4
3.60 < 4.00	3.80	2	4	8
			35	−7

$$\overline{X} \cong m_a + \frac{\left(\sum_{j=1}^{g} U_j f_j\right)}{n} C = 3.00 + \left(\frac{-7}{35}\right)(.40) = 2.920$$

It is of course noted that the approximation for $\overline{X}$ is identical, regardless of whether Equation (4.10) or (4.11) is used.

4.6.3 Approximating the Median

The median for the frequency distribution can be approximated by first finding the particular class interval in which the $n/2$ ordered observation lies and then determining, through interpolation within that class interval, the numerical value associated with the $n/2$ ordered observation. Thus,

$$\text{Median} \cong B_M + \left(\frac{\frac{n}{2} - f_{BM}}{f_M}\right) C \qquad (4.12)$$

where B_M = lower boundary of the class interval containing the median
f_M = number of observations in the class interval containing the median
f_{BM} = total number of observations before the class interval containing the median
C = width of the class interval containing the median
$\frac{n}{2}$ = median observation

From Table 4.6 it is seen that the $n/2$ ordered observation, 17.5, is one of the 13 observations contained in the class interval whose lower boundary is 2.80 and in addition, that the total number of observations prior to this class interval is 14. That is, there are five observations contained in the first class interval; a total of 14 observations are contained in the first and second class intervals; and a total of 27 observations are found in the first three class intervals. Thus 17.5, the $n/2$ ordered observation, must be the 3.5th ordered observation out of the 13 observations in the class interval whose lower boundary is 2.80. Hence through interpolation, the median grade-point index for the 35 accounting majors is approximated as

$$\text{Median} \cong B_M + \left(\frac{\frac{n}{2} - f_{BM}}{f_M}\right) C = 2.80 + \left(\frac{17.5 - 14}{13}\right)(.40) = 2.908$$

Table 4.6
APPROXIMATING THE MEDIAN, MODE, AND QUANTILES

	Grade-Point Index	Number of Students f_j	Cumulated Totals
	2.00 < 2.40	5	5
Q_1 class	2.40 < 2.80	9	14
Median class, Q_3 class	2.80 < 3.20	13*	27
	3.20 < 3.60	4	31
	3.60 < 4.00	4	35
		35	

* "Most typical" class: 2.80 < 3.20.

A distinct advantage of using the median over the mean as a measure of central tendency with grouped data is that the median may be approximated from Equation (4.12) regardless of whether or not the frequency distribution has equal class-interval widths or open-ended classes.

4.6.4 Approximating the Mode

While several methods exist for approximating the mode from grouped data (References 1 and 2), it is sufficient here to merely scan the frequency distribution (Table 4.6) and report the class interval containing the most observations as the "most typical" or modal class. Thus for the 35 accounting students, the modal class contains grade-point indexes from 2.80 to under 3.20.

4.6.5 Approximating the Quartiles

Like the median, the quartiles are approximated by locating the positioning point of the particular ordered observation of interest and then interpolating in that class interval. Using Table 4.6 the first and third quartiles are calculated from Equations (4.13) and (4.14), respectively, which, as will be noted, are quite similar to the formula for the median, Equation (4.12), that is,

$$Q_1 \cong B_{Q_1} + \left(\frac{\frac{n}{4} - f_{BQ_1}}{f_{Q_1}}\right) C \tag{4.13}$$

where Q_1 = first quartile
B_{Q_1} = lower boundary of the class interval containing the first quartile
f_{Q_1} = number of observations in the class interval containing the first quartile
f_{BQ_1} = total number of observations before the class interval containing the first quartile
C = width of the class interval containing the first quartile
$\frac{n}{4}$ = first quartile observation

and

$$Q_3 \cong B_{Q_3} + \left(\frac{\frac{3n}{4} - f_{BQ_3}}{f_{Q_3}}\right) C \tag{4.14}$$

where Q_3 = third quartile
B_{Q_3} = lower boundary of the class interval containing the third quartile
f_{Q_3} = number of observations in the class interval containing the third quartile
f_{BQ_3} = total number of observations prior to the class interval containing the third quartile
C = width of the class interval containing the third quartile
$\frac{3n}{4}$ = third quartile observation

Hence,

$$Q_1 \cong B_{Q_1} + \left(\frac{\frac{n}{4} - f_{BQ_1}}{f_{Q_1}}\right) C = 2.40 + \left(\frac{8.75 - 5}{9}\right)(.40) = 2.567$$

and

$$Q_3 \cong B_{Q_3} + \left(\frac{\frac{3n}{4} - f_{BQ_3}}{f_{Q_3}}\right) C = 2.80 + \left(\frac{26.25 - 14}{13}\right)(.40) = 3.177$$

The first and third quartiles for the 35 accounting majors' grade-point indexes are, respectively, approximated by 2.567 and 3.177.

4.6.6 Approximating the Range and Interquartile Range

With grouped data the range may be approximated as the difference between the upper boundary in the largest class and the lower boundary in the smallest class. Therefore from Table 4.6 the range is approximated by $4.0 - 2.0 = 2.0$.

The interquartile range, however, is still defined as the difference (or spread) between the first and third quartiles. Using Equation (4.4), the interquartile range for the 35 accounting majors' grade-point indexes may be approximated from the frequency distribution (Table 4.6) by

$$\text{Interquartile Range} = Q_3 - Q_1 \cong 3.177 - 2.567 = 0.610$$

4.6.7 Approximating S^2 and S: Definitional Method

To calculate the variance and standard deviation from grouped data, definitional formulas analogous to Equations (4.5) and (4.6) for ungrouped data may be developed. Since the variance "sort of" measures the average of the squared differences between each observation and its mean and since the midpoints of each class of a frequency distribution are used to represent the observations in the classes, the variance may be approximated from a frequency distribution by:

1. Taking the squares of the differences between each midpoint and the mean
2. Multiplying or "weighting" each of the squared differences by the respective number of observations pertaining to each class
3. Summing these products
4. Dividing this total by $n - 1$

The standard deviation is readily approximated as the square root of the variance. Thus,

$$S^2 \cong \frac{\sum_{j=1}^{g} (m_j - \bar{X})^2 f_j}{n - 1} \tag{4.15}$$

where $\bar{X}$ = approximated mean
S^2 = variance of the sample
n = number of observations in the sample
g = number of groups or classes in the frequency distribution
m_j = midpoint or class mark of the jth class
f_j = number of observations tallied into the jth class

Table 4.7
CALCULATING S^2 BY THE DEFINITIONAL METHOD

Grade-Point Index	Class Marks m_j	$(m_j - \bar{X})$	$(m_j - \bar{X})^2$	Number of Students f_j	$(m_j - \bar{X})^2 f_j$
2.00 < 2.40	2.20	−0.72	0.5184	5	2.5920
2.40 < 2.80	2.60	−0.32	0.1024	9	0.9216
2.80 < 3.20	3.00	+0.08	0.0064	13	0.0832
3.20 < 3.60	3.40	+0.48	0.2304	4	0.9216
3.60 < 4.00	3.80	+0.88	0.7744	4	3.0976
				35	7.6160

Also,[6]

$$S \cong \sqrt{\frac{\sum_{j=1}^{g} (m_j - \bar{X})^2 f_j}{n - 1}} \tag{4.16}$$

Thus it is seen from Table 4.7 that the variance and standard deviation for the sample of 35 accounting students' grade-point indexes are

$$S^2 \cong \frac{\sum_{j=1}^{g} (m_j - \bar{X})^2 f_j}{n - 1} = \frac{7.6160}{34} = 0.2240$$

and

$$S \cong \sqrt{0.2240} = 0.473$$

4.6.8 Approximating S^2 and S: Coded-Computational Method

Although Equations (4.15) and (4.16) may be used for calculating S^2 and S for grouped data regardless of whether or not the class intervals are of equal width, it is observed from Table 4.7 that such computations may be quite laborious and cumbersome—even with a calculator possessing several memories.[7] Therefore, as with the mean, for distributions having equal class-interval widths a more practical coded-computational formula may be employed. Using a sequence of consecutive integer values U_j as the coded replacements for m_j, the class midpoints (as described in Section 4.6.2),

[6] It should be noted here that the two mathematical properties of the arithmetic mean, which were presented in Section 4.3.1, also hold for grouped data. Thus,

$$\sum_{j=1}^{g} (m_j - \bar{X}) f_j = 0 \qquad \text{and} \qquad \sum_{j=1}^{g} (m_j - \bar{X})^2 f_j = \text{Minimum Total}$$

[7] With open-ended classes the variance and standard deviation cannot be calculated (unless the particular observations in the open-ended groups are known).

Table 4.8 demonstrates the computation of the variance and standard deviation by the coded-computational formulas given by

$$S^2 \cong C^2 \left[\frac{\sum_{j=1}^{g} U_j^2 f_j - \frac{\left(\sum_{j=1}^{g} U_j f_j\right)^2}{n}}{n-1} \right] \tag{4.17}$$

where

S^2 = variance of the sample
C = width of each class interval
U_j = integer code assigned to the jth class midpoint
f_j = number of observations tallied into the jth class
n = number of observations in the sample
g = number of groups or classes in the frequency distribution

Also,

$$S \cong C \sqrt{\frac{\sum_{j=1}^{g} U_j^2 f_j - \frac{\left(\sum_{j=1}^{g} U_j f_j\right)^2}{n}}{n-1}} \tag{4.18}$$

As depicted in Table 4.8, the third class mark is selected as the arbitrary origin and coded zero. Hence,

$$S^2 \cong C^2 \left[\frac{\sum_{j=1}^{g} U_j^2 f_j - \frac{\left(\sum_{j=1}^{g} U_j f_j\right)^2}{n}}{n-1} \right] = (.40)^2 \left[\frac{49 - \frac{(-7)^2}{35}}{34} \right]$$

$$\cong \frac{7.6160}{34} = 0.2240$$

and

$$S \cong \sqrt{0.2240} = 0.473$$

The results for S^2 (and S) are identical—regardless of whether the definitional formulas or coded-computational formulas are used. While at a glance the latter look more imposing, they are indeed much simpler to use than the definitional formulas and are suggested whenever the class intervals of the frequency distribution are of equal width. In fact the only additional term

Table 4.8
CALCULATING S^2 BY THE CODED-COMPUTATIONAL METHOD

Grade-Point Index	Class Marks m_j	Coded Values U_j	U_j^2	Number of Students f_j	U_jf_j	$U_j^2f_j$
2.00 < 2.40	2.20	−2	4	5	−10	20
2.40 < 2.80	2.60	−1	1	9	− 9	9
2.80 < 3.20	3.00 = m_a	0	0	13	0	0
3.20 < 3.60	3.40	1	1	4	4	4
3.60 < 4.00	3.80	2	4	4	8	16
				35	−7	49

needed in Equation (4.17) that was not already obtained from Table 4.5 when computing the mean is $\sum_{j=1}^{g} U_j^2 f_j$. This is easily found as the summation of the products of the squares of the coded values in each class and the number of observations in the class.

4.6.9 Approximating the Coefficient of Variation

Whether or not the data are grouped, the coefficient of variation is defined as a measure of relative scatter around the mean and may be computed as the ratio of the standard deviation to the mean. Thus using Equation (4.9), the coefficient of variation CV for the 35 accounting students' grade-point indexes may be approximated from the frequency distribution (Tables 4.5 and 4.8) as

$$\text{CV} = \left(\frac{S}{\overline{\overline{X}}}\right) 100\% \cong \left(\frac{0.473}{2.920}\right) 100\% = 16.2\%$$

4.6.10 Describing the Shape of the Distribution

From the frequency distribution the mean is approximated as 2.920 and the median is approximated as 2.908. Since the mean exceeds the median, should it be concluded that the data on grade-point indexes for accounting students are positive or right-skewed rather than symmetrical? Relative to the amount of spread in the data (as measured by the standard deviation), however, the difference by which the mean exceeds the median appears to be negligible. Hence for descriptive purposes, is it not more appropriate to conclude that the data are approximately symmetrical rather than slightly skewed? The best way to judge the shape for descriptive purposes is to examine the polygon of the particular distribution of data. In the following section both the polygon and the ogive for the distribution of grade-point indexes of accounting students will be reconstructed, and the properties of location, dispersion, and shape will be reexamined from the charts.

4.7 GRAPHICAL INTERPRETATIONS OF DESCRIPTIVE MEASURES FROM GROUPED DATA

In the previous section approximations for various descriptive measures of location, dispersion, and shape were obtained for data grouped into a frequency distribution. Since these results are only approximations to the actual results that would be obtained from the ungrouped data, it often suffices for descriptive purposes to obtain, where possible, approximations of these descriptive measures from charts.

4.7.1 The Polygon

Figure 4.8 reproduces the percentage polygon of grade-point indexes of 35 accounting students (Figure 3.12).

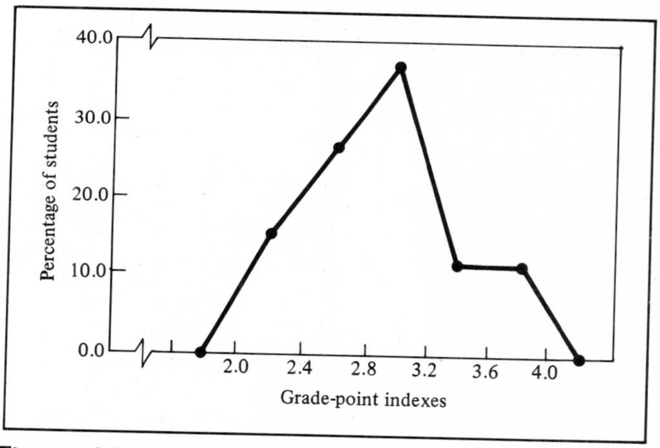

Figure 4.8 Percentage polygon of grade-point indexes of 35 accounting students.

SOURCE: Data are taken from Table 3.6.

From the polygon it appears that the shape of the data is approximately symmetrical. Moreover, since we may recall that the polygon is plotted at the midpoint of each class interval, it is observed that the most typical or modal class contains grade-point indexes that range from 2.80 to under 3.20.

Had the data been perfectly symmetrical, as is represented by the hypothetical bell-shaped normal curve of Figure 4.9, all the measures of location —mean, median, and mode—would have been identical. The mean acts as a "balancing" point, the median cuts the data into two equal areas, and the mode is the most typical value. For other kinds of distributions, however, the presence of symmetry is dependent upon the equality of the mean and the median. The mode may differ or not exist. As examples, Figure 4.10 represents a hypothetical U-shaped curve which is symmetrical and bimodal and

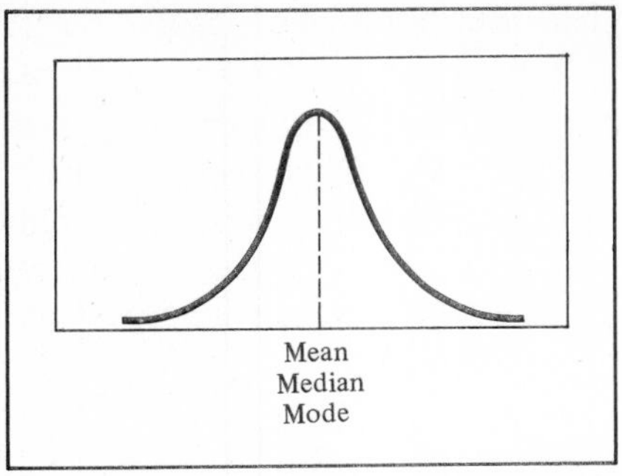

Figure 4.9 Hypothetical bell-shaped curve.

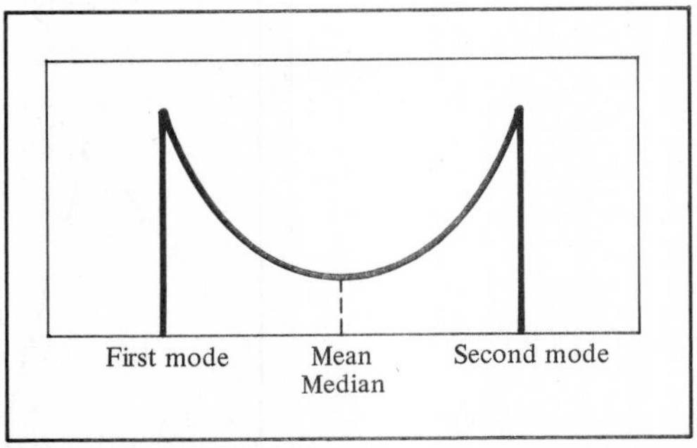

Figure 4.10 Hypothetical U-shaped curve.

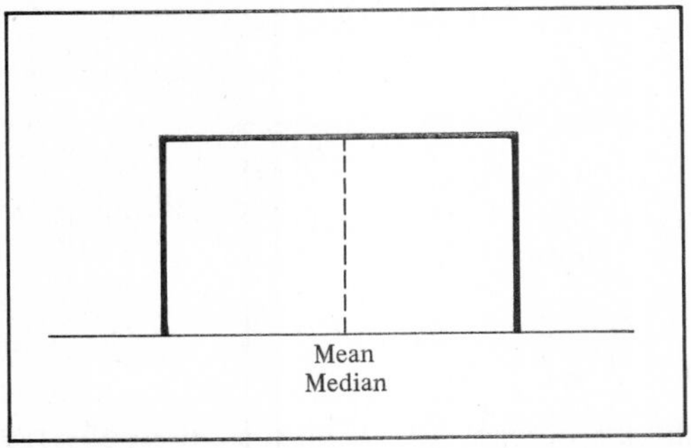

Figure 4.11 Hypothetical rectangularly shaped curve.

Figure 4.11 depicts a hypothetical rectangularly shaped curve which is symmetrical but possesses no mode. In each distribution, though, the mean and median are identical.

For asymmetrical distributions, however, the direction of the skewness depends upon the location of the extreme values. If the extreme values are the larger observations, the mean will be the measure of location most greatly distorted toward the upward direction. Since the mean exceeds the median and the mode, such distributions are said to be positive or right-skewed. This

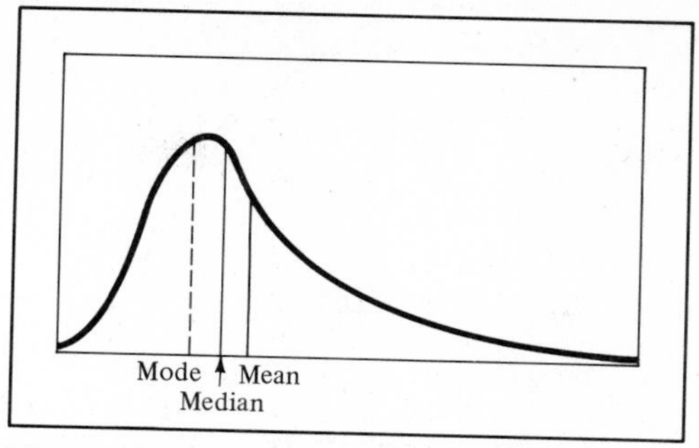

Figure 4.12 Hypothetical right-skewed distribution.

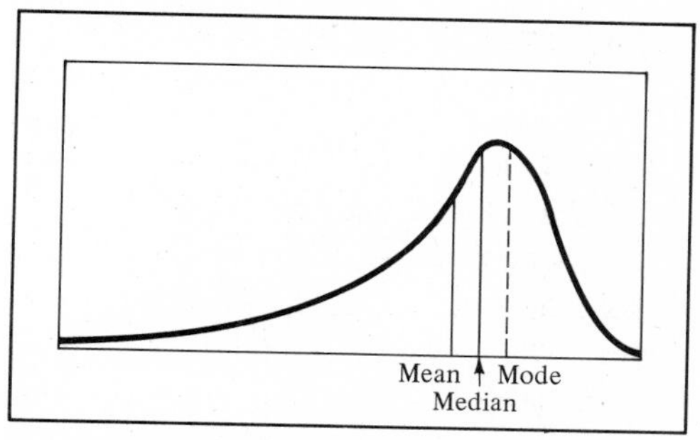

Figure 4.13 Hypothetical left-skewed distribution.

is depicted in Figure 4.12. Note that the tail of the distribution is distorted to the right. On the other hand, if the extreme values are the smaller observations, the mean will be the measure of location most greatly reduced. Since the mean is exceeded by the median and the mode, such distributions are considered negative or left-skewed. As shown in Figure 4.13, the tail of the distribution is distorted to the left.

4.7.2 The Ogive

Figure 4.14 reproduces the cumulative percentage polygon or ogive of the grade-point indexes of 35 accounting students (Figure 3.14). As indicated in Figure 4.14 it is only necessary to use the "less than" curve to obtain approximations for the median, the first and third quartiles, the interquartile range, and the range. Horizontal lines at 25.0%, 50.0%, and 75.0% are extended to the "less than" curve, and then perpendicular lines are constructed to the horizontal axis from the respective points of intersection. Thus the following approximations are obtained:

$$Q_1 \cong 2.57$$

$$\text{Median} \cong 2.91$$

$$Q_3 \cong 3.18$$

$$\text{Interquartile Range} = Q_3 - Q_1 \cong 0.61$$

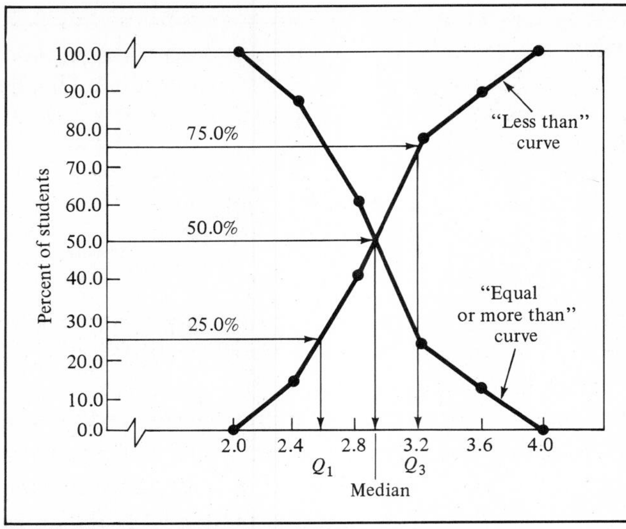

Figure 4.14 Cumulative percentage polygon (ogive) of grade-point indexes of 35 accounting majors.

SOURCE: Data are taken from Table 3.8.

In addition, since the ogive is plotted at the boundaries of each class, the range may be approximated as the difference in the upper boundary of the last class and the lower boundary of the first class so that

$$\text{Range} \cong 4.00 - 2.00 = 2.00$$

4.7.3 Graphical Interpretations: An Overview

Using the polygon of Figure 4.8 and the ogive of Figure 4.14 it is observed that most of the descriptive measures that have been developed in this chapter can be interpreted graphically. If the data appear to be approximately symmetrical, the mean can be considered equal to the median. On the other hand, however, if the data appear to be skewed, the mean cannot be approximated graphically. In addition, the variance, standard deviation, and coefficient of variation cannot be directly interpreted graphically, regardless of the shape of the data.

4.8 DESCRIPTIVE MEASURES: AN OVERVIEW AND THE ROLE OF THE COMPUTER

In this chapter we have developed a variety of descriptive summary measures which may be used to examine the properties of location, dispersion, and shape for a set of data obtained from a sample. With ungrouped data we have computed actual values for these descriptive measures. Moreover, when the numbers of observations were sufficiently large to warrant their grouping into a frequency distribution, we have discussed analogous formulas for approximating the statistics. Finally, we have also demonstrated how most of the descriptive measures can be approximated graphically.

In conclusion, Table 4.9 presents, in summary, the values for the various descriptive measures:

1. Computed from ungrouped data
2. Approximated from grouped data
3. Interpreted from charts

for both the 35 accounting students and the 59 non-accounting students. By scanning these results it will be clear that both grouped-data calculations and graphical interpretations, which are much less laborious, yield fairly close approximations to the actual results obtained from ungrouped-data calculations. These approximations would even be better if the sample sizes were to be larger. The reader will be asked to verify these results (see Problems 4.32 and 4.33). In so doing, the reader will gain an appreciation for the amount of labor saved by employing grouped-data approximations or graphical interpretations.

Table 4.9
SUMMARY OF DESCRIPTIVE RESULTS BY THREE METHODS

Descriptive Measure	Grade-Point Indexes of 35 Accounting Majors Obtained from			Grade-Point Indexes of 59 Non-Accounting Majors Obtained from		
	Ungrouped Data	*Grouped Data*	*Charts*	*Ungrouped Data*	*Grouped Data*	*Charts*
$\bar{X}$	2.881	2.920	*	2.882	2.932	*
Median	2.91	2.908	2.91	2.84	2.936	2.94
Mode	2.00, 3.00, 3.20	2.80 < 3.20	2.80 < 3.20	3.00	2.80 < 3.20	2.80 < 3.20
Q_1	2.56	2.567	2.57	2.60	2.607	2.60
Q_3	3.18	3.177	3.18	3.20	3.210	3.22
Range	1.86	2.000	2.00	2.06	2.400	2.40
Interquartile range	0.62	0.610	0.61	0.60	0.603	0.62
S^2	0.2309	0.2240	†	0.2081	0.2104	†
S	0.480	0.473	†	0.456	0.459	†
CV	16.7%	16.2%	†	15.8%	15.7%	†
Shape	Approximately symmetrical	Approximately symmetrical	Approximately symmetrical	Approximately symmetrical	Approximately symmetrical	Approximately symmetrical

* May be considered equal to median since data are observed to be approximately symmetrical.
† Cannot be directly interpreted graphically.

On the other hand, however, once the reader gains familiarity with both the techniques of mass data presentation developed in the previous chapter and the usefulness of descriptive summary measures discussed in this chapter, much of the drudgery can usually be avoided if one of the several "computer packages" available for descriptive statistical methods is utilized. Among the more widely known software packages are BMDP, SPSS, SAS, and STATPACK (References 3, 6, 8, and 9). As examples, Figures 4.15 and 4.16 are STATPACK computer printouts representing the frequency histogram of the

```
FREQUENCY      5    9   13    4    4
----------+----+----+----+----+----+----+----+----+----+----+-
   13                    H
   12                    H
   11                    H
   10                    H
    9               H    H
    8               H    H
    7               H    H
    6               H    H
    5          H    H    H
    4          H    H    H    H    H
    3          H    H    H    H    H
    2          H    H    H    H    H
    1          H    H    H    H    H
----------+----+----+----+----+----+----+----+----+----+----+-
        2.000     2.800     3.600
             2.400     3.200     4.000
```

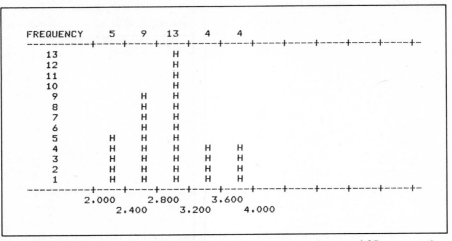

Figure 4.15 Frequency histogram of grade-point indexes of 35 accounting students obtained by STATPACK.

grade-point indexes attained by the 35 accounting majors (see Figure 3.11) and some of the descriptive summary measures calculated from the raw data (see Table 4.9), respectively.

```
ANALYSIS
?* ELEMENTARY STATISTICS

NEW DATA
?* NO

VARIABLE          TOTAL          MEAN        MAXIMUM        MINIMUM
       1        100.830         2.881          3.860          2.000

VARIABLE          RANGE      VARIANCE       STD.DEV.      STD.ERROR
       1          1.860         0.231          0.481          0.081
```

Figure 4.16 Some descriptive summary measures for the grade-point indexes of 35 accounting students obtained by STATPACK.

The reader must be cautioned, however, to use such packaged programs with care and to become thoroughly familiar with their capabilities. Because of the recent widespread application of such statistical packages in business research and social science investigations, the American Statistical Association has established a prestigious committee to investigate the features of the various packages in order to prevent statistical errors from entering into their designs.

PROBLEMS

4.1 It is stated in Section 4.3.1 that one important property of the arithmetic mean is

$$\sum_{i=1}^{n} (X_i - \bar{X}) = 0$$

(a) Using the sample of six senior bookkeepers' hourly wage rates, verify that this property holds.

(b) Using the sample of six section managers' hourly wage rates, verify that this property holds.

4.2 A track coach must decide on which one of two sprinters to select for the 100-meter dash at an upcoming meet. The coach will base the decision on the results of five races between the two men, run over a one-hour period with 15 minute rest intervals between. The following times (in seconds) were recorded for the five races:

	Race				
Athlete	*1*	*2*	*3*	*4*	*5*
Jones	11.1	11.0	11.0	15.8	11.1
Smith	11.3	11.4	11.4	11.5	11.4

(a) Based on the above data, which of the two sprinters should the coach select? Why?

(b) Should the selection be different if the coach knew that Jones had fallen at the start of the fourth race?

(c) Discuss the differences in the concepts of the mean and the median as measures of central tendency and how this relates to (a) and (b).

4.3 Using the data corresponding to the samples of hourly wage rates of senior bookkeepers and of section managers, verify that the computation of the standard deviation is identical regardless of whether the definitional formula (4.6) or the computational formula (4.8) is used.

4.4 Using the data corresponding to the samples of hourly wage rates of senior bookkeepers and of section managers, verify the computations displayed in Table 4.3.

4.5 Using the data in Table 4.7, verify that, with grouped data,

$$\sum_{j=1}^{g} (m_j - \bar{X})f_j = 0$$

4.6 Using the data in Table 4.7, verify that, with grouped data,

$$\sum_{j=1}^{g} (m_j - \bar{X})^2 f_j = \text{Minimum Total}$$

by showing that the calculation 7.6160 for the numerator of Equation (4.15) must be a smaller total than if the median or any other value were used in lieu of $\bar{X}$.

4.7 The following data are the ages of a sample of 25 automobile salesmen in the United States and a sample of 25 automobile salesmen in Western Europe.

United States					Western Europe				
23	63	25	22	32	43	26	30	27	40
56	30	34	56	30	35	48	36	47	41
25	48	44	27	26	34	45	30	38	33
38	26	30	39	30	35	44	24	33	40
36	32	36	38	33	31	23	29	37	28

(a) For each set of **ungrouped** data compute the mean, median, Q_1, Q_3, range interquartile range, standard deviation, and coefficient of variation.

(b) For each set of data form the frequency distribution, percentage distribution, and cumulative percentage distribution. Plot the percentage polygons and ogives.

(c) Graphically interpret the range, median, quartiles, interquartile range, and shape of each data set.

4.8 The following data are the monthly rental prices for a sample of 10 one-bedroom unfurnished apartments in Manhattan and a sample of 10 one-bedroom unfurnished apartments in Brooklyn:

Manhattan	\$355, \$400, \$385, \$380, \$340, \$375, \$365, \$399, \$647, \$519
Brooklyn	\$250, \$275, \$225, \$205, \$194, \$225, \$190, \$245, \$ 75, \$300

(a) For each set of data compute the mean, median, range, standard deviation, and coefficient of variation.

(b) What can be said about one-bedroom unfurnished apartments renting in Brooklyn versus those renting in Manhattan?

* 4.9 Sales tax receipts in a particular community are collected quarterly. The following data represent the receipts (in thousands of dollars) collected for the first quarter from a sample of nine retail outlets in the community: 16, 18, 11, 17, 13, 10, 22, 15, 16.

(a) Compute the mean, median, and standard deviation.

(b) If there are a total of 300 retail outlets in this community, estimate the total amount of sales tax receipts that will be collected this quarter and estimate the standard deviation of the total amount.

(c) Other things being equal, estimate the total annual amount of sales tax receipts that will be collected from all the retail outlets for the year and estimate the standard deviation.

4.10 Upon examining the monthly billing records of a mail-order book company, the auditor takes a sample of 10 of its unpaid accounts. The amounts owed the company were: $4, $18, $11, $7, $7, $10, $5, $33, $9, $12.

(a) Compute:

(1) Mean	(6) Range
(2) Median	(7) Interquartile range
(3) Mode	(8) Variance
(4) Q_1	(9) Standard deviation
(5) Q_3	(10) Coefficient of variation

(b) Are these data skewed? If so, how?

(c) If a total of 250 bills were still outstanding, use the mean and the standard deviation to estimate the total amount owed to the company and the standard deviation of the total amount.

*4.11 The following data are the price-to-earnings ratios (rounded to the nearest value) for a random sample of 50 stocks from the New York Stock Exchange:

7	9	8	6	12	6	9	15	9	16
8	5	14	8	7	6	10	8	11	4
10	6	16	5	10	12	7	10	15	7
10	8	8	10	18	8	10	11	7	10
7	8	15	23	13	9	8	9	9	13

(a) For the ungrouped data compute the mean, median, and standard deviation.

(b) Group the data into a frequency distribution and approximate the mean, median, and standard deviation.

(c) How would you express the shape of the data?

(d) Based on the Bienaymé–Chebyshev rule, between what two values would we estimate that at least 75% of the data are contained?

(e) What percentage of the data are actually contained within ± 2 standard deviations of the mean? Compare the results with those from (d).

4.12 For the following random variables indicate the shape that the polygon of its distribution is likely to have:

Random Variable	Shape
______ 1. Number of issues of *Atlantic Monthly* read per year	a. Bell-shaped and symmetrical
______ 2. Numerical outcome of a roll of a die	b. Right-skewed
______ 3. Life of 100-watt bulbs (in hours)	c. Left-skewed
______ 4. Income of urban residents	d. U-shaped
______ 5. SAT mathematics scores by math majors	e. Rectangularly shaped
______ 6. Heights of 10-year-old girls	f. "Reverse J" shaped

The following problems refer to the sample data collected in the survey described in Chapter 2.

4.13 Professor V. Ulyanov of the Administrative Sciences Department met with the researcher and the dean to discuss future tuition policy. Upon examining the responses to question 8 of the student survey (see Figures 2.2 and 2.6) regarding the maximum amount of annual tuition the student would be willing (and able) to pay, the professor casually remarked that "in a truly Socialist state the student would have to pay according to what he could afford." Under such a system, discuss how one might estimate for the coming year the total amount of revenue funded through tuition and the standard deviation of the total amount if the population size, $N = 2{,}202$ students, were expected to remain stable (that is, admissions and retentions balance out graduates and reductions)?

4.14 From the responses to questions 4 and 8 of the student survey (see Figures 2.2 and 2.6) calculate the mean, median, Q_1, Q_3, standard deviation, and coefficient of variation of the amount of tuition that accounting majors and non-accounting majors claim they are able to pay. Discuss.

4.15 From the responses to questions 4 and 6 of the student survey (see Figures 2.2 and 2.6) calculate the mean, median, standard deviation, and coefficient of variation of the high-school averages attained by students majoring in accounting versus those not majoring in accounting. Discuss.

4.16 From the responses to questions 4 and 7 of the student survey (see Figures 2.2 and 2.6) calculate the mean, median, Q_1, Q_3, standard deviation, and coefficient of variation of the expected starting annual salaries for accounting versus non-accounting majors. Discuss.

4.17 From the responses to questions 2 and 7 of the student survey (see Figures 2.2 and 2.6) calculate the mean, median, Q_1, Q_3, standard deviation, and coefficient of variation of the expected starting annual salaries for juniors versus seniors. Discuss your results in light of the fact that economic forecasts indicate a 6% rate of inflation over the coming year.

4.18 From the responses to questions 4 and 13 of the student survey (see Figures 2.2 and 2.6) calculate the mean, median, Q_1, Q_3, standard deviation, and coefficient of variation of the number of pairs of jeans owned by accounting versus non-accounting majors. Discuss.

4.19 From the responses to questions 4 and 16 of the student survey (see Figures 2.2 and 2.6) calculate the mean, median, Q_1, Q_3, standard deviation, and coefficient of variation of the heights of accounting versus non-accounting majors. For each group determine the percentage of students whose heights are between $\bar{X} \pm 1S$ and $\bar{X} \pm 2S$. Discuss.

4.20 From the responses to questions 2 and 16 of the student survey (see Figures 2.2 and 2.6) do Problem 4.19 for juniors versus seniors.

4.21 From the responses to questions 4 and 17 of the student survey (see Figures 2.2 and 2.6) calculate the mean, median, Q_1, Q_3, standard deviation, and coefficient of variation of the weights of accounting versus non-accounting majors.

For each group determine the percentage of the students whose weights are between $\overline{X} \pm 1S$ and $\overline{X} \pm 2S$. Discuss.

4.22 From the responses to questions 2 and 17 of the student survey (see Figures 2.2 and 2.6) do Problem 4.21 for juniors versus seniors.

4.23 From the responses to questions 2 and 5 of the student survey (see Figures 2.2 and 2.6) calculate the mean, median, Q_1, Q_3, standard deviation, and coefficient of variation of grade-point indexes of juniors versus seniors. Discuss.

4.24 From the responses to questions 3 and 5 of the student survey (see Figures 2.2 and 2.6) calculate the mean, median, Q_1, Q_3, standard deviation, and coefficient of variation of grade-point indexes of students who plan to attend graduate school versus those that do not plan to attend graduate school. Discuss.

4.25 From the responses to questions 3 and 7 of the student survey (see Figure 2.2 and 2.6) calculate the mean, median, Q_1, Q_3, standard deviation, and coefficient of variation of the annual starting salaries expected after the baccalaureate for students planning to attend graduate school versus those that do not plan to attend graduate school. Discuss.

Questions 3 + 7

4.26 From the responses to questions 16 and 19 of the student survey (see Figures 2.2 and 2.6) calculate the mean, median, Q_1, Q_3, standard deviation, and coefficient of variation of the heights of students who smoke versus those that do not smoke. Discuss.

4.27 From the responses to questions 17 and 19 of the student survey (see Figures 2.2 and 2.6) calculate the mean, median, Q_1, Q_3, standard deviation, and coefficient of variation of the weights of students who smoke versus those that do not smoke. Discuss.

4.28 From the responses to questions 5 and 21 of the student survey (see Figures 2.2 and 2.6) calculate the mean, median, Q_1, Q_3, standard deviation, and coefficient of variation of the grade-point indexes of students favoring twelfth-grade admission standards versus those who favor no minimum standards or eighth-grade standards (combined). Discuss.

4.29 From the responses to questions 5 and 22 of the student survey (see Figures 2.2 and 2.6) calculate the mean, median, Q_1, Q_3, standard deviation, and coefficient of variation of the grade-point indexes of students favoring more stringent retention standards versus those who do not. Discuss.

4.30 Use your own sample data (Problem 3.12) to select problems from among Problems 4.14 through 4.29.

4.31 For your own class (Problem 3.13) select problems from among Problems 4.14 through 4.29.

4.32 Using the data in Figures 3.1, 3.13, 3.15, and Table 3.7 for the sample of 59 non-accounting majors verify:
(a) The ungrouped data calculations given in Table 4.9.
(b) The grouped data calculations given in Table 4.9.
(c) The graphical interpretations given in Table 4.9.

4.33 Using the data in Figures 3.1, 3.13, 3.15, and Table 3.4 for the sample of 35 accounting majors verify:
(a) The ungrouped data calculations given in Table 4.9.
(b) The grouped data calculations given in Table 4.9.
(c) The graphical interpretations given in Table 4.9.

4.34 Referring to Problem 3.26, use the relative frequency ogives depicted in Figure 3.21 to construct the frequency distributions of family income in Sample A and in Sample B. Using the two frequency distributions, approximate the mean, median, variance, and standard deviation of family income in each sample. Compare the two sets of results.

4.35 Referring to Problem 3.27, use the frequency distributions of Table 3.17 to approximate the mean, median, variance, and standard deviation of the accounts receivable for Month 1 and for Month 2. Compare the two sets of results.

4.36 Referring to Problem 3.28, use the frequency distributions of Table 3.18 to approximate the mean, median, variance, and standard deviation of the number of days to pay oil bills in each suburb. Compare the two sets of results.

Note: *Definitional formulas* [Equations (4.10), (4.15), and (4.16)] must be used for your computations in Suburb B.

REFERENCES

1. ARKIN, H., AND R. COLTON, *Statistical Methods* 5th ed. (New York: Barnes and Noble, 1970).
2. CROXTON, F., D. COWDEN, AND S. KLEIN, *Applied General Statistics,* 3rd ed. (Englewood Cliffs, N.J.: Prentice-Hall, 1967).
3. DIXON, W. J., AND M. P. BROWN, eds., *BMDP* (Berkeley: University of California Press, 1977).
4. KENDALL, M. G., AND A. STUART, *The Advanced Theory of Statistics,* vol. 1 (London: Charles Griffin, 1958).
5. NETER, J., W. WASSERMAN, AND G. WHITMORE, *Applied Statistics* (Boston: Allyn and Bacon, 1978).
6. NIE, N., C. HULL, J. JENKINS, K. STEINBRENNER, AND D. BENT, *Statistical Package for the Social Sciences,* 2nd ed. (New York: McGraw-Hill, 1975).
7. RICHMOND, S., *Statistical Analysis* 2nd ed. (New York: Ronald Press, 1964).
8. SERVICE, J., *A User's Guide to the Statistical Analysis System* (Raleigh: North Carolina State University Press, 1972).
9. *STATPACK: Statistical Package,* 2nd ed., developed by IBM, February 1970.

CHAPTER 5

BASIC PROBABILITY

5.1 INTRODUCTION

In the past three chapters we have examined methods of collecting, reducing, presenting, and describing a set of data. In this chapter we begin to study various rules of basic probability that can be used to evaluate the chance of occurrence of different phenomena. These rules will be expanded so that probability theory can be utilized in making inferences about a population based only on sample statistics.

5.2 OBJECTIVE AND SUBJECTIVE PROBABILITY

What is meant by the word **probability?** It refers to the relative chance that a particular event will occur. It could mean the chance of picking a black card from a deck of cards, the chance that a student selected at random in the dean's survey has an average below B, or the chance that a new product on the market will be successful.

Each of the previous examples refers to one of three definitions of the probability of occurrence of a particular event. The first definition is often called **classical probability.** In this instance, the probability of success is based upon a prior knowledge of the process involved. The probability of success is defined as the number of successful outcomes divided by the total number of outcomes:

$$\text{Probability of Success} = \frac{\text{Number of Favorable Outcomes}}{\text{Total Number of Outcomes}}$$

For example, in finding the probability of picking a black card from a deck of cards, the correct answer would be 26/52 or 1/2, since there are 26 black cards in a standard deck of cards. What does this probability of 1/2 of picking a black card tell us? If we replace each card after it is [illegible], does it mean that one out of the next two cards will be black? On the contrary, we cannot say for sure what will happen on the next several selections. However, we can say that in the long run, if this selection process is [illegible] on a "large" number of trials, the proportion of black cards selected will approach .50.

In the first example, the number of successes and the number of outcomes were known from the process involved, the deck of cards. This brings us to the second type of probability called **empirical classical probability.** The probability of success is still defined as the number of favorable outcomes divided by the total number of outcomes. However, in the case of empirical classical probability these outcomes are based upon observed data, not on a prior knowledge of a process such as a deck of cards. In our second example, the dean's study, the probability that a student has an average below B can be found by selecting a random sample of students from the entire population of students. In Chapter 2 we may recall that a sample of 94 students was selected. Of these 94 students, 57 had an average below B. Therefore the probability that a student has an average below B is 57/94 or .606. In this case the "success" refers to the student having an average below B. This illustrates that the favorable event can be defined in whatever manner is appropriate for the particular problem involved.

The previous example involved the examination of a fairly large number of outcomes. However, in many cases only a small number of past outcomes of an event may be available, or there may not be any past outcomes to examine (such as in the marketing of a new product).

The third type of probability then is called **subjective probability.** Whereas in the previous two definitions the probability of a favorable event was computed objectively—either from a prior probability or from actual (past) data—subjective probability refers to the chance of occurrence placed upon an event by a particular individual. This chance may be quite different from the subjective probability assigned by a different individual. For example, the inventor of a new toy may place quite a different probability on the chance of success for the toy than the president of the company that is considering marketing the toy. The assignment of probabilities to various events is usually based upon a combination of an individual's past experience, personal opinion, and analysis of a particular situation. Subjective probability is especially used in making decisions (see Chapter 11) in situations in which the probability of various events cannot be determined empirically.

5.3 BASIC PROBABILITY CONCEPTS

5.3.1 Sample Spaces and Events

The basic elements of probability theory are the outcomes of the process or phenomenon under study. Each possible type of occurrence is referred to as an event. **A simple event is an event that can be described by a single characteristic. The collection of all the possible events is called the sample space.**

We can achieve a better understanding of these terms by referring to several examples. First, let us examine the standard deck of 52 playing cards (Figure 5.1) in which there are four suits (hearts, clubs, diamonds, and spades) each of which has 13 different cards (ace, king, queen, jack, ten, nine, eight, seven, six, five, four, three, two).

If we are to randomly select a card from the deck, we might like to know various probabilities, such as:

1. What is the probability the card is black?
2. What is the probability the card is an ace?
3. What is the probability the card is a black ace?
4. What is the probability the card is black or an ace?
5. If we knew that the card selected was black, what is the probability that it is also an ace?

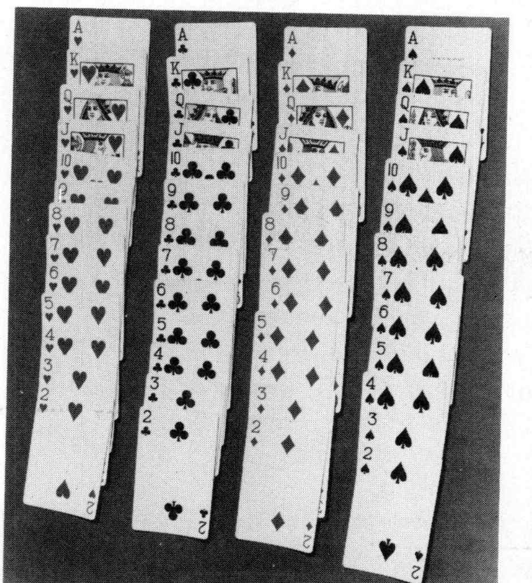

Figure 5.1 Standard deck of 52 playing cards.

We will also examine a second example that refers to the data collected as part of the survey in Chapter 2. Suppose that from the total sample of 94 students we pick a student at random.

1. What is the probability that the student is an accounting major?
2. What is the probability that the student has a B average or above?
3. What is the probability that the student is an accounting major who has a B average or above?
4. What is the probability that the student is an accounting major or has a B average or above?
5. If we knew that the student selected had a B average or above, what then would be the probability that he or she was an accounting major?

In the case of the deck of cards, the sample space consists of the entire deck of 52 cards, made up of various events, depending on how they are classified. For example, if the events are classified by suit, then there are four events: heart, club, diamond, and spade. If the events are classified by card value, there would be 13 events: ace, king, . . . , and two.

In the example from the survey, the sample space is made up of the 94 students in the sample who filled out the questionnaire. If the results are cross-classified according to the two criteria, accounting versus non-accounting major and B average or above versus below B average, there are four simple events that can occur: accounting major, non-accounting major, B average or above, below B average.

The manner in which the sample space is subdivided depends on the type of probabilities that are to be determined. With this in mind, it is of interest to define the complement of an event as follows: **The complement of event A includes all events that are not part of event A.** It is given by the symbol A'.

The complement of the event "black" would consist of all those cards that were not black (that is, all the red cards). The complement of spade would contain all cards that were not spades (that is, hearts, clubs, and diamonds). The complement of accounting majors is, of course, non-accounting majors since the students have been subdivided into only two categories.

A joint event is an event that has two or more characteristics. The event "black ace" is a joint event since the card must be both black *and* ace in order to qualify as a black ace. In a similar manner, the event "accounting major and B or above average" is a joint event since the student must be an accounting major *and* have a B or above average.

5.3.2 Contingency Tables and Venn Diagrams

As in the case of descriptive data, there are several alternative ways in which a particular sample space can be viewed. The first method involves cross-classifying the appropriate events in a cross-classification table called a **contingency table** (see Chapter 3).

Table 5.1
CONTINGENCY TABLE FOR COLOR-ACE VARIABLES

	Red	Black	Totals
Ace	2	2	4
Non-ace	24	24	48
Totals	26	26	52

If the two variables of interest for the card example were "color of card" and "presence of ace," the contingency table (Table 5.1) would look as shown at the top of this page.

The values in each cell of the table were obtained by subdividing the sample space of 52 cards according to the number of aces and the color of the card. It can be noted that if the totals in the margins are known, only one cell entry in this 2 × 2 table is needed in order to obtain the entries in the remaining three cells.

The contingency table for the 94 students is developed by referring back to Figure 2.6 and entering each student in the appropriate cross classification and tallying the entries. These values are summarized in Table 5.2.

The contingency table provides a clear presentation of the number of possible outcomes of the relevant variables. This is especially true if there are more than two events, as in Table 3.1, or if there are more than two variables being considered simultaneously, as in Table 3.3.

The second way of presenting the sample space is by using a Venn diagram. This diagram graphically represents the various events as "unions" and "intersections" of circles.

Figure 5.2 represents a typical Venn diagram for a two-variable (A and B) problem with each variable having only two events (A and A', B and B').

The circle on the left (the lighter one) represents all events that are part of A. The circle on the right (the darker one) represents all events that are part of B. The area contained within circle A *and* circle B (center area) is considered the **intersection** of A and B ($A \cap B$), since this area is part of A *and* is also part of B. The total area of the two circles is the **union** of A and B ($A \cup B$) and contains all outcomes that are part of event A, part of event B, or part of both A and B. The area in the diagram outside $A \cup B$ contains those outcomes that are neither part of A nor part of B.

Table 5.2
CONTINGENCY TABLE FOR ACCOUNTING–B AVERAGE VARIABLES

	Accounting	Non-accounting	Totals
Below B average	20	37	57
B average or above	15	22	37
Totals	35	59	94

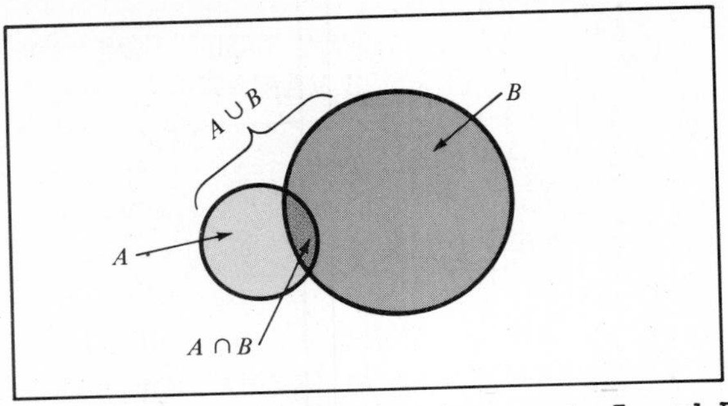

Figure 5.2 Venn diagram for events *A* and *B*.

In order to develop Venn diagrams for our two examples, *A* and *B* must be defined in each case. It does not matter which event is defined as *A* or *B*, as long as we are consistent in evaluating the various events.

Therefore for the first example, the events can be defined as follows:

A = ace B = black

A' = non-ace B' = red

In drawing the Venn diagram (Figure 5.3), the value of the intersection of *A and B* ($A \cap B$) must be determined so that the entire circle can be divided into its parts. $A \cap B$ consists of all black aces in the deck (that is, two outcomes).

Since there are two black aces, the remainder of event *A* (ace) consists of the red aces (there are two). The remainder of event *B* (black cards) consists of all black cards that are not aces (there are 24). The remaining cards are those that are neither black nor ace (there are also 24).

The second example can be presented in a Venn diagram (Figure 5.4) by defining the events as follows:

A = accounting major B = B average or above

A' = non-accounting major B' = below B average

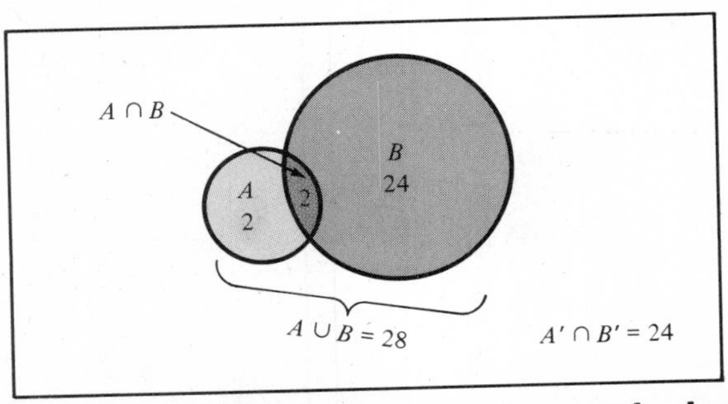

Figure 5.3 Venn diagram for card deck example.

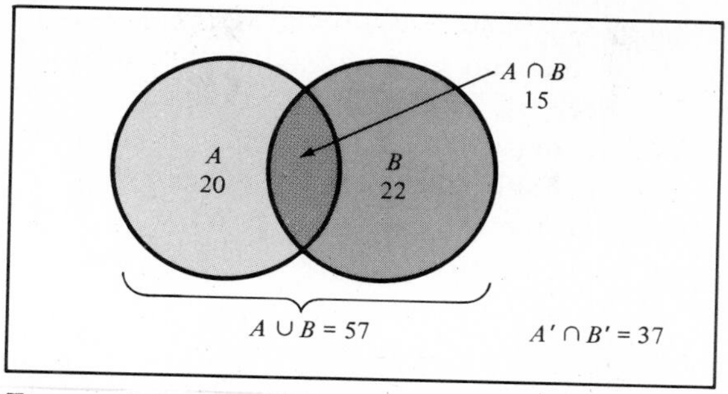

Figure 5.4 Venn diagram for accounting—B average example.

Since there are a total of 35 accounting majors among whom there are 15 accounting majors with a B average or above ($A \cap B$), there must be 20 accounting majors who have below B averages. Since there are a total of 37 students with B averages or above, there must be 22 students who are not accounting majors and have B averages or above. Finally, there happen to be 37 students who are not accounting majors nor have a B average or above.

5.4 SIMPLE (MARGINAL) PROBABILITY

Thus far we have focused upon the meaning of probability and on defining and illustrating several sample spaces. We shall now begin to answer some of the questions posed in the previous sections by developing rules for different types of probability.

Simple probability refers to the probability of occurrence of a simple event, $P(A)$, an event described by a single characteristic such as the probability of black, the probability of an ace, the probability of an accounting major, the probability that a student has an average below B.

We have already noted that the probability of selecting a black card on the draw of a card was 26/52 or 1/2 since there were 26 black cards in the 52-card deck.

How would we find the probability of picking an ace from the deck? We would find the number of aces in the deck by totaling the black aces and the red aces in the deck:

$$P(\text{Ace}) = \frac{\text{Number of Aces in Deck}}{\text{Number of Cards in Deck}}$$

$$= \frac{\text{Number of Red Aces} + \text{Number of Black Aces}}{\text{Total Number of Cards}}$$

$$= \frac{2 + 2}{52} = \frac{4}{52}$$

Simple probability is also called **marginal probability** since the total number of successes (aces in this case) can be obtained from the appropriate margin of the contingency table. The probability of an ace, $P(A)$, could be obtained from the Venn diagram (Figure 5.3) by looking at the number of outcomes contained in circle A. There are four outcomes in circle A, two contained in $A \cap B$ and two outside of $A \cap B$. This, of course, gives us the same result as analyzing the contingency table (Table 5.1).

Let us refer to the second example. We want to find the probability that a randomly selected student

1. Is an accounting major
2. Has a B average or above

The probability that the student is an accounting major can be determined by referring to the contingency table (Table 5.2):

$$P(\text{Accounting}) = \frac{\text{Number of Accounting Majors}}{\text{Total Number in Sample}}$$

$$= \frac{35}{94} = .372$$

Let us obtain from the Venn diagram $P(B)$, the probability that the student has a B average or above. Looking at Figure 5.4 we can see that circle B contains 37 outcomes. Therefore,

$$P(\text{B or above}) = \frac{37}{94} = .394$$

5.5 JOINT PROBABILITY

Marginal probability referred to the occurrence of simple events such as the probability of selecting an ace or the probability that a student is an accounting major. **Joint probability** refers to phenomena containing two or more events such as the probability of a black ace, a red queen, or an accounting major with B average or above.

The joint event "*A and B*" ($A \cap B$) means that both event A *and* event B must occur simultaneously. The event "black ace" consists only of those cards that are black and are also aces. Referring to the contingency table (Table 5.1) those cards that are black *and* ace consist only of the outcomes in the single cell, "black ace." Since there are two black aces, the probability of picking a card that is a black ace is

$$P(\text{Black } \textit{and} \text{ Ace}) = \frac{\text{Number of Black Aces}}{\text{Number of Cards in Deck}}$$

$$= \frac{2}{52}$$

This result can also be obtained by examining the Venn diagram of Figure 5.3. The joint event "*A and B*" (black ace) consists of the intersection ($A \cap B$) of events A (ace) and B (black) which contains two outcomes. Therefore the probability of a black ace is equal to 2/52.

The probability of choosing an accounting major who has a B average or above would be obtained from Figure 5.4 in the following manner:

$$P(\text{Accounting } \textit{and} \text{ B average}) = \frac{15}{94}$$

since there are 15 students who are accounting majors *and* have B averages or above.

Now that we have discussed the concept of joint probability, the marginal probability of a particular event can be viewed in an alternative manner. That is, the marginal probability of an event actually contains several joint probabilities. For example, if B consists of two events, B_1 and B_2, then we can observe that the probability of event A, $P(A)$, consists of the joint probability of event A occurring with event B_1 and the joint probability of event A occurring with event B_2. Thus, in general

$$P(A) = P(A \textit{ and } B_1) + P(A \textit{ and } B_2) + \cdots + P(A \textit{ and } B_k) \tag{5.1}$$

where $B_1, B_2, \ldots, B_k$ are k mutually exclusive events.

Therefore the probability of an ace can be expressed as follows:

$$\begin{aligned} P(\text{Ace}) &= P(\text{Ace } \textit{and} \text{ Red}) + P(\text{Ace } \textit{and} \text{ Black}) \\ &= \frac{2}{52} + \frac{2}{52} \\ &= \frac{4}{52} \end{aligned}$$

This is precisely what we did when we added up the number of favorable outcomes that made up the simple event.

5.6 ADDITION RULE

Now that we have developed a means of finding the probability of event A and the probability of event "*A and B*," we should like to examine a rule (the addition rule) that is used for finding the probability of event "*A or B*" ($A \cup B$). This rule, called **union**, refers to the occurrence of either event A, or event B, or both A and B.

The event "black *or* ace" would include all cards that were black, were aces, or were black aces. The event "accounting major *or* B average" would include all students who were accounting majors, or had B averages and above, or had both these characteristics.

Each cell of the contingency table (Table 5.2) can be examined to determine whether it is part of the event (accounting major *or* B average). The cell "accounting major *and* below B average" is part of the event since it includes accounting majors. The cell "non-accounting major *and* B average" is included because it contains students with B averages. Finally, the cell "accounting *and* B average" is included because it contains students who are accounting majors and have B averages. Therefore the probability can be obtained as follows:

$$\begin{aligned} P(\text{Accounting } \textit{or} \text{ B Average}) &= P(\text{Accounting } \textit{and} \text{ B Average}) \\ &\quad + P(\text{Accounting } \textit{and} \text{ below B Average}) \\ &\quad + P(\text{Non-accounting } \textit{and} \text{ B Average}) \\ &= \frac{15}{94} + \frac{20}{94} + \frac{22}{94} \\ &= \frac{57}{94} \end{aligned}$$

This result can be seen from the Venn diagram. The union of events A and B consists of all outcomes contained either in circle A or in circle B. As we can see, this adds up to 57 outcomes $(20 + 15 + 22 = 57)$.

The computation of the probability of the event A *or* B, $P(A \cup B)$ can be expressed in the following **general addition rule:**

$$P(A \cup B) = P(A \text{ or } B) = P(A) + P(B) - P(A \text{ and } B) \qquad (5.2)$$

Applying this addition rule to the previous example we obtain the following result:

$$\begin{aligned} P(\text{Accounting } \textit{or} \text{ B Average}) &= P(\text{Accounting}) + P(\text{B Average}) \\ &\quad - P(\text{Accounting } \textit{and} \text{ B Average}) \\ &= \frac{35}{94} + \frac{37}{94} - \frac{15}{94} \\ &= \frac{57}{94} \end{aligned}$$

The addition rule takes the probability of A and adds it to the probability of B. The intersection of A *and* B must be subtracted from this total because it has already been included twice in computing the probability of A and of B. This can be clearly demonstrated by referring to the Venn diagram. If the outcomes of the event accounting major (circle A) are added to those of the event B average (circle B), then the event "Accounting major *and* B average" (the intersection of A *and* B) has been included in each of these events. Therefore, since this has been "double counted," it must be subtracted to provide the correct result.

On the other hand, how would we calculate the probability of picking a card that was a heart or a spade? Using the addition rule, we have the following:

$$P(\text{Heart } or \text{ Spade}) = P(\text{Heart}) + P(\text{Spade}) - P(\text{Heart } and \text{ Spade})$$
$$= \frac{13}{52} + \frac{13}{52} - \frac{0}{52}$$
$$= \frac{26}{52}$$

The probability that a card will be both a heart *and* a spade is zero, since each individual card can be of one and only one suit. The Venn diagram for this example is shown in Figure 5.5.

The intersection in this case is nonexistent (called the **null set**) because it contains no outcomes since heart and spade cannot occur simultaneously in the same card. Whenever the intersection of events A *and* B has zero probability of occurrence, these events are called **mutually exclusive** since the occurrence of one event precludes the occurrence of the other. The **addition rule for mutually exclusive events** reduces to the following:

$$P(A \text{ or } B) = P(A) + P(B) \qquad (5.3)$$

assuming A and B are mutually exclusive.

Let us examine one additional example. What would be the probability of picking a card that was red *or* black. Since red and black are mutually exclusive events, the addition rule would be as follows:

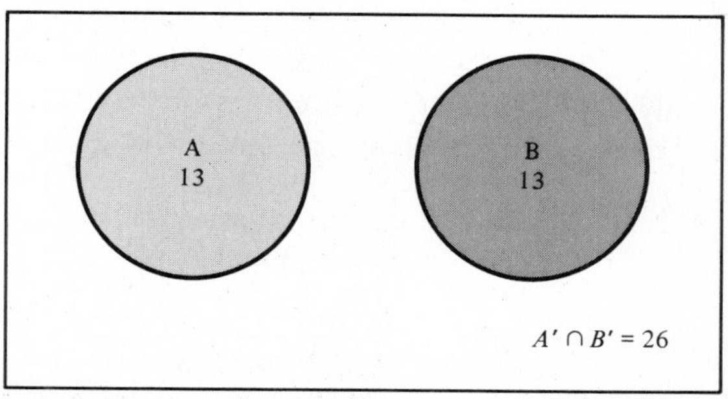

Figure 5.5 Venn diagram for heart and spade examples. A = heart, B = spade.

$$P(\text{Red } or \text{ Black}) = P(\text{Red}) + P(\text{Black})$$

Since red and black are mutually exclusive,

$$P(\text{Red } or \text{ Black}) = \frac{26}{52} + \frac{26}{52} = \frac{52}{52} = 1.0$$

The probability of red *or* black adds up to 1.0. This means that the card selected must be either red or black. Whenever the probability for a set of events add up to 1.0, the events are said to be **collectively exhaustive.**

5.7 CONDITIONAL PROBABILITY

Each of the cases that we have examined thus far has involved the probability of a particular event when sampling from the entire sample space. However, how would we find various probabilities if certain information about the events involved was already known? For example, if we were told that the card was black, what would be the probability that the card was an ace? If we were told that the student selected had a B average or above, what would be the probability that he or she is an accounting major?

When we are computing the probability of a particular event (A), given information about the occurrence of another event (B), this probability is referred to as **conditional probability** $(A|B)$. The conditional probability $(A|B)$ can be defined as follows:

$$P(A|B) = \frac{P(A \text{ and } B)}{P(B)} \tag{5.4}$$

where $P(A \text{ and } B)$ = joint probability of A and B

$P(B)$ = marginal probability of B

Rather than utilizing Equation (5.4) for finding conditional probability, the contingency table or Venn diagram can be evaluated. In the first example, we wish to find $P(\text{Ace}|\text{Black})$. Here the information is given that the card is black. Therefore the sample space does not consist of all 52 cards in the deck; it only consists of the black cards. Of the 26 black cards, two are aces. Therefore the probability of an ace, given that we know the card is black is

$$P(\text{Ace}|\text{Black}) = \frac{\text{Number of Black Aces}}{\text{Number of Black Cards}}$$

$$= \frac{2}{26}$$

This result (2/26) can also be obtained by using Equation (5.4) as illustrated on the next page.

$$P(A|B) = \frac{P(A \text{ and } B)}{P(B)}$$

If

$$\text{event } A = \text{ace}$$
$$\text{event } B = \text{black}$$

then

$$P(\text{Ace}|\text{Black}) = \frac{P(\text{Ace } \textit{and} \text{ Black})}{P(\text{Black})}$$
$$= \frac{2/52}{26/52}$$
$$= \frac{2}{26}$$

Let us now examine the second example mentioned, that of determining P(Accounting|B or above). Since the information given is that the student has a B average or above, the sample space is reduced to those students who have B averages or above (37 students). Of these 37 students it can be seen that 15 are accounting majors. Therefore the probability that the student is an accounting major given that he or she has a B average or above can be computed as follows (see Table 5.2):

$$P(\text{Accounting}|\text{B or above}) = \frac{\text{Number of Accounting Majors with B or above Averages}}{\text{Number of B or above Averages}}$$
$$= \frac{15}{37}$$

Again, Equation (5.4) would provide the same answer as follows:

$$P(A|B) = \frac{P(A \text{ and } B)}{P(B)}$$

If

$$\text{event } A = \text{accounting major}$$
$$\text{event } B = \text{B or above average}$$

then

$$P(\text{Accounting}|\text{B or above}) = \frac{P(\text{Accounting } \textit{and} \text{ B or above})}{P(\text{B or above})}$$
$$= \frac{15/94}{37/94}$$
$$= \frac{15}{37}$$

It has been observed that in the first example the probability that the card picked is an ace, given that we know it is black, is 2/26. We may remember that the probability of picking an ace out of the deck, *P*(Ace), was 4/52, which reduces to 2/26. This result reveals some important information. The prior knowledge that the card was black did not affect the probability that the card was an ace. This characteristic is called **statistical independence** and can be defined as follows:

$$P(A|B) = P(A) \qquad (5.5)$$

where $P(A|B)$ = conditional probability of A given B
$P(A)$ = marginal probability of A

Thus we may note that two events A and B are statistically independent if and only if $P(A|B) = P(A)$.

Here, "color of the card" and "being an ace" are statistically independent events. Knowledge of one event in no way affects the probability of the second event.

We should also like to determine whether being an accounting major is independent of whether the student has a B average or above. The proportion of students with B or above averages who are accounting majors (Accounting|B or above) is 15/37 = .405 while the proportion of all students who are accounting majors is 35/94 = .372. Since $P(\text{Accounting}|\text{B or above}) \neq P(\text{Accounting})$, being an accounting major and having a B average are not statistically independent. The proportion of students who have B averages is not the same for accounting and non-accounting majors. Therefore, knowledge of a student's average does give information about the probability of being an accounting major, and these two events are not independent.

5.8 MULTIPLICATION RULE

The formula for conditional probability can be manipulated algebraically so that the joint probability (A *and* B) can be determined from the conditional probability of an event. From the formula for conditional probability we have

$$P(A|B) = \frac{P(A \text{ and } B)}{P(B)}$$

Solving for the joint probability (A *and* B), we have **the general multiplication rule:**

$$P(A \text{ and } B) = P(A \mid B)P(B) \qquad (5.6)$$

The use of this multiplication rule can be examined with the following example. Let us say that a display of 15 T-shirts in a department store contains three different sizes: small, medium, and large, and that of the 15 T-shirts, three are small, six are medium, and six are large. We are to randomly select two T-shirts from the 15 T-shirts. What would be the probability that both T-shirts selected are small? Here the multiplication rule could be used in the following way:

$$P(A \text{ and } B) = P(A|B)P(B)$$

Therefore if

$$A_s = \text{second T-shirt selected is small}$$
$$B_s = \text{first T-shirt selected is small}$$

we have

$$P(A_s \text{ and } B_s) = P(A_s|B_s)P(B_s)$$

The probability that the first T-shirt is small would be 3/15 since three of the 15 T-shirts are small. However, the probability that the second T-shirt is also small depends upon the result of the first selection. If the first T-shirt is not returned to the display after its size is determined (sampling *without* replacement), then the number of T-shirts remaining will be 14. If the first T-shirt selected was small, the probability that the second also is small is 2/14 since there would be two small T-shirts remaining in the display. Therefore, we have the following:

$$P(A_s \text{ and } B_s) = \left(\frac{2}{14}\right)\left(\frac{3}{15}\right)$$
$$= \frac{6}{210} = .029$$

However, what if the first T-shirt selected was returned to the display after its size had been determined? Then the probability of picking a small T-shirt on the second selection is the same as on the first selection (sampling *with* replacement) since there are three small T-shirts out of 15 in the display. Therefore, we have the following:

$$P(A_s \text{ and } B_s) = P(A_s|B_s)P(B_s)$$
$$= \left(\frac{3}{15}\right)\left(\frac{3}{15}\right)$$
$$= \frac{9}{225} = .04$$

This example of sampling *with* replacement illustrates that the second selection is independent of the first since the second probability was not influenced by the first selection. Therefore, the **multiplication rule for independent events** can be expressed as follows (by substituting $P(A)$ for $P(A|B)$):

$$P(A \text{ and } B) = P(A)P(B) \tag{5.7}$$

If this rule holds for two events, A and B, then A and B are statistically independent. Therefore, there are two ways to determine statistical independence.

1. Events A and B are statistically independent if and only if $P(A|B) = P(A)$.
2. Events A and B are statistically independent if and only if $P(A \text{ and } B) = P(A)P(B)$.

It should be noted that for a 2 × 2 contingency table, if this is true for one joint event, it will be true for all joint events.

Now that we have discussed the multiplication rule, the formula for marginal probability (5.1) can be written as follows. If

$$P(A) = P(A \text{ and } B_1) + P(A \text{ and } B_2) + \cdots + P(A \text{ and } B_k)$$

then using the multiplication rule, we have

$$P(A) = P(A|B_1)P(B_1) + P(A|B_2)P(B_2) + \cdots + P(A|B_k)P(B_k) \tag{5.8}$$

where $B_1, B_2, \ldots, B_k$ are k mutually exclusive events.

The use of this formula will be illustrated by examining the randomized response technique, a way of anonymously obtaining responses to highly controversial questions.

5.9 THE RANDOMIZED RESPONSE TECHNIQUE: AN APPLICATION OF CONDITIONAL PROBABILITY

In Chapter 2 (Section 2.6.4) we discussed how the randomized response technique could be used to obtain answers to highly personal questions. Depending on the result of a random event, such as the flip of a coin, the respondent answers either the personal question (question A) or the control question (question B). The question that is being answered is unknown to anyone but the respondent. However, from the results observed for all respondents and from the rules of conditional probability, we can estimate the proportion of all respondents who are responding "yes" to the personal question.

The marginal probability of an event has been defined in Equation (5.8). In this case there are two events, question A and question B. There is a known probability that each of these questions will be answered. Therefore, if

$$
\begin{aligned}
P(Y) &= \text{probability of a "yes" answer} \\
P(A) &= \text{probability that the question answered is question A} \\
P(B) &= \text{probability that the question answered is question B} \\
P(Y|A) &= \text{probability of a "yes" answer given question A is answered} \\
P(Y|B) &= \text{probability of a "yes" answer given question B is answered}
\end{aligned}
$$

then

$$P(Y) = P(Y|A)P(A) + P(Y|B)P(B) \qquad (5.9)$$

Since we can determine the probability of answering "yes" to the control question, and the total probability of any "yes" answer can be found by totaling up the number of "yes" answers in the sample, the probability of a "yes" answer to the personal question, $P(Y|A)$, can be found as follows:

$$P(Y|A) = \frac{P(Y) - P(Y|B)P(B)}{P(A)} \qquad (5.10)$$

where $P(Y) = \dfrac{\text{Number of "Yes" Answers}}{\text{Sample Size}}$

Since $P(Y|A)$ must be a number between 0 and 1, we can estimate $P(Y|A)$ whenever $P(Y)$ satisfies the following inequality:

$$P(Y|B)P(B) \leq P(Y) \leq P(A) + P(Y|B)P(B)$$

Referring to the data collected by the researcher and the dean (Figure 2.2), in question 26 we wanted to determine the proportion of students who are receiving any scholarship, stipend, or financial assistance to attend college. The question set was worded with the following type of instructions:

> For the following set of questions flip a coin; if the coin lands on "heads" only answer question A; if the coin lands on "tails" only answer question B.

Do not divulge the result of the coin flip or the question you are answering. Place answer here: YES ☐ NO ☐

A. Are you receiving any scholarship, stipend, or financial assistance to attend college?
B. Was your mother born in a year ending in an odd number (1, 3, 5, 7, 9)?

From the responses to question 26 of the survey, as summarized in Figure 2.6, we can see that out of a sample of 94 respondents, 50 answered "yes" to the personal question (concerning financial assistance). This sample information can be used to help determine the proportion of students receiving financial assistance. Since a fair coin is being flipped, the probability of answering question A, $P(A)$, is 1/2 and that of answering question B, $P(B)$, is also 1/2. If it is assumed that an individual is equally likely to be born in an odd- or even-numbered year, the probability of answering "yes" to question B, $P(Y|B)$, is 1/2. With all this information, we can solve for the probability of answering "yes" to question A. Using Equation (5.10), we have

$$P(Y|A) = \frac{P(Y) - P(Y|B)P(B)}{P(A)}$$

If

$$P(A) = 1/2$$
$$P(B) = 1/2$$
$$P(Y|B) = 1/2$$

$$P(Y|A) = \frac{P(Y) - (1/2)(1/2)}{1/2}$$
$$= \frac{P(Y) - 1/4}{1/2}$$

We will be able to estimate $P(Y|A)$ as long as $P(Y)$ satisfies

$$1/4 \leq P(Y) \leq 1/2 + 1/4$$

or

$$.25 \leq P(Y) \leq .75$$

Since

$$P(Y) = \frac{\text{Number of "Yes" Answers}}{\text{Sample Size}}$$

we have

$$P(Y) = \frac{50}{94} = .532$$

Therefore we will be able to estimate $P(Y|A)$ as

$$P(Y|A) = \frac{.532 - .25}{.50} = .564$$

Our estimate, based on the sample results, is that the proportion of students receiving financial assistance is .564.

We have seen that we could estimate the proportion who answered "yes" to question A, $P(Y|A)$, only if the proportion of "yes" answers, $P(Y)$, was within certain limits. If we have prior information that can be used to approximate $P(Y|A)$, then the randomized response question can be set up using appropriate values of $P(A)$ and $P(Y|B)$.

These restrictions are placed upon the probabilities because of the nature of the randomized response technique. The responses of each individual participating in the survey are anonymous, since they are known only to each respondent. Therefore, the techniques of conditional probability must be utilized as a means of measuring these probabilities.

5.10 BAYES' THEOREM

Conditional probability takes into account information about the occurrence of one event to predict the probability of another event. This concept can be extended to "revise" probabilities based on new information and to determine the probability that a particular effect was due to a specific cause. The procedure for revising these probabilities is known as **Bayes' theorem** [since it was originally developed by Rev. Thomas Bayes (1702–1761) in an attempt to prove the existence of God].

Bayes' theorem can be applied to the following problem: The marketing manager of a toy manufacturing firm is planning to introduce a new toy into the market. In the past, 40% of the toys introduced by the company have been successful, 60% have not been successful. Before the toy is actually marketed, market research is conducted and a report, either favorable or unfavorable, is compiled. In the past, 80% of the successful toys received favorable reports and 30% of the unsuccessful toys also received favorable reports. The marketing manager would like to know the probability that the new toy will be successful if it receives a favorable report.

Bayes' theorem can be developed from the definitions of conditional and marginal probability in the following manner:

$$P(A \text{ and } B) = P(A|B)P(B) \tag{5.11a}$$

but also

$$P(A \text{ and } B) = P(B|A)P(A) \tag{5.11b}$$

From Equations (5.11*a*) and (5.11*b*) we have

$$P(B|A)P(A) = P(A|B)P(B)$$

so that

$$P(B|A) = \frac{P(A|B)P(B)}{P(A)} \tag{5.11c}$$

Since from Equation (5.8)

$$P(A) = P(A|B_1)P(B_1) + P(A|B_2)P(B_2) + \cdots + P(A|B_k)P(B_k)$$

Bayes' theorem is:

$$P(B_i|A) = \frac{P(A|B_i)P(B_i)}{P(A|B_1)P(B_1) + P(A|B_2)P(B_2) + \cdots + P(A|B_k)P(B_k)} \tag{5.11d}$$

where B_i is the ith event out of k mutually exclusive events.

Now we can use Bayes' theorem to solve the toy manufacturer's problem. Let

event S = successful toy $\quad S'$ = unsuccessful toy

event F = favorable report $\quad F'$ = unfavorable report

and

$$P(S) = .40 \qquad P(F|S) = .80$$
$$P(S') = .60 \qquad P(F|S') = .30$$

Then

$$\begin{aligned} P(S|F) &= \frac{P(F|S)P(S)}{P(F|S)P(S) + P(F|S')P(S')} \\ &= \frac{(.80)(.40)}{(.80)(.40) + (.30)(.60)} \\ &= \frac{.32}{.32 + .18} \\ &= .64 \end{aligned}$$

The probability of a successful toy given that a favorable report was received is .64. Thus the probability of an unsuccessful toy given that a favorable report was received is .36 since there are only two possible events.

$$P(S'|F) = 1 - P(S|F)$$

The denominator of Bayes' theorem represents the marginal probability of event F, in this case a favorable market research report. Therefore the proportion of toys that receive favorable market research reports is .50.

Table 5.3
BAYES' THEOREM CALCULATIONS FOR TOY MANUFACTURER'S PROBLEM

Events S_i	Prior Probability $P(S_i)$	Conditional Probability $P(F\|S_i)$	Joint Probability $P(F\|S_i)P(S_i)$	Revised Probability $P(S_i\|F)$
S = successful toy	.40	.80	.32	$.32/.50 = .64 = P(S\|F)$
S' = unsuccessful toy	.60	.30	.18	$.18/.50 = .36 = P(S'\|F)$
			.50	

The computation of the probabilities can also be demonstrated by Table 5.3. This illustrative table is particularly helpful when there are more than two events that can occur, as in the following example. The quality control manager of a tire factory would like to determine which production shift produced a tire that has blown out. There are three shifts of workers in the factory: day, evening, and night. Based on past data, of the tires produced by the factory 40% were by the day shift, 40% by the evening shift, and 20% by the night shift. 5% of the tires produced by the day shift blew out, 10% of the tires produced by the evening shift blew out, and 20% of the tires produced by the night shift blew out. What is the probability that the tire that blew out was produced by the day shift? the evening shift? the night shift?

To solve this, we use the following information:

event B = tire blew out
event B' = tire did not blow out

event S_1 = day shift
event S_2 = evening shift
event S_3 = night shift

and

$$P(S_1) = .40 \qquad P(B|S_1) = .05$$
$$P(S_2) = .40 \qquad P(B|S_2) = .10$$
$$P(S_3) = .20 \qquad P(B|S_3) = .20$$

For example, for the day shift,

$$\begin{aligned} P(S_1|B) &= \frac{P(B|S_1)P(S_1)}{P(B|S_1)P(S_1) + P(B|S_2)P(S_2) + P(B|S_3)P(S_3)} \\ &= \frac{(.05)(.40)}{(.05)(.40) + (.10)(.40) + (.20)(.20)} \\ &= \frac{.02}{.02 + .04 + .04} \\ &= \frac{.02}{.10} = .20 \end{aligned}$$

Since the denominator is .10, it should be noted that the total probability of a blow out, $P(B)$, over all shifts is .10 (see Table 5.4).

Table 5.4
BAYES' THEOREM CALCULATIONS FOR TIRE COMPANY PROBLEM

Events S_i	Prior Probability $P(S_i)$	Conditional Probability $P(B\|S_i)$	Joint Probability $P(B\|S_i)P(S_i)$	Revised Probability $P(S_i\|B)$
S_1 = day shift	.40	.05	.02	$.02/.10 = .20 = P(S_1\|B)$
S_2 = evening shift	.40	.10	.04	$.04/.10 = .40 = P(S_2\|B)$
S_3 = night shift	.20	.20	.04	$.04/.10 = .40 = P(S_3\|B)$
			.10	

Referring to the data collected in the dean's survey, Bayes' theorem can also be utilized to determine the conditional probability that an accounting major will have a B average or above, $P(B$ or above|Accounting), based upon information about the proportion of students with B or above averages who are accounting majors, P(Accounting|B or above).

In Sections 5.4 and 5.7 we have already computed the following:

event A = accounting major   event B = B average or above

event A' = non-accounting major   event B' = below B average

From Table 5.2 we have

$$P(B) = 37/94 \qquad P(A|B) = 15/37$$
$$P(B') = 57/94 \qquad P(A|B') = 20/57$$

Therefore,

$$P(B|A) = \frac{P(A|B)P(B)}{P(A|B)P(B) + P(A|B')P(B')}$$
$$= \frac{(15/37)(37/94)}{(15/37)(37/94) + (20/57)(57/94)}$$

Thus,

$$P(B|A) = \frac{15/94}{35/94} = 15/35 = .429$$

In these three problems Bayes' theorem has been used:

1. To revise the probability of marketing a successful toy in the light of new information (a favorable market research report)
2. To determine the probability that a certain effect (a blown-out tire) resulted from a specific shift of workers
3. To compute a conditional probability given information about other conditional probabilities

Bayes' theorem will be further discussed in Chapter 11 (Bayesian decision making) where it will be used to help choose between alternative courses of action.

5.11 COUNTING RULES

Each rule of probability has involved the counting of the number of favorable outcomes and the total number of outcomes. However, in many instances, because of the large number of possible outcomes that may occur, it is not feasible to list all the possible outcomes. For these types of cases, rules for counting have been developed.

We shall examine five different counting rules. First of all, suppose that a coin was being flipped 10 times. How would we determine the number of different possible outcomes (the sequences of heads and tails)?

Counting Rule 1: If any one of k different mutually exclusive and collectively exhaustive events can occur on each of n trials, the number of possible outcomes is equal to

$$k^n \tag{5.12}$$

If a coin (having two sides) is tossed 10 times, the number of outcomes would be $2^{10} = 1{,}024$. If a die (having six sides) is rolled twice, the number of different outcomes would be $6^2 = 36$.

Second, let us say that the number of possible events is different on some of the trials. For example, a state motor vehicle department would like to know how many license plate numbers would be available if the license plate consisted of three digits followed by two letters. The fact that three values are digits (each having 10 possible outcomes) while two positions are letters (each having 26 outcomes) leads to the second rule of counting.

Counting Rule 2: If there are k_1 events on the first trial, k_2 events on the second trial, $\cdots$, and k_n events on the nth trial, then the number of possible outcomes is

$$(k_1)(k_2) \cdots (k_n) \tag{5.13}$$

Thus if a license plate consisted of three digits followed by two letters, the total numbers of possible outcomes would be $(10)(10)(10)(26)(26) = 676{,}000$. Taking another example, if a restaurant menu had a choice of four appetizers, ten entrees, three beverages, and six desserts, the total number of possible dinners would be $4 \times 10 \times 3 \times 6 = 720$.

A third counting rule, one that is extremely important in computing probabilities, involves the computation of the number of ways that a set of objects can be arranged in order. If a set of six textbooks are to be arranged on a shelf, how can we determine the number of ways in which the six books may be arranged? The number of ways could be determined in the following manner. Any of the six books could occupy the first position on the shelf. Once the first position is filled, there are five books to choose from in filling the second. This procedure is continued until all the positions are occupied. The total number of arrangements would be equal to (6)(5)(4)(3)(2)(1) = 720. This concept can be generalized in counting rule 3.

Counting Rule 3: The number of ways that all n objects can be arranged in order is

$$n! = n(n-1) \cdots (1) \tag{5.14}$$

where $n!$ is called n **"factorial"** and 0! is defined as 1.

The number of ways that six books could be arranged is

$$\begin{aligned} n! &= (6)(5)(4)(3)(2)(1) \\ &= 720 \end{aligned}$$

In many instances it is important to know the number of ways that X objects, selected from n objects ($X \leq n$), can be arranged in order. For example, modifying the previous problem, if six textbooks are involved, but there is room for only four books on the shelf, how many ways can these books be arranged on the shelf?

This rule is called the rule of **permutations.**

Counting Rule 4: Permutations: The number of ways of arranging X objects selected from n objects in order is

$$\frac{n!}{(n-X)!} \tag{5.15}$$

Therefore, the number of ordered arrangements of four books selected from six books is equal to

$$\frac{n!}{(n-X)!} = \frac{6!}{(6-4)!} = \frac{6!}{2!} = \frac{6(5)(4)(3)(2)(1)}{2(1)} = 360$$

Finally, in many problems we are not interested in the *order* of the outcomes, but only in the number of ways that X objects can be selected out of n objects, *irrespective* of order. This rule is called the rule of **combinations.**

Counting Rule 5: Combinations: The number of ways of selecting X objects out of n objects, irrespective of order is equal to

$$\frac{n!}{X!(n-X)!} \tag{5.16}$$

This expression may be denoted by the symbol $\binom{n}{X}$.

Therefore, the number of combinations of four books selected from six books is expressed by $\binom{6}{4}$. This is equal to

$$\frac{n!}{X!(n-X)!} = \frac{6!}{4!\,(6-4)!} = \frac{6!}{4!2!} = \frac{6(5)(4)(3)(2)(1)}{4(3)(2)(1)(2)(1)} = 15$$

5.12 SUMMARY

In this chapter we have examined various rules of probability as well as applications of these rules to a variety of problems. Probability theory is the foundation of statistical inference. These concepts will be extended to more complicated situations in subsequent chapters so that methodology can be developed to use descriptive statistics in order to make inferences about populations.

PROBLEMS

* 5.1 A manager of a women's store wishes to determine the relationship between the type of customer and the type of payment. She has collected the following data:

	Payment	
Customer	Credit	Cash
Regular	70	50
Nonregular	40	40

What is the probability that if a customer is selected at random:

(a) The customer is regular?

(b) The customer is regular **and** buys on credit?

(c) The customer is regular **or** pays cash?

(d) Assume we know that the customer is regular. What is the probability then that she will buy on credit?

(e) Are the two events being a regular customer **and** buying on credit independent? Explain.

5.2 The purchase of many consumer goods may involve various degrees of preplanning. Items such as a pair of slacks or a suit may be the result of a few days' or weeks' planning. Other more expensive purchases such as a car or a house may require planning ahead for a year or more before actual buying takes place.

Numerous intensive studies have been conducted of consumer planning for the purchase of durable goods such as television sets, refrigerators, washing machines, stoves, and automobiles. In one such study a randomly selected sample of 1,000 individuals were asked whether they were planning to buy a new television in the next 12 months. A year later the same persons were interviewed again to find out whether they actually bought a new television. The response to both interviews (data hypothetical) is cross-tabulated in the table below:

	B (Buyers)	B' (Nonbuyers)	Total
P (planned to buy)	200	50	250
P' (did not plan to buy)	100	650	750
Total	300	700	1,000

If an individual is selected at random, what is the probability that in the last year he or she

(a) Has bought a new television?

(b) Planned to buy a new television?

(c) Planned to buy **and** actually bought a new television?

(d) If the respondent planned to buy a new television, what is the probability that he or she actually bought one?

(e) If the respondent did not plan to buy a new television, what is the probability that he or she did not buy a new television?

(f) Is planning to buy a new television **and** actually buying one statistically independent?

5.3 When rolling a die once, what is the probability that

(a) The face of the die is odd?

(b) The face is even **or** odd?

(c) The face is even **or** a one?

(d) The face is odd **or** a one?

(e) The face is both even **and** a one?

(f) Given the face is odd, it is a one?

5.4 The 500 credit customers of the Upay Credit Co. are categorized according to the number of years they have held a credit account with Upay Credit Co. and by their average credit balance. 210 of the customers have credit balances below \$100. 260 of the customers have held a credit account for at least five years. 80 of the customers have balances over \$100 **and** have had credit accounts for less than five years. If one credit customer is selected at random, what is the probability that he or she

(a) Has a credit balance above \$100?

(b) Has a credit balance above \$100 **or** has had a credit account for at least five years?

(c) Has a credit balance below \$100 **and** has had a credit account for less than five years?

(d) Assume we know that a customer has had a credit account for at least five years. What then is the probability that he has a cash balance below \$100?

(e) Show whether having a credit balance above $100 **and** having a credit account for at least five years are statistically independent.

Hint: Set up a 2 × 2 table or a Venn diagram to evaluate the probabilities.

* 5.5 A computer dating company currently has on its files the names (and addresses) of 200 women. Of these 200, a total of 35 are 5 feet 2 inches or under; 60 are blondes; 12 of the blondes are 5 feet 2 inches or under. Paul Smith mails in his application.

(a) What is the probability that he receives the name of a blonde woman?

(b) What is the probability that he receives the name of a woman who is blonde **and** is taller than 5 feet 2 inches?

(c) What is the probability that he receives the name of a blonde woman **or** a woman shorter than 5 feet 2 inches?

(d) He calls the woman and makes a date. They agree to meet at Lew's Place. As he approaches, she is sitting at the counter and he sees that she has blonde hair. What then is the probability that she is taller than 5 feet 2 inches?

(e) Determine whether the two events of having blonde hair **and** being taller than 5 feet 2 inches are statistically independent.

Hint: Set up a 2 × 2 table or a Venn diagram to evaluate the probabilities.

5.6 Of 250 employees of a company a total of 130 smoke cigarettes. There are 150 males working for this company. 85 of the males smoke cigarettes. What is the probability that an employee chosen at random

(a) Does not smoke cigarettes?

(b) Is female **and** smokes cigarettes?

(c) Is male **or** smokes cigarettes?

(d) Let us say we meet a female employee of the company. What then is the probability she does not smoke cigarettes?

(e) Determine whether cigarette smoking **and** sex are statistically independent.

Hint: Set up a 2 × 2 table or a Venn diagram to evaluate the probabilities.

5.7 Suppose a survey has been undertaken to determine if there is a relationship between the size of a city in which someone lives and the company preference in the purchase of an automobile. A random sample of 200 car owners from large cities, 150 from suburbs, and 150 from rural areas were selected with the following results:

	Company					
Type of Area	GM	Ford	Chrysler	American	Foreign	Total
Large city	64	40	26	8	62	200
Suburb	53	35	24	6	32	150
Rural	53	45	30	6	16	150
Total	170	120	80	20	110	500

If a car owner is selected at random, what is the probability that he or she

(a) Owns a GM car?

(b) Lives in a suburb?

(c) Lives in a large city **and** owns a foreign car?

(d) Lives in a rural area **and** owns a Ford?

(e) Lives in a suburb **or** owns a Chrysler?

(f) Assume that we know that the person selected lives in a large city. What is the probability that she owns a foreign car?

(g) Is the size of the place in which someone lives statistically independent of company preference in automobile ownership? Explain.

5.8 The Statistics Association at a large state university would like to determine whether there is a relationship between a student's interest in statistics and his or her ability in mathematics. A random sample of 200 students is selected and they are asked whether their ability in mathematics and interest in statistics is low, average, or high. The results were as follows:

	Ability in Mathematics			
Interest in Statistics	*Low*	*Average*	*High*	*Total*
Low	60	15	15	90
Average	15	45	10	70
High	5	10	25	40
Total	80	70	50	200

If a student is selected at random, what is the probability that he or she

(a) Has a high ability in mathematics?

(b) Has an average interest in statistics?

(c) Has a low ability in mathematics **and** a low interest in statistics?

(d) Has a high interest in statistics **or** a high ability in mathematics?

(e) Assume that we know the person selected has a high ability in mathematics. What is the probability that she has a high interest in statistics?

(f) Are interest in statistics and ability in mathematics statistically independent? Explain.

* 5.9 A nationwide market research study was undertaken to determine the preferences of various age groups of males for different sports. A random sample of 1,000 men was selected and each individual was asked to indicate his favorite sport. The results were as follows:

	Sport				
Age Group	*Baseball*	*Football*	*Basketball*	*Hockey*	*Total*
Under 20	26	47	41	36	150
20–29	38	84	80	48	250
30–39	72	68	38	22	200
40–49	96	48	30	26	200
50 and over	134	44	18	4	200
Total	366	291	207	136	1,000

If a respondent is selected at random, what is the probability that he

(a) Prefers baseball?

(b) Is between 20 and 29 years old?

(c) Is between 20 and 29 years old **and** prefers basketball?

(d) Is at least 50 years of age **or** prefers baseball?

P(Hockey | under 20) = P[Hockey · Under 20] / P(under 20) = 36/150

(e) Assume that the person selected is under 20 years old. What is the probability that he prefers hockey?

5.10 For the cross-classification table developed in Problem 3.14, what is the probability that a student selected at random

(a) Is a senior?
(b) Is planning to go to graduate school?
(c) Is a sophomore **or** is not planning to go to graduate school?
(d) Is a freshman **and** is planning to go to graduate school?
(e) If we selected a senior, what is the probability that he is planning to go to graduate school?
(f) Are class designation **and** graduate school intentions statistically independent? Explain.

5.11 For the sample of 94 students (Figure 2.6) set up a cross-classification table for sex (question 1) and the attitude toward legislation prohibiting Sunday shopping (question 10). Determine the probability that a student selected at random

(a) Is a male.
(b) Is opposed to legislation prohibiting shopping.
(c) Is a male **and** favors the legislation.
(d) Is a female **or** favors the legislation.
(e) If a female is selected, what is the probability that she favors legislation prohibiting Sunday shopping?
(f) Are the two variables (sex **and** attitude toward shopping) independent? Explain.

5.12 For the sample of 94 students (Figure 2.6) set up a cross-classification table for the amount of cigarette smoking (question 19) and the attitude toward smoking in the classroom (question 20). Determine the probability that a student selected at random

(a) Does not smoke.
(b) Thinks smoking should be permitted in classrooms.
(c) Does not smoke **or** does not think smoking should be permitted in classrooms.
(d) Smokes two or more packs per day **and** thinks smoking should be permitted in classrooms.
(e) If a student does not smoke, what is the probability that he or she thinks that smoking should be permitted in classrooms?
(f) Is amount of cigarette smoking statistically independent of attitude toward smoking in the classroom? Explain.

5.13 For the sample of 94 students (Figure 2.6) set up a cross-classification table for sex (question 1) and class designation (question 2). Determine the probability that a randomly selected student

(a) Is a female sophomore.
(b) Is a junior.
(c) Is a male **or** a senior.
(d) Is a male given you know he is a freshman.
(e) Is sex independent of class designation? Explain.

5.14 For the sample of 94 students (Figure 2.6) set up a cross-classification table for class designation (question 2) and employment (question 9). Determine the probability that a randomly selected student

(a) Is unemployed.
(b) Is an unemployed senior.
(c) Is unemployed **or** a senior.
(d) Given that a senior is selected. What then is the probability that the student is unemployed?
(e) Are class designation **and** employment status statistically independent? Explain.

5.15 For the sample of 94 students (Figure 2.6) set up a cross-classification table for sex (question 1) and blood pressure status (question 18). Find the probability that a student selected at random

(a) Has had high blood pressure.
(b) Is a male having had high blood pressure.
(c) Is either a male **or** has had high blood pressure.
(d) If the student is male, what then is the probability he has had high blood pressure?
(e) Do sex **and** blood pressure status appear to be independent? Explain.

5.16 For the sample of 94 students (Figure 2.6) set up a cross-classification table for class designation (question 2) and attitude toward retention standards (question 22). Determine the probability that a randomly selected student

(a) Desires more stringent retention standards at the college.
(b) Is a junior **and** desires more stringent retention standards.
(c) Is a junior **or** desires more stringent retention standards.
(d) If a student is a junior, what then is the probability that he or she desires more stringent retention standards?
(e) Are class designation **and** attitude toward retention standards statistically independent? Explain the above in light of proposed policy which will require students below the junior level to "qualify" for retention at the college through special second-year examinations.

5.17 Use your own sample data (Problem 3.12) to select problems from among Problems 5.10 through 5.16.

5.18 **(Class Project)** For your own class (Problem 3.13) select problems from among Problems 5.10 through 5.16.

5.19 From among the qualitative type questions in the questionnaire (Figure 2.2) develop a set of five other tables of cross classifications that might be of interest.

5.20 A standard deck of cards is being used to play a game. There are four suits (hearts, diamonds, clubs, and spades), each having thirteen cards (ace, 2, 3, 4, 5, 6, 7, 8, 9, 10, jack, queen, and king), making a total of 52 cards. This complete deck is thoroughly mixed, and you will receive the first two cards from the deck **without** replacement.

(a) What is the probability that both cards are queens?

(b) What is the probability that the first card is a 10 **and** the second card is a 5 or 6?

(c) If we were sampling **with** replacement, what would be the answer in (a)?

(d) In the game of blackjack the picture cards (jack, queen, king) count as 10 points while the ace counts as either 1 or 11 points. All other cards are counted at their face value. Blackjack is achieved if your two cards sum to 21 points. What is the probability of getting blackjack in this problem?

* 5.21 A bin contains two defective tubes and five good tubes. Two tubes are randomly selected from the bin **without** replacement.

(a) What is the probability that both tubes are defective?

(b) What is the probability that the first tube selected is defective **and** the second tube is good?

5.22 A box of nine baseball gloves contains two left-handed gloves and seven right-handed gloves. If two gloves are randomly selected from the box **without** replacement, what is the probability that

(a) Both gloves selected will be right-handed?

(b) There will be one right-handed glove **and** one left-handed glove selected?

(c) If three gloves were selected, what is the probability that all three will be left-handed?

(d) If we were sampling **with** replacement, what would be the answers to (a) and (c)?

* 5.23 Referring to the medical survey discussed in Section 2.6.4, if a sample of 100 respondents were selected and 10 responded "yes" to the question set:

A. Have you ever had an abortion?
B. Was your father born in the month of January?

determine the proportion of women who have had at least one abortion.

5.24 Discuss whether the randomized response technique is or is not a good method of obtaining responses to highly personal questions.

* 5.25 A television station would like to measure the ability of its weather forecaster. Past data have been collected that indicate the following:

1. The probability the forecaster predicted sunshine on sunny days is .80.
2. The probability the forecaster predicted sunshine on rainy days is .40.
3. The probability of a sunny day is .60.

Find the probability that

(a) It will be sunny given that the forecaster has predicted sunshine.

(b) The forecaster will predict sunshine.

5.26 An advertising executive is studying television viewing habits of married men and women during prime time hours. Based on past viewing records he has determined that during prime time husbands are watching television 60% of the time. It has also been determined that when the husband is watching television, 40% of the time the wife is also watching. When the husband is not watching television, 30% of the time the wife is watching television. Find the probability that

(a) If the wife is watching television, the husband is also watching television.

(b) The wife is watching television during prime time.

* 5.27 The Olive Construction Co. is determining whether it should submit a bid for the construction of a new shopping center. In the past, Olive's main competitor, Base Construction Co., has submitted bids 70% of the time. If Base Co. does not bid on a job, the probability that the Olive Co. will get the job is .50; if Base Co. does bid on a job, the probability that the Olive Co. will get the job is .25.

(a) If the Olive Construction Co. gets the job, what is the probability that the Base Construction Co. did not bid?

(b) What is the probability that Olive Construction Co. will get the job?

5.28 The editor of a major textbook publishing company is trying to decide whether to publish a proposed new business statistics textbook. Previous textbooks published indicate that 10% are huge successes, 20% are modest successes, 40% break even, and 30% are losers. However, before a publishing decision is made, the book will be reviewed. In the past, 99% of the huge successes received favorable reviews, 70% of the moderate successes received favorable reviews, 40% of the break-even books received favorable reviews, and 20% of the losers received favorable reviews.

(a) If the proposed text receives a favorable review, how should the editor revise the probabilities of the various outcomes to take this information into account?

(b) What proportion of textbooks receive favorable reviews?

5.29 A municipal bond rating service has three rating categories (A, B, and C). In the past year, of the municipal bonds issued throughout the country, 70% have been rated A, 20% have been rated B, and 10% have been rated C. Of the municipal bonds rated A, 50% were issued by cities, 40% by suburbs, and 10% by rural areas. Of the municipal bonds rated B, 60% were issued by cities, 20% by suburbs, and 20% by rural areas. Of the municipal bonds rated C, 90% were issued by cities, 5% by suburbs, and 5% by rural areas.

(a) If a new municipal bond is to be issued by a city, what is the probability that it will receive an A rating?

(b) What proportion of the municipal bonds are issued by cities?

(c) What proportion of the municipal bonds are issued by suburbs?

5.30 A lock on a bank vault consists of three dials, each with 30 positions. In order for the vault to open when closed, each of the three dials must be in the correct position.

(a) How many different possible lock combinations are there for this lock?

(b) What is the probability that, if you randomly select a position on each dial, you will be able to open the bank vault?

* 5.31 (a) If a coin is tossed seven times, how many different outcomes are possible?

(b) If a die is tossed seven times, how many different outcomes are possible?

(c) Discuss the differences in your answer to (a) and (b).

5.32 A famous fast-food restaurant has a menu that consists of ten entrees, two vegetables, four beverages, and three desserts. How many different meals (consisting of one entree, vegetable, beverage, and dessert) can be ordered at this restaurant?

* 5.33 The Daily Double at the local racetrack consists of picking the winners of the first two races. If there are ten horses in the first race and 13 horses in the second race, how many Daily Double possibilities are there?

5.34 There are six teams in the Western Division of the National League: Cincinnati, Los Angeles, San Francisco, Houston, San Diego, and Atlanta. How many different orders of finish are there in which the teams can place? Do you really believe that all these orders are equally likely? Discuss.

* 5.35 A gardener has seven rows available in his vegetable garden to place tomatoes, eggplant, peppers, cucumbers, beans, lettuce, and squash. Each vegetable will be allotted one and only one row. How many ways are there to position these vegetables in his garden?

5.36 A basketball team must schedule a game with each of three different teams. There are five different dates available for games. How many different schedules can be made?

* 5.37 The Big Triple at the local racetrack consists of picking the correct order of finish of the first three horses in the ninth race. If there are 12 horses entered in today's ninth race, how many Big Triple outcomes are there?

5.38 The Quinella at the local racetrack consists of picking the horses that will place first and second in a race irrespective of order. If eight horses are entered in a race, how many Quinella combinations are there?

* 5.39 A student has seven books that she would like to place in an attaché case. However, only four books can fit into the attaché case. How many ways are there of placing four books into the attaché case?

REFERENCES

1. BIERMAN, H., C. BONINI, AND W. HAUSMAN, *Quantitative Analysis for Business Decisions,* 3rd ed., (Homewood, Ill.: Richard D. Irwin, 1974).
2. HAYS, W. L., *Statistics for the Social Sciences* (New York: Holt, Rinehart and Winston, 1973).
3. KIRK, R. E., ed., *Statistical Issues: A Reader for the Behavioral Sciences* (Belmont, Cal.: Wadsworth, 1972).
4. LAPIN, L., *Statistics: Meaning and Method* (New York: Harcourt, Brace, Jovanovich, 1975).
5. MOSTELLER, F., R. ROURKE, AND G. THOMAS, *Probability with Statistical Applications,* 2nd ed., (Reading, Mass.: Addison-Wesley, 1970).
6. NOETHER, G., *Introduction to Statistics: A Nonparametric Approach,* 2nd ed., (Boston: Houghton Mifflin, 1976).

CHAPTER 6

BASIC PROBABILITY DISTRIBUTIONS

6.1 INTRODUCTION

In the previous chapter we established various rules of probability and examined some counting techniques. In this chapter we will utilize this information as we explore various probability models which represent certain phenomena of interest.

6.1.1 The Probability Distribution for a Random Variable

As was discussed in Section 2.5, quantitative random variables or phenomena of interest are those whose responses or outcomes may be expressed numerically. Such random variables may also be classified as discrete or continuous—the former arising from a counting process and the latter arising from a measuring process. As examples, from the questionnaire developed by the dean and researcher in Chapter 2, the responses of the students to the questions about the number of pairs of jeans each student owned or the number of clubs, groups, teams, etc., to which each student belonged are representative of probability distributions for discrete random variables while the responses of students to the questions regarding age, height, weight, anticipated starting salary, tuition, etc., are representative of probability dis-

Table 6.1
THEORETICAL PROBABILITY DISTRIBUTION OF THE RESULTS OF ROLLING ONE FAIR DIE

Face of Outcome	Probability
1 ⚀	1/6
2 ⚁	1/6
3 ⚂	1/6
4 ⚃	1/6
5 ⚄	1/6
6 ⚅	1/6
Totals	1

tributions for continuous random variables. With this in mind we define the probability distribution for a discrete random variable as follows: **A probability distribution for a discrete random variable is a mutually exclusive listing of all possible numerical outcomes for that random variable such that a particular probability of occurrence is associated with each outcome.**

Assuming that a fair six-sided die will not stand on edge or roll out of sight (null events), Table 6.1 represents the probability distribution for the outcomes of a single roll of the fair die. Since all possible outcomes are included, this listing is complete (or collectively exhaustive) and thus the probabilities must sum up to 1.

Using the addition rule for mutually exclusive events, the probability of a face ⚂ is

$$P(⚂) = 1/6$$

furthermore, the probability of an odd face is

$$P(\text{Odd}) = P(⚀) + P(⚂) + P(⚄)$$

$$= 1/6 + 1/6 + 1/6 = 3/6$$

moreover, the probability of a face of ⚁ or less is

$$P(\text{⚁ or less}) = P(\text{⚀}) + P(\text{⚁}) = 1/6 + 1/6 = 2/6$$

and the probability of a face larger than ⚅ is

$$P(> \text{⚅}) = 0$$

For continuous random variables, however, the exact probability of any particular value cannot be computed; we may only determine the probability that a value of the random variable of interest is contained within a particular interval. If our nonoverlapping (mutually exclusive) listing contains all possible intervals (is collectively exhaustive) the probabilities will again sum up to 1. This is demonstrated in Table 6.2 which theoretically represents the thickness of a batch of 10,000 brass washers of a certain type manufactured by a large company. Such a probability distribution may be considered as a relative frequency distribution as described in Section 3.3 where, except for the two open-ended classes, the class mark or midpoint of each (remaining) class interval represents the data in that interval.

Regardless of whether the random variable is discrete or continuous, the probability distribution for the random variable may be

1. A theoretical listing of outcomes and probabilities (as in Tables 6.1 and 6.2) which can be obtained from a mathematical model or function representing some phenomenon of interest.
2. An empirical listing of outcomes and their observed relative frequencies.
3. A subjective listing of outcomes associated with their subjective or

Table 6.2
THICKNESS OF 10,000 BRASS WASHERS

Thickness (inches)	Relative Frequency or Probability
Under .0180	48/10,000
.0180 < .0182	122/10,000
.0182 < .0184	325/10,000
.0184 < .0186	695/10,000
.0186 < .0188	1198/10,000
.0188 < .0190	1664/10,000
.0190 < .0192	1896/10,000
.0192 < .0194	1664/10,000
.0194 < .0196	1198/10,000
.0196 < .0198	695/10,000
.0198 < .0200	325/10,000
.0200 < .0202	122/10,000
.0202 or above	48/10,000
Total	1

"contrived" probabilities representing the degree of conviction of the decision maker as to the likelihood of the possible outcomes (as discussed in Section 5.2).

In this chapter we will be mainly concerned with the first kind of probability distribution—the listing obtained from a mathematical model or function representing some phenomenon of interest.

6.1.2 Mathematical Models of Discrete Random Variables: The Probability Distribution Function

A model may be generally considered as a miniature representation of some underlying phenomenon. In particular, **a mathematical model is a mathematical expression representing some underlying phenomenon.** For discrete random variables, this mathematical expression is known as a **probability distribution function.** When such mathematical expressions are available, the exact probability of occurrence of any particular outcome or value of the random variable can be computed. In such cases, then, the entire probability distribution can be obtained and listed. For example, the probability distribution represented in Table 6.1 is one in which the discrete random variable of interest is said to follow the **uniform probability distribution function.** This and other types of mathematical models for discrete random variables will be discussed in Section 6.3.

6.1.3 Mathematical Models of Continuous Random Variables: The Probability Density Function

On the other hand, for continuous random variables the mathematical expression representing some underlying phenomenon is known as a **probability density function.** When these mathematical expressions are available, the probabilities that particular values of the random variable occur within certain ranges or intervals may be calculated. As an example, the probability distribution represented in Table 6.2 is one in which the continuous random phenomenon of interest is said to follow the bell-shaped **normal probability density function.** This and other types of mathematical models for continuous random variables will be discussed in Section 6.8. In the next section, however, we consider ways in which we may describe the major characteristics or properties of a probability distribution.

6.2 MATHEMATICAL EXPECTATION

In order to summarize a probability distribution we shall compute its major characteristics—the mean and the standard deviation. While our discussion pertains strictly to discrete phenomena, it should be noted that analogous formulas exist for obtaining the mean and the standard deviation pertaining to continuous phenomena.

6.2.1 Expected Value of a Discrete Random Variable

The expected value of some discrete random phenomenon may be considered as its average over all possible values. This summary measure can be obtained by calculating the arithmetic mean of all possible values in the probability distribution weighted by their respective probabilities. Thus $E(X)$ or μ, the expected value of the random variable X, may be expressed as

$$\mu = E(X) = \sum_{i=1}^{N} X_i P(X_i) \tag{6.1}$$

where $E(X)$ = expected value of X

X = discrete random variable of interest

X_i = ith value of X

$P(X_i)$ = probability of occurrence of the ith value of X

$i = 1, 2, \ldots, N$

For the theoretical probability distribution of the results of rolling one fair die (Table 6.1) the expected value of the roll may be computed as

$$\begin{aligned}\mu = E(X) = \sum_{i=1}^{N} X_i P(X_i) &= (1)(1/6) + (2)(1/6) + (3)(1/6) \\ &\quad + (4)(1/6) + (5)(1/6) + (6)(1/6) \\ &= 1/6 + 2/6 + 3/6 + 4/6 + 5/6 + 6/6 \\ &= 21/6 = 3.5\end{aligned}$$

It is noted that the expected value of the results of rolling a fair die is not "literally meaningful" since we can never obtain a face of 3.5. However, we can expect to observe the six different faces with equal likelihood so that, in the long run, over many rolls, the average value would be 3.5. To make this particular situation meaningful, however, we introduce the following carnival game: How much money should we be willing to put up in order to have the opportunity of rolling a fair die if we were to be paid, in dollars, the amount on the face of the die? Since the expected value of a roll of a fair die is 3.5, the expected long-run payoff is \$3.50 per roll, that is, on any particular roll our payoff will be \$1, \$2, . . . , or \$6, but over many, many rolls the payoff can be expected to average out to \$3.50 per roll. Now if we want the game to be fair, neither we nor our opponent (the "house") should have an advantage. Thus we should be willing to pay \$3.50 per roll to play. If the house wants to charge us \$4.00 per roll we can expect to lose from such gambling, on the average, \$.50 per roll over time, and unless we derive some intrinsic satisfaction (costing on the average \$.50 per roll), we should refrain from participating in such a game. Usually though, in any casino or carnival-type game the expected long-run payoff to the participant is negative, otherwise the house would not be in business (Reference 8). Such games as Craps, Chuck-a-luck, Under-or-over-seven, or Roulette (see Problems 6.1, 6.2, 6.3, and 6.4) attract large numbers of participants and, in each case, the expected return over time favors the house. Since a decision maker who acts rationally would want to maximize gains or minimize losses, the only way rational persons would freely partake in such transactions is if something other than expected monetary value is the ultimate criterion. The concept of the **expected utility of money** will be developed in Chapter 11. It is the criterion which rational participants are considering, implicitly or explicitly, when they partake in such games. On the other hand, however, it should be realized that it is the house which uses the expected monetary value criterion when it participates in such games.

6.2.2 Variance of a Discrete Random Variable

The variance of a discrete random variable may be considered the average squared deviation about the mean—taken over all possible values. The variance can be computed as the summation of the product of each squared deviation about the mean in the probability distribution weighted by the respective probabilities. Hence, the variance σ^2 is computed as

$$\sigma^2 = \text{Var}(X) = \sum_{i=1}^{N} (X_i - \mu)^2 P(X_i) \tag{6.2}$$

where $\text{Var}(X)$ = variance of X

X = discrete random variable of interest

X_i = ith value of X

$P(X_i)$ = probability of occurrence of the ith value of X

$i = 1, 2, \ldots, N$

Moreover, the standard deviation of the probability distribution is given by

$$\sigma = \sqrt{\text{Var}(X)} = \sqrt{\sum_{i=1}^{N} (X_i - \mu)^2 P(X_i)} \tag{6.3}$$

For the theoretical probability distribution of the results of rolling one fair die (Table 6.1) the variance and the standard deviation may be computed by

$$\sigma^2 = \text{Var}(X) = \sum_{i=1}^{N} (X_i - \mu)^2 P(X_i)$$
$$= (1 - 3.5)^2(1/6) + (2 - 3.5)^2(1/6)$$
$$+ (3 - 3.5)^2(1/6) + (4 - 3.5)^2(1/6)$$
$$+ (5 - 3.5)^2(1/6) + (6 - 3.5)^2(1/6)$$
$$= 2.9166$$

and

$$\sigma = \sqrt{\text{Var}(X)} = 1.71$$

In terms of our carnival game, the mean payoff per roll is \$3.50 with a standard deviation of \$1.71.

6.3 DISCRETE DISTRIBUTIONS

Many different types of mathematical models have been developed to represent various discrete phenomena which occur in the social and natural sciences, in medical research, and in business. Among the more useful of these represent data characterized by the uniform probability distribution, the binomial probability distribution, the hypergeometric probability distribution, and the Poisson probability distribution. Each of these discrete distributions will be discussed in the following sections.

6.4 UNIFORM DISTRIBUTION

A uniform probability distribution reflecting the possible outcomes of a roll of a single fair die was presented in Table 6.1. The essential characteristic of the uniform distribution is that all outcomes of the random variable are equally likely to occur. Thus the probability that the face  of the fair die turns up is the same as that for any other result—1/6—since there are six possible outcomes.

The mathematical expression representing the probability that a discrete random variable X, which follows the uniform distribution, takes on a particular value from among a series of consecutive integer values is given by

$$P(X) = \frac{1}{(b - a) + 1} \tag{6.4}$$

where b = largest possible outcome of X

a = smallest possible outcome of X

Although the mean and the standard deviation for any discrete random variable could be computed from Equations (6.1) and (6.3), simpler methods exist for making these summary computations when the random variable is considered to be uniformly distributed. In such cases the mean or expected value of the random variable X can be shown to be

$$\mu = E(X) = (a + b)/2 \tag{6.5}$$

and the standard deviation is obtained from

$$\sigma = \sqrt{\text{Var }(X)} = \sqrt{\frac{[(b - a) + 1]^2 - 1}{12}} \tag{6.6}$$

Applications of uniformly distributed random variables are found in the development of lotteries and other forms of gaming; in the generation of random numbers for engineering and/or simulation experiments; and, as will be discussed in Chapter 11, in the assessment of one's "prior probabilities" or "prior beliefs" regarding the outcome of some future event for decision-making purposes.

6.5 BINOMIAL DISTRIBUTION

One discrete probability distribution which is extremely useful for describing many phenomena is the binomial distribution. The binomial model also plays a particularly important role as an approximation to the hypergeometric distribution (which we shall consider in Section 6.6.2). The binomial distribution possesses four essential properties:

1. Each item may be considered as having been selected from an infinite population without replacement or from a finite population with replacement.
2. Each item or observation may be classified into one of two mutually exclusive and collectively exhaustive categories—success or failure.
3. The probability of an observation being classified as success, p, is constant from observation to observation. Thus, the probability of an item being classified as failure, $1 - p$, is constant over all observations.
4. The outcome (that is, success or failure) of any observation is independent of the outcome of any other observation.

The discrete random variable or phenomenon of interest which follows the binomial distribution is the number of successes obtained in a sample of n observations. Thus the binomial distribution has enjoyed numerous applications—from games of chance (what is the probability that red will come up 12 or more times in 19 spins of the roulette wheel?) to product quality control (what is the probability that in a sample of 20 tires of the same type none will be defective if 8% of all such tires produced at a particular plant are defective?).

In both of these problems the four properties of the binomial distribution are clearly satisfied. For the roulette example, a particular set of spins may be construed as the sample taken from an infinite population of spins without replacement. When spinning the roulette wheel, each observation is categorized as red (success) or not red (failure). The probability of spinning red, p, on an American roulette wheel is 18/38 and is assumed to remain stable over all observations. Thus the probability of failure (spinning black or green), $1 - p$, is 20/38 each and every time the roulette wheel spins. Moreover, the roulette wheel has no memory—the outcome of any one spin is independent of preceding or following spins so that, for example, the probability of obtaining red on the 32nd spin, given that the previous 31 spins were all red, remains equal to p, 18/38, if the roulette wheel is a fair one[1] (Figure 6.1).

In the product quality control example, the sample of tires is also selected without replacement from an ongoing production process, an infinite population of manufactured tires.[2] As each tire in the sample is inspected, it is cate-

Figure 6.1 Roulette wheel.

[1] It has been reported (Reference 9) that one time at Monte Carlo red came up on 32 consecutive spins. The probability of such an occurrence is indeed a very small number—$(18/37)^{32} = .0000000000969$. The Monte Carlo roulette wheel, like other European wheels, has 37 equal-size sectors—18 of which are red, 18 are black, and one of which is green.

[2] As an example of a binomial random variable arising from sampling with replacement from a finite population, consider the probability of obtaining two clubs in five draws from a random bridge deck where the selected card is replaced and the deck shuffled after each draw.

gorized as defective (success) or nondefective (failure).[3] Over the entire sample of tires the probability of any particular tire being classified as defective, p, is .08 so that the probability of any tire being categorized as nondefective, $1 - p$, is .92. The production process is assumed to be stable. This would be the case if:

1. The machinery producing the tires does not wear down
2. The raw materials used are uniform
3. The labor is consistent

Moreover, for such a production process, the probability of one tire being classified as defective or nondefective is independent of the classification for any other tire.

What distinguishes the two examples of binomial probability models described above are the parameters n and p. Each time a set of parameters—the number of observations in the sample, n, and the probability of success, p—is specified, a particular binomial probability distribution can be generated.

6.5.1 Development

As another example of a phenomenon which satisfies the conditions of the binomial distribution, and one which is convenient for intuitively deriving an expression for the probabilities which arise in binomial problems, we shall return to the rolling of a fair die which was discussed in Section 6.4. Here, however, we consider success to be the outcome face ⚄ and failure to be any other outcome. Suppose we are now interested in three rolls of this same die in order to determine how frequently the face ⚄ is obtained.[4] What might occur? None of the rolls might land on ⚄; one of the rolls may be a ⚄; two of the rolls may land on ⚄; or all three rolls may land on ⚄. Can the binomial random variable, the number of ⚄ faces occurring on three rolls of a fair die, take on any other value? Obviously not! If we roll the same die three times and are interested in how often a particular value (face ⚄) occurs, that value cannot exceed the number of rolls

[3] Note that when we are looking for defective tires, the discovery of such an event is deemed a success. Thus for statistical purposes, a success may refer to business failures, deaths due to a particular illness, and other phenomena which, in nonstatistical terminology, would be deemed unsuccessful.

[4] Three rolls of the same die is equivalent to one roll of each of three dice. See Problem 6.2 on the game of Chuck-a-luck.

n, nor can it be lower than zero. Hence the range of a binomial random variable is from 0 to n.

Suppose then, for example, we roll a fair die three times and observe the following result:

First Roll	Second Roll	Third Roll
⚄	Not 5	⚄

We now wish to determine the probability of this occurrence, that is, what is the probability of obtaining two successes (face ⚄) in three rolls *in the particular sequence* above? Since it may be assumed that rolling dice is a stable process, the probability of each roll occurring as above is

First Roll	Second Roll	Third Roll
$p = 1/6$	$1 - p = 5/6$	$p = 1/6$

Since each outcome is independent of the others, the probability of obtaining the given sequence is

$$p(1-p)p = p^2(1-p)^1 = p^2(1-p) = (1/6)^2(5/6) = 5/216$$

Thus out of 216 possible and equally likely outcomes from rolling a fair die three times, five will have the face ⚄ as the first and last roll, with a face other than ⚄ (that is, ⚀, ⚁, ⚂, ⚃, or ⚅) as the middle roll and the particular sequence above will be obtained.

Now, however, we may ask how many different sequences are there for obtaining two faces of ⚄ out of $n = 3$ rolls of the die? Using Equation (5.16) from Section 5.11 we have

$$\binom{n}{X} = \frac{n!}{X!\,(n-X)!} = \frac{3!}{2!\,(3-2)!} = 3$$

such sequences. These three possible sequences are

Sequence 1 = ⚄ Not 5 ⚄
with Probability $p(1-p)p = p^2(1-p)^1 = 5/216$

Sequence 2 = ⚄ ⚄ Not 5
with Probability $pp(1-p) = p^2(1-p)^1 = 5/216$

Sequence 3 = Not 5  

with Probability $(1 - p)pp = p^2(1 - p)^1 = 5/216$

Therefore, the probability of obtaining exactly two faces of  from three rolls of a die is equal to

(Number of Possible Sequences) × (Probability of a Particular Sequence)

$$(3) \times (5/216) = 15/216 = .0694$$

A similar, intuitive derivation can be obtained for the other three possible outcomes of the random variable—no face , one face , or all three faces . However, as n, the number of observations, gets large, this type of intuitive approach becomes quite laborious, and a mathematical model is more appropriate. In general, the following mathematical model represents the binomial probability distribution for obtaining the number of successes (X), given a knowledge of the parameters n and p:

$$P(X \mid n, p) = \frac{n!}{X!\,(n - X)!}\, p^X (1 - p)^{n-X} \qquad (6.7)$$

where n = sample size

p = probability of success

$1 - p$ = probability of failure

X = number of successes in the sample

We note, however, that the generalized form shown in Equation (6.7) is merely a restatement of what we had intuitively derived above. The product $p^X(1 - p)^{n-X}$ tells us the probability of obtaining exactly X successes out of n observations in a *particular sequence* while the term

$$\frac{n!}{X!\,(n - X)!}$$

tells us *how many* sequences or arrangements of the X successes out of n observations are possible. Hence, given the number of observations n and the probability of success p, we may determine the probability of $X = 0, 1, 2, \ldots, n$ successes by

$$P(X|n,p) = \text{(Number of Possible Sequences)} \times \text{(Probability of a Particular Sequence)}$$
$$= \frac{n!}{X!\,(n - X)!}\, p^X(1 - p)^{n-X}$$

Such computations may be quite tedious, especially as n gets large. However, we may obtain the probabilities directly from Appendix E, Table E.7, and thereby avoid any computational drudgery. Appendix E, Table E.7 provides, for various combinations of the parameters n and p, the probabilities that the binomial random variable takes on values of $X = 0, 1, 2, \ldots, n$. However, the reader should be cautioned that the p values in Appendix E, Table E.7, are taken to only two decimal places and thus, in some circumstances, due to rounding errors, the probabilities will only be approximations to the true result. As a case in point, for our dice rolling experiment, we first find in Appendix E, Table E.7, the combination $n = 3$ with p rounded to .17. To obtain the probability of exactly two successes, we read the probability corresponding to the row $X = 2$ and the result is .0720 (as demonstrated in Table 6.3).

Table 6.3
OBTAINING A BINOMIAL PROBABILITY

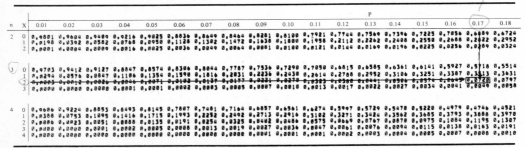

		p																	
n	X	0.01	0.02	0.03	0.04	0.05	0.06	0.07	0.08	0.09	0.10	0.11	0.12	0.13	0.14	0.15	0.16	0.17	0.18
2	0	0,9801	0,9604	0,9409	0,9216	0,9025	0,8836	0,8649	0,8464	0,8281	0,8100	0,7921	0,7744	0,7569	0,7396	0,7225	0,7056	0,6889	0,6724
	1	0,0198	0,0392	0,0582	0,0768	0,0950	0,1128	0,1302	0,1472	0,1638	0,1800	0,1958	0,2112	0,2262	0,2408	0,2550	0,2688	0,2822	0,2952
	2	0,0001	0,0004	0,0009	0,0016	0,0025	0,0036	0,0049	0,0064	0,0081	0,0100	0,0121	0,0144	0,0169	0,0196	0,0225	0,0256	0,0289	0,0324
3	0	0,9703	0,9412	0,9127	0,8847	0,8574	0,8306	0,8044	0,7787	0,7536	0,7290	0,7050	0,6815	0,6585	0,6361	0,6141	0,5927	0,5718	0,5514
	1	0,0294	0,0576	0,0847	0,1106	0,1354	0,1590	0,1816	0,2031	0,2236	0,2430	0,2614	0,2788	0,2952	0,3106	0,3251	0,3387	0,3513	0,3631
	2	0,0003	0,0012	0,0026	0,0046	0,0071	0,0102	0,0137	0,0177	0,0221	0,0270	0,0323	0,0380	0,0441	0,0506	0,0574	0,0645	0,0720	0,0797
	3	0,0000	0,0000	0,0000	0,0001	0,0001	0,0002	0,0003	0,0005	0,0007	0,0010	0,0013	0,0017	0,0022	0,0027	0,0034	0,0041	0,0049	0,0058
4	0	0,9606	0,9224	0,8853	0,8493	0,8145	0,7807	0,7481	0,7164	0,6857	0,6561	0,6274	0,5997	0,5729	0,5470	0,5220	0,4979	0,4746	0,4521
	1	0,0388	0,0753	0,1095	0,1416	0,1715	0,1993	0,2252	0,2492	0,2713	0,2916	0,3102	0,3271	0,3424	0,3562	0,3685	0,3793	0,3888	0,3970
	2	0,0006	0,0023	0,0051	0,0088	0,0135	0,0191	0,0254	0,0325	0,0402	0,0486	0,0575	0,0669	0,0767	0,0870	0,0975	0,1084	0,1195	0,1307
	3	0,0000	0,0000	0,0001	0,0002	0,0005	0,0008	0,0013	0,0019	0,0027	0,0036	0,0047	0,0061	0,0076	0,0094	0,0115	0,0138	0,0163	0,0191
	4	0,0000	0,0000	0,0000	0,0000	0,0000	0,0000	0,0000	0,0000	0,0001	0,0001	0,0001	0,0002	0,0003	0,0004	0,0005	0,0007	0,0008	0,0010

SOURCE: Extracted from Appendix E, Table E.7.

Thus Appendix E, Table E.7, has given us an approximate answer to the true probability, .0694, obtained from Equation (6.7) using the fraction $1/6 = p$ rather than the rounded decimal value .17.

6.5.2 Characteristics

We may recall that each time a set of parameters—n and p—is specified, a particular binomial probability distribution can be generated. This can be readily seen by examining Appendix E, Table E.7, for various combinations of n and p. We note that a binomial distribution may be symmetric or skewed. Whenever $p = .5$, the binomial distribution will be symmetric regardless of how large or small the value of n. However, when $p \neq .5$, the distribution will be skewed. The closer p is to .5 and the larger the number of observations, n, the less skewed the distribution will be. Thus the distribution of the number of occurrences of red in 19 spins of the roulette wheel is only slightly skewed to the right since $p = 18/38$. On the other hand, with small p the distribution will be highly right-skewed—as is observed for the distribution of the number of defective tires in a sample of 20 where $p = .08$. For very large p, the distribution would be highly left-skewed. We leave it to the reader to verify the effect of n and p on the shape of the distribution by plotting the histograms in Problems 6.9(c) and 6.10(d). However, to summarize the above characteristics, three binomial distributions are depicted in Figure 6.2.

Panel A represents the probability of obtaining the face 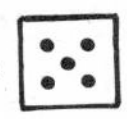 on three rolls of a fair die; panel B represents the probability of obtaining "heads" on three tosses of a fair coin; and panel C represents the probability of obtaining "heads" on four tosses of a fair coin. Thus a comparison of panel A with panel B demonstrates the effect on shape when the sample sizes are the same but the probabilities for success differ. Moreover, a comparison of panel B with C shows the effect on shape when the probabilities for success are the same but the sample sizes differ.

The mean of the binomial distribution can be readily obtained as the product of its two parameters n and p, that is, instead of using Equation (6.1), for data that are binomially distributed we compute

$$\mu = E(X) = np \tag{6.8}$$

Intuitively, this makes sense. For example, if we spin the roulette wheel 19 times, how frequently should we "expect" the color red to come up? On the average, over the long run, we would theoretically expect $\mu = E(X) = np = (19)(18/38) = 9$ occurrences of red in 19 spins. The standard deviation of the binomial distribution is obtained from

$$\sigma = \sqrt{\text{Var}(X)} = \sqrt{np(1-p)} \tag{6.9}$$

In this section we have developed the binomial model as a useful discrete probability distribution in its own right. The binomial distribution, however, plays an even more important role when it is used to approximate the hypergeometric probability distribution, as will be described in the next section. It is in this latter capacity that the binomial is used in statistical inference problems regarding estimating or testing proportions (as will be discussed in Chapters 7 through 9).

Panel A

Three Rolls of a Fair Die

X = *Number of* ⚄

$$P\left(X = 0 | n = 3, p = \frac{1}{6}\right) = \frac{3!}{0!3!}\left(\frac{1}{6}\right)^0\left(\frac{5}{6}\right)^3 = \frac{125}{216}$$

$$P\left(X = 1 | n = 3, p = \frac{1}{6}\right) = \frac{3!}{1!2!}\left(\frac{1}{6}\right)^1\left(\frac{5}{6}\right)^2 = \frac{75}{216}$$

$$P\left(X = 2 | n = 3, p = \frac{1}{6}\right) = \frac{3!}{2!1!}\left(\frac{1}{6}\right)^2\left(\frac{5}{6}\right)^1 = \frac{15}{216}$$

$$P\left(X = 3 | n = 3, p = \frac{1}{6}\right) = \frac{3!}{3!0!}\left(\frac{1}{6}\right)^3\left(\frac{5}{6}\right)^0 = \frac{1}{216}$$

$$\overline{\quad 1 \quad}$$

The probability of getting at least two face ⚄ is $\frac{15}{216} + \frac{1}{216} = \frac{16}{216}$

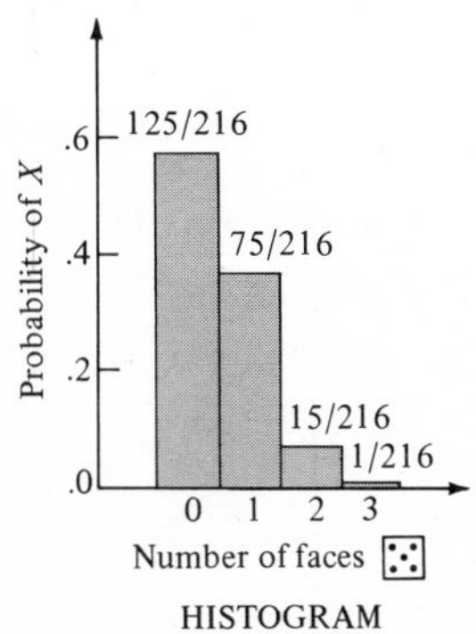

Panel B

Three Tosses of a Fair Coin

X = *Number of "Heads"*

$$P\left(X = 0 | n = 3, p = \frac{1}{2}\right) = \frac{3!}{0!3!}\left(\frac{1}{2}\right)^0\left(\frac{1}{2}\right)^3 = \frac{1}{8}$$

$$P\left(X = 1 | n = 3, p = \frac{1}{2}\right) = \frac{3!}{1!2!}\left(\frac{1}{2}\right)^1\left(\frac{1}{2}\right)^2 = \frac{3}{8}$$

$$P\left(X = 2 | n = 3, p = \frac{1}{2}\right) = \frac{3!}{2!1!}\left(\frac{1}{2}\right)^2\left(\frac{1}{2}\right)^1 = \frac{3}{8}$$

$$P\left(X = 3 | n = 3, p = \frac{1}{2}\right) = \frac{3!}{3!0!}\left(\frac{1}{2}\right)^3\left(\frac{1}{2}\right)^0 = \frac{1}{8}$$

$$\overline{\quad 1 \quad}$$

The probability of getting at least two heads is $\frac{3}{8} + \frac{1}{8} = \frac{4}{8}$

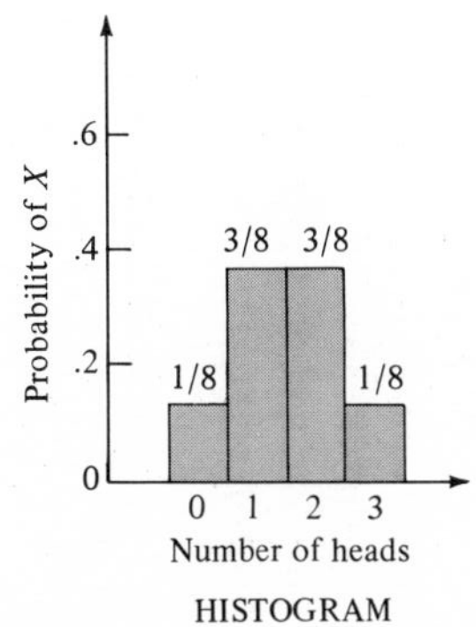

Panel C

Four Tosses of a Fair Coin

X = *Number of "Heads"*

$$P\left(X = 0 | n = 4, p = \frac{1}{2}\right) = \frac{4!}{0!4!}\left(\frac{1}{2}\right)^0\left(\frac{1}{2}\right)^4 = \frac{1}{16}$$

$$P\left(X = 1 | n = 4, p = \frac{1}{2}\right) = \frac{4!}{1!3!}\left(\frac{1}{2}\right)^1\left(\frac{1}{2}\right)^3 = \frac{4}{16}$$

$$P\left(X = 2 | n = 4, p = \frac{1}{2}\right) = \frac{4!}{2!2!}\left(\frac{1}{2}\right)^2\left(\frac{1}{2}\right)^2 = \frac{6}{16}$$

$$P\left(X = 3 | n = 4, p = \frac{1}{2}\right) = \frac{4!}{3!1!}\left(\frac{1}{2}\right)^3\left(\frac{1}{2}\right)^1 = \frac{4}{16}$$

$$P\left(X = 4 | n = 4, p = \frac{1}{2}\right) = \frac{4!}{4!0!}\left(\frac{1}{2}\right)^4\left(\frac{1}{2}\right)^0 = \frac{1}{16}$$

$$\overline{\quad 1 \quad}$$

The probability of getting at least two heads is $\frac{6}{16} + \frac{4}{16} + \frac{1}{16} = \frac{11}{16}$

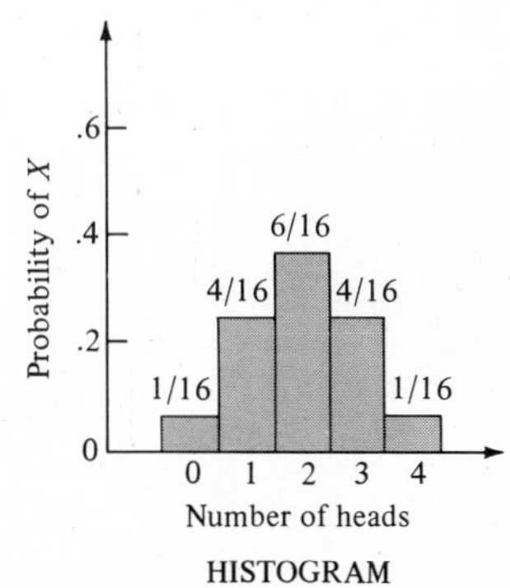

Figure 6.2 Comparison of three binomial distributions.

where the expression $\sqrt{\frac{N-n}{N-1}}$ is a **finite population correction factor** which arises because of the process of sampling without replacement from finite populations. (This correction factor will be discussed in greater detail in Section 7.4).

6.6.2 Using the Binomial Distribution to Approximate the Hypergeometric Distribution

Figure 6.5 permits us to compare the theoretical results of an experiment obtained from the binomial model against those from the hypergeometric distribution. Panel A represents the binomial probability distribution generated by considering the number of clubs obtained in a sample of five cards selected with replacement, while panel B depicts the hypergeometric distribution generated when sampling without replacement. The comparative results are fairly similar because the sample size $n = 5$ represents but a small portion (9.6%) of the population size $N = 52$. As a rule of thumb then, when sampling without replacement, whenever the sample size is less than 5% of the population size (that is, $n < .05N$) we may use the binomial probability distribution to approximate the hypergeometric distribution, since the results will be quite similar and the binomial model is less cumbersome. The importance of this application of the binomial distribution should not be overlooked. Much sample survey work (such as that undertaken by the dean and researcher in developing the questionnaire presented in Chapter 2) deals with hypergeometric–binomial type data. Estimation and hypothesis testing for the true proportion of success will be studied in Chapters 8 and 9.

6.7 POISSON DISTRIBUTION

Another discrete probability distribution which should be considered because of its usefulness is the Poisson distribution. Not only is the Poisson model applicable to many types of discrete phenomena, but it also may be used to approximate the binomial distribution for those situations in which n is very large and p is very small.[5] From Equation (6.7) it is clearly seen that as n gets large, the computations for the binomial distribution become tedious. However, for those situations in which p is also very small, the following mathematical expression for the Poisson model may be used to facilitate the computations and give a good approximation to the true (binomial) result:

[5] But what constitutes large n and small p? A rule of thumb used by some writers (see, for example, References 1 and 4) is that the Poisson approximation to the binomial distribution is appropriate for combinations of $n \geq 20$ with $p \leq .05$. However, we will see from Figure 6.6 that rather good approximations are obtained for $n = 20$ and $p = .08$. In addition, we should note here that when n is large but p is not very small, appropriate approximations to the binomial distribution can be obtained from the normal distribution. This will be discussed in Sections 6.10 and 6.11.

Panel A	Panel B
Sampling with Replacement (Binomial)	*Sampling without Replacement (Hypergeometric)*
$P(X=0 \mid n=5, p=13/52) = \binom{5}{0}\left(\frac{13}{52}\right)^0\left(\frac{39}{52}\right)^5 = .23730$	$P(X=0 \mid n=5, N=52, A=13) = \frac{\binom{13}{0}\binom{39}{5}}{\binom{52}{5}} = .22153$
$P(X=1 \mid n=5, p=13/52) = \binom{5}{1}\left(\frac{13}{52}\right)^1\left(\frac{39}{52}\right)^4 = .39551$	$P(X=1 \mid n=5, N=52, A=13) = \frac{\binom{13}{1}\binom{39}{4}}{\binom{52}{5}} = .41142$
$P(X=2 \mid n=5, p=13/52) = \binom{5}{2}\left(\frac{13}{52}\right)^2\left(\frac{39}{52}\right)^3 = .26367$	$P(X=2 \mid n=5, N=52, A=13) = \frac{\binom{13}{2}\binom{39}{3}}{\binom{52}{5}} = .27428$
$P(X=3 \mid n=5, p=13/52) = \binom{5}{3}\left(\frac{13}{52}\right)^3\left(\frac{39}{52}\right)^2 = .08789$	$P(X=3 \mid n=5, N=52, A=13) = \frac{\binom{13}{3}\binom{39}{2}}{\binom{52}{5}} = .08154$
$P(X=4 \mid n=5, p=13/52) = \binom{5}{4}\left(\frac{13}{52}\right)^4\left(\frac{39}{52}\right)^1 = .01465$	$P(X=4 \mid n=5, N=52, A=13) = \frac{\binom{13}{4}\binom{39}{1}}{\binom{52}{5}} = .01073$
$P(X=5 \mid n=5, p=13/52) = \binom{5}{5}\left(\frac{13}{52}\right)^5\left(\frac{39}{52}\right)^0 = .00098$	$P(X=5 \mid n=5, N=52, A=13) = \frac{\binom{13}{5}\binom{39}{0}}{\binom{52}{5}} = .00050$
1	1

Panel A

The probability of getting two or fewer clubs out of five cards is .23730 + .39551 + .26367 = .89648

Mean: $\mu = np = (5)\left(\frac{13}{52}\right) = 1.25$

Standard deviation: $\sigma = \sqrt{np(1-p)} = \sqrt{(5)\left(\frac{13}{52}\right)\left(\frac{39}{52}\right)}$
$= .97$

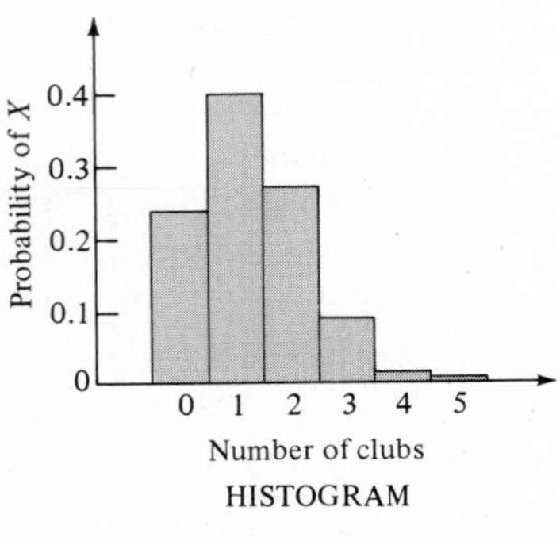

Panel B

The probability of getting two or fewer clubs out of five cards is .22153 + .41142 + .27428 = .90723.

Mean: $\mu = np = (5)\left(\frac{13}{52}\right) = 1.25$

Standard deviation: $\sigma = \sqrt{np(1-p)} \cdot \sqrt{\frac{N-n}{N-1}}$
$= \sqrt{(5)\left(\frac{13}{52}\right)\left(\frac{39}{52}\right)} \cdot \sqrt{\frac{52-5}{52-1}}$
$= .93$

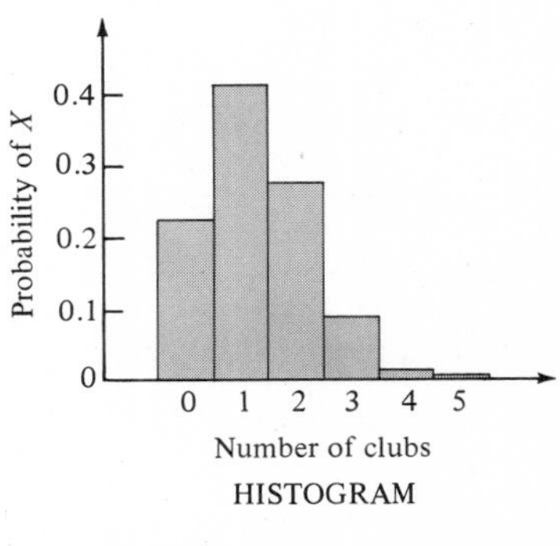

Figure 6.5 Comparing the binomial and hypergeometric distributions.

$$P(X|n,p) \cong \frac{e^{-np}(np)^X}{X!} \tag{6.15}$$

where n = sample size

p = true probability of success

e = base of the Naperian (natural) logarithmic system—a mathematical constant approximated by 2.71828

X = number of successes in the sample

Theoretically it should be noted that the Poisson random variable may range from 0 to ∞. However, when used as an approximation to the binomial distribution, the Poisson random variable—the number of successes out of n observations—clearly cannot exceed the sample size n. Moreover, with large n and small p, Equation (6.15) implies that the probability of observing a large number of successes becomes small and approaches zero quite rapidly. Hence because of the severe degree of right-skewness in such a probability distribution, no difficulty arises when employing the Poisson approximation to the binomial.

It is interesting to note that the Poisson distribution has but one parameter, μ, that is, the mean and the variance of a Poisson model are equal. Hence when using the Poisson distribution to approximate the binomial model we may compute

$$\mu = E(X) = np \tag{6.16}$$

and we may approximate

$$\sigma = \sqrt{\text{Var}(X)} \cong \sqrt{np} \tag{6.17}$$

Now it may be recalled from Equation (6.9) that the standard deviation of the binomial model is $\sqrt{np(1-p)}$. Why then is $\sigma \cong \sqrt{np}$ in Equation (6.17)? What happens to the $(1 - p)$? Intuitively, it should be clear that if p is very small and approaching zero, then its complement, $1 - p$, must be very large and approaching 1. If, at the same time, n is very large (approaching ∞) so that the product np could be thought of as a constant, then

$$\sigma = \sqrt{np(1-p)} \cong \sqrt{np1} = \sqrt{np}$$

6.7.1 Using the Poisson Distribution to Approximate the Binomial Distribution

To illustrate the use of the Poisson approximation for the binomial, we compute the probability of obtaining exactly one defective tire from a sample of 20 if 8% of the tires manufactured at a particular plant are defective. Thus from Equation (6.15) we have

$$P(X = 1|n = 20, p = .08) \cong \frac{e^{-(20)(.08)}[(20)(.08)]^1}{1!} = \frac{e^{-1.6}(1.6)^1}{1}$$

However, rather than having to use the natural logarithmic system to determine this probability, tables of the Poisson distribution (Appendix E, Table E.6) can be employed. Referring to these tables, the only values necessary are the parameter μ and the desired number of successes X. Since in the above example $\mu = 1.6$ and $X = 1$, we have from Appendix E, Table E.6,

$$P(X = 1|\mu = 1.6) = .3230$$

This is shown in Table 6.4.

Table 6.4
OBTAINING A POISSON PROBABILITY

	μ									
X	1.1	1.2	1.3	1.4	1.5	1.6	1.7	1.8	1.9	2.0
0	.3329	.3012	.2725	.2466	.2231	.2019	.1827	.1653	.1496	.1353
1	~~.3662~~	~~.3614~~	~~.3543~~	~~.3452~~	~~.3347~~ →	.3230	.3106	.2975	.2842	.2707
2	.2014	.2169	.2303	.2417	.2510	.2584	.2640	.2678	.2700	.2707
3	.0738	.0867	.0998	.1128	.1255	.1378	.1496	.1607	.1710	.1804
4	.0203	.0260	.0324	.0395	.0471	.0551	.0636	.0723	.0812	.0902

SOURCE: Extracted from Appendix E, Table E.6.

Had the true distribution, the binomial, been employed instead of the approximation, we would compute

$$P(X = 1|n = 20, p = .08) = \binom{20}{1}(.08)^1(.92)^{19} = .3282$$

This computation, though, is tedious on most calculators. Clearly, with Appendix E, Table E.7 available, however, one could argue that we should look up the binomial probability directly for $n = 20$, $p = .08$, and $X = 1$ and not bother calculating it or using the Poisson approximation. On the other hand, Appendix E, Table E.7 only shows binomial probabilities for n from 2 through 20, so that for $n > 20$ the Poisson approximation should certainly be used if p is very small.

To summarize our findings, Figure 6.6 compares the binomial distribution

Panel A

*Binomial Distribution**

$P(X = 0|n = 20, p = .08) = .1887$

$P(X = 1|n = 20, p = .08) = .3282$

$P(X = 2|n = 20, p = .08) = .2711$

$P(X = 3|n = 20, p = .08) = .1414$

$P(X = 4|n = 20, p = .08) = .0523$

$P(X = 5|n = 20, p = .08) = .0145$

$P(X = 6|n = 20, p = .08) = .0032$

$P(X = 7|n = 20, p = .08) = .0005$

$P(X = 8|n = 20, p = .08) = .0001$

$P(X = 9|n = 20, p = .08) = .0000$

$\vdots \qquad \vdots$

$P(X = 20|n = 20, p = .08) = \underline{.0000}$

1

The probability of discovering two or more defective tires is
$1 - [P(X = 0) + P(X = 1)]$.
$1 - [.1887 + .3282] = .4831$

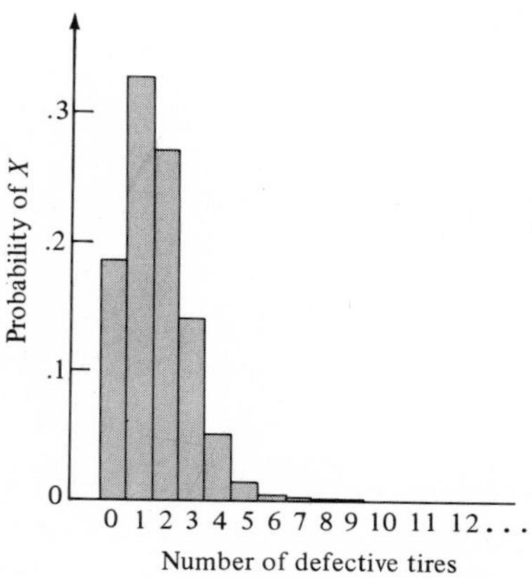

HISTOGRAM

Binomial

Panel B

Poisson Distribution†

$$P(X = 0|n = 20, p = .08) = \frac{e^{-1.6}(1.6)^0}{0!} = .2019$$

$$P(X = 1|n = 20, p = .08) = \frac{e^{-1.6}(1.6)^1}{1!} = .3230$$

$$P(X = 2|n = 20, p = .08) = \frac{e^{-1.6}(1.6)^2}{2!} = .2584$$

$$P(X = 3|n = 20, p = .08) = \frac{e^{-1.6}(1.6)^3}{3!} = .1378$$

$$P(X = 4|n = 20, p = .08) = \frac{e^{-1.6}(1.6)^4}{4!} = .0551$$

$$P(X = 5|n = 20, p = .08) = \frac{e^{-1.6}(1.6)^5}{5!} = .0176$$

$$P(X = 6|n = 20, p = .08) = \frac{e^{-1.6}(1.6)^6}{6!} = .0047$$

$$P(X = 7|n = 20, p = .08) = \frac{e^{-1.6}(1.6)^7}{7!} = .0011$$

$$P(X = 8|n = 20, p = .08) = \frac{e^{-1.6}(1.6)^8}{8!} = .0002$$

$$P(X = 9|n = 20, p = .08) = \frac{e^{-1.6}(1.6)^9}{9!} = .0000$$

$$P(X = 10|n = 20, p = .08) = \frac{e^{-1.6}(1.6)^{10}}{10!} = .0000$$

$$\vdots$$

$$P(X = 20|n = 20, p = .08) = \frac{e^{-1.6}(1.6)^{20}}{20!} = \underline{.0000}$$

1

The probability of discovering two or more defective tires is approximately $1 - [P(X = 0) + P(X = 1)]$
$1 - [.2019 + .3230] = .4751$

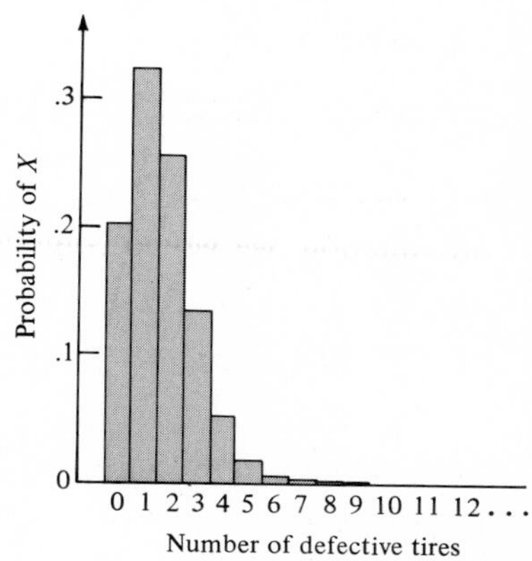

HISTOGRAM

Poisson

Figure 6.6 Binomial distribution and its Poisson approximation.

* The binomial probabilities are taken from Appendix E, Table E.7.
† The Poisson probabilities are taken from Appendix E, Table E.6.

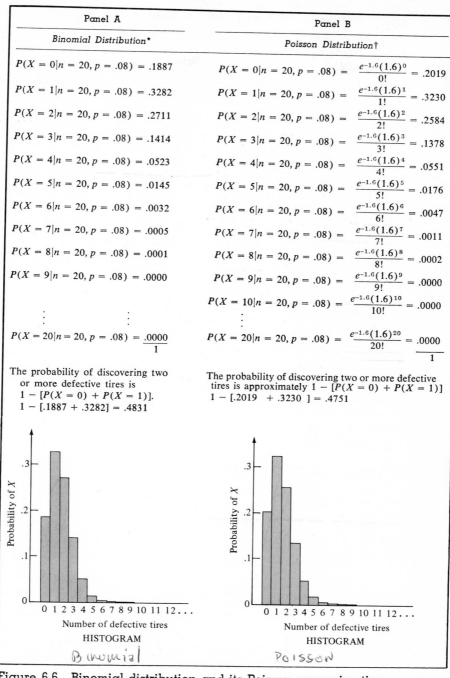

(panel A) and its Poisson approximation (panel B) for the number of defective tires in a sample of 20. The similarities of the two results are clearly evident, thus demonstrating the usefulness of the Poisson approximation.

6.7.2 Applications of the Poisson Model

It may now be recalled that the Poisson probability distribution is also important because of the numerous random phenomena which seem to follow it (see References 2 and 5 for a more mathematically rigorous development of the Poisson process). As examples of data which are Poisson distributed, we may think of the number of business failures per week in a particular state, the number of major industrial strikes per week in the United Kingdom, the number of runs per inning of a baseball game, the number of arrivals per minute at a toll bridge, and, of course, the classic example—the number of deaths due to being kicked in the head by a horse per year per corps in the Prussian cavalry.[6] Referring back to the student survey described in Chapter 2, our dean and researcher were concerned with the number of groups, teams, clubs, organizations, etc., that each student belonged to. In each case that we have described, the discrete random variable refers to the number of successes per unit (that is, per week, inning, minute, etc.) and the parameter μ refers to the average or expected number of successes per unit. As mentioned previously, an interesting characteristic about a Poisson distribution is that its mean and variance are equal. Thus,

$$E(X) = \mu \tag{6.18}$$

and

$$\sigma = \sqrt{\text{Var}(X)} = \sqrt{\mu} \tag{6.19}$$

Substituting the parameter μ for np in Equation (6.15), the mathematical expression for the Poisson distribution for obtaining X successes, given that μ successes are expected, is

$$P(X|\mu) = \frac{e^{-\mu}\mu^X}{X!} \tag{6.20}$$

[6] Ehrenberg (Reference 2) demonstrates quite clearly how the data obtained by the military historian von Bortkewitsch (1898) are approximated (fitted) by the Poisson distribution (see Problem 10.36).

where μ = expected number of successes

e = mathematical constant approximated by 2.71828

X = number of successes per unit

As an example we may consider the following problem. If, on the average, a switchboard receives $\mu = 3.0$ phone calls per minute, what is the probability that in any given minute more than two phone calls will be received? Now,

$$P(X > 2|\mu = 3.0) = P(X = 3|\mu = 3.0) + P(X = 4|\mu = 3.0) + \cdots + P(X = \infty|\mu = 3.0)$$

But the terms on the right can be expressed as

$$1 - P(X \leq 2|\mu = 3.0)$$

Thus,

$$P(X > 2|\mu = 3.0) = 1 - \{P(X = 0|\mu = 3.0) + P(X = 1|\mu = 3.0) + P(X = 2|\mu = 3.0)\}$$

Now substituting Equation (6.20), we have

$$P(X > 2|\mu = 3.0) = 1 - \left\{\frac{e^{-3.0}(3.0)^0}{0!} + \frac{e^{-3.0}(3.0)^1}{1!} + \frac{e^{-3.0}(3.0)^2}{2!}\right\}$$

From Appendix E, Table E.6, we can readily obtain the probabilities of 0, 1, or 2 successes, given a mean of 3.0 successes. Therefore,

$$P(X > 2|\mu = 3.0) = 1 - \{.0498 + .1494 + .2240\}$$
$$= 1 - .4232 = .5768$$

Thus we see that there is roughly a 42.3% chance that two or fewer calls will be received by the switchboard per minute. Therefore, a 57.7% chance exists that three or more calls will be received.

6.8 CONTINUOUS PROBABILITY DENSITY FUNCTIONS

Now that we have studied several discrete probability distribution functions, we may turn our attention to the continuous probability density functions—those that arise due to some measuring process on various phenomena of interest. Several such continuous models have important applications in engineering and the physical sciences as well as in business and the social sciences. Examples of continuous random phenomena are: time between arrivals (of telephone calls into a switchboard, customers in a supermarket, or cars at a toll bridge); customer servicing; as well as life testing or time to failure (Reference 6). However, with continuous phenomena many of the necessary mathematical expressions require a knowledge of integral calculus and are beyond the scope of this book. Nevertheless, one continuous probability density function which we shall focus upon has been deemed so important for applications that special probability tables (Appendix E, Table E.2) have been devised in order to eliminate the need for what otherwise would require laborious mathematical computations. This particular continuous probability density function is known as the Gaussian or **normal** distribution.

6.9 NORMAL DISTRIBUTION

The normal probability density function is vitally important in statistics because of three main reasons. First, numerous continuous phenomena seem to follow it or can be approximated by it. As examples, from the survey conducted by the dean and the researcher, the populations of heights of female students, weights of male students, grade-point indexes, and high-school averages are all continuous random variables that may be approximated by a normal distribution. Second, the normal distribution can be used to approximate various discrete probability distributions and thereby avoid much otherwise needed computational drudgery. Third, but certainly not last, as we shall indicate in Chapter 7 (and emphasize in Chapters 8, 9, 10, 12, and 13), the normal distribution provides the basis for classical statistical inference because of its relationship to the central limit theorem.

Figure 6.7 depicts the relative frequency histogram and polygon which theoretically represents the thickness of a batch of 10,000 brass washers that was given in Table 6.2. From this we see that the normal distribution is bell-shaped and symmetrical.

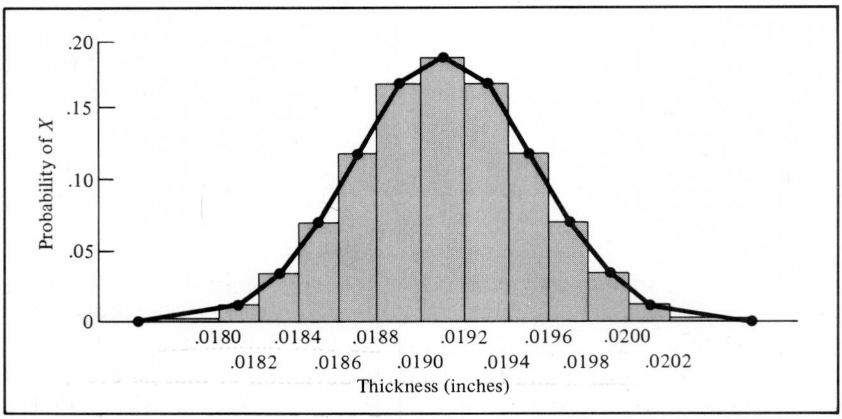

Figure 6.7 Relative frequency histogram and polygon of the thickness of 10,000 brass washers.

SOURCE: Data are taken from Table 6.2.

6.9.1 Properties of the Normal Distribution

Theoretically, as depicted in Figure 4.9, the normal distribution has two interesting properties: the measures of central tendency (mean, median, and mode) are identical and the continuous random variable of interest has an infinite range ($-\infty < X < +\infty$). Nevertheless, from a practical point of view, for some phenomenon which may be approximated by the normal model, slight differences may exist between these measures of central tendency. Moreover, for data which follow or are approximated by a normal model, the practical range falls about 3 standard deviations (distances) above and below the mean. As a case in point, we consider the data on the thickness of brass washers. The random variable of interest, thickness, cannot possibly take on values of zero or below, nor can a washer be so thick that it becomes unusable. From Table 6.2 we note that only 48 out of every 10,000 brass washers manufactured can be expected to have a thickness of .0202 inch or more, while an equal number can be expected to have a thickness under .0180 inch. Thus the chance of randomly obtaining a washer so thin or so thick is .0048 + .0048 = .0096—or almost 1 in 100. Hence we shall leave it to the reader to verify (see Problem 6.24) that 99.04% of these manufactured washers can be expected to have a thickness between .0180 and .0202 inch, that is, 2.59 standard deviations (distances) above and below the mean.

The mathematical expression representing the probability density function, denoted by the symbol $f(X)$, for the normal distribution is

$$f(X) = \frac{1}{\sqrt{2\pi}\,\sigma} e^{\frac{-(X-\mu)^2}{2\sigma^2}} \tag{6.21}$$

where e = mathematical constant approximated by 2.71828

π = mathematical constant approximated by 3.14159

μ = true mean

σ = true standard deviation

X = values of the continuous random variable where $-\infty < X < +\infty$

Since e and π are mathematical constants, it is clear from Equation (6.21) that the probabilities of occurrences within particular ranges of the random variable X are dependent upon the two parameters of the normal distribution—the mean μ and the standard deviation σ, that is, every time we specify a particular combination of μ and σ, a different normal probability distribution may be generated. This may be illustrated from Figure 6.8 where three different normal distributions are depicted. Distributions A and B have the same mean μ, but differ in terms of dispersion (that is, the standard devia-

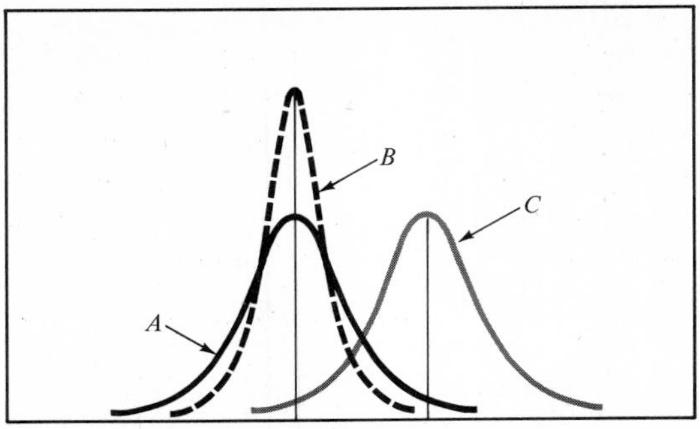

Figure 6.8 Three hypothetical normal distributions having differing parameters μ and σ.

tion). On the other hand, distributions A and C have the same standard deviation σ, but differ in terms of central tendency (that is, the mean). Furthermore, distributions B and C depict two normal probability density functions which differ with respect to both the μ and σ.

Unfortunately, the form of the mathematical expression (6.21) used for generating a normal probability density function, and/or its particular probabilities, is computationally tedious. To avoid such computations, it would be nice if a set of tables were to exist so that we could merely look up the desired probabilities. But alas—it is impossible to construct such a set of tables! An infinite number of combinations of the parameters μ and σ exist, and each such combination would be required to generate a particular normal probability density function such as the three depicted in Figure 6.8. This of course would require a set of tables of infinite size (or of an infinite number of pages, assuming one page for each generated distribution based on the combinations of μ and σ)!

6.9.2 Standardizing the Normal Distribution

Fortunately, the problem is alleviated by *standardizing* the data, that is, regardless of whatever units of measurement the normal random variable X has (inches, kilograms, dollars, minutes, etc.,) and regardless of whatever combinations of parameters μ and σ the data possess, we may convert these data onto a standardized scale by using the following transformation formula:

$$Z = \frac{X - \mu}{\sigma} \qquad (6.22)$$

Hence in this manner, any normal random variable X can be converted to a standardized normal random variable Z. While the original data for the

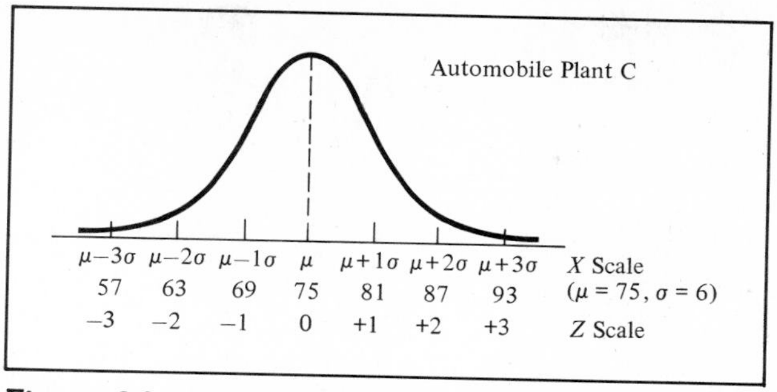

Figure 6.9 Transformation of scales.

random variable X had mean μ and standard deviation σ, the standardized random variable Z will always have mean $\mu_Z = 0$ and standard deviation $\sigma_Z = 1$. Therefore since the standardized data will always have mean 0 and standard deviation 1, it will only be necessary to generate and then tabulate but one such distribution (see Appendix E, Table E.2) and, by using the transformation formula (6.22), we will always be able to convert any set of normally distributed data and determine its probabilities.

To illustrate the transformation of scale for units of measurement X to a standardized form Z, let us consider the following problem. Suppose a production manager was investigating the time it took factory workers to assemble a particular part in Automobile Plant C and determined the data (time in seconds) to be normally distributed with a mean μ of 75 seconds and a standard deviation σ of 6 seconds. We see from Figure 6.9 that every measurement X has a corresponding standardized measurement Z obtained from the conversion formula (6.22). Hence from Figure 6.9 it is clear that a time of 81 seconds required for a factory worker to complete the task is equivalent to 1 standardized unit (that is, **standard deviation unit**) above the mean $[Z = (81 - 75)/6 = +1]$ while a time of 57 seconds required for a worker to assemble the part is equivalent to 3 standardized units below the mean $[Z = (57 - 75)/6 = -3]$. Clearly then, a transformation from X to Z is analogous to converting such scales as yards to meters, pounds to kilograms, or degrees Fahrenheit to degrees Celsius.

Suppose now that the production manager conducted the same time-and-motion study at Automobile Plant B where the workers were trained to assemble the part by a different method. This time, however, suppose she determined that the time to perform the task was normally distributed with mean μ of 60 seconds and standard deviation σ of 3 seconds. The data are depicted in Figure 6.10. We note, for example, in comparison with the aforementioned results at Automobile Plant C, that at this automobile plant a time of 57 seconds to complete the task is only 1 standardized unit below the mean for the group $[Z = (57 - 60)/3 = -1]$. Moreover, a time of 63 seconds is 1 standardized unit above the mean time for assemblage

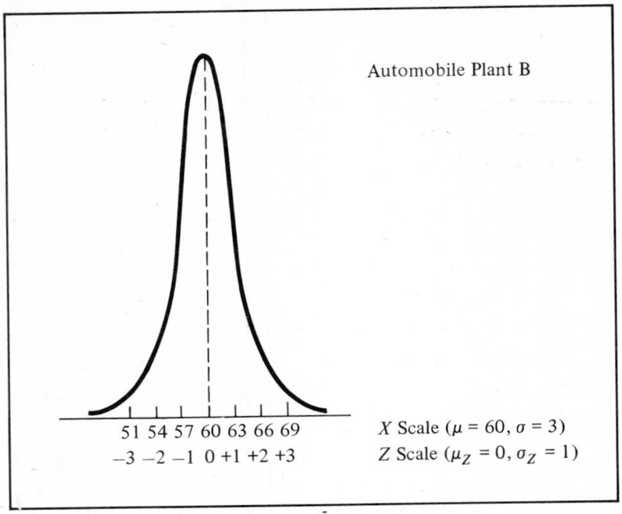

Figure 6.10 A different transformation of scales.

$[Z = (63 - 60)/3 = +1]$, while a time of 51 seconds is 3 standardized units below the group mean $[Z = (51 - 60)/3 = -3]$.

6.9.3 Using the Normal Probability Tables

The two bell-shaped curves in Figures 6.9 and 6.10 depict the relative frequency or percentage polygons for the normal distributions representing the time trials for all factory workers to assemble a part at two particular automobile plants—B and C. Since at each plant the time trials to assemble the part are known for every factory worker, the set of time trial data represent the entire population at a particular plant, and therefore the *probabilities* or proportion of area under the curve (that is, the relative frequency polygon) must add up to 1. Clearly then, the area under the curve between any two reported time trials represents but a portion of the total area possible. Suppose now that we wish to inquire, from Figure 6.9, as to how likely or possible it is that a factory worker selected at random from Automobile Plant C should require between 75 and 81 seconds to complete the task, that is, from the mean time at Automobile Plant C up to 1 standardized unit above the particular plant mean? From Appendix E, Table E.2, we may easily obtain that answer.

Appendix E, Table E.2, represents the probabilities or areas under the normal curve calculated from the mean μ to the particular value of interest X. Using Equation (6.22), this corresponds to the probability or area under the standardized normal curve from the mean ($\mu_Z = 0$) to the transformed value of interest Z. Only positive entries for Z are listed in the table since it is clear that for such a symmetrical distribution with zero mean, the area from the mean to $+Z$ (that is, Z standardized units above the mean) must be identical to the area from the mean to $-Z$ (that is, Z standardized units below the mean).

To use Appendix E, Table E.2, we note that all Z values must first be recorded to two decimal places. Thus our particular Z value of interest is recorded as + 1.00. To read the probability or area under the curve from the mean to $Z = +1.00$, we scan down the Z column from Appendix E, Table E.2, until we locate the Z value of interest (in tenths). Hence we stop in the row $Z = 1.0$. Next we read across this row until we intersect the column that contains the appropriate Z value in hundredths. Therefore, in the body of the table the tabulated probability for $Z = 1.00$ corresponds to the intersection of the row $Z = 1.0$ with the column $Z = .00$ as shown in Table 6.5. From Appendix E, Table E.2, this probability is .3413, that is, as depicted in Figure 6.11, there is a 34.13% chance that a factory worker selected at random from Automobile Plant C will require between 75 and 81 seconds to assemble the part.

On the other hand, we know from Figure 6.10 that at Automobile Plant B

Table 6.5
OBTAINING AN AREA UNDER THE NORMAL CURVE

Z	.00	.01	.02	.03	.04	.05	.06	.07	.08	.09
0.0	.0000	.0040	.0080	.0120	.0160	.0199	.0239	.0279	.0319	.0359
0.1	.0398	.0438	.0478	.0517	.0557	.0596	.0636	.0675	.0714	.0753
0.2	.0793	.0832	.0871	.0910	.0948	.0987	.1026	.1064	.1103	.1141
0.3	.1179	.1217	.1255	.1293	.1331	.1368	.1406	.1443	.1480	.1517
0.4	.1554	.1591	.1628	.1664	.1700	.1736	.1772	.1808	.1844	.1879
0.5	.1915	.1950	.1985	.2019	.2054	.2088	.2123	.2157	.2190	.2224
0.6	.2257	.2291	.2324	.2357	.2389	.2422	.2454	.2486	.2518	.2549
0.7	.2580	.2612	.2642	.2673	.2704	.2734	.2764	.2794	.2823	.2852
0.8	.2881	.2910	.2939	.2967	.2995	.3023	.3051	.3078	.3106	.3133
0.9	.3159	.3186	.3212	.3238	.3264	.3289	.3315	.3340	.3365	.3389
1.0	.3413	.3438	.3461	.3485	.3508	.3531	.3554	.3577	.3599	.3621
1.1	.3643	.3665	.3686	.3708	.3729	.3749	.3770	.3790	.3810	.3830

SOURCE: Extracted from Appendix E, Table E.2.

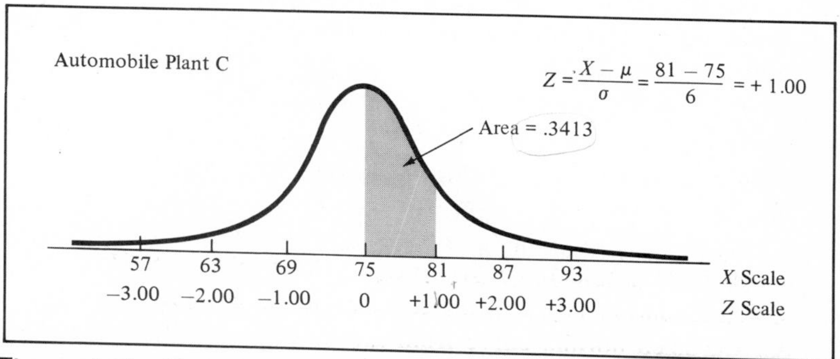

Figure 6.11 Determining the area between the mean and Z from a standardized normal distribution.

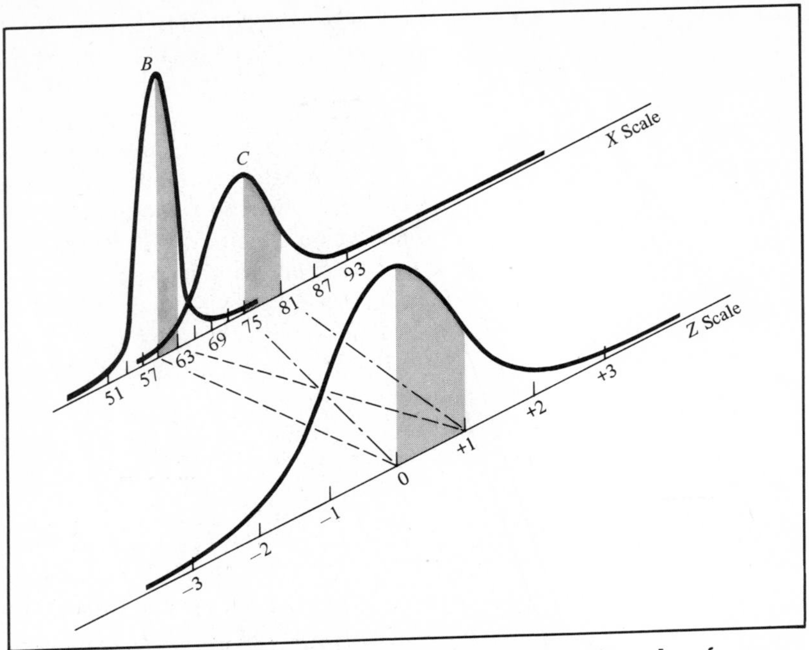

Figure 6.12 Demonstrating a transformation of scales for corresponding portions under two curves.

a time of 63 seconds is 1 standardized unit above the mean time of 60 seconds. Thus the likelihood of a randomly selected factory worker at Automobile Plant B completing the assemblage in between 60 and 63 seconds is also .3413.[7] These results are clearly illustrated in Figure 6.12, which demonstrates that regardless of the mean μ and standard deviation σ of a particular set of data that are normally distributed, a transformation to a standardized scale can always be made from Equation (6.22), and, by using Appendix E, Table E.2, any probability or portion of area under the curve can be obtained. Clearly, from Figure 6.12 it is evident that the probability or area under the curve from 60 to 63 seconds at Automobile Plant B is identical to the probability or area under the curve from 75 to 81 seconds at Automobile Plant C.

6.9.4 Applications

Now that we have learned to use Appendix E, Table E.2, in conjunction with formula (6.22), many different types of probability questions pertaining to the normal distribution can be resolved. For instance, at Automobile Plant C, where the mean time for assemblage is 75 seconds and the standard deviation is 6 seconds, how can we determine the probability that a randomly selected factory worker will perform the task in under 75 seconds or over 81 seconds, $P(X < 75 \text{ or } X > 81)$? Since we have already determined the probability

[7] Mathematically this may be expressed as $P(60 \leq X \leq 63) = P(0 \leq Z \leq 1) = .3413$.

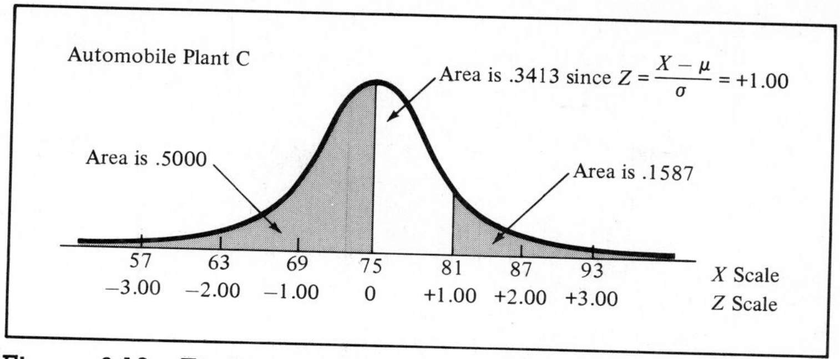

Figure 6.13 Finding $P(X < 75 \text{ or } X > 81)$.

that a randomly selected factory worker will need between 75 and 81 seconds to complete the task, it is obvious from Figure 6.11 that our desired probability must be its complement, that is, $1 - .3413 = .6587$. Another way of viewing this problem, however, is to separately obtain both the probability of completing the part in under 75 seconds and the probability of completing the part in over 81 seconds and then use the addition rule for mutually exclusive events [Equation (5.3)] to obtain the desired result. This is depicted in Figure 6.13.

Since the mean and median are theoretically the same for normally distributed data, it is clear that 50% of the workers can complete the task in under 75 seconds.[8] To show this, from Equation (6.22) we have

$$Z = \frac{X - \mu}{\sigma} = \frac{75 - 75}{6} = 0.00$$

Using Appendix E, Table E.2, we see that the area under the normal curve from the mean to $Z = 0.00$ is .0000. Hence the area under the curve less than $Z = 0.00$ must be $.5000 - .0000 = .5000$ (which happens to be the area for the entire left side of the distribution, from the mean to $Z = -\infty$).

Now we wish to obtain the probability of completing the part in over 81 seconds. But Equation (6.22) only gives the areas under the curve from the mean to Z, not from Z to $+\infty$. Thus we find the probability from the mean to Z and subtract this result from .5000 to obtain the desired answer. Since we know that the area or portion of the curve from the mean to $Z = +1.00$ is .3413, the area from $Z = +1.00$ to $Z = +\infty$ must be $.5000 - .3413 = .1587$. Hence the probability that a randomly selected factory worker will

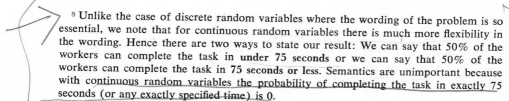

[8] Unlike the case of discrete random variables where the wording of the problem is so essential, we note that for continuous random variables there is much more flexibility in the wording. Hence there are two ways to state our result: We can say that 50% of the workers can complete the task in **under 75 seconds** or we can say that 50% of the workers can complete the task in **75 seconds or less**. Semantics are unimportant because with continuous random variables the probability of completing the task in exactly 75 seconds (or any exactly specified time) is 0.

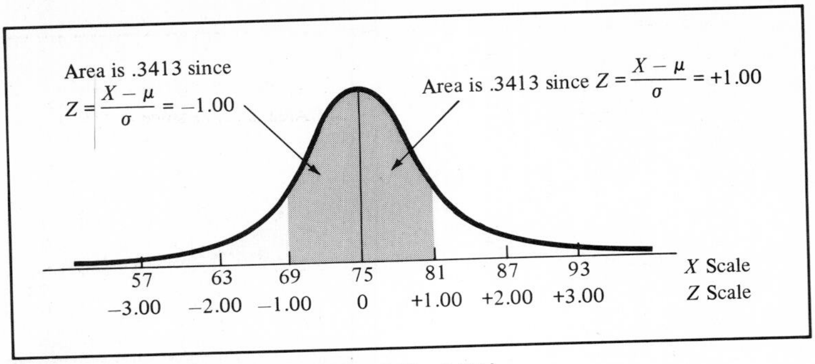

Figure 6.14 Finding $P(69 \leq X \leq 81)$.

perform the task in under 75 or over 81 seconds, $P(X < 75 \text{ or } X > 81)$, is $.5000 + .1587 = .6587$.

Suppose that we are now interested in determining the probability that a randomly selected factory worker can complete the part in 69 to 81 seconds, that is, $P(69 \leq X \leq 81)$. We note from Figure 6.14 that one of the values of interest is above the mean assemblage time of 75 seconds while the other value is below it. Since our conversion formula (6.22) only permits us to find probabilities from a *particular* value of interest to the mean, we see from Figure 6.14 that this type of problem must be solved in three steps: First, we can find the probability or portion of the curve from the mean to 81 seconds; then we can find the probability from the mean to 69 seconds; finally, we can sum up the two mutually exclusive results.

However, we already know that the area under the normal curve from the mean to 81 seconds is .3413. To find the area from the mean to 69 seconds, we have

$$Z = \frac{X - \mu}{\sigma} = \frac{69 - 75}{6} = -1.00$$

Since Appendix E, Table E.2, only shows positive entries for Z (because of symmetry), it is clear that the area from the mean to $Z = -1.00$ must be identical to the area from the mean to $Z = +1.00$. Discarding the negative sign then, we look up in Appendix E, Table E.2, the value $Z = 1.00$ and find the probability to be .3413. Hence the probability that the task can be completed in between 69 and 81 seconds is $.3413 + .3413 = .6826$.

This is a rather important finding for if we may generalize for a moment we can see that for any normal distribution there is a .6826 chance that a randomly selected item will fall within ± 1 standardized unit above or below the mean. Moreover, we will leave it to the reader to verify from Appendix E, Table E.2 (see Problem 6.23) that there is a .9544 chance that any randomly selected normally distributed observation will fall within ± 2 standardized units above or below the mean and that there is a .9973 chance that the observation will fall between ± 3 standardized units above or below the

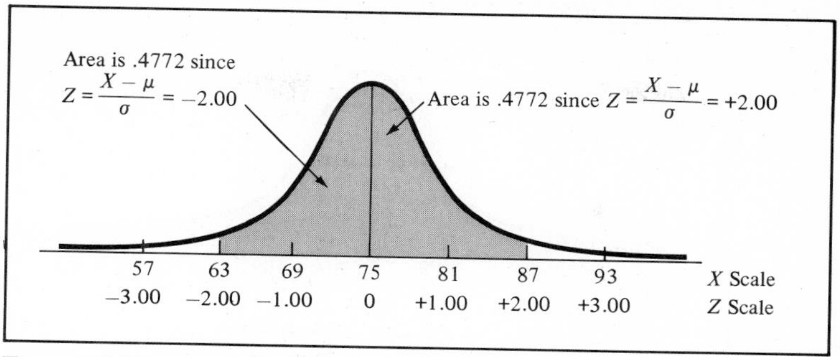

Figure 6.15 Finding $P(63 \leq X \leq 87)$.

mean. For our particular automobile plant (Plant C) this tells us that slightly more than two out of every three factory workers (68.26%) can be expected to complete the task within ±1 standard deviation (standardized unit) from the mean. Moreover, from Figure 6.15, slightly more than 19 out of every 20 factory workers (95.44%) can be expected to complete the assemblage within ±2 standard deviations (standardized units) from the mean (that is, between 63 and 87 seconds), and, from Figure 6.16, practically all factory workers (99.73%) can be expected to complete the part within ±3 standard deviations (standardized units) from the mean (that is, between 57 and 93 seconds). From Figure 6.16 it is indeed quite unlikely (.0027 or only 27 factory workers in 10,000) that a randomly selected factory worker will be so fast or so slow that he or she could be expected to complete the assemblage of the part in under 57 seconds or over 93 seconds. Thus it is clear why 6σ (that is, 3 standard deviations above the mean to 3 standard deviations below the mean) is often used as a practical approximation of the range for normally distributed data.

As another illustration of determining probabilities from the normal curve, suppose we wish to find how likely it is that a randomly selected factory worker can complete the task in 62 to 69 seconds. Since both values of interest are below the mean, it is clear from Figure 6.17 that the desired prob-

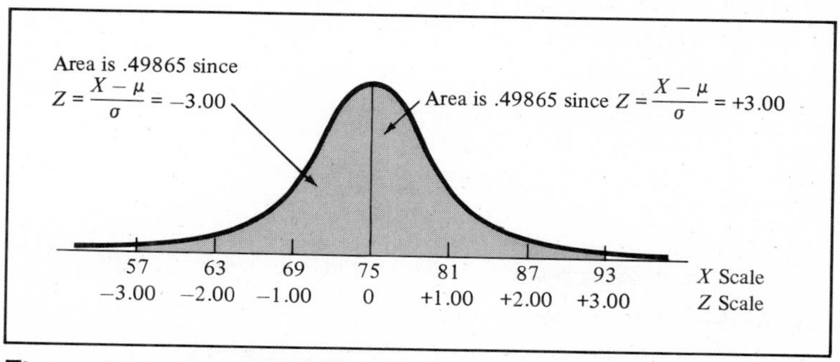

Figure 6.16 Finding $P(57 \leq X \leq 93)$.

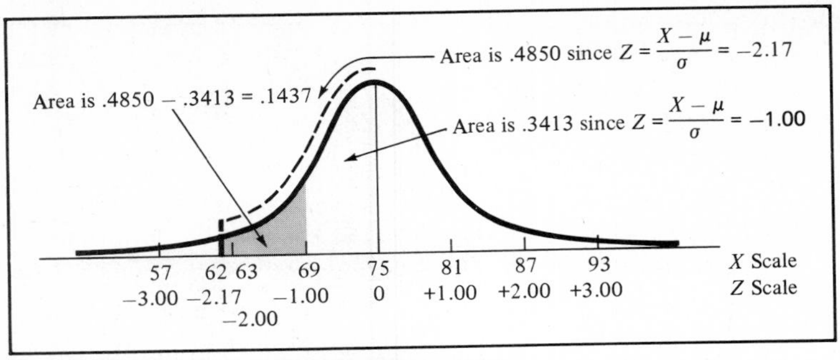

Figure 6.17 Finding $P(62 \leq X \leq 69)$.

ability (or area under the curve between these two values) is less than .5000. Since our conversion formula (6.22) only permits us to find probabilities from a particular value of interest to the mean, we see from Figure 6.17 that such a problem may be solved as follows: First, we can find the probability or portion of the curve from the mean to 62 seconds; then we can find the probability or area under the curve from the mean to 69 seconds; finally, we can subtract the smaller portion from the larger to obtain our desired result.

However, we have already determined that the area under the normal curve from the mean to 69 seconds is .3413 since $Z = (69 - 75)/6 = -1.00$. To determine the area under the curve from the mean to 62 seconds, we have

$$Z = \frac{X - \mu}{\sigma} = \frac{62 - 75}{6} = \frac{-13}{6} = -2.17$$

Neglecting the negative sign, we look up the Z value of 2.17 in Appendix E, Table E.2, by matching the appropriate Z row (2.1) with the appropriate Z column (.07) as shown in Table 6.6.

Table 6.6
OBTAINING AN AREA UNDER THE NORMAL CURVE

Z	.00	.01	.02	.03	.04	.05	.06	.07	.08	.09
0.0	.0000	.0040	.0080	.0120	.0160	.0199	.0239	.0279	.0319	.0359
0.1	.0398	.0438	.0478	.0517	.0557	.0596	.0636	.0675	.0714	.0753
0.2	.0793	.0832	.0871	.0910	.0948	.0987	.1026	.1064	.1103	.1141
0.3	.1179	.1217	.1255	.1293	.1331	.1368	.1406	.1443	.1480	.1517
0.4	.1554	.1591	.1628	.1664	.1700	.1736	.1772	.1808	.1844	.1879
⋮	⋮	⋮	⋮	⋮	⋮	⋮	⋮	⋮	⋮	⋮
2.0	.4772	.4778	.4783	.4788	.4793	.4798	.4803	.4808	.4812	.4817
2.1	.4821	.4826	.4830	.4834	.4838	.4842	.4846	.4850	.4854	.4857
2.2	.4861	.4864	.4868	.4871	.4875	.4878	.4881	.4884	.4887	.4890
2.3	.4893	.4896	.4898	.4901	.4904	.4906	.4909	.4911	.4913	.4916
2.4	.4918	.4920	.4922	.4925	.4927	.4929	.4931	.4932	.4934	.4936

SOURCE: Extracted from Appendix E, Table E.2.

Therefore, the resulting probability or area under the curve from the mean to 2.17 standardized units below it is .4850. Hence by subtracting the smaller area from the larger one we determine that there is only a .1437 probability of randomly selecting a factory worker who could be expected to complete the task in between 62 and 69 seconds.

In our previous applications regarding normally distributed data we have sought to determine from Appendix E, Table E.2, various probabilities or areas under the normal curve. Suppose, however, that we now switch the problem around and put the "cart before the horse"! That is, suppose now that we are interested in determining a particular value of the random variable for a given probability of occurrence. As a case in point, in how many seconds can a factory worker be expected to assemble the part if only 10% of all factory workers at Automobile Plant C can assemble the part in less time? From our previous information at Automobile Plant C we know the parameters $\mu = 75$ seconds and $\sigma = 6$ seconds. We wish to find the time trial X (in seconds) such that the proportion of workers who could be expected to finish the task in less time than X is .1000. This situation is depicted in Figure 6.18.

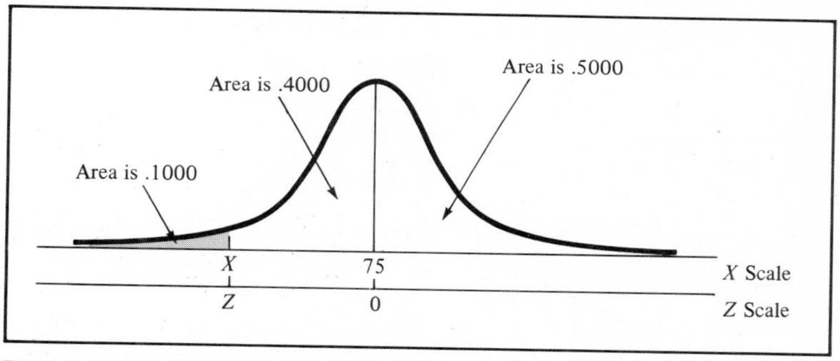

Figure 6.18 Finding Z to determine X.

Since 10% of the factory workers are expected to complete the task in under X seconds, then 90% of the workers would be expected to require X seconds or more to do the job. Nevertheless, from Figure 6.18 it is clear that this 90% can be broken down into two parts—time trials slower than (above) the mean (that is, 50%) and time trials between the mean and the desired value X (that is, 40%). While we do not know X, we can easily determine the corresponding standardized value Z since the area under the normal curve from the standardized mean 0 to this Z must be .4000. Using the body of Appendix E, Table E.2, we search for the area or probability .4000. The closest result is .3997 as shown in Table 6.7. Working from this area to the margins of Appendix E, Table E.2, we see that the Z value corresponding to the particular Z row (1.2) and Z column (.08) is 1.28. However, from Figure 6.18 the Z value must be recorded as a negative (that is, $Z = -1.28$) since it is below the standardized mean of 0.

Table 6.7
OBTAINING A Z VALUE CORRESPONDING TO A PARTICULAR AREA UNDER THE NORMAL CURVE

Z	.00	.01	.02	.03	.04	.05	.06	.07	.08	.09
0.0	.0000	.0040	.0080	.0120	.0160	.0199	.0239	.0279	.0319	.0359
0.1	.0398	.0438	.0478	.0517	.0557	.0596	.0636	.0675	.0714	.0753
0.2	.0793	.0832	.0871	.0910	.0948	.0987	.1026	.1064	.1103	.1141
0.3	.1179	.1217	.1255	.1293	.1331	.1368	.1406	.1443	.1480	.1517
0.4	.1554	.1591	.1628	.1664	.1700	.1736	.1772	.1808	.1844	.1879
⋮	⋮	⋮	⋮	⋮	⋮	⋮	⋮	⋮	⋮	⋮
1.0	.3413	.3438	.3461	.3485	.3508	.3531	.3554	.3577	.3599	.3621
1.1	.3643	.3665	.3686	.3708	.3729	.3749	.3770	.3790	.3810	.3830
1.2	.3849	.3869	.3888	.3907	.3925	.3944	.3962	.3980	.3997	.4015
1.3	.4032	.4049	.4066	.4082	.4099	.4115	.4131	.4147	.4162	.4177
1.4	.4192	.4207	.4222	.4236	.4251	.4265	.4279	.4292	.4306	.4319

SOURCE: Extracted from Appendix E, Table E.2.

Once Z is determined, we can now use the transformation formula (6.22) to determine the value of interest, X, that is, since

$$Z = \frac{X - \mu}{\sigma} \qquad (6.22)$$

then

$$X = \mu + Z\sigma \qquad (6.22a)$$

Substituting, we have $X = 75 + (-1.28)(6) = 67.32$ seconds. Thus we could expect that 10% of the workers will be able to complete the task in less than 67.32 seconds.

In this section we have studied, through numerous illustrations, one important aspect of the normal distribution—how to determine various probabilities of occurrence for data which are either normally distributed or which can be approximated by the normal distribution. In each case it was helpful to draw the normal curve so that we could more intuitively understand how large or small these probabilities of interest were likely to be. Therefore, at this time we leave it to the reader to examine the various panels in Figure 6.19 to verify the simple probability extensions to problems we have already considered regarding the times to assemble a part at Automobile Plant C.

6.10 THE NORMAL DISTRIBUTION AS AN APPROXIMATION TO VARIOUS DISCRETE PROBABILITY DISTRIBUTIONS

In the previous section we demonstrated the importance of the normal probability density function because of the numerous phenomena which seem to follow it or whose distributions can be approximated by it. In this and the

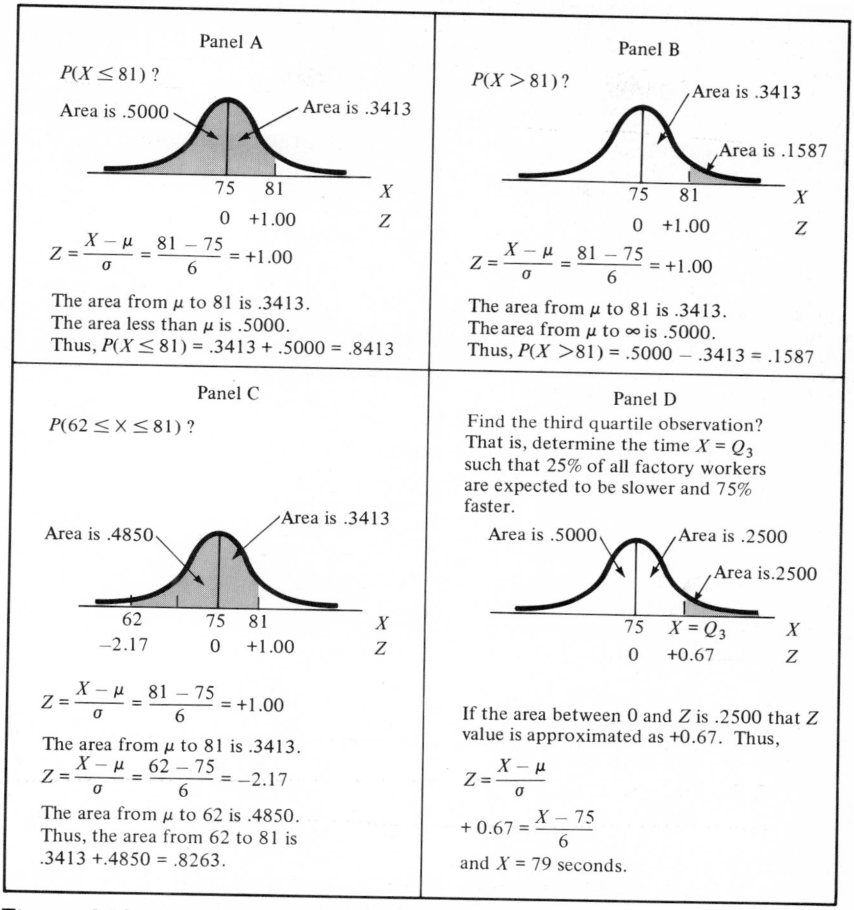

Figure 6.19 Determining probabilities from the normal distribution.

following section we shall demonstrate another important aspect of the normal distribution—how it may be used to approximate various important discrete probability distribution functions such as the binomial, the hypergeometric, and the Poisson.

Specifically, in this section we shall discuss these approximation methods in their simplest form. On the other hand, it has been argued (Reference 3) that whenever a continuous distribution (such as the normal) is used to approximate a discrete distribution, a more accurate approximation can be had if an adjustment or **correction for continuity** is employed. We shall develop these more accurate but computationally cumbersome approximation methods in Section 6.11.

6.10.1 Approximating the Binomial Distribution

In Section 6.5 we stated that the binomial distribution will be symmetric (like the normal distribution) whenever $p = .5$. When $p \neq .5$ the binomial distribution will not be symmetric. However, the closer p is to .5 and the larger the number of sample observations n, the more symmetric the distribution becomes. On the other hand, the larger the number of observations in the sample, the more tedious it is to compute the exact probabilities of success by use of Equation (6.7). Fortunately though, whenever the sample size is large, the normal distribution can be used to approximate the exact probabilities of success that otherwise may have to be attained through laborious computations. As one rule of thumb this normal approximation can be used whenever the following two conditions are met: (1) the product of the two parameters n and p equals or exceeds 5; and (2) the product $n(1 - p)$ equals or exceeds 5.

Using the transformation formula (6.22),

$$Z = \frac{X - \mu}{\sigma} \tag{6.22}$$

we have

$$Z \cong \frac{X - np}{\sqrt{np(1 - p)}} \tag{6.23}$$

where $\mu = np$, mean of the binomial distribution

$\sigma = \sqrt{np(1 - p)}$, standard deviation of the binomial distribution

$X =$ number of successes

and the approximate probabilities of success are obtained from Appendix E, Table E.2.

To illustrate this, suppose, in the product quality control example, that a sample of $n = 1{,}600$ tires of the same type are obtained at random from an on-going production process in which 8% of all such tires produced are defective. What is the probability that in such a sample *not more than* 150 tires will be defective?

Since both $np = (1{,}600)(.08) = 128$ and $n(1 - p) = (1{,}600)(.92) = 1{,}472$ exceed 5, we may use the normal distribution [Equation (6.23)] to approximate the binomial:

$$Z \cong \frac{X - np}{\sqrt{np(1 - p)}} = \frac{150 - 128}{\sqrt{(1{,}600)(.08)(.92)}} = \frac{22}{10.85} = +2.03$$

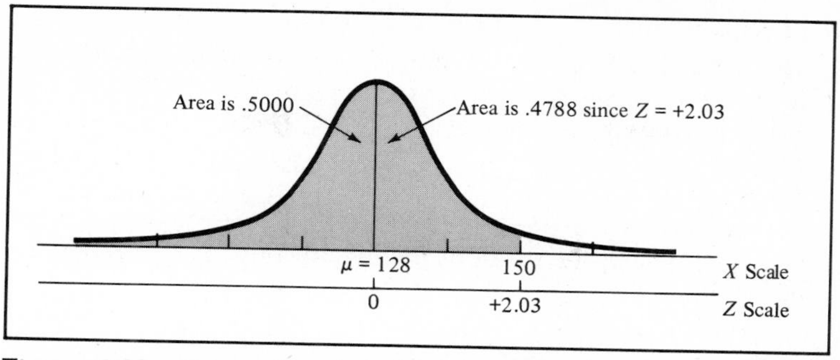

Figure 6.20 Approximating the binomial distribution.

Thus from Figure 6.20 the probability of not more than $X = 150$ successes corresponds, on the standardized Z scale, to a value of not more than $+2.03$. Using Appendix E, Table E.2, the area under the curve between the mean and $Z = +2.03$ is .4788 so that the approximate probability is given by $.5000 + .4788 = .9788$.

To appreciate the amount of labor saved by using the normal approximation to the binomial model in lieu of the exact probability computations, just imagine making the following 151 computations from Equation (6.7) before summing up the results:

$$\binom{1{,}600}{0}(.08)^{0}(.92)^{1{,}600} + \binom{1{,}600}{1}(.08)^{1}(.92)^{1{,}599} + \cdots + \binom{1{,}600}{150}(.08)^{150}(.92)^{1{,}450}$$

6.10.2 Approximating the Hypergeometric Distribution

The normal approximation to the hypergeometric distribution is also used whenever both $np \geq 5$ and $n(1-p) \geq 5$. In fact we may recall from Sections 6.5 and 6.6 that the only distinction in the application of the hypergeometric distribution and the binomial distribution is that the latter deals with problems in which sampling is accomplished with replacement from a finite population or without replacement from an infinite population, while the hypergeometric distribution deals with problems in which sampling is achieved without replacement from a finite population. Thus to approximate the exact probabilities of success that otherwise would be obtained through tedious computations, we have

$$Z = \frac{X - \mu}{\sigma} \tag{6.22}$$

so that

$$Z \cong \frac{X - np}{\sqrt{np(1-p)} \cdot \sqrt{\dfrac{N-n}{N-1}}} \tag{6.24}$$

where $\mu = np$, the mean of the hypergeometric distribution

$\sigma = \sqrt{np(1-p)} \cdot \sqrt{\dfrac{N-n}{N-1}}$, the standard deviation of the hypergeometric distribution

X = number of successes

and the approximate probabilities of success are obtained from Appendix E, Table E.2.

We note here though that such a computation for the standard deviation in Equation (6.24) is only necessary when the sample size, selected without replacement from a finite population, is large in comparison to the population size. In Section 6.6.2 we stated as a rule of thumb that this is considered to occur if the sample size equals or exceeds 5% of the population size. For much sample survey work, however, when sampling without replacement from a finite population, the sample size is usually small when compared to the population (that is, $n < .05N$). In these cases we may neglect the expression $\sqrt{\dfrac{N-n}{N-1}}$, the **finite population correction factor,** and thereby consider the problem identical to that of the normal approximation to the binomial distribution given by Equation (6.23).

To illustrate this using the survey data obtained by the dean and the researcher, we may consider the following: What is the probability of obtaining 35 *or more* accounting majors in a sample of 94 students if, in the population of 2,202 students, 43.6% of them (959 students) are accounting majors? Here $n = 94$ and $N = 2{,}202$. Since $n < .05N$, the expression $\sqrt{\dfrac{N-n}{N-1}}$ may be neglected. Using Equation (6.23) in lieu of Equation (6.24), we compute:

$$Z \cong \frac{X - np}{\sqrt{np(1-p)}} = \frac{35 - (94)(.436)}{\sqrt{(94)(.436)(.564)}} = \frac{35 - 40.984}{4.808} = -1.24$$

From Figure 6.21 and Appendix E, Table E.2, the probability of obtaining 35 or more accounting majors in a sample of 94 students is approximated by $.3925 + .5000 = .8925$.

6.10.3 Approximating the Poisson Distribution

The normal distribution may also be used to approximate the Poisson model whenever the parameter μ, the expected number of successes, equals or ex-

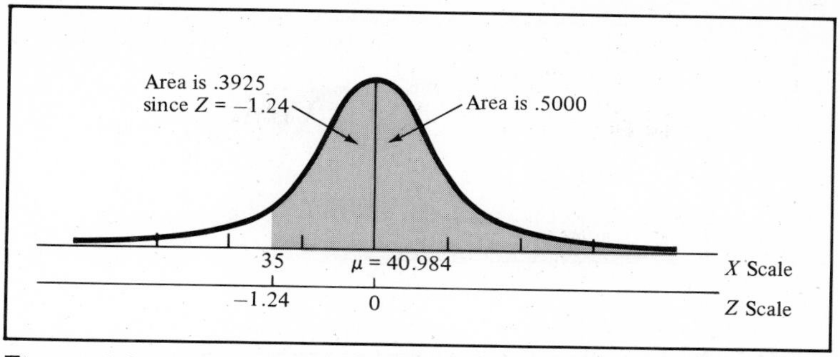

Figure 6.21 Approximating the hypergeometric distribution.

ceeds 5. Since the value of the mean and the variance of a Poisson distribution are the same, we have the following transformation formula:

$$Z = \frac{X - \mu}{\sigma} \tag{6.22}$$

and thus,

$$Z \cong \frac{X - \mu}{\sqrt{\mu}} \tag{6.25}$$

where μ = expected number of successes or mean of the Poisson distribution

$\sigma = \sqrt{\mu}$, the standard deviation of the Poisson distribution

X = number of successes

The approximate probabilities of success are obtained from Appendix E, Table E.2.

To illustrate this let us suppose that at Automobile Plant C the average number of work stoppages per day due to equipment problems during the production process is 12.0. What then is the approximate probability of having 15 *or fewer* work stoppages due to equipment problems on any given day? From Equation (6.25) we have

$$Z \cong \frac{X - \mu}{\sqrt{\mu}} = \frac{15 - 12.0}{\sqrt{12.0}} = +0.87$$

From Figure 6.22 and Appendix E, Table E.2, the approximate probability of having 15 or fewer work stoppages due to equipment problems on any given day is .5000 + .3078 = .8078.

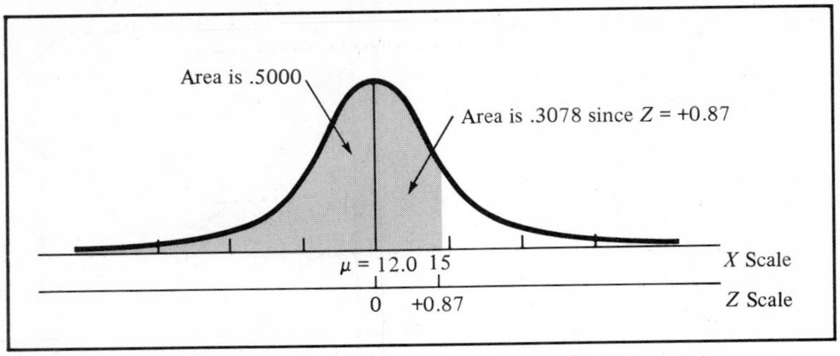

Figure 6.22 Approximating the Poisson distribution.

6.11 THE NORMAL DISTRIBUTION AS AN APPROXIMATION TO VARIOUS DISCRETE PROBABILITY DISTRIBUTIONS: USING THE CORRECTION FOR CONTINUITY ADJUSTMENT

In this section we shall introduce the **correction for continuity** adjustment which, when used with the normal distribution, will provide more accurate approximations to such discrete probability distributions as the binomial, the hypergeometric, or the Poisson.[9] Moreover, by revising the approximation formulas [Equations (6.23), (6.24), and (6.25)] so as to include the **continuity correction,** we shall also be able to obtain probability approximations for *individual* values of the random variable—which otherwise could not be estimated.

6.11.1 Obtaining Individual Probability Approximations

We may recall that with a continuous distribution such as the normal, the probability of obtaining a particular value of a random variable cannot be determined. On the other hand, when the normal distribution is used to approximate a discrete distribution, a correction for continuity adjustment can be used so that we may approximate the probability of a specific value. For example, in the binomial problem presented in Section 6.10.1 a random sample of n = 1,600 tires of the same type was selected from an on-going production process in which 8% of all such tires produced are defective. Suppose that we now want to compute the probability of obtaining *exactly* 150 defectives. In such a problem, the correction for continuity defines the integer value of interest to range from one-half unit below it to one-half unit above it. In our problem, the probability of obtaining 150 defective tires would be defined as the area (under the normal curve) between 149.5 and 150.5. Thus by adjusting Equation (6.23), this probability can be approximated as follows:

[9] We may recall from Sections 2.5 and 6.1 that a discrete random variable can only take on a specified value while a continuous random variable used to approximate it could take on any values whatsoever within a continuum or interval around that specified value. The correction for continuity adjustment provides more accuracy by taking into account these differences.

$$Z \cong \frac{150.5 - 128}{\sqrt{(1,600)(.08)(.92)}} = +2.07$$

and

$$Z \cong \frac{149.5 - 128}{\sqrt{(1,600)(.08)(.92)}} = +1.98$$

From Appendix E, Table E.2, we note that the area under the normal curve from the mean to $X = 150.5$ is .4808 while the area under the curve from the mean to $X = 149.5$ is .4761. Thus as depicted in Figure 6.23, the approximate probability of obtaining 150 defective tires is the difference in the two areas, .0047.

6.11.2 Obtaining More Accurate Probability Approximations

Suppose that we now wish to use the correction for continuity adjustment to determine the probability that in such a sample of $n = 1,600$ tires *not more than* 150 tires would be defective. This is the same problem raised in Section 6.10.1. Here, however, since the correction for continuity will be utilized, our approximation will be more accurate. Under the binomial distribution the probability of obtaining not more than 150 defective tires consists of all events up to and including 150 defectives, that is, $P(X \leq 150) = P(X = 0) + P(X = 1) + \cdots + P(X = 150)$ and the true probability may be laboriously computed from

$$\sum_{X=0}^{150} \binom{1,600}{X} (.08)^X (.92)^{1,600-X}$$

Since the correction for continuity defines the value $X = 150$ as lying in the interval from 149.5 to 150.5, in order to approximate the probability of not more than 150 defective tires, we need to compute the area under the normal curve up to 150.5. Referring to Figure 6.23, the area under the normal curve from the mean to 150.5 is .4808. Thus the area up to 150.5 is approximately

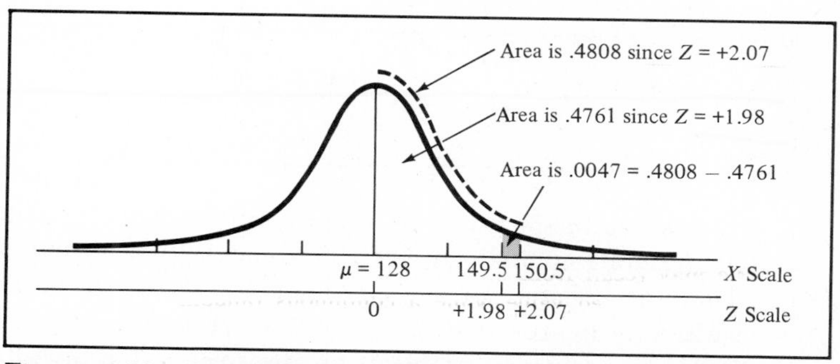

Figure 6.23 Approximating an exact binomial probability.

.5000 + .4808 = .9808. This approximation of the true probability is more accurate than the approximation .9788 computed in Section 6.10.1.

Now let us further suppose that we wanted to approximate the probability that *fewer than* 150 tires would be defective. Since this probability consists of all events up to and including 149, we need to compute the area under the normal curve up to 149.5. Referring to Figure 6.23, the area under the normal curve from the mean to 149.5 is .4761. Thus the area up to 149.5 is .5000 + .4761 = .9761.

In the previous two problems we have obtained approximate probabilities for values up to and including a particular point as well as for values less than a particular point. In these two problems the correction for continuity adjustment was used to either add .5 to 150 or to subtract .5 from 150, depending on the type of probability desired. Thus when approximating the binomial distribution, Equation (6.23) can be rewritten to incorporate this correction for continuity as follows:

$$Z \cong \frac{(X - np) \pm .5}{\sqrt{np(1-p)}} \tag{6.26}$$

where ± .5, the correction for continuity, is added or subtracted as needed.

In a similar manner, when approximating the hypergeometric or Poisson distribution, Equations (6.24) and (6.25), respectively, can be revised to include the correction for continuity adjustment. Thus, for approximating the hypergeometric distribution, we have

$$Z \cong \frac{(X - np) \pm .5}{\sqrt{np(1-p)} \cdot \sqrt{\dfrac{N-n}{N-1}}} \tag{6.27}$$

and for approximating the Poisson distribution, we have

$$Z \cong \frac{(X - \mu) \pm .5}{\sqrt{\mu}} \tag{6.28}$$

To illustrate the fact that the correction for continuity adjustment provides a more accurate approximation to the true result, let us return to the problem involving the Poisson distribution presented in Section 6.10.3. We recall that we wished to approximate the probability of having 15 *or fewer* work stoppages due to equipment problems on any given day if at Automobile Plant C the average number of such work stoppages per day is 12.0. From Equation (6.25), without the correction for continuity adjustment, we approximated this result to be .8078. Now, however, using the correction for continuity adjustment in Equation (6.28), we obtain[10]

$$Z \cong \frac{(15 - 12.0) + .5}{\sqrt{12.0}} = +1.01$$

From Figure 6.24 and Appendix E, Table E.2, we note that the area under the normal curve from the mean to 15.5 is .3438. Hence the area up to 15.5 is .5000 + .3438 = .8438. Therefore, the approximate probability of having 15 or fewer work stoppages due to equipment problems on any given day is .8438. This approximation compares quite favorably to the exact Poisson probability, .8445, obtained from Equation (6.20) or from Appendix E, Table E.6. In addition, we note that it is more accurate than the approximation (.8078) computed in Section 6.10.3 without using the correction for continuity adjustment.

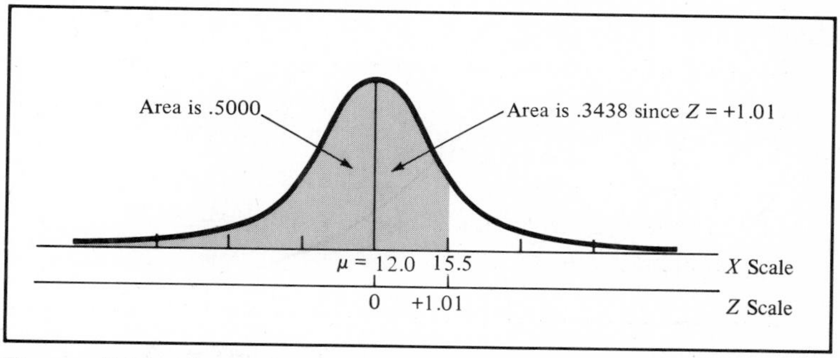

Figure 6.24 Approximating the Poisson distribution using correction for continuity adjustment.

We have seen from the example in Section 6.11.1 that if we are interested in obtaining probability approximations for individual values of the random variable, then it is, of course, necessary to use the correction for continuity adjustment. On the other hand, for other types of probability approximations, no hard and fast rule exists for using the correction for continuity adjustment. Since it is known that the advantages of increased accuracy become minimal with larger sample sizes and since the employment of the correction for continuity adjustment increases the computational complexity, researchers who favor the simpler approximation methods of Section 6.10 question

[10] This result is obtained since the correction for continuity adjustment defines the value $X = 15$ as lying in the interval from 14.5 to 15.5. Therefore, in order to approximate the probability of 15 or fewer work stoppages, we need to compute the area under the normal curve up to 15.5.

whether its use is worth the efforts involved. For simplicity, the correction for continuity will not be used in the remainder of this text. In most instances our sample sizes will be large enough so that the differences in the two approximation methods will be negligible. The reader should be aware, however, that in the appropriate sections of Chapters 8, 9, 10, and 13, the correction for continuity adjustment may be included in the equations if desired.

6.12 SUMMARY

In this chapter we have examined a variety of discrete probability distributions and one particularly important continuous distribution, the normal probability density function. The normal distribution was shown to be useful in its own right and also useful as an approximation of various discrete models. In the next chapter, however, we shall investigate how the normal distribution provides the basis for classical statistical inference. In subsequent chapters pertaining to inference we shall see that the data (that is, the phenomena of interest) follow or can be approximated by the various probability distribution functions considered here.

PROBLEMS

* 6.1 Let us consider rolling a pair of six-sided dice. The random variable of interest represents the sum of the two numbers (that is, faces) that occur when the pair of fair dice are rolled. The probability distribution is given below:

X	$P(X)$
2	1/36
3	2/36
4	3/36
5	4/36
6	5/36
7	6/36
8	5/36
9	4/36
10	3/36
11	2/36
12	1/36
	1

(a) Determine the mean or expected sum from rolling a pair of fair dice.
(b) Compute the variance and the standard deviation.

The game of Craps deals with the rolling of a pair of fair dice. A **field bet** in the game of Craps is a one-roll bet and is based on the outcome of the pair of dice. For every $1 bet you make: you can lose the $1 if the sum is 5, 6, 7, or 8; you can win $1 if the sum is 3, 4, 9, 10, or 11; or you can win $2 if the sum is either 2 or 12.

(c) Form the probability distribution function representing the different outcomes that are possible in a field bet.
(d) Determine the mean of this probability distribution.
(e) What is the player's expected long-run profit (or loss) from a $1 field bet? Interpret.

(f) What is the expected long-run profit (or loss) from a \$1 field bet to the house? Interpret.

(g) Would you play this game and make a field bet?

6.2 In the carnival game of Chuck-a-luck three fair dice are rolled after the player has placed a bet on the occurrence of a particular face of the dice, say ⚄. For every \$1 bet that you place: you can lose the \$1 if none of the three dice show the face ⚄; you can win \$1 if one die shows the face ⚄; you can win \$2 if two of the dice show the face ⚄; or you can win \$3 if all three dice show face ⚄.

(a) Form the probability distribution function representing the different monetary values (winnings or losses) that are possible (from one roll of the three dice).

Hint: Review Section 6.5.1 and see Figure 6.2 (panel A).

(b) Determine the mean of this probability distribution.

(c) What is the player's expected long-run profit (or loss) from a \$1 bet? Interpret.

(d) What is the expected long-run profit (or loss) to the house? Interpret.

(e) Would you play Chuck-a-luck and make a bet?

6.3 In the carnival game Under-or-over-seven a pair of fair dice are rolled once and the resulting sum determines whether or not the player wins or loses his or her bet. For example, the player can bet \$1 on the sum being under 7—that is, 2, 3, 4, 5, or 6. For such a bet the player will lose the \$1 if the outcome equals or exceeds 7 or will win a \$1 if the result is under 7. Similarly, the player can bet the \$1 on the sum being over 7—8, 9, 10, 11, or 12. Here the player wins \$1 if the result is over 7 but loses the \$1 if the result is 7 or under. A third method of play is to bet the \$1 on the outcome 7. For this bet the player will win \$4 if the result of the roll is 7 and lose the \$1 otherwise.

(a) Form the probability distribution function representing the different outcomes that are possible for a \$1 bet on being under 7.

(b) Form the probability distribution function representing the different outcomes that are possible for a \$1 bet on being over 7.

(c) Form the probability distribution function representing the different outcomes that are possible for a \$1 bet on 7.

(d) Prove that the expected long-run profit (or loss) to the player is the same—no matter which method of play is used.

(e) Would you prefer to play Under-or-over-seven or Chuck-a-luck (Problem 6.2) or make a field bet in Craps (Problem 6.1)? Why?

6.4 Roulette as played in American casinos is characterized by a wheel containing 38 equisized sectors with numbers 0, 00, 1, 2, . . . , 36 (see Figure 6.1). 18 of these 38 sectors are colored red, 18 are colored black, while the sectors 0 and 00 are colored green. Each time the wheel spins the resulting outcome determines whether the player wins or loses the bet. While many methods of play are possible, two such popular methods are described here. The player may bet \$1, say, on a color; or the player may bet the \$1 on a particular number. If the player bets \$1 on the color red, for example, the player will win \$1 if the color red turns up and lose the \$1 if either black or green turns up. On the other hand, if the player chooses to bet a particular number, \$35 is won if that number turns up or the \$1 is lost if otherwise.

(a) Form the probability distribution function representing the different outcomes that are possible for a \$1 bet on the color red.

(b) Form the probability distribution function representing the different outcomes that are possible for a \$1 bet on the number 29.

(c) Prove that the expected long-run profit (or loss) to the player is the same—no matter which method of play is used.

6.5 For the 50-year period of 1928 through 1977 the World Series in baseball has lasted four games on nine occasions, five games on 11 occasions, six games on 10 occasions, and seven games on 20 occasions.

(a) Form the probability distribution function for the discrete random variable X, the number of games played per World Series.

(b) Compute the mean or expected number of games played.

(c) Compute the standard deviation.

(d) From (a), what is the (historical) probability that any given World Series will last longer than five games?

6.6 The distribution of the number of home runs hit per baseball game for all games played this season in a well-known college athletic conference is

Number of Home Runs X	$P(X)$
0	.20
1	.30
2	.25
3	.10
4	.10
5	.04
6	.01

(a) Is this a probability distribution? Explain fully.

(b) Compute the expected number of home runs per game.

(c) Compute the variance and the standard deviation.

6.7 Using the summation rules (see Appendix B, Section B.3) show that the expression for σ^2 given in Equation (6.2) can also be written as

$$\sigma^2 = \sum_{i=1}^{N} X_i^2 P(X_i) - \mu^2$$

Verify your results using the data presented in Table 6.1.

6.8 It is known that 30% of the defective parts produced in a certain operation can be made satisfactory by rework.

(a) What is the probability that in a batch of six such defective parts at least three can be satisfactorily reworked?

(b) What is the probability that none of them can be reworked?

(c) What is the probability that all of them can be reworked?

* 6.9 The probability that a patient fails to recover from a particular operation is .1.

(a) What is the probability that exactly two of the next eight patients having this operation will not recover?

(b) What is the probability that at most one patient of the eight will not recover?

(c) Using the probabilities from Appendix E, Table E.7, plot the histogram for this binomial distribution.

6.10 An antitank gunner has a 70% chance of hitting his target each time he fires from within 200 meters.

(a) What is the probability that he fires at his target 10 times and misses all 10 times?

(b) How many times can he be expected to hit the target?

(c) What is the probability that he hits his target at least nine times in 10 shots?

(d) Using the probabilities extracted from Appendix E, Table E.7, plot the histogram for this binomial distribution.

6.11 Based upon past experience, 15% of the bills of a large mail-order book company are incorrect. If a random sample of three current bills is selected, what is the probability that

(a) Exactly two bills are incorrect?

(b) No more than two bills are incorrect?

(c) At least two bills are incorrect?

(d) What assumptions about the probability distribution are necessary to solve this problem?

6.12 The probability that an insurance salesman will make a sale on his first visit to a new customer is .25. If a salesman has three new customers to visit today, what is the probability that he makes a sale to

(a) Exactly one new customer?

(b) At most one new customer?

(c) What assumptions about making a sale to a new customer on the first visit are needed to evaluate the probabilities?

6.13 A well-known basketball player has an 80% chance of making his foul shots each time he shoots from the free throw line.

(a) What is the probability that he shoots four times in a game and hits all four times?

(b) What is the probability that he makes at most three foul shots out of his four attempts?

6.14 The typing pool at a large law firm contains 25 secretaries, 10 of whom had been with the firm for more than five years. If the executive officer of the firm wishes to randomly select three secretaries for assignment to a new case, what is the probability that

(a) None of the secretaries will have over five years of experience?
(b) One of the secretaries will have over five years of experience?
(c) Two of the secretaries will have over five years of experience?
(d) All three of the secretaries will have over five years of experience?
(e) Why is this a probability distribution?
(f) What discrete model does this represent?
(g) Compute the mean and the standard deviation from Equations (6.1) and (6.3) and then verify your computations by applying the direct formulas peculiar to this discrete model.

* 6.15 An auditor for the Internal Revenue Service is selecting a sample of six tax returns from persons in a particular profession for possible audit. If two or more of these indicate "improper" deductions, the entire group (population) of 100 tax returns will be audited. What is the probability of a more detailed audit if the percentage of improper returns is

(a) 25?
(b) 30?
(c) Discuss the differences in your results depending on the true percentage of improper returns.

6.16 The dean of our business school wishes to form an executive committee of five from among the 40 tenured faculty members at the school. If the selection is to be random and at the school there are eight tenured faculty members in accounting, what is the probability that the committee will contain

(a) None of them?
(b) At least one of them?
(c) Not more than one of them?
(d) How many tenured faculty members in accounting would be expected?

* 6.17 The average number of calls per minute received by a well-known television repair service is 1.2. What is the probability that in any given minute

(a) Fewer than two calls will be received?
(b) More than three calls will be received?
(c) Fewer than two calls or more than three calls will be received?
(d) Either two or three calls will be received?

6.18 The average number of claims per hour made to an insurance company for damages or losses incurred in moving is 3.1. What is the probability that in any given hour

(a) Fewer than three claims will be made?
(b) Exactly three claims will be made?
(c) Three or more claims will be made?

6.19 Based upon past records, the average number of two-car accidents within a New York City police precinct is 3.4 per day. What is the probability that there will be

(a) At least six such accidents in this precinct on any given day?
(b) Not more than two such accidents in this precinct on any given day?
(c) Fewer than two such accidents in this precinct on any given day?

6.20 Based upon past experience, 2% of the bills of a large mail-order book company are incorrect. If a sample of 20 bills is selected, find the probability that at least one bill will be incorrect. Do this using two probability distributions and briefly compare and explain your results.

* 6.21 Only 10% of all insects exposed to a certain insecticide under laboratory conditions survive.

(a) If a sample of five insects are exposed to this insecticide, what is the probability that

(1) All five insects will survive?
(2) All five insects will fail to survive?
(3) Not more than two insects will survive?

(b) Solve a(1), a(2), and a(3) using the Poisson distribution as an approximation of the binomial distribution and briefly compare your results.

(c) If a sample of 30 insects are exposed to this insecticide, what is the **approximate** probability that

(1) Exactly one insect will survive?
(2) All 30 insects will fail to survive?
(3) Not more than two insects will survive?

6.22 One out of every 100 light bulbs produced by the Starr Lighting Co. fail before the end of a one-week period when left burning continuously. One bulb is installed on each of the 50 floors of a large apartment building in New York City. What is the **approximate** probability that

(a) One bulb will be burned out at the end of the week?
(b) More than three of the bulbs will be burned out at the end of the week?
(c) Fewer than three of the bulbs will be burned out at the end of the week?
(d) Three of the bulbs will be burned out at the end of the week?

6.23 Verify the following:

(a) The area under the normal curve between the mean and 2 standard deviations above and below it is .9544.

(b) The area under the normal curve between the mean and 3 standard deviations above and below it is .9973.

6.24 The thickness of a batch of 10,000 brass washers of a certain type manufactured by a large company is normally distributed with a mean of .0191 inch and with a standard deviation of .000425 inch. Verify that 99.04% of these washers can be expected to have a thickness between .0180 and .0202 inch.

* 6.25 The length of time needed to service a car at a gas station is normally distributed with mean $\mu = 4.5$ minutes and standard deviation $\sigma = 1.1$ minutes. What is the probability that a randomly selected car will require

(a) More than 6 minutes of service or under 5 minutes of service?
(b) Between 3.5 and 5.6 minutes of service?
(c) At most 3.5 minutes of service?
(d) What must the servicing time be if only 5% of all cars require more than this amount of time?

6.26 The average (mean) toll for telephone calls to South America is $19; the standard deviation is $6.50. Assuming that the tolls are normally distributed,
(a) What percentage of these tolls are less than $29?
(b) What percentage of these tolls are between $10 and $22?
(c) What percentage of these tolls are between $10 and $17?
(d) What is the toll (in dollars and cents) that is exceeded by 90% of the callers?

* 6.27 A statistical analysis of 1,000 long-distance telephone calls made from a large business office indicates that the length of these calls is normally distributed with $\mu = 129.5$ seconds and $\sigma = 30.0$ seconds.
(a) What percentage of these calls lasted less than 180 seconds?
(b) What is the probability that a particular call lasted between 89.5 and 169.5 seconds?
(c) How many calls lasted less than 60 seconds or more than 150 seconds?
(d) What percentage of the calls lasted between 100 and 120 seconds?
(e) What must the length of a particular call be if only 1% of all calls are shorter?
(f) If the researcher could not assume that the data were normally distributed, what then would be the probability that a particular call lasted between 89.5 and 169.5 seconds?
Hint: Recall the Bienaymé–Chebyshev rule in Section 4.4.3.
(g) Discuss the differences in your answers to (b) and (f).

6.28 The price per gallon of unleaded gasoline at service stations in a large city is normally distributed with a population mean of 80.7 cents and a population standard deviation of 3 cents. What proportion of the service stations have a price per gallon for unleaded gasoline
(a) Between 78.7 and 82 cents per gallon?
(b) Between 77.1 and 78.7 cents per gallon?
(c) Less than 75.9 cents per gallon?
(d) Below what value will 2% of the prices fall?
(e) Between what two values symmetrically distributed around the population mean will 75% of the prices fall?

6.29 Monthly food expenditures average $280 for families of size four in a large city with a standard deviation of $80. Assuming that the monthly food expenditures are normally distributed
(a) What percentages of these expenditures are less than $350?
(b) What percentages of these expenditures are between $200 and $250?
(c) What percentages of these expenditures are less than $200 or greater than $350?
(d) Determine Q_1 and Q_3 from the normal curve.

6.30 Plastic bags used for packaging produce are manufactured so that the breaking strength of the bag is normally distributed with a mean of 5 pounds per square inch and a standard deviation of 1 pound per square inch. What proportion of the bags produced have a breaking strength

(a) Between 5 and 5.5 pounds per square inch?
(b) Between 3.2 and 4.2 pounds per square inch?
(c) At least 3.6 pounds per square inch?
(d) Less than 3.17 pounds per square inch?
(e) Between what two values symmetrically distributed around the mean will 95% of the plastic bags fall?

6.31 The reaction time to a certain psychological experiment is normally distributed with mean $\mu = 20$ seconds and standard deviation $\sigma = 4$ seconds.

(a) What is the probability that a subject has a reaction time between 14 and 30 seconds?
(b) What is the probability that a subject has a reaction time between 25 and 30 seconds?
(c) What percentage of subjects have reaction times over 14 seconds?
(d) What is the reaction time for which only 1% of all subjects are faster?

6.32 A set of final examination grades in an introductory statistics course was found to be normally distributed with a mean of 73 and a standard deviation of 8.

(a) What is the probability of getting at most a grade of 91 on this exam?
(b) What percentage of students scored between 65 and 89?
(c) What percentage of students scored between 81 and 89?
(d) What must the final exam grade be if only 5% of the students taking the test scored higher?
(e) If the professor "curves" (gives A's to the top 10% of the class regardless of the score), are you better off with a grade of 81 on this exam or a grade of 68 on a different exam where the mean is 62 and the standard deviation is 3? Show statistically and explain.

6.33 At a well-known business school (which competes with the one our dean and researcher are associated with) the grade-point indexes of its 1,000 undergraduates are approximately normally distributed with mean $\mu = 2.83$ and standard deviation $\sigma = 0.38$.

(a) What is the probability that a randomly selected student has a grade-point index between 2.00 and 3.00?
(b) What percentage of the student body are on probation, that is, have grade-point indexes below 2.00?
(c) How many students at this school are expected to be on the dean's list, that is, have grade-point indexes equal to or exceeding 3.20?
(d) What grade-point index will be exceeded by 15% of the student body?

6.34 The number of days between billing and payment of charge accounts of a large department store is approximately normally distributed with a mean of 18 days and standard deviation of 4 days. What proportion of the bills will be paid

(a) Between 12 and 18 days?
(b) Between 20 and 23 days?

(c) In less than 8 days?

(d) In 12 or more days?

(e) Within how many days will 99.5% of the bills be paid?

(f) Between what two values symmetrically distributed around the mean will 98% of the bills fall?

6.35 A particular woman is 67 inches tall and weighs 135 pounds. If the heights of women are normally distributed with $\mu_H = 65$ inches and $\sigma_H = 2.5$ inches and if the weights of women are normally distributed with $\mu_W = 125$ pounds and $\sigma_W = 10$ pounds, determine whether this woman's more unusual characteristic is her height or her weight.

6.36 The heights of 12-year-old boys follows the normal distribution with mean $\mu = 61.8$ inches. Find the standard deviation σ if 98% of the boys in the population are shorter than 65.4 inches.

6.37 For overseas flights, an airline has three different choices on its dessert menu—ice cream, apple pie, and chocolate cake. Based on past experience the airline feels that each dessert is equally likely to be chosen.

(a) If a random sample of four passengers is selected, what is the probability that at least two will choose ice cream for dessert?

(b) If a random sample of 30 passengers is selected, what is the **approximate** probability that at least two will choose ice cream for dessert?

6.38 Based upon past experience, 40% of all customers at a certain automobile service station pay for their purchases with a credit card. If a random sample of three customers is selected, what is the probability that

(a) None pay with a credit card?

(b) Two pay with a credit card?

(c) At least two pay with a credit card?

(d) Not more than two pay with a credit card?

If a random sample of 200 customers is selected, what is the **approximate** probability that

(e) At least 75 pay with a credit card?

(f) Not more than 70 pay with a credit card?

(g) Between 70 and 75 customers, inclusive, pay with a credit card?

6.39 Servicing records indicate that 50% of all new automobiles purchased of a particular brand will require some type of repair within the 90-day warranty period. For a random sample of $n = 12$ such new automobiles, use the binomial distribution to determine

(a) The probability that eight or nine of them will require repair within the 90-day warranty period.

(b) The probability that not more than two of them will require repair within the 90-day warranty period.

Use the normal approximation to the binomial distribution to determine

(c) The **approximate** probability that eight or nine of them will require repair within the 90-day warranty period.

(d) The **approximate** probability that not more than two of them will require repair within the 90-day warranty period.

(e) Discuss the differences in your answers in (a) versus (c) and in (b) versus (d).

6.40 It is known that one out of every three people entering a particular department store will make at least one purchase. If a random sample of $n = 5$ persons is selected, what is the probability that

(a) Two or more of them will make at least one purchase?

(b) At most four of them will make at least one purchase?

If a random sample of $n = 81$ persons is selected, what is the **approximate** probability that

(c) 30 or more of them will make at least one purchase?

(d) At most 40 of them will make at least one purchase?

* 6.41 The dean of our business school wishes to form a faculty senate of 20 members from among the faculty of 100. If the selection is to be random and at the school 25 of the faculty members are in accounting, what is the **approximate** probability that the faculty senate body will contain

(a) At least two accounting faculty?

(b) Between two and six accounting faculty?

(c) How many accounting faculty would be expected in the faculty senate?

6.42 A public health officer in a certain area suspects that 10% of the children are severely undernourished. 2,000 children live in this area.

(a) If a sample of four children is selected, what is the probability that at least one will be undernourished?

(b) If a sample of 80 children is selected, what is the **approximate** probability that at least five will be undernourished?

* 6.43 The number of cars arriving per minute at a toll booth on a particular bridge is Poisson distributed with $\mu = 2.5$.

(a) What is the probability that in any given minute

(1) No cars arrive?

(2) Not more than two cars arrive?

(b) If μ, the expected number of cars arriving at the toll booth per 10-minute interval, is 25.0, what is the **approximate** probability that in any given 10-minute period

(1) Not more than 20 cars arrive?

(2) Between 20 and 30 cars arrive?

6.44 Cars arrive at a car wash at a rate of nine per half-hour.

(a) What is the probability that in any given half-hour period at least three cars arrive?

(b) What is the **approximate** probability that in any given half-hour period at least three cars arrive?

(c) Compare your results in (a) and (b).

REFERENCES

1. Daniel, W. W., and J. C. Terrell, *Business Statistics: Basic Concepts and Methodology* (Boston: Houghton Mifflin, 1975).
2. Ehrenberg, A. S. C., *Data Reduction: Analyzing and Interpreting Statistical Data* (London: Wiley, 1975).
3. Freund, J. E., and F. J. Williams, *Elementary Business Statistics: The Modern Approach,* 3rd ed., (Englewood Cliffs, N.J.: Prentice-Hall, 1977).
4. Hamburg, M., *Statistical Analysis for Decision Making,* 2nd ed., (New York: Harcourt, Brace, Jovanovich, 1977).
5. Larson, H. J., *Introduction to Probability Theory and Statistical Inference* (New York: Wiley, 1969).
6. Miller, I., and J. E. Freund, *Probability and Statistics for Engineers,* 2nd ed., (Englewood Cliffs, N.J.: Prentice-Hall, 1977).
7. Scarne, J., *Scarne's Complete Guide to Gambling* (New York: Simon and Schuster, 1961).
8. Thorp, E. O., *Beat the Dealer* (New York: Random House, 1962).
9. Weaver, W., "Probability," in A. Shuchman, ed., *Scientific Decision Making in Business* (New York: Holt, Rinehart and Winston, 1963).

CHAPTER 7

SAMPLING DISTRIBUTIONS

7.1 THE NEED FOR SAMPLING DISTRIBUTIONS

One of the major goals of statistical analysis is to use those statistics (such as the sample average, the sample standard deviation, and the sample proportion) which are obtained from sample data in order to estimate their *true* value in the population. The process of generalizing these sample results to the population is referred to as **statistical inference.** In the previous two chapters we have examined basic rules of probability and investigated various probability distributions such as the uniform, binomial, hypergeometric, Poisson, and normal distributions. In this chapter we shall use these rules of probability along with the knowledge of certain probability distributions to begin focusing on how certain statistics (such as the mean) can be utilized in making inferences about the true population parameters.

We should realize that a survey researcher is concerned with drawing conclusions about a population, not about a sample. For example, a political pollster would be interested in the sample results only as a way of estimating the actual proportion of the votes that each candidate will receive from the population of voters. Likewise, the auditor, in selecting a sample of vouchers, is only interested in using the sample total for estimating the actual total amount in the population. Moreover, our dean and researcher would only utilize a sample survey as a way of drawing inferences about the population of students. Thus in each of these situations, the sample is the "vehicle" for drawing conclusions about the population.

In practice, a single sample of a predetermined size is selected at random from the population. The values that are to be included in the sample are determined through the use of a random number generator, such as a table of random numbers (see Section 2.9). Hypothetically, in order to be able to use the sample mean to estimate the population mean, we should examine every possible sample (and its mean) that could have occurred in the process of selecting one sample of a certain size. If this selection of all possible samples actually were to be done, the distribution of the results would be referred to as a **sampling distribution.** Although in practice only one such sample is actually selected, the concept of a sampling distribution must be examined so that probability theory can be used in making inferences about the population values.

7.2 SAMPLING DISTRIBUTION OF THE MEAN

7.2.1 Why the Arithmetic Mean?

In Chapter 4 we examined several measures including the arithmetic mean, median, and mode in discussing the central tendency of quantitative variables. Why, then, should we now focus upon the arithmetic mean in making inferences? There are several properties (Reference 6) that make the arithmetic mean the best estimator for drawing inferences about the population mean. Three of the more important properties are:

1. Unbiasedness
2. Efficiency
3. Consistency

The first property, **unbiasedness,** involves the average sample value of the mean. If all possible samples of a certain size n were selected from the population, the mean of all these sample means would be the population mean, μ.

This property can be demonstrated empirically by looking at the following example: A population of four typists were asked to type the same page of a manuscript. The number of errors made by each typist were:

Typist	Number of Errors
A	3
B	2
C	1
D	4

This population distribution is shown in Figure 7.1

The mean of this population μ is equal to

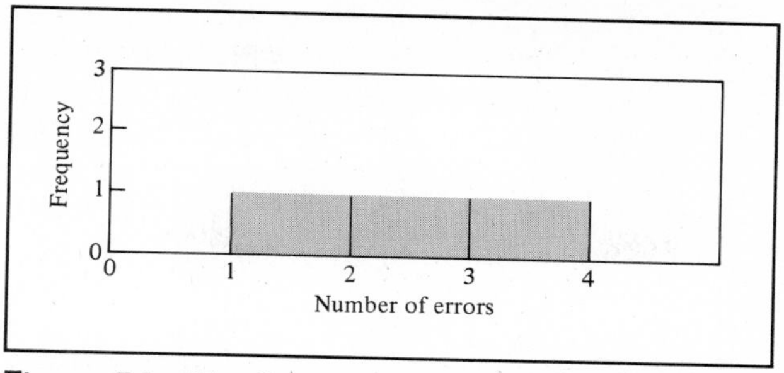

Figure 7.1 Number of errors made by a population of four typists.

$$\mu = \frac{3 + 2 + 1 + 4}{4} = 2.5 \text{ Errors}$$

If samples of two typists are selected *with* replacement from this population, there are 16 possible samples that could be selected ($N^n = 4^2 = 16$). These possible sample outcomes are shown in Table 7.1.

If all these 16 sample means are averaged, the mean of these values ($\mu_{\bar{X}}$) is equal to 2.5, which is the mean of the population, μ.

On the other hand, if sampling was being done *without* replacement, there would be six possible samples of two typists

$$\frac{N!}{n!(N-n)!} = \frac{4!}{2!2!} = 6$$

These six possible samples are listed in Table 7.2.

Table 7.1
ALL 16 SAMPLES OF $n = 2$ TYPISTS FROM A POPULATION OF $N = 4$ TYPISTS WHEN SAMPLING WITH REPLACEMENT

Sample	Typists	Sample Outcomes	Sample Mean $\bar{X}_i$
1	A, A	3, 3	$\bar{X}_1 = 3$
2	A, B	3, 2	$\bar{X}_2 = 2.5$
3	A, C	3, 1	$\bar{X}_3 = 2$
4	A, D	3, 4	$\bar{X}_4 = 3.5$
5	B, A	2, 3	$\bar{X}_5 = 2.5$
6	B, B	2, 2	$\bar{X}_6 = 2$
7	B, C	2, 1	$\bar{X}_7 = 1.5$
8	B, D	2, 4	$\bar{X}_8 = 3$
9	C, A	1, 3	$\bar{X}_9 = 2$
10	C, B	1, 2	$\bar{X}_{10} = 1.5$
11	C, C	1, 1	$\bar{X}_{11} = 1$
12	C, D	1, 4	$\bar{X}_{12} = 2.5$
13	D, A	4, 3	$\bar{X}_{13} = 3.5$
14	D, B	4, 2	$\bar{X}_{14} = 3$
15	D, C	4, 1	$\bar{X}_{15} = 2.5$
16	D, D	4, 4	$\bar{X}_{16} = 4$
			$\mu_{\bar{X}} = 2.5 = \mu$

Table 7.2
ALL SIX POSSIBLE SAMPLES OF $n = 2$ TYPISTS FROM A POPULATION OF $N = 4$ TYPISTS WHEN SAMPLING WITHOUT REPLACEMENT

Sample	Typists	Sample Outcome	Sample Mean $\bar{X}_i$
1	A, B	3, 2	$\bar{X}_1 = 2.5$
2	A, C	3, 1	$\bar{X}_2 = 2$
3	A, D	3, 4	$\bar{X}_3 = 3.5$
4	B, C	2, 1	$\bar{X}_4 = 1.5$
5	B, D	2, 4	$\bar{X}_5 = 3$
6	C, D	1, 4	$\bar{X}_6 = 2.5$
			$\mu_{\bar{X}} = 2.5 = \mu$

In this case, also, the average of all sample means ($\mu_{\bar{X}}$) is equal to the population mean, 2.5. Therefore we have shown that the sample arithmetic mean is an unbiased estimator of the population mean.

The second property, **efficiency,** refers to how well a sample statistic estimates a population parameter. Generally, the sample mean is considered an efficient estimator of the population mean since it is more stable from sample to sample than either the median or the mode and, therefore, is likely to provide a closer estimate of the true mean than either the median or the mode. The median, as the middle-ranked value in the ordered array, seems to vary somewhat more than the mean. The mode, the value that occurs most often, is rather unstable, particularly for small samples, often fluctuating drastically from sample to sample.

The third property, **consistency,** refers to the effect of the sample size on the usefulness of an estimator. As the sample size increases, the variation of the sample mean from the population mean becomes smaller and smaller so that the sample arithmetic mean becomes a better estimate of the population mean.

7.2.2 Standard Error of the Mean

The fluctuation in the average number of typing errors which was obtained when sampling with replacement from all 16 possible samples is illustrated in Figure 7.2.

In this small example, although we can observe a good deal of fluctuation in the sample mean—depending on which typists were selected—there is not nearly as much fluctuation as in the actual population itself. The fact that the sample means are less variable than the population data follows logically from an understanding of the averaging process. A particular sample mean averages together all the values in the sample. A population may consist of individual outcomes that can take on a wide range of values from extremely small to extremely large. However, if an extreme value falls into the sample, although it will have an effect on the mean, the effect will be reduced since it is being averaged in with all the other values in the sample. Moreover, as the sample size increases, the effect of a single extreme value gets even smaller, since it is

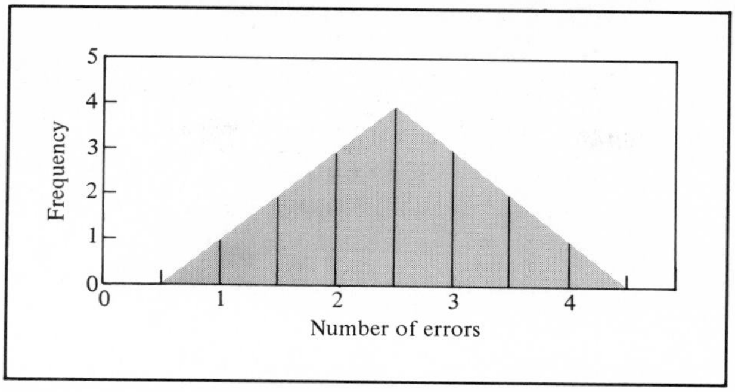

Figure 7.2 Sampling distribution of the average number of errors for samples of two typists.

being averaged with more observations. This phenomenon is expressed statistically in the value of the standard deviation of the sample mean. This is the measure of variability of the mean from sample to sample and is referred to in statistics as the standard deviation of the sample mean or the **standard error of the mean,** $\sigma_{\bar{X}}$. When sampling with replacement, the standard error of the mean is equal to

$$\sigma_{\bar{X}} = \frac{\sigma}{\sqrt{n}} \tag{7.1}$$

the standard deviation of the population divided by the square root of the sample size. Therefore, as the sample size increases, the standard error of the mean will decrease by a factor equal to the square root of the sample size. This relationship between the standard error of the mean and the sample size will be further examined in Chapter 8 when the sample size will be determined to meet specific criteria desired in estimating the mean.

7.2.3 Normal Populations

Now that we know what the average and the standard deviation of the sample mean will be, what distribution will the sample mean follow?

It can be shown that if we sample with replacement from a population that is normally distributed with mean μ and standard deviation σ, the sampling distribution of the mean, $\bar{X}$, for any size n will also be normally distributed with mean μ and have a standard error of the mean $\sigma_{\bar{X}} = \sigma/\sqrt{n}$.

In the most trivial case, if we draw samples of size $n = 1$, each possible sample mean is a particular observation from the population since

$$\overline{X} = \frac{\sum_{i=1}^{n} X_i}{n} = \frac{X}{1} = X$$

If we know that the population follows a normal distribution with mean μ and standard deviation σ, then, of course, the sampling distribution of $\overline{X}$ for samples of size $n = 1$ must also follow the normal distribution with mean μ and standard error of the mean $\sigma_{\overline{X}} = \sigma/\sqrt{n} = \sigma/\sqrt{1} = \sigma$.

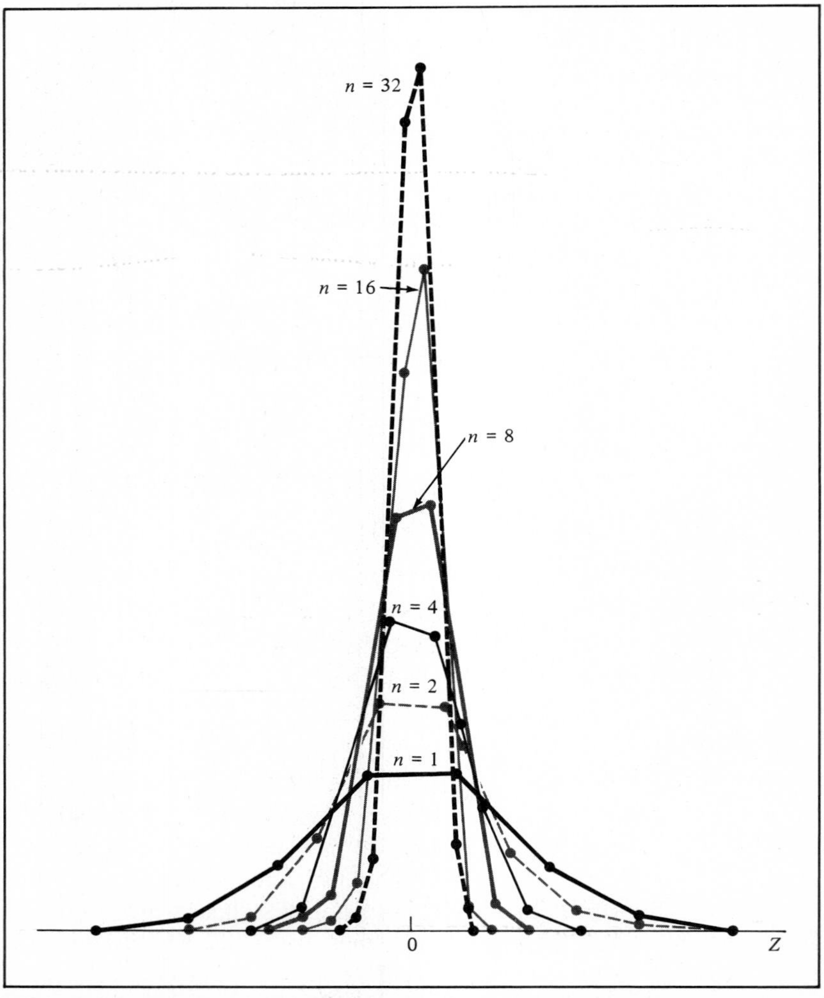

Figure 7.3 Sampling distributions of the mean from 500 samples of size n = 1, 2, 4, 8, 16, and 32 selected from a normal population.

Moreover, as the sample size increases, the sampling distribution of the mean still follows a normal distribution with mean μ. However, with increased sample size, the standard error becomes smaller, so that more sample means are closer to the population mean. This can be observed by referring to Figure 7.3. In this figure, 500 samples of size 1, 2, 4, 8, 16, and 32 were randomly selected, using a computer program, from a normally distributed population. We can see clearly from the polygons in Figure 7.3 that while the sampling distribution of the mean is approximately[1] normal for each sample size, the sample means are distributed closer to the population mean as the sample size is increased.

In order to understand how the sampling distribution of the mean may be used, let us look at the following example. A tire company produces 10,000 tires per week at a large factory. The quality control of the process is strictly maintained so that the tread life of the tires produced is normally distributed with an average of 25,000 miles and a standard deviation of 5,000 miles.

If a sample of 100 tires is randomly selected for analysis, and the average tire life is computed for this sample, what type of results could be expected?

For example, do you think that the sample mean would be 25,000 miles? 50,000 miles? 24,000 miles? The sample acts as a "miniature representation" of the population so that if the values in the population were normally distributed, the values in the sample should be approximately normally distributed. Moreover, if the population mean is 25,000 miles, the sample mean has a good chance of being "close to" 25,000 miles.

To explore this problem even further, how can we determine the probability that the sample of 100 tires will have a mean between 24,000 and 25,000? We know from our study of the normal distribution (Section 6.9) that the area between any value X and the population mean μ can be found by converting to standardized Z units

$$Z = \frac{X - \mu}{\sigma} \qquad (7.2)$$

and finding the appropriate value in the table of the normal distribution (Appendix E, Table E.2). In those situations we were studying how any particular value X deviates from its mean. Now, in the tire example, the value of interest is actually a sample mean $\overline{X}$, and we wish to study how a particular value $\overline{X} = 24{,}000$ deviates from the population mean. The standard deviation of $\overline{X}$ is, of course, the standard error of the mean. Thus we have the following formula when working with the sampling distribution of the mean:

[1] We must remember that "only" 500 samples out of an infinite number of samples have been selected, so that the sampling distributions shown are only approximations of the true distributions.

$$Z = \frac{\overline{X} - \mu}{\sigma_{\overline{X}}} = \frac{\overline{X} - \mu}{\frac{\sigma}{\sqrt{n}}} \tag{7.3}$$

To find the area between 24,000 and 25,000 miles (Figure 7.4) we have

$$Z = \frac{\overline{X} - \mu}{\frac{\sigma}{\sqrt{n}}} = \frac{24{,}000 - 25{,}000}{\frac{5{,}000}{\sqrt{100}}}$$

$$= \frac{-1{,}000}{500} = -2.0$$

Looking up 2.0 in Appendix E, Table E.2, we find an area of .4772. Therefore, 47.72% of all the possible samples of size 100 would have means between 24,000 and 25,000 miles.

Now this is clearly not the same as saying that a certain percentage of tires will last between 24,000 and 25,000 miles. In fact the percentage of tires that will last between 24,000 and 25,000 miles can be computed as follows:

$$Z = \frac{X - \mu}{\sigma} = \frac{24{,}000 - 25{,}000}{5{,}000}$$

$$= \frac{-1{,}000}{5{,}000} = -.20$$

The area corresponding to $Z = -.20$ in Appendix E, Table E.2, is .0793. Therefore 7.93% of the tires are expected to last between 24,000 and 25,000 miles. By comparing the two results we can see that many more sample means than actual individual values are expected to fall between 24,000 and 25,000 miles. Each sample mean consists of the average of 100 tires. Some extremely

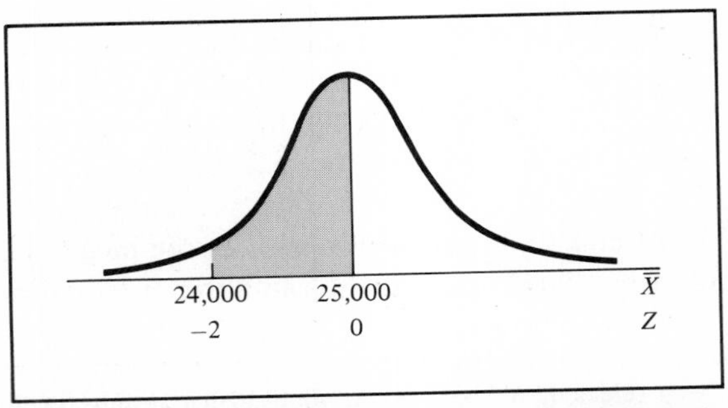

Figure 7.4 Diagram of normal curve needed to find area between 24,000 and 25,000 miles.

small and/or extremely large individual tire values have probably been averaged out in computing this mean value. In order for the mean of a sample of 100 tires to be extremely small, a large number of the individual values would have to be extremely small, not just a few values.

How would these results have been affected by using a different sample size? Let us see what the effect would be of decreasing the sample size to 25. Here we would have the following:

$$Z = \frac{\bar{X} - \mu}{\frac{\sigma}{\sqrt{n}}} = \frac{24{,}000 - 25{,}000}{\frac{5{,}000}{\sqrt{25}}}$$

$$= \frac{-1{,}000}{1{,}000} = -1.0$$

The area corresponding to $Z = -1.0$ is .3413. Therefore only 34.13% of all the samples of size 25 would be expected to have means between 24,000 and 25,000 miles, as compared to 47.72% for samples of size 100.

Instead of determining the proportion of sample means that are expected to fall within a certain interval, we might be more interested in finding out the interval within which a fixed proportion of the samples (means) would fall. For example, let us find the interval around the population mean that will include 95% of the sample means. The 95% could be divided into two equal parts, half below the mean and half above the mean (see Figure 7.5). Analogous to Section 6.9, we are determining a distance below and above the population mean containing a specific area of the normal curve. From Equation (7.3) we have

$$Z_L = \frac{\bar{X}_L - \mu}{\frac{\sigma}{\sqrt{n}}}$$

where $Z_L = -Z$.

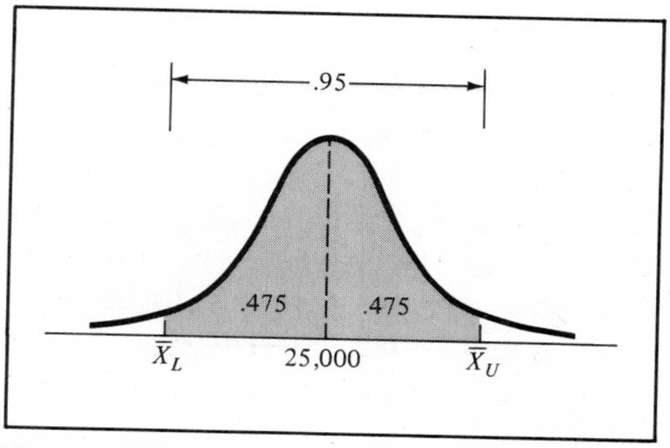

Figure 7.5 Diagram of normal curve needed to find upper and lower limits to include 95% of sample means.

and

$$Z_U = \frac{\bar{X}_U - \mu}{\frac{\sigma}{\sqrt{n}}}$$

where $Z_U = +Z$. Therefore, the lower value of $\bar{X}$ is:

$$\bar{X}_L = \mu - Z\frac{\sigma}{\sqrt{n}} \qquad (7.4a)$$

while the upper value of $\bar{X}$ is:

$$\bar{X}_U = \mu + Z\frac{\sigma}{\sqrt{n}} \qquad (7.4b)$$

Since $\sigma = 5{,}000$ and $n = 100$ and the value of Z corresponding to an area of .475 under the normal curve is 1.96, the lower and upper values can be found as follows:

$$\bar{X}_L = 25{,}000 - (1.96)\frac{5{,}000}{\sqrt{100}} \qquad \bar{X}_L = 25{,}000 - 980 = 24{,}020$$

$$\bar{X}_U = 25{,}000 + (1.96)\frac{5{,}000}{\sqrt{100}} \qquad \bar{X}_U = 25{,}000 + 980 = 25{,}980$$

Our conclusion would be that 95% of all sample means of size 100 should fall between 24,020 and 25,980 miles.

7.2.4 Sampling from Nonnormal Populations

In the previous section we examined the sampling distribution of the mean when we knew that the population was normally distributed. However, it is unrealistic to think that the population of interest will always be normally distributed. In many cases, either we do not have any real knowledge as to the shape of the population or we know that the population is not normally distributed. What will the sampling distribution of the mean look like for populations that are not normal? This question brings us to probably the most important theorem in basic statistics, the central limit theorem.

Central Limit Theorem: As the sample size (number of observations in each sample) gets large enough, the sampling distribution of the mean can be approximated by the normal distribution. This is true regardless of the distribution of the individual values in the population.

Now the question that surely arises at this point is what sample size is "large enough?" A great deal of statistical research has gone into this particular question. Statisticians have found that regardless of how nonnormal the population distribution is, once the sample size is at least 30, the sampling distribution of the mean will be approximately normal. However, if we have some knowledge of the population, we may be able to apply the central limit theorem for even smaller sample sizes.

The application of the central limit theorem to different populations can be illustrated by referring to Figures 7.6 to 7.9. Each of the depicted sampling distributions has been obtained by using the computer to select 500 different samples from their respective population distributions. These samples were selected for varying sizes ($n = 2, 4, 8, 16, 32$) from four different continuous distributions (normal, uniform, U-shaped, and exponential).

The first figure, Figure 7.6, illustrates the sampling distribution of the mean selected from a normal population. In the previous section we stated that if the population is normally distributed, the sampling distribution of the mean will be normally distributed regardless of the sample size. An examination of the sampling distributions shown in Figure 7.6 gives empirical evidence to this statement. For each sample size studied, the sampling distribution of the mean is approximately normally distributed.[1]

The second figure, Figure 7.7, presents the sampling distribution of the mean based upon a population that follows a continuous uniform (rectangular) distribution. As depicted in panel A, for samples of size $n = 1$, each value in the population is equally likely. However, when samples of only two are selected, there is a "peaking" or "central limiting" effect already working. Thus in this case we can observe somewhat more values "close to" the mean of the population than far out at the extremes. Moreover, as the sample size increases, the sampling distribution of the mean rapidly approaches a normal distribution. Here, once there are samples of at least eight observations, the sample mean approximately follows a normal distribution.

The third figure, Figure 7.8, refers to the sampling distribution of the mean obtained from a U-shaped population. From panel A we can observe that this population, although symmetrical, has its low frequency in the center and its high frequency at the lower and upper extremes of the distribution. Although the population has very few values near the center, even when the sample size is only two, many of the sample means are already clustered around the center of the distribution (at the population mean μ). An examination of the other panels of Figure 7.8 reveals that once the sample size reaches eight, the sampling distribution of the mean is approximately normally distributed.

Finally, the fourth figure, Figure 7.9, exemplifies the sampling distribution of the mean obtained from a highly right-skewed population, called the exponential distribution (Reference 6). From Figure 7.9 we note that as the

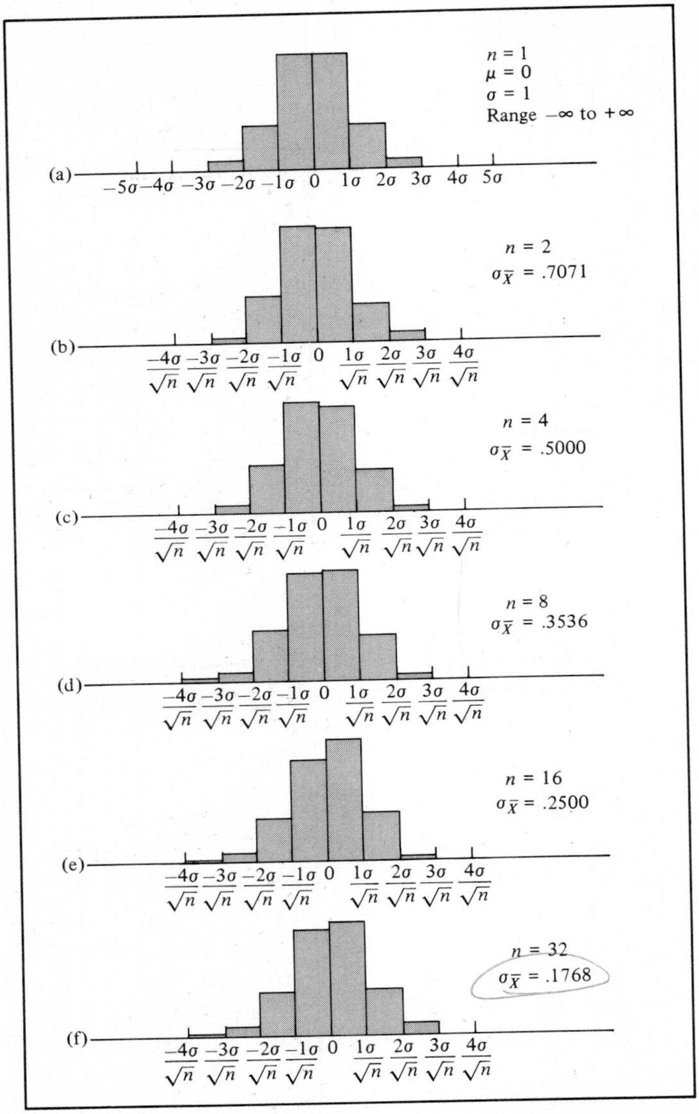

Figure 7.6 Normal distribution and the sampling distribution of the mean from 500 samples of size $n = 2$, 4, 8, 16, 32.

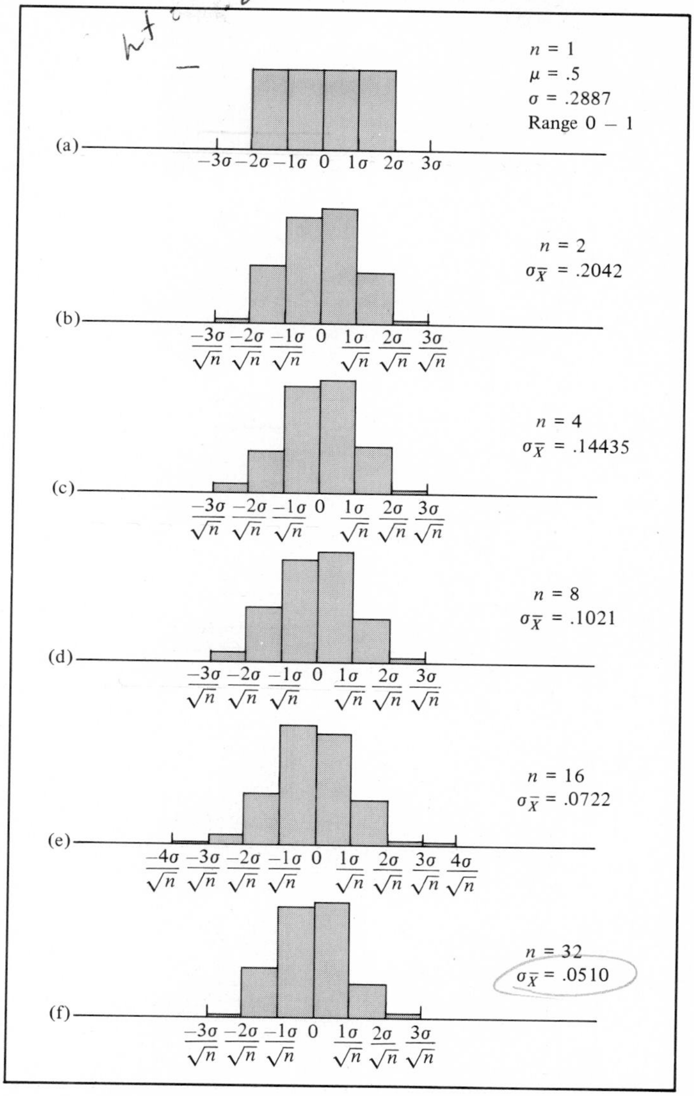

Figure 7.7 Continuous uniform (rectangular) distribution and the sampling distribution of the mean from 500 samples of size $n = 2, 4, 8, 16, 32$.

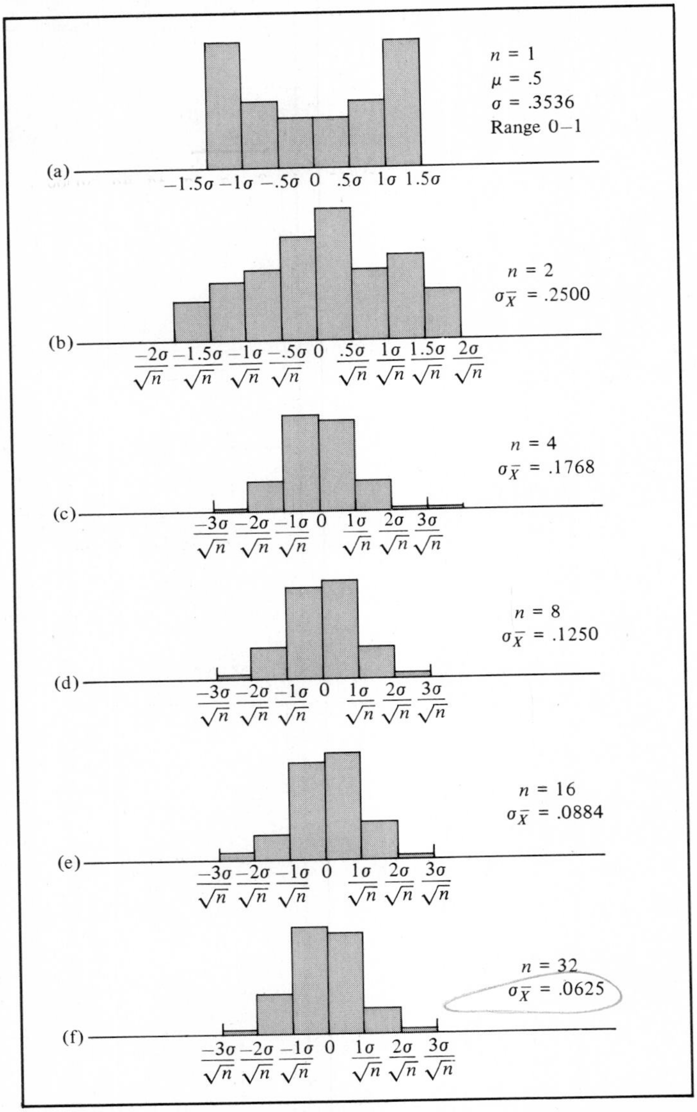

Figure 7.8 U-shaped distribution and the sampling distribution of the mean from 500 samples of size $n = 2, 4, 8, 16, 32$.

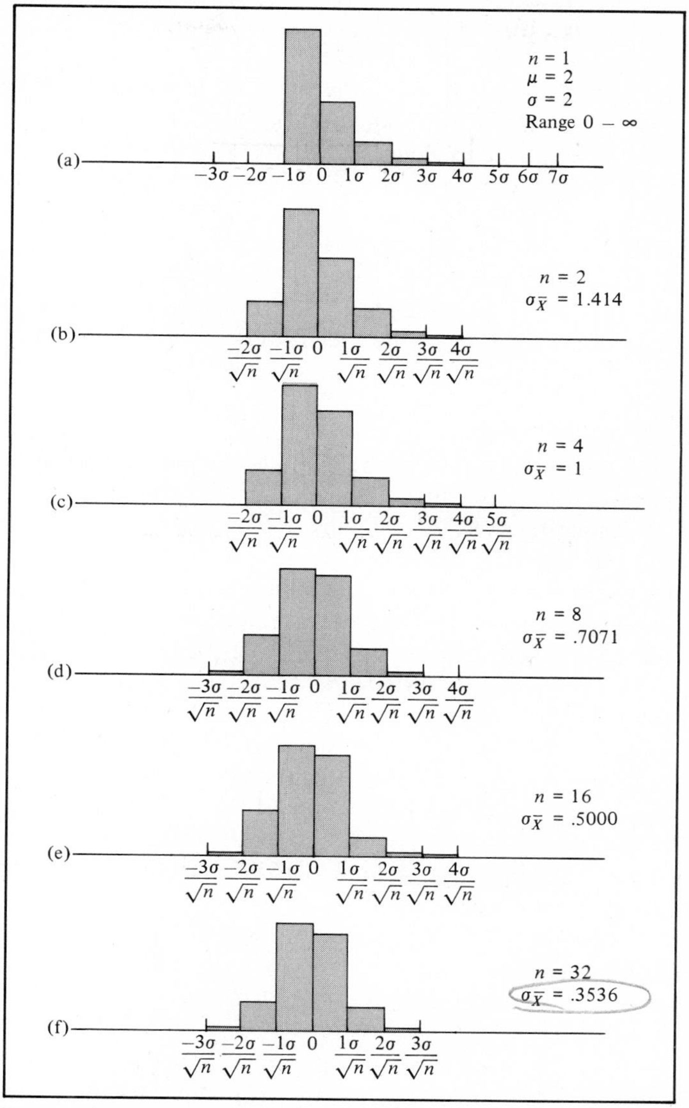

Figure 7.9 Exponential distribution and the sampling distribution of the mean from 500 samples of size $n = 2$, 4, 8, 16, 32.

sample size increases, the sampling distribution becomes less skewed. When samples of size 16 are taken, the distribution of the mean is slightly skewed, while for samples of size 32 the sampling distribution of the mean appears to be normally distributed.

In Figures 7.6 to 7.9 we have examined several well-known statistical distributions. Before drawing further conclusions about the central limit theorem, however, we could also examine the empirical distributions of some of the quantitative variables collected by the dean and the researcher. As an example, let us focus upon two of them—the maximum tuition that a student is willing or able to pay (question 8) and the grade-point index (question 5) that he or she has achieved. Five hundred different samples of each size (2, 4, 8, 16, 32, and 94) were selected with replacement by the computer from the entire population of 2,202 students. Figure 7.10 illustrates the sampling distribution of the mean for the variable—tuition. Panel A of Figure 7.10 depicts the entire empirical population distribution of tuition for all $N = 2{,}202$ students. Figure 7.11 presents the sampling distribution of the mean for the variable—grade-point index. Panel A of Figure 7.11 depicts the entire empirical population distribution of grade-point index for all $N = 2{,}202$ students.

Referring to panel A of Figure 7.10, we observe that the population distribution of tuition is right-skewed with a large amount of variation. As was the case with the exponential distribution, as the sample size increases, the sampling distribution of the mean becomes less skewed. Once samples of at least 16 are selected, the sampling distribution of the average tuition is approximately normally distributed.

On the other hand, from Figure 7.11, an examination of panel A reveals that the distribution of grade-point index appears to be fairly symmetrical. Thus from panels B and C of Figure 7.11 we can see that for samples as small as two and four, the distribution of the sample average grade-point index clusters closely to the population mean. Once samples of at least eight are selected, the sampling distribution of the mean grade-point index approximates a normal distribution.

We may now use the results obtained from our well-known statistical distributions (normal, uniform, U-shaped, exponential) and from the empirical distributions of the dean's survey to summarize our conclusions as follows:

1. Once samples of at least 30 observations are selected, the sampling distribution of the mean will be approximately normally distributed regardless of the population distribution.
2. If the population distribution is fairly symmetric, then once samples of at least 15 observations are selected, the sampling distribution of the mean will be approximately normal.
3. If the population is normally distributed, the sampling distribution of the mean will be normally distributed regardless of the sample size.

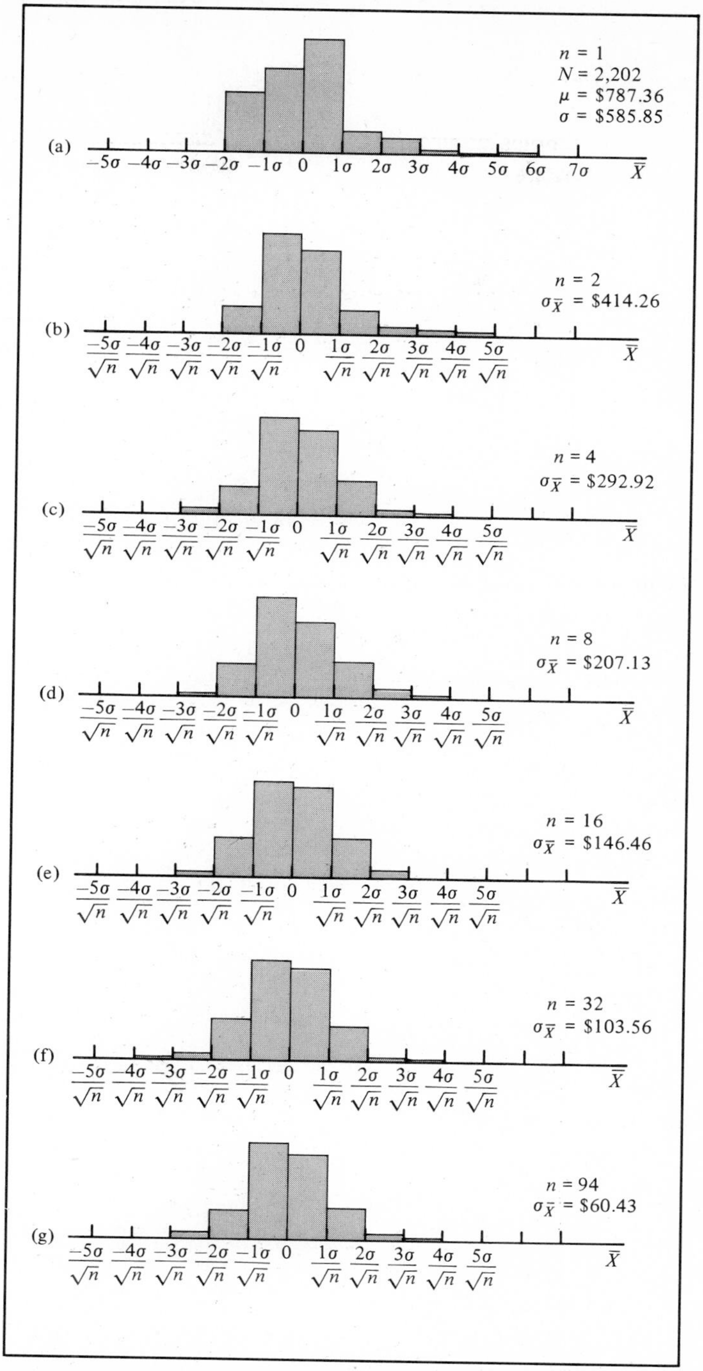

Figure 7.10 Distribution of tuition from the population of $N = 2{,}202$ students and the sampling distribution of the mean from 500 samples of size $n = 2, 4, 8, 16, 32, 94$.

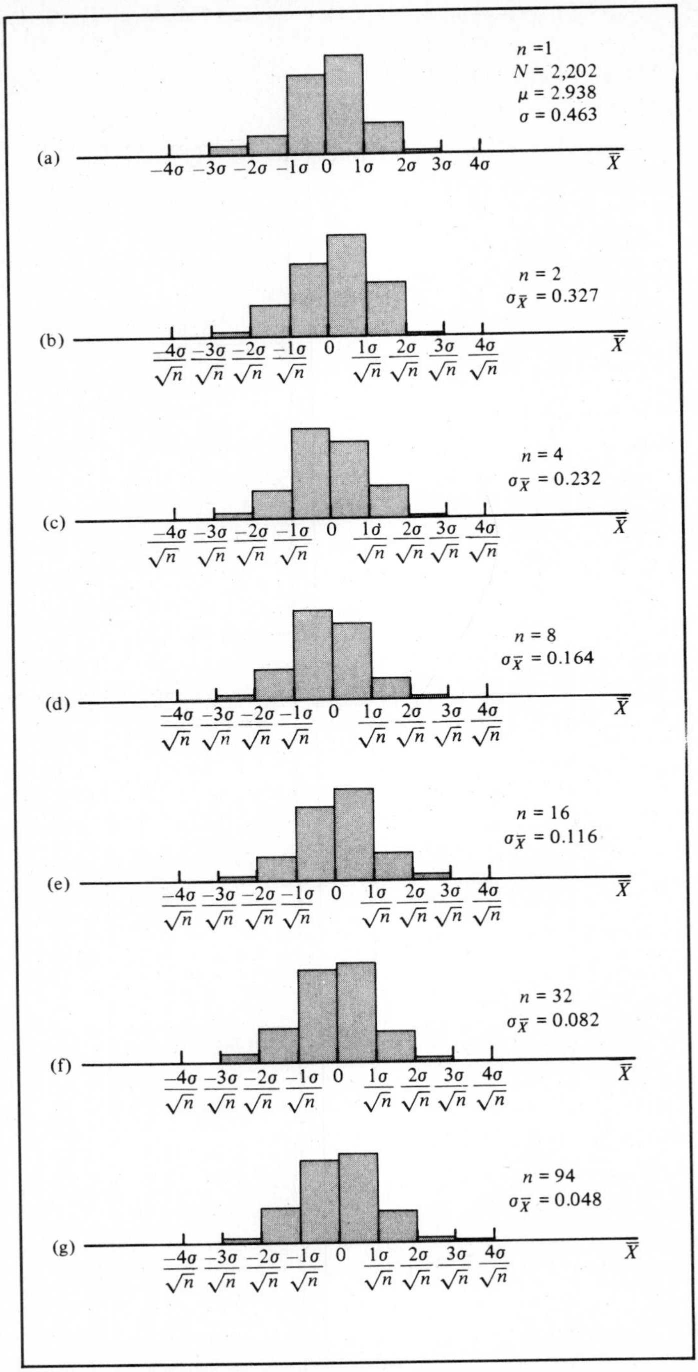

Figure 7.11 Distribution of grade-point index from the population of $N = 2{,}202$ students and the sampling distribution of the mean from 500 samples of size $n = 2, 4, 8, 16, 32, 94$.

The central limit theorem then is of crucial importance in using statistical inference to draw conclusions about a population. It allows the researcher to make inferences about the population mean without having to know the specific shape of the population distribution. This means (as we shall investigate in subsequent chapters) that the sample statistics computed, such as the sample mean, the standard deviation, and the proportion, provide the necessary information for estimating the true values in the population.

7.3 SAMPLING DISTRIBUTION OF THE PROPORTION

In our discussion of sampling distributions thus far in this chapter we have been concerned with the distribution of the mean of quantitative variables. On the other hand, when examining qualitative variables, the characteristic that is usually considered is the proportion of successes. As examples, in our survey the dean and the researcher might be interested in estimating the proportion of students who are majoring in accounting. Moreover, a political pollster would be interested in estimating the true proportion of the votes that will be obtained by a particular candidate. Finally, an auditor would like to determine the true rate of occurrence of a particular type of error.

In several previous chapters we have briefly discussed qualitative variables. In Chapter 5 we defined the proportion of successes p as

$$p = \frac{X}{n} = \frac{\text{Number of Successes}}{\text{Sample Size}}$$

while in Chapter 6 we saw that the number of successes followed a binomial distribution expressed as follows:

$$P(X \text{ successes}) = \frac{n!}{X!(n-X)!}p^X(1-p)^{n-X}$$

where the average number of successes μ was equal to np and the standard deviation of the number of successes σ was equal to $\sqrt{np(1-p)}$.

Now, instead of expressing the variable in terms of number of successes X, we can readily convert the variable to proportion of successes by dividing by n, the sample size. Thus the average or expected proportion of successes is p, while the standard deviation of the proportion of successes σ_p is equal to $\sqrt{\frac{p(1-p)}{n}}$.

It was noted in Section 6.10 that as the sample size increased, the binomial distribution could be approximated by the normal distribution. The general rule of thumb was that if np and $n(1-p)$ each were at least 5, the normal distribution provides a good approximation of the binomial distribution. In most instances in which inferences are being made about the proportion, the sample size is quite large so that the normal distribution yields a good approximation to the binomial distribution. Therefore, the normal distribution can be used in investigating the sampling distribution of the proportion. We can use this information in referring to the following example.

The manager of a furniture store has determined that 20% of the sales in the past year included delivery of the furniture within 30 days of purchase. If a random sample of 400 sales is selected, what is the probability that the sample proportion of orders delivered within 30 days will be between .20 and .25?

Since the sampling distribution of the proportion is approximately normally distributed,[2] we have the following:

$$Z = \frac{\bar{X} - \mu}{\sigma_{\bar{X}}}$$

and because we are dealing with sample proportions (not sample means) and

$$p_s = \text{sample proportion}$$

$$p = \text{population proportion}$$

$$\sigma_p = \sqrt{\frac{p(1-p)}{n}}$$

we have

$$Z \cong \frac{p_s - p}{\sqrt{\dfrac{p(1-p)}{n}}} \qquad (7.5)$$

Substituting,

$$Z \cong \frac{p_s - p}{\sqrt{\dfrac{p(1-p)}{n}}} = \frac{.25 - .20}{\sqrt{\dfrac{.20(.80)}{400}}} = \frac{.05}{\sqrt{\dfrac{.16}{400}}}$$

$$= \frac{.05}{.02} = 2.50$$

Using Appendix E, Table E.2, the area under the normal curve from $Z = 0$ to $Z = 2.50$ is .4938. Therefore the probability of obtaining a sample proportion between .20 and .25 is .4938. This means that if the true proportion of successes in the population is .20, then 49.38% of the samples of size 400 would be expected to have sample proportions between .20 and .25.

[2] When working with the sampling distribution of the proportion for very large samples, the continuity correction factor (see Section 6.11) is usually omitted since it will have minimal effect on the results.

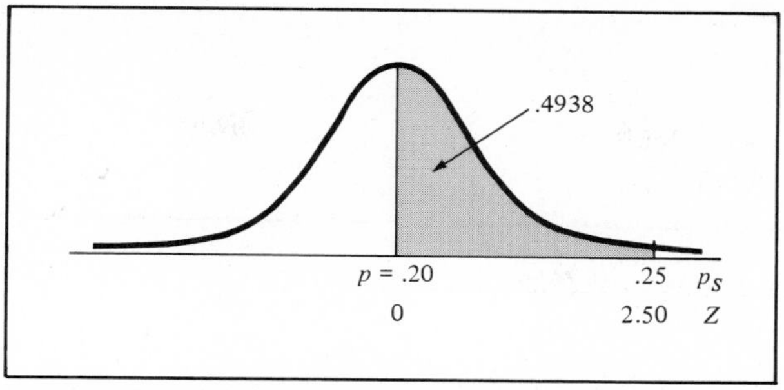

Figure 7.12 Diagram of normal curve needed to find the area between the proportions .20 and .25.

As with quantitative data, knowledge of the sampling distribution will allow inferences to be made about a population proportion based only upon the proportion of successes in a single sample. These concepts of inference will be developed further in the next two chapters.

7.4 SAMPLING FROM FINITE POPULATIONS

The central limit theorem and the standard error of the mean were based upon the premise that the samples selected were chosen with replacement. However, in survey research and in business, sampling is conducted without replacement from populations that are of a finite size N. In these cases, particularly when the sample size n is not small as compared to the population size N, a **finite population correction factor (fpc)** should be used in developing the particular sampling distribution.

The finite population correction factor may be expressed as[3]

$$\text{fpc} = \frac{N-n}{N-1} \tag{7.6}$$

Therefore

$$\sigma_{\bar{X}}^2 = \frac{\sigma^2}{n}\left(\frac{N-n}{N-1}\right) \tag{7.7a}$$

[3] The finite population correction factor essentially expresses the proportion of observations that have not been included in the sample $\left(1 - \frac{n}{N} = \frac{N-n}{N}\right)$. This result is approximately equal to $\frac{N-n}{N-1}$ when N is large.

so that

$$\sigma_{\bar{X}} = \frac{\sigma}{\sqrt{n}}\sqrt{\frac{N-n}{N-1}} \qquad (7.7b)$$

correction factor

where n = sample size
N = population size

If we are referring to proportions, we have the following:

$$\sigma_p^{\ 2} = \frac{p(1-p)}{n}\,\frac{N-n}{N-1} \qquad (7.8a)$$

and

$$\sigma_p = \sqrt{\frac{p(1-p)}{n}} \cdot \sqrt{\frac{N-n}{N-1}} \qquad (7.8b)$$

In the tire company example (Section 7.2) a sample of 100 tires was selected from a population of 10,000 tires. Using the finite population correction factor, we have

$$\sigma_{\bar{X}} = \frac{\sigma}{\sqrt{n}}\sqrt{\frac{N-n}{N-1}}$$

$$= \frac{5{,}000}{\sqrt{100}}\sqrt{\frac{10{,}000-100}{10{,}000-1}} = 500\sqrt{.99} = 500(.995) = 497.50$$

Thus $Z = -1{,}000/497.5 = -2.02$, and from Appendix E, Table E.2, the appropriate area under the normal curve is .4783.

It is evident from this example that the use of the finite population correction factor had little effect on the standard error of the mean or on the normal curve area, since the sample size is small relative to the size of the population ($n/N = .01$).

In the example with the furniture store (Section 7.3) suppose that the proportion of .20 of the orders being delivered within 30 days of purchase is based upon the fact that out of a population of 5,000 orders, 1,000 were delivered within 30 days of purchase. The sample of size 400 out of this finite population results in the following:

$$\sigma_p = \sqrt{\frac{p(1-p)}{n}} \cdot \sqrt{\frac{N-n}{N-1}}$$

$$= \sqrt{\frac{.2(.8)}{400}} \cdot \sqrt{\frac{5{,}000 - 400}{5{,}000 - 1}} = \sqrt{\frac{.16}{400}} \cdot \sqrt{\frac{4{,}600}{4{,}999}} = \sqrt{\frac{.16}{400}} \cdot \sqrt{.92}$$

$$= .02(.959) = .0192$$

Thus $Z = .05/.0192 = 2.60$, and from Appendix E, Table E.2, the appropriate area under the normal curve is .4953. In this example the finite population correction factor has reduced the previously computed standard error by about 4%.

7.5 SUMMARY AND OVERVIEW

In this chapter we have examined the distribution of the sample mean and the sample proportion. The importance of the normal distribution in statistics has been further emphasized by examining the central limit theorem. We have seen that knowledge of a population distribution is not always necessary in drawing conclusions about a sampling distribution of the mean or proportion. These concepts are central to the development of statistical inference. The main objective of statistical inference is to take information, based only upon a sample, and use this information to draw conclusions and make decisions about various population values. The statistical techniques developed to achieve these objectives are discussed in the next three chapters (confidence intervals and tests of hypotheses).

PROBLEMS

* 7.1 Plastic bags used for packaging produce are manufactured so that the breaking strength of the bag is normally distributed with a mean of 5 pounds per square inch and a standard deviation of 1 pound per square inch.

(a) What proportion of the bags produced have a breaking strength between 5 and 5.5 pounds per square inch?

(b) What proportion of the bags produced have a breaking strength between 4 and 4.1 pounds per square inch?

(c) If many random samples of 16 bags are selected, what would the mean and the standard error of the mean be expected to equal?

(d) What distribution would the sample means follow?

(e) What proportion of the sample means would be between 5 and 5.5 pounds per square inch?

(f) What proportion of the sample means would be between 4 and 4.1 pounds per square inch?

(g) Compare the answers to (a) and (e), to (b) and (f), and explain the differences in the answers.

7.2 Long-distance telephone calls are normally distributed with $\mu = 8$ minutes and $\sigma = 2$ minutes. If random samples of 25 calls were selected,

(a) Compute $\sigma_{\bar{X}}$.

(b) What proportion of the sample means would be between 7.8 and 8.2 minutes?

(c) What proportion of sample means would be between 7.5 and 8 minutes?

If random samples of 100 calls were selected,

(d) What proportion of the sample means would be between 7.8 and 8.2 minutes?

(e) Explain the difference in the results of (b) and (d).

* 7.3 A soft-drink machine is regulated so that the amount dispensed is normally distributed with μ = 7 ounces and σ = .5 ounces. If samples of nine cups are taken, what value will be exceeded by 95% of the sample means?

7.4 The number of hours of life of a type of transistor battery is normally distributed with μ = 100 hours and σ = 20 hours.

(a) What proportion of the batteries will last between 100 and 125 hours?

If random samples of 16 batteries are selected, what proportion of the sample means will be

(b) Between 100 and 125 hours?

(c) More than 90 hours?

(d) Within what limits around the population mean will 90% of sample means fall?

(e) Is the central limit theorem necessary to answer (b), (c), and (d)? Explain.

* 7.5 The luncheon expense vouchers of the executives of a large advertising firm have a population mean μ = \$10 per person and a σ = \$4 per person. If random samples of 16 accounts are selected,

(a) Below what dollar value will 99% of the sample means fall?

(b) What proportion of the sample means will be between \$8 and \$12?

(c) What assumption must be made in order to solve (a) and (b)?

If random samples of 64 accounts are selected,

(d) Below what dollar value will 99% of the sample means fall?

(e) What assumption must be made in order to solve (d)?

7.6 The number of customers per week at each store of a supermarket chain has a population mean μ = 5,000 and σ = 500. If a random sample of 25 stores is selected,

(a) What is the probability that the sample mean will be below 5,075 customers per week?

(b) Within what limits around the population mean can we be 95% certain that the sample mean will fall?

* 7.7 A large chain of home improvement centers stocks a nationally known brand of portable electric drills. In order to achieve maximum volume discount, the drill will be reordered for all stores at the same time. The inventory reorder decision is to reorder when the average inventory at a sample of stores is below 25 drills. Based on past data, the standard deviation is assumed to be 10 drills. If a random sample of 25 stores is selected, what is the probability that the drill will be reordered

(a) When the true average inventory of all stores is 20 drills?
(b) When the true average inventory of all stores is 30 drills?
(c) What assumption must be made in (a) and (b)?
(d) What would be your answer to (a) and (b) if the sample size was increased to 36?

7.8 (**Class Project**) The table of random numbers is an example of a uniform distribution since each digit is equally likely to occur. Starting in the row corresponding to the day of the month in which you were born, using the table of random numbers (Appendix E, Table E.1) taking **one-digit** numbers, select samples of size $n = 2$, $n = 5$, $n = 10$. Compute the sample mean $\overline{X}$ of each sample. For each sample size, each student should select five different samples so that a frequency distribution of the sample means can be developed for the results of the entire class. What can be said about the shape of the sampling distribution for each of these sample sizes?

7.9 (**Class Project**) A coin having one side "heads" and the other side "tails" is to be flipped 10 times and the number of heads obtained is to be recorded. If each student performs this experiment five times, a frequency distribution of the number of "heads" can be developed from the results of the entire class. Does this distribution seem to approximate the normal distribution?

7.10 (**Class Project**) The number of cars waiting in line at a car wash is distributed as follows:

Length of Waiting Line (cars)	Probability
0	.25
1	.40
2	.20
3	.10
4	.04
5	.01

The table of random numbers can be used to select samples from this distribution by assigning numbers as follows:

1. **Two-digit** random numbers are to be selected.
2. If a random number between 00 and 24 is selected, record a length of 0; if between 25 and 64, record a length of 1; if between 65 and 84, record a length of 2; if between 85 and 94, record a length of 3; if between 95 and 98, record a length of 4; if it is 99, record a length of 5.

Select samples of size $n = 2$, $n = 10$, $n = 25$. Compute the sample mean for each sample. For example, if a sample size 2 results in random numbers 18 and 46, these would correspond to lengths of 0 and 1, respectively, producing a sample mean of 0.5. If each student selects five different samples for each sample size, a frequency distribution of the sample means (for each sample size) can be developed from the results of the entire class. What conclusions can you draw about the sampling distribution of the mean as the sample size is increased?

7.11 (Class Project) Appendix A is a data base consisting of the responses of the population of 2,202 students to the questionnaire shown in Figure 2.2. Each student is to select five samples of size $n = 2, 5, 10$, and 25 and compute the sample mean $\overline{X}$ of each sample for the following quantitative variables to be selected by the instructor:

(a) Expected starting annual salary (question 7)
(b) Number of pairs of jeans owned (question 13)
(c) Grade-point index (question 5)
(d) High-school average (question 6)
(e) Age (question 15)
(f) Number of clubs affiliated with (question 23)

The results of the entire class can be used to develop sampling distributions for the chosen variable for each of the given sample sizes. What can be said about the shape of the sampling distribution for each sample size of each chosen variable?

7.12 (Class Project) Appendix A is a data base consisting of the responses of the population of 2,202 students to the questionnaire shown in Figure 2.2. Each student is to select five samples of size $n = 10$, $n = 25$, and $n = 100$ and compute the proportion of successes p in each sample for the following qualitative variables to be selected by the instructor:

(a) Proportion of accounting majors (question 4, code 2)
(b) Proportion of students who are unemployed (question 9, code 3)
(c) Proportion of students who own calculators (question 14, code 1)
(d) Proportion of students who have high blood pressure (question 18, code 1)
(e) Proportion of students who would place their money in a savings bank (question 11, code 1)
(f) Proportion of students who would be primarily interested in purchasing a standard or intermediate size car (question 12, code 2)
(g) Proportion of students who plan to attend graduate school (question 3, code 1)

The results of the entire class can be used to develop sampling distributions for the qualitative variable chosen for each of the given sample sizes. What can be said about the shape of the sampling distribution for each sample size of each chosen qualitative variables?

* 7.13 Historically, 10% of a large shipment of machine parts are defective. If random samples of 400 parts are selected, what proportion of the samples will have

(a) Between 9% and 10% defective parts?
(b) Less than 8% defective parts?
(c) If a sample size of only 100 was selected, what would your answers have been in (a) and (b)?

7.14 A political pollster is conducting an analysis of sample results in order to make predictions on election night. Assuming a two-candidate election, if a specific candidate receives at least 55% of the vote in the sample, then that candidate will be forecast as the winner of the election. If a random sample of 100 voters is selected, what is the probability that a candidate will be forecasted as the winner when

(a) The true percentage of his vote is 50.1%?
(b) The true percentage of his vote is 60%?
(c) The true percentage of his vote is 49%?
(d) If the sample size was increased to 400, what would be your answer to (a), (b), and (c)? Discuss.

7.15 Based on past data, 30% of the credit card purchases at a large department store are for amounts above $100. If random samples of 100 purchases are selected,

(a) What proportion of samples are likely to have between 20% and 30% of the purchases over $100?

(b) Within what symmetrical limits of the population percentage will 95% of the sample percentages fall?

7.16 Referring to Problem 7.5, if there is a population of 500 vouchers, what would be your answers to (a) and (b) of that problem?

* 7.17 Referring to Problem 7.7, if the chain has a total of 250 stores, what would be your answers to (a) and (b) of that problem?

7.18 Referring to Problem 7.13, if the shipment included 5,000 machine parts, what would be your answers to (a) and (b) of that problem?

REFERENCES

1. COCHRAN, W. G., *Sampling Techniques,* 3rd ed. (New York: Wiley, 1977).
2. DYCKMAN, T. R., AND L. J. THOMAS, *Fundamental Statistics for Business and Economics* (Englewood Cliffs, N.J.: Prentice-Hall, 1977).
3. FREUND, J., AND F. WILLIAMS, *Elementary Business Statistics: The Modern Approach,* 3rd ed. (Englewood Cliffs, N.J.: Prentice-Hall, 1977).
4. LAPIN, L., *Statistics: Meaning and Method* (New York: Harcourt, Brace, Jovanovich, 1975).
5. LI, W. G., *Statistical Inference,* vol. I (Ann Arbor, Mich.: Edwards Brothers, 1969).
6. MOOD, A. M., F. A. GRAYBILL, AND D. C. BOES, *Introduction to the Theory of Statistics,* 3rd ed. (New York: McGraw-Hill, 1974).

CHAPTER 8

ESTIMATION

8.1 POINT AND CONFIDENCE INTERVAL ESTIMATES

Statistical inference is the process of using sample results to estimate or draw conclusions about the characteristics or parameters of a population. In this chapter we shall examine estimation procedures which attempt to measure particular characteristics of a population such as the mean or the proportion. There are two major types of estimates, **point estimates** and **interval estimates.** A point estimate uses a single sample value to estimate the population parameter involved. For example, the sample mean $\overline{X}$ is a point estimate of the population mean μ. The sample variance S^2 is a point estimate of the population variance σ^2. In order for the point estimate to be a good estimator of the population parameter, several mathematical properties are desirable (Reference 6). For example, we stated in Section 7.2 that the sample arithmetic mean was an unbiased estimator of the population mean. This meant that if the average of all the sample means was taken, it would be equal to the population mean. This property is highly desirable since it indicates that the average sample statistic is equal to the population parameter.[1]

[1] It is for this reason that the denominator of the sample variance is $n - 1$ instead of n, so that S^2 will be an unbiased estimator of σ^2, that is, if

However, the value of the point estimate will fluctuate from sample to sample since only a portion of the population is being selected in each sample. But in the last chapter we saw that the distribution of the sample mean and the sample proportion for large enough sample sizes followed a normal distribution. This information about the distribution of the sample statistic is taken into account in developing an interval estimate of the population parameter. Instead of having an estimate based on a single value, an interval is used for estimating the population parameter. This interval has a specified *confidence* or probability of correctly estimating the true value of the population parameter.

8.2 CONFIDENCE INTERVAL ESTIMATION OF THE MEAN (σ KNOWN)

In the discussion of the sampling distribution of the mean in Chapter 7 we noted that either from the central limit theorem or from a knowledge of the population we could determine the proportion of sample means that fell within certain limits of the population mean. For example, in Section 7.2, for the distribution of tire mileage having $\mu = 25{,}000$, $\sigma = 5{,}000$, and a sample size of 100, we saw that when sampling with replacement, 95% of the sample means would fall between 24,020 and 25,980 miles. Although this statement is useful in characterizing the sampling distribution, it is not the type of conclusion needed in making inferences from the sample to the population. This statement takes the population mean and standard deviation and only draws a conclusion about the sample mean. However, what is needed, is to take a *single sample mean* and estimate an *unknown population mean* (using either a population or a sample standard deviation) with a certain degree of confidence or probability.

Sample ⟶ Population

Inference

$\bar{X} \longrightarrow \mu$

In a real situation the population mean is unknown and is the quantity to be estimated. The population standard deviation is very rarely known and is usually estimated from the results of the sample. Suppose, however, that we were sampling with replacement and the population standard deviation σ was

---

$$S^2 = \frac{\sum_{i=1}^{n}(X_i - \bar{X})^2}{n-1} \quad \text{and} \quad \sigma^2 = \frac{\sum_{i=1}^{N}(X_i - \mu)^2}{N}$$

then $E(S^2) = \sigma^2$, and therefore S^2 is an unbiased estimator of σ^2.

known. We saw that if we took $(\mu \pm (1.96)\sigma/\sqrt{n})$, 95% of all sample means would be included within that interval. Let us see what results would occur if we used $\overline{X} \pm (1.96)\sigma/\sqrt{n}$ to estimate μ, the population mean. Since $\sigma = 5{,}000$ and $n = 100$, this leads to $\overline{X} \pm (1.96)(5{,}000)/\sqrt{100}$ or $\overline{X} \pm 980$.

If, for example, the mean of our sample was 24,100, the interval developed to estimate μ would be 24,100 ± 980, that is,

$$23{,}120 \leq \mu \leq 25{,}080$$

Since μ is actually equal to 25,000, the estimate made of μ based upon this sample result would be a correct estimate, since μ is located within this interval (see Figure 8.1).

What if the sample mean was 25,500 miles? Then, the interval would be 25,500 ± 980, that is,

$$24{,}520 \leq \mu \leq 26{,}480$$

Observe that the population mean of 25,000 is included within this interval, and hence the estimate of μ is a correct statement.

Surely our question at this point is: Will μ always be included within the sample interval? What if the sample mean were 23,000 miles? The interval would then be 23,000 ± 980, that is,

$$22{,}020 \leq \mu \leq 23{,}980$$

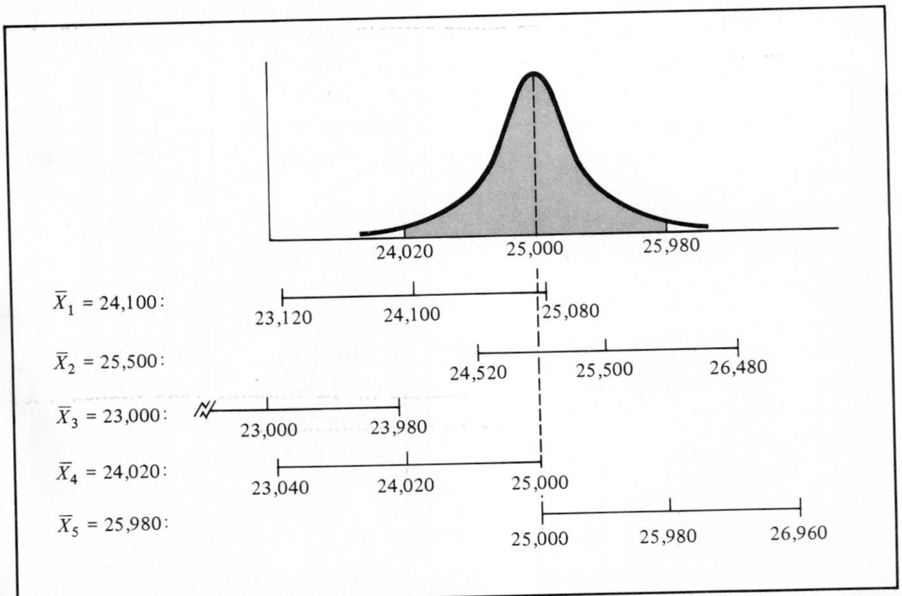

Figure 8.1 Confidence interval estimates for five different samples of size $n = 100$ taken from a population where $\mu = 25{,}000$ and $\sigma = 5{,}000$.

We should realize that this estimate of μ would not be a correct statement since the true population mean of 25,000 miles is not included in the interval. Therefore, we have a paradoxical situation. For some samples the estimate of μ will be correct, while for others the estimate will not be correct. We therefore must determine what proportion of samples will result in correct estimates. In order to find this, let us examine two other possible sample results, $\bar{X} = 24{,}020$ and $\bar{X} = 25{,}980$. If $\bar{X} = 24{,}020$ miles, the interval will be 24,020 ± 980, that is,

$$23{,}040 \leq \mu \leq 25{,}000$$

Since the population mean of 25,000 miles is at the upper limit of the interval, the estimate is a correct one. Finally, if $\bar{X} = 25{,}980$, we would have 25,980 ± 980, that is,

$$25{,}000 \leq \mu \leq 26{,}960$$

In this case the population mean of 25,000 is included at the lower limit of the interval. Therefore based upon the last two examples, if the sample mean falls between 24,020 and 25,980 miles, the population mean will be included *somewhere* within the interval. However, we know from the sampling distribution of the mean that 95% of the sample means will fall between 24,020 and 25,980 miles. Therefore, 95% of all sample means will include the population mean within the interval developed. This is what is meant by **95% confidence.** We would have 95% confidence that the sample selected is one where the true mean is located within the interval developed. Since μ is usually unknown, we never actually know whether the specific interval does or does not include the population mean. For a particular sample we can state the level of confidence that the interval does include the population mean.

The meaning of a confidence interval estimate can be better understood by studying the responses to a particular question taken from the dean's survey. We may recall that a random sample of 94 respondents had been selected from a population of 2,202 students. One of the variables studied was the maximum amount of yearly tuition that a student could afford to pay or was willing to pay (question 8). If we desired to estimate the average tuition in the population, what would be the meaning of a 95% confidence interval? The average tuition in the sample of 94 students (see Figure 2.6) was computed to be \$1,046.17. A 95% confidence interval estimate for this sample[2] can be found as follows:

$$1{,}046.17 \pm (1.96)\frac{\sigma}{\sqrt{94}}$$

[2] It is assumed in this example that the sample was drawn with replacement. As was indicated in Chapter 2, survey sampling is conducted without replacement from finite populations. A finite population correction factor which can be utilized when sampling without replacement from finite populations is discussed in Section 8.7.

It is interesting to note that unlike most actual situations the standard deviation of the tuition variable is available from the entire population of 2,202 students (see Appendix A). Since the standard deviation of the population is \$585.85, the actual 95% confidence interval estimate based upon this sample would be $1{,}046.17 \pm (1.96)\,(585.85)/\sqrt{94} = 1{,}046.17 \pm 118.43$, that is,

$$\$927.64 \leq \mu \leq \$1{,}164.60$$

This interval estimates the true mean tuition to be between \$927.64 and \$1,164.60. In general a 95% confidence interval estimate can be interpreted to mean that if all possible samples were taken, 95% of them would include the true population mean *somewhere* within their interval, while only 5% of them would fail to estimate the true mean correctly. We can illustrate this concept empirically by selecting 100 samples, each containing 94 respondents from the student population, and determining how many[3] include the population mean within their interval. From the population of 2,202 students the true mean can be computed as \$787.36. Table 8.1 indicates the sample mean and the 95% confidence interval estimate from each of the 100 samples containing 94 observations.

From the empirical results indicated in Table 8.1 we note that 95 out of the 100 samples include the population mean within their confidence intervals. In an actual business situation, of course, the true population mean is unknown and is the quantity to be estimated. However, if a single sample were selected, the chance would be 95% that it was one which correctly estimated the true mean μ.

In the previous examples the intervals had 95% confidence of including the true mean. In general we might desire either more or less confidence than 95%. In some problems we might desire a higher degree of assurance (such as 99%) of including the population mean within the interval. In other cases we might be willing to accept less assurance (such as 90%) of correctly estimating the true population mean. The level of confidence of 95% leads to a Z value of $\pm$ 1.96 (see Figure 8.2).

Therefore, to obtain the confidence interval estimate of the mean with σ known, we have

$$\bar{X} \pm Z\frac{\sigma}{\sqrt{n}}$$

or

$$P\left[\bar{X} - Z\frac{\sigma}{\sqrt{n}} \leq \mu \leq \bar{X} + Z\frac{\sigma}{\sqrt{n}}\right] = 1 - \alpha \qquad (8.1)$$

where $1 - \alpha$ = level of confidence.

[3] We should realize that 95% confidence does **not** mean that exactly 95 out of every 100 samples will include the population mean within the confidence interval, only that on the average out of **all** possible samples, 95% will include the true mean.

Table 8.1
95% CONFIDENCE INTERVAL ESTIMATES OF THE AVERAGE TUITION FROM 100 SAMPLES OF 94 STUDENTS

Sample Number	Sample Mean $\bar{X}$	Lower Limit	Upper Limit	Is μ = \$787.36 included in the interval?
*1	\$1,046.170	\$927.640	\$1,164.600	No
2	725.234	606.799	843.668	Yes
3	731.734	613.299	850.168	Yes
4	818.223	699.789	936.658	Yes
5	858.766	740.331	977.200	Yes
6	731.479	613.044	849.913	Yes
7	736.638	618.204	855.073	Yes
8	792.479	674.044	910.913	Yes
9	747.287	628.853	865.721	Yes
10	661.904	543.469	780.338	No
11	870.755	752.321	989.189	Yes
12	879.234	760.799	997.668	Yes
13	801.638	683.204	920.073	Yes
14	781.808	663.374	900.243	Yes
15	835.638	717.204	954.073	Yes
16	866.127	747.693	984.562	Yes
17	793.489	675.055	911.924	Yes
18	807.021	688.587	925.456	Yes
19	781.032	662.597	889.466	Yes
20	734.234	615.799	852.668	Yes
21	816.245	697.810	934.679	Yes
22	748.787	630.353	867.221	Yes
23	802.447	684.012	920.881	Yes
24	801.681	683.246	920.115	Yes
25	798.638	680.204	917.073	Yes
26	813.138	694.704	931.573	Yes
27	747.394	628.959	865.828	Yes
28	763.170	644.736	881.604	Yes
29	812.627	694.193	931.062	Yes
30	736.734	618.299	855.168	Yes
31	743.510	625.076	861.945	Yes
32	747.702	629.267	866.136	Yes
33	788.117	669.682	906.551	Yes
34	867.085	748.650	985.519	Yes
35	821.957	703.523	940.392	Yes
36	859.349	740.959	977.828	Yes
37	753.883	635.448	872.317	Yes
38	758.713	640.278	877.147	Yes
39	711.251	592.916	829.785	Yes
40	743.096	624.661	861.530	Yes
41	789.925	671.491	908.360	Yes
42	751.957	633.523	870.392	Yes
43	814.468	696.033	932.902	Yes
44	743.670	625.236	862.104	Yes
45	821.702	703.267	940.136	Yes
46	915.000	796.565	1,033.434	No
47	833.276	714.842	951.711	Yes
48	741.085	622.650	859.519	Yes
49	878.617	760.182	997.051	Yes
50	865.521	747.087	983.956	Yes
51	776.298	657.863	894.732	Yes
52	821.245	702.810	939.679	Yes
53	778.872	660.438	897.307	Yes
54	865.000	746.565	938.434	Yes
55	800.521	682.087	918.956	Yes
56	763.064	644.629	881.498	Yes
57	807.319	688.885	925.753	Yes
58	791.619	673.257	910.126	Yes
59	751.872	633.438	870.307	Yes
60	757.021	638.587	875.456	Yes

Table 8.1 (*Continued*)

Sample Number	Sample Mean $\bar{X}$	Lower Limit	Upper Limit	Is μ = \$787.36 included in the interval?
61	714.149	595.714	832.583	Yes
62	745.979	627.544	864.413	Yes
63	793.596	675.161	912.030	Yes
64	899.085	780.650	1,017.519	Yes
65	733.936	615.501	852.370	Yes
66	745.245	626.810	863.679	Yes
67	903.202	784.767	1,021.636	Yes
68	824.234	705.799	942.668	Yes
69	859.225	740.821	977.689	Yes
70	685.702	567.267	804.136	Yes
71	772.734	654.299	891.168	Yes
72	779.596	661.161	898.030	Yes
73	796.436	678.001	914.870	Yes
74	800.042	681.608	918.477	Yes
75	710.691	592.257	829.126	Yes
76	907.596	789.161	1,026.030	No
77	797.564	679.129	915.998	Yes
78	785.819	667.385	904.253	Yes
79	864.394	745.959	982.828	Yes
80	749.521	631.087	867.956	Yes
81	926.191	807.757	1,044.626	No
82	856.830	738.395	975.264	Yes
83	862.425	743.991	980.860	Yes
84	858.000	739.565	976.434	Yes
85	875.340	756.906	993.775	Yes
86	866.372	747.938	984.807	Yes
87	852.979	734.544	971.413	Yes
88	832.181	713.746	950.615	Yes
89	764.159	645.725	882.594	Yes
90	829.776	711.342	948.211	Yes
91	737.308	618.874	855.743	Yes
92	823.159	704.725	941.594	Yes
93	708.596	590.161	827.030	Yes
94	825.319	706.885	943.753	Yes
95	710.223	591.789	828.658	Yes
96	704.383	585.948	822.817	Yes
97	853.202	734.767	971.636	Yes
98	797.510	679.076	915.945	Yes
99	846.447	728.012	964.881	Yes
100	842.149	723.714	960.583	Yes

* Sample 1 is the one obtained by the dean and the researcher (see page 234).

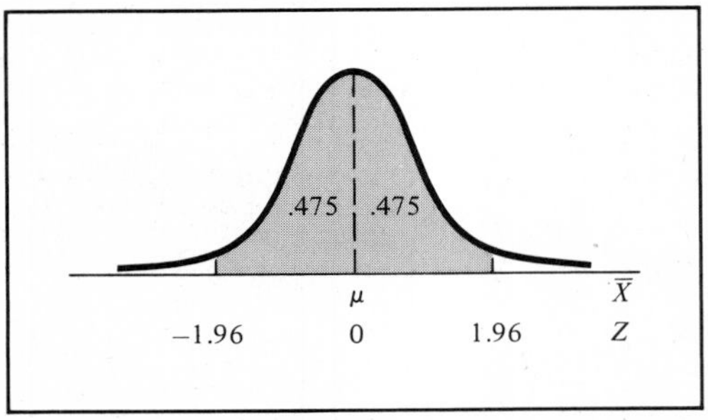

Figure 8.2 Normal curve for determining the Z value needed for 95% confidence.

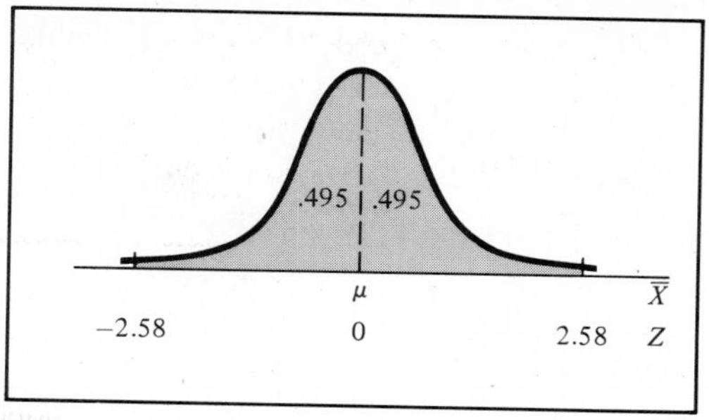

Figure 8.3 Normal curve for determining the Z value needed for 99% confidence.

If 99% confidence was desired, the area of .99 would be divided in half, leaving .495 between each limit and μ (see Figure 8.3). The Z value corresponding to an area of .495 under the normal curve is 2.58. Therefore, the Z value for 99% confidence is $\pm$ 2.58.

Since the population standard deviation of the amount of tuition paid is known to be \$585.85, the 99% confidence interval estimate for average tuition could be obtained from the sample of 94 students by using a Z value of 2.58 in the confidence interval formula [see Equation (8.1)]. The 99% confidence interval estimate would be

$$\overline{X} \pm Z\frac{\sigma}{\sqrt{n}} = 1{,}046.17 \pm (2.58)\frac{(585.85)}{\sqrt{94}} = 1{,}046.17 \pm 155.90$$

$$\$890.27 \leq \mu \leq \$1{,}202.07$$

The question might logically be raised at this point as to why we would not increase our confidence to make it as large as possible. We should realize, however, that any increase in confidence is achieved only after paying a "price." The price that is paid is having a wider interval in the estimate of the population mean. The increase of confidence from 95% to 99% has resulted in a Z value of 2.58 rather than 1.96. Therefore, the interval for 99% confidence will be less precise. We would have more confidence that the true population mean is within a broader range. This trade-off between the size of the confidence interval and the level of confidence will be discussed in greater depth when we investigate the methods of determining sample size (Section 8.5).

8.3 CONFIDENCE INTERVAL ESTIMATION OF THE MEAN (σ UNKNOWN)

As previously stated, just as the mean of the population μ is usually not known, the actual standard deviation of the population σ is also not likely to be known. Therefore, we need to obtain a confidence interval estimate of μ by using only the sample statistics of $\bar{X}$ and S.

The distribution that has been developed to be applied to this situation is **Student's t distribution.**[4]

If our variable X is normally distributed, then

$$\frac{\bar{X} - \mu}{\frac{S}{\sqrt{n}}}$$

is distributed as t with $n - 1$ degrees of freedom.

The t distribution is a symmetrical distribution that is similar to the normal distribution although it has more area in the tails and less in the center than does the normal distribution (see Figure 8.4). However, as the number of degrees of freedom increases, the t distribution gradually approaches the normal distribution until the two are practically identical.

The concept of a degree of freedom was briefly mentioned previously in Chapter 4. In computing the sample variance S^2, since the numerator of $\sum_{i=1}^{n} (X_i - \bar{X})^2$ required the computation of the sample arithmetic mean, only $(n - 1)$ values were "free" to vary. Therefore the statistic t, that is, $t = (\bar{X} - \mu)/(S/\sqrt{n})$, which uses S instead of σ, has $(n - 1)$ degrees of freedom. The confidence interval for the mean with σ unknown is expressed as follows:

$$\bar{X} \pm t_{n-1}\frac{S}{\sqrt{n}}$$

$$P\left[\bar{X} - t_{n-1}\frac{S}{\sqrt{n}} \leq \mu \leq \bar{X} + t_{n-1}\frac{S}{\sqrt{n}}\right] = 1 - \alpha \qquad (8.2)$$

where $1 - \alpha$ = level of confidence.

In practice, as long as the sample size is not too small and the population is not very skewed, the t distribution can be used in estimating the population mean when σ is unknown. The critical values of the t distribution, for the appropriate degrees of freedom, can be obtained from the tables of the t distribution presented in Appendix E, Table E.3. The value at the top of each column of the t table indicates the area in one tail of the distribution (either lower or upper tail), while each row represents the particular value at each degree of freedom. For example, with 24 degrees of freedom, if 95% confidence were desired, the appropriate value of t would be found in the follow-

[4] Student was the pen name used by W. S. Gosset who developed the t distribution at the turn of this century while working for Guinness Breweries in Dublin.

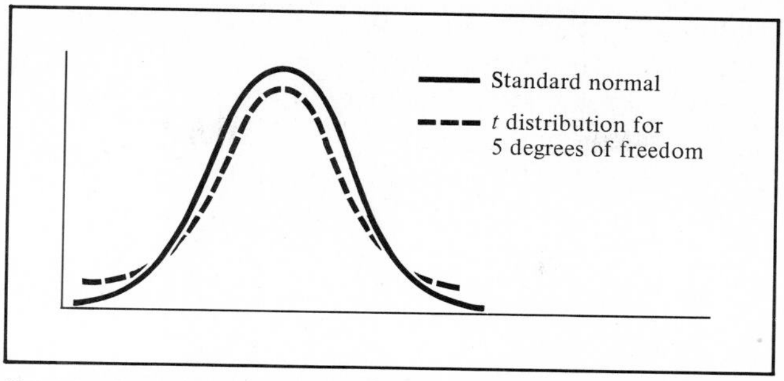

Figure 8.4 Standard normal distribution and t distribution for 5 degrees of freedom.

ing manner (as shown in Table 8.2). The 95% confidence level indicates that there would be an area of .025 in each tail of the distribution. Looking in the column for an upper tail area of .025 and in the row corresponding to 24 degrees of freedom results in a value of t of 2.0639, which for convenience is rounded to 2.064. Since t is a symmetrical distribution, if the upper tail value is +2.064, the value for the lower tail (lower .025) would be −2.064. A t value of 2.064 means that the probability that t would exceed +2.064 is .025 or 2.5% (see Figure 8.5).

In order to see how confidence intervals for a mean can be developed when the population standard deviation is unknown, we shall examine the following problems.

In the first problem, a department store chain has 10,000 credit card holders who are billed monthly for their purchases. The controller of the

Table 8.2
DETERMINING THE CRITICAL VALUE FROM THE t TABLE FOR AN AREA OF .025 IN EACH TAIL WITH 24 DEGREES OF FREEDOM

Degrees of Freedom	Upper Tail Areas					
	.25	.10	.05	.025	.01	.005
1	1.0000	3.0777	6.3138	12.7062	31.8207	63.6574
2	0.8165	1.8856	2.9200	4.3027	6.9646	9.9248
3	0.7649	1.6377	2.3534	3.1824	4.5407	5.8409
4	0.7407	1.5332	2.1318	2.7764	3.7469	4.6041
5	0.7267	1.4759	2.0150	2.5706	3.3649	4.0322
⋮	⋮	⋮	⋮	⋮	⋮	⋮
21	0.6864	1.3232	1.7207	2.0796	2.5177	2.8314
22	0.6858	1.3212	1.7171	2.0739	2.5083	2.8188
23	0.6853	1.3195	1.7139	2.0687	2.4999	2.8073
24	0.6848	1.3178	1.7109	2.0639	2.4922	2.7969
25	0.6844	1.3163	1.7081	2.0595	2.4851	2.7874

SOURCE: Extracted from Appendix E, Table E.3.

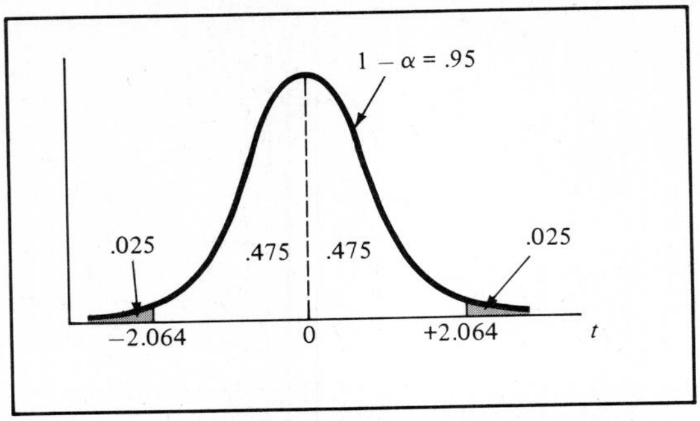

Figure 8.5 *t* distribution with 24 degrees of freedom.

store would like to take a sample of credit card holders in order to determine the average amount spent each month by all individuals having credit cards at this department store. A random sample of 25 credit card holders was selected, and the sample average was \$75 with a sample standard deviation of \$20.

The controller would like to estimate the true average amount spent by credit card holders at the department store. He would like to have 95% confidence that the interval obtained includes the true population average. In this problem, since there are 24 degrees of freedom, the critical value of t for 95% confidence is 2.064 (see Figure 8.5). Therefore, we have $\bar{X} = \$75$, $S = \$20$, $n = 25$, and $t = 2.064$. Thus,

$$\bar{X} \pm t_{n-1}\frac{S}{\sqrt{n}} = 75 \pm (2.064)\frac{(20)}{\sqrt{25}} = 75 \pm 8.256$$

$$\$66.74 \leq \mu \leq \$83.26$$

We would conclude with 95% confidence that the average monthly purchases of credit card holders is between \$66.74 and \$83.26. The 95% confidence interval states that we are 95% sure that the sample we have selected is one in which the true population mean is located within the interval. This 95% confidence actually means that if all possible samples of size 25 were selected (something that would never be done in practice), 95% of the intervals developed would include the true population mean *somewhere* within the interval.

In the second problem, let us refer to the survey of the dean and the researcher. The dean would like to estimate the expected annual starting salary of accounting majors. In the random sample of 94 students, 35 have indicated that they are accounting majors. Questions 4 and 7 of the questionnaire (see Figure 2.2) can be used along with the sample data of Figure 2.6 to determine that the sample average expected annual salary is \$13,100 with a sample standard deviation of \$3,925.41. A 99% confidence interval estimate of the true population average expected starting salary of accounting majors would be computed as follows: $\bar{X}$ = \$13,100.00, S = \$3,925.41, and $n = 35$.

Since there are $35 - 1 = 34$ degrees of freedom in this problem, from the t distribution (Appendix E, Table E.3) the critical value for 99% confidence is 2.728. Therefore, using Equation (8.2), we have

$$\bar{X} \pm t_{n-1}\frac{S}{\sqrt{n}} = 13{,}100 \pm (2.728)\frac{(3{,}925.41)}{\sqrt{35}} = 13{,}100 \pm 1{,}810.07$$

$$\$11{,}289.93 \le \mu \le \$14{,}910.07$$

We would conclude with 99% confidence that the average expected annual starting salary of accounting majors is between \$11,289.93 and \$14,910.07.

We have previously noted that as the number of degrees of freedom increases, the t distribution approaches the normal distribution. For large sample sizes the various percentage points will be virtually the same for the t and normal distributions. For example, the critical value on the normal distribution for 99% confidence is 2.58. The corresponding value on the t distribution is 2.75 for 30 degrees of freedom, 2.66 for 60 degrees of freedom, 2.63 for 100 degrees of freedom and 2.61 for 150 degrees of freedom. Therefore, we might say that when there are at least 150 degrees of freedom, the normal distribution can be used interchangeably with the t distribution.

Referring back to the first problem, what if the controller also wanted to estimate the total monthly amount spent by the credit card holders? In Chapter 4 we estimated the total [Equation (4.2)]:

$$\text{Total} = N\bar{X} \qquad (8.3)$$

Now that the confidence interval for the mean has been developed, the same procedures can be used to develop a confidence interval estimate of the population total τ.

Confidence Interval Estimate for the Population Total τ

Since, from Equation (8.2), the confidence interval estimate of the population mean μ is obtained from $\bar{X} \pm t_{n-1}\frac{S}{\sqrt{n}}$, then the confidence interval estimate for the population total τ is

$$N\bar{X} \pm t_{n-1}\frac{NS}{\sqrt{n}}$$

$$P\left[N\bar{X} - t_{n-1}\frac{NS}{\sqrt{n}} \le \tau \le N\bar{X} + t_{n-1}\frac{NS}{\sqrt{n}}\right] = 1 - \alpha \qquad (8.4)$$

Since in this example we have $N = 10{,}000$, $\bar{X} = \$75$, $S = \$20$, and $n = 25$, then,

$$\text{Total} = N\bar{X} = 10{,}000(75) = \$750{,}000$$

and since at 95% confidence, 24 degrees of freedom, $t = 2.064$, we have

$$750{,}000 \pm (2.064)(10{,}000)\frac{(20)}{\sqrt{25}} = 750{,}000 \pm 82{,}560$$

$$\$667{,}440 \leq \tau \leq \$832{,}560$$

The 95% confidence interval estimates that the total amount spent by credit card holders in the last month is between \$667,440 and \$832,560.

8.4 CONFIDENCE INTERVAL ESTIMATION FOR THE PROPORTION

The concept of estimation can be extended to qualitative data to estimate the proportion of successes in the population based only upon sample data. We noted in Chapters 6 and 7 that when np and $n(1-p)$ were at least 5, the binomial distribution generally could be approximated by the normal distribution. If we desire to estimate the population proportion p from the sample proportion p_s we could set up the following confidence interval estimate for the population proportion p.

Confidence Interval Estimate for the Population Proportion

$$p_s \pm Z\sqrt{\frac{p_s(1-p_s)}{n}} \qquad (8.5)$$

$$P\left[p_s - Z\sqrt{\frac{p_s(1-p_s)}{n}} \leq p \leq p_s + Z\sqrt{\frac{p_s(1-p_s)}{n}}\right] = 1 - \alpha$$

The sample proportion p_s is used as an estimate of p [in $p(1-p)/n$] since the true value of p is unknown and in fact is the quantity to be estimated.

In order to see how this confidence interval estimate of the proportion can be utilized, we will examine two problems. First, returning to the previous example concerning the department store, suppose that the controller also wanted to determine the proportion of credit card customers who would shop at the store if it were open on Sundays. In order to estimate this, a random sample of 100 credit card holders was selected and 60 indicated that they would shop at the store on Sundays. The controller wanted to obtain a 99% confidence interval estimate of the true proportion who would shop on Sunday. A confidence level as high as 99% was chosen because the controller wanted a greater chance of correctly estimating the true proportion before making this major marketing decision.

The confidence interval for the true population proportion of credit card customers who would shop on Sunday would be computed as follows: $p_s = 60/100 = .60$, and with 99% confidence $Z = 2.58$. Using Equation (8.5), we have

$$p_s \pm Z\sqrt{\frac{p_s(1 - p_s)}{n}} = .60 \pm (2.58)\sqrt{\frac{.6(.4)}{100}}$$

$$= .60 \pm (2.58)(.049) = .60 \pm .1264$$

$$.4736 \leq p \leq .7264$$

Therefore, the controller would estimate with 99% confidence that between 47.36% and 72.64% of the credit card holders would shop at this department store if it were open on Sunday.

In the second problem the dean would like to determine the proportion of students who believe that smoking should not be permitted in the classroom (question 20). From the sample of 94 students selected (see Figure 2.6) it can be determined that 69 believe that smoking should not be permitted in classrooms. A 95% confidence interval estimate of the true population proportion of students who are opposed to smoking in classrooms can be developed as follows: $p_s = 69/94 = .734$ and for 95% confidence, $Z = 1.96$. Again, from Equation (8.5),

$$p_s \pm Z\sqrt{\frac{p_s(1 - p_s)}{n}} = .734 \pm (1.96)\sqrt{\frac{.734(.264)}{94}}$$

$$= .734 \pm (1.96)(.0454) = .734 \pm .089$$

$$.645 \leq p \leq .823$$

Therefore, the dean can conclude with 95% confidence that between 64.5% and 82.3% of the students in the population do not believe that smoking should be permitted in classrooms.

We should note that in these cases since the number of successes (and failures) are 60 and 69 (and 40 and 25), the normal distribution provides an excellent approximation for the binomial distribution. However, if the sample size is not large or the percentage of successes is either very low or very high, then the binomial distribution should be used rather than the normal distribution (Reference 1). The exact confidence intervals for various sample sizes and proportions of successes have been tabled by Fisher and Yates (Reference 3) and are discussed in Huntsburger *et al.* (Reference 5).

8.5 SAMPLE SIZE DETERMINATION FOR THE MEAN

In each of our examples concerning confidence interval estimation, the sample size was arbitrarily determined without regard to the size of the confidence interval. In the business world the determination of the proper sample size is a complicated procedure which is subject to the constraints of budget, time, and ease of selection. For example, if the controller wished to estimate the average monthly amount spent by credit card holders, he would try to determine in advance how "good" an estimate would be required. This would mean that he would decide how much error he was willing to accept in estimating the population average monthly amount spent by credit card customers. Was accuracy required to be within ± \$1, ± \$5, ± \$10, ± \$20, etc.? The controller would also determine in advance how sure (confident) he wanted to be of correctly estimating the true population parameter. In determining the sample size for estimating the mean, these requirements must be kept in mind along with information about the standard deviation.

If σ was known, the confidence interval estimate for the population mean is obtained from Equation (8.1).

$$\overline{X} \pm Z\frac{\sigma}{\sqrt{n}}$$

Recall Equation (7.3), $Z = (\overline{X} - \mu)/\sigma_{\overline{X}}$; thus we have $(\overline{X} - \mu) = Z\sigma/\sqrt{n}$. The **sampling error** is equal to the difference between the estimate from our sample $\overline{X}$ and the parameter to be estimated μ. This sampling error e can be defined as

$$e = \frac{Z\sigma}{\sqrt{n}} \quad (8.6a)$$

Solving this equation for n, we have

$$n = \frac{Z^2\sigma^2}{e^2} \quad (8.6b)$$

Therefore, to determine the sample size, three factors must be known:

1. The confidence level desired, Z
2. The sampling error permitted, e
3. The standard deviation, σ

First of all, the Z value can be determined once the desired level of confidence is known. Second, the sampling error e is the amount of error (plus or minus) that we are willing to accept in using the sample statistic to estimate the population parameter. Third, an estimate of the standard deviation must be available in order to determine the required sample size. In some cases the standard deviation of the variable is known. In other instances, past (historical) data may be available that can be extrapolated to estimate the current standard deviation. If the standard deviation cannot be estimated from past data, a pilot study can be taken, and results can be used to obtain an estimate of the standard deviation.

Returning to the previous example, the controller would like to estimate the population mean to within ±$5 of the true value. He would like to be 95% confident of correctly estimating the true mean. Based upon previous studies undertaken in the past year, the standard deviation is estimated as $21. With this information the sample size can be determined in the following manner for $e = \pm\$5$, $\sigma = \$21$, and 95% confidence ($Z = 1.96$):

$$n = \frac{Z^2\sigma^2}{e^2} = \frac{(1.96)^2(21)^2}{5^2} = \frac{3.8416(441)}{25} = 67.8$$

Therefore, $n = 68$.

We have chosen a sample of size 68 because the rule of thumb used in determining sample size is to always round up to the next highest integer value in order to slightly oversatisfy the criteria desired.

Therefore, if the controller had utilized these criteria, a sample of 68 should have been taken, not a sample of 100. However, it should be noted that the standard deviation has been estimated at $21 based upon past experience. If the actual sample standard deviation is very different from this value, the computed sampling error will be affected proportionately.

8.6 SAMPLE SIZE DETERMINATION FOR A PROPORTION

In the previous section we discussed the determination of sample size needed for the estimation of a quantitative parameter (the mean). Now suppose that the controller wishes to determine the sample size necessary for estimating the true proportion of credit card holders who would shop on Sunday. The methods of sample size determination that are utilized in estimating a true proportion are similar to those employed in estimating a mean.

The confidence interval estimate of the true proportion p is obtained from

$$p_s \pm Z\sqrt{\frac{p_s(1-p_s)}{n}} \tag{8.5}$$

Recall from Equation (7.5) that since $Z \cong \dfrac{p_s - p}{\sqrt{\dfrac{p\,(1-p)}{n}}}$,

we have

$$p_s - p = Z\sqrt{\frac{p(1-p)}{n}}$$

The sampling error is equal to the difference between the estimate from the sample p_s and the parameter to be estimated p. This sampling error can be defined as

$$e = Z\sqrt{\frac{p(1-p)}{n}} \tag{8.7a}$$

Solving for n, we obtain

$$n = \frac{Z^2p(1-p)}{e^2} \tag{8.7b}$$

In determining the sample size for estimating a proportion, three factors are needed:

1. The level of confidence desired, Z
2. The sampling error permitted, e
3. The estimated true proportion of success, p

The level of confidence desired in estimating the true value of the proportion will enable us to obtain the appropriate Z value on the normal distribution. The sampling error is the amount of error we are willing to accept in estimating the true proportion. The actual (true) proportion of success in this population p is the quantity that we would like to estimate by taking the sample. In this case we have two alternatives available. If the true proportion of success can be estimated based on past data or experience, this estimate can be used for p. However, what if no information were available? What can be used to estimate p? In such a case we would try to be as "conservative" as possible in estimating p. From Equation (8.7b), we would like to use the value of p that makes the quantity $p(1-p)$ as large as possible. It can be shown empirically that when $p = .5$, then $p(1-p)$ is at its maximum value. Several values of p along with the accompanying products of $p(1-p)$ are given below:

$$
\begin{aligned}
p &= .5 & p(1-p) &= .5(.5) = .25\\
p &= .4 & p(1-p) &= .4(.6) = .24\\
p &= .7 & p(1-p) &= .7(.3) = .21\\
p &= .1 & p(1-p) &= .1(.9) = .09\\
p &= .99 & p(1-p) &= .99(.01) = .0099
\end{aligned}
$$

Therefore, when we have no prior knowledge or estimate of the true proportion p we should use $p = .5$ as the most conservative way of determining the sample size. However, the use of $p = .5$ may result in an overestimate of the sample size. Since the actual sample proportion is utilized in the confidence interval, if it is very different from .50, the width of the confidence interval may be substantially narrower than originally intended.

In our example the controller wanted to estimate the proportion of credit card holders who would shop at the department store if it were open on Sunday. Since an important business decision is involved, the sample size should be carefully determined according to the requirements of the controller. In this case the controller would like a high degree of confidence, such as 99%, that the true proportion is being correctly estimated. The controller also would like to have a fairly small sampling error (such as $\pm.025$) in estimating the true proportion. Since the question involved is a new store policy, no information about the true proportion is available from past data. Therefore, p is set equal to .5. With these criteria in mind, the sample size needed can be determined in the following manner with 99% confidence ($Z = 2.58$), $e = .025$, $p = .5$, and

$$n = \frac{Z^2p(1-p)}{(e)^2} = \frac{(2.58)^2(.5)(.5)}{(.025)^2} = \frac{6.6564(.25)}{.000625} = 2{,}662.56$$

Thus, $n = 2{,}663$.

Therefore, in order to be 99% confident of estimating the proportion to within $\pm$ .025 of its true value, a sample size of 2,663 would be needed.

Since this is a rather large sample size, probably beyond the budget of the controller, a smaller sample size would be taken. However, the smaller sample size would only be achieved at the price of reducing the level of confidence and/or increasing the sampling error. Recall that in the original example the controller selected a sample of only 100 credit customers. This substantially smaller sample size had the same level of confidence (99%), but had about five times as much sampling error ($\pm.1264$) in estimating the true proportion.

8.7 ESTIMATION AND SAMPLE SIZE DETERMINATION FOR FINITE POPULATIONS

In the last chapter (Section 7.4) we saw that when sampling without replacement from finite populations, the **finite population correction (fpc) factor** served to reduce the standard error by a factor equal to $(N - n)/(N - 1)$. When estimating population parameters from such samples without replacement, the finite population correction factor should be used for developing confidence interval estimates.

Therefore, the confidence interval estimate for the mean would become

$$\bar{X} \pm t_{n-1}\frac{S}{\sqrt{n}}\sqrt{\frac{N-n}{N-1}} \qquad (8.8)$$

In the controller's example, a sample of 25 accounts was selected from a population of 10,000 accounts. Using the finite population correction factor, we would have, with $\bar{X} = \$75$, $S = \$20$, $n = 25$, $N = 10{,}000$, and $t = 2.064$,

$$\bar{X} \pm t_{n-1}\frac{S}{\sqrt{n}}\sqrt{\frac{N-n}{N-1}} = 75 \pm 2.064\frac{(20)}{\sqrt{25}}\sqrt{\frac{10{,}000-25}{10{,}000-1}}$$
$$= 75 \pm 8.256(.9988) = 75 \pm 8.246$$
$$\$66.754 \leq \mu \leq \$83.246$$

In this case, since the sample was a very small fraction of the population, the correction factor had little effect on the confidence interval estimate (as was computed in Section 8.3).

The confidence interval estimate of the total when sampling without replacement would be

$$N\bar{X} \pm t_{n-1}\frac{NS}{\sqrt{n}}\sqrt{\frac{N-n}{N-1}} \tag{8.9}$$

In the same example, the total would be estimated as follows, with $N = 10{,}000$, $S = \$20$, $n = 25$, $\bar{X} = \$75$, and $t = 2.064$:

$$N\bar{X} \pm t_{n-1}\frac{NS}{\sqrt{n}}\sqrt{\frac{N-n}{N-1}}$$
$$= 10{,}000(75) \pm (2.064)(10{,}000)\frac{(20)}{\sqrt{25}}\sqrt{\frac{10{,}000-25}{10{,}000-1}}$$
$$= 750{,}000 \pm 82{,}560(.99881) = 750{,}000 \pm 82{,}460.93$$
$$\$667{,}539.07 \leq \tau \leq \$832{,}460.93$$

Again, since the sample was a small fraction of the population, the correction factor did not greatly affect the estimate of the total (as was previously computed in Section 8.3).

The confidence interval estimate of the proportion, when sampling without replacement, would be

$$p_s \pm Z\sqrt{\frac{p_s(1-p_s)}{n}}\sqrt{\frac{N-n}{N-1}} \tag{8.10}$$

In the controller's example a sample of 100 customers was selected from a population of 10,000 customers. The confidence interval estimate would be determined in the following manner when sampling without replacement. We have $p_s = 60/100 = .60$, $Z = 2.58$, $n = 100$, and $N = 10{,}000$. Thus,

$$p_s \pm Z\sqrt{\frac{p_s(1-p_s)}{n}}\sqrt{\frac{N-n}{N-1}} = .60 \pm 2.58\sqrt{\frac{.6(.4)}{100}}\sqrt{\frac{10{,}000-100}{10{,}000-1}}$$

$$= .60 \pm (2.58)(.049)\sqrt{.99} = .60 \pm (.1264)(.995) = .60 \pm .1258$$

$$.4742 \le p \le .7258$$

Once again, because the sample is a small fraction of the population, the correction factor has a negligible effect on the confidence interval estimate (as was previously computed in Section 8.4).

Just as the correction factor was used in developing confidence interval estimates, it also should be used in determining sample size when sampling without replacement. For example, in estimating proportions, the sampling error would be

$$e = Z\sqrt{\frac{p(1-p)}{n}}\sqrt{\frac{N-n}{N-1}} \qquad (8.7c)$$

while in estimating means, the sampling error would be

$$e = Z\frac{\sigma}{\sqrt{n}}\sqrt{\frac{N-n}{N-1}} \qquad (8.6c)$$

The sample size needed can then be determined in a two-stage procedure. First, the sample size would be determined as in Sections 8.5 and 8.6 without regard to the correction factor. Then the correction factor would be applied to obtain the final sample size.

In determining the sample size in estimating the mean, we would have, from Equation (8.6*b*),

$$n_0 = \frac{Z^2\sigma^2}{e^2}$$

Applying the correction factor to this results in

$$n = \frac{n_0}{\dfrac{n_0 + (N-1)}{N}} \qquad (8.11)$$

In the controller's example the sample size needed in order to be 95% confident of being correct to within $\pm\$5$ (assuming a standard deviation of \$21) was 68 since n_0 was computed as 67.8. Using the correction factor leads to the following:

$$n = \frac{(67.8)}{\dfrac{67.8 + (10{,}000 - 1)}{10{,}000}} = 67.35$$

Thus, $n = 68$.

In this case the use of the correction factor made no difference in the sample size selected. However, in general this may not be the case. For example, in order for the controller to estimate the true proportion of credit card holders who would shop on Sunday with 99% confidence to within $\pm .025$, a sample size of 2,663 would be required since n_0 was computed as 2,662.56. This sample size was determined from Equation (8.7*b*), $n_0 = Z^2p(1 - p)/e^2$. Applying the correction factor to this leads to

$$n = \frac{n_0}{\dfrac{n_0 + (N - 1)}{N}} = \frac{2{,}662.56}{\dfrac{2{,}662.56 + (10{,}000 - 1)}{10{,}000}} = 2{,}102.87$$

Thus, $n = 2{,}103$.

Here, since more than 25% of the population was to be sampled, the finite population correction factor had a substantial effect on the sample size—reducing it from 2,663 to 2,103.

We may recall that in Chapter 2 we determined that a sample of 94 students had to be selected in order to meet the desired requirements which were based on the answers to question 8 (tuition) and question 22 (attitude toward retention standards). Since the random variable "tuition" is quantitative, in order to determine the sample size required we use Equation (8.6*b*) and (8.11). Three quantities are needed—the desired confidence level Z, the sampling error e, and the standard deviation σ. After considerable thought and consultation, the dean had decided that he would like to have 95% confidence that the estimate of the average tuition that students were willing or able to pay is correct to within $\pm\$100$ of the true value. Based upon a similar study undertaken at another business school, the standard deviation of the tuition is estimated as \$505. With this information, the sample size can be determined in the following manner, with $e = \pm\$100$, $\sigma = \$505$ (estimated), and 95% confidence ($Z = 1.96$),

$$n_0 = \frac{Z^2\sigma^2}{e^2} = \frac{(1.96)^2(505)^2}{(100)^2} = \frac{(3.8416)(255025)}{10{,}000} = 97.97$$

Thus,

$$n = \frac{n_0}{\dfrac{n_0 + (N - 1)}{N}} = \frac{97.97}{\dfrac{97.97 + (2{,}202 - 1)}{2{,}202}} = 93.83$$

Therefore, $n = 94$.

However, before we can conclude that a sample of 94 respondents is needed, we must evaluate the sample size required for the qualitative variable "attitude toward retention standards." This can be found by using Equations (8.7*b*) and (8.11) after determining three quantities—the confidence level desired Z, the sampling error e, and an estimate of the true proportion of students who favor more stringent retention standards p. Once again, as with the quantitative variable, considerable thought was given to determining the desired values. However, the dean concluded that he would like 90% confidence that the estimate of the true proportion of students favoring more stringent retention standards is correct to within ± 9%. Since the dean had no prior estimate of the true proportion, the value of $p = .5$ is used. With this information, the sample size can be determined in the following manner with, $e = \pm.09$, $p = .5$, and 95% confidence ($Z = 1.96$):

$$n_0 = \frac{Z^2p(1-p)}{e^2} = \frac{(1.96)^2(.5)(.5)}{(.09)^2} = \frac{(3.8416)(.25)}{.0081} = 83.51$$

Thus,

$$n = \frac{n_0}{\dfrac{n_0 + (N-1)}{N}} = \frac{83.51}{\dfrac{83.51 + (2{,}202 - 1)}{2{,}202}} = 80.49$$

Therefore, $n = 81$.

We have seen that a sample of 94 respondents is needed to satisfy our requirements for the tuition variable, while a sample of 81 respondents is required to study the attitudes toward retention standards. However, since we must satisfy both requirements simultaneously with one sample, the larger sample size of 94 respondents must be utilized for the dean's survey.

PROBLEMS

* 8.1 The quality control manager of an electric light bulb factory wishes to estimate the average life of a shipment of light bulbs. A random sample of 64 light bulbs is selected. The results indicated a sample average life of 540 hours with a standard deviation of 120 hours. Set up a 95% confidence interval estimate of the true average life of light bulbs in this shipment.

8.2 A new breakfast cereal is to be test marketed for 1 month at stores of a large supermarket chain. The results for a sample of 36 stores indicated average sales of $1,200 with a standard deviation of $180.

(a) Set up a 99% confidence interval estimate of the true average sales of this new breakfast cereal.

(b) If the supermarket chain has 200 stores, set up a 99% confidence interval estimate of the total sales of this product at the supermarket chain.

* 8.3 The alumni association of a large state university would like to estimate the average annual salaries of graduates of the class of 1968. A random sample of 100 individuals revealed an average salary of $18,210 with a standard deviation of

$1,450. Set up a 95% confidence interval estimate of the true average annual salary of graduates of the class of 1968.

8.4 Referring to the data of Problem 4.8,

(a) Set up a 90% confidence interval estimate of the average monthly rental of one-bedroom apartments in Brooklyn.

(b) Set up a 90% confidence interval estimate of the average monthly rental of one-bedroom apartments in Manhattan.

* 8.5 Referring to the data of Problem 4.9,

(a) Set up a 95% confidence interval estimate of the average quarterly sales tax receipts of the retail outlets.

(b) Set up a 95% confidence interval estimate of the total sales tax receipts that are to be collected this quarter.

8.6 Referring to the data of Problem 4.10,

(a) Set up a 99% confidence interval estimate of the average amount in unpaid accounts.

(b) Set up a 99% confidence interval estimate of the total amount owed the book company.

8.7 Referring to Problem 4.11, set up a 95% confidence interval estimate of the population average price-to-earnings ratio.

8.8 The manager of a bank in a small city would like to determine the proportion of its depositors who are paid on a weekly basis. A random sample of 100 depositors is selected in which 30 state that they are paid weekly. Set up a 90% confidence interval estimate of the true proportion of the bank's depositors who are paid weekly.

* 8.9 A suburban bus company is considering the institution of a commuter bus route from a particular suburb into the central business district of the city. A random sample of 50 commuters is selected, and 18 indicate that they would use this bus route. Set up a 95% confidence interval estimate of the true proportion of commuters who would utilize this new bus route.

8.10 An auditor for a consumer protection agency would like to determine the proportion of claims that are paid by a health insurance company within 2 months of receipt of the claim. A random sample of 200 claims is selected, and it is determined that 80 were paid out within 2 months of the receipt of the claim. Set up a 99% confidence interval estimate of the true proportion of the claims paid within 2 months.

* 8.11 An auditor for a wholesale metal parts supplier would like to estimate the rate of occurrence of errors in the billing prices charged to customers. A random sample of 300 vouchers transacted in the last month indicated that 45 contained errors in the billing prices charged to customers. Set up a 95% confidence interval estimate of the true proportion of errors in the billing prices.

8.12 A stationery supply store receives a shipment of a certain brand of inexpensive ball point pens from the manufacturer. The owner of the store wishes to estimate the proportion of pens that are defective. A random sample of 300 pens is tested, and 30 are found to be defective. Set up a 90% confidence interval estimate of the proportion of defective pens in the shipment. If the shipment can be returned if it is more than 5% defective, based on the sample results, can the owner return this shipment?

* 8.13 A survey is planned to determine the average annual family medical expenses of employees of a large company. The management of the company wishes to be 95% confident that the sample average is correct to within ±$50 of the true average family expense. A pilot study indicates that the standard deviation can be estimated as $400. How large a sample size is necessary?

8.14 Referring to Problem 8.1, if the process standard deviation is 100 hours and the manager wishes to estimate the average life to within ±20 hours of the true average with 95% confidence, what sample size is needed?

* 8.15 Referring to Problem 8.2, if the average amount of sales is to be estimated to within ±$100 with 99% confidence and the standard deviation is assumed to be $200, what sample size is needed?

8.16 Referring to Problem 8.3, if the average salary of graduates of the class of 1968 is to be estimated to within ±$500 of the true average with 95% confidence and the standard deviation is assumed to be $1,000 based upon past data, what sample size is needed?

* 8.17 A consumer group would like to estimate the average monthly electric bills for the month of July of one-family homes in a large city. Based upon studies conducted in other cities, the standard deviation is assumed to be $20. The group would like to estimate the average bill for July to within ±$5 of the true average with 99% confidence. What sample size is needed?

8.18 A political pollster would like to estimate the proportion of voters who will vote for the Democratic candidate in a presidential campaign. The pollster would like 90% confidence that her prediction is correct to within ±.04 of the true proportion. What sample size is needed?

* 8.19 A cable television company would like to estimate the proportion of its customers that would purchase a cable television program guide. The company would like to have 95% confidence that its estimate is correct to within ±.05 of the true proportion. Past experience in other areas indicates that 30% of the customers will purchase the program guide. What sample size is needed?

8.20 Referring to Problem 8.8, if the manager wishes to be 90% confident of being correct to within ±.05 of the true proportion of customers who are paid weekly, what sample size is needed?

* 8.21 Referring to Problem 8.9, if the bus company wishes to be 95% confident of being correct to within ±.02 of the true proportion of commuters who would utilize the bus service, and based on experience with other routes the true proportion is assumed to be .40, what sample size is needed?

8.22 Referring to Problem 8.11, if the auditor wishes to be 95% confident of being correct to within ±2.5% of the true error rate of billing prices, what sample size is needed? Based upon past experience, the error rate of occurrence is assumed to be 10%.

* 8.23 Referring to Problems 8.1 and 8.14, if the shipment contains a total of 2,000 light bulbs,

(a) Set up a 95% confidence interval estimate of the true average life of light bulbs in this shipment.

(b) Determine the sample size needed to estimate the average life to within ±20 hours with 95% confidence.

8.24 Referring to Problem 8.13, what sample size is necessary if the company has 3,000 employees?

* 8.25 Referring to Problems 8.8 and 8.20, if the bank has 1,000 depositors,

(a) Set up a 90% confidence interval estimate of the true proportion of depositors who are paid weekly.

(b) Determine the sample size needed to estimate the true proportion to within ±.05 with 90% confidence.

8.26 Referring to Problems 8.11 and 8.22, if a total of 5,000 vouchers were transacted in the last month,

(a) Set up a 95% confidence interval estimate of the true proportion of errors in the billing prices.

(b) Determine the sample size needed to estimate the true proportion to within ±.025 with 95% confidence.

8.27 Referring to Problem 8.1, set up 99% and 90% confidence interval estimates of the true average life of light bulbs in the shipment. Compare and discuss the meaning of the three confidence interval estimates.

8.28 Referring to Problem 8.11, set up 99% and 90% confidence interval estimates of the true proportion of errors in the billing prices charged to customers. Compare and discuss the meaning of the three confidence interval estimates.

8.29 Referring to the results shown in Section 5.9, set up a 95% confidence interval estimate of the true proportion of students who are receiving any scholarship, stipend, or financial assistance to attend college.

The following problems refer to the sample data obtained from the questionnaire (Figure 2.2) and presented in Figure 2.6.

8.30 Set up a 95% confidence interval estimate of the expected annual starting salary (question 7) of non-accounting majors (question 4).

8.31 Set up a 90% confidence interval estimate of the average grade-point index (question 5) in the population.

8.32 Set up a 95% confidence interval estimate of the average age (in years) of students (question 15).

8.33 Set up a 99% confidence interval estimate of the average number of clubs (question 23) that students are affiliated with.

8.34 Set up a 90% confidence interval estimate of the average amount of tuition (question 8) that accounting majors (question 4) claim they are able to pay.

8.35 Set up a 99% confidence interval estimate of the high-school average (question 6) of non-accounting majors (question 4).

8.36 Set up a 95% confidence interval estimate of the average number of pairs of jeans owned (question 13) by accounting majors (question 4).

8.37 Set up a 99% confidence interval estimate of the average weight (question 17) of students who smoke (question 19, codes 2, 3, and 4).

8.38 Set up a 95% confidence interval estimate of the proportion of students who are female (question 1, code 2).

8.39 Set up a 90% confidence interval estimate of the proportion of students who are employed full time (question 9, code 1).

8.40 Set up a 99% confidence interval estimate of the proportion of students that do not smoke cigarettes (question 19, code 1).

8.41 Set up a 95% confidence interval estimate of the proportion of students that favor more stringent retention standards (question 22, code 1).

8.42 Set up a 95% confidence interval estimate of the proportion of students who are planning to go to graduate school (question 3, code 1).

8.43 Set up a 99% confidence interval estimate of the true proportion of students who are accounting majors (question 4, code 2).

8.44 Set up a 95% confidence interval estimate of the true proportion of students at this school who own a calculator (question 14, code 1).

8.45 Set up a 90% confidence interval estimate of the true proportion of students at this school who would invest in stocks (question 11, code 5).

8.46 Use your own sample data (Problem 3.12) to solve selected problems from among Problems 8.30 through 8.45.

8.47 For your own class (Problem 3.13) select problems from among Problems 8.30 through 8.45.

REFERENCES

1. COCHRAN, W. G., *Sampling Techniques,* 3rd ed. (New York: Wiley, 1977).
2. DANIEL, W. W., AND J. C. TERRELL, *Business Statistics: Basic Concepts and Methodology* (Boston: Houghton Mifflin, 1975).
3. FISHER, R. A., AND F. YATES, *Statistical Tables for Biological, Agricultural and Medical Research,* 5th ed. (Edinburgh: Oliver and Boyd, 1957).
4. FREUND, J. E., AND F. J. WILLIAMS, *Elementary Business Statistics: The Modern Approach,* 3rd ed. (Englewood Cliffs, N.J.: Prentice-Hall, 1977).
5. HUNTSBURGER, D., P. BILLINGSLEY, AND D. J. CROFT, *Statistical Inference for Management and Economics* (Boston: Allyn and Bacon, 1975).
6. MOOD, A. M., F. A. GRAYBILL, AND D. C. BOES, *Introduction to the Theory of Statistics,* 3rd ed. (New York: McGraw-Hill, 1974).
7. NOETHER, G. *Introduction to Statistics: A Nonparametric Approach,* 2nd ed. (Boston: Houghton Mifflin, 1976).
8. SNEDECOR, G. W., AND W. G. COCHRAN, *Statistical Methods,* 6th ed. (Ames, Iowa: Iowa State University Press, 1967).

CHAPTER 9

HYPOTHESIS TESTING I: INTRODUCTION AND CONCEPTS

9.1 INTRODUCTION

In Chapters 7 and 8, Sampling Distributions and Estimation, the concept that a sample statistic such as a mean (or a proportion) would follow a particular distribution under various circumstances was used to develop the confidence interval as a way of estimating the true value of a mean (or proportion). In this chapter we will begin to focus on another phase of statistical inference—hypothesis testing.

In many circumstances, decisions must be made based only on sample information. A quality control manager must determine whether a process is working properly. A marketing manager must determine whether a new marketing strategy will increase sales. An auditor must be able to determine the propriety of a firm's books. In drawing these types of conclusions, the decision maker would like to be as certain as possible that the correct conclusion is being reached.

9.2 THE HYPOTHESIS TESTING PROCEDURE

In Chapter 8 several examples were discussed in developing confidence interval estimates. Here we will focus on two of these problems, one based on a tire company case study and the other based upon various questions that the researcher and the dean wish to answer.

In the first problem the tire company has two shifts of workers, a day shift and an evening shift. A random sample of 100 tires produced by each shift is selected in order to assist the manager in drawing conclusions about each of the following questions:

1. Is the average life of tires produced by the day shift equal to 25,000 miles?
2. Is the average life of tires produced by the evening shift less than 25,000 miles?
3. Do more than 8% of the tires produced by the day shift blow out before 10,000 miles?

In the second problem the dean and the researcher wish to use the results of the survey to draw conclusions about each of the following:

1. Have the average grade-point indexes of accounting majors changed in the last 10 years if a census of all accounting majors at that time indicated a population average of 2.705?
2. Has there been an "inflation" of B or better averages if the records of 10 years ago indicate that 27% of all students at that time had averages of B or above?

In order to answer each of these questions, a decision based upon sample information must be made. In the first problem, let us focus initially on the question of whether the average life of tires produced by the day shift is equal to 25,000 miles. The manager can draw one of two conclusions:

1. The average tire life is 25,000 miles.
2. The average tire life is not 25,000 miles; either it is less than 25,000 miles or it is more than 25,000 miles. If this latter conclusion is reached, corrective action will be taken to determine why below standard or above standard tires are being produced.

From the point of view of statistical hypothesis testing, these two conclusions would be represented as follows:

$$H_0 \text{ (null hypothesis)}: \quad \mu = 25{,}000 \text{ miles}$$

$$H_1 \text{ (alternate hypothesis)}: \quad \mu \neq 25{,}000 \text{ miles}$$

In the second problem, let us examine the question of whether the average grade-point index of accounting majors is equal to 2.705? One of two conclusions can be reached:

1. The average grade-point index has not changed from 2.705.
2. The average grade-point index has changed from 2.705.

Again, from the hypothesis testing point of view, these two conclusions would be expressed as

$$H_0 \text{ (null hypothesis)}: \quad \mu = 2.705$$
$$H_1 \text{ (alternative hypothesis)}: \quad \mu \neq 2.705$$

The **null hypothesis** represents the conclusion that we would draw if the process were operating properly. It is analogous to the presumption of innocence until guilt is proven in the American legal system. In our two problems, we assume that the process is working properly ($\mu = 25{,}000$ miles) or that the grade-point index has not changed ($\mu = 2.705$) unless evidence of "guilt" can be found, which shows that the process is not working properly or that the grade-point index has changed. The **alternative hypothesis,** which is usually the opposite of the null hypothesis, represents the conclusion that would be drawn if evidence of guilt is found. In our cases, "guilt" either refers to the average tire life being different from 25,000 miles or to the average grade-point index changing from 2.705. Therefore the alternative hypothesis would be that the average tire life is not 25,000 miles or the average grade-point index is not 2.705.

In developing the hypothesis testing procedure, let us refer to the tire example. We would begin by assuming that the average tire life is 25,000 miles. If a sample was taken and the sample mean was "close to" 25,000 miles, it is reasonable to conclude that the true mean is actually 25,000 miles. If the sample mean was very different from 25,000 miles, it is reasonable to conclude that the true mean is not 25,000 miles. Rather than arbitrarily saying "close to" or "different from" 25,000 miles, statistical hypothesis testing quantifies the decision-making process.

For each type of hypothesis testing procedure, an appropriate **test statistic** can be computed. This test statistic measures how closely the sample value (such as an average) has come to the null hypothesis. The test statistic either follows a well-known statistical distribution (such as normal, *t*, etc.), or a distribution can be developed for the particular test statistic.

The appropriate distribution of the test statistic is divided into two regions, a **region of rejection** and a **region of nonrejection.** If the test statistic falls into the latter region, the null hypothesis cannot be rejected, and the manager would conclude that the process is working properly while the dean would conclude that the average had not changed from 2.705. If the test statistic falls into the rejection region, the null hypothesis would be rejected and the manager would conclude that the true mean was not 25,000 miles, while the dean would conclude that the average had changed from 2.705 (see Figure 9.1).

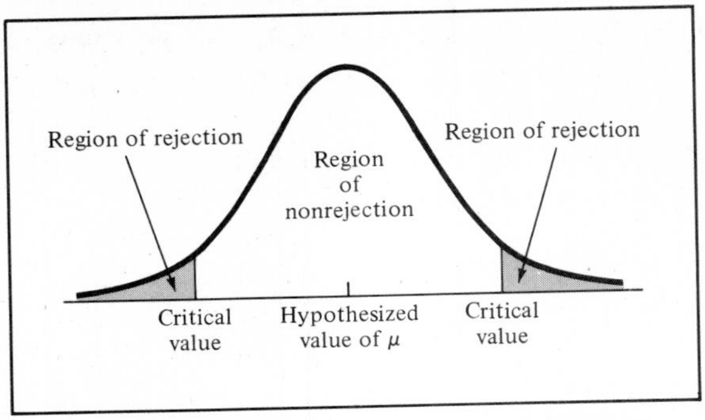

Figure 9.1 Regions of rejection and nonrejection in hypothesis testing.

In making this decision concerning the null hypothesis, we must determine the **critical value** on the statistical distribution that divides the nonrejection region (in which the null hypothesis cannot be rejected) from the rejection region. The critical value, however, depends upon the size of the rejection region. As we will see in the next section, the decision concerning the size of the rejection region affects the risks of making different types of incorrect decisions.

9.3 TYPE I AND TYPE II ERRORS

In using a sample to draw inferences about the population, the decision maker is taking a risk that the incorrect conclusion will be reached. There are two types of errors that can occur in the hypothesis testing procedure.

The first error, called Type I error (α), is the probability that the null hypothesis H_0 will be rejected when, in fact, it is true. In our tire example this error would occur if we concluded (based on our sample data) that the average tire life was not 25,000 miles when it was actually 25,000 miles. In the survey example the Type I error would occur if we concluded (based on the sample data) that the grade-point average had changed when actually there had been no change.

The Type I error α is also called the *level of significance.* Traditionally, the statistician controls the Type I error by establishing the risk level he or she is willing to tolerate in terms of rejecting a true null hypothesis. The selection of the particular risk level α is, of course, dependent upon the importance (significance) of the problem. Once the value for α is specified, the size of the rejection region is known since α is the probability of rejection under the null hypothesis. From this fact the upper and lower critical values that divide the rejection and nonrejection regions can be determined.

The second error, called Type II error (β), is the probability that the null hypothesis H_0 will not be rejected when it is false and should be rejected. In our tire example the Type II error would occur if we have concluded that the process is working properly (producing tires with a life of 25,000 miles) when in fact the average tire life is different from 25,000 miles. In the survey example the Type II error would occur if we concluded (based on the sample data) that the grade-point average had not changed when actually the grade-point average was no longer equal to 2.705.

Unlike the Type I error α which is set at a specific value, the value of the Type II error β depends upon the way in which the null hypothesis is not true. For example, with our tire study it was claimed (our null hypothesis) that the average tread life of the tires is 25,000 miles. If, in fact, the actual average tire life were only 10,000 miles, the chance (β) would indeed be quite small that, based upon the sample evidence, we would erroneously conclude that the average tire life was the hypothesized 25,000 miles. On the other hand, if the actual average tire life were 24,900 miles (very close to the null hypothesis), the chance (β) would be fairly high that, based upon the sample evidence, we would erroneously conclude that the average tire life was the hypothesized 25,000 miles.

The two types of errors can be illustrated in Table 9.1.

The complement ($1 - \beta$) of the Type II error, which is the chance of rejecting the null hypothesis when it is false, is called the **power of a statistical test.** The power of the test is a measure of the sensitivity of the hypothesis testing procedure since it determines the chance of correctly rejecting the null hypothesis in different circumstances. In the tire example we could compute the chance of concluding that the process is not working properly (rejecting the null hypothesis) for differing values of the true mean (10,000, 20,000, 24,900 miles, etc.). A more detailed discussion of the power of a statistical test is presented in Section 9.10.

For a given sample size the decision maker must balance the two types of errors. If α is to be decreased, then β will be increased. If β is to be decreased, then α will be increased. The values for α and β depend on the importance of each risk in a particular problem. The α risk in the tire case involves taking corrective action when none is necessary. The β risk involves *not* taking corrective action when the tire production process is not working properly. The importance of α and β depend on the costs inherent in each type of error. For example, if it were very costly to take corrective action (such as closing the factory), then the α risk would be most important, and perhaps β could be allowed to increase (computations involving α and β are presented in Section 9.10).

Table 9.1
HYPOTHESIS TESTING

Statistical Decision	Actual Situation	
	H_0 *True*	H_0 *False*
Do not reject H_0	$1 - \alpha$	Type II error (β)
Reject H_0	Type I error (α)	Power $= 1 - \beta$

9.4 STEPS OF HYPOTHESIS TESTING

1. State the null hypothesis.
2. State the alternative hypothesis.
3. Specify the level of significance α.
4. Determine the sample size n.
5. Set up the critical values that divide the rejection and nonrejection regions.
6. Determine the test statistic.
7. Collect the data and compute the sample value of the appropriate test statistic.
8. Determine whether the test statistic has fallen into the rejection or the nonrejection region.
9. Determine the statistical decision.
10. Express the statistical decision in terms of the problem.

Steps 1 and 2: The null and alternative hypotheses must be stated in statistical terms. If we were testing whether the average tire life was 25,000 miles, the null hypothesis would be that μ equals 25,000 miles, while the alternative hypothesis would be that μ was not equal to 25,000 miles.

Step 3: The level of significance α is specified according to the importance of α and β in the problem.

Step 4: The sample size is determined by taking into account the importance of α and β (see Sections 8.5 and 9.10) and by considering budgeting constraints in carrying out the study.

Step 5: Once the null and alternative hypotheses are known, and the level of significance and sample size decided upon, the critical values of the appropriate distribution can be set up to indicate the rejection and nonrejection regions.

Step 6: The technique to be used to determine whether the sample statistic has fallen into the rejection or nonrejection region must be determined. The appropriate test statistic must be indicated along with how the sample statistic is to be compared to the hypothesized parameter. For example, in the case of means, we find how far the statistic deviates from the hypothesized parameter in standard deviation units:

$$\frac{\text{Statistic} - \text{Parameter } (H_0)}{\text{Standard Error of Statistic}}$$

Step 7: The data are collected and the actual value of the test statistic is computed.

Step 8: The value of the test statistic is compared to the critical value on the appropriate distribution to determine whether it falls in the rejection or nonrejection region.

Step 9: The hypothesis testing decision is determined. If the test statistic falls into the nonrejection region, the null hypothesis H_0 cannot be rejected.

If the test statistic falls into the rejection region, the null hypothesis H_0 is rejected.

Step 10: Once the decision is made, its consequences must be expressed in terms of the particular problem. For example, in the tire problem, if the null hypothesis were rejected it would mean that the population average was believed to be different from 25,000 at a specific level of significance. The conclusion could then be made that corrective action should be taken to determine why the process was not working properly. In the survey problem, not rejecting the null hypothesis would result in the conclusion that there was no evidence that the grade-point average of accounting majors had changed in the last 10 years.

9.5 RESULTS OF TWO CASE STUDIES: THE TIRE COMPANY AND THE DEAN'S SURVEY

At the beginning of this chapter we developed two case studies, one based on a tire company, and the other based upon the survey that has been discussed throughout this text. The results of samples of 100 tires from the day shift and the evening shift are summarized in Table 9.2.

Table 9.2
SUMMARY TABLE FOR TIRE STUDY

Day Shift	Evening Shift
$\bar{X}_{day} = 25{,}430$ miles	$\bar{X}_{eve} = 23{,}310$ miles
$S_{day} = 4{,}000$ miles	$S_{eve} = 3{,}000$ miles
5 tires blew out before 10,000 miles	10 tires blew out before 10,000 miles
$n_{day} = 100$	$n_{eve} = 100$

The results necessary to answer the various questions asked by the dean are available from Tables 4.9 and 5.2. These are summarized in Table 9.3.

Table 9.3
SUMMARY TABLE FOR DEAN'S STUDY

Accounting	Non-Accounting
$\bar{X}_{acc} = 2.881$	$\bar{X}_{n\text{-}acc} = 2.882$
$S_{acc} = 0.48$	$S_{n\text{-}acc} = 0.456$
15 accounting majors have B averages or above	22 non-accounting majors have B averages or above
$n_{acc} = 35$	$n_{n\text{-}acc} = 59$

SOURCE: Data are taken from Tables 4.9 and 5.2.

9.6 TEST OF HYPOTHESIS FOR THE MEAN (ONE SAMPLE)

In the tire company problem the null and alternative hypotheses were set up as follows:

$$H_0: \quad \mu = 25{,}000$$
$$H_1: \quad \mu \neq 25{,}000$$

If the standard deviation σ of the tires produced by the day shift was known, then based on the central limit theorem, the sampling distribution of the mean would follow the normal distribution, and the test statistic which is based upon the difference between the sample mean $\bar{X}$ and the hypothesized mean μ would be found as follows:

$$Z = \frac{\bar{X} - \mu}{\frac{\sigma}{\sqrt{n}}} \tag{9.1}$$

If the size of the rejection region α were set at 5%, then the critical values of the normal distribution could be determined. Since the rejection region is divided into the two tails of the distribution, the 5% is divided into two equal parts of 2.5%.

Since we have the normal distribution, the critical values can be expressed in standard deviation units. A rejection region of .025 in each tail of the normal distribution results in an area of .475 between the hypothesized mean and the critical value. Looking up this area in the normal distribution (Appendix E, Table E.2), we find that the critical values that divide the rejection and nonrejection regions are +1.96 and −1.96 (see Figure 9.2).

Therefore the decision rule would be

$$\begin{aligned} \text{Reject } H_0 \text{ if } Z &> +1.96 \\ \text{or if } Z &< -1.96 \\ \text{otherwise do not reject } H_0 & \end{aligned}$$

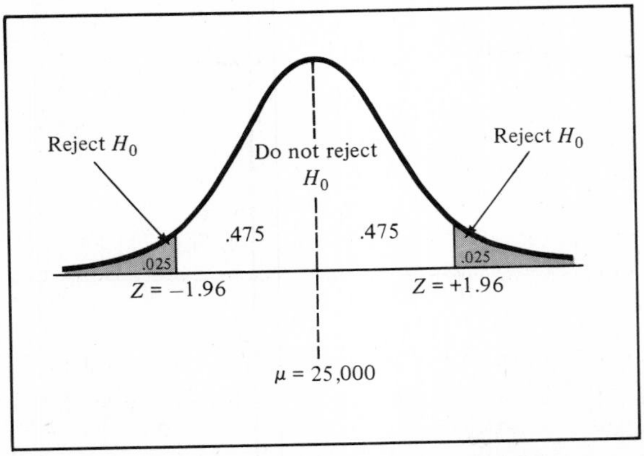

Figure 9.2 Testing a hypothesis about the mean (σ known) at the 5% level of significance.

However, in most cases the standard deviation σ of the population is unknown. The standard deviation is estimated by computing S, the standard deviation of the sample. If the population is assumed to be normal, the sampling distribution of the mean will follow a t distribution with $n - 1$ degrees of freedom. In practice it has been found that as long as the sample size is not very small and the population is not very skewed, the t distribution gives a good approximation to the sampling distribution of the mean. The test statistic for determining the difference between the sample mean $\bar{X}$ and the population mean μ when the sample standard deviation S is used, is given by

$$t_{n-1} = \frac{\bar{X} - \mu}{\frac{S}{\sqrt{n}}} \tag{9.2}$$

For a sample of 100, if a level of significance α of .05 is selected, the critical values of the t distribution with $100 - 1 = 99$ degrees of freedom can be obtained (as shown in Table 9.4 and Figure 9.3).

Table 9.4
DETERMINING THE CRITICAL VALUE FROM THE t TABLE FOR AN AREA OF .025 IN EACH TAIL WITH 99 DEGREES OF FREEDOM

Degrees of Freedom	Upper Tail Areas					
	.25	.10	.05	.025	.01	.005
91	0.6772	1.2909	1.6618	1.9864	2.3680	2.6309
92	0.6772	1.2908	1.6616	1.9861	2.3676	2.6303
93	0.6771	1.2907	1.6614	1.9858	2.3671	2.6297
94	0.6771	1.2906	1.6612	1.9855	2.3667	2.6291
95	0.6771	1.2905	1.6611	1.9853	2.3662	2.6286
96	0.6771	1.2904	1.6609	1.9850	2.3658	2.6280
97	0.6770	1.2903	1.6607	1.9847	2.3654	2.6275
98	0.6770	1.2902	1.6606	1.9845	2.3650	2.6269
99	0.6770	1.2902	1.6604	→ **1.9842**	2.3646	2.6264
100	0.6770	1.2901	1.6602	1.9840	2.3642	2.6259

SOURCE: Extracted from Appendix E, Table E.3.

Since this is a two-tailed test, the rejection region of .05 is again divided into two equal parts of .025 each. Using the t tables given in Appendix E, Table E.3, the critical values are -1.984 and $+1.984$. The decision rule is

Reject H_0 if $t_{99} > +1.984$
or if $t_{99} < -1.984$
otherwise do not reject H_0

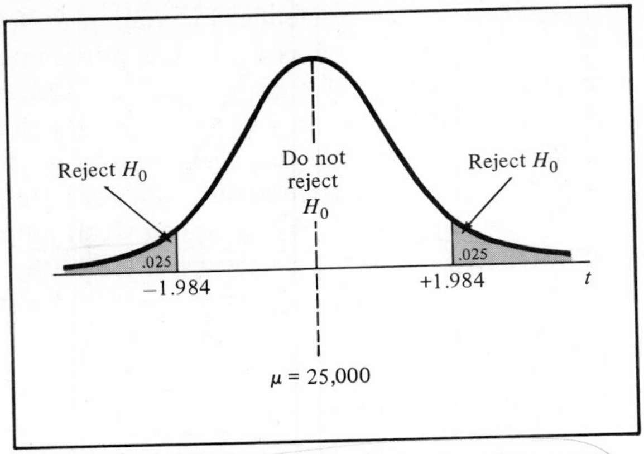

Figure 9.3 Testing a hypothesis about the mean (σ unknown) at the 5% level of significance with 99 degrees of freedom.

The sample results for the day shift were $\overline{X}_{\text{day}} = 25{,}430$ miles, $S_{\text{day}} = 4{,}000$ miles, and $n_{\text{day}} = 100$. Since we are testing whether the mean is different from 25,000 miles, we have, from Equation (9.2),

$$t_{n-1} = \frac{\overline{X} - \mu}{\dfrac{S}{\sqrt{n}}}$$

$$t_{100-1} = \frac{25{,}430 - 25{,}000}{\dfrac{4{,}000}{\sqrt{100}}}$$

$$= \frac{+430}{400} = +1.075$$

Since $t_{99} = +1.075$, we see that $-1.984 < +1.075 < +1.984$, so do not reject H_0.

Therefore the decision is not to reject the null hypothesis H_0. The conclusion drawn is that the average tire life is 25,000 miles. In order to take into account the possibility of a Type II error, this statement can be phrased as "there is no evidence that the average tire life is different from 25,000 miles for tires produced by the day shift."

For the first question raised by the dean, the null and alternative hypotheses would be

$$H_0: \quad \mu = 2.705$$

$$H_1: \quad \mu \neq 2.705$$

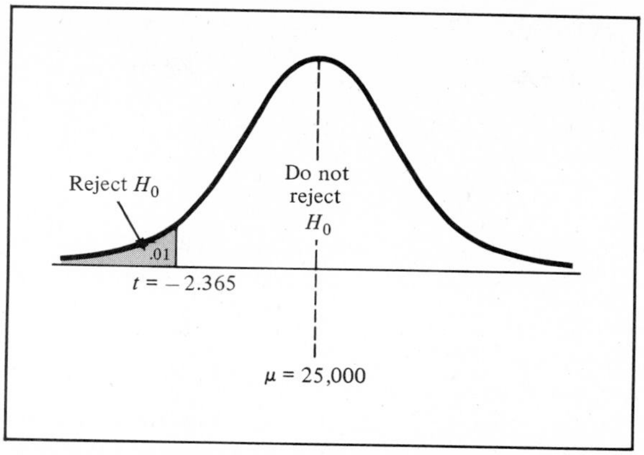

Figure 9.6 One-tailed test of hypothesis for a mean (σ unknown) at the 1% level of significance with 99 degrees of freedom.

Once again, since a single population mean is involved, the test statistic is

$$t_{n-1} = \frac{\overline{X} - \mu}{\frac{S}{\sqrt{n}}}$$

and for our data,

$$t_{99} = \frac{23{,}310 - 25{,}000}{\frac{3{,}000}{\sqrt{100}}} = \frac{-1{,}690}{300} = -5.63$$

Since $t_{99} = -5.63 < -2.365$, reject H_0.

The decision would be to reject the null hypothesis H_0. The conclusion drawn at the .01 level of significance is that the average life of tires produced by the evening shift is below 25,000 miles. Therefore the manager of the factory would conclude that the process is not working properly and would take the appropriate corrective action.

We have previously noted that as the number of degrees of freedom increases, the t distribution approaches the normal distribution. For large sample sizes the critical values of the t distribution will be virtually the same as the normal distribution. For example, the critical value at the .01 level of significance (two-tailed) of the normal distribution is 2.58. The corresponding value of the t distribution is 2.75 for 30 degrees of freedom, 2.66 for 60 degrees of freedom, 2.63 for 100 degrees of freedom, and 2.61 for 150 degrees of freedom. Therefore, we might say that when there are at least 150 degrees of freedom, the normal distribution can be used interchangeably with the t distribution to obtain the appropriate critical values.

9.8 TEST OF HYPOTHESIS FOR A PROPORTION (ONE SAMPLE)

In the previous section we used hypothesis testing procedures for quantitative data (means). The concept of hypothesis testing can also be used to test hypotheses about qualitative data. In our two studies, for example, the manager of the tire company wanted to determine the proportion of tires that blew out before 10,000 miles, while the dean wanted to know whether the proportion of students with B or above averages had increased. Each of these is an example of a qualitative variable, since we wish to draw conclusions about the proportion of values that have a particular characteristic.

In the first case the manager of the tire factory wants the quality of the tires produced to be high enough so that very few tires blow out before 10,000 miles. If more than 8% of the tires blow out before 10,000 miles, the conclusion will be reached that the process is not working properly. The null and alternative hypotheses for this problem can be stated as follows:

$$H_0: \quad p \leq .08 \qquad \text{(working properly)}$$

$$H_1: \quad p > .08 \qquad \text{(not working properly)}$$

The number of successes follows a binomial process. However, as we have seen previously when developing confidence intervals, if the sample size is large enough (np and $n(1-p) \geq 5$), the normal distribution gives a good approximation to the binomial distribution. The test statistic can be stated in two forms, in terms of either the proportion of successes [see Equation (9.3)] or the number of successes [see Equation (9.4)]:

$$Z \cong \frac{p_s - p}{\sqrt{\dfrac{p(1-p)}{n}}} \tag{9.3}$$

where $p_s = \dfrac{X}{n} = \dfrac{\text{Number of Successes in Sample}}{\text{Sample Size}}$

p = proportion of successes from the null hypothesis

or

$$Z \cong \frac{X - np}{\sqrt{np(1-p)}} \tag{9.4}$$

Both formulas will result in the same exact answer for the problem.

Let us determine whether the process is working properly for tires produced by the day shift. The results for the day shift indicated that five tires in a

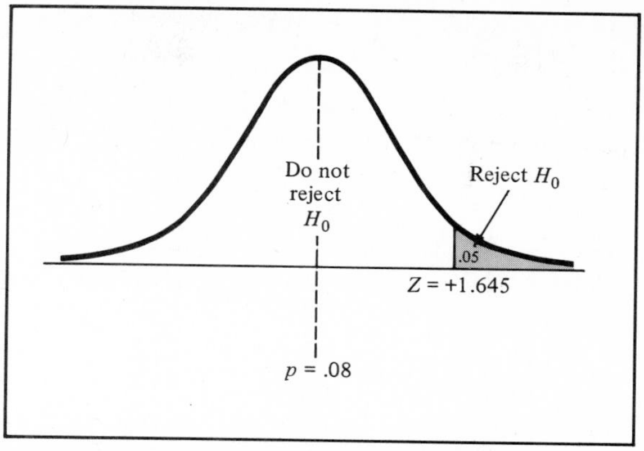

Figure 9.7 One-tailed test of hypothesis for a proportion at the 5% level of significance.

sample of 100 blew out before 10,000 miles. For this problem, if a level of significance α of .05 is selected, the rejection and nonrejection regions would be set up as shown in Figure 9.7, and the decision rule would be

Reject H_0 if $Z > +1.645$; otherwise do not reject H_0

From our data,

$$p_s = \frac{5}{100} = .05$$

and thus,

$$Z \cong \frac{p_s - p}{\sqrt{\dfrac{p(1-p)}{n}}} = \frac{.05 - .08}{\sqrt{\dfrac{.08(.92)}{100}}} = \frac{-.03}{\sqrt{.000736}} = \frac{-.03}{.0271} = -1.107$$

$Z \cong -1.107 < +1.645$; therefore, do not reject H_0

The null hypothesis would not be rejected since the test statistic has not fallen into the rejection region. The conclusion would be drawn that there is no evidence that more than 8% of the tires produced by the day shift blow out before 10,000 miles. The manager has not found any evidence that an excessive number of blowouts are occurring on tires produced by the day shift.

In the second problem we may recall that the dean wanted to determine whether there had been an "inflation" of grades in the past 10 years. Based on past records, 27% of the students had B averages or above at that time. Therefore the null and alternative hypotheses would be

H_0: $p \leq .27$ (no grade inflation)

H_1: $p > .27$ (grades have inflated)

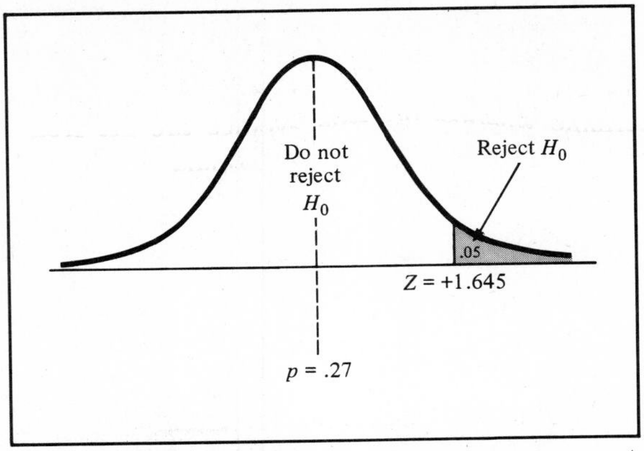

Figure 9.8 One-tailed test of hypothesis for a proportion at the .05 level of significance.

Since we are drawing a conclusion about all students, the accounting and non-accounting majors can be combined into one group. Out of a total of 94 students, 37 have averages of B or above.

Using the second formula for testing a hypothesis about a proportion and selecting the .05 level of significance, we have the following decision rule (see Figure 9.8):

$$\text{Reject } H_0 \text{ if } Z > 1.645; \text{ otherwise do not reject } H_0$$

For our data,

$$Z \cong \frac{X - np}{\sqrt{np(1-p)}} = \frac{37 - 94(.27)}{\sqrt{94(.27)(.73)}} = \frac{37 - 25.38}{\sqrt{18.5274}} = \frac{+11.62}{4.304} = +2.70$$

$$Z \cong +2.70 > +1.645; \text{ therefore reject } H_0$$

Since the test statistic has fallen into the rejection region, the null hypothesis has been rejected. We can conclude that the proportion of students with B or above averages is greater than .27. Therefore the dean can say that grade inflation may have occurred over the past 10 years, since the proportion of the students with B or above averages has increased significantly.

9.9 A CONNECTION BETWEEN CONFIDENCE INTERVALS AND HYPOTHESIS TESTING

In the last two chapters we have examined the two major areas of statistical inference: confidence intervals and hypothesis testing. They have been based on the same set of concepts but have been used for different purposes. Confidence intervals were used to estimate parameters while hypothesis testing has been utilized in making decisions about specified population parameters. In many cases, confidence intervals and tests of hypothesis can be used interchangeably. This can be illustrated in the case of the test of hypothesis for a mean. Referring back to the first question of the tire study, we attempted to determine whether the true average tire life from the day shift was 25,000 miles. This was tested with the formula

$$t_{n-1} = \frac{\bar{X} - \mu}{\frac{S}{\sqrt{n}}} \tag{9.2}$$

This problem could also have been solved by obtaining a confidence interval estimate of μ. Had the hypothesized value of μ (25,000 miles) fallen into the interval, the null hypothesis would not be rejected. On the other hand, if it did not fall into the interval, the null hypothesis would be rejected. Using Equation (8.2), the confidence interval estimate could be set up from the following data: $\bar{X} = 25{,}430$, $S = 4{,}000$, and $n = 100$. $\bar{X} \pm t_{n-1} S/\sqrt{n}$ covers μ with $(1 - \alpha)\%$ confidence.

$$P\left[\bar{X} - t_{n-1}\frac{S}{\sqrt{n}} < \mu < \bar{X} + t_{n-1}\frac{S}{\sqrt{n}}\right] = (1 - \alpha)\% \text{ confidence}$$

For a confidence level of 95%,

$$25{,}430 \pm (1.984)\frac{(4{,}000)}{\sqrt{100}} = 25{,}430 \pm 793.6$$

$$24{,}636.4 \leq \mu \leq 26{,}223.6$$

Since the interval includes the hypothesized value of 25,000 miles, the null hypothesis would not be rejected. This, of course, was the same decision reached by using the hypothesis testing technique.

9.10 POWER OF A TEST

9.10.1 Development

In our initial discussions of statistical hypothesis testing we defined the two types of risks that are taken when decisions are made about population parameters based only upon sample evidence. The first type of error (α) occurred when the null hypothesis was rejected when, in fact, it was true and should not have been rejected. The second type of error (β) occurred when the null hypothesis was not rejected when, in fact, it was false and should have been rejected. The power of the statistical test $(1 - \beta)$ indicated the sensitivity of the statistical procedure in detecting changes which have occurred by measuring the probability of rejecting the null hypothesis when it is false and should be rejected. The power of the statistical test is dependent upon how different the true mean really is from the value being hypothesized (under H_0). If there is a large difference between the true mean and the hypothesized mean, the power of the test will be much greater than if the difference between the population mean and the hypothesized mean is small.

In this section the concept of the power of a statistical test will be further developed with an example taken from our tire study. In Section 9.7 we previously noted that the production manager was interested in determining whether the process was operating properly and producing tires that lasted an average of at least 25,000 miles. The null and alternative hypotheses for this problem were set up as follows:

$$H_0: \quad \mu \geq 25{,}000 \qquad \text{(process working properly)}$$

$$H_1: \quad \mu < 25{,}000 \qquad \text{(process not working properly)}$$

Let us suppose that, based upon past experience, the standard deviation is assumed to be 3,500 miles. If a level of significance (α risk) of 5% is selected and a random sample of 100 tires is obtained, the value of $\bar{X}$ that will enable us to reject the null hypothesis can be found from Equation (7.4*a*) as follows:

$$\bar{X}_L = \mu_0 - Z\frac{\sigma}{\sqrt{n}}$$

where μ_0 is the hypothesized mean.

Since we have a one-tailed test with a level of significance of 5%, the value of Z equal to 1.645 can be obtained from Appendix E, Table E.2 (see Figure 9.9), and for this problem we have

$$\bar{X}_L = 25{,}000 - (1.645)\frac{(3{,}500)}{\sqrt{100}} = 25{,}000 - 575.75 = 24{,}424.25$$

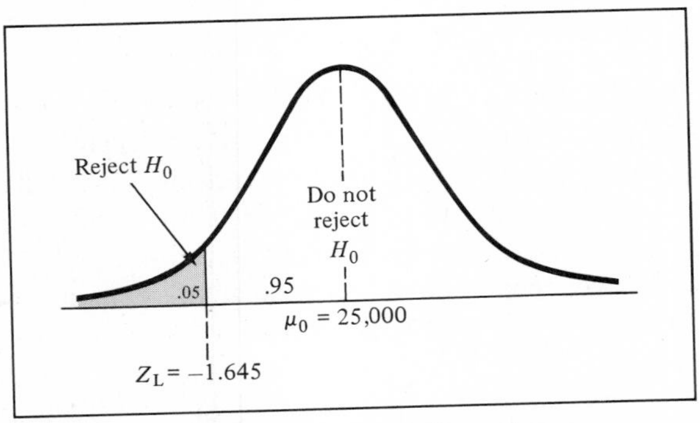

Figure 9.9. Determining the lower critical value for a one-tailed test for a population mean at the 5% level of significance.

The decision rule for this problem then will be

$$\text{Reject } H_0 \text{ if } \bar{X} < 24{,}424.25\text{; otherwise do not reject } H_0$$

This decision rule states that if a random sample of 100 tires reveals a sample mean of less than 24,424.25 miles, the null hypothesis will be rejected and the production manager will conclude that the process is not working properly. If in fact this is the case, the power of the test measures the probability of concluding that the process is not working properly for differing values of the true population mean.

Suppose, for example, we would like to determine the chance of rejecting the null hypothesis when the true population mean is actually 24,000 miles. Based upon our decision rule, we need to determine the probability or area under the normal curve below 24,424.25 miles. From the central limit theorem we can assume that the sampling distribution of the mean follows a normal distribution. Therefore this area under the normal curve below 24,424.25 miles can be expressed in standard deviation units since we are finding the probability of rejecting the null hypothesis when the true mean has shifted to 24,000 miles (see Figure 9.10). Using Equation (9.1) we have

$$Z = \frac{\bar{X} - \mu_1}{\frac{\sigma}{\sqrt{n}}}$$

where μ_1 = true population mean.

Thus,

$$Z = \frac{24{,}424.25 - 24{,}000}{\frac{3{,}500}{\sqrt{100}}} = +1.21$$

From Appendix E, Table E.2 (table of the normal distribution), there is a 38.69% chance of observing a Z value between the mean and $+1.21$ stan-

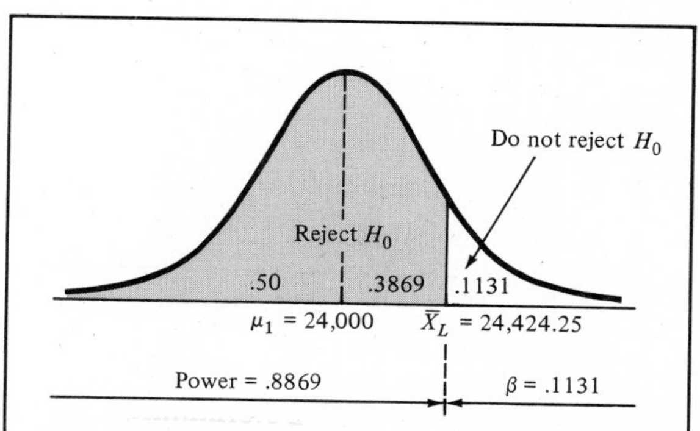

Figure 9.10 Determining the power of the test and the Type II error for μ_1 = 24,000 miles

dard deviations. Since we wish to determine the area below 24,424.25, the area under the curve below the mean (50%) must be added to this value, and the power of the test is found to be 88.69%. The Type II error β, representing the chance that the null hypothesis ($\mu = 25{,}000$) will not be rejected, is 1 − .8869, or .1131 (11.31%).

Now that we have determined the power of the test if the population mean were really equal to 24,000, we can also calculate the power for any other value that μ could attain. For example, what would be the power of the test if the population mean were really equal to 20,000 miles? Assuming the same standard deviation, sample size, and level of significance, the decision rule would still be:

Reject H_0 if $\overline{X} < 24{,}424.25$; otherwise do not reject H_0

Once again, since we are testing a hypothesis for a mean, from Equation (9.1) we have

$$Z = \frac{\overline{X} - \mu_1}{\frac{\sigma}{\sqrt{n}}}$$

If the population mean shifts down to 20,000 miles (see Figure 9.11), then

$$Z = \frac{24{,}424.25 - 20{,}000}{\frac{3{,}500}{\sqrt{100}}} = +12.64$$

The critical value is more than 12 standard deviation units above the mean, thereby indicating that the power of the test is approximately 100%, with virtually no chance of committing a Type II error.

In the previous two cases we have found that the power of the test was

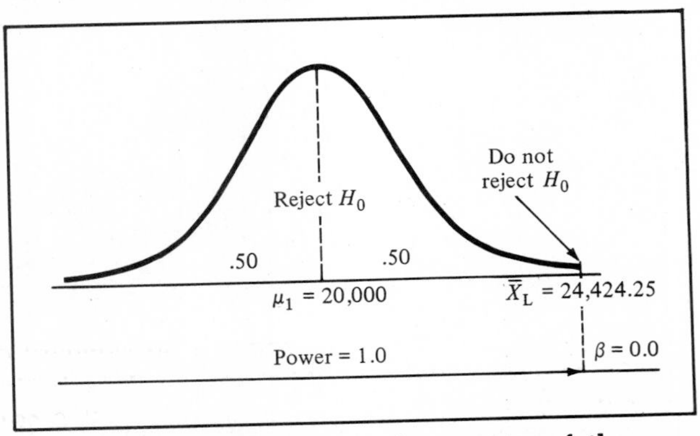

Figure 9.11 Determining the power of the test and the Type II error for $\mu_1 = 20{,}000$ miles.

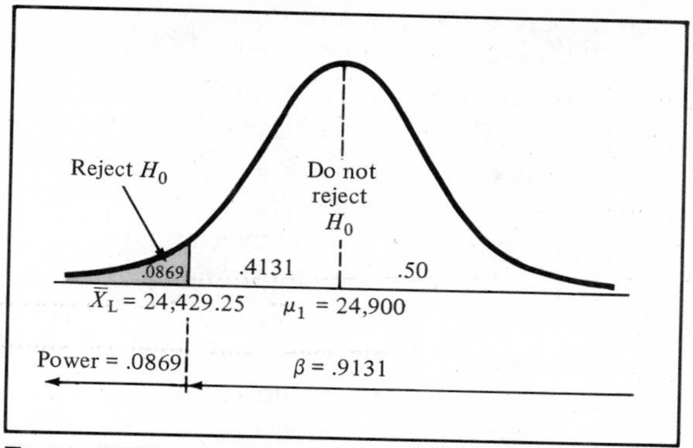

Figure 9.12 Determining the power of the test and the Type II error for $\mu_1 = 24{,}900$ miles.

quite high, while conversely, the chance of committing a Type II error was quite low. In our next example we shall compute the power of the test if the population mean were really equal to 24,900 miles—a value which is very close to the hypothesized mean of 25,000 miles.

Once again, from Equation (9.1), since we are testing a hypothesis about a mean, we have

$$Z = \frac{\bar{X} - \mu_1}{\dfrac{\sigma}{\sqrt{n}}}$$

If the population mean were really equal to 24,900 miles (see Figure 9.12), then

$$Z = \frac{24{,}424.25 - 24{,}900}{\dfrac{3{,}500}{\sqrt{100}}} = -1.36$$

so that

$$\begin{array}{r} .5000 \\ -\ .4131 \\ \hline .0869 \end{array} = \text{Power} = 1 - \beta$$

From Appendix E, Table E.2, we can see that the probability (area under the curve) between the mean and -1.36 standard deviation units is .4131 (that is, 41.31%). Since, in this instance, the rejection region is in the lower tail of the distribution, the power of the test is 8.69%, while the chance of making a Type II error is 91.31%.

The results for these three cases are plotted in Figure 9.13 while the computations are summarized in Figure 9.14. Figure 9.13 illustrates the power of the test for various possible values of μ_1 (including, of course, the ones that we have examined). This is called a **power curve.**

From Figure 9.13 we observe that the power of this one-tailed test increases sharply (and approaches 100%) as the tire population mean takes on values farther below the hypothesized 25,000 miles. Clearly, for this one-tailed test the smaller the true mean μ_1 is when compared to the hypothesized mean, the greater will be the power to detect this disparity.[1] On the other hand, for values of μ_1 close to 25,000 miles the power is rather small since the test is unable to effectively detect small differences between the true population mean and the hypothesized value of 25,000 miles. If the population mean were actually 25,000 miles, then the power of the test would be equal to α,

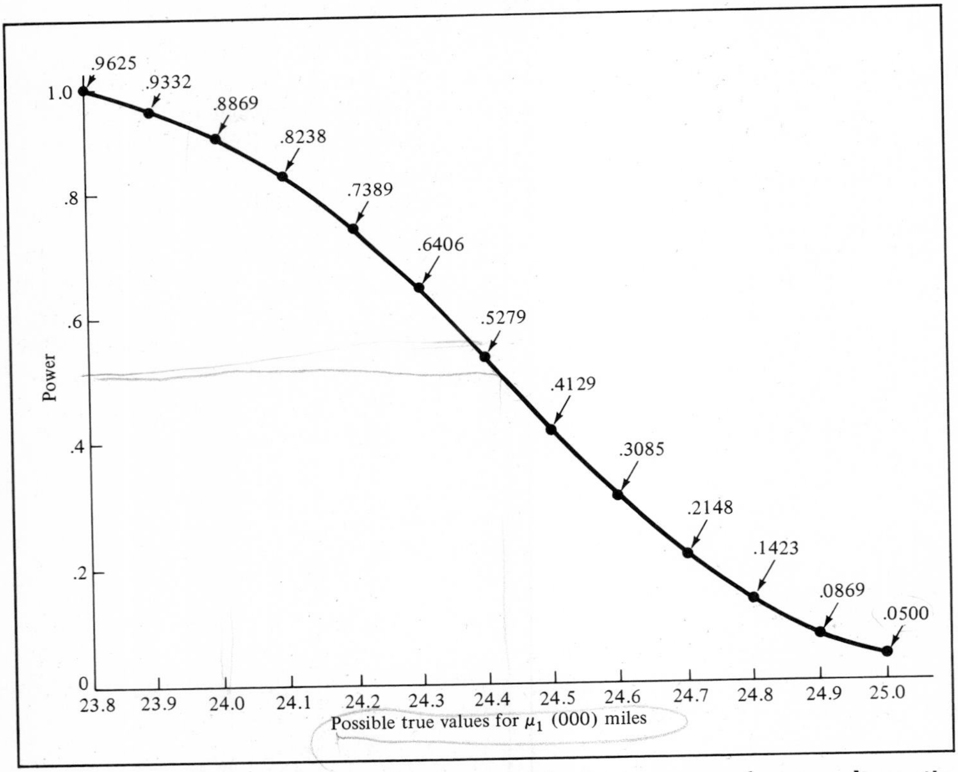

Figure 9.13 Power curve of the tire production process for an alternative hypothesis of $\mu_1 \leq 25{,}000$ miles.

[1] Of course, for situations involving one-tailed tests in which the actual mean μ_1 really exceeds the hypothesized mean, the converse would be true. The larger the actual mean μ_1 would be when compared to the hypothesized mean, the greater would be the power. On the other hand, for two-tailed tests, the greater the **distance** between the actual mean μ_1 and the hypothesized mean, the greater the power of the test.

the level of significance (which is 5% in this problem), since the null hypothesis would actually be true.

The drastic changes in the power of the test for differing values of the true population means can be observed by reviewing the different panels of Figure 9.14. From panels A and B, we can see that when the population mean does not greatly differ from 25,000 miles, the chance of rejecting the null hypothesis, based upon the decision rule involved, is not large. However, once the true population mean shifts substantially below the hypothesized 25,000 miles, the power of the test greatly increases, approaching its maximum value of 100%.

In our discussion of the power of a statistical test we have utilized a one-tailed test, a level of significance of 5%, and a sample size of 100 tires. With this in mind we can determine the effect on the power of the test by varying

1. The type of statistical test—one-tailed versus two-tailed
2. The level of significance α
3. The sample size n

9.10.2 The Effects on Power of Changes in the Type of Statistical Test

In formulating the null hypothesis in Section 9.10.1, we noted that the production manager desired to reject the null hypothesis only when the mean tire life was significantly below 25,000 miles (one-tailed test). The question that we now wish to answer is what would be the effect on the power of the test if the manager wanted to reject the null hypothesis when the mean tire life was significantly *different* from 25,000 miles? In this case we would have a two-tailed test that would be set up as follows:

$$H_0: \quad \mu = 25{,}000 \text{ miles}$$
$$H_1: \quad \mu \neq 25{,}000 \text{ miles}$$

If we assume that the level of significance is still 5%, the standard deviation is still 3,500 miles, and the sample size is still 100 tires, then from Equations (7.4*a* and 7.4*b*) we may determine the values of $\bar{X}_{\mathrm{L}}$ and $\bar{X}_{\mathrm{U}}$ which would cause us to reject the null hypothesis:

$$\bar{X}_{\mathrm{L}} = \mu - Z\frac{\sigma}{\sqrt{n}}$$

$$\bar{X}_{\mathrm{U}} = \mu + Z\frac{\sigma}{\sqrt{n}}$$

Here, however, the total level of significance of 5% is contained in both rejection regions so that, from Appendix E, Table E.2, the appropriate Z value

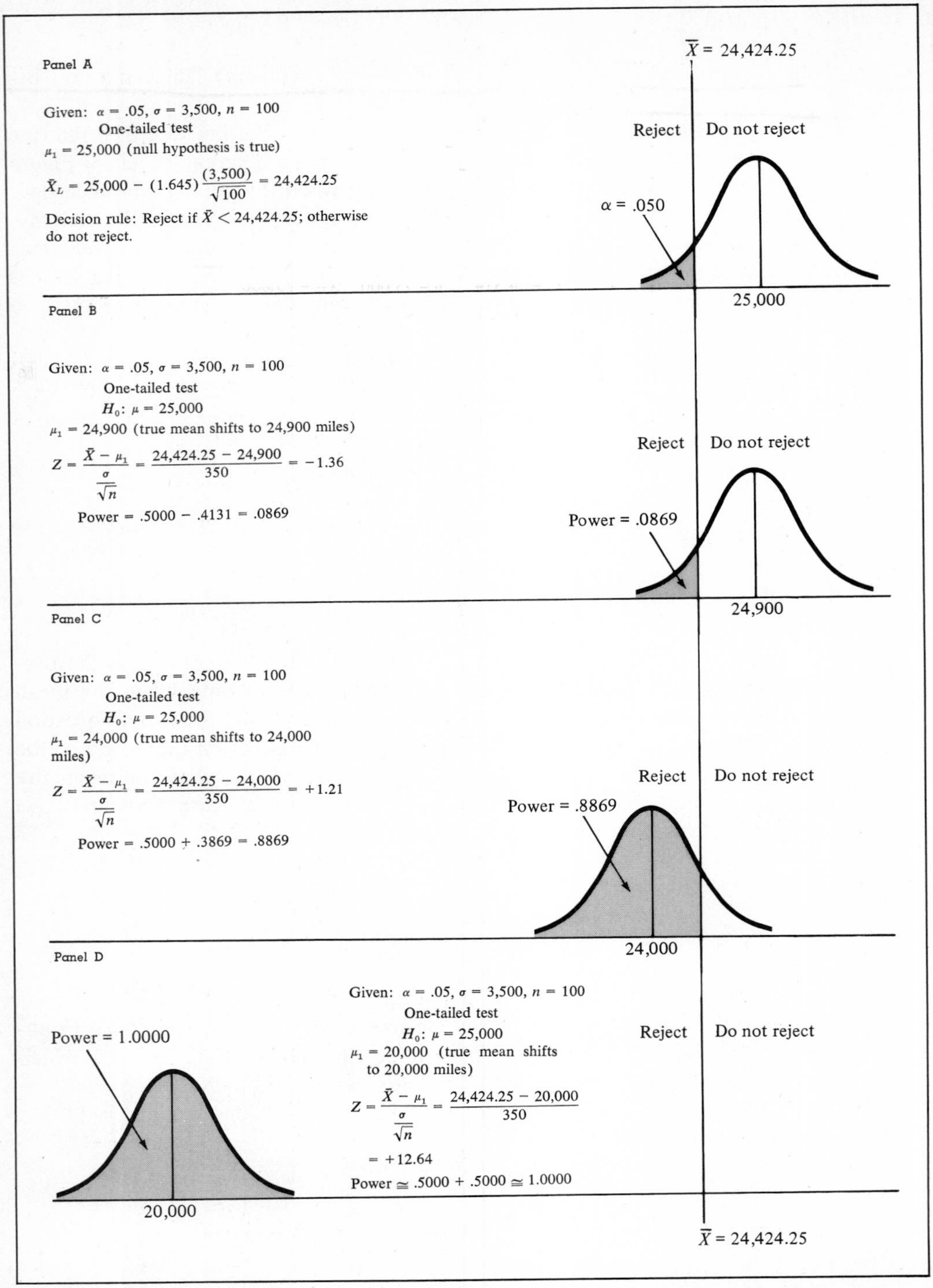

Figure 9.14 Determining statistical power for varying values of the true population mean.

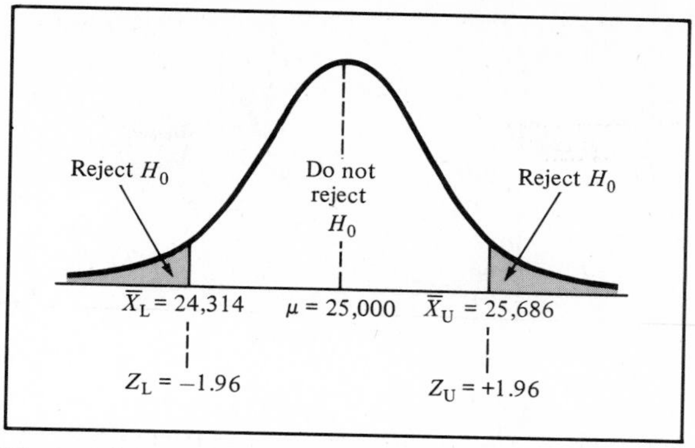

Figure 9.15 Determining the lower and upper critical values for a two-tailed test for a population mean at 5% level of significance.

is 1.96 (see Figure 9.15). $\overline{X}_L$, the lower critical value of $\overline{X}$, would be obtained from

$$\overline{X}_L = 25,000 - (1.96)\frac{(3,500)}{\sqrt{100}} = 25,000 - 686 = 24,314 \text{ miles}$$

while $\overline{X}_U$, the upper critical value of $\overline{X}$, would be obtained from

$$\overline{X}_U = 25,000 + (1.96)\frac{(3,500)}{\sqrt{100}} = 25,000 + 686 = 25,686 \text{ miles}$$

Therefore the decision rule for the two-tailed test would be

Reject H_0 if $\overline{X} > 25,686$
or if $\overline{X} < 24,314$
otherwise do not reject H_0

Now that the decision rule has been established, we can find the power of the test for the case in which the population mean is really 24,000 miles. We may recall from panel C of Figure 9.14 that for the particular one-tailed test the power was 88.69%. Using our decision rule for the two-tailed test we now need to find the probability that a sample mean will fall below 24,314 miles or above 25,686 miles if the true mean is really 24,000 miles. Using Equation (9.1) for these data we would have

$$Z_U = \frac{\overline{X}_U - \mu_1}{\frac{\sigma}{\sqrt{n}}} \quad \text{and} \quad Z_L = \frac{\overline{X}_L - \mu_1}{\frac{\sigma}{\sqrt{n}}}$$

Thus (see Figure 9.16),

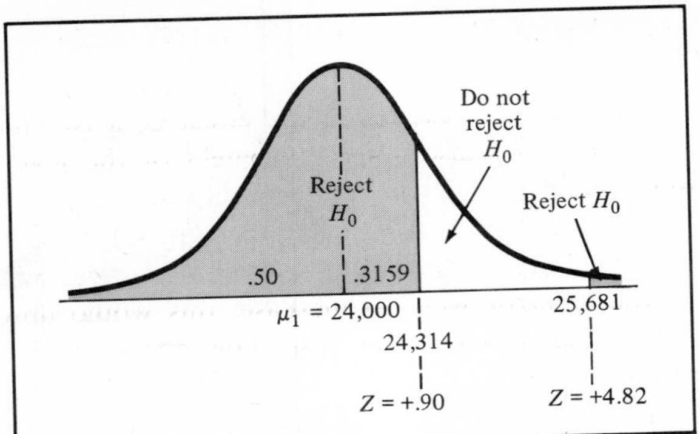

Figure 9.16 Determining the power of a two-tailed test and the Type II error for $\mu_1 = 24{,}000$ miles at the 5% level of significance.

$$Z_U = \frac{25{,}686 - 24{,}000}{\frac{3{,}500}{\sqrt{100}}} = \frac{1{,}686}{350} = +4.82 \quad \rightarrow \quad .5000 \text{ (Area)}$$

$$Z_L = \frac{24{,}314 - 24{,}000}{\frac{3{,}500}{\sqrt{100}}} = \frac{314}{350} = +.90 \quad \rightarrow \quad .3159$$

$$.1841$$

$$\text{Power} = 1 - .1841 = .8159$$

Since 25,686 miles is 4.82 standard deviation units above the population mean of 24,000, from Appendix E, Table E.2, we observe that the area under the normal curve (probability) between 25,686 and 24,000 miles is approximately .50. In a similar manner, the area between 24,314 and 24,000 miles is found to be .3159 (since 24,314 miles is .90 standard deviation unit above 24,000 miles). Therefore the power of the test is equal to .8159 (or 81.59%), since the null hypothesis will be rejected if the sample mean is below 24,314 miles or above 25,686 miles. Comparing this result with the one obtained in panel C of Figure 9.14 illustrates that, for given values of α, σ, and n, a one-tailed test is more powerful than a two-tailed test for a specified population mean. This result occurs because the one-tailed test places the rejection region entirely in one tail of the distribution. Thus we may conclude that if we have prior information that leads us to test the null hypothesis against a specifically directed alternative, then a one-tailed test will provide a more powerful test than a two-tailed test. On the other hand, we should realize that if we are only interested in *differences* from the null hypothesis, not in the *direction* of the difference, the two-tailed test is the appropriate procedure to utilize.

9.10.3 Effects on Power of Changes in the Level of Significance

Now that we have examined the effect of a two-tailed test on the power of the test, we may also study the effect of the level of significance α on the power of the test. In Section 9.3 we mentioned that if the Type I error α (the level of significance) is decreased, then, for a fixed sample size, the chance of a Type II error β would increase. Since β and power are complementary, if the Type II error were to increase, this would unfortunately result in a decrease in the power of the test. This effect can be observed by once again referring to panel C of Figure 9.14, where a one-tailed test was utilized with a sample size of 100 tires. Now, however, let us select a level of significance of 1% rather than 5%. Using Equation (7.4a) we have

$$\bar{X}_L = \mu - Z\frac{\sigma}{\sqrt{n}}$$

and, for these data (see Figure 9.17),

$$\bar{X}_L = 25{,}000 - (2.33)\frac{(3{,}500)}{\sqrt{100}} = 25{,}000 - 815.5 = 24{,}184.5$$

Therefore the decision rule would be

Reject H_0 if $\bar{X} < 24{,}184.5$ miles; otherwise do not reject H_0

Now that the decision rule has been determined, the power of this one-tailed test can be found for the case in which the population mean is really 24,000 miles. Based on our decision rule, we would need to find the prob-

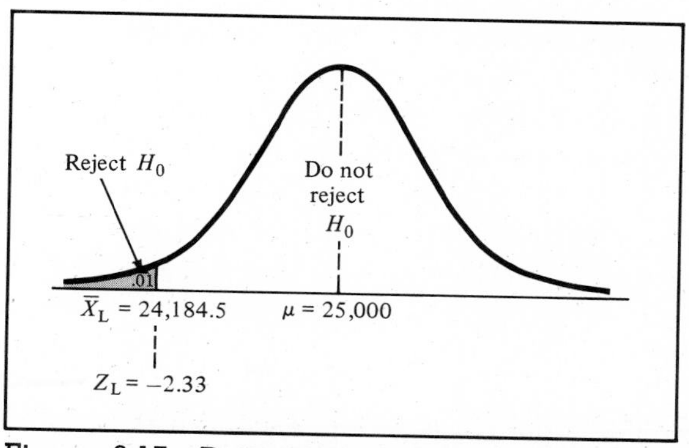

Figure 9.17 Determining the lower critical value for a one-tailed test for a population mean at the 1% level of significance.

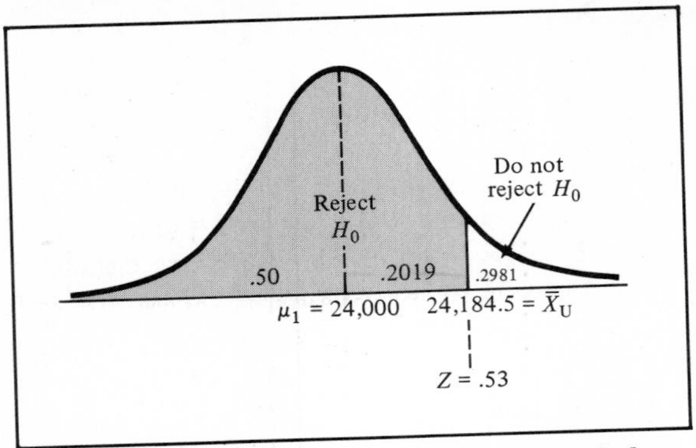

Figure 9.18 Determining the power of the test and Type II error for $\mu_1 = 24{,}000$ miles at the 1% level of significance.

ability that the sample mean falls below 24,184.5 miles. Using Equation (9.1) with a population mean of 24,000 miles, we have (see Figure 9.18)

$$Z = \frac{24{,}184.5 - 24{,}000}{\frac{3{,}500}{\sqrt{100}}} = \frac{184.5}{350} = .53$$

$$\text{Power} = .50 + .2019 = .7019$$

Since the critical value for 24,184.5 miles is .53 standard deviation unit above the population mean of 24,000 miles, from Appendix E, Table E.2, we may determine the probability to be .2019. Since the null hypothesis will be rejected when the sample mean is below 24,184.5 miles, and since 50% of the values will be below the population mean of 24,000 miles, the power of the test will be 70.19%. Thus if we compare the resulting statistical power (70.19%) to the power of a test using a 5% level of significance (88.69%), we can see that a reduction in the level of significance from 5% to 1% produces a substantial reduction in the power of the test.

9.10.4 Effects on Power of Changes in Sample Size

Now that the effects on power of a two-tailed test and changes in the level of significance have each been studied using a fixed sample size, it becomes necessary to vary the sample size to determine its effect on statistical power. This effect can be seen by reducing the sample size from 100 to 25 tires and determining the power of the one-tailed test at the 5% level of significance when the population mean is really 24,000 miles. Using Equation (7.4*a*), we have

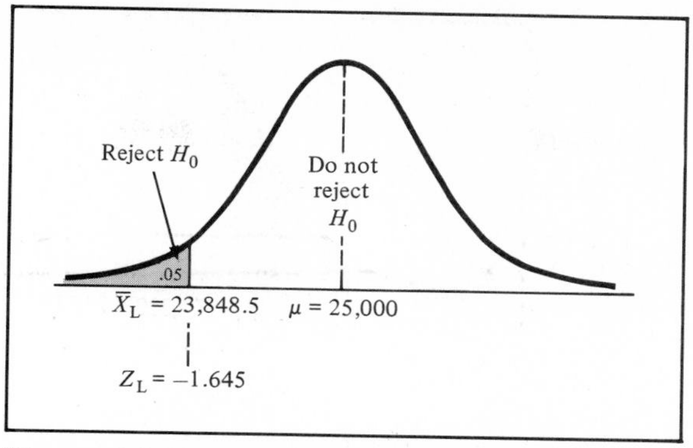

Figure 9.19 Determining the lower critical value for a one-tailed test for a population mean using a sample of size 25.

$$\bar{X}_L = \mu - Z\frac{\sigma}{\sqrt{n}}$$

and thus, for a sample of 25 tires (see Figure 9.19),

$$\bar{X}_L = 25{,}000 - (1.645)\frac{(3{,}500)}{\sqrt{25}} = 25{,}000 - 1{,}151.5 = 23{,}848.5$$

Therefore the decision rule for this problem containing a sample of 25 tires would be

Reject H_0 if $\bar{X} < 23{,}848.5$ miles; otherwise do not reject H_0

The power of the test can be determined by finding the probability that the sample mean falls below 23,848.5 miles when the population mean is actually 24,000 miles. Using Equation (9.1) for a true population mean of 24,000 miles, we have (see Figure 9.20)

$$Z = \frac{23{,}848.5 - 24{,}000}{\dfrac{3{,}500}{\sqrt{25}}} = -.22$$

Since 23,848.5 is .22 standard deviation unit below the population mean of 24,000 miles, from Appendix E, Table E.2, we may determine the probability of such an occurrence to be .0871. Since 50% of the mean values will be below 24,000 miles, the probability of obtaining a sample mean $\bar{X}$ below 23,848.5 miles (which represents the power of the test) will be .4129 (or 41.29%). Comparing this power to that which was obtained for a sample of

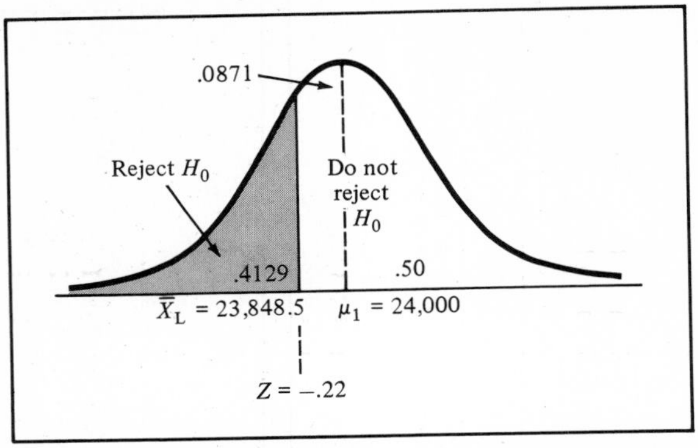

Figure 9.20 Determining the power of the test and the Type II error for $\mu_1 = 24{,}000$ miles with a sample of size 25.

100 tires (88.69%), as shown in panel C of Figure 9.14, indicates that, other things remaining equal, a reduction in the sample size may seriously reduce the power of the test.

9.10.5 Effects on Power: Summary of Findings

In examining the power of a statistical test we have observed that the type of test used (one-tailed versus two-tailed), the level of significance selected, and the sample size chosen each can seriously affect the results. A summary of our findings is presented in Figure 9.21.

9.10.6 Determining Sample Size Based on α and β Risks

In planning a statistical study we have already seen in Section 8.5 that the desired sample size can be determined for a specified confidence level and sampling error. In a decision-making procedure such as hypothesis testing, however, we may determine the sample size needed for a desired level of significance α and the desired power of a test $(1 - \beta)$ as follows:

$$n = \frac{\sigma^2 (Z_\alpha - Z_\beta)^2}{(\mu_0 - \mu_1)^2} \tag{9.5}$$

where σ^2 = variance of the population

Z_α = Z value for a given α level of significance

Z_β = Z value for a given β risk of a Type II error

μ_0 = value of the population mean under the null hypothesis

μ_1 = value of the population mean under the alternative hypothesis

Conclusion 1—For a one-tailed test having a specified α, n, and σ, the greater the distance between the actual mean μ_1 and the hypothesized mean, the greater is the power of the test (provided the correct direction of the alternative has been stated). **Note:** For a two-tailed test having a specified α, n, and σ, the greater the distance between the actual mean μ_1 and the hypothesized mean, the greater is the power of the test.	Given: H_0: $\mu = 25{,}000$ $\alpha = .05$ $n = 100$ $\sigma = 3{,}500$ One-tailed test If $\mu_1 = 24{,}000$, then power = .8869 (and $\beta = .1131$). If $\mu_1 = 24{,}900$, then power = .0869 (and $\beta = .9131$).
Conclusion 2—For a specified α, n, σ, and actual mean μ_1, a one-tailed test is more powerful than a two-tailed test and, therefore, should be used whenever the researcher is able to specify the direction of the alternative hypothesis.	Given: H_0: $\mu = 25{,}000$ $\alpha = .05$ $n = 100$ $\sigma = 3{,}500$ $\mu_1 = 24{,}000$ If one-tailed test is used, then power = .8869 (and $\beta = .1131$). If two-tailed test is used, then power = .8159 (and $\beta = .1841$).
Conclusion 3—For a specified n, σ, type of test, and actual mean μ_1, the larger the level of significance α that is chosen, the smaller is the risk of Type II error β and the greater is the power of the test.	Given: H_0: $\mu = 25{,}000$ $n = 100$ $\sigma = 3{,}500$ one-tailed test $\mu_1 = 24{,}000$ If $\alpha = .05$, then power = .8869 (and $\beta = .1131$). If $\alpha = .01$, then power = .7019 (and $\beta = .2981$).
Conclusion 4—For a specified α, σ, type of test, and actual mean μ_1, the larger the sample size n that is chosen, the greater is the power of the test.	Given: H_0: $\mu = 25{,}000$ $\alpha = .05$ $\sigma = 3{,}500$ One-tailed test $\mu_1 = 24{,}000$ If $n = 100$, then power = .8869 (and $\beta = .1131$). If $n = 25$, then power = .4129 (and $\beta = .5871$).

Figure 9.21 Studying the power of a test.

To demonstrate this sample size determination procedure for a specified Type I (α) and Type II (β) risk, we may refer once again to our tire study. If we may assume that we wish to have an 80% chance (power) of rejecting the null hypothesis (of 25,000 miles) when the population mean is really equal to 24,000 miles and we are willing to permit a Type I error of 5%, what sample size is required? Using Equation (9.5) we have

$$n = \frac{\sigma^2(Z_\alpha - Z_\beta)^2}{(\mu_0 - \mu_1)^2}$$

and, for the tire study,

$$\sigma = 3{,}500 \text{ miles}$$

$$\mu_0 = 25{,}000 \text{ miles}$$

$$\mu_1 = 24{,}000 \text{ miles}$$

Using a level of significance of $\alpha = .05$ for a one-tailed test, the rejection region can be established as follows (see Figure 9.22).

The Z_α value obtained from Appendix E, Table E.2, is equal to -1.645 because the rejection region contains 5% of the area under the normal curve (so that the area between the lower critical value and the null hypothesized mean of 25,000 miles is 45%).

If a power of 80% is desired when the tire population mean is 24,000 miles, the value of Z_β can also be obtained from Appendix E, Table E.2 (see Figure 9.23). Since we wish to have a power of 80% ($\beta = 20\%$) of rejecting the null hypothesis, this results in an area of .30 between the true population mean of 24,000 miles and the critical value (which corresponds to .84 standard deviation unit above the tire population mean).

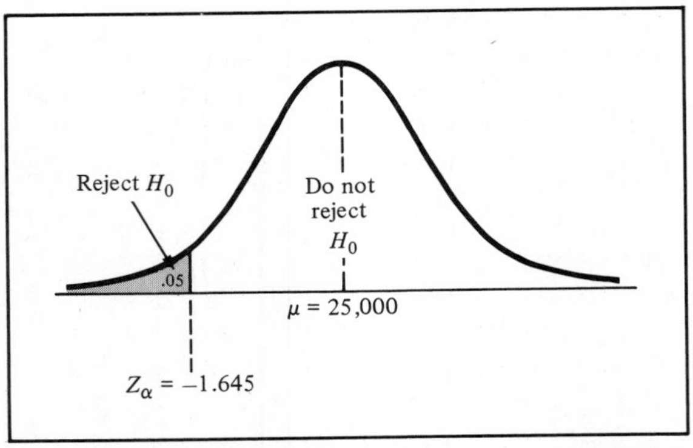

Figure 9.22 Determining the lower critical value in a one-tailed test for population mean when the sample size is unknown.

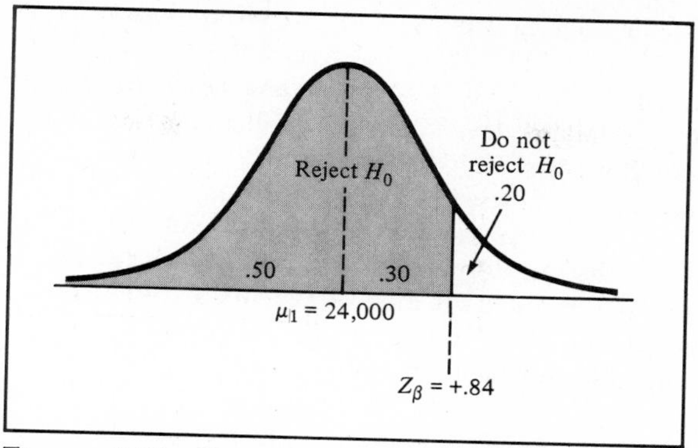

Figure 9.23 Determining the critical value for μ_1 = 24,000 miles where the sample size is unknown.

Therefore from Equation (9.5) the sample size would be found as follows:

$$n = \frac{(3{,}500)^2\,(-1.645 - .84)^2}{(25{,}000 - 24{,}000)^2} = \frac{(12{,}250{,}000)\,(-2.485)^2}{(1{,}000)^2} = 75.645$$

Hence, $n = 76$.

Thus, a sample size of 76 tires would be required if the production manager was willing to have a 5% risk of making a Type I error and an 80% chance of rejecting the null hypothesis (of 25,000 miles) and detecting that the true population mean has actually shifted to 24,000 miles.

PROBLEMS

9.1 Distinguish between
- (a) Null hypothesis and alternative hypothesis
- (b) Type I error and Type II error
- (c) One-tailed test and two-tailed test
- (d) Region of nonrejection and region of rejection

9.2 The editor of a textbook publishing firm must decide whether to publish a textbook written by a particular professor. Based upon publication costs, the editor has arrived at the following conclusion. If there is evidence that more than 15% of the colleges in the country would consider adopting this textbook, then the textbook will be published. If no evidence can be demonstrated, the textbook will not be published. A random sample of 100 colleges will be selected.
- (a) Explain the meaning of the Type I and Type II errors in this problem.
- (b) Which error would be more important to the editor? Why?
- (c) Which error would be more important to the professor? Why?
- (d) Is this problem a one-tailed test or a two-tailed test? Explain.

* 9.3 The credit manager for an oil company claims that the average monthly balance of credit card holders is \$30. To test this claim, an auditor selects a random sample of 100 accounts and finds that the average owed is \$35 with a sample standard deviation of \$12.50. At the .01 level of significance what should the auditor conclude?

9.4 The manufacturer of Dillco auto batteries claims that Dillco auto batteries last an average of 40,000 miles. A random sample of 81 batteries were tested and resulted in an average life of 40,800 miles with a sample standard deviation of 5,400 miles. At the .05 level of significance is the claim of the manufacturer of Dillco batteries valid?

* 9.5 The Dollar Bill Steel Company manufactures steel bars. The production process turns out steel bars with an average length of at least 2.8 feet when the process is working properly. A sample of 25 bars is selected from the production line. The sample indicates an average length of 2.43 feet and a standard deviation of .20 foot. The company wishes to determine whether the machine needs any adjustment.

(a) State the null and alternative hypotheses.

(b) If the company wishes to test the hypothesis at the .05 level of significance, what decision would it make?

9.6 The owner of a wholesale appliance dealership must estimate her average accounts receivable at the end of a monthly period. Based upon past experience, she estimates the average accounts receivable at \$10,000. Rather than accept this estimate of the owner, the accountant for this firm decides to select a random sample of 36 accounts. The results were as follows: $\overline{X} = \$12{,}300$ and $S = \$3{,}300$. At the .10 level of significance should the estimate of the owner be accepted?

9.7 The marketing manager of a large retail supermarket chain believes that there will be a large demand for vegetable seedlings at suburban stores during the month of May. If sales of vegetable seedlings exceed \$100 per week, they will be sold at all of the chain's suburban stores. A random sample of 16 stores is selected and the results of the store test indicated an average sale of \$120 with a sample standard deviation of \$25.

(a) At the .01 level of significance should vegetable seedlings be sold at all of the chain's suburban stores?

(b) What assumption is necessary to perform this test?

9.8 Refer to Problem 4.10.

(a) Is there evidence at the .05 level of significance that the average unpaid balance is greater than \$10?

(b) What assumption is necessary to perform this test?

9.9 Refer to Problem 4.8.

(a) Is there evidence at the .01 level of significance that the average rental for a one-bedroom unfurnished apartment in Brooklyn is greater than \$200?

(b) What assumption is necessary to perform this test?

9.10 A large nationwide chain of home improvement centers is having an end-of-season clearance of lawnmowers. The number of lawnmowers sold during this sale at a sample of 10 stores was as follows:

8 11 0 4 7 8 10 5 8 3

(a) At the .05 level of significance, is there evidence that an average of more than 5 lawnmowers per store have been sold during this sale?
(b) What assumption is necessary to perform this test?

* 9.11 The personnel department of a large company would like to determine the amount of time it takes for employees to arrive at work. A random sample of 12 employees is selected and the time in minutes to arrive at work is recorded with the following results:

15 30 50 60 25 65 45 90 75 50 50 20

(a) At the .01 level of significance, is there evidence that the average travel time of employees is less than 60 minutes?
(b) What assumption is necessary to perform this test?

9.12 The owner of a baseball club claims that at least 50% of all sports fans have watched a baseball game in the past year. A random sample of 400 indicated that 160 had watched a baseball game in the past year. At the .05 level of significance, is the claim of the owner justified?

* 9.13 A television manufacturer claims that 90% of his television sets do not require any repair during the first 2 years of operation. The Consumer Protection Agency selects a sample of 100 sets and finds that 14 sets required some repair within the first 2 years of operation. At the .01 level of significance, what conclusion should be reached by the Consumer Protection Agency?

9.14 The owner of a wholesale distributing firm would like to know the proportion of accounts receivable that are more than 60 days past due. The owner estimates that 15% of the accounts receivable are more than 60 days past due. A random sample of 200 accounts receivable revealed that 44 were more than 60 days past due. At the .05 level of significance is the estimate of the owner valid?

* 9.15 A home heating oil company estimates that 60% of its customers pay their oil bills within 15 days of delivery. A random sample of 400 bills indicated that 180 were paid within 15 days of delivery. At the .10 level of significance, is the estimate of the company correct?

9.16 Referring to Problem 9.2, if the random sample of 100 colleges indicated that 25 would consider adopting this textbook, should the editor publish the textbook (use a level of significance of .05)?

* 9.17 A coin-operated soft-drink machine was designed to discharge, when it is operating properly, at least 7 ounces of beverage per cup with a standard deviation of 0.2 ounce. If a random sample of 16 cupfuls is selected by a statistician for a

consumer testing service, and the statistician is willing to have a Type I (α) risk of 5%, compute the power of the test and the probability of a Type II (β) error if the population average amount dispensed is:

(a) 6.9 ounces per cup

(b) 6.8 ounces per cup

(c) If the statistician wishes to have a power of 99% of detecting a shift of the population mean from 7.0 ounces to 6.9 ounces, what sample size must be selected?

9.18 Refer to Problem 9.17. If the statistician wishes to have an α risk of 1%, compute the power of the test and the probability of a Type II error (β) if the population average amount dispensed is:

(a) 6.9 ounces

(b) 6.8 ounces

(c) Compare the results in (a) and (b) of this problem and Problem 9.17.

* 9.19 Refer to Problem 9.17. If the statistician selected a random sample of 25 cupfuls, and used an α risk of 5%, compute the power of the test and the probability of a Type II (β) error if the population average amount dispensed is:

(a) 6.9 ounces

(b) 6.8 ounces

(c) Compare the results in (a) and (b) of this problem and Problem 9.17.

9.20 A machine that fills cereal boxes places 368 grams of cereal in each box when it is working properly. The amount placed in the cereal box is normally distributed with a standard deviation of 30 grams. The production manager will stop filling boxes only if there is evidence that the average amount of cereal placed in each box is less than 368 grams. If a random sample of 25 boxes is selected, and the production manager is willing to have a Type I (α) risk of 5%, compute the power of the test and the probability of a Type II (β) error if the population average amount placed in the cereal box is:

(a) 360 grams

(b) 355 grams

(c) If the production manager wishes to have a 90% power of detecting a shift of the population mean from 368 grams to 360 grams, what sample size must be selected?

9.21 Refer to Problem 9.20. If the statistician wishes to have a α risk of 10%, compute the power of the test and the probability of a Type II error (β) if the population average amount is:

(a) 360 grams

(b) 355 grams

(c) Compare the results in (a) and (b) of this problem and Problem 9.20.

9.22 Refer to Problem 9.20. If the production manager selected a random sample of 100 cereal boxes and used an α risk of 5%, compute the power of the test and the probability of a Type II (β) error if the population average amount is:

(a) 360 grams

(b) 355 grams
(c) Compare the results in (a) and (b) of this problem and Problem 9.20.

9.23 Refer to Problem 9.20. If the production manager will stop filling boxes when there is evidence that the average amount of cereal placed in each box is **different** from 368 grams (either less than **or** greater than) and a random sample of 25 boxes is selected with an α risk of 5%, compute the power of the test and the probability of a Type II error (β) if the population average amount placed in the cereal box is:

(a) 360 grams
(b) 355 grams
(c) Compare the results in (a) and (b) of this problem and Problem 9.20.

9.24 A businessman was considering the establishment of a Sunday morning bagel delivery service in a local suburb. Based upon the cost of this service and the profits to be made, he has arrived at the following conclusion: If there is evidence that the average order will be more than 14 bagels per household in this suburban area, then the delivery service will be instituted. If no evidence can be demonstrated, the delivery service will not be instituted. Based on past experience with other suburbs, the standard deviation is estimated to be 3 bagels. A random sample of 36 households is to be surveyed. The businessman is willing to have a 1% risk that the service will be instituted when the average demand is below 14 bagels per household.

(a) Compute the probability of instituting the bagel delivery service when the average demand is 15 bagels per household.

(b) Compute the probability of instituting the bagel delivery service when the average demand is 17 bagels per household.

(c) If the businessman wishes to have a 90% chance of instituting the bagel delivery service when the population average demand is 17 bagels, what sample size should be selected?

9.25 Refer to Problem 9.24. If the businessman is willing to have a 5% risk (rather than a 1% risk) that the service will be instituted when the average demand is below 14 bagels per household, compute the probability of instituting the bagel delivery service when the average demand is:

(a) 15 bagels per household
(b) 17 bagels per household
(c) Compare the results in (a) and (b) of this problem and Problem 9.24.

9.26 Refer to Problem 9.24. If the businessman selected a random sample of 64 households and is willing to have an α risk of 1%, compute the probability of instituting the bagel delivery service when the average demand is:

(a) 15 bagels per household
(b) 17 bagels per household
(c) Compare the results in (a) and (b) of this problem and Problem 9.24.

* 9.27 A large chain of discount toy stores would like to determine whether a certain toy should be marketed. Based upon past experience with similar toys, the

marketing director of the chain has decided that the toy will be marketed only if there is evidence that more than an average of 100 toys per month will be sold at each store. A random sample of 25 stores is selected for a test marketing period of 1 month. Based on past experience, the standard deviation is estimated to be 10 toys. The marketing director is willing to have a 5% risk that the toys will be marketed when the average sale is no more than 100 toys per month. Compute the probability that the toy will be marketed when the population average number of toys sold is:

(a) 105

(b) 108

(c) If the marketing director wishes to have a 98% chance of marketing the toy when the population average demand is 110 per month, how large a sample of stores must be selected?

9.28 Refer to Problem 9.27. If the marketing director is willing to have a 10% risk (rather than a 5% risk) of marketing the toy when the average sale is no more than 100 toys per month, compute the probability that the toy will be marketed when the population average number of toys sold is:

(a) 105

(b) 108

(c) Compare the results in (a) and (b) of this problem and Problem 9.27.

* 9.29 Refer to Problem 9.27. If the marketing director could only select a sample of 16 stores in which to test market the toy, and is willing to have an α risk of 5%, compute the probability of marketing the toy when the population average number of toys sold is:

(a) 105

(b) 108

(c) Compare the results in (a) and (b) of this problem and Problem 9.27.

The following problems refer to the sample data obtained from the questionnaire of Figure 2.2 and presented in Figure 2.6.

9.30 Using a level of significance of .05, test the claim that the true mean grade-point average of seniors is 2.80 (see Figure 2.6, questions 2 and 5). ☐

9.31 Using a level of significance of .10, test the claim that the true mean high-school average attained by freshmen is 85 (see Figure 2.6, questions 2 and 6). ☐

9.32 Using a level of significance of .01, test the claim that the true mean anticipated starting salaries of accounting majors is at least $14,000 (see Figure 2.6, questions 4 and 7). ☐

9.33 Using a level of significance of .05, test the claim that the true proportion of students intending to go to graduate school is .40 (see Figure 2.6, question 3). ☐

9.34 Using a level of significance of .01, test the claim that the true proportion of students favoring stringent retention standards is .50 (see Figure 2.6, question 22). ☐

9.35 Using a level of significance of .10, test the claim that at most one third of all students smoke (see Figure 2.6, question 19).

9.36 Using a level of significance of .05, test the claim that at least 10% of all students favor Sunday shopping legislation (see Figure 2.6, question 10).

9.37 Using a level of significance of .01, test the claim that at most one student in nine has had a diagnosed case of high blood pressure (see Figure 2.6, question 18).

9.38 Using a level of significance of .10, test the claim that at least half the student body are not affiliated with any clubs, groups, teams, or other campus organizations (see Figure 2.6, question 23).

9.39 Use your own sample data (Problem 3.12) and select problems from among Problems 9.30 through 9.38.

9.40 For your own class (Problem 3.13), select problems from among Problems 9.30 through 9.38.

REFERENCES

1. BRADLEY, J. V., *Distribution Free Statistical Tests* (Englewood Cliffs, N.J.: Prentice-Hall, 1968).
2. DANIEL, W., *Introductory Statistics with Applications* (Boston: Houghton Mifflin, 1977).
3. HAMBURG, M., *Basic Statistics: A Modern Approach* (New York: Harcourt Brace Jovanovich, 1974).
4. HUNTSBERGER, D., P. BILLINGSLEY, AND D. J. CROFT, *Statistical Inference for Management and Economics* (Boston: Allyn and Bacon, 1975).
5. NETER, J., W. WASSERMAN, AND G. A. WHITMORE, *Applied Statistics* (Boston: Allyn and Bacon, 1978).
6. PFAFFENBERGER, R. C., AND J. H. PATTERSON, *Statistical Methods for Business and Economics* (Homewood, Ill.: Richard Irwin, 1977).
7. SNEDECOR, G. W., AND W. G. COCHRAN, *Statistical Methods*, 6th ed. (Ames, Iowa: Iowa State University Press, 1967).

CHAPTER 10

HYPOTHESIS TESTING II: ADDITIONAL PROCEDURES

10.1 INTRODUCTION

In the previous chapter we examined hypothesis testing procedures pertaining to whether a mean or a proportion was equal to some specified value. These cases usually are referred to as **one-sample tests,** since a single sample is selected from a population of interest and a computed statistic from the sample is compared to a hypothesized value. In this chapter, we shall extend our discussion of hypothesis testing to consider additional procedures pertaining to quantitative and qualitative data.

We may recall from Section 9.2 that we considered various questions of interest to the production manager of a tire company and the dean of a business school. In this chapter we shall discuss other questions of concern to these people.

Referring to the tire company, suppose that the manager also wanted to draw conclusions about each of the following questions:

1. Is there a difference in the average life of tires produced by the day shift as compared to the evening shift?

2. Is the tread life lower for tires driven at a highway speed of 65 miles per hour than for tires driven at 55 miles per hour?
3. Is there a difference between the day and evening shifts in the proportion of tires that blow out before 10,000 miles?
4. Is there a difference between the day, evening, and night shifts in the proportion of tires that blow out before 10,000 miles?
5. Is there a relationship between the work shift of an employee and his or her experience?
6. Is there evidence that the standard deviation of the life of tires produced on Friday is above 3,500 miles (and thereby out of control)?
7. Is the variability of the life of tires produced on Friday afternoon greater than those produced on Friday morning?

Referring to the survey, suppose the dean and the researcher also wanted to draw conclusions about each of the following:

1. Is there a difference in average grade-point index between accounting and non-accounting majors?
2. Is there a difference between actual and reported grade-point index among accounting majors who are juniors?
3. Is there a difference between accounting and non-accounting majors in the proportion of students who have a B or above average?
4. Is there a difference in the proportion of students who own calculators among the various majors?
5. Is there a relationship between a student's major and the intention to attend graduate school?
6. Does the number of clubs or organizations a student is affiliated with follow a Poisson distribution?
7. Is the population standard deviation of the grade-point index of economics and finance majors equal to 0.40?
8. Is there any difference in the variability in the grade-point index of economics and finance majors and management majors?

Statistical hypothesis testing procedures necessary for answering these questions will be developed in the following sections.

10.2 TESTING FOR THE DIFFERENCE BETWEEN THE MEANS OF TWO POPULATIONS: INDEPENDENT SAMPLES

Let us first extend the hypothesis testing concepts developed in the previous chapter to situations in which we would like to determine whether there is any difference between the means of two independent populations. Suppose then we consider two independent populations, each having a mean and standard deviation (symbolically represented as follows):

Population I	Population II
μ_1, σ_1	μ_2, σ_2

The test to be performed can be either two-tailed or one-tailed, depending on whether we are testing if the two population means are merely *different* or if one mean is *greater than* the other mean.

Two-Tailed Test	One-Tailed Test	One-Tailed Test
H_0: $\mu_1 = \mu_2$ H_1: $\mu_1 \neq \mu_2$	H_0: $\mu_1 \geq \mu_2$ H_1: $\mu_1 < \mu_2$	H_0: $\mu_1 \leq \mu_2$ H_1: $\mu_1 > \mu_2$

where μ_1 = mean of population 1
μ_2 = mean of population 2

The statistic used to determine the difference between the population means is based upon the difference between the sample means $(\bar{X}_1 - \bar{X}_2)$. This statistic, $(\bar{X}_1 - \bar{X}_2)$, because of the central limit theorem, will follow the normal distribution for a large enough sample size. The test statistic is

$$Z = \frac{\bar{X}_1 - \bar{X}_2}{\sqrt{\dfrac{\sigma_1^2}{n_1} + \dfrac{\sigma_2^2}{n_2}}} \qquad (10.1)$$

However, as we mentioned previously, in most cases we do not know the standard deviation of either of the two populations (σ_1, σ_2). The only information usually available are the sample means and sample standard deviations $(\bar{X}_1, \bar{X}_2; S_1, S_2)$. If the assumptions are made that each population is normally distributed and that the **population variances are equal** $(\sigma_1^2 = \sigma_2^2)$, the t distribution with $n_1 + n_2 - 2$ degrees of freedom can be used to test for the difference between the means of the two populations.

If a two-tailed test is to be used to determine whether there is any difference between the means, the null and alternate hypotheses will be (see Figure 10.1)

$$H_0: \quad \mu_1 = \mu_2$$
$$H_1: \quad \mu_1 \neq \mu_2$$

Since we have assumed equal variances in the two populations, the variances of the two samples (S_1^2, S_2^2) can be pooled together or combined to form one estimate (S_p^2) of the population variance. The test statistic will be

$$t_{n_1+n_2-2} = \frac{\bar{X}_1 - \bar{X}_2}{\sqrt{S_p^2\left(\dfrac{1}{n_1} + \dfrac{1}{n_2}\right)}} \qquad (10.2)$$

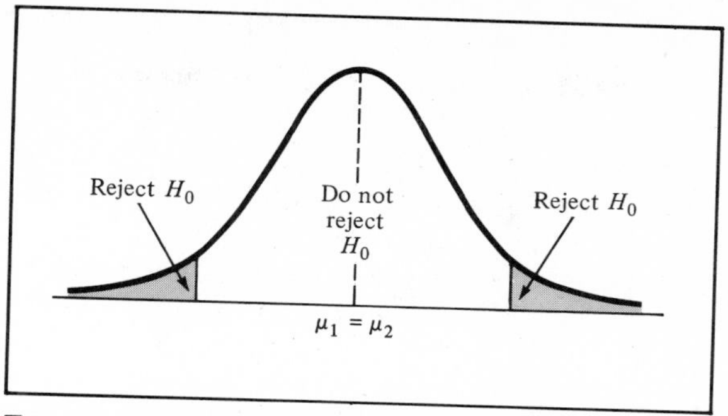

Figure 10.1 Rejection regions for the two-tailed test for the difference between two means.

where

$$S_p^2 = \frac{(n_1 - 1)S_1^2 + (n_2 - 1)\ S_2^2}{n_1 + n_2 - 2}$$

where S_p^2 = pooled variance of the two groups
$\overline{X}_1$ = sample mean in population 1
S_1^2 = sample variance in population 1
n_1 = sample size for population 1
$\overline{X}_2$ = sample mean in population 2
S_2^2 = sample variance in population 2
n_2 = sample size for population 2

In our two studies we may recall that we wanted to know whether there was any difference in the average life of tires produced by the day and evening shifts and whether there was any difference in the grade-point index between accounting and non-accounting majors. In the tire company problem the results can be summarized as follows:

Day Shift	Evening Shift
$\overline{X}_{day}$ = 25,430	$\overline{X}_{eve}$ = 23,310
S_{day} = 4,000	S_{eve} = 3,000
n_{day} = 100	n_{eve} = 100

The null and alternative hypotheses for this example would be

$$H_0: \quad \mu_{day} = \mu_{eve}$$

$$H_1: \quad \mu_{day} \neq \mu_{eve}$$

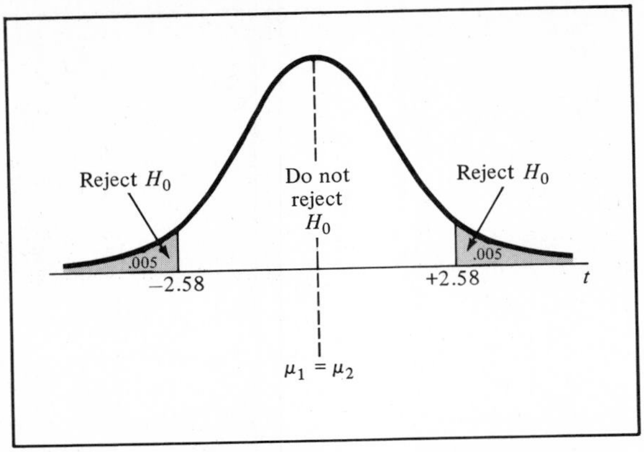

Figure 10.2 Two-tailed test of hypothesis for the difference between the means at the 1% level of significance.

If the test were conducted at the .01 level of significance, the t distribution with $100 + 100 - 2 = 198$ degrees of freedom would be utilized. Since there are more than 150 degrees of freedom, we can use the normal distribution interchangeably with the t distribution. Hence the critical values are $+2.58$ and -2.58 (see Figure 10.2), and the decision rule is

$$\begin{aligned} &\text{Reject } H_0 \text{ if } t_{198} > +2.58 \\ &\text{or if } t_{198} < -2.58 \\ &\text{otherwise do not reject } H_0 \end{aligned}$$

For our data we have

$$t_{198} = \frac{\bar{X}_1 - \bar{X}_2}{\sqrt{S_p^2\left(\frac{1}{n_1} + \frac{1}{n_2}\right)}}$$

where[1]

[1] In this example the two groups had equal sample sizes. For this special case ($n_1 = n_2$) the formula for the pooled variance can be simplified. Thus if $n_1 = n_2$, then

$$S_p^2 = \frac{S_1^2 + S_2^2}{2}$$

for

$$t_{n_1+n_2-2} = \frac{\bar{X}_1 - \bar{X}_2}{\sqrt{S_p^2\left(\frac{1}{n_1} + \frac{1}{n_2}\right)}}$$

$$S_p^2 = \frac{(n_1 - 1)S_1^2 + (n_2 - 1)S_2^2}{n_1 + n_2 - 2}$$

$$= \frac{99(4,000)^2 + 99(3,000)^2}{100 + 100 - 2} = \frac{1,584,000,000 + 891,000,000}{198}$$

$$= \frac{2,475,000,000}{198} = 12,500,000$$

so that

$$t_{198} = \frac{25,430 - 23,310}{\sqrt{12,500,000\left(\frac{1}{100} + \frac{1}{100}\right)}} = \frac{2,120}{\sqrt{250,000}} = \frac{2,120}{500} = 4.24$$

Since we see that $4.24 > 2.58$, we reject H_0.

The null hypothesis has been rejected because the test statistic has fallen into the rejection region. The sample difference between the day and evening shifts is much larger than what could have occurred by chance if the two populations had equal means. Therefore the manager can draw the conclusion that there is a significant difference in the average life of tires produced by the day and evening shifts.

In the second problem we were attempting to determine whether there was any difference in grade-point index between accounting and non-accounting majors. The data can be summarized as below:

Accounting Majors	Non-accounting Majors
$\bar{X}_{acc} = 2.881$	$\bar{X}_{n\text{-}acc} = 2.882$
$S_{acc} = 0.48$	$S_{n\text{-}acc} = 0.456$
$n_{acc} = 35$	$n_{n\text{-}acc} = 59$

The null and alternative hypotheses for this problem would be

$$H_0: \mu_{acc} = \mu_{n\text{-}acc}$$

$$H_1: \mu_{acc} \neq \mu_{n\text{-}acc}$$

If this test was being conducted at the .01 level of significance, the critical values from the t tables (Appendix E, Table E.3) would be $+2.63$ and -2.63 since it is a two-tailed test with $94 - 2 = 92$ degrees of freedom (see Figure 10.3), and the decision rule is

Reject H_0 if $t_{92} > +2.63$
or if $t_{92} < -2.63$
otherwise do not reject H_0

Since the test statistic for determining whether there is a difference between the means is

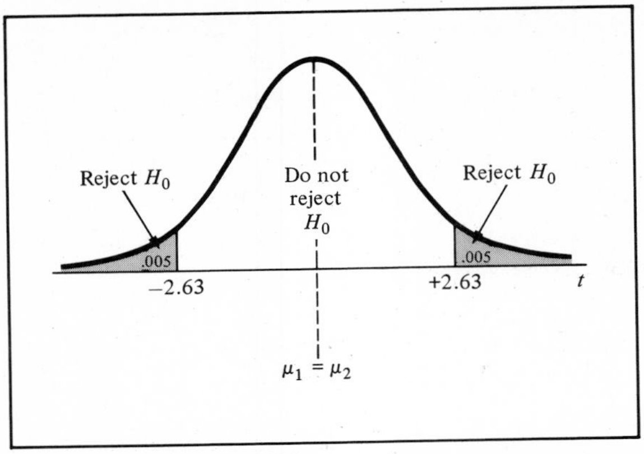

Figure 10.3 Testing a hypothesis for the difference between two means at the 1% level of significance with 92 degrees of freedom.

$$t_{n_1+n_2-2} = \frac{\bar{X}_1 - \bar{X}_2}{\sqrt{S_p^2\left(\frac{1}{n_1} + \frac{1}{n_2}\right)}}$$

where

$$S_p^2 = \frac{(n_1 - 1)S_1^2 + (n_2 - 1)S_2^2}{n_1 + n_2 - 2}$$

we have the following:

$$S_p^2 = \frac{34(.48)^2 + 58(.456)^2}{92} = \frac{7.8506 + 12.0698}{92} = \frac{19.9204}{92} = .2165$$

and

$$t_{92} = \frac{2.881 - 2.882}{\sqrt{.2165\left(\frac{1}{35} + \frac{1}{59}\right)}} = \frac{-.001}{\sqrt{.0098}} = \frac{-.001}{.099} = -.01$$

Since $t_{92} = -.01$, we see that $-2.63 < -.01 < +2.63$; therefore, do not reject H_0.

Since the test statistic clearly has not fallen into the rejection region, the null hypothesis cannot be rejected. The conclusion can be reached that there is no evidence of a difference in the grade-point index between accounting and non-accounting majors.

In testing for the difference between the means, we have assumed a normal distribution and equality of variances of the two populations. We must examine the consequences on the t test of departures from each of these assumptions. With respect to the assumption of normality, the t test is a "robust" test in that it is not sensitive to modest departures from normality. As long as the sample sizes are not extremely small, the assumption of normality can be violated without serious effect on the power of the test. The second assumption, equality of variances (see Section 10.10), creates what is called in statistics the **Behrens-Fisher problem** when it is violated. When σ_1^2 is significantly different from σ_2^2, an approximation to the t distribution can be utilized (Reference 9).

10.3 THE PAIRED DIFFERENCE t TEST FOR RELATED SAMPLES

10.3.1 Rationale

In Section 10.2 we discussed the test for the difference between the means of two independent populations. In this section, we shall develop a procedure for analyzing the difference between the means of two groups when the data are obtained from samples that are **related;** that is, the results of the first group are not independent of the second group. This dependent characteristic of the two groups occurs either because the items or individuals are **paired** or **matched** according to some characteristic, or because **repeated measurements** are obtained from the same set of items or individuals. In either case, the variable of interest becomes the *difference* between the values of the observations rather than the observations themselves.

The first approach to the related samples problem involves matching of items or individuals according to some characteristic of interest. For example, if the tire company wanted to study the effect of different driving conditions on the tread life of tires, a control for differences in the quality of the types of tires involved should be established. In this situation two tires of each type can be tested, with one tire assigned to condition 1 and the other tire assigned to condition 2.

The second approach to the related samples problem involves taking repeated measurements on the same items or individuals. Under the theory that the same items or individuals will behave alike if treated alike, the objective of the analysis is to show that any differences between two measurements of the same items or individuals are due to different treatment conditions. For example, in our survey, the dean and the researcher wished to study the accuracy of the responses to question 5, grade-point index. For those students included in their sample, the actual grade-point index could have been obtained from college records filed in the registrar's office. If the dean wanted to determine whether there was any difference between the reported and the actual grade-point index, it would be appropriate to obtain *both* these measurements from the same students rather than taking two independent samples of students as discussed in Section 10.2. Such an approach would reduce the variability among the grade-point indexes of the students themselves and thereby focus upon differences between the reported and actual grade-point index of each student. Therefore, regardless of whether matched (paired) samples or repeated measurements are utilized, the objective is to study the difference between two measurements by reducing the effect of the variability due to the items or individuals themselves.

10.3.2 Development

In order to determine whether any difference exists between two related samples, the individual values for each group must be obtained (as shown in Table 10.1).

Table 10.1
DETERMINING THE DIFFERENCE BETWEEN TWO RELATED GROUPS

	Group		Difference
Observation	1	2	(D)
1	X_{11}	X_{21}	$D_1 = X_{11} - X_{21}$
2	X_{12}	X_{22}	$D_2 = X_{12} - X_{22}$
3			
⋮	⋮	⋮	⋮
i	X_{1i}	X_{2i}	$D_i = X_{1i} - X_{2i}$
⋮	⋮	⋮	⋮
n	X_{1n}	X_{2n}	$D_n = X_{1n} - X_{2n}$

where X_{1i} = ith value in group 1
X_{2i} = ith value in group 2
$D_i = X_{1i} - X_{2i}$ = difference between the ith value in group 1 and the ith value in group 2

From the central limit theorem, the average difference $\bar{D}$ follows a normal distribution when the population standard deviation of the difference σ_D is known and the sample size is large enough. Since only the sample standard deviation of the difference S_D is usually available, the test statistic is

$$t_{n-1} = \frac{\bar{D}}{\dfrac{S_D}{\sqrt{n}}} \qquad (10.3)$$

with $(n - 1)$ degrees of freedom, where

$$\bar{D} = \frac{\sum_{i=1}^{n} D_i}{n}$$

$$S_D = \sqrt{\frac{n\sum_{i=1}^{n} D_i^2 - \left(\sum_{i=1}^{n} D_i\right)^2}{n(n-1)}}$$

n = sample size

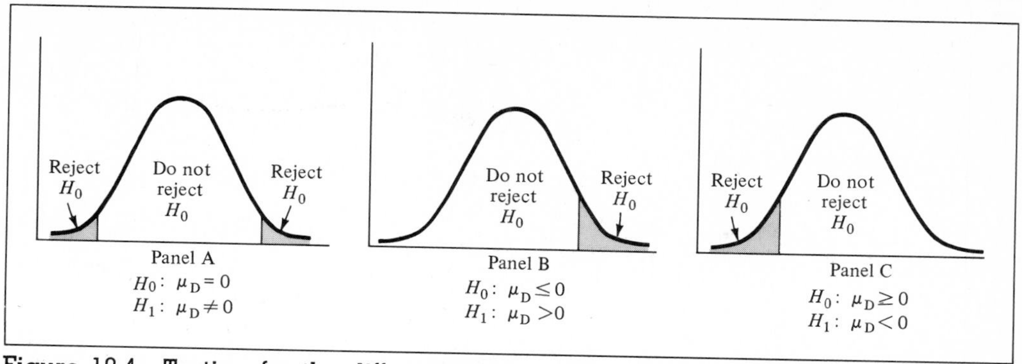

Figure 10.4 Testing for the difference between the means in related samples. Panel A, two-tailed test. Panel B, one-tailed test. Panel C, one-tailed test.

The three panels of Figure 10.4 indicate the null and alternative hypothesis and rejection regions for the possible one-tailed and two-tailed tests. If, as shown in panel A, the test of the hypothesis is two-tailed, the rejection region is split into the lower and upper tails of the *t* distribution. However, if the test is one-tailed, the rejection region is either in the upper tail (panel B of Figure 10.4) or in the lower tail (panel C of Figure 10.4) of the *t* distribution, depending on the direction of the alternative hypothesis.

10.3.3 Applications

To apply the test for the difference between the means of two related samples, let us refer back to the example mentioned in Section 10.3.1. The tire company wished to determine whether the tread life was lower for tires driven at a highway speed of 65 miles per hour than for tires driven at a highway speed of 55 miles per hour. In order to reduce the influence of tire variability, a pair of tires from each of eight different quality types was selected. One tire of each type was driven at 65 miles per hour and the other was driven at 55 miles per hour. The results for the eight types are shown in Table 10.2.

Table 10.2
TIRE LIFE (THOUSANDS OF MILES) FOR A RANDOM SAMPLE OF EIGHT TYPES OF TIRES UNDER TWO DRIVING CONDITIONS

	Driving Condition		
Tire Type	*I* 65 mph	*II* 55 mph	$(X_{1i} - X_{2i})$ Difference (D_i)
a	24.31	26.42	−2.11
b	31.27	33.77	−2.50
c	30.71	35.42	−4.71
d	28.64	30.32	−1.68
e	23.60	22.85	+0.75
f	36.41	42.71	−6.30
g	21.46	25.09	−3.63
h	30.62	31.76	−1.14

For these data,

$$\sum_{i=1}^{n} D_i = -21.32 \qquad \sum_{i=1}^{n} D_i^2 = 90.4376 \qquad n = 8$$

thus

$$\bar{D} = \frac{\sum_{i=1}^{n} D_i}{n} = \frac{-21.32}{8} = -2.665$$

and

$$S_D^2 = \frac{n\sum_{i=1}^{n} D_i^2 - \left(\sum_{i=1}^{n} D_i\right)^2}{n(n-1)} = \frac{8(90.4376) - (-21.32)^2}{8(7)} = \frac{268.9584}{56}$$

$$= 4.803$$

so that

$$S_D = 2.192$$

Since the tire company wishes to determine whether the tire life will be lower at a highway speed of 65 miles per hour compared to 55 miles per hour, we have a one-tailed test in which the null and alternative hypotheses can be stated as follows:

$$H_0: \quad \mu_D \geq 0 \qquad \text{(that is, } \mu_{65} \geq \mu_{55}\text{)}$$

$$H_1: \quad \mu_D < 0 \qquad \text{(that is, } \mu_{65} < \mu_{55}\text{)}$$

Since samples of eight tire types have been taken, if a level of significance of 1% is selected, the decision rule can be stated as (see Figure 10.5)

Reject H_0 if $t_7 < -2.998$; otherwise do not reject H_0

From Equation (10.3) we recall that the test statistic is

$$t_{n-1} = \frac{\bar{D}}{\frac{S_D}{\sqrt{n}}}$$

and thus for this example we have

$$t_7 = \frac{-2.665}{\frac{2.192}{\sqrt{8}}} = -3.439$$

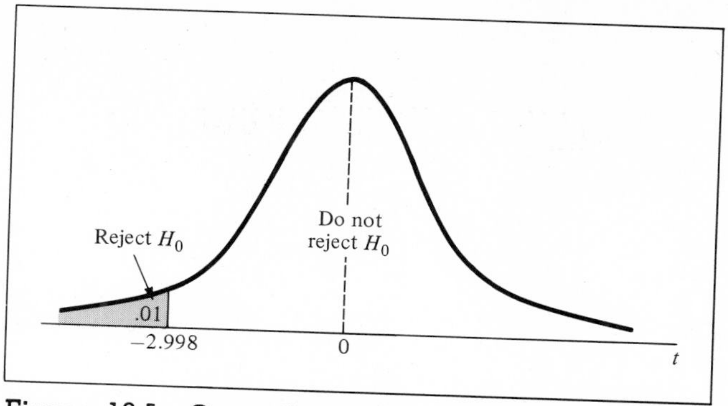

Figure 10.5 One-tailed test for the paired difference at the 1% level of significance with 7 degrees of freedom.

Since $-3.439 < -2.998$, we reject H_0.

Since the null hypothesis has been rejected, we would conclude that the tire life is lower for cars driven at a highway speed of 65 miles per hour than it is for 55 miles per hour.

A second application of the paired difference test can be illustrated by referring to the survey of the dean. In Section 10.3.1 we may recall that the dean was interested in determining the accuracy of the responses to question 5, grade-point index. In order to accomplish this objective, the actual grade-point index of a portion of the sample, juniors who are accounting majors, was obtained from the college records and compared with those reported. The results are shown in Table 10.3.

For these data,

$$\sum_{i=1}^{n} D_i = +.38 \qquad \sum_{i=1}^{n} D_i^2 = .074 \qquad n = 11$$

thus

$$\bar{D} = \frac{\sum_{i=1}^{n} D_i}{n} = \frac{+.38}{11} = +.0345$$

and

$$S_D^2 = \frac{n\sum_{i=1}^{n} D_i^2 - \left(\sum_{i=1}^{n} D_i\right)^2}{n(n-1)} = \frac{11(.074) - (.38)^2}{11(10)} = \frac{.6696}{110} = \; = .0061$$

Table 10.3
REPORTED GRADE-POINT INDEX AND ACTUAL GRADE-POINT INDEX OF A SAMPLE OF 11 JUNIORS WHO ARE ACCOUNTING MAJORS

Student	Reported Grade-Point Index (X_{1i})	Actual Grade-Point Index (X_{2i})	Difference (D_i) $D_i = X_{1i} - X_{2i}$
1	2.80	2.69	+.11
2	2.00	1.96	+.04
3	2.92	2.80	+.12
4	2.62	2.48	+.14
5	2.76	2.83	−.07
6	2.91	2.92	−.01
7	3.75	3.75	0
8	2.55	2.65	−.10
9	2.73	2.70	+.03
10	3.12	3.10	+.02
11	3.00	2.90	+.10

and therefore,

$$S_D = .078$$

Since the dean is interested in determining whether there is a difference in the reported and actual grade-point indexes, we have a two-tailed test in which the null and alternative hypothesis can be stated as follows:

H_0: $\mu_D = 0$ ($\mu_1 = \mu_2$ there is no difference in reported and actual grade-point index)

H_1: $\mu_D \neq 0$ ($\mu_1 \neq \mu_2$ there is a difference in reported and actual grade-point index)

Since a sample of 11 juniors who are accounting majors was involved, if a level of significance of 5% is selected, the decision rule can be stated as (see Figure 10.6):

Reject H_0 if $t_{10} > +2.228$
or $t_{10} < -2.228$
otherwise do not reject H_0

From Equation (10.3) the test statistic is

$$t_{n-1} = \frac{\bar{D}}{\frac{S_D}{\sqrt{n}}}$$

Therefore, for the dean's example we have

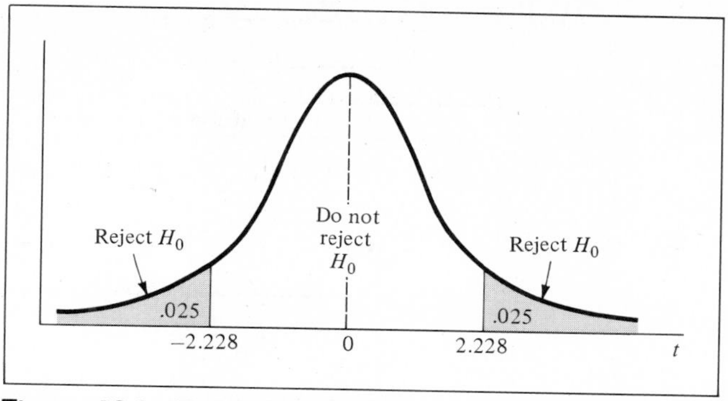

Figure 10.6 Two-tailed test for the paired difference at the 5% level of significance with 10 degrees of freedom.

$$t_{10} = \frac{+\,.0345}{\dfrac{.078}{\sqrt{11}}} = +\,1.467$$

Since we can see that $-\,2.228 < +\,1.467 < +\,2.228$, do not reject H_0.

Since the null hypothesis cannot be rejected, the dean should conclude that, for the types of students studied, there is no evidence of a difference between the actual and reported grade-point index.

10.4 TESTING FOR THE DIFFERENCE BETWEEN TWO PROPORTIONS USING THE NORMAL APPROXIMATION

Rather than being concerned with the differences between two populations in terms of a quantitative variable, we could be interested in differences in some qualitative characteristic. A test for the difference between two proportions based upon independent samples[2] can be performed using two different methods, but the results will be equivalent.

The first method of testing for the difference between two proportions involves the use of the normal distribution. This test is based on the difference between the two sample proportions which may be approximated by a normal distribution for large sample sizes.

For the two populations involved, we are interested in either determining whether there is any *difference* in the proportion of successes in the two groups (two-tailed test) or whether one group had a *higher* proportion of successes than the other group (one-tailed test).

[2] A test based upon the difference in the proportions of two related samples has been developed by McNemar (see Reference 1).

Two-Tailed Test	One-Tailed Test	One-Tailed Test
H_0: $p_1 = p_2$	H_0: $p_1 \geq p_2$	H_0: $p_1 \leq p_2$
H_1: $p_1 \neq p_2$	H_1: $p_1 < p_2$	H_1: $p_1 > p_2$

where p_1 = proportion of successes in population 1
p_2 = proportion of successes in population 2

The test statistic would be

$$Z \cong \frac{p_{s_1} - p_{s_2}}{\sqrt{\bar{p}(1 - \bar{p})\left(\frac{1}{n_1} + \frac{1}{n_2}\right)}} \tag{10.4}$$

with

$$\bar{p} = \frac{X_1 + X_2}{n_1 + n_2}$$

$$p_{s_1} = \frac{X_1}{n_1} \qquad p_{s_2} = \frac{X_2}{n_2}$$

where p_{s_1} = sample proportion in population 1
p_{s_2} = sample proportion in population 2
X_1 = number of successes in sample 1
X_2 = number of successes in sample 2
n_1 = sample size for population 1
n_2 = sample size for population 2
$\bar{p}$ = pooled estimate of the population proportion

The estimate for the population proportion that we shall use is based upon the null hypothesis. Under the null hypothesis it is assumed that the two population proportions are equal. Therefore we may obtain an overall estimate of the population proportion by pooling together the two sample proportions. The estimate $\bar{p}$ is simply the number of successes in the two samples combined $(X_1 + X_2)$ divided by the total sample size $(n_1 + n_2)$.

The test of the difference between two proportions can be applied in our first example to determine whether there is any difference between the day and evening shifts in the proportion of tires that blow out before 10,000 miles and in our second example to determine if there is any difference in the proportion of accounting and non-accounting majors who have B or above averages. The data for the first example, the tire company study, is summarized as follows:

Day Shift	Evening Shift
$n_{\text{day}} = 100$ 5 tires blew out before 10,000 miles	$n_{\text{eve}} = 100$ 10 tires blew out before 10,000 miles

The null and alternative hypotheses for this problem are

$$H_0: \quad p_{\text{day}} = p_{\text{eve}}$$
$$H_1: \quad p_{\text{day}} \neq p_{\text{eve}}$$

If the test was to be carried out at the .10 level of significance, the critical values would be -1.645 and $+1.645$ (see Figure 10.7), and our decision rule would be

Reject H_0 if $Z > +1.645$
or if $Z < -1.645$
otherwise do not reject H_0

For our data,

$$p_{s\,\text{day}} = \frac{X_{\text{day}}}{n_{\text{day}}} = \frac{5}{100} = .05 \qquad p_{s\,\text{eve}} = \frac{X_{\text{eve}}}{n_{\text{eve}}} = \frac{10}{100} = .10$$

Since we are testing for the difference between two proportions, from Equation (10.4) the test statistic would be

$$\bar{p} = \frac{5 + 10}{100 + 100} = 0.075$$

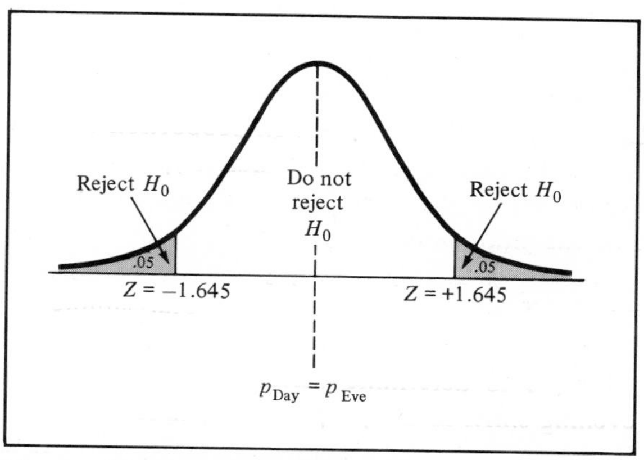

Figure 10.7 Testing a hypothesis about the difference between two proportions at the 10% level of significance.

and

$$Z \cong \frac{p_{s_1} - p_{s_2}}{\sqrt{\bar{p}(1-\bar{p})\left(\frac{1}{n_1}+\frac{1}{n_2}\right)}}$$

$$= \frac{.05 - .10}{\sqrt{(.075)(.925)\left(\frac{1}{100}+\frac{1}{100}\right)}} = \frac{-.05}{\sqrt{.069375\left(\frac{2}{100}\right)}} = \frac{-.05}{\sqrt{.0013875}}$$

$$= \frac{-.05}{.03724} = -1.34$$

Since $Z = -1.34$, we see that $-1.645 < -1.34 < +1.645$ and the decision would be to not reject H_0. The conclusion reached is that there is no evidence of any difference between the day and evening shifts in the proportion of tires that blow out before 10,000 miles.

In the above problem, the rejection region has been divided up into both tails of the distribution. Of course, if the alternative hypothesis had been a specific direction (such as $p_1 > p_2$), then the rejection region would be entirely contained in one tail of the normal distribution.

10.5 THE CHI-SQUARE TEST FOR THE DIFFERENCE BETWEEN TWO PROPORTIONS

For the problem of the previous section, we may also view the data in terms of the frequency of blowouts for each of the two shifts. A two-way table, called a **contingency table** (see Sections 3.2.2 and 5.3.2) can be developed to indicate the frequency of successes and failures for each group (Table 10.4).

The "cells" in the table indicate the frequency for each possible combination of outcome (tire blowout–tire did not blow out) and shift of workers (day–evening). For example, the frequency of 95 indicated that 95 tires produced by the day shift did *not* blow out before 10,000 miles. This value was arrived at by subtracting the five tires that did blow out from the 100 tires produced by the day shift. The totals in the margins of the table result from the sum of the values in the appropriate row or column of the table.

Table 10.4
CONTINGENCY TABLE FOR DAY–EVENING SHIFT PROBLEM

	Day	Evening	Totals
Number of tires that blew out before 10,000 miles	5	10	15
Number of tires that did not blow out before 10,000 miles	95	90	185
Totals	100	100	200

10.5.1 Test for the Difference Between Two Proportions

This contingency table is the basis for testing the difference between the two proportions using a second method of analysis, the **chi-square test.** The method begins from the null hypothesis that the proportion of successes is the same in the two populations ($p_1 = p_2$). If this null hypothesis were true, the frequencies that theoretically should be found in each cell could be computed. For example, we can determine how many blowouts before 10,000 miles can theoretically be expected from tires produced by the day shift and by the evening shift. A total of 15 out of 200 tires in the combined samples (7.5%) are blowouts. If there was no difference between the day and evening shifts, then theoretically 7.5% of each sample should blow out. This would mean a theoretical frequency of $.075 \times 100 = 7.5$ blowouts for the day shift and $.075 \times 100 = 7.5$ blowouts for the evening shift. The number of tires that theoretically should not blow out can be found by subtracting the theoretical number of blowouts from the sample size in each shift. The actual (observed) frequencies and theoretical frequencies (circled numbers) are shown in Table 10.5.

The theoretical frequencies f_t for each cell can also be computed by using a simple formula:

$$f_t = \frac{n_R n_C}{n} \tag{10.5}$$

where n_R = total number in row

n_C = total number in column

n = total sample size

Therefore the theoretical frequency of 7.5 for day shift blowouts can be arrived at by taking the total number in the appropriate row (15), multiplying by the total number in the appropriate column (100), and dividing by the total sample size (200). The computation of the theoretical frequencies for each cell is illustrated on the next page:

Table 10.5
OBSERVED AND THEORETICAL FREQUENCIES FOR THE TIRE BLOWOUT PROBLEM

Tires	Day		Evening		Totals
Blowouts	5	(7.5)	10	(7.5)	15
Not blowouts	95	(92.5)	90	(92.5)	185
Totals	100		100		200

$$f_t = \frac{15(100)}{200} = 7.5 \qquad f_t = \frac{15(100)}{200} = 7.5$$

$$f_t = \frac{185(100)}{200} = 92.5 \qquad f_t = \frac{185(100)}{200} = 92.5$$

The chi-square method of analysis depends on the squared difference between the observed frequency f_o and the theoretical frequency f_t in each cell. If there is no "real" difference in the proportion of successes in the two groups, then the squared difference between the observed and theoretical frequencies should be small. If the proportions for the two groups are significantly different, then the squared difference between the observed and theoretical frequencies should be large. The test statistic is $(f_o - f_t)^2/f_t$ summed up over all cells of the table. This statistic approximately follows a chi-square (χ^2) distribution with the degrees of freedom equal to the number of rows in the contingency table minus 1 $(R - 1)$ times the number of columns in the contingency table minus 1 $(c - 1)$:

$$\chi^2_{(R-1)(c-1)} \cong \sum_{\substack{\text{all} \\ \text{cells}}} \frac{(f_o - f_t)^2}{f_t} \tag{10.6}$$

where f_o = observed frequency in each cell
f_t = theoretical frequency in each cell
R = number of rows in the contingency table
c = number of columns in the contingency table

The null hypothesis of equality between the two proportions ($p_1 = p_2$) will be rejected *only* when the computed value of the test statistic is greater than the critical value of the chi-square distribution with the appropriate number of degrees of freedom (see Figure 10.8).

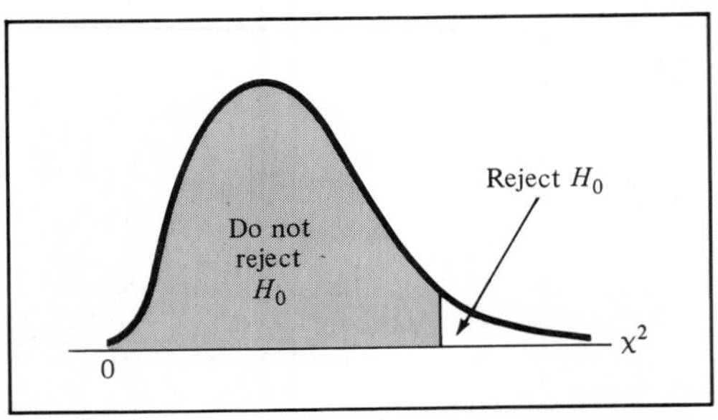

Figure 10.8 Testing a hypothesis for the difference between two proportions using the chi-square test.

Table 10.6
OBTAINING THE CRITICAL VALUE FROM THE CHI-SQUARE DISTRIBUTION FOR $\alpha = .05$ AND 1 DEGREE OF FREEDOM

Degrees of Freedom	Lower Tail Areas (Percentiles)									
	.005	.01	.025	.05	.10	.25	.75	.90	.95	.975
1			0.001	0.004	0.016	0.102	1.323	2.706 →	3.841	5.024
2	0.010	0.020	0.051	0.103	0.211	0.575	2.773	4.605	5.991	7.378
3	0.072	0.115	0.216	0.352	0.584	1.213	4.108	6.251	7.815	9.348
4	0.207	0.297	0.484	0.711	1.064	1.923	5.385	7.779	9.488	11.143
5	0.412	0.554	0.831	1.145	1.610	2.675	6.626	9.236	11.071	12.833

SOURCE: Extracted from Appendix E, Table E.4.

The chi-square distribution is a skewed distribution whose shape depends solely on the number of degrees of freedom. As the number of degrees of freedom increases, the chi-square distribution becomes more symmetrical. Appendix E, Table E.4 contains various percentile points of the chi-square distribution for different degrees of freedom (see Table 10.6).

The value at the top of each column indicates the area (percentile) in the lower portion (or left side) of the chi-square distribution. For example, with 1 degree of freedom the value of the 95th percentile point of the distribution is 3.841 (see Figure 10.9).

This means that for 1 degree of freedom the probability of exceeding a chi-square value of 3.841 is .05. Therefore once we determine the level of significance and the degrees of freedom, the critical value of chi-square can be found.

Let us return to the computation of the chi-square value for the tire problem. Here there are two rows and two columns in the table. Therefore there will be $(2 - 1) \times (2 - 1) = 1$ degree of freedom. The test statistic is:

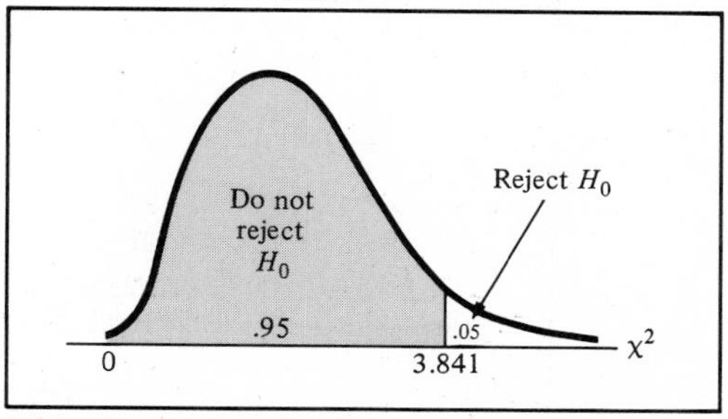

Figure 10.9 Finding the 95th percentile of the chi-square distribution with 1 degree of freedom.

Table 10.7
COMPUTATION OF CHI-SQUARE TEST STATISTIC FOR TIRE BLOWOUT PROBLEM

f_o	f_t	$(f_o - f_t)$	$(f_o - f_t)^2$	$(f_o - f_t)^2/f_t$
5	7.5	−2.5	6.25	.8333
95	92.5	+2.5	6.25	.0675
10	7.5	+2.5	6.25	.8333
90	92.5	−2.5	6.25	.0675
				1.8016

$$\chi_1^2 \cong \sum_{\substack{\text{all}\\\text{cells}}} \frac{(f_o - f_t)^2}{f_t} \tag{10.6}$$

The data are given in Table 10.7.

If the test was being conducted at the .10 level of significance, the critical value of chi-square would be 2.706 since the .10 in the upper tail means that .90 would be below (to the left of) the critical value. Looking in Appendix E, Table E.4, the value of chi-square for the 90th percentile when there is 1 degree of freedom is 2.706. In this problem, $\chi_1^2 = 1.8016 < 2.706$ (see Figure 10.10).

Therefore the null hypothesis would not be rejected. The conclusion would be reached that there is no evidence of a difference between the day and evening shifts in the proportion of tires that blow out before 10,000 miles.

The chi-square test assumes that there are at least five theoretical frequencies in each cell of the contingency table. This assumption is important primarily for the 2 × 2 contingency table which has only 1 degree of freedom. Other procedures, such as Fisher's exact test (Reference 1), can be utilized if this assumption is not met.

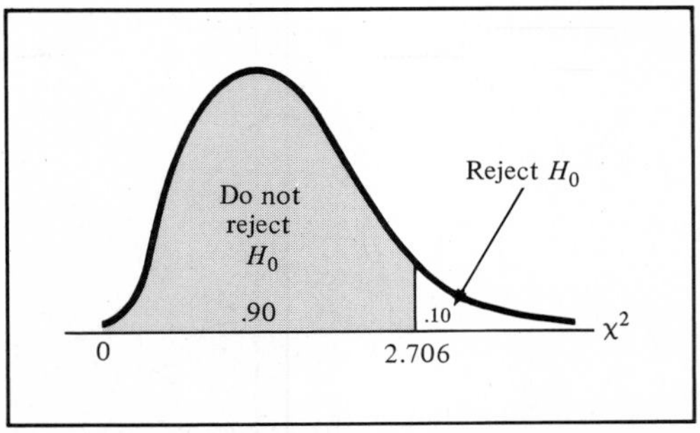

Figure 10.10 Finding the critical value of chi-square at the 10% level of significance with 1 degree of freedom.

10.5.2 Test for Independence

The null and alternative hypotheses of the chi-square test can be stated in another form. The contingency table is a two-way table that is cross classifying two qualitative variables (shifts and blowouts). The statement of no difference in the proportion of blowouts between the day and evening shifts can be viewed in an alternative manner. The statement also means that the two variables are independent (that is, there is no relationship between the proportion of blowouts and the shift in which the tire is produced). When the conclusions are stated in this manner, the chi-square test is called a **test of independence.** The null and alternative hypotheses would be stated as follows:

H_0: There is independence (no relationship) between shift and proportion of blowouts

H_1: There is a relationship (dependence) between shift and proportion of blowouts

If the null hypothesis cannot be rejected, the conclusion could be drawn that there was no evidence of a relationship between the two variables. If the null hypothesis is rejected, the conclusion would be drawn that there is evidence of a relationship between the two variables. Regardless of whether the chi-square test is considered a test for the difference between two proportions or a test of independence, the computations and results are exactly the same.

10.5.3 Testing for Equality of Proportions by Z and by χ^2: A Comparison of Results

The methods used in our tire study can also be applied to the survey data obtained by the researcher and the dean to determine whether any difference exists between accounting and non-accounting majors in the proportion of students who have averages of B or above. The relevant information is summarized below:

Accounting	Non-accounting
$n_{acc} = 35$ 15 students have B or above averages	$n_{n\text{-}acc} = 59$ 22 students have B or above averages

The null and alternative hypotheses for this problem would be

$$H_0: \quad p_{acc} = p_{n\text{-}acc}$$
$$H_1: \quad p_{acc} \neq p_{n\text{-}acc}$$

If this test were to be performed using the normal distribution and a level of significance of .05 was selected, the critical values would be $+1.96$ and -1.96 (see Figure 10.11), and our decision rule would be

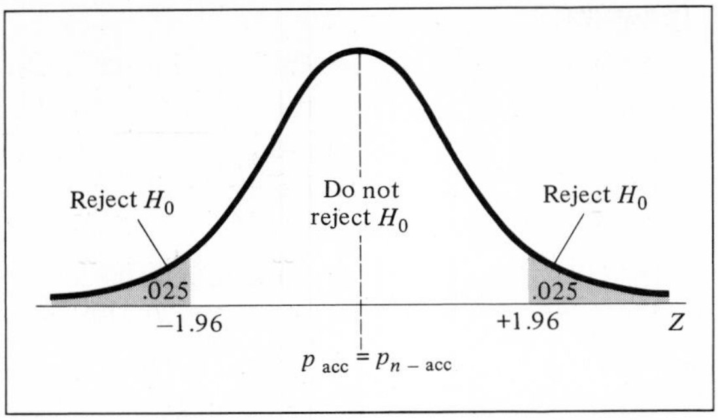

Figure 10.11 Testing for the difference between two proportions at the 5% level of significance using the normal approximation.

$$\begin{aligned}\text{Reject } H_0 \text{ if } \quad & Z > +1.96 \\ \text{or if } \quad & Z < -1.96\end{aligned}$$
otherwise do not reject

Therefore, for our data we would have

$$p_{s_{\text{acc}}} = \frac{X_{\text{acc}}}{n_{\text{acc}}} = \frac{15}{35} = .4286 \qquad p_{s\,\text{n-acc}} = \frac{X_{\text{n-acc}}}{n_{\text{n-acc}}} = \frac{22}{59} = .3729$$

Since we are testing for the difference between two proportions, from Equation (10.4) the test statistic would be

$$Z \cong \frac{p_{s_1} - p_{s_2}}{\sqrt{\bar{p}\,(1 - \bar{p})\left(\frac{1}{n_1} + \frac{1}{n_2}\right)}}$$

where

$$\bar{p} = \frac{15 + 22}{35 + 59} = \frac{37}{94} = .3936$$

so that

$$Z \cong \frac{.4286 - .3729}{\sqrt{(.3936)\ (.6064)\left(\frac{1}{35} + \frac{1}{59}\right)}} = \frac{.0554}{\sqrt{.01086}} = \frac{.0554}{.1042} = +0.532$$

Since $Z = +0.532$, we see that $-1.96 < +.532 < +1.96$; therefore, do not reject H_0.

The conclusion reached is that there is no evidence of any difference be-

Table 10.8
EVALUATION OF STUDENT PERFORMANCE

Grade-point Index	Major: Accounting	Major: Non-accounting	Totals
Below B average	20	37	57
B average or above	15	22	37
Totals	35	59	94

SOURCE: Data are taken from Table 5.2.

tween accounting and non-accounting majors in the proportion of students who have B or above averages.

This problem can be solved (see Table 10.8) in an alternative way by using the chi-square test based on the contingency table developed as Table 5.2.

The theoretical frequencies can be obtained from Equation (10.5)

$$f_t = \frac{n_R n_C}{n}$$

so that

$$f_t = \frac{57\ (35)}{94} = 21.2234 \qquad f_t = \frac{57\ (59)}{94} = 35.7766$$

$$f_t = \frac{37\ (35)}{94} = 13.7766 \qquad f_t = \frac{37\ (59)}{94} = 23.2234$$

Since there are two rows and two columns in this table, there will be $(2 - 1)(2 - 1) = 1$ degree of freedom. Using a level of significance of .05 results in a critical value of 3.841 (see Figure 10.12).

The chi-square test statistic, equal to

$$\sum_{\substack{\text{all}\\ \text{cells}}} \frac{(f_o - f_t)^2}{f_t}$$

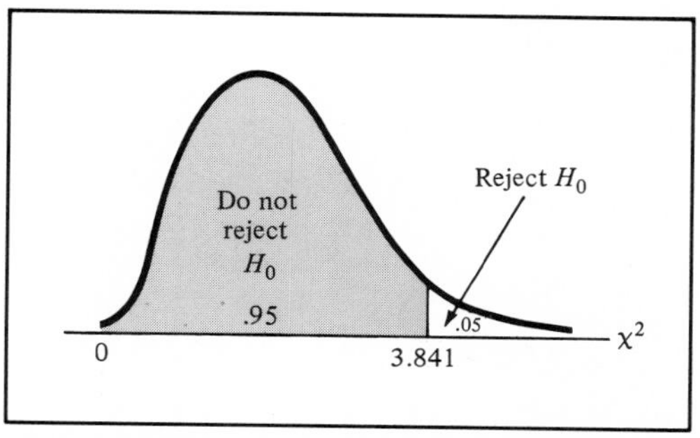

Figure 10.12 Testing for the difference between two proportions at the 5% level of significance using the chi-square test.

Table 10.9
COMPUTATION OF THE CHI-SQUARE TEST STATISTIC FOR DATA FROM TABLE 10.8

f_o	f_t	$(f_o - f_t)$	$(f_o - f_t)^2$	$(f_o - f_t)^2/f_t$
20	21.2234	−1.2234	1.4967	.07052
15	13.7766	+1.2234	1.4967	.10864
37	35.7766	+1.2234	1.4967	.04183
22	23.2234	−1.2234	1.4967	.06445
				.28544

would be computed as shown in Table 10.9. Since $\chi_1^2 = .28544 < 3.841$, do not reject H_0.

Again, the null hypothesis could not be rejected and the dean would conclude that there is no evidence of a difference between accounting and nonaccounting majors in the proportion of students having B or above averages. Therefore the dean would conclude that there is no evidence of a relationship between being an accounting major and having a B average or above.

We have seen in these problems that both the normal distribution and chi-square tests have not rejected the null hypothesis. The two methods are equivalent for testing the same hypothesis. This can be explained by the interrelationship between the value for the normal distribution and that of the chi-square distribution with 1 degree of freedom. The chi-square statistic will be the square of the test statistic based upon the normal distribution. For instance, with the problem concerning the researcher and the dean the computed Z value was 0.532 while the computed chi-square value was $(0.532)^2 \cong 0.28544$. Also, if we compare the critical values of the two distributions, we can see that at the .05 level of significance the chi-square value of 3.841 is the square of the normal value of 1.96 (that is, $\chi_1^2 = Z^2$).

Therefore it is clear that in testing for the *difference* between two proportions, the normal distribution and the chi-square test provide equivalent methods. However, if we are interested in determining a *directional difference,* such as $p_1 > p_2$, then the normal distribution test must be used with the entire rejection region located in one tail of the distribution.

In these examples we have seen that the chi-square test has 1 degree of freedom since there are two rows and two columns. The 1 degree of freedom can be derived from the nature of the 2 × 2 contingency table. The theoretical frequencies are determined from the totals in the margins of the table. Once we have determined the theoretical frequency for one of the cells, the theoretical frequencies for the other three cells will be fixed. In the tire study, for example, once the theoretical frequency of 7.5 for blowouts in the day shift is found, we know that the theoretical number of nonblowouts for the day shift must be 92.5, since the total must add to 100. We also know that the theoretical number of blowouts in the evening shift must be 7.5 since the combined total of blowouts must add to 15. Finally, for the evening shift, if the theoretical number of blowouts is 7.5, then the number of nonblowouts must be 92.5, since the total must add to 100. Therefore since all other theoretical frequencies will be known once the theoretical frequency of one cell is known, the contingency table is said to have 1 degree of freedom.

10.6 TESTING FOR THE DIFFERENCE BETWEEN THE PROPORTIONS OF c POPULATIONS

In the previous sections we have studied differences between two populations, first for quantitative variables (means) and then for qualitative variables (proportions). The chi-square method of testing for the difference between two proportions can be extended to the general case in which there are c independent populations to be compared. The null and alternative hypotheses would be

$$H_0: \quad p_1 = p_2 = p_3 = \cdots = p_c$$

H_1: At least one proportion is different from the others

The contingency table would have 2 rows and c columns, so that there would be $c - 1$ degrees of freedom in the chi-square test. Referring to the tire example, let us assume that the tire company added a third shift of workers, a night shift, and that a random sample of 50 tires from this shift indicated that 10 tires blew out before 10,000 miles. The hypothesis to be tested concerns the question of whether there is any difference in the proportion of blowouts among the tires produced by the three shifts. The null and alternative hypotheses would be

$$H_0: \quad p_{\text{day}} = p_{\text{eve}} = p_{\text{night}}$$

H_1: At least one shift has a different proportion of blowouts from the others

The data are given in Table 10.10.

The theoretical frequencies in each cell can be computed by using the

Table 10.10
TESTING $p_1 = p_2 = \cdots = p_c$

	Shift			
	Day	*Evening*	*Night*	Totals
Number of tires that blow out before 10,000 miles	5	10	10	25
Number of tires that do not blow out before 10,000 miles	95	90	40	225
Totals	100	100	50	250

Table 10.11
COMPUTATION OF THE CHI-SQUARE TEST STATISTIC FOR A 2 × 3 CONTINGENCY TABLE

f_o	f_t	$(f_o - f_t)$	$(f_o - f_t)^2$	$(f_o - f_t)^2/f_t$
5	10	−5	25	2.5
90	90	0	0	0
95	90	+5	25	.2777
10	5	+5	25	5.0
10	10	0	0	0
40	45	−5	25	.5555
				8.3332

formula $f_t = n_R n_c/n$. Therefore, the theoretical frequencies for the six cells will be

$$f_t = \frac{25(100)}{250} = 10 \qquad f_t = \frac{25(100)}{250} = 10 \qquad f_t = \frac{25(50)}{250} = 5$$

$$f_t = \frac{225(100)}{250} = 90 \qquad f_t = \frac{225(100)}{250} = 90 \qquad f_t = \frac{225(50)}{250} = 45$$

Since this is a 2 × 3 contingency table, the chi-square value can be computed from Equation (10.6) where

$$\chi^2_{(2-1)(3-1)} \cong \sum_{\substack{\text{all}\\ \text{cells}}} \frac{(f_o - f_t)^2}{f_t}$$

as shown in Table 10.11.

At the .05 level of significance the critical value of chi-square with $(2 - 1)(3 - 1) = 2$ degrees of freedom is 5.991 (see Appendix E, Table E.4). This is depicted in Figure 10.13. Therefore the decision rule is to reject H_0 if $\chi_2^2 > 5.991$.

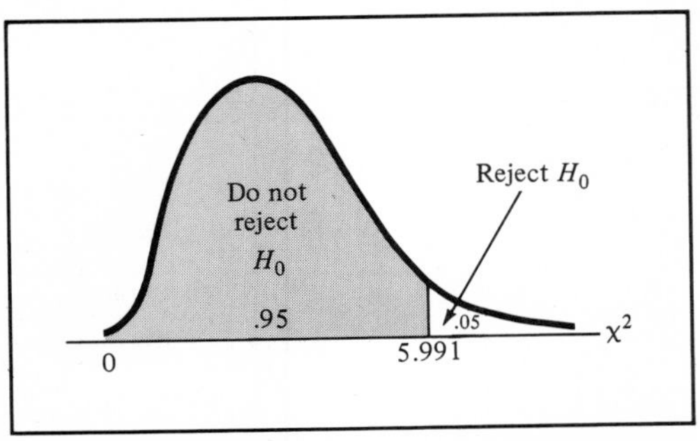

Figure 10.13 Testing for the difference between the proportions in three groups at the 5% level of significance with 2 degrees of freedom.

Table 10.12
CROSS CLASSIFICATION OF CALCULATOR OWNERSHIP AND MAJOR

Major	Own a Calculator		
	Yes	No	Totals
Undecided	5	1	6
Accounting	28	7	35
Economics or finance	8	0	8
Management	7	6	13
Marketing	13	10	23
Statistics or computer systems	1	1	2
Other	5	2	7
Totals	67	27	94

Since our computed $\chi_2^2 = 8.3332 > 5.991$, our decision is to reject H_0. Thus we may conclude that the proportion of tires that blow out before 10,000 miles is not the same for all three shifts. This conclusion also implies that the shift of workers and the proportion of blowouts are not independent variables. The various shifts have produced different proportions of blowouts.

As a second illustration, we can also apply this chi-square test for the difference between the proportions of c populations to an example from the dean's survey. One of the questions that the dean sought to determine related to calculator ownership. The dean wanted to know whether the proportion of students who owned calculators was the same for the various majors. This information can be collected by cross classifying major (question 4) and calculator ownership (question 14) in the above contingency table (Table 10.12).

Since a theoretical frequency of at least five in each cell is advisable in order to use the chi-square test (Reference 3), the majors economics or finance, management, statistics or computer systems, other, and undecided can be grouped into a category called "other than accounting or marketing." The revised cross classifications can be expressed in the contingency table, Table 10.13.

Since there are now three groups remaining, the null and alternative hypotheses can be stated as follows:

Table 10.13
REVISED CROSS CLASSIFICATION TABLE

Major	Own a Calculator		
	Yes	No	Totals
Accounting	28	7	35
Marketing	13	10	23
Other than accounting or marketing	26	10	36
Totals	67	27	94

H_0: $p_{\text{acc}} = p_{\text{mkt}} = p_{\text{other}}$ (no relationship between major and calculator ownership)

H_1: At least one major has a different proportion of calculator ownership than the others (there is a relationship between major and calculator ownership)

If these hypotheses were to be tested at the .01 level of significance, the critical value (from Appendix E, Table E.4) would be 9.21 (since there would be $(3 - 1)(2 - 1) = 2$ degrees of freedom and the critical value is at the 99th percentile point of the distribution). This is depicted in Figure 10.14.

Since the theoretical frequencies would be computed from Equation (10.5), $f_t = n_R n_C/n$, we would have for the six cells:

$$f_t = \frac{(35)(67)}{94} = 24.9468 \qquad f_t = \frac{(35)(27)}{94} = 10.0532$$

$$f_t = \frac{(23)(67)}{94} = 16.3936 \qquad f_t = \frac{(23)(27)}{94} = 6.6064$$

$$f_t = \frac{(36)(67)}{94} = 25.6596 \qquad f_t = \frac{(36)(27)}{94} = 10.3404$$

The chi-square test statistic would be computed as shown in Table 10.14 since

$$\chi_2^2 \cong \sum_{\substack{\text{all}\\ \text{cells}}} (f_o - f_t)^2/f_t$$

Since $\chi_2^2 = 3.76239 < 9.21$, the null hypothesis could not be rejected and we would conclude that there is no evidence of a difference between the various majors in the proportion of students who own calculators. This con-

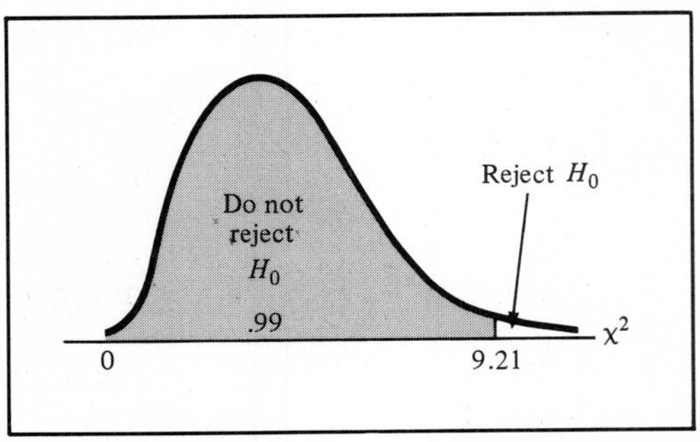

Figure 10.14 Testing for a relationship between calculator ownership and major at the 1% level of significance with 2 degrees of freedom.

Table 10.14
COMPUTATION OF THE CHI-SQUARE TEST STATISTIC FOR THE CALCULATOR OWNERSHIP–MAJOR PROBLEM

f_o	f_t	$(f_o - f_t)$	$(f_o - f_t)^2$	$(f_o - f_t)^2/f_t$
28	24.9468	3.0532	9.32203	.37367
13	16.3936	−3.3936	11.51652	.70250
26	25.6596	.3404	.11587	.00452
7	10.0532	−3.0532	9.32203	.92726
10	6.6064	3.3936	11.51652	1.74323
10	10.3404	− .3404	.11587	.01121
				3.76239

clusion can also be expressed as "there is no evidence of a relationship between major and calculator ownership."

10.7 CHI-SQUARE TEST OF INDEPENDENCE IN THE $R \times c$ TABLE

We have just seen that the chi-square test can be used to determine differences between the proportion of successes in any number of populations. For a contingency table that has R rows and c columns, the chi-square test can be generalized as a test of independence. In the first example the manager of the tire company would like to know the factors which could possibly explain the differences in the proportion of blowouts between the three shifts of workers. He feels that experience may be a factor in determining the quality of the tires produced. Therefore the personnel records of the company are sampled to determine the length of time an employee has worked for the company. Random samples of 100 day-shift, 50 evening-shift, and 50 night-shift employees were selected with the results shown in Table 10.15.

The null and alternative hypotheses would be

H_0: No relationship (independence) between shift and experience

H_1: There is a relationship between shift and experience

The theoretical frequencies would be computed from Equation (10.5), $f_t = n_R n_c / n$, as follows:

Table 10.15
CROSS CLASSIFICATION OF SHIFT AND EXPERIENCE

Employees	Shift			
Years of Experience	*Day*	*Evening*	*Night*	Totals
Under 2 years	50	30	40	120
2 to under 5 years	35	15	10	60
5 to under 10 years	15	5	0	20
Totals	100	50	50	200

Table 10.16
COMPUTATION OF THE CHI-SQUARE TEST STATISTIC FOR THE SHIFT-EXPERIENCE PROBLEM

f_o	f_t	$(f_o - f_t)$	$(f_o - f_t)^2$	$(f_o - f_t)^2/f_t$
50	60	−10	100	1.6667
35	30	5	25	.8333
15	10	5	25	2.5
30	30	0	0	0
15	15	0	0	0
5	5	0	0	0
40	30	10	100	3.3333
10	15	− 5	25	1.6667
0	5	− 5	25	5.0
				15.0000

$$f_t = \frac{120(100)}{200} = 60 \qquad f_t = \frac{120(50)}{200} = 30 \qquad f_t = \frac{120(50)}{200} = 30$$

$$f_t = \frac{60(100)}{200} = 30 \qquad f_t = \frac{60(50)}{200} = 15 \qquad f_t = \frac{60(50)}{200} = 15$$

$$f_t = \frac{20(100)}{200} = 10 \qquad f_t = \frac{20(50)}{200} = 5 \qquad f_t = \frac{20(50)}{200} = 5$$

The chi-square test statistic for this problem can be computed from the data given in Table 10.16.

Since this is a 3×3 contingency table there are $(3-1)(3-1) = 4$ degrees of freedom. From Appendix E, Table E.4 with $\alpha = .05$, the decision rule would be to reject H_0 if $\chi_4^2 > 9.488$ (since this is the critical value at the 95th percentile with 4 degrees of freedom). This is depicted in Figure 10.15.

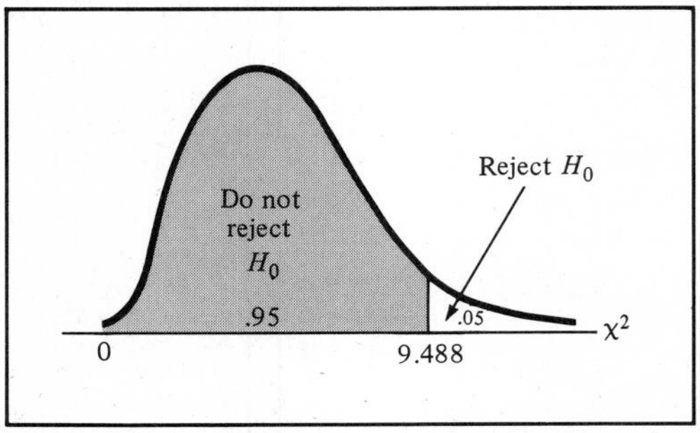

Figure 10.15 Testing for a relationship between shift and experience at the 5% level of significance with 4 degrees of freedom.

Table 10.17
CROSS CLASSIFICATION OF GRADUATE SCHOOL INTENTIONS AND MAJOR

Major	Are You Planning to Attend Graduate School? Yes	No	Not Sure	Totals
Undecided	2	1	3	6
Accounting	11	8	16	35
Economics or finance	2	1	5	8
Management	6	2	5	13
Marketing	14	4	5	23
Statistics or computer systems	1	0	1	2
Other	2	1	4	7
Totals	38	17	39	94

Since the computed value of $\chi_4^2 = 15.0 > 9.488$, our decision is to reject H_0. We would conclude that there is a relationship between shift and amount of experience. An examination of the data seems to suggest that the night shift has a higher proportion of inexperienced workers while the day shift has a higher proportion of experienced workers.

This general test of independence can also be used by the researcher and the dean to determine whether there is a relationship between a student's major and intention to attend graduate school. The contingency table for these two variables can be set up by cross classifying questions 3 and 4 of the questionnaire (Figure 2.2) as shown in Table 10.17.

Since several cells in this 7×3 contingency table have theoretical frequencies substantially below 5, the various majors will again be reduced to three categories: accounting, marketing, and "other than accounting or marketing." It should be noted that one of the remaining cells (marketing—not planning to attend graduate school) has a theoretical frequency below 5. However, in a large contingency table (such as the resulting 3×3 table) the existence of a single cell with a theoretical frequency below 5 does not seriously affect the chi-square test unless that theoretical frequency is below 1 (Reference 3). Therefore the resulting cross classification table of major and graduate school intention is shown in Table 10.18.

Table 10.18
REVISED CROSS CLASSIFICATION TABLE

Major	Are You Planning to Attend Graduate School Yes	No	Not Sure	Totals
Accounting	11	8	16	35
Marketing	14	4	5	23
Other than accounting or marketing	13	5	18	36
Totals	38	17	39	94

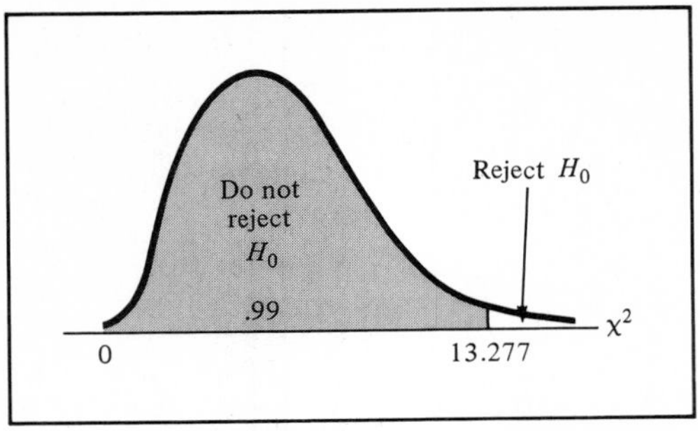

Figure 10.16 Testing for a relationship between major and graduate school intention at the 1% level of significance with 4 degrees of freedom.

The null and alternative hypotheses for this condensed table would be

H_0: There is no relationship between major and graduate school intentions

H_1: There is a relationship between major and graduate school intentions

If this hypothesis were to be tested at the .01 level of significance, the critical value would be 13.277 (since there are $(3-1)(3-1)=4$ degrees of freedom). This is shown in Figure 10.16.

The theoretical frequencies for the nine cells, computed from Equation (10.5), are

$$f_t = \frac{35(38)}{94} = 14.1489 \quad f_t = \frac{35(17)}{94} = 6.3298 \quad f_t = \frac{35(39)}{94} = 14.5213$$

$$f_t = \frac{23(38)}{94} = 9.2979 \quad f_t = \frac{23(17)}{94} = 4.1596 \quad f_t = \frac{23(39)}{94} = 9.5425$$

$$f_t = \frac{36(38)}{94} = 14.5532 \quad f_t = \frac{36(17)}{94} = 6.5106 \quad f_t = \frac{36(39)}{94} = 14.9362$$

The chi-square test statistic $\left(\sum_{\substack{\text{all}\\\text{cells}}} (f_o - f_t)^2/f_t\right)$ for these data would then be computed as shown in Table 10.19.

Since $\chi_4^2 = 6.98978 < 13.277$, the null hypothesis could not be rejected and we would conclude that there is no evidence of a relationship between major field of study and the graduate school intentions of the students.

Table 10.19
COMPUTATION OF THE CHI-SQUARE TEST STATISTIC FOR THE MAJOR–GRADUATE SCHOOL INTENTION PROBLEM

f_o	f_t	$(f_o - f_t)$	$(f_o - f_t)^2$	$(f_o - f_t)^2/f_t$
11	14.1489	−3.1489	9.91557	.70080
14	9.2979	4.7021	22.10974	2.37793
13	14.5532	−1.5532	2.41243	.16577
8	6.3298	1.6702	2.78957	.44070
4	4.1596	− .1596	.02547	.00612
5	6.5106	−1.5106	2.28191	.35049
16	14.5213	1.4787	2.18655	.15714
5	9.5425	−4.5425	20.63431	2.16236
18	14.9362	3.0638	9.38687	.62846
				6.98978

10.8 THE CHI-SQUARE GOODNESS-OF-FIT TEST FOR PROBABILITY DISTRIBUTIONS

In the last several sections we have been utilizing the chi-square test to determine whether or not the proportion of successes was equal in each of several groups as well as to test for an association between two qualitative variables. In each of these applications of the chi-square test, from Equation (10.6), we had, for a particular number of degrees of freedom (df)

$$\chi_{df}^2 \cong \sum_{\substack{\text{all} \\ \text{cells}}} \frac{(f_o - f_t)^2}{f_t}$$

This test statistic measured the squared difference between the theoretical (f_t) and observed (f_o) frequencies in a cell divided by the theoretical frequency in each cell. The theoretical frequency in the cell was based on a null hypothesis of either no difference in the proportion of successes over all groups or no association between the variables.

In this section we shall examine still another important application of the chi-square test, that of testing the **goodness of fit** of a set of data to a specific probability distribution. In testing the goodness of fit of a set of data, we compare the actual frequencies in each category (or class interval) to the frequencies that theoretically would be expected to occur if the data followed a specific probability distribution. In performing a chi-square goodness-of-fit test, several steps must be followed. First of all we must hypothesize the probability distribution which is to be fitted to the data. Second, once the appropriate hypothesized probability distribution has been determined, the values of each parameter of the distribution (such as the mean) must either be hypothesized or estimated from the actual data. Next the specific hypothesized probability distribution is used to determine both the probability and then the theoretical frequency for each category or class interval. Finally the chi-square test statistic can be employed to test whether the specific distribution is a "good fit" to the data.

Table 10.20
FREQUENCY DISTRIBUTION OF "NUMBER OF CLUBS AFFILIATED WITH" FOR A RANDOM SAMPLE OF 94 STUDENTS

Number of Clubs	Frequency
0	45
1	22
2	16
3	7
4	3
5	1
6 or more	0
Totals	94

SOURCE: Data are taken from Figure 2.6.

The use of the chi-square goodness-of-fit test can be seen by referring to the data collected by the dean and the researcher. We may recall from Section 6.7 (discussing the Poisson distribution) that the question pertaining to the number of clubs a student is affiliated with (Figure 2.2, question 23) might follow the Poisson distribution. In order to determine whether this is true, we have extracted the information concerning "number of clubs affiliated with" from the sample of 94 students and presented it in Table 10.20.

In order to determine whether this variable, "number of clubs," follows a Poisson distribution, we may set up the following hypotheses:

H_0: The number of clubs follows a Poisson distribution

H_1: The number of clubs follows a probability distribution other than a Poisson distribution

In this example, since the Poisson distribution has one parameter, its mean μ, we can either include a specified value for the mean as part of the null hypothesis or we can estimate the mean from the sample results. For this latter case, which will be used in our example, we would be hypothesizing

Table 10.21
COMPUTATION OF THE SAMPLE AVERAGE NUMBER OF CLUBS FROM THE FREQUENCY DISTRIBUTION

Number of Clubs m_j	Frequency f_j	$m_j f_j$
0	45	0
1	22	22
2	16	32
3	7	21
4	3	12
5	1	5
		92

only that the data follow the Poisson distribution without specifying a population mean μ.

In order to compute an estimate of the mean of the Poisson distribution we must refer back to Section 4.6.1 in which we computed the sample mean from a set of data grouped into a frequency distribution. In Table 10.20 each integer value may be considered the midpoint of a class. So, from Equation (4.10), $\overline{X} = \sum_{j=1}^{g} m_j f_j / n$, we would obtain Table 10.21. For this example

$$\overline{X} = \frac{\sum_{j=1}^{g} m_j f_j}{n} = \frac{92}{94} = 0.98$$

Therefore, $\overline{X} \cong 1.0$.

Since the value of the sample mean is very close to 1.0, for the purposes of finding the probability from the table of the Poisson distribution (Appendix E, Table E.6), a value of $\mu = 1.0$ would be utilized. Therefore from Appendix E, Table E.6, for $\mu = 1.0$, the probability of X successes ($X = 0, 1, 2, 3, 4, 5, 6$ or more) can be determined. The theoretical frequency for each category is obtained by multiplying the appropriate probability by the sample size n. These results are summarized in Table 10.22.

Table 10.22
ACTUAL AND THEORETICAL EXPECTED FREQUENCIES FOR "NUMBER OF CLUBS AFFILIATED WITH"

Number of Clubs	Actual Frequency f_o	$p(X)$ Probability for Poisson with $\mu = 1.0$	Theoretical Frequency $f_t = n \cdot p(X)$
0	45	.3679	94 × .3679 = 34.58
1	22	.3679	94 × .3679 = 34.58
2	16	.1839	94 × .1839 = 17.29
3	7	.0613	94 × .0613 = 5.76
4	3	.0153	94 × .0153 = 1.44
5	1	.0031	94 × .0031 = 0.29
6 or more	0	.0006	94 × .0006 = 0.06
Totals	94	1.0000	94.00

We can see from Table 10.22 that the theoretical frequency of belonging to either 4, 5, or 6 or more clubs is below 5. However, we may recall that in our previous discussion of the chi-square test, we assumed that each cell contained at least 5 theoretical frequencies. Since the classes at the extreme upper portion of the distribution have theoretical frequencies below 5, they are combined together or with the adjacent categories for the purpose of performing the analysis.

The chi-square test for determining whether the data follow a specific probability distribution is computed from

$$\chi^2_{K-p-1} \cong \sum_K \frac{(f_o - f_t)^2}{f_t} \tag{10.7}$$

where f_o = observed frequency

f_t = theoretical frequency

K = number of categories or classes that remain after collapsing classes

p = number of parameters estimated from the data

In our example, after the collapsing of classes, there are four classes that remain (0, 1, 2, 3 or more clubs). Since the mean of the Poisson distribution had been computed from the actual data, the number of degrees of freedom would equal

$$K - p - 1 = 4 - 1 - 1 = 2 \text{ degrees of freedom}$$

On the other hand, it should be mentioned here that had a specific value of the mean been included as part of the null hypothesis, then the 1 degree of freedom p would not be lost and there would be 3 degrees of freedom $(K - 1)$ for obtaining the critical value.

If a level of significance of .05 is selected, the critical value of chi-square with 2 degrees of freedom is 5.991 (from Appendix E, Table E.4). This is shown in Figure 10.17.

Therefore the decision rule is

$$\text{Reject } H_0 \text{ if } \chi_2^2 > 5.991; \text{ otherwise do not reject } H_0$$

From Table 10.22 the value of chi-square can be computed as shown in Table 10.23.

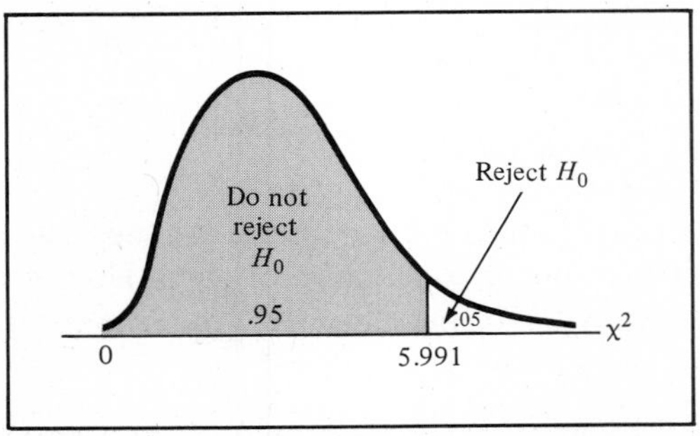

Figure 10.17 Testing the goodness of fit of the Poisson distribution at the 5% level of significance with 2 degrees of freedom.

Table 10.23
COMPUTATION OF THE CHI-SQUARE TEST STATISTIC FOR THE NUMBER OF CLUBS PROBLEM

Number of Clubs X	f_o	f_t	$(f_o - f_t)$	$(f_o - f_t)^2$	$(f_o - f_t)^2/f_t$
0	45	34.58	10.42	108.5764	3.1399
1	22	34.58	−12.58	158.2564	4.5765
2	16	17.29	− 1.29	1.6641	0.0962
3 or more	11	7.55	3.45	11.9025	1.5765
					9.3891

Since 9.3891 $>$ 5.991, reject H_0.

Since the null hypothesis has been rejected, the dean must conclude that the distribution of the number of clubs students are affiliated with does not follow a Poisson distribution.

In this section we have limited our discussion of the chi-square goodness-of-fit test to the Poisson distribution. Other well-known probability distributions such as the uniform, binomial, and normal distributions can also be "fitted" to sets of data to determine the goodness of fit. In each case the parameters of the distributions involved (such as the mean and the standard deviation) must either be hypothesized or estimated from the data so that the theoretical expected frequencies may be determined.

10.9 TESTING A HYPOTHESIS ABOUT A POPULATION VARIANCE

In addition to tests for goodness of fit, in our discussion of hypothesis testing thus far in this chapter we have focused upon decisions about qualitative variables (proportions) and quantitative variables (means). However, when analyzing quantitative variables, it is often important to draw conclusions about the variability as well as the average of a characteristic of interest. For example, in our tire study the manager of the factory would probably be interested in determining whether or not the variability in the life of tires is within acceptable limits before concluding that the production process is working properly. Here, then, we would be interested in drawing conclusions about either the population standard deviation or variance.

In attempting to draw conclusions about the variability in the population, we first must determine what statistical test can be used to represent the distribution of the variability in the sample data. If the variable (tire tread life) is assumed to be normally distributed, then the statistic $(n - 1)S^2/\sigma^2$ will follow the chi-square (χ^2) distribution with $(n - 1)$ degrees of freedom.

Therefore the test statistic for testing whether or not the population variance is equal to a specified value is

$$\chi^2_{n-1} = \frac{(n - 1)S^2}{\sigma^2} \qquad (10.8)$$

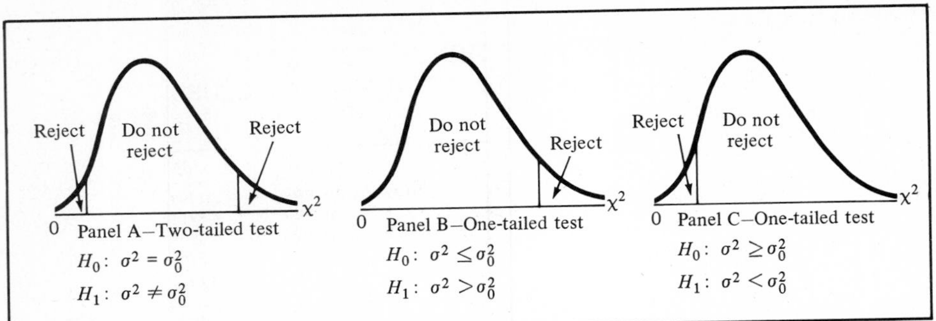

Figure 10.18 Testing a hypothesis about a population variance: One-tailed and two-tailed tests. Panel A, two-tailed test. Panel B, one-tailed test, Panel C, one-tailed test.

where n = sample size
S^2 = sample variance
σ^2 = hypothesized population variance

If, as shown in panel A of Figure 10.18, the test of hypothesis is two-tailed, the rejection region is split into both the lower and the upper tails of the chi-square distribution. However, if the test is one-tailed, the rejection region is either in the upper tail (panel B of Figure 10.18) or in the lower tail (panel C of Figure 10.18) of the chi-square distribution, depending on the direction of the alternative hypothesis.

To apply the test of a hypothesis for a population variance, let us again refer back to the tire production study. Based upon past experience, the production manager has determined that if the population standard deviation of tire life is no more than 3,500 miles, the process will be considered to be working properly (in control). The manager is particularly interested in determining whether there is too much variability in the life of tires produced on Fridays. Thus a random sample of 16 tires produced on Friday is selected with the results shown in Table 10.24.

Table 10.24
TIRE LIFE FOR A RANDOM SAMPLE OF 16 TIRES PRODUCED ON A FRIDAY

Tire Life (Thousands of Miles)	
22.42	20.62
18.36	26.47
21.46	19.75
19.20	20.30
23.40	17.84
27.38	26.34
23.46	28.27
23.51	24.73

For these data

$$\sum_{i=1}^{n} X_i = 363.51 \qquad \sum_{i=1}^{n} X_i^2 = 8{,}420.0585 \qquad n = 16$$

The sample variance can now be computed from Equation (4.7*a*) as follows:

$$S^2 = \frac{n\sum_{i=1}^{n} X_i^2 - \left(\sum_{i=1}^{n} X_i\right)^2}{n(n-1)}$$

$$= \frac{16(8{,}420.0585) - (363.51)^2}{16(15)} = \frac{2{,}581.4159}{240} = 10.7559$$

Thus,

$$S = 3.27962 \text{ (thousands of miles)}$$

so that

$$S = 3{,}279.62 \text{ miles}$$

Since the primary concern of the production manager is that the variability is not too large, we have a one-tailed test, and the null and alternative hypotheses can be stated as follows:

H_0: $\sigma \leq 3{,}500$ miles (process is working properly)
or $\sigma^2 \leq 12{,}250{,}000$

H_1: $\sigma > 3{,}500$ miles (process is not working properly)
or $\sigma^2 > 12{,}250{,}000$

Hence the null hypothesis would be rejected only if the test statistic were greater than the critical value on the chi-square distribution.

Since there are 15 degrees of freedom (16 − 1), if a level of significance of .05 is selected, the decision rule can be stated as (see Figure 10.19)

Reject H_0 if $\chi_{15}^2 > 24.996$; otherwise do not reject H_0

For these data we have, from Equation (10.8),

$$\chi_{n-1}^2 = \frac{(n-1)S^2}{\sigma^2} = \frac{(16-1)(10{,}755{,}900)}{(3{,}500)^2} = 13.17$$

Since $13.17 < 24.996$, do not reject H_0.

Since the sample statistic (13.17) is less than the critical value on the chi-square distribution (24.996), the null hypothesis cannot be rejected. The pro-

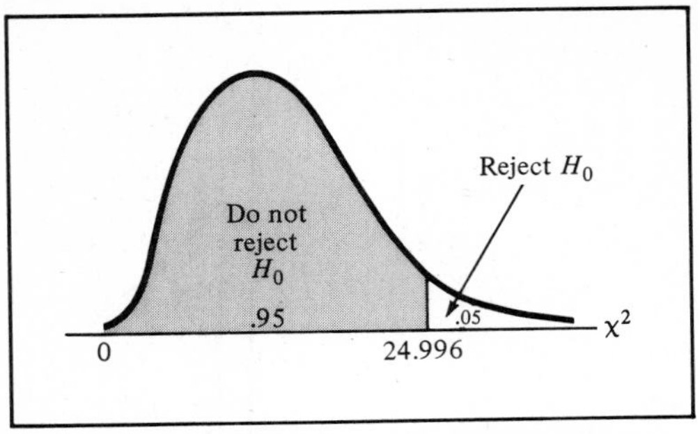

Figure 10.19 One-tailed test of hypothesis for a population variance at the 5% level of significance with 15 degrees of freedom.

duction manager should conclude that there is no evidence that the process is not working properly.

A second application of the test of hypothesis for a population variance can be illustrated by referring to the data collected by the dean and the researcher and presented in Figure 2.6. For this particular case the dean would like to know whether the population standard deviation in the grade-point index of economics and finance majors is equal to 0.40. From Figure 2.6 we can extract the grade-point index for each of the eight economics and finance majors included in the sample. This information is presented in Table 10.25.

For these data,

$$\sum_{i=1}^{n} X_i = 23.35 \qquad \sum_{i=1}^{n} X_i^2 = 68.3961 \qquad n = 8$$

Computing the sample variance from Equation (4.7a), we have

$$S^2 = \frac{n \sum_{i=1}^{n} X_i^2 - \left(\sum_{i=1}^{n} X_i\right)^2}{n(n-1)} = \frac{8(68.3961) - (23.35)^2}{8(7)} = .0347$$

Table 10.25
GRADE-POINT INDEX OF THE EIGHT ECONOMICS AND FINANCE MAJORS

Grade-Point Index	
2.82	3.21
3.00	3.01
2.75	3.07
2.64	2.85

SOURCE: Data are taken from Figure 2.6.

and

$$S = .1863$$

Since the dean wishes to know whether the standard deviation equals 0.40 (the variance equals .16), we have a two-tailed test, and the null and alternative hypothesis can be stated as follows:

$$\begin{aligned} H_0&: \quad \sigma = 0.40 \\ \text{or}& \quad \sigma^2 = 0.16 \\ H_1&: \quad \sigma \neq 0.40 \\ \text{or}& \quad \sigma^2 \neq 0.16 \end{aligned}$$

In this example, then, the null hypothesis would be rejected if the test statistic fell into either the lower or upper tails of the chi-square distribution (Figure 10.20).

Since there are 7 degrees of freedom ($8 - 1 = 7$), if a level of significance of 5% was selected, the lower (χ_L^2) and upper (χ_U^2) critical values could be obtained from the table of the chi-square distribution (Appendix E, Table E.4). We may recall from Section 10.5 (see Table 10.6) that the value at the top of the table of the chi-square distribution indicates the lower tail area of the distribution. Therefore the lower critical value χ_L^2 of 1.69 can be obtained from Appendix E, Table E.4, by looking in the column labeled ".025" for 7 degrees of freedom. The upper critical value χ_U^2 of 16.013 can be obtained by first determining that the percentage below χ_U^2 is 97.5% (since there are 2.5% in the upper tail) and looking in the column labeled ".975" for 7 degrees of freedom.

Therefore, the decision rule would be

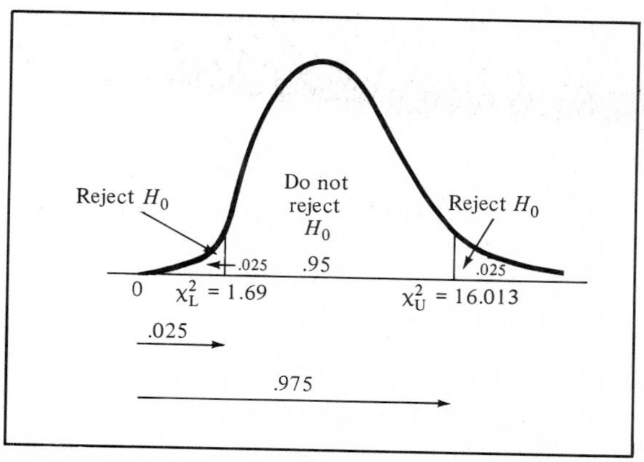

Figure 10.20 Two-tailed test of hypothesis about a population variance at the 5% level of significance with 7 degrees of freedom.

Reject H_0 if $\chi_7^2 > 16.013$
or if $\chi_7^2 < 1.690$
otherwise do not reject H_0

From Equation (10.8) we recall that the test statistic for a population variance is

$$\chi^2_{n-1} = \frac{(n-1)S^2}{\sigma^2}$$

and thus, for this example we have

$$\chi_7^2 = \frac{(8-1)(.0347)}{(.16)} = 1.518$$

Since $1.518 < 1.69$, we reject H_0.

Since the null hypothesis has been rejected, the dean would conclude that the population variance of the grade-point index for economics and finance majors is not equal to 0.16. This is equivalent to saying that the standard deviation is not equal to 0.40.

In testing a hypothesis about a variance, we should be aware that we have assumed a normally distributed population. Unfortunately this test is sensitive to departures from this assumption so that if the population is not normally distributed, particularly for small sample sizes, the accuracy of the test can be seriously affected. (Reference 1).

10.10 TESTING OF A HYPOTHESIS FOR THE EQUALITY OF VARIANCES FROM TWO POPULATIONS

In the previous section we discovered that a hypothesis concerning the value of the variance or standard deviation of a population can be tested by using the chi-square distribution. In a manner analogous to those already discussed for means and proportions, we may also be interested in determining whether two populations have the same variability. Either we may be interested in testing the assumption of equal variances that we had made for the t test in Section 10.2, or we may be interested in studying the variances of two populations as an end in itself.

In order to compare the equality of the variances of two independent populations, a statistical procedure has been devised that is based upon the ratio of the two sample variances. If the data from each population are assumed to be normally distributed, then the ratio S_1^2/S_2^2 follows a distribution called the F distribution (see Appendix E, Table E.5), which was named after the famous statistician R. A. Fisher. From Appendix E, Table E.5, we can see that the critical values of the F distribution depends upon *two* sets of degrees of freedom: the degrees of freedom in the numerator and in the

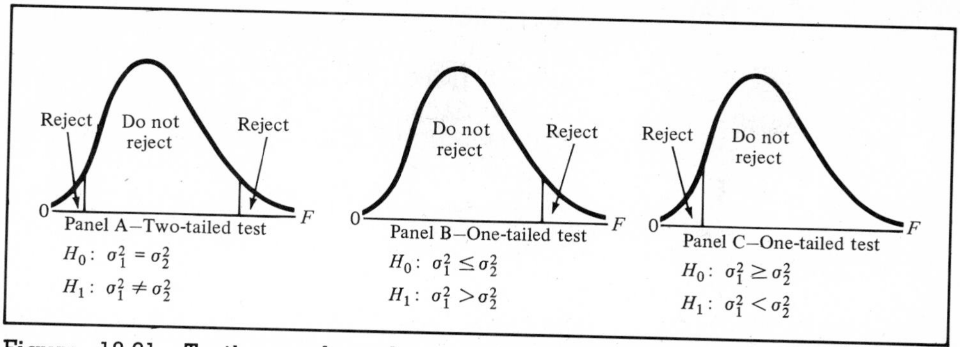

Figure 10.21 Testing a hypothesis about the equality of two population variances: one and two-tailed tests. Panel A, two-tailed test. Panel B, one-tailed test. Panel C, one-tailed test.

denominator. The test statistic for testing the ratio between two variances would be

$$F_{(n_1-1),\,(n_2-1)} = \frac{S_1^2}{S_2^2} \tag{10.9}$$

where
n_1 = sample size in group 1
n_2 = sample size in group 2
$n_1 - 1$ = degrees of freedom in group 1
$n_2 - 1$ = degrees of freedom in group 2
S_1^2 = sample variance in group 1
S_2^2 = sample variance in group 2

In testing the ratio of two variances, either one-tailed or two-tailed tests can be employed as indicated in Figure 10.21.

In order to demonstrate how we may test for the equality of two variances, we can return to the tire study from the previous section. We may recall that a random sample of 16 tires produced on a Friday was selected, and the life of these tires was presented in Table 10.24. In addition to making a decision concerning the variability in the life of all tires produced on Friday, the production manager would also like to determine whether the variability in the life of tires produced on Friday afternoon is *greater than* that of those produced Friday morning. The random sample of 16 tires (Table 10.24) is subdivided into those produced in the morning and those produced in the afternoon as shown in Table 10.26.

Table 10.26
RANDOM SAMPLE OF THE LIFE OF TIRES PRODUCED ON FRIDAY SUBDIVIDED INTO MORNING AND AFTERNOON

Tire Life (Thousands of Miles)	
Morning	*Afternoon*
22.42	23.51
18.36	20.62
21.46	26.47
19.20	19.75
23.40	20.30
27.38	17.84
23.46	26.34
	28.37
	24.73

SOURCE: Data are taken from Table 10.24.

For these data,

Morning	Afternoon
$\sum_{i=1}^{n} X_i = 155.68$	$\sum_{i=1}^{n} X_i = 207.83$
$\sum_{i=1}^{n} X_i^2 = 3{,}516.5136$	$\sum_{i=1}^{n} X_i^2 = 4{,}903.5449$
$n_m = 7$	$n_a = 9$
$df_m = n_m - 1 = 7 - 1 = 6$	$df_a = n_a - 1 = 9 - 1 = 8$

Once again, from Equation (4.7*a*), we have

$$S^2 = \frac{n \sum_{i=1}^{n} X_i^2 - \left(\sum_{i=1}^{n} X_i\right)^2}{n(n-1)}$$

Therefore,

$$S_m^2 = \frac{7(3{,}516.5136) - (155.68)^2}{7(6)} = 9.0317$$

$$S_a^2 = \frac{9(4{,}903.5449) - (207.83)^2}{9(8)} = 13.036$$

Since the production manager wishes to determine whether or not there is more variability in the afternoon than in the morning, a one-tail test can be set up as follows:

$$H_0: \quad \sigma_a^2 \leq \sigma_m^2$$

$$H_1: \quad \sigma_a^2 > \sigma_m^2$$

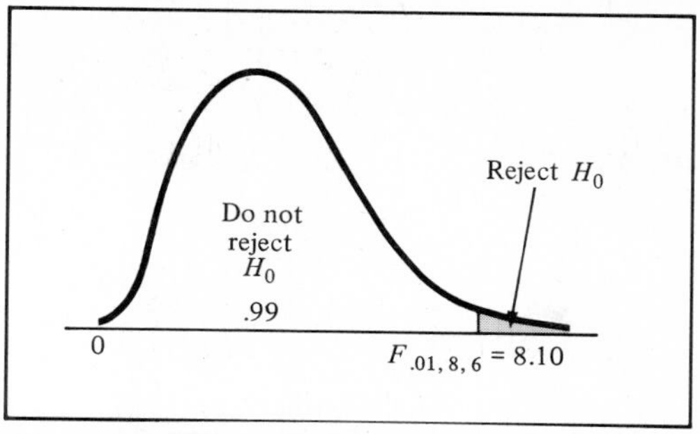

Figure 10.22 One-tailed test for the equality of two variances at the 1% level of significance with 8 and 6 degrees of freedom.

For this example, since group 1 consists of the tires produced in the afternoon, and group 2 consists of tires produced in the morning, the rejection region is located in the upper tail of the distribution (see Figure 10.22).

Using Equation (10.9),

$$F_{(n_1-1),\ (n_2-1)} = \frac{S_1^2}{S_2^2}$$

we have

$$F_{(9-1),\ (7-1)} = \frac{S_a^2}{S_m^2}$$

Since there are 8 degrees of freedom in the numerator and 6 degrees of freedom in the denominator, if a level of significance of .01 is selected, the critical value on the F distribution can be obtained from Appendix E, Table E.5, by looking in the column labeled "8" and the row labeled "6" and obtaining the value for the upper 1% of this F distribution, which is 8.10 (see Table 10.27). Therefore the decision rule is

Reject H_0 if $F_{8,6} > 8.10$; otherwise do not reject H_0

For our data we can compute

$$F_{8,6} = \frac{13.036}{9.0317} = 1.443$$

Since $1.443 < 8.10$, do not reject H_0.

Since the null hypothesis cannot be rejected, the production manager can conclude that there is no evidence that the variability in the life of tires produced on Friday afternoon is higher than for tires produced on Friday morning.

Table 10.27
OBTAINING THE CRITICAL VALUE OF F WITH 8 AND 6 DEGREES OF FREEDOM AT THE .01 LEVEL OF SIGNIFICANCE

Denominator df_2	Numerator df_1: 1	2	3	4	5	6	7	8	9	10	12
1	4052	4999.5	5403	5625	5764	5859	5928	5982	6022	6056	6106
2	98.50	99.00	99.17	99.25	99.30	99.33	99.36	99.37	99.39	99.40	99.42
3	34.12	30.82	29.46	28.71	28.24	27.91	27.67	27.49	27.35	27.23	27.05
4	21.20	18.00	16.69	15.98	15.52	15.21	14.98	14.80	14.66	14.55	14.37
5	16.26	13.27	12.06	11.39	10.97	10.67	10.46	10.29	10.16	10.05	9.89
6	13.75	10.92	9.78	9.15	8.75	8.47	8.26	8.10	7.98	7.87	7.72
7	12.25	9.55	8.45	7.85	7.46	7.19	6.99	6.84	6.72	6.62	6.47
8	11.26	8.65	7.59	7.01	6.63	6.37	6.18	6.03	5.91	5.81	5.67

SOURCE: Extracted from Appendix E, Table E.5.

Using the data collected by the dean and the researcher, we may recall that in the previous section we had tested whether or not the variance of economics and finance majors was equal to 0.16. However, the dean would also like to know whether or not there is any difference in the variability of the grade-point index of economics and finance majors and management majors. Table 10.28 depicts the grade-point index of the 13 management majors (code 4, question 4) included in the sample which are extracted from Figure 2.6.

For these data,

$$\sum_{i=1}^{n} X_i = 41.10 \qquad \sum_{i=1}^{n} X_i^2 = 132.655 \qquad n_{\text{mgt}} = 13$$

Once again, from Equation (4.7*a*), we have

$$S^2_{\text{mgt}} = \frac{n\sum_{i=1}^{n} X_i^2 - \left(\sum_{i=1}^{n} X_i\right)^2}{n(n-1)}$$

Table 10.28
GRADE-POINT INDEX OF 13 MANAGEMENT MAJORS

Grade-Point Index	
3.43	3.86
3.51	2.80
3.32	3.75
2.80	2.69
3.82	2.83
2.45	2.84
3.00	

SOURCE: Data are taken from Figure 2.6.

Therefore,

$$S^2_{\text{mgt}} = \frac{13(132.655) - (41.1)^2}{13(12)} = 0.2263$$

Since the dean wishes to determine *only whether there is a difference* in the variability between economics and finance majors and management majors, a two-tail test can be set up as follows:

$$H_0: \quad \sigma^2_{\text{e\&f}} = \sigma^2_{\text{mgt}}$$

$$H_1: \quad \sigma^2_{\text{e\&f}} \neq \sigma^2_{\text{mgt}}$$

Because this is a two-tailed test, the rejection region is split into the lower and upper tails of the F distribution. If a level of significance of .05 were selected, each rejection region would contain 2.5% of the distribution (see Figure 10.23).

The upper critical value can be obtained directly from Appendix E, Table E.5, the tables of the F distribution. Looking up 12 degrees of freedom in the numerator and 7 degrees of freedom in the denominator, the upper tail 2.5% value is 4.67. On the other hand, the lower critical value cannot be obtained directly from Appendix E, Table E.5, the tables of the F distribution. However, any lower tail critical value on the F distribution can be found by using the following formula:

$$F_{L(a,b)} = \frac{1}{F_{U(b,a)}} \qquad (10.10)$$

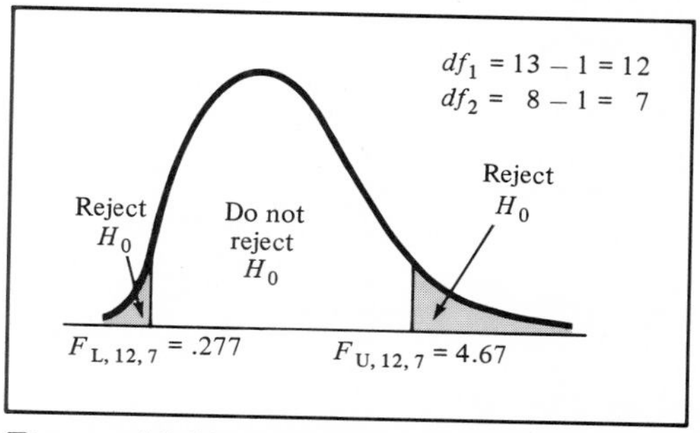

Figure 10.23 Two-tailed test for the equality of two variances at the 5% level of significance with 12 and 7 degrees of freedom.

where F_L = lower critical value of F
F_U = upper critical value of F
a = number of degrees of freedom in group A
b = number of degrees of freedom in group B

Therefore, in this problem we have

$$F_{L(12,7)} = \frac{1}{F_{U(7,12)}}$$

To compute our desired lower tail critical value, we need to obtain the upper .025 value of F with 7 degrees of freedom in the numerator and 12 degrees of freedom in the denominator and take its reciprocal.

Thus from Appendix E, Table E.5, we find that this upper critical value $F_{U(7,12)}$ is 3.61. Therefore, from Equation (10.10),

$$F_{L(12,7)} = \frac{1}{F_{U(7,12)}} = \frac{1}{3.61} = 0.277$$

The decision rule for this example can then be stated as:

Reject H_0 if $F_{12,7} > 4.67$
or if $F_{12,7} < 0.277$
otherwise do not reject H_0

Since, from Equation (10.9),

$$F_{(n_1-1),\ (n_2-1)} = \frac{S_1^2}{S_2^2}$$

we have, for this example,

$$F_{(13-1),\ (8-1)} = \frac{S^2_{\text{mgt}}}{S^2_{\text{e\&f}}}$$

and since $S^2_{\text{e\&f}} = .0347$ and $S^2_{\text{mgt}} = .2263$,

$$F_{12,7} = \frac{.2263}{.0347} = 6.52 > 4.67$$

therefore we reject H_0.

Thus the dean would reject the null hypothesis and conclude that there is a difference in the variability of the grade-point index for economics and finance majors and management majors.

In testing for the ratio of two population variances, as was the case with the variance of a single population, we should be aware that the test assumes that each of the two populations is normally distributed. Unfortunately this test is not robust for departures from this assumption (Reference 1), particularly when the sample sizes in the two groups are not equal. Therefore if the populations are not at least approximately normally distributed, the accuracy of the procedure can be seriously affected (References 1 and 3 present other procedures for testing the equality of two variances).

10.11 INFERENCE: AN OVERVIEW AND THE ROLE OF THE COMPUTER

In Chapters 8, 9, and 10 we have developed a variety of methods which are useful for making inferences based on sample information. In Chapter 8 we considered confidence interval estimation procedures, while here and in Chapter 9 we have been concerned with methods of hypothesis testing as a way of drawing conclusions from sample data. As the reader gains familiarity with these techniques, much time and effort can be saved when attempting to analyze and interpret large-scale problems if one of the several "computer packages" available for statistical inference is utilized (References 2, 5, 7, 8, 10). As an example, the dean and the researcher would surely have desired to create a file or record of the responses for each of the 94 students in the sample so that the information could later be retrieved and used for various types of statistical analyses. To accomplish this they might have used, for example, an SPSS program (References 5 and 7) to create a file on computer tape containing all the information for each subject. They could then have used a variety of SPSS subprograms which perform statistical analyses. Among these are CONDESCRIPTIVE; FREQUENCIES; T-TEST; and CROSSTABS. The CONDESCRIPTIVE subprogram computes various descriptive summary measures for quantitative random variables (see Chapter 4). The FREQUENCIES subprogram (1) creates bar charts for qualitative

```
ANALYSIS OF STUDENT QUESTIONNAIRES.                                                        11/22/76

FILE   STDNTSVR (CREATION DATE = 11/22/76)   SURVEY OF ATTITUDES, INTERESTS, & CHARACTERISTICS

- - - - - - - - - - - - - - - - - - - - - - - - - - - - - - T - T E S T - - - - - - - - - - - - - - - -

GROUP 1 - MAJOR2   EQ        2.= ACCOUNTING
GROUP 2 - MAJOR2   EQ        8.= NON-ACCOUNTING
                                                                       * POOLED VARIANCE ESTIMATE *
                                                          *            *                          *
VARIABLE          NUMBER                STANDARD   STANDARD *   F   2-TAIL *   T   DEGREES OF 2-TAIL *
                 OF CASES     MEAN     DEVIATION    ERROR   * VALUE  PROB. * VALUE   FREEDOM   PROB. *
--------------------------------------------------------------------------------------------------------
GPI        GRADE POINT INDEX                                *              *                          *
      GROUP 1=ACC   35       2.8809       0.481      0.081  *              *                          *
                                                            *  1.11  0.714 *  -0.01     92      0.993 *
      GROUP 2=N-ACC 59       2.8817       0.456      0.059  *              *                          *
                                                            *    ↑         *    ↑                     *
--------------------------------------------------------------------------------------------------------
                                                            Computed            Computed
                                                            F statistic         t statistic
```

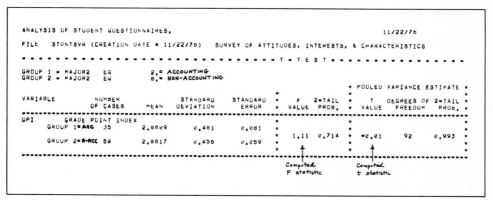

Figure 10.24 SPSS output for *t*-test of differences in grade-point indexes of accounting and non-accounting students.

```
ANALYSIS OF STUDENT QUESTIONNAIRES,                                                11/22/76        PAGE    4

FILE   STDNTSVR (CREATION DATE = 11/22/76)   SURVEY OF ATTITUDES, INTERESTS, & CHARACTERISTICS       OF UNDERG

* * * * * * * * * * * * * * * * * * *  C R O S S T A B U L A T I O N   O F  * * * * * * * * * * * * * * * * * *
    MAJOR1     RECODED FOR ACCTG,MKTG, & OTHER                     BY  CALCOWN   OWNERSHIP OF CALCULATOR
* * * * * * * * * * * * * * * * * * * * * * * * * * * * * * * * * * * * * * * * * * * * * * * * * * PAGE  1 OF

                    CALCOWN
            COUNT  I
                   IYES      NO          ROW
                   I                     TOTAL
                   I      1.I      2.I
MAJOR1     --------I--------I--------I
               2.  I    28  I     7  I    35
  ACCOUNTING       I        I        I  37.2%
                  -I--------I--------I
               5.  I    13  I    10  I    23
  MARKETING        I        I        I  24.5%
                  -I--------I--------I
               8.  I    26  I    10  I    36
  OTHER OR UNDECID I        I        I  38.3%
                  -I--------I--------I
            COLUMN     67       27        94
             TOTAL    71.3%    28.7%  .100.0%

CHI SQUARE =     3.76243 WITH   2 DEGREES OF FREEDOM   SIGNIFICANCE =  0.1524
```

Figure 10.25 SPSS output for χ^2 test of differences in calculator ownership according to major.

random variables; (2) creates histograms and frequency distributions for quantitative random variables; and (3) computes descriptive summary measures (see Chapters 3 and 4). The T-TEST subprogram computes the means and standard deviations for two independent samples and then performs the *t* test discussed in Section 10.2. Moreover, the T-TEST subprogram also performs the test for the equality of two population variances presented in Section 10.10. The CROSSTABS subprogram forms tables of cross classifications (see Sections 3.2 and 5.3) and performs the chi-square tests discussed in Sections 10.5 through 10.7. As examples, Figure 10.24 presents the SPSS

```
ANALYSIS OF STUDENT QUESTIONNAIRES,                                                11/22/76        PAGE    5

FILE   STDNTSVR (CREATION DATE = 11/22/76)   SURVEY OF ATTITUDES, INTERESTS, & CHARACTERISTICS       OF UNDERG

* * * * * * * * * * * * * * * * * * *  C R O S S T A B U L A T I O N   O F  * * * * * * * * * * * * * * * * * *
    MAJOR1     RECODED FOR ACCTG,MKTG, & OTHER                     BY  GRDSCHL   GRADUATE SCHOOL INTENTION
* * * * * * * * * * * * * * * * * * * * * * * * * * * * * * * * * * * * * * * * * * * * * * * * * * PAGE  1 OF

                    GRDSCHL
            COUNT  I
                   IYES      NO       NOT SURE   ROW
                   I                             TOTAL
                   I      1.I      2.I      3.I
MAJOR1     --------I--------I--------I--------I
               2.  I    11  I     8  I    16  I    35
  ACCOUNTING       I        I        I        I  37.2%
                  -I--------I--------I--------I
               5.  I    14  I     4  I     5  I    23
  MARKETING        I        I        I        I  24.5%
                  -I--------I--------I--------I
               8.  I    13  I     5  I    18  I    36
  OTHER OR UNDECID I        I        I        I  38.3%
                  -I--------I--------I--------I
            COLUMN     38       17       39        94
             TOTAL    40.4%    18.1%    41.5%   100.0%

CHI SQUARE =     6.98334 WITH   4 DEGREES OF FREEDOM   SIGNIFICANCE =  0.1368
```

Figure 10.26 SPSS output for χ^2 test of independence between major and graduate school intention.

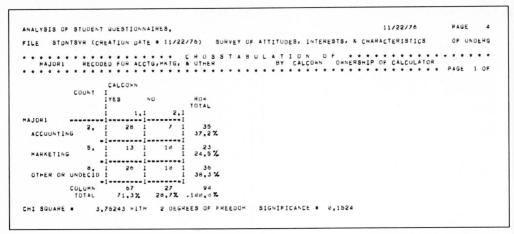

output for the T-TEST subprogram for testing whether there were any differences in the mean grade-point indexes of accounting versus non-accounting students (see Section 10.2), while Figure 10.25 presents the SPSS output for the CROSSTABS subprogram for testing whether or not there were any differences in calculator ownership according to major area of study (see Table 10.13). Lastly, Figure 10.26 presents the SPSS output for the CROSSTABS subprogram for determining whether or not there is a relationship between one's major area of study and one's intention to attend graduate school (see Table 10.18).

The reader, however, must again be cautioned to be thoroughly familiar with the capabilities of any such selected package in order to prevent errors in the results.

PROBLEMS

* 10.1 Management of the Dollar Bill Steel Co. wishes to determine if there is any difference in performance between the day shift of workers and the evening shift of workers. A sample of 100 day shift workers reveals an average output of 74.3 parts per hour with a standard deviation of 16 parts per hour. A sample of 100 evening shift workers reveals an average output of 69.7 parts per hour with a standard deviation of 18 parts per hour. At the .10 level of significance is there a difference in output between the day shift and the evening shift?

10.2 An independent testing agency has been contracted to determine whether there is any difference in gasoline mileage output of two different gasolines on the same model automobile. Gasoline A was tested on 200 cars and produced a sample average of 18.5 miles per gallon with a sample standard deviation of 4.6 miles per gallon. Gasoline B was tested on a sample of 100 cars and produced a sample average of 19.34 miles per gallon with a sample standard deviation of 5.2 miles per gallon. At the .05 level of significance is there a difference in performance of the two gasolines?

* 10.3 The budget director of a large retailing firm wanted to determine if there was any difference in the luncheon expense vouchers for executives of two departments of the firm. A random sample of 15 vouchers from Department 1 and 15 vouchers from Department 2 were selected with the following results:

Department 1	Department 2
$\bar{X}_1$ = \$27.50	$\bar{X}_2$ = \$23.25
S_1 = \$ 6.00	S_2 = \$ 7.50

At the .01 level of significance is the average luncheon expense voucher higher in Department 1?

10.4 The advertising manager of a breakfast cereal company wanted to know whether a new package shape would increase sales of the product. A random sample of 20 stores with the new package shape revealed weekly sales that averaged 130 pounds of this cereal with a standard deviation of 10 pounds. A random sample of 20 stores with the old package shape revealed weekly sales that averaged 117 pounds with a standard deviation of 12 pounds. At the .05 level of significance, did the new package shape result in increased sales?

10.5 A medical researcher was studying birth weights of babies. A random sample of five boys and six girls is selected from all babies born at the hospital in the past year. The birth weight (in pounds) of the babies was the following:

Boys	5.3	2.8	6.4	6.8	7.4	
Girls	8.0	4.7	7.3	6.2	3.4	5.5

(a) At the .01 level of significance is there a difference in birth weight of boys and girls born at the hospital in the past year?

(b) What assumptions are necessary to perform this test?

10.6 A consumer reporting agency wished to determine whether an "unknown brand" calculator sells at a lower price than the "famous brand" calculator of the same type. A random sample of eight stores was selected, and the prices at the stores of each of the two calculators was recorded with the following results:

Store	Unknown Brand	Famous Brand
1	10	11
2	8	11
3	7	10
4	9	12
5	11	11
6	10	13
7	9	12
8	8	10

(a) At the .01 level of significance does the unknown brand sell for a lower price?

(b) What assumption is necessary to perform this test?

* 10.7 A systems analyst is testing the feasibility of using a new computer system. The analyst will switch the processing to the new system only if there is evidence that the new system uses less processing time than the old system. In order to make a decision, a sample of seven jobs was selected and the processing time in seconds was recorded on the two systems with the following results:

Job	Old	New
1	8	6
2	4	3
3	10	7
4	9	8
5	8	5
6	7	8
7	12	9

(a) At the .01 level of significance does the new system use less processing time?

(b) What assumption is necessary to perform this test?

10.8 The data tabulated below represent the closing prices over a two-day period of a random sample of ten stocks listed on the New York Stock Exchange.

	Stock									
Closing Prices	*A*	*B*	*C*	*D*	*E*	*F*	*G*	*H*	*I*	*J*
Monday	84¼	109½	82½	57½	93	70	63	28½	49¼	66¼
Tuesday	83½	109¼	83	58¼	94½	69½	63¼	28¼	50½	68

(a) At the .05 level of significance is there evidence to enable a security analyst to conclude that stock market prices have significantly increased over the two-day period?

(b) What assumption is necessary to perform this test?

* 10.9 A financial analyst wishes to know whether there had been a significant change in earnings per share from one period of time to another among the largest industrial corporations in the United States. A random sample of 15 companies selected from the *Fortune* 500 yielded the following results:

	Earnings per Share ($)	
Company	*Year 1*	*Year 2*
1	4.12	4.79
2	2.85	3.20
3	2.81	3.34
4	3.39	1.94
5	2.03	−2.86
6	4.91	3.69
7	2.28	2.50
8	4.10	4.30
9	6.39	7.16
10	0.52	1.78
11	2.44	0.80
12	−2.25	−1.31
13	5.01	5.06
14	1.85	2.15
15	1.95	2.07

(a) At the .01 level of significance is there a difference in earnings per share between the two years?

(b) What assumption is necessary to perform this test?

10.10 In order to measure the effect of a storewide sales campaign on other nonsaleable items, the research director of a national supermarket chain took a random sample of 13 pairs of stores which were matched according to average weekly sales volume. One store of each pair (the experimental group) was exposed to the sales campaign, while the other member of the pair (the control group) was not. The following data indicate the results over a weekly period:

	Sales ($000) of Nonsaleable Items	
Store	*With Sales Campaign*	*Without Sales Campaign*
1	67.2	65.3
2	59.4	54.7
3	80.1	81.3
4	47.6	39.8
5	97.8	92.5
6	38.4	37.9
7	57.3	52.4
8	75.2	69.9
9	94.7	89.0
10	64.3	58.4
11	31.7	33.0
12	49.3	41.7
13	54.0	53.6

(a) At the .05 level of significance can the research director conclude that the sales campaign has increased the sales of nonsaleable items?

(b) What assumption is necessary to perform this test?

10.11 We wish to determine if there is any difference in the popularity of football between college-educated males and non-college-educated males. A sample of 100 college-educated males revealed 55 who considered themselves football fans. A sample of 200 non-college educated males revealed 125 who considered themselves football fans. Is there any difference in football popularity between college-educated and non-college-educated males at the .01 level of significance?

10.12 A marketing study conducted in Detroit showed that in a random sample of 100 married women who work full-time, 64 prefer to buy instant coffee. However, in a random sample of 100 married women who work part-time, only 56 prefer to buy instant coffee. Using a .05 level of significance, is there any difference between the proportion of the two groups of married women that prefer instant coffee?

* 10.13 A housing survey is to be taken in New York and San Francisco to determine the proportion of housing units occupied by high-income families. A random sample of 600 housing units in New York revealed 150 units occupied by high-income families. A sample of 300 units in San Francisco revealed 120 units occupied by high-income families.

(a) Use two different statistical techniques to determine if there is any difference between New York and San Francisco in the proportion of housing units occupied by high-income families.

(b) Compare the results obtained from the two methods in (a).

(c) If you wanted to know whether San Francisco had a higher proportion of housing units occupied by high income families than New York, what method would you use to perform the statistical test?

Note: Use a level of significance of .01 throughout the problem.

10.14 An accountant was studying the readability of the annual reports of two major companies. A random sample of 100 certified public accountants was selected. 50 were randomly assigned to read the annual report of Company A, and the other 50 were to read the annual report of company B. Based upon a standard measure of readability 17 found Company A's annual report "understandable" and 23 found Company B's annual report "understandable." At the .10 level of significance is there any difference between the two companies in the proportion of CPAs that found the annual report understandable?

10.15 We wish to determine whether there is a sex difference in preference for margarine versus butter. A sample of 80 males indicated that 28 preferred margarine to butter. A sample of 120 females indicated that 52 preferred margarine to butter. Is there any difference in preference for margarine versus butter between males and females at the .05 level of significance?

10.16 A product testing organization was interested in studying Neveready and Pennysonic transistor radio batteries to determine the number of hours that the batteries lasted. Random samples of 25 Neveready batteries and 25 Pennysonic batteries indicated the following results:

Neveready	Pennysonic
$\bar{X}$ = 110.6 hours	$\bar{X}$ = 103.8 hours
S = 10 hours	S = 12 hours
16 lasted more than 100 hours	14 lasted more than 100 hours

Use hypothesis testing to draw conclusions about **each** of the following:

(a) Is the proportion of Neveready batteries that last more than 100 hours equal to .50?

(b) Do Pennysonic batteries last more than an average of 95 hours?

(c) Is there a difference in average hours between Neveready and Pennysonic batteries?

(d) Is there a difference between Neveready and Pennysonic in the proportion of batteries that last more than 100 hours?

Note: Use a level of significance of .10 in this entire problem.

* 10.17 The R & M department store has two charge plans available for its credit account customers. The management of the store wishes to collect information about each plan and study the differences between the two plans. It is interested in the average monthly balance and percentage of monthly balances above $100. A random sample of 25 accounts of plan A and 50 accounts of plan B was selected with the following results:

Plan A	Plan B
$n_A = 25$	$n_B = 50$
$\bar{X}_A = \$75$	$\bar{X}_B = \$110$
$S_A = \$15$	$S_B = \$14.14$
5 monthly balances above $100	25 monthly balances above $100

Use statistical inference (confidence intervals or tests of hypothesis) to draw conclusions about **each** of the following:

(a) Average monthly balance of plan B accounts.

(b) Proportion of plan A accounts with monthly balances above $100.

(c) Is the average monthly balance of plan A accounts equal to $105?

(d) Is the proportion of plan B accounts with monthly balances above $100 equal to .40?

(e) Is there a difference in average monthly balance between plan A and plan B?

(f) Is there any difference between plan A and plan B in the proportion of accounts with monthly balances above $100?

Note: Use a level of significance of .01 (99% confidence) throughout the problem.

10.18 The alumni association of a large urban school of business has selected a random sample of 100 graduates from the class of 1966. In the sample 36 were accounting majors and 64 were non-accounting majors. Summary information concerning average annual income and number of graduates earning $30,000 per year revealed the following:

Accounting	Non-accounting
$n_1 = 36$	$n_2 = 64$
$\bar{X}_1 = \$24{,}000$	$\bar{X}_2 = \$21{,}000$
$S_1 = \$3{,}000$	$S_2 = \$4{,}000$
15 earn at least $30,000 per year	15 earn at least $30,000 per year

Use hypothesis testing to draw conclusions about each of the following (.05 level of significance):

(a) Is the average salary of non-accounting majors equal to $20,000?

(b) Is the proportion of accounting majors earning at least $30,000 per year equal to .50?

(c) Is the average annual income higher for accounting majors than for non-accounting majors?

(d) Is there a difference in the proportion of accounting and non-accounting majors earning at least $30,000 per year?

10.19 The superintendent of a school district wished to study absenteeism of teachers in the school district during the past academic year. A random sample of 50 teachers in the district was selected—25 teachers from the primary schools and 25 teachers from the secondary schools—with the following results:

Primary	Secondary
$n_1 = 25$	$n_2 = 25$
$\bar{X}_1 = 8.6$ days	$\bar{X}_2 = 9.7$ days
$S_1 = 3.0$ days	$S_2 = 4.0$ days
8 were absent more than 10 days	12 were absent more than 10 days

Use statistical inference (confidence intervals and hypothesis testing) to draw conclusions about each of the following:

(a) Average days absent of primary school teachers.

(b) Proportion of secondary school teachers who are absent more than 10 days.

(c) Is the average days absent less than 9 days for primary school teachers?

(d) Is the proportion of secondary school teachers that are absent more than 10 days equal to .50?

(e) Is there a difference in average days absent between primary and secondary school teachers?

(f) Is there a difference in the proportion of primary and secondary school teachers that are absent more than 10 days per year?

Note: Use a .05 level of significance (95% confidence) throughout the problem.

10.20 An auditor would like to know if there is a difference in the proportion of improper travel expense vouchers in three different departments of a company. A random sample of 25 vouchers from department A, 25 vouchers from department B and 50 vouchers from department C revealed the following information:

	Department		
	A	*B*	*C*
Improper vouchers	6	5	9
Proper vouchers	19	20	41

Is there any difference in the proportion of improper vouchers between departments A, B, and C at the .05 level of significance?

* 10.21 The faculty council of a large university would like to determine the opinion of various groups toward a proposed trimester academic calendar. A random sample of 100 undergraduate students, 50 graduate students, and 50 faculty members is selected with the following results:

Opinion	Undergraduates	Graduate	Faculty
Favor trimester	63	27	30
Oppose trimester	37	23	20
Totals	100	50	50

At the .01 level of significance is there evidence of a difference in attitude toward the tri-semester between the various groups?

10.22 An agronomist is studying three different varieties of tomato to determine whether there is a difference in the proportion of seeds that germinate. Random samples of 100 seeds of each variety (beefsteak, plum, and cherry) are subjected to the same starting conditions with the following results:

	Tomato Variety		
Number of Seeds	*Beefsteak*	*Plum*	*Cherry*
Germinated	82	70	58
Did not germinate	18	30	42
Totals	100	100	100

At the .10 level of significance is there a difference between the varieties of tomatoes in the proportion of seeds that germinate?

* 10.23 The director of a large shopping center would like to know if there are differences in the proportion of women shoppers at various times during the

week. Random samples of 300 weekday shoppers, 300 weeknight shoppers, and 400 weekend shoppers were selected with the following results:

	Weekday	Weeknight	Weekend
Male	90	125	185
Female	210	175	215
Totals	300	300	400

At the .05 level of significance is there a difference in the proportion of women shoppers at the various times of the week?

10.24 The quality control manager of an automobile parts factory would like to know if there is a difference in the proportion of defective parts produced on different days of the work week. Random samples of 100 parts produced on each day of the week were selected with the following results:

	Mon.	Tues.	Wed.	Thurs.	Fri.
Number of defective parts	12	7	7	10	14
Number of acceptable parts	88	93	93	90	86
Totals	100	100	100	100	100

At the .05 level of significance is there a difference in the proportion of defective parts produced on the various days of the week?

* 10.25 The Statistics Association at a large state university would like to determine whether there is a relationship between student interest in statistics and ability in mathematics. A random sample of 200 students is selected and they are asked whether their ability in mathematics and interest in statistics are low, average, or high. The results were as follows:

Interest in Statistics	Ability in Mathematics: Low	Average	High	Totals
Low	60	15	15	90
Average	15	45	10	70
High	5	10	25	40
Totals	80	70	50	200

At the .01 level of significance is there a relationship between interest in statistics and ability in mathematics?

10.26 A manufacturer of automobile batteries wishes to determine whether there are any differences on three different media (TV, radio, magazine) in terms of recall of an ad. The results of an advertising study were as follows:

Recall Ability	Media: Magazine	TV	Radio	Totals
Number of persons remembering ad	25	10	7	42
Number of persons not remembering ad	73	93	108	274
Totals	98	103	115	316

At the .10 level of significance determine whether there is a relationship between media and ability to recall the ad.

10.27 A nationwide market research study was undertaken to determine the preferences of various age groups of males for different sports. A random sample of 1,000 men was selected, and each individual was asked to indicate his favorite sport. The results were as follows:

Age Group	Sport: Baseball	Football	Basketball	Hockey	Totals
Under 20	26	47	41	36	150
20–29	38	84	80	48	250
30–39	72	68	38	22	200
40–49	96	48	30	26	200
50 and over	134	44	18	4	200
Totals	366	291	207	136	1,000

At the .01 level of significance is there a relationship between age of men and preference for sports?

10.28 Suppose a survey has been undertaken to determine if there is a relationship between the size of a city in which someone resides and the company preference in the purchase of an automobile. A random sample of 200 car owners from large cities, 150 from suburbs, and 150 from rural areas were selected with the following results:

Type of Area	Company: GM	Ford	Chrysler	American	Foreign	Totals
Large city	64	40	26	8	62	200
Suburb	53	35	24	6	32	150
Rural	53	45	30	6	16	150
Totals	170	120	80	20	110	500

At the .05 level of significance is there a relationship between type of area of residence and company preference in an automobile purchase?

* 10.29 The owner of a home heating oil delivery firm would like to investigate how fast bills are paid in three different suburban areas. Random samples of 100 accounts are selected in each area. The number of days between delivery of the oil and payment of the bill is recorded with the following results:

Days before Payment	Area: I	II	III
1–15	34	42	40
16–30	48	50	46
More than 30	18	8	14
Totals	100	100	100

At the .10 level of significance is there a difference between the three areas in how quickly the oil bills are paid?

10.30 The manager of a computer facility has collected data on the number of times that service to users was interrupted (usually due to machine failure) in each day over the past 500 days.

Interruptions per Day	Number of Days
0	160
1	175
2	86
3	41
4	18
5	12
6	8
	500

Does the distribution of service interruptions come from a Poisson distribution at the .01 level of significance?

10.31 A statistician for a professional hockey team collected data on the number of goals scored per game by the team in an 80-game season. The results from last year indicated the following:

Number of Goals Scored	Frequency
0	5
1	10
2	15
3	18
4	14
5	11
6	3
7	2
8	0
9	1
10	1
	80

Does the distribution of number of goals scored follow a Poisson distribution at the .05 level of significance?

10.32 Referring to Problem 10.30, at the .01 level of significance does the distribution of service interruptions follow a Poisson distribution with a population mean of 1.5 interruptions per day?

10.33 Referring to Problem 10.31, at the .05 level of significance does the distribution of goals scored follow a Poisson distribution with a population mean of 3 goals scored per game?

10.34 The manager of the commercial mortgage department of a large bank has collected data during the past 2 years concerning the number of commercial mortgages approved per week. The results from these 2 years (104 weeks) indicated the following:

Number of Commercial Mortgages Approved	Frequency
0	13
1	25
2	32
3	17
4	9
5	6
6	1
7	1
	104

Does the distribution of the number of commercial mortgages approved per week in the last 2 years follow a Poisson distribution at the .01 level of significance?

* 10.35 The number of customers waiting for service on the express (checkout counter) line of a large supermarket is examined at random on 50 occasions during a 12-hour period. The results are as follows:

3	5	1	4	0	6	4	3	4	2
0	4	8	3	5	5	6	5	7	4
4	2	6	9	3	4	1	2	3	3
4	3	3	5	1	3	4	5	2	5
7	1	4	2	12	2	7	4	6	3

Fit a Poisson distribution to the number of customers waiting for service per occasion and, using a level of significance of .01, test for goodness of fit.

10.36 The military historian Von Bortkewitsch (1898) obtained the following data on the number of deaths (per year per corps) due to being kicked in the head by a horse in the Prussian cavalry (Reference 2 of Chapter 6). Fit a Poisson distribution to these data and, using a level of significance of .05, test for goodness of fit.

Number of Deaths (per Year per Corps)	Frequency
0	109
1	65
2	22
3	3
4	1
	200

10.37 Roll a single die 60 times and, for each roll, make a tally of the face which appears. Fit a uniform distribution to these data (see Table 6.1) and, using a level of significance of .05, test for goodness of fit.

Hint: The appropriate degrees of freedom for data fitted by a uniform distribution are the number of (remaining) groups or classes (after possible condensing) minus 1.

10.38 To check the uniformity of the table of random numbers (Appendix E, Table E.1) make a tally of the 400 digits 0, 1, 2, . . . , 9 which appear in rows 1 through 10. Fit a uniform distribution to these data and, using a level of significance of .01, test for goodness of fit.

Hint: The appropriate degrees of freedom for data fitted by a uniform distribution are the number of (remaining) groups or classes (after possible condensing) minus 1.

10.39 Roll a pair of dice 180 times and, for each roll, make a tally of the sum of the two numbers that appear. Fit a triangular distribution to these data (see Problem 6.1) and, using a level of significance of .05, test for goodness of fit.

Hint: The appropriate degrees of freedom for data fitted by this triangular distribution are the number of (remaining) groups or classes (after possible condensing) minus 1.

10.40 The manager of computer operations of a large company wishes to study computer usage of two departments within the company, the accounting department and the research department. A random sample of 5 jobs from the accounting department in the last month and 6 jobs from the research department in the last month were selected, and the processing time (in seconds) for each job was recorded with the following results:

Accounting	9	3	8	7	12	
Research	4	13	10	9	9	6

Draw conclusions about each of the following:

(a) Is the standard deviation of the processing time of the accounting department greater than 1 second?

(b) Is there a difference in processing time between the accounting department and the research department?

(c) What assumptions must be made in order to do (b)?

(d) Is there a difference in the variance between the accounting department and the research department?

(e) What assumption is needed to do (d)?

Note: Use a .05 level of significance throughout the problem.

* 10.41 A consumer reporting agency wished to compare the price of a particular brand of a calculator in two different cities. A random sample of six stores in one city and eight stores in a second city were selected with the following results:

City I	\$10	12	9	14	12	10		
City II	\$13	16	8	12	14	13	11	14

Draw conclusions about each of the following:

(a) Is the standard deviation in city I equal to \$2?

(b) Is there a difference in the price of the calculator in the two cities?

(c) What assumptions must be made in order to do (b)?

(d) Is there a difference in the variances in the two cities?

(e) What assumption is needed to do (d)?
Note: Use a .05 level of significance throughout this problem.

10.42 Referring to Problem 9.5, based upon the sample of steel bars, at the .05 level of significance is there evidence that the population standard deviation is greater than 0.10 foot?

* 10.43 Referring to Problem 9.6, based upon the sample of accounts, at the .01 level of significance, does the standard deviation differ from $3,000?

10.44 Referring to Problem 9.7, based upon the sample of stores, at the .05 level of significance is there evidence that the population standard deviation is greater than $15?

* 10.45 Referring to Problem 10.3, at the .05 level of significance is the variance in luncheon expense vouchers in department 2 greater than in department 1?

10.46 Referring to Problem 10.5, at the .01 level of significance is there a difference in the variances of birth weights for boys and girls?

10.47 Referring to Problem 10.19, at the .10 level of significance is the variance in absenteeism of secondary school teachers higher than the variance for primary school teachers?

The following problems refer to the sample data obtained from the questionnaire of Figure 2.2 and presented in Figure 2.6.

10.48 Using a level of significance of .05, determine whether there is a difference in high-school averages attained by accounting versus non-accounting students (see Problem 4.15).

10.49 Using a level of significance of .01, determine whether there is a difference in anticipated starting salaries of accounting versus non-accounting majors (see Problem 4.16).

10.50 Using a level of significance of .01, determine whether there is a difference in anticipated starting salaries of juniors versus seniors (see Problem 4.17).

10.51 Using a level of significance of .10, determine whether there is a difference in the number of pairs of jeans owned by accounting versus non-accounting students (see Problem 4.18).

10.52 Using a level of significance of .05, determine whether there is a difference in the grade-point averages achieved by juniors versus seniors (see Problem 4.23).

10.53 Using a level of significance of .05, determine whether there is a difference in the grade-point indexes of students planning to attend graduate school versus those who do not intend to go (see Problem 4.24).

10.54 Using a level of significance of .01, determine whether there is a difference in the anticipated starting salaries of students planning to attend graduate school versus those who do not plan to go (see Problem 4.25).

10.55 Using a level of significance of .05, determine whether there is a difference in the grade-point indexes attained by students favoring a twelfth-year English and mathematics standard of admissions versus those who favor open admissions for all high school graduates.

10.56 Using a level of significance of .05, determine whether there is a difference in the grade-point indexes attained by students desiring stringent retention standards versus those who do not (see Problem 4.29).

10.57 Using a level of significance of .05, determine whether there is a difference in the heights of males who smoke versus those males who do not smoke at all (see Figure 2.6, questions 1, 16, and 19).

10.58 Using a level of significance of .05, determine whether there is a difference in the weights of female students who smoke versus female students who do not smoke at all (see Figure 2.6, questions 1, 17, and 19).

10.59 Using a level of significance of .10, determine whether there is a difference in the ages of male versus female students (see Figure 2.6, questions 1 and 15).

10.60 Using a level of significance of .01, determine whether there is a difference in the anticipated starting salaries of male versus female students (see Figure 2.6, questions 1 and 7).

10.61 Using a level of significance of .05, determine whether there is a difference in the amount of tuition unemployed students versus full-time employed students claim they are willing to pay (see Figure 2.6, questions 8 and 9).

10.62 Using a level of significance of .05, determine whether there is a relationship between class designation and graduate school intention (see Problem 5.10(f)).

10.63 Using a level of significance of .10, determine whether there is a relationship between sex and attitude toward Sunday shopping (see Problem 5.11(f)).

10.64 Using a level of significance of .01, determine whether there is a relationship between smoking and attitude toward smoking in the classroom (see Problem 5.12(f)).

10.65 Using a level of significance of .05, determine whether there is a relationship between sex and class designation (see Problem 5.13(e)).

10.66 Using a level of significance of .01, determine whether there is a relationship between class designation and employment status (see Problem 5.14(e)).

10.67 Using a level of significance of .05, determine whether there is a relationship between sex and blood pressure (see Problem 5.15(e)).

10.68 Using a level of significance of .05, determine whether there is a relationship between class designation and attitude toward retention standards (see Problem 5.16(e)).

10.69 Using a level of significance of .05, determine whether there is a relationship between attitude toward admission standards and attitude toward retention standards (see Figure 2.6, questions 21 and 22).

10.70 Using a level of significance of .10, determine whether there is a relationship between graduate school intentions and calculator ownership (see Figure 2.6, questions 3 and 14).

10.71 Using a level of significance of .05, determine whether there is a relationship between sex and investment intent (see Figure 2.6, questions 1 and 11).

10.72 Using a level of significance of .05, determine whether there is a relationship between sex and automobile preference (see Figure 2.6, questions 1 and 12).

10.73 Using a level of significance of .01, determine whether there is a relationship between sex and attitude toward admission standards (see Figure 2.6, questions 1 and 21).

10.74 Using a level of significance of .01, determine whether there is a relationship between sex and attitude toward retention standards (see Figure 2.6, questions 1 and 22).

10.75 Using a level of significance of .10, determine whether there is a relationship between employment status and graduate school intent (see Figure 2.6, questions 3 and 9).

10.76 Use your own sample data (Problem 3.12) and select problems from among Problems 10.48 through 10.75.

10.77 For your own class (Problem 3.13), select problems from among Problems 10.48 through 10.75.

10.78 For the sample of $n = 94$ students, fit a Poisson distribution to the data on the number of pairs of jeans owned (see Figure 2.6, question 13) and, using a level of significance of .01, test for goodness of fit.

10.79 For the sample of $n = 94$ students, fit a uniform distribution to the data on the ratings of Student Personnel Services (see Figure 2.6, question 24) and, using a level of significance of .05, test for goodness of fit.

Hint: The appropriate degrees of freedom for data fitted by a uniform distribution are the number of (remaining) groups or classes (after possible condensing) minus 1.

10.80 For the sample of $n = 94$ students, fit a uniform distribution to the data on the ratings of Library Services (see Figure 2.6, question 25) and, using a level of significance of .05, test for goodness of fit.

Hint: The appropriate degrees of freedom for data fitted by a uniform distribution are the number of (remaining) groups or classes (after possible condensing) minus 1.

REFERENCES

1. BRADLEY, J. V., *Distribution-Free Statistical Tests* (Englewood Cliffs, N.J.: Prentice-Hall, 1968).
2. DIXON, W. J., AND M. B. BROWN, *BMDP,* (Berkeley, Calif.: University of California Press, 1977).
3. GIBBONS, J. D., *Nonparametric Methods for Quantitative Analysis* (New York: Holt, Rinehart and Winston, 1976).
4. HAMBURG, M., *Basic Statistics: A Modern Approach* (New York: Harcourt Brace Jovanovich, 1974).
5. KLECKA, W., N. NIE, AND C. HULL, *SPSS Primer* (New York: McGraw-Hill, 1975).
6. NETER, J., W. WASSERMAN, AND G. WHITMORE, *Fundamental Statistics for Business and Economics,* 4th ed. (Boston: Allyn and Bacon, 1973).
7. NIE, N., C. HULL, J. JENKINS, K. STEINBRENNER, AND D. BENT, *Statistical Package for the Social Sciences,* 2nd ed. (New York: McGraw-Hill, 1975).
8. SERVICE, J., *A User's Guide to the Statistical Analysis System* (Raleigh: North Carolina State University Press, 1972).
9. SNEDECOR, G. W., AND W. G. COCHRAN, *Statistical Methods,* 6th ed. (Ames, Iowa: Iowa State University Press, 1967).
10. *STATPACK: Statistical Package,* 2nd ed., developed by IBM, February 1970.

CHAPTER 11

BAYESIAN DECISION MAKING

11.1 INTRODUCTION

In the last three chapters we have investigated what is usually referred to as the classical approach to decision making; that is, we have examined procedures for estimating parameters, such as means and proportions, and we have explored methods of decision making through hypothesis testing. In contrast to this approach, the past 20 years have witnessed the rapid development of alternative methods to decision making, called the Bayesian approach. Whereas the classical approach does not directly take into account the subjective or otherwise "prior" probability of various events, the Bayesian approach revises any prior information available to take into account sample data that have been collected.

Any decision-making situation contains four basic features:

1. **Alternative courses of action**—The decision maker must have several possible choices to evaluate prior to selecting one course of action. For example, the dean can use the survey results to help decide whether to reor-

ganize student personnel or library services; a consumer must decide which brand of a product to purchase; a marketing manager must determine whether a particular toy will be profitable.

2. **Events or states of the world**—If only one possible event (such as one brand always being preferred) can occur, the decision-making process can be quite simple. However, in most instances there are several events that can occur. These events are sometimes referred to as "states of the world." Sometimes the probability of occurrence of each of these events can be estimated. In other cases no information is available to determine the likelihood of each event.

3. **Payoffs**—In order to evaluate each possible course of action, the result of each event with each course of action must have a value (or payoff) placed upon it. In business problems this value is often expressed in terms of profits or costs, although other payoffs such as units of satisfaction or *utility* are sometimes utilized.

4. **Decision criteria**—The decision maker must determine the manner in which the best course of action will be determined. One criterion that is often utilized in business is to choose the alternative that results in the largest average profit. However, other criteria are available for making decisions, particularly when little or no information is available concerning the probability of occurrence of the various events.

Thus, in this chapter various decision-making problems in business will be developed in detail. We will examine the effect of sample information on the decision process and look at several criteria used to make business decisions.

11.2 THE PAYOFF TABLE

In order to see how Bayesian decision making can be utilized, let us examine two different problems.

First, let us return to the problem faced by the marketing manager of the toy company which was presented in Section 5.10. The marketing manager of this toy company must determine whether a new toy should be introduced onto the market. He is aware that there are risks of making an incorrect decision depending on the ultimate success of this toy. For example, the toy could be marketed and turn out to be unsuccessful. On the other hand, the manager could decide not to market a toy that would have been a success.

In order to evaluate these possibilities, a payoff table can be constructed. This table is a two-way table that contains each possible **event** (state of the world) that can occur for each **alternative course of action.** For each combination of an event with a particular course of action, the **payoff** must be available. For our purposes we shall primarily consider payoffs in terms of profits or costs. In the toy problem the payoff table, assuming only two possible events (a successful toy or an unsuccessful toy), is indicated in Table 11.1.

Table 11.1
PAYOFF TABLE FOR TOY MARKETING PROBLEM

	Alternative Courses of Action, A_i	
Event, E_j	Market, A_1	Do Not Market, A_2
Successful toy, E_1	X_{11}	X_{21}
Unsuccessful toy, E_2	X_{12}	X_{22}

X_{ij} is the payoff that occurs when course of action i is selected and event j occurs. The marketing manager would have to determine the payoffs or profits for each action–event combination. Let us say that there is a fixed cost of $3,000 incurred prior to making the decision to market the toy. Based on past experience, the marketing manager determines that if the toy is successful, a profit of $45,000 ($48,000 minus $3,000 in fixed costs) will be obtained. If the toy is not successful, there will be a loss of $36,000 ($33,000 loss in the marketing of the product plus $3,000 in fixed costs). These payoffs can now be inserted into the original payoff table as shown in Table 11.2.

As depicted in Table 11.2, a loss of $3,000 will occur if the toy is not marketed, regardless of whether the toy actually would have been successful. The $3,000 represents the fixed cost incurred regardless of the marketing decision.

In this first problem the decision structure contained only two possible alternative courses of action and two possible events. In general, however, there can be any number of alternative courses of action and events. This can be observed by examining a second problem.

The manufacturer of swimming pools must determine (in advance of the summer season) the number of swimming pools that should be produced for the coming year. Each swimming pool costs $500 to produce and is sold for $1,000. Any swimming pool not sold by the end of the season can be disposed of for $300 each at an end of season clearance sale. In order to determine the correct production level, the manufacturer must obtain information pertaining to the demand for swimming pools. To simplify the problem, based upon past experience, the following three levels of demand are postulated:

1. Low demand—considered to mean that 1,000 pools would be demanded

Table 11.2
COMPLETED PAYOFF TABLE FOR TOY MARKETING PROBLEM

	Alternative Courses of Action, A_i	
Event, E_j	Market, A_1	Do Not Market, A_2
Successful toy, E_1	+$45,000	−$3,000
Unsuccessful toy, E_2	−$36,000	−$3,000

2. Moderate demand—considered to mean that 5,000 pools would be demanded
3. High demand—considered to mean that 10,000 pools would be demanded

Since we have condensed the actual demand function into three categories (low, moderate, high), we can view our alternative courses of action in terms of these categories. Therefore our production levels could be set at 1,000 pools or 5,000 pools or 10,000 pools, corresponding to the possible levels of demand. Once these possibilities are set up, the profits for each alternative–event combination can be computed.

If 1,000 pools were produced, and 1,000 pools were demanded, the profit would be computed by taking

$$(\text{Profit per Pool Sold})(\text{Number of Pools Sold}) + (\text{Loss per Pool Not Sold})(\text{Number of Unsold Pools})$$

Since all 1,000 pools would be sold, we have

$$\text{Profit per Pool Sold} = (\$1{,}000 - \$500) = \$500$$
$$\text{Loss per Pool Unsold} = (\$300 - \$500) = -\$200$$

so that

$$(\$500)(1{,}000) + (-\$200)(0) = \$500{,}000$$

The profit for this combination would be \$500,000. If 5,000 pools were demanded, the profit would still be \$500,000, since only 1,000 pools would have been produced. However, there would be an undefined "opportunity" cost (not included in this problem) of failing to satisfy the demand of 4,000 customers. In a similar manner, if 10,000 pools were demanded, the profit would still be \$500,000, although 9,000 customers would not be satisfied and might pursue alternative producers in the future.

If 5,000 pools were produced we can again determine the profit if the demand were 1,000, 5,000, and 10,000 pools, respectively. If 1,000 pools were demanded, the profit would be

$$\begin{aligned}(+\$500)(1{,}000) + (-\$200)(4{,}000) &= \$500{,}000 + (-\$800{,}000)\\ &= -\$300{,}000\end{aligned}$$

Therefore, in this case there would be a loss of \$300,000. However, if 5,000 pools were demanded, there would be a profit of \$2,500,000 (\$500 × 5,000 pools), since all 5,000 pools produced would be sold. Likewise, if 10,000 pools were demanded, the profit would still be \$2,500,000, since the producer only has 5,000 pools to sell. We should realize however that the demand of the additional 5,000 customers would not be satisfied.

Table 11.3
PAYOFF TABLE FOR POOL PRODUCTION PROBLEM

	Alternative Courses of Action, A_i		
	Production Levels		
Event, E_j	1,000	5,000	10,000
Low demand, 1,000 pools	+$500,000	−$300,000	−$1,300,000
Moderate demand, 5,000 pools	+$500,000	+$2,500,000	+$1,500,000
High demand, 10,000 pools	+$500,000	+$2,500,000	+$5,000,000

If 10,000 pools were produced, the profits for the various levels of demand would be as follows:

1. 1,000 pools demanded, 10,000 pools produced:

$$\begin{aligned}\text{Profit} &= 1{,}000(+\$500) + 9{,}000(-\$200)\\ &= +\$500{,}000 + (-\$1{,}800{,}000)\\ &= -\$1{,}300{,}000\end{aligned}$$

2. 5,000 pools demanded, 10,000 pools produced:

$$\begin{aligned}\text{Profit} &= 5{,}000(+\$500) + 5{,}000(-\$200)\\ &= +\$2{,}500{,}000 + -\$1{,}000{,}000\\ &= +\$1{,}500{,}000\end{aligned}$$

3. 10,000 pools demanded, 10,000 pools produced:

$$\begin{aligned}\text{Profit} &= 10{,}000(+\$500) + 0(-\$200)\\ &= +\$5{,}000{,}000\end{aligned}$$

The completed payoff table for this example is presented in Table 11.3.

11.3 DECISION MAKING USING EXPECTED MONETARY VALUE

Now that the profit for each event under each alternative course of action has been indicated in the payoff table, we need to consider how the best course of action will be chosen. In many cases no information is available about the probability of occurrence of the various events. However, in other instances the probabilities of the events can be estimated in several ways. First, information may be available from past experience that can be used to estimate the probabilities. For example, the marketing manager may be able to determine the proportion of similar toys that have been successful in the past and use that information to estimate the chance of success for the new toy. Second, the manager can combine this information with his own opinion, to subjectively assess the likelihood of the various events. For example, the producer of swimming pools can combine the experience of what has occurred in previous years with his own assessment of the demand for this year to develop the subjective probabilities of the events. These probabilities are subjective because each individual would evaluate information differently, thereby producing different sets of probabilities for the events. Third, the probabilities of the events could follow a particular distribution such as the normal, binomial, or Poisson distribution.

If the probabilities of the various events can be obtained, this information, along with the payoffs for each event–action combination, can be used to determine the best course of action. We have already used such a procedure in Section 6.2.1 to determine whether to play the carnival game. In that case we were deciding whether it would be profitable to enter this game of chance. The expected value, $E(X)$, of the game was computed by multiplying the return on each outcome, X, by the probability of occurrence of each outcome, $P(X)$, and then summing the results. Since the expected value was less than the cost of playing the game, we decided not to play the game, since we could expect that the average loss would be \$.50 each time we played the game. When this criterion is applied to decision-making problems, it is referred to as **expected monetary value** (EMV), since the expected profit of each alternative is being computed. The expected monetary value indicates the average profit that would be gained if a particular alternative was selected in many similar decision-making situations. Since decisions are made on a one-time basis, the decision criterion is to choose the alternative course of action that maximizes the expected monetary value.

The expected monetary value for a course of action i is the profit for each combination of event j and action $i(X_{ij})$ times the probability of occurrence of the event P_j summed over all events. This expression is given by

$$\text{EMV}_i = \sum_j X_{ij} P_j \tag{11.1}$$

where EMV_i = expected monetary value of action i

X_{ij} = payoff of action i for event j

P_j = probability of occurrence of event j

We may recall (see Section 5.10) that based upon past experience, the marketing manager of the toy company has subjectively assigned a probability of 0.40 to the event "successful toy" and a probability of .60 to the event "unsuccessful toy." Once these probabilities are available, the expected monetary value for alternative A_i can be determined by multiplying the payoff for each combination of event j and alternative i by its corresponding probability (P_j) and summing over all the events. These computations are illustrated for the toy example in Table 11.4.

Table 11.4
EXPECTED MONETARY VALUE FOR EACH ALTERNATIVE FOR THE TOY MARKETING PROBLEM

		Alternative Courses of Action, A_i			
P_j	Event, E_j	Market, A_1	$X_{1j}P_j$	Do Not Market, A_2	$X_{2j}P_j$
.4	Successful toy	+\$45,000	(\$45,000)(.4) = +\$18,000	−\$3,000	(−\$3,000)(.4) = −\$1,200
.6	Unsuccessful toy	−\$36,000	(−\$36,000)(.6) = −\$21,600	−\$3,000	(−\$3,000)(.6) = −\$1,800
			EMV(A_1) = −\$3,600		*EMV(A_2) = −\$3,000 (Optimal Decision)

We have seen in this example that the expected profit for marketing this toy is −\$3,600 (an average loss of \$3,600) while the expected profit of not marketing the product is −\$3,000 (a loss of \$3,000 that represents the fixed cost incurred prior to the final decision). In this case the alternative that maximizes the expected monetary value (minimizes the loss) is not to market the product, since −\$3,000 is less negative than −\$3,600.

Turning to the second case, the manufacturer of the swimming pools must obtain an estimate of the demand for swimming pools for the next summer season. The manufacturer decides to estimate the probability of the various demand levels by utilizing sales data for the past 5 years and subjectively taking into account his assessment of the prevailing economic conditions for the coming year. Using these criteria, the probability of the various levels of demand are estimated as the following:

$$P(\text{low demand}) = 0.20$$
$$P(\text{moderate demand}) = 0.50$$
$$P(\text{high demand}) = 0.30$$

Once these demand levels have been assigned subjective probabilities, the expected monetary value for each alternative course of action can be determined. The computations for this problem are illustrated in Table 11.5.

The expected monetary value is highest when the action is taken to produce 10,000 swimming pools. This expected monetary value of +\$1,990,000 is slightly higher than the amount for producing 5,000 swimming pools (+\$1,940,000) and is substantially greater than the expected monetary value achieved for the production of 1,000 pools (\$500,000). Therefore, based upon the *expected monetary value* criterion, the optimal action for the manufacturer would be to produce 10,000 swimming pools for the coming summer season. One should realize, however, that a prudent businessman, upon examining these results, may not choose to produce as many as 10,000 swimming pools. If 10,000 pools are produced, there is a 20% chance that the manufacturer will lose \$1,300,000 if this action is taken. Therefore the manufacturer possibly would choose to produce 5,000 pools, an action that will result

Table 11.5
EXPECTED MONETARY VALUE FOR EACH ALTERNATIVE FOR THE SWIMMING POOL PROBLEM

		Alternative Courses of Action, A_i					
		Production Levels					
P_j	Event, E_j	1,000	$X_{1j}P_j$	5,000	$X_{2j}P_j$	10,000	$X_{3j}P_j$
.20	Low demand (1,000 pools)	+\$500,000	(\$500,000)(.2) = \$100,000	−\$300,000	(−\$300,000)(.2) = −\$60,000	−\$1,300,000	(−\$1,300,000)(.2) = −\$260,000
.50	Moderate demand (5,000 pools)	+\$500,000	(\$500,000)(.5) = \$250,000	+\$2,500,000	(\$2,500,000)(.5) = +\$1,250,000	+\$1,500,000	(\$1,500,000)(.5) = +\$750,000
.30	High demand (10,000 pools)	+\$500,000	(\$500,000)(.3) = \$150,000	+\$2,500,000	(\$2,500,000)(.3) = \$750,000	+\$5,000,000	(\$5,000,000)(.3) = \$1,500,000
			EMV(1,000) = +\$500,000		EMV(5,000) = +\$1,940,000		*EMV(10,000) = +\$1,990,000 (Optimal Decision)

in a smaller expected profit, but one that will guard against the chance of an extremely heavy loss. This type of decision-making criterion, based on the concept of the *utility* of money, will be further explored in Section 11.7.

11.4 THE OPPORTUNITY LOSS TABLE

In the previous section we have determined the expected monetary value for each alternative. This value took into account the fact that different alternatives would be most preferred depending upon the event that actually occurred. However, we can view the payoff table from a different perspective by determining the best action to take if the event that was going to occur was known. For example, in the toy problem, if we knew that the toy was going to be successful, of course we would market it; if we knew it was not going to be successful, we would not market it. This concept can be expanded to determine the amount of profit that is lost when the best alternative is not selected for a particular event. **The opportunity loss is defined as the difference between the highest possible profit for an event and the actual profit obtained for the particular action taken.**

For example, in the toy problem, for the event "successful toy," the maximum profit is achieved when the product is marketed (\$45,000). The opportunity that would be lost by not marketing the toy would be the difference between \$45,000 and −\$3,000, which is \$48,000. On the other hand, if the toy were unsuccessful, the best action would be not to market the product (−\$3,000 profit). The opportunity that would be lost in making the incorrect decision results in a loss of −\$3,000 − (−\$36,000) = \$33,000. It should be noted that the opportunity loss will always be a *nonnegative* number since it represents the difference between the profit under the best action and any other alternative courses of action for a particular event. The complete opportunity loss table is summarized in Table 11.6.

In the second problem the opportunity loss table would involve three demand levels as well as three alternative courses of action. For each demand

Table 11.6
OPPORTUNITY LOSS TABLE FOR TOY PROBLEM

Event, E_j	Optimum Action	Profit of Optimum Action	Actions A_i	
			Market, A_1	Do Not Market, A_2
Successful toy	Market	+\$45,000	\$45,000 − \$45,000 = \$0	\$45,000 − (−\$3,000) = 48,000
Unsuccessful toy	Do not market	−\$ 3,000	−\$3,000 − (−\$36,000) = \$33,000	−\$3,000 − (−\$3,000) = \$0

level of swimming pools we need to determine the difference between the profit under the best action and the profits under the other actions. For example, if 1,000 pools are demanded, the best action is to produce 1,000 pools at a profit of \$500,000. The action of producing 5,000 pools results in an opportunity loss of \$500,000 − (−\$300,000) = \$800,000. The complete set of opportunity losses for this problem is summarized in Table 11.7.

11.5 EXPECTED VALUE OF PERFECT INFORMATION (EVPI)

Once the opportunity loss table has been developed, the alternative courses of action can be evaluated by determining the expected opportunity loss of each alternative. This is similar to the expected monetary value concept, except that opportunity losses rather than profits are being evaluated. Therefore if l_{ij} represents the opportunity loss of action i for event j, the expected opportunity loss (EOL) of action i is equal to

$$\text{EOL}(A_i) = \sum_j l_{ij}P_j \tag{11.2}$$

where P_j = probability of occurrence of event j

The computation of the expected opportunity loss for the toy problem is illustrated in Table 11.8.

Since we are looking at losses rather than profits, the decision criterion is to choose the alternative that has the *smallest* expected opportunity loss. As one would expect, the optimal act in this problem is action A_2 (do not market), since its expected opportunity loss is below that of marketing the

Table 11.7
OPPORTUNITY LOSS TABLE FOR SWIMMING POOL PROBLEM

Event, E_j	Optimum Action	Profit of Optimum Action	Production Levels		
			1,000	5,000	10,000
Low demand (1,000 pools)	Produce 1,000	+\$500,000	\$500,000 − \$500,000 = \$0	\$500,000 − (−\$300,000) = \$800,000	\$500,000 − (−\$1,300,000) = \$1,800,000
Moderate demand (5,000 pools)	Produce 5,000	+\$2,500,000	\$2,500,000 − \$500,000 = \$200,000	\$2,500,000 − \$2,500,000 = \$0	\$2,500,000 − \$1,500,000 = \$1,000,000
High demand (10,000 pools)	Produce 10,000	+\$5,000,000	\$5,000,000 − \$500,000 = \$4,500,000	\$5,000,000 − \$2,500,000 = \$2,500,000	\$5,000,000 − \$5,000,000 = \$0

Table 11.8
EXPECTED OPPORTUNITY LOSS FOR EACH ALTERNATIVE IN THE TOY PROBLEM

		Actions, A_i			
P_j	Event, E_j	*Market*, A_1	$l_{1j}P_j$	*Do Not Market*, A_2	$l_{2j}P_j$
.4	Successful toy	$0	$0(.4) = $0	$48,000	$48,000(.4) = $19,200
.6	Unsuccessful toy	$33,000	$33,0000(.6) = $19,800	$0	$0(.6) = $0
			EOL(A_1) = $19,800		*EOL(A_2) = $19,200 (Optimal Decision)

product. The decision reached using this procedure will be exactly the same as that arrived at through the expected monetary value criterion since they represent alternative ways of viewing the same payoff table.

However, the computation of the expected opportunity losses enables us to obtain additional information about the decision problem. If the marketing manager could *always* predict the future, he could determine the proper action that should be taken. Thus, if he knew that the toy was going to be successful, then it would be marketed at a profit of $45,000. On the other hand, if he knew that the toy was not going to be successful, it would not be marketed and a loss of $3,000 (profit of −$3,000) would be incurred. Since 40% of the toys are successful and 60% are not, the **expected profit under certainty** could be computed in the following manner:

$$\begin{aligned}\text{Expected Profit under Certainty} &= .4(\$45{,}000) + .6(-\$3{,}000)\\ &= \$18{,}000 - \$1{,}800\\ &= \$16{,}200\end{aligned}$$

This value, $16,200, represents the profit that would be made if the market manager knew with *certainty* whether the toy was going to be successful. However, we realize that the marketing manager does not always know whether or not the toy will be successful. The question could be raised here as to what value there is in having perfect information. It is in this context that the expected opportunity loss can be interpreted. The expected opportunity loss of the best alternative represents the expected value of perfect information (EVPI). This is the *maximum* amount that the marketing manager would be willing to pay for obtaining perfect information. It also represents the difference between the expected profit under certainty and the expected monetary value of the best alternative. The relationship is such that:

$$\begin{aligned}\text{EVPI} = &\ (\text{Expected Profit under Certainty})\\ &- (\text{Expected Monetary Value of Best Alternative}) \qquad (11.3)\end{aligned}$$

In this example we have

$$\text{Expected Profit under Certainty} = \$16{,}200$$
$$\text{Expected Monetary Value of Best Alternative} = -\$3{,}000$$
$$\text{EVPI} = \$19{,}200$$

The expected monetary value of the best alternative (do not market) was −$3,000. Since an expected profit of $16,200 could be achieved if perfect information were available, the marketing manager would be willing to pay up to $19,200 to obtain perfect information. Therefore the expected value of perfect information (EVPI) is $19,200.

This concept of expected value of perfect information can be developed for any number of events and any number of alternative courses of action. In the swimming pool problem there are three demand levels and three alternative levels of production. By referring to the values in Table 11.9, the expected opportunity losses for each production level can be computed.

The optimal action is to produce 10,000 pools with an expected opportunity loss of $860,000. Therefore the maximum amount that the manufacturer would be willing to pay for perfect information is $860,000. The expected profit under certainty would be

$$\begin{aligned}\text{Expected Profit under Certainty} \\ &= \$500{,}000(.2) + \$2{,}500{,}000(.5) + \$5{,}000{,}000(.3) \\ &= \$100{,}000 + \$1{,}250{,}000 + \$1{,}500{,}000 \\ &= \$2{,}850{,}000\end{aligned}$$

If perfect information were available, the expected profit would be $2,850,000. Since the expected monetary value of producing 10,000 pools is $1,990,000, the difference of $860,000 represents the maximum amount that the manufacturer would pay for perfect information.

Table 11.9
EXPECTED OPPORTUNITY LOSS FOR EACH ALTERNATIVE IN THE SWIMMING POOL PROBLEM

		Production Levels					
P_j	Event, E_j	*1,000*	$l_{1j}P_j$	*5,000*	$l_{2j}P_j$	*10,000*	$l_{3j}P_j$
.2	Low demand (1,000 pools)	$0	0(.2) = $0	$800,000	($800,000)(.2) = $1,600,000	$1,800,000	($1,800,00)(.2) = $360,000
.5	Moderate demand (5,000 pools)	$2,000,000	($2,000,000)(.5) = $1,000,000	$0	0(.5) = $0	$1,000,000	($1,000,000)(.5) = $500,000
.3	High demand (10,000 pools)	$4,500,000	($4,500,000)(.3) = $1,350,000	$2,500,000	($2,500,000)(.3) = $750,000	$0	$0(.3) = $0
			EOL(1,000) = $2,350,000		EOL(5,000) = $910,000		*EOL(10,000) = $860,000 (Optimal Decision) EVPI = $860,000

11.6 DECISION MAKING WITH SAMPLE INFORMATION

In the previous sections of this chapter we have developed the framework for analyzing a decision problem that contains several alternative courses of action. The expected monetary value (EMV) criterion was utilized as a means of choosing between various alternatives, and the opportunity loss table was developed in order to determine the expected value of perfect information (EVPI). Rather than relying solely on past experience and subjective probability, the decision maker has the option of collecting sample information prior to making a final decision. For example, the marketing manager of the toy company could ask the market research department to prepare a report that would estimate the sales potential of this particular toy. If the report was favorable, the marketing manager would be more inclined to market the product. On the other hand, if the report was unfavorable, the marketing manager would be more inclined not to market the product:

In Section 5.10 (Bayes' theorem) we developed a quantitative way of incorporating this sample information into the decision problem. Bayes' theorem revises the subjective probability of each event to take into account the results of the sample. In the toy example we have seen that the marketing manager has estimated the probability of a successful toy as 40% and the probability of an unsuccessful toy as 60%. Referring back to the example in Section 5.10, the marketing manager knows that in the past 80% of the successful toys received favorable marketing research reports, while only 30% of the unsuccessful toys received favorable marketing research reports. Suppose that the marketing research department, after evaluating the sales potential of this toy, issues a favorable sales potential report. The marketing research manager can use this new information to revise his estimate of whether the toy will be successful. Referring to Bayes' theorem [Equation (5.11d)], in this problem we have

Event S = successful toy S' = unsuccessful toy

Event F = favorable report F' = unfavorable report

$P(S) = .40$ $P(F|S) = .80$

$P(S') = .60$ $P(F|S') = .30$

Therefore, using Bayes theorem, we have

$$P(S|F) = \frac{P(F|S)P(S)}{P(F|S)P(S) + P(F|S')P(S')}$$

$$= \frac{(.80)(.40)}{(.80)(.40) + (.30)(.60)}$$

$$= 0.64$$

Table 11.10
EXPECTED MONETARY VALUE USING REVISED PROBABILITIES FOR EACH ALTERNATIVE IN THE TOY MARKETING PROBLEM

P_j	Event, E_j	Market, A_1	$X_{1j}P_j$	Do Not Market, A_2	$X_{2j}P_j$
.64	Successful toy	+\$45,000	(\$45,000)(.64) = +\$28,000	−\$3,000	(−\$3,000)(.64) = −\$1,920
.36	Unsuccessful toy	−\$36,000	(−\$36,000)(.36) = \$12,960	−\$3,000	(−\$3,000)(.36) = −\$1,080
			*EMV(A_1) = \$15,840 (Optimal Decision)		EMV(A_2) = −\$3,000

Since there are only two events, S and S',

$$P(S'|F) = 1 - .64 = .36$$

Since the original subjective probabilities were used to determine the best decision based upon the expected monetary value criterion, the expected profit of each alternative must be reevaluated by using these revised probabilities. The revised expected monetary value computations are illustrated in Table 11.10.

In this case the optimal decision is to market the product, since an average profit of \$15,840 could be expected as compared to a loss of \$3,000 if the toy was not marketed. This decision is different from the one that was optimal prior to the collection of the sample information (the market research report). The information contained in the report had a substantial effect on the prior assessment of the marketing manager, so that the best decision became one in which the product would be marketed.

Just as the marketing manager of the toy company was able to utilize sample information prior to making a decision, the manufacturer of swimming pools would most likely desire to obtain a forecast of the demand for swimming pools in the upcoming season prior to making a commitment to produce a specific number of pools. Sample information probably would be particularly crucial in formulating a decision, since either large profits or large losses could occur, depending upon the alternative chosen and the event that actually occurs. Suppose the manufacturer would like to obtain a forecast of the economic conditions in the next 3 months since the level of sales may depend heavily upon the state of the economy. Based upon past experience, the manufacturer estimates that when there has been low demand D_1, 90% of the time there has been a recession or economic lull E_1 (10% of the time there has been a prosperous economy, E_2). When there has been moderate demand D_2, 60% of the time there has been a prosperous economy (40% of the time there has been a recession or economic lull). When there has been high demand D_3, 95% of the time there has been a prosperous economy (5% of the time there has been a recession or economic lull). This information can be summarized as follows:

$P(\text{Recession}|\text{Low}) = .90$ $\quad$ $P(\text{Prosperity}|\text{Low}) = .10$

$P(\text{Recession}|\text{Moderate}) = .40$ $\quad$ $P(\text{Prosperity}|\text{Moderate}) = .60$

$P(\text{Recession}|\text{High}) = .05$ $\quad$ $P(\text{Prosperity}|\text{High}) = .95$

If the economic forecast was for prosperity, Bayes' theorem could be used to revise the prior probabilities in the following way:

$$P(D_1) = P(\text{Low}) = .20$$
$$P(D_2) = P(\text{Moderate}) = .50$$
$$P(D_3) = P(\text{High}) = .30$$

$$P(\text{Low}|\text{Prosperity}) = \frac{P(\text{Prosperity}|\text{Low})P(\text{Low})}{P(\text{Prosperity}|\text{Low})P(\text{Low}) + P(\text{Prosperity}|\text{Moderate})P(\text{Moderate}) + P(\text{Prosperity}|\text{High})P(\text{High})}$$

$$P(D_1|E_2) = \frac{P(E_2|D_1)P(D_1)}{P(E_2|D_1)P(D_1) + P(E_2|D_2)P(D_2) + P(E_2|D_3)P(D_3)}$$

$$= \frac{(.10)(.20)}{(.10)(.20) + (.60)(.50) + .95(.30)} = \frac{.02}{.02 + .30 + .285}$$

$$= \frac{.02}{.605} = .033$$

Moreover, we may compute

$$P(D_2|E_2) = \frac{.30}{.02 + .30 + .285} = \frac{.30}{.605} = .496$$

and

$$P(D_3|E_2) = \frac{.285}{.02 + .30 + .285} = \frac{.285}{.605} = .471$$

Therefore, comparing these probabilities to the original prior estimates, the forecast of a prosperous economy has increased the probability of having a high demand, while it has decreased the probability of low demand. Since the original probabilities of the manufacturer have changed, the payoff table needs to be reevaluated in the light of this sample information. The revised expected monetary values of the alternatives are indicated in Table 11.11.

Therefore, the optimal action is to produce 10,000 pools since this decision results in an expected monetary value of $3,056,100. In this instance the sample information has not altered the decision reached using prior probabilities. However, the forecast of a prosperous economy has caused the action of producing 10,000 pools to appear substantially better than producing only 5,000 pools.

Table 11.11
EXPECTED MONETARY VALUE USING REVISED PROBABILITIES FOR EACH ALTERNATIVE IN THE SWIMMING POOL PROBLEM

		Production Levels					
P_j	Event, E_j	1,000	$X_{1j}P_j$	5,000	$X_{2j}P_j$	10,000	$X_{3j}P_j$
.033	Low demand (1,000 pools)	+$500,000	$500,000(.033) = $16,500	−$300,000	(−$300,000)(.033) = −$9,900	−$1,300,000	(−$1,300,000)(.033) = −$42,900
.496	Moderate demand (5,000 pools)	+$500,000	$500,000(.496) = $248,000	+$2,500,000	($2,500,000)(.496) = $1,240,000	+$1,500,000	($1,500,000)(.496) = $744,000
.471	High demand (10,000 pools)	+$500,000	$500,000(.471) = $235,500	+$2,500,000	($2,500,000)(.471) = $1,171,500	+$5,000,000	($5,000,000)(.471) = $2,355,000
			EMV(1,000) = $500,000		EMV(5,000) = +$2,407,600		*EMV(10,000) = $3,056,100 (Optimal Decision)

11.7 UTILITY

In Section 11.3 we observed that a businessman might not wish to rigidly follow the expected monetary value criterion when large losses of money are involved. For example, in the swimming pool problem the selection of a 10,000-pool production level included the chance that a large sum of money could be lost even though this alternative had the highest expected monetary value. Therefore it is clear that the actual criterion for decision is not based solely upon monetary value, but on the usefulness or **utility** of a particular amount of money. For example, suppose we were faced with the following choice:

1. A fair coin is to be flipped; if it lands on heads we will receive $.60, if it lands on tails we will pay $.40
2. Not playing the game

What decision should we choose? The expected value of playing this game would be .60(.50) − .40(.50) = +$.10, while the expected value of not playing the game is 0.

Most people would decide to play the game since its expected value is positive and there are only small amounts of money involved. However, if the game was formulated with a payoff of $60,000 when the coin lands on heads and a loss of $40,000 when the coin lands on tails, the expected value of playing the game would be +$10,000. With these payoffs, even though the expected value is positive, most individuals (or businessmen) would not play the game because of the severe negative consequences of losing $40,000. Each additional dollar amount of either profit or loss does not have the same utility as the previous dollar. Large negative amounts (for most individuals) have severely negative utility, while the extra value of each incremental dollar of profit decreases once high enough profit levels are reached.

An important part of the decision problem, which is beyond the scope of this textbook, is to develop a utility curve that represents the actual value of each specified dollar amount in the problem. There are three basic types of

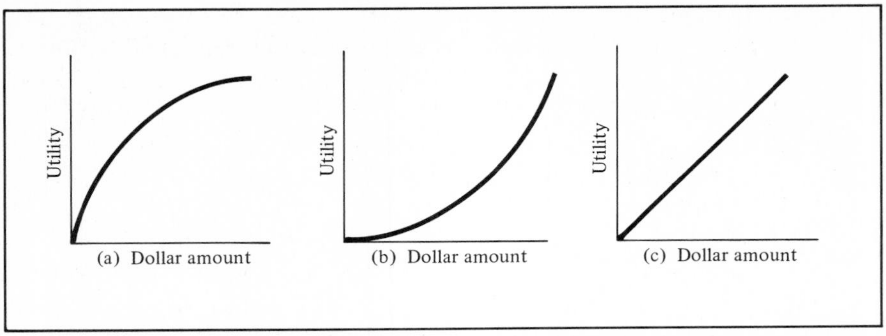

Figure 11.1 Three types of utility curves. (a) Risk averter. (b) Risk seeker. (c) Risk neutral.

utility curves: that of the risk averter, the risk seeker, and the risk neutral person. These curves are illustrated in Figure 11.1.

The first type of curve, the risk averter, shows a rapid increase in utility for initial amounts of money, followed by a gradual leveling off of the increased utility for increasing dollar amounts. This curve would be appropriate for most individuals or businesses since the value of each additional dollar is not as great once large amounts have already been earned.

On the other hand, the risk seeker curve represents the utility of one who enjoys taking risks. The utility of increasing amounts becomes greater for larger dollar amounts. This curve could perhaps represent an individual who is interested only in "striking it rich" and is willing to take large risks to obtain the opportunity of making large profits.

The third curve, the risk neutral curve, represents the utility curve of the expected monetary value approach. For this curve, each additional dollar of profit has the same value as the previous dollar.

Once a utility curve has been developed for a particular decision maker in a specific situation, the dollar amounts would be converted to utilities, and the expected utility of each alternative course of action would be determined. The alternative with the highest expected utility would be the action selected.

11.8 A COMPARISON OF CLASSICAL AND BAYESIAN DECISION MAKING

In Chapters 8, 9, 10, and 11 we have considered two types of decision making: classical and Bayesian. In Chapter 8 we developed interval estimates of population parameters such as the mean and the proportion using only results obtained from sample data. In Chapters 9 and 10 decisions about population values and comparisons between the values of two populations were made based upon the outcome of sample data. In Chapter 11 decisions were made between alternative courses of action by directly using economic information. Subjective prior probabilities were taken into account along with sample results to make decisions between competing alternatives.

In comparing these two approaches, there are two fundamental differences between them. The first major distinction concerns the use of economic information. The Bayesian decision-making procedure bases its decision upon a direct evaluation of the payoffs of each event for each alternative course of action. On the other hand, classical decision making uses economic information only indirectly in evaluating the Type I (α) and Type II (β) errors and in setting up the null and alternative hypotheses. The consequences of incorrect decisions are taken into account in determining these two types of risks. Once these risks are determined, a decision rule is developed given a specific sample size.

The second major distinction relates to the use of subjective probability. The Bayesian inferential approach takes prior probabilities and revises them based upon sample results. The prior probabilities often consist of the subjective assessment of the different events made by an individual. This assessment could vary greatly from person to person. The classical inferential procedure considers only the sample results in drawing conclusions about a population.

PROBLEMS

* 11.1 A vendor at a local baseball stadium must determine whether to sell ice cream or soda at today's game. The vendor believes that the profit made will depend upon the weather. The payoff table is as follows:

	Action	
Event	*Sell Soda*	*Sell Ice Cream*
Cool weather	+$40	+$20
Warm weather	+$55	+$80

Based upon her past experience at this time of the year, the vendor estimates the probability of warm weather as .60.

(a) Determine whether the vendor should sell soda or sell ice cream based upon the expected monetary value (EMV) criterion.

(b) Set up the opportunity loss table.

(c) Compute the expected opportunity loss (EOL) of each alternative course of action.

(d) Determine the expected value of perfect information (EVPI).

(e) What is the interpretation of EVPI?

Prior to making her decision, the vendor decides to hear the forecast of the local weatherman. In the past, when it has been cool the weatherman has forecast cool weather 80% of the time. When it has been warm, the weatherman has forecast warm weather 70% of the time. If today's forecast is for cool weather,

(f) Revise the prior probabilities of the vendor in light of this new information.

(g) Use these revised probabilities along with the expected monetary value criterion to determine whether she should sell ice cream or soda today.

11.2 An investor has a certain amount of money available to invest now. Three alternative portfolio selections are available. The estimated profits of each portfolio under each economic condition is indicated in the following payoff table:

	Portfolio Selection		
Event	*A*	*B*	C
Economy declines	+$500	−$2,000	− $7,000
No change	+$1,000	+$2,000	− $1,000
Economy expands	+$2,000	+$5,000	+$20,000

Based upon his own past experience, the investor assigns the following probabilities to each economic condition:

$$P(\text{Economy Declines}) = .30$$
$$P(\text{No Change}) = .50$$
$$P(\text{Economy expands}) = .20$$

(a) Determine the best portfolio selection for the investor according to the expected monetary value criterion.

(b) Set up the opportunity loss table.

(c) Compute the expected opportunity loss (EOL) of each alternative course of action.

(d) Determine the expected value of perfect information (EVPI).

(e) What is the interpretation of the EVPI?

Prior to making his investment decision, the investor has decided to consult with his stock broker. In the past, when the economy has declined, the stock broker has given a rosy forecast 20% of the time (with a gloomy forecast 80% of the time). When there has been no change in the economy, the stock broker has given a rosy forecast 40% of the time. When there has been an expanding economy, the stock broker has given a rosy forecast 70% of the time. The stock broker in this case gives a gloomy forecast for the economy.

(f) Revise the prior probabilities of the investor in light of this economic forecast by the stock broker.

(g) Use these revised probabilities along with the expected monetary value criterion to determine the best portfolio selection for the investor.

* 11.3 The Dollar Bill Steel Co. must determine whether it should build a large factory or a small factory. The profit for steel sold is $10 per ton. A small factory can be built for a cost of $200,000 and would have a production capacity of 50,000 tons. A large factory can be built for a cost of $400,000, but would have a production capacity of 100,000 tons. The probability distribution of sales of steel is as follows:

Sales (tons)	Probability
10,000	.1
20,000	.4
50,000	.2
100,000	.3

(a) Determine which type of factory should be built according to the expected monetary value criterion.

(b) Set up the opportunity loss table.

(c) Compute the expected opportunity loss (EOL).

(d) Determine the expected value of perfect information (EVPI) and interpret its meaning.

11.4 An author is trying to choose between two publishing companies that are competing for the marketing rights to her new novel. Company A has offered the author $10,000 plus $2 for each book sold. Company B has offered the author $2,000 plus $4 for each book sold. The author estimates the distribution of demand for this book as follows:

Number of Books Sold	Probability
1,000	.45
2,000	.20
5,000	.15
10,000	.10
50,000	.10

(a) Using the expected monetary value criterion, determine whether the author should sell the marketing rights to Company A or Company B.

(b) Set up the opportunity loss table.

(c) Compute the expected opportunity loss (EOL).

(d) Determine the expected value of perfect information (EVPI) and interpret its meaning.

Prior to making a final decision, the author has decided to have an experienced reviewer examine her novel. This reviewer has an outstanding reputation for predicting success for a novel. In the past, for novels that sold 1,000 copies, only 1% received favorable reviews. Of novels that sold 2,000 copies, only 5% received favorable reviews. Of novels that sold 5,000 copies, 25% received favorable reviews. Of novels that sold 10,000 copies, 60% received favorable reviews. Finally, of novels that sold 50,000 copies, 99% received favorable reviews.

After examining the author's novel, the reviewer gives it an unfavorable review.

(e) Revise the prior probabilities of number of books sold in light of this unfavorable review.

(f) Using these revised probabilities, according to the expected monetary value criterion, should the author sell the marketing rights to Company A or Company B?

* 11.5 The Islander Fishing Co. purchases clams for $1.00 per pound from Peconic Bay fishermen for sale to various New York restaurants for $1.50 per pound. Any clams not sold to the restaurants by the end of the week can be sold to a local soup company for $0.25 per pound. The probability of various levels of demand are as follows:

Demand (pounds)	Probability
500	.2
1,000	.4
2,000	.4

(a) Using the expected monetary value criterion, determine the optimal number of pounds of clams that the company should purchase from the fishermen.

(b) Set up the opportunity loss table.

(c) Compute the expected opportunity loss (EOL).

(d) Determine the expected value of perfect information (EVPI) and interpret its meaning.

11.6 Shop-Quik Supermarkets purchases large quantities of white bread for sale during a week. The bread is purchased for 35 cents per loaf and is sold for 60 cents per loaf. Any loaves of bread not sold by the end of the week can be sold to a local thrift shop for 20 cents per loaf. Based on past demand, the probability of various levels of demand is as follows:

Demand (loaves)	Probability
6,000	.1
8,000	.5
10,000	.3
12,000	.1

(a) Using the expected monetary value criterion, determine the optimal number of loaves of bread that should be purchased.

(b) Set up the opportunity loss table.

(c) Compute the expected opportunity loss (EOL).

(d) Determine the expected value of perfect information (EVPI) and interpret its meaning.

11.7 The owner of a large home heating oil delivery company would like to determine whether to offer a solar heating installation service to its customers. The owner of the company has determined that a start-up cost of $150,000 would be necessary, but a profit of $2,000 can be made on each solar heating system installed. The owner estimates the probability of various demand levels as follows:

Number of Units Installed	Probability
50	.4
100	.3
200	.3

(a) Using the expected monetary value criterion, determine whether the company should offer this solar heating installation service.

(b) Set up the opportunity loss table.

(c) Compute the expected opportunity loss (EOL).

(d) Determine the expected value of perfect information (EVPI) and interpret its meaning.

11.8 The LeFleur Garden Center chain purchases Christmas trees from a supplier for sale during the holiday season. The trees are purchased for $6 each and are sold for $12 each. Any trees not sold can be disposed of for $3 each. The probability of various levels of demand is as follows:

Demand (number of trees)	Probability
100	.2
200	.6
500	.2

(a) Using the expected monetary value criterion, determine the number of trees that the chain should purchase from the supplier.

(b) Set up the opportunity loss table.

(c) Compute the expected opportunity loss (EOL).

(d) Determine the expected value of perfect information (EVPI) and interpret its meaning.

11.9 A bakery wishes to make orange Halloween cakes. Each cake costs $2 to make and is sold for $3.50. Any cakes not sold during the day **cannot** be sold at a later time. Based on past experience, the baker estimates the demand for these cakes will be either 5, 10, or 15 with probabilities of .6, .3, and .1, respectively.

(a) Using the expected monetary value criterion, determine the number of cakes that the baker should make.

(b) Set up the opportunity loss table.

(c) Compute the expected opportunity loss (EOL).

(d) Determine the expected value of perfect information (EVPI) and interpret its meaning.

11.10 The producer of a nationally distributed brand of potato chips would like to determine the feasibility of changing the product package from a cellophane bag to an unbreakable container. The product manager believes that there would be three possible national market responses to a change in product package: weak, moderate, and strong. The payoff, in increased or decreased profit as compared to the current product package, is the following:

	Action	
Event	*Use New Package*	*Keep Old Package*
Weak national response	−$4,000,000	0
Moderate national response	+$1,000,000	0
Strong national response	+$5,000,000	0

Based upon past experience, the product manager assigns the following probabilities to the different levels of national response:

$$P(\text{Weak National Response}) = .30$$
$$P(\text{Moderate National Response}) = .60$$
$$P(\text{Strong National Response}) = .10$$

(a) Using the expected monetary value (EMV) criterion, determine whether the new product package should be adopted.

(b) Set up the opportunity loss table.

(c) Compute the expected opportunity loss (EOL).

(d) Determine the expected value of perfect information (EVPI) and interpret its meaning.

Prior to making a final decision, the product manager would like to test market the new package in a selected city. In this selected city the new package is substituted for the old package and a determination is made as to whether sales have increased, decreased, or stayed the same during a specified period of time. In previous test marketing of other products, when there has been a subsequent weak national response, sales in the test city have decreased 60% of the time, stayed the same 30% of the time, and increased only 10% of the time. When there has been a moderate national response, sales in the test city have decreased 20% of the time, stayed the same 40% of the time, and increased 40% of the time. When there has been a strong national response, sales in the test city have decreased 5% of the time, stayed the same 35% of the time, and increased 60% of the time.

(e) If sales in the test city stayed the same, revise the prior probabilities of the product manager in light of this new information.

(f) Use these revised probabilities along with the expected monetary value criterion to determine whether the new product package should be adopted.

(g) If sales in the test city decreased, revise the prior probabilities of the product manager in light of this new information.

(h) Use the revised probabilities obtained in (g) along with the expected monetary value criterion to determine whether the new product package should be adopted.

* 11.11 A businessman would like to determine whether it would be profitable to establish a gardening service in a local suburb. The businessman believes that there are four possible levels of demand for this gardening service:

1. Very low demand—1% of the households would use this gardening service
2. Low demand—5% of the households would use this gardening service
3. Moderate demand—10% of the households would use this gardening service
4. High demand—25% of the households would use this gardening service

Based upon past experience in other suburbs, the businessman assigns the following probabilities to the various demand levels:

$$P(\text{Very Low Demand}) = .20$$
$$P(\text{Low Demand}) = .50$$
$$P(\text{Moderate Demand}) = .20$$
$$P(\text{High Demand}) = .10$$

The businessman has calculated the following profits or losses of this garden service for each demand level (over a period of 1 year):

	Action	
Demand	*Provide Garden Service*	*No Garden Service*
Very low ($p = .01$)	−$10,000	0
Low ($p = .05$)	−$ 1,000	0
Moderate ($p = .10$)	+$ 8,000	0
High ($p = .25$)	+$20,000	0

(a) Using the expected monetary value criterion, determine whether the gardening service should be instituted in this suburb.

The businessman decides that prior to a final decision, a survey of households in this suburb should be taken to determine demand for this gardening service. If a random sample of 20 households is selected and three would use this gardening service:

(b) Revise the prior probabilities of the businessman in light of this sample information. **Hint:** Use the binomial distribution to determine the conditional probability of this outcome given a particular level of demand.

(c) Use these revised probabilities along with the expected monetary value criterion to determine whether the businessman should institute this gardening service.

11.12 The manufacturer of a brand of inexpensive felt tip pens maintains a production process that produces 10,000 pens per day. In order to maintain the highest quality of this product, the manufacturer guarantees free replacement of any defective pen sold. It has been calculated that each defective pen produced costs 20¢ for the manufacturer to replace. Based upon past experience, four rates of producing defective pens are possible:

1. Very low—1% of the pens produced are defective
2. Low—5% of the pens produced are defective
3. Moderate—10% of the pens produced are defective
4. High—20% of the pens produced are defective

The manufacturer can reduce the rate of defective pens produced by having a mechanic fix the machines at the end of the day. This mechanic can reduce the defective rate to 1%, but it will cost $80 for the mechanic's services.

A payoff table based upon the daily production of 10,000 pens, indicating the replacement costs for each of two alternatives (calling in the mechanic and not calling in the mechanic) is indicated below:

	Action	
Defective Rate	*Do Not Call Mechanic*	*Call Mechanic*
Very low (1%)	$ 20	$100
Low (5%)	$100	$100
Moderate (10%)	$200	$100
High (20%)	$400	$100

Based upon past experience, each defective rate is assumed to be equally likely to occur. At the end of a particular day's production, a random sample of 15 pens is selected of which 2 are defective.

(a) If the manufacturer were to use the Bayesian decision-making approach, revise the prior probabilities to take into account the sample information. Use the expected monetary value criterion to determine whether to call the mechanic. **Hint:** Use the binomial distribution to determine the conditional probability of this sample outcome given a particular defective rate.

(b) In the examination of this payoff table, the manufacturer notes that when the defective rate is 5% there is no difference in the cost of calling in the mechanic and not calling in the mechanic. Therefore the manufacturer has decided that a mechanic will be called in only if there is evidence that the defective rate is greater than 5%. If the manufacturer were to use a classical decision-making procedure (see Chapter 9), at the .01 level of significance, should the mechanic be called?

(c) Briefly compare and contrast the Bayesian decision-making approach of (a) with the classical decision-making approach of (b).

REFERENCES

1. BIERMAN, H., C. P. BONINI, AND W. H. HAUSMAN, *Quantitative Analysis for Business Decisions,* 4th ed. (Homewood, Ill.: Richard D. Irwin, 1973).
2. GORDON, G., AND I. PRESSMAN, *Quantitative Decision Making for Business* (Englewood Cliffs, N.J.: Prentice-Hall, 1978).
3. GREEN, P., AND D. TULL, *Research for Marketing Decisions,* 4th ed.: (Englewood Cliffs, N.J.: Prentice-Hall, 1978).
4. LAPIN, L., *Quantitative Methods for Business Decisions* (New York: Harcourt Brace Jovanovich, 1976).
5. SCHLAIFER, R., *Introduction to Statistics for Business Decisions* (New York: McGraw-Hill, 1959).
6. VALINSKY, D., "Statistics" in C. Heyel, ed., *The Encyclopedia of Management* (New York: Reinhold Publishing, 1963).
7. WINKLER, R. L., *Introduction to Bayesian Inference and Decision* (New York: Holt, Rinehart and Winston, 1972).

CHAPTER 12

THE ANALYSIS OF VARIANCE

12.1 INTRODUCTION

In Chapter 10 we used statistical inference to draw conclusions about differences between two groups based on either a quantitative variable (means) or a qualitative variable (proportions). The chi-square technique, which was used to examine differences in the proportion of successes in two groups, was then extended (see Section 10.6) so that we were able to draw conclusions about the difference in the proportion of successes in more than two groups. In this chapter we will study the analogous situation when the data are quantitative. That is, we shall briefly examine methods that have been developed to test for differences between the means of several groups. This methodology is classified under the general title of "the analysis of variance." The procedures covered, however, only pertain to the "one-way" analysis of variance, since only one *factor* (with several groups) is to be considered. If the researcher wished to draw conclusions about a study in which several factors are simultaneously considered, the experimental design required would necessarily be more complex and beyond the scope of this text (References 2, 4, 6, and 8).

12.2 SEVERAL MEASURES OF VARIATION

In Chapter 10 two examples pertaining to quantitative data were investigated. One involved a comparison of the average life (in miles) of tires produced by the day and evening shifts at a tire factory. The second study referred to the survey of the dean and the researcher in which the grade-point indexes of accounting and non-accounting majors were compared.

Referring to the tire company problem, we previously noted that there was a third shift of workers, a night shift, at the factory. Suppose that the production manager would like to determine whether there is any difference between the three shifts in the average life of tires produced. Random samples of five tires produced by each shift were selected, and the results (in thousands of miles of tire life) are indicated in Table 12.1.

Table 12.1
TIRE LIFE (IN THOUSANDS OF MILES) OF SAMPLES OF FIVE TIRES FROM THE DAY, EVENING, AND NIGHT SHIFTS

	Day	Evening	Night
	25.40	23.40	20.00
	26.31	21.80	22.20
	24.10	23.50	19.75
	23.74	22.75	20.60
	25.10	21.60	20.40
Mean	$\bar{X}_{day} = 24.93$	$\bar{X}_{eve} = 22.61$	$\bar{X}_{night} = 20.59$

In order to answer this kind of question, we must test whether the various groups all have the same population average. The null and alternative hypotheses would be stated as follows:

$$H_0: \quad \mu_1 = \mu_2 = \mu_3 = \cdots = \mu_c$$

H_1: Not all the means are equal

For the tire company study, since there are three shifts, the null and alternative hypotheses would be

$$H_0: \quad \mu_{day} = \mu_{eve} = \mu_{night}$$

H_1: Not all the shifts have equal means

We have seen in Table 12.1 that there are differences in the sample means of the three shifts. Tires produced in the day shift have an average of 24,930 miles, the evening shift tires have an average of 22,610 miles, and the night shift tires last an average of 20,590 miles. The question that must be answered is whether these sample results are sufficiently different for us to conclude that the population averages of the shifts are not all equal, or whether the sample differences can reasonably be explained by random variation.

Since under the null hypothesis the population means of the three shifts are presumed equal, a measure of the **total variation** among all the tires can be obtained by summing up the squared differences between each observation (tire) and an overall mean, based upon all the tires. The total variation would be computed as

$$\text{Total Variation} = \sum_{i=1}^{c}\sum_{j=1}^{n_i}(X_{ij} - \bar{\bar{X}})^2 \qquad (12.1)$$

where X_{ij} = jth observation in group i

$\bar{\bar{X}} = \dfrac{\sum_{i=1}^{c}\sum_{j=1}^{n_i} X_{ij}}{n}$ is called the "grand mean"

n_i = number of observations in group i

n = total number of observations in sample

c = number of groups

Since this total variation is actually the sum of squared deviations, it is also referred to as the **total sum of squares** (total SS). In the tire study the total variation would be computed as follows:

$$\bar{\bar{X}} = \frac{\sum_{i=1}^{c}\sum_{j=1}^{n_i} X_{ij}}{n} = \frac{(25.40 + 26.31 + \cdots + 23.40 + \cdots + 20.40)}{15}$$

$$= \frac{340.65}{15} = 22.71$$

$$\text{Total Variation} = \sum_{i=1}^{c}\sum_{j=1}^{n_i}(X_{ij} - \bar{\bar{X}})^2$$

$$= \begin{pmatrix} (25.40 - 22.71)^2 \\ +(26.31 - 22.71)^2 \\ +(24.10 - 22.71)^2 \\ +(23.74 - 22.71)^2 \\ +(25.10 - 22.71)^2 \end{pmatrix} + \begin{pmatrix} (23.40 - 22.71)^2 \\ +(21.80 - 22.71)^2 \\ +(23.50 - 22.71)^2 \\ +(22.75 - 22.71)^2 \\ +(21.60 - 22.71)^2 \end{pmatrix} + \begin{pmatrix} (20.00 - 22.71)^2 \\ +(22.20 - 22.71)^2 \\ +(19.75 - 22.71)^2 \\ +(20.60 - 22.71)^2 \\ +(20.40 - 22.71)^2 \end{pmatrix}$$

$$= \begin{pmatrix} 7.2361 \\ +12.9600 \\ +\ 1.9321 \\ +\ 1.0609 \\ +\ 5.7121 \end{pmatrix} + \begin{pmatrix} +\ .4761 \\ +\ .8281 \\ +\ .6241 \\ +\ .0016 \\ +1.2321 \end{pmatrix} + \begin{pmatrix} +7.3441 \\ +\ .2601 \\ +8.7616 \\ +4.4521 \\ +5.3361 \end{pmatrix}$$

$$= 58.2172$$

This total variation, which we have just computed, measures differences between each value X_{ij} and the overall (grand) mean $\overline{\overline{X}}$. The total variation, however, can be subdivided into two separate components: one part consists of variation **between** the groups (since the sample means of the groups are not necessarily equal); the second part consists of variation **within** the groups (since within each group—shift of workers—all the values—lives of tires—are not the same). It is always true that

$$\text{Total Variation} = (\text{Between Group Variation}) + (\text{Within Group Variation}) \tag{12.2}$$

The **between group variation** measures the differences between the sample mean of each group $\overline{X}_i$ and the grand mean $\overline{\overline{X}}$, weighted by the number of observations in each group. The between group variation would be computed as

$$\text{Between Group Variation} = \sum_{i=1}^{c} n_i(\overline{X}_i - \overline{\overline{X}})^2 \tag{12.3}$$

where n_i = number of observations in group i

$\overline{X}_i$ = sample mean of group i

$\overline{\overline{X}}$ = grand mean

In the tire study the between group variation would be computed as follows:

$$\begin{aligned}\text{Between Group Variation} &= 5(24.93 - 22.71)^2 + 5(22.61 - 22.71)^2 \\ &\quad + 5(20.59 - 22.71)^2 \\ &= 5(2.22)^2 + 5(-.10)^2 + 5(-2.12)^2 \\ &= 24.642 + .05 + 22.472 = 47.164\end{aligned}$$

The **within group variation** measures the variation of each value from the mean of its own group and cumulates these squared differences over all groups. In the tire study the within group variation would be computed as:

$$\text{Within Group Variation} = \sum_{i=1}^{c} \sum_{j=1}^{n_i} (X_{ij} - \overline{X}_i)^2 \tag{12.4}$$

where $\overline{X}_i$ = mean of group i

X_{ij} = jth observation in group i

In the tire study the within group variation would be computed as follows:

Within Group Variation

$$= \begin{pmatrix} (25.40 - 24.93)^2 \\ +(26.31 - 24.93)^2 \\ +(24.10 - 24.93)^2 \\ +(23.74 - 24.93)^2 \\ +(25.10 - 24.93)^2 \end{pmatrix} + \begin{pmatrix} (23.40 - 22.61)^2 \\ +(21.80 - 22.61)^2 \\ +(23.50 - 22.61)^2 \\ +(22.75 - 22.61)^2 \\ +(21.60 - 22.61)^2 \end{pmatrix} + \begin{pmatrix} (20.00 - 20.59)^2 \\ +(22.20 - 20.59)^2 \\ +(19.75 - 20.59)^2 \\ +(20.60 - 20.59)^2 \\ +(20.40 - 20.59)^2 \end{pmatrix}$$

$$= \begin{pmatrix} .2209 \\ +1.9044 \\ +\ .6889 \\ +1.4161 \\ +\ .0289 \end{pmatrix} + \begin{pmatrix} .6241 \\ +\ .6561 \\ +\ .7921 \\ +\ .0196 \\ +1.0201 \end{pmatrix} + \begin{pmatrix} .3481 \\ +2.5921 \\ +\ .7056 \\ +\ .0001 \\ +\ .0361 \end{pmatrix}$$

$$= 11.0532$$

Referring back to the previous computations for the total variation and the between group variation, we check to determine that

$$\text{Total Variation} = (\text{Between Group Variation}) + (\text{Within Group Variation})$$

$$58.2172 = 47.164 + 11.0532 = 58.2172$$

12.3 THE *F* DISTRIBUTION

In order to determine whether or not the means of the various groups are all equal, we can examine two different **variances,** one based on differences between groups and the other based upon differences within groups. We may recall from Chapter 4 that a variance is computed by dividing the sum of squared deviations by its appropriate degrees of freedom. In the analysis of variance, the sum of squared deviations is represented by the respective measures of variation. The variance within groups S_w^2 measures variability around the mean of each group. Since this variability is not affected by group differences, it can be considered a measure of the random variation of values within a group. On the other hand, the variance between groups S_B^2 takes into account not only the random fluctuations from observation to observation, but it also measures differences from one group to another. If there is no real difference from group to group, any sample differences will be explainable by random variation, and the variance between groups S_B^2 should be close to the variance within groups S_w^2. However, if there really is a difference between the groups, the variance between groups S_B^2 will be significantly larger than the variance within groups S_w^2.

Since there are two variances involved, the test statistic is based upon the ratio of the two variances, S_B^2/S_W^2 (see Section 10.10). The ratio of two variances follows a distribution called the F distribution (see Appendix E, Table E.5), named after the famous statistician R. A. Fisher; that is,

$$F_{df_1, df_2} = \frac{S_B^2}{S_W^2} \tag{12.5}$$

From Appendix E, Table E.5, we can see that the shape of the F distribution depends upon two sets of degrees of freedom: the degrees of freedom between group means df_1 in the numerator and the degrees of freedom within groups df_2 in the denominator.

Referring to our analysis of variance example, the variance between groups S_B^2 is equal to

$$S_B^2 = \frac{\text{Between Group Variation}}{\text{Degrees of Freedom Between Groups*}} = \frac{\sum_{i=1}^{c} n_i(\bar{X}_i - \bar{\bar{X}})^2}{c - 1} \tag{12.6}$$

where c = number of groups. The variance within groups S_W^2 is equal to

$$S_W^2 = \frac{\text{Within Group Variation}}{\text{Degrees of Freedom Within Groups†}} = \frac{\sum_{i=1}^{c}\sum_{j=1}^{n_i} (X_{ij} - \bar{X}_i)^2}{n - c} \tag{12.7}$$

As we previously stated, if there actually was a difference between groups, the variance between the groups S_B^2 should be significantly greater than the variance within the groups S_W^2. Therefore the decision rule would be to reject the null hypothesis of no difference between the groups if

$$F_{(c-1),(n-c)} = \frac{S_B^2}{S_W^2} > F_{\alpha,(c-1),(n-c)}$$

where α = level of significance (see Figure 12.1).

* There are $c - 1$ degrees of freedom because there were c groups from which the overall mean was determined.

† Within each group there are $(n_i - 1)$ degrees of freedom. When these are summed over all c groups, there are $n - c$ degrees of freedom within groups.

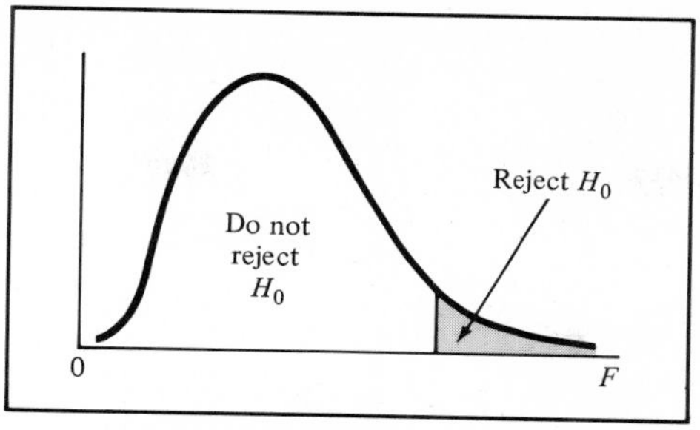

Figure 12.1 Regions of rejection and nonrejection in the analysis of variance.

Since there are $c - 1$ degrees of freedom between groups and $n - c$ degrees of freedom within groups, the F test would have $(c - 1)$ degrees of freedom in the numerator and $(n - c)$ in the denominator.

Returning to the tire study of the previous section, since between group variation = 47.164, within group variation = 11.0532, $n = 15$, and $c = 3$, we have

$$S_B^2 = \frac{\text{Between Group Variation}}{\text{Degrees of Freedom Between Groups}} = \frac{\sum_{i=1}^{c} n_i(\bar{X}_i - \bar{\bar{X}})^2}{c - 1}$$

$$= \frac{47.164}{3 - 1} = \frac{47.164}{2} = 23.582$$

$$S_w^2 = \frac{\text{Within Group Variation}}{\text{Degrees of Freedom Within Groups}} = \frac{\sum_{i=1}^{c}\sum_{j=1}^{n_i}(X_{ij} - \bar{X}_i)^2}{n - c}$$

$$= \frac{11.0532}{15 - 3} = \frac{11.0532}{12} = .9211$$

Therefore we have

$$F_{2,12} = \frac{S_B^2}{S_w^2} = \frac{23.582}{.9211} = 25.60$$

If a significance level of .01 was decided upon, the critical value of the F distribution could be determined from Appendix E, Table E.5. The values in the body of this table refer to selected *upper* percentage points of the F distribution. In this problem, since there are 2 degrees of freedom in the numerator and 12 degrees of freedom in the denominator, the critical value of F at the .01 level of significance is 6.93 (see Table 12.2).

Table 12.2
OBTAINING THE CRITICAL VALUE OF F WITH 2 AND 12 DEGREES OF FREEDOM AT THE .01 LEVEL

Denominator df_2	Numerator df_1 1	2	3	4	5	6	7	8	9	10	12
6	13.75	10.92	9.78	9.15	8.75	8.47	8.26	8.10	7.98	7.87	7.72
7	12.25	9.55	8.45	7.85	7.46	7.19	6.99	6.84	6.72	6.62	6.47
8	11.26	8.65	7.59	7.01	6.63	6.37	6.18	6.03	5.91	5.81	5.67
9	10.56	8.02	6.99	6.42	6.06	5.80	5.61	5.47	5.35	5.26	5.11
10	10.04	7.56	6.55	5.99	5.64	5.39	5.20	5.06	4.94	4.85	4.71
11	9.65	7.21	6.22	5.67	5.32	5.07	4.89	4.74	4.63	4.54	4.40
12	9.33	6.93	5.95	5.41	5.06	4.82	4.64	4.50	4.39	4.30	4.16
13	9.07	6.70	5.74	5.21	4.86	4.62	4.44	4.30	4.19	4.10	3.96

SOURCE: Extracted from Appendix E, Table E.5.

The decision rule would then be to reject the null hypothesis (H_0: $\mu_1 = \mu_2 = \mu_3$) if the calculated F equals or exceeds 6.93 (see Figure 12.2). In our example, since $F_{2,12} = 25.60 > 6.93$, we may reject H_0 and conclude that there is a significant difference in the average life of tires produced by the three shifts.

12.4 THE ANALYSIS OF VARIANCE TABLE

Since there are several steps involved in the computation of both the between and the within group variances, the entire set of results may be organized into an analysis of variance (ANOVA) table. This table, which is presented as Table 12.3, includes the sources of variation, the sums of squares (variations), the degrees of freedom, the variances, and the calculated F value.

In the tire company problem that we have been discussing, the computations just completed can be summarized in the following analysis of variance table (Table 12.4).

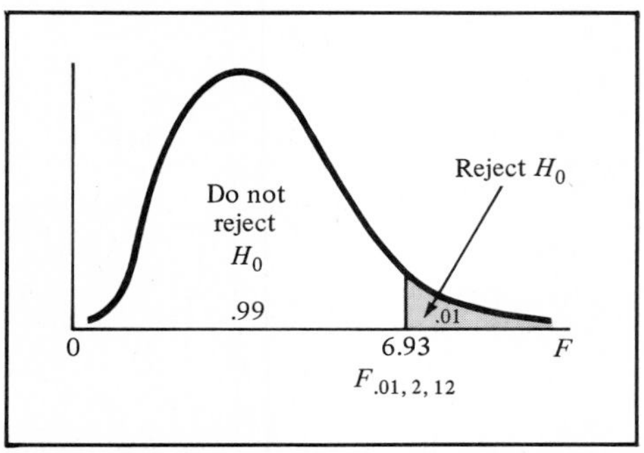

Figure 12.2 Regions of rejection and nonrejection for the analysis of variance at the 1% level of significance with 2 and 12 degrees of freedom.

Table 12.3
ANALYSIS OF VARIANCE TABLE

Source	Sum of Squares (SS)	Degrees of Freedom	(Mean Square) Variance	F
Between groups	$SS_B = \sum_{i=1}^{c} n_i(\bar{X}_i - \bar{\bar{X}})^2$	$c - 1$	$S_B^2 = \frac{SS_B}{c-1}$	$F = \frac{S_B^2}{S_W^2}$
Within groups	$SS_W = \sum_{i=1}^{c}\sum_{j=1}^{n_i} (X_{ij} - \bar{X}_i)^2$	$n - c$	$S_W^2 = \frac{SS_W}{n-c}$	
Total	$\sum_{i=1}^{c}\sum_{j=1}^{n_i} (X_{ij} - \bar{\bar{X}})^2$	$n - 1$		

12.5 COMPUTATIONAL METHODS

In the last several sections we have developed the concept of having several independent measures of variation, and we have used this approach to determine whether there were any differences in the means of several groups. The formulas utilized in computing the between group sum of squares, the within group sum of squares, and the total sum of squares were conceptual (or definitional). These formulas most certainly involved a large amount of tedious computations. However, as we may recall from Section 4.4, when the variance was computed, simple *computational* formulas were derived from the definitional formula. These computational formulas greatly simplify the calculations of the variances. The computational formulas for the analysis of variance require the calculation of the following quantities:

1. $\sum_{i=1}^{c}\sum_{j=1}^{n_i} X_{ij}^2$—the sum of the squared values of each observation in the data
2. T_i—the total (sum) of the values in group i
3. Grand total (GT) $= \sum_{i=1}^{c}\sum_{j=1}^{n_i} X_{ij}$—the grand total (sum) of all values

Table 12.4
ANALYSIS OF VARIANCE TABLE
FOR THE TIRE PROBLEM

Source	Sum of Squares (SS)	Degrees of Freedom	Variance	F
Between groups (shifts)	$\sum_{i=1}^{c} n_i(\bar{X}_i - \bar{\bar{X}})^2 = 47.164$	$c - 1$ $3 - 1 = 2$	$S_B^2 = \frac{47.164}{2} = 23.582$	$F = \frac{S_B^2}{S_W^2}$
Within groups (shifts)	$\sum_{i=1}^{c}\sum_{j=1}^{n_i} (X_{ij} - \bar{X}_i)^2 = 11.0532$	$n - c$ $15 - 3 = 12$	$S_W^2 = \frac{11.0532}{12} = .9211$	$= \frac{23.582}{.9211} = 25.60$
Total	$\sum_{i=1}^{c}\sum_{j=1}^{n_i} (X_{ij} - \bar{\bar{X}})^2 = 58.2172$	$n - 1$ $15 - 1 = 14$		

4. $\sum_{i=1}^{c} T_i^2$—the summation of the squared total of the values in each group i

Using computational formulas, we can then compute the various sums of squares in the following way:

$$\text{Between Group Variation} = \text{SS}_B = \frac{\sum_{i=1}^{c} T_i^2}{n_i} - \frac{(\text{GT})^2}{n} \qquad (12.3a)$$

$$\text{Within Group Variation} = \text{SS}_W = \sum_{i=1}^{c}\sum_{j=1}^{n_i} X_{ij}^2 - \frac{\sum_{i=1}^{c} T_i^2}{n_i} \qquad (12.4a)$$

$$\text{Total Variation} = \text{SS}_{\text{total}} = \sum_{i=1}^{c}\sum_{j=1}^{n_i} X_{ij}^2 - \frac{(\text{GT})^2}{n} \qquad (12.1a)$$

The format for the analysis of variance table using the computational formulas is given in Table 12.5.

Returning to the tire company problem, the difference between the shifts had been determined by using the definitional formulas. In order to use the simpler computational methods, the following four quantities must be computed:

$$\sum_{i=1}^{c}\sum_{j=1}^{n_i} X_{ij}^2 \qquad T_i \qquad \sum_{i=1}^{c}\frac{T_i^2}{n_i} \qquad \frac{(\text{GT})^2}{n}$$

Table 12.5
ANALYSIS OF VARIANCE TABLE USING COMPUTATIONAL FORMULAS

Source	Sum of Squares (SS)	Degrees of Freedom	Variance	F
Between groups	$\text{SS}_B = \frac{\sum_{i=1}^{c} T_i^2}{n_i} - \frac{(\text{GT})^2}{n}$	$c - 1$	$S_B^2 = \frac{\text{SS}_B}{c-1}$	$F = \frac{S_B^2}{S_W^2}$
Within groups	$\text{SS}_W = \sum_{i=1}^{c}\sum_{j=1}^{n_i} X_{ij}^2 - \frac{\sum_{i=1}^{c} T_i^2}{n_i}$	$n - c$	$S_W^2 = \frac{\text{SS}_W}{n-c}$	
Total	$\sum_{i=1}^{c}\sum_{j=1}^{n_i} X_{ij}^2 - \frac{(\text{GT})^2}{n}$	$n - 1$		

Referring back to Table 12.1 there were three shifts of workers (day, evening, and night). The results, giving the total and the mean for each shift, are summarized below:

	Day	Evening	Night
Sample size n_i	$n_1 = 5$	$n_2 = 5$	$n_3 = 5$
Total T_i	$T_1 = 124.65$	$T_2 = 113.05$	$T_3 = 102.95$
Sample mean $\bar{X}_i$ (thousands of miles)	$\bar{X}_1 = 24.93$	$\bar{X}_2 = 22.61$	$\bar{X}_3 = 20.59$

$$\text{GT} = \sum_{i=1}^{c}\sum_{j=1}^{n_i} X_{ij} = 124.65 + 113.05 + 102.95 = 340.65$$

$$\frac{(\text{GT})^2}{n} = \frac{(340.65)^2}{15} = 7{,}736.1615$$

$$\sum_{i=1}^{c}\sum_{j=1}^{n_i} X_{ij}^2 = 25.40^2 + 26.31^2 + \cdots + 20.00^2 + \cdots + 20.95^2$$

$$= 7{,}794.3787$$

$$\sum_{i=1}^{c}\frac{T_i^2}{n_i} = \frac{(124.65)^2 + (113.05)^2 + (102.95)^2}{5} = 7{,}783.3255$$

With this information the analysis of variance table for the tire problem can be set up using the computational formulas (Table 12.6).

If we compare the results given in Table 12.4 with those of Table 12.6, we can see, of course, that the two methods (definitional and computational) have produced exactly the same results. Therefore it is recommended that the computational method be utilized for all calculations needed for the analysis of variance table since these computations are less tedious and simpler to perform.

Table 12.6
ANALYSIS OF VARIANCE TABLE FOR THE TIRE PROBLEM USING COMPUTATIONAL FORMULAS

Source	Sum of Squares (SS)	Degrees of Freedom	Variance	F
Between shifts	$SS_B = \sum_{i=1}^{c}\frac{T_i^2}{n_i} - \frac{(\text{GT})^2}{n}$ $7{,}783.3225 - 7{,}736.1615$ $= 47.164$	$c - 1$ $3 - 1 = 2$	$S_B^2 = \frac{47.164}{2}$ $= 23.582$	$F = \frac{23.582}{.9211}$ $= 25.60$
Within shifts	$SS_W = \sum_{i=1}^{c}\sum_{j=1}^{n_i} X_{ij}^2 - \sum_{i=1}^{c}\frac{T_i^2}{n_i}$ $7{,}794.3787 - 7{,}783.3255$ $= 11.0532$	$n - c$ $15 - 3 = 12$	$S_W^2 = \frac{11.0532}{12}$ $= .9211$	
Total	$\sum_{i=1}^{c}\sum_{j=1}^{n_i} X_{ij}^2 - \frac{(\text{GT})^2}{n}$ $7{,}794.3787 - 7{,}736.1615$ $= 58.2172$	$n - 1$ $15 - 1 = 14$		

Now that we have developed this computational method for the analysis of variance, we can more easily analyze the problem of concern to the researcher and the dean as to whether there is any difference in the grade-point indexes of students preferring various admission standards. We may recall from our student survey that in question 21 of the questionnaire (see Figure 2.2) students were asked to indicate their preference for one of four minimum standards of admission:

1. No minimum standards; open to all with high school degree
2. Eighth-grade mathematics and English levels
3. Tenth-grade mathematics and English levels
4. Twelfth-grade mathematics and English levels

If the dean would like to determine whether there is any difference in the average grade-point index of students who prefer the four different admission standards, this information can be obtained by cross classifying questions 5 and 21 of the questionnaire. These results are presented in Table 12.7.

The null and alternative hypotheses for this problem would be stated as follows.

H_0: $\mu_{no} = \mu_8 = \mu_{10} = \mu_{12}$ (there is no difference in the grade-point index of students preferring different admission standards)

H_1: Not all means are equal (there is a difference in the grade-point index of students according to their preference for different admission standards)

From Table 12.7 it is clear that the sample sizes of the four groups are quite different—ranging from a low of 7 values to a high of 45 values. Although it is desirable (for reasons of power and simplicity) to have equal or approximately equal sample sizes, it is not a requirement for the one-factor analysis of variance.

In order to analyze the data presented in Table 12.7, various sums, needed for the analysis of variance, must be computed. The calculations are indicated below:

	No Minimum Standards	8th Grade	10th Grade	12th Grade
Sample size n_i	7	10	45	32
Total T_i	19.64	25.4	130.86	94.95
Sample mean $\bar{X}_i$	2.806	2.54	2.908	2.967

$$\text{GT} = \sum_{i=1}^{c} \sum_{j=1}^{n_i} X_{ij} = 19.64 + 25.4 + 130.86 + 94.95 = 270.85$$

$$\frac{(\text{GT})^2}{n} = \frac{(270.85)^2}{94} = \frac{73{,}359.7225}{94} = 780.42257$$

$$\sum_{i=1}^{c}\sum_{j=1}^{n_i} X_{ij}^2 = 2.30^2 + \cdots + 2.50^2 + 2.36^2 + \cdots + 2.00^2 + 2.80^2$$
$$+ \cdots + 2.00^2 + 2.00^2 + \cdots + 2.93^2 = 800.3469$$

$$\sum_{i=1}^{c}\frac{T_i^2}{n_i} = \frac{(19.64)^2}{7} + \frac{(25.40)^2}{10} + \frac{(130.86)^2}{45} + \frac{(94.95)^2}{32}$$
$$= 55.104228 + 64.516 + 380.54088 + 281.73445 = 781.89555$$

Table 12.7
GRADE-POINT INDEX OF STUDENTS PREFERRING EACH OF FOUR DIFFERENT MINIMUM ADMISSION STANDARDS

None	8th Grade	10th Grade	12th Grade
2.00	2.55	2.80	2.00
2.30	3.20	3.17	2.76
3.00	2.00	2.70	2.91
3.75	2.36	2.90	3.20
3.25	2.45	3.20	3.75
2.84	2.80	2.56	3.09
2.50	2.69	2.28	3.54
	2.86	2.92	3.18
(n_{no} = 7)	2.66	3.00	3.00
	1.83	3.86	2.93
		2.51	2.00
	(n_8 = 10)	2.62	3.51
		2.91	3.32
		3.60	2.80
		2.73	2.90
		3.12	3.00
		2.48	2.75
		2.64	2.60
		3.62	3.34
		2.10	2.50
		3.00	3.20
		3.43	3.00
		2.40	3.01
		3.82	2.25
		2.82	2.80
		3.00	3.07
		2.64	2.81
		2.45	3.08
		2.82	3.53
		3.86	2.75
		2.50	3.61
		2.30	2.76
		3.20	(n_{12} = 32)
		2.64	
		3.21	
		2.94	
		3.39	
		2.85	
		2.89	
		2.51	
		2.83	
		3.89	
		2.85	
		2.90	
		2.00	
		(n_{10} = 45)	

SOURCE: Data are taken from Figure 2.6.

Table 12.8
ANALYSIS OF VARIANCE TABLE FOR THE ADMISSION STANDARDS PROBLEM

Source	Sum of Squares (SS)	Degrees of Freedom	Variance	F_{calc}
Between admission standard levels	$\dfrac{\sum_{i=1}^{c} T_i^2}{n_i} - \dfrac{(GT)^2}{n}$ $781.89555 - 780.42257$ $= 1.47298$	$c - 1$ $4 - 1 = 3$	$S_B^2 = \dfrac{1.47298}{3}$ $= .491$	$F_{3,90} = \dfrac{.491}{.205}$ $= 2.395$
Within admission standard levels	$\sum_{i=1}^{c}\sum_{j=1}^{n_i} X_{ij}^2 - \dfrac{\sum_{i=1}^{c} T_i^2}{n_i}$ $800.3469 - 781.89555$ $= 18.45135$	$n - c$ $94 - 4 = 90$	$S_w^2 = \dfrac{18.45135}{90}$ $= .205$	
Total	$\sum_{i=1}^{c}\sum_{j=1}^{n_i} X_{ij}^2 - \dfrac{(GT)^2}{n}$ $800.3469 - 780.42257$ $= 19.92433$	$n - 1$ $94 - 1 = 93$		

With these results the above analysis of variance table can be set up using the computational formulas (Table 12.8).

If a level of significance of .05 were chosen, we would need to determine the critical value of the F distribution with 3 degrees of freedom in the numerator and 90 degrees of freedom in the denominator. Referring to the F distribution (Appendix E, Table E.5), since there is no value given for 90 degrees of freedom in the denominator, we can roughly interpolate between the values given for (3,60) and (3,120), that is, we would obtain an approximate value of 2.72 by interpolating between 2.76 ($F_{.05,3,60}$) and 2.68 ($F_{.05,3,120}$). Since $F_{3,90} = 2.395 < 2.72$ (see Figure 12.3), we cannot reject the null hypothesis. Thus based upon the sample data, we would conclude that there is no evidence of a difference in the grade-point index of students preferring different standards of admission.

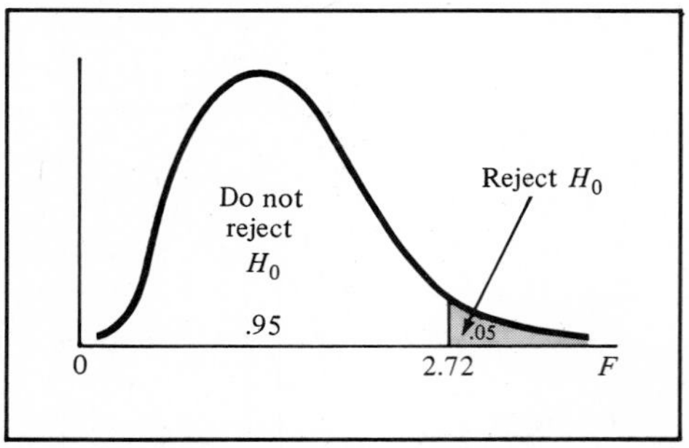

Figure 12.3 Regions of rejection and nonrejection for the analysis of variance at the 5% level of significance with 3 and 90 degrees of freedom.

12.6 ASSUMPTIONS OF THE ANALYSIS OF VARIANCE

In Chapters 9 and 10, we mentioned the assumptions that were made in applying each particular statistical procedure to a set of data and the consequences of departures from these assumptions. The analysis of variance technique also makes certain assumptions about the data being investigated. There are three major assumptions in the analysis of variance:

1. Normality
2. Homogeneity of variance
3. Independence of error

The first assumption, normality, states that the values in each group are normally distributed. Just as in the case of the t test, the analysis of variance test is "robust" against departures from the normal distribution; that is, as long as the distributions are not extremely different from a normal distribution, the level of significance of the analysis of variance test is not greatly affected by lack of normality, particularly for large samples.

The second assumption, homogeneity of variance, states that the variance within each group should be equal for all groups ($\sigma_1^2 = \sigma_2^2 = \cdots = \sigma_c^2$). This assumption is needed in order to combine or pool the variances within the groups into a single "within groups" source of variation. If there are equal sample sizes in each group, inferences based upon the F distribution may not be seriously affected by unequal variances. However, if there are unequal sample sizes in different groups, unequal variances from group to group can have serious effects on drawing inferences made from the analysis of variance. Thus from the point of view of both computational simplicity and robustness there should be equal sample sizes in all groups whenever possible.

The third assumption, independence of error, states that the error (variation of each value around its own group mean) should be independent for each value. This assumption often refers to data that are collected over a period of time in which the variation of values around the mean is not the same for each time period. Departures from this assumption can seriously affect inferences from the analysis of variance. These problems are discussed more thoroughly in Reference 4.

12.7 COMPARISON OF SUBSETS OF GROUPS

In the two examples discussed in this chapter the analysis of variance was used to determine whether there was any difference in the average value over several groups. An additional portion of this analysis would consist of determining which particular groups (or subsets of groups) are different. Of course, this would be particularly appropriate once a significant difference between the groups has been found through the F test.

Such questions can be examined in reference to the two problems that have been discussed in this chapter. In the tire problem, for example, it was determined that there was evidence of a significant difference in the life of tires produced by the three shifts. Based upon this result, the production manager would like to determine the shifts that are different. Is the evening shift different from the night shift? Is the day shift different from the other shifts? Nevertheless, in making comparisons between groups or subsets of groups, the key question as to the statistical approach is whether a particular hypothesis was formulated before the data were collected (a priori) or whether it was suggested by the data themselves (post hoc). For example, had the dean and the researcher wanted to know (prior to the actual data collection) whether there was a difference between those students who preferred at least a tenth-grade admission standard (groups 3 and 4) versus those who preferred less than a tenth-grade admission standard (groups 1 and 2), an a priori statistical procedure (such as orthogonal contrasts—see Reference 4) should have been utilized. However, if the comparison is suggested after inspection of the data, such as the comparison of the night shift to the other shifts, then a post hoc procedure (such as Scheffé's test—see Reference 4) should be chosen. A more detailed discussion of such specific multiple-comparison procedures is beyond the scope of this text. These topics are discussed in greater depth in References 2, 3, 4, and 6.

12.8 COMPUTERS AND THE ANALYSIS OF VARIANCE

In previous discussions of our survey data base we mentioned the role of the computer in the computation of descriptive statistics and in hypothesis testing

```
                    - - - -  A N A L Y S I S   O F   V A R I A N C E  - - - -


VALUE LABEL                          SUM        MEAN      STD DEV    SUM OF SQ              N

OPEN ADMISSION                     19.640      2.806      0.595        2.126       (       7)
8TH YEAR MATH & ENG                25.400      2.540      0.485        1.479       (      10)
10TH YEAR MATH & ENG              130.860      2.908      0.462        9.412       (      45)
12TH YEAR MATH & ENG               94.950      2.967      0.419        5.435       (      32)
                              ------------  ---------  -----------  ------------
  TOTAL                           270.850      2.881      0.463       18.452       (      94)


* * * * * * * * * * * * * * * * * A N O V A   T A B L E * * * * * * * * * * * * * * * * * *
*                                                                                          *
*                        SUM OF SQUARES     DEGREES OF FREEDOM      MEAN SQUARE            *
*                                                                                          *
*  BETWEEN GROUPS            1.4728              (       3)            0.4909              *
*                                                                                          *
*  WITHIN GROUPS            18.4517              (      90)            0.2050              *
*                                                                                          *
*    TOTAL                  19.9246              (      93)                                *
*                                                                                          *
* * * * * * * * * * * * * * * * * * * * * * * * * * * * * * * * * * * * * * * * * * * * * *
*                                                                                          *
*    F =     2.3946                                                                        *
*                                                                                          *
* * * * * * * * * * * * * * * * * * * * * * * * * * * * * * * * * * * * * * * * * * * * * *
```

Figure 12.4 Computer output from the SPSS subprogram BREAKDOWN for the data of Table 12.7.

of both quantitative and qualitative variables. A computer package such as SPSS (Reference 5) employed such subprograms as CONDESCRIPTIVE, FREQUENCIES, CROSSTABS, and T-TEST to perform these analyses (see Sections 4.8 and 10.11). In a similar fashion, a computer package could be used to perform the analysis of variance. This is particularly true if more than one factor is being considered or if a large sample size is involved (such as was the case in our example with 94 students). If the SPSS package is utilized, subprograms called BREAKDOWN, ANOVA (for general analysis of variance problems), or ONEWAY (one-way analysis of variance problems only) could be applied. The application of the BREAKDOWN subprogram to the data of Table 12.7 is illustrated in Figure 12.4.

PROBLEMS

* 12.1 A consumer magazine was interested in determining whether any difference existed in the average life of four different brands of transistor radio batteries. A random sample of four batteries of each brand was tested with the following results (in hours):

Brand 1	Brand 2	Brand 3	Brand 4
12	14	21	14
15	17	19	21
18	12	20	25
10	19	23	20

At the .05 level of significance is there a difference in the average life of these four brands of transistor radio batteries?

12.2 An industrial psychologist would like to determine the effect of alcoholic consumption on the typing ability of a group of secretaries. 15 secretaries were randomly assigned to one of three consumption levels (0 ounces, 1 ounce, and 2 ounces). Each secretary was instructed to type the same standard page. The number of errors made by the secretary was recorded with the following results:

Alcoholic Consumption (ounces)		
0	*1*	*2*
2	7	10
5	5	6
3	6	10
6	3	12
4	9	12

At the .01 level of significance does the amount of alcohol consumed affect the number of errors made by typists?

12.3 The retailing manager of a food chain wishes to determine whether product location has any effect on the sale of pet toys. Three different aisle locations are to be considered: front, middle, and rear. A random sample of 18 stores was selected with 6 stores randomly assigned to each aisle location. At the end of a 1-week trial period, the sales of the product in each store were as follows:

Aisle Location		
Front	*Middle*	*Rear*
86	20	46
72	32	28
54	24	60
40	18	22
50	14	28
62	16	40

(a) At the .01 level of significance is there a difference in sales between the various aisle locations?

(b) Based upon these results, what comparisons between groups or sets of groups are suggested by the data?

12.4 An auditor for a school district would like to study teacher absenteeism rates at different grade levels. Random samples of four teachers are respectively selected from elementary schools, junior high schools, and high schools, and the number of days absent in the past year was as follows:

Elementary School	Junior High School	High School
7	13	7
4	14	2
10	9	6
6	8	9

(a) At the .05 level of significance is there any difference in absenteeism at the various grade levels?

(b) Based upon these results, what comparisons between groups or sets of groups are suggested by the data?

12.5 The owner of a home heating oil delivery company would like to investigate the speed with which bills are paid in three suburban areas. Random samples of five accounts are selected in each area, and the number of days between delivery and payment of the bill is recorded with the following results:

Area		
I	*II*	*III*
8	10	32
18	16	8
14	28	16
20	25	27
12	7	17

At the .01 level of significance is there any difference in payment (as measured by days) between the three suburban areas?

12.6 An agronomist wishes to study the yield (in pounds) of four different types of squash. A field was divided into 16 plots with 4 plots randomly assigned to each variety. The four varieties of squash were

1. Butternut (winter) squash
2. Acorn (winter) squash

3. Zucchini (summer) squash
4. Scallop (summer) squash

The results of the experiment (yield in pounds) were as follows:

1	2	3	4
86	40	30	48
74	48	36	54
88	54	42	42
76	46	34	56

(a) At the .05 level of significance is there any difference in yield between the four varieties of squash?

(b) What comparisons might the agronomist want to make prior to collecting the data, and what comparisons should be made after examining the data?

* 12.7 A professor of computer science wanted to conduct an experiment to investigate the relative efficiency of the three computer languages FORTRAN, COBOL, and PL/1 in solving a large-scale problem. A random sample of 15 computer science majors, equally proficient in all three languages, was selected. The 15 students were randomly assigned to a particular language and told to count the number of hours of work needed to solve the problem. The results were as follows:

FORTRAN	COBOL	PL/I
20	23	20
17	20	23
26	27	19
19	30	24
24	26	27

At the .01 level of significance is there any difference in efficiency in man-hours between the three computer languages?

12.8 An auditor for the Internal Revenue Service would like to compare the efficiency of four regional tax processing centers. A random sample of five returns was selected at each center, and the number of days between receipt of the tax return and final processing was determined. The results (in days) were as follows:

Regional Centers			
East	*Midwest*	*South*	*West*
49	47	39	52
54	56	55	42
40	40	48	57
60	51	43	46
43	55	50	50

At the .05 level of significance is there evidence of a difference in processing time between the four regional centers?

* 12.9 A statistics professor wanted to study the difference in four different strategies of playing the game of Blackjack (Twenty-One). The four strategies were:

1. Dealer's strategy
2. Five-count strategy
3. Basic ten-count strategy
4. Advanced ten-count strategy

The game of Blackjack was programmed into a digital computer, and data from five sessions of each strategy were collected. The profits (or losses) from each session were as follows:

Strategy			
Dealer's	Five Count	Basic Ten Count	Advanced Ten Count
−$56	−$26	+$16	+$60
−$78	−$12	+$20	+$40
−$20	+$18	−$14	−$16
−$46	−$ 8	+$ 6	+$12
−$60	−$16	−$25	+$ 4

At the .01 level of significance is there any difference in profitability of the four gambling strategies?

12.10 A new-car dealer would like to study the amount of money spent on optional equipment purchased for full-sized cars. A random sample of 20 purchases is selected. The respondents were divided into the following age classifications: 18–24, 25–29, 30–39, 40–59, 60 and over. The amount of optional equipment purchased (in hundreds of dollars) was organized by age groups as follows:

Age				
18–24	25–29	30–39	40–59	60 and over
6.31	7.64	8.37	11.23	6.74
4.27	5.36	9.26	10.64	7.36
5.75	3.85	10.16	8.32	5.12
	6.24	6.48	9.00	
		7.86	7.53	

At the .05 level of significance is there any difference in the amount of money spent on optional equipment purchased by the various age groups?

The following problems refer to the sample data obtained from the survey (Figure 2.2) and presented in Figure 2.6.

12.11 At the .01 level of significance is there any difference in grade-point index (question 5) of seniors (code 4, question 2) based upon plans to attend graduate school (question 3)?

12.12 At the .01 level of significance is there any difference in high-school average (question 6) of seniors (code 4, question 2) based upon plans to attend graduate school (question 3)?

12.13 At the .05 level of significance is there any difference between accounting, economics, management, and marketing majors (question 4) in the maximum tuition (question 8) that sophomores (code 2, question 2) are willing to pay? **Hint:** In order to simplify computations, code the values into thousands of dollars.

12.14 At the .05 level of significance is there any difference between accounting, economics, management, and marketing majors (question 4) in the grade-point index (question 5) of sophomores (code 2, question 2)?

12.15 At the .05 level of significance is there any difference in the number of pairs of jeans owned (question 13) by sophomore, junior, and senior (question 2) accounting majors (code 2, question 4)?

12.16 At the .05 level of significance is there any difference in the number of pairs of jeans owned (question 13) by accounting, economics, management, and marketing majors (question 4)?

12.17 At the .01 level of significance is there a difference in the height (question 16) of women (code 2, question 1) based on smoking habits (question 19)?

12.18 At the .01 level of significance is there a difference in the weight (question 17) of males (code 1, question 1) based on automobile preference (question 12)?

12.19 At the .05 level of significance is there any difference between accounting, economics, management, and marketing majors (question 4) in the annual starting salary (question 7) that juniors (code 3, question 2) expect to obtain? **Hint:** In order to simplify computations, code the values into thousands of dollars.

12.20 At the .05 level of significance is there any difference in the age (question 15) of students planning to attend graduate school, not planning to attend graduate school, or not sure of attending graduate school (question 3)?

12.21 At the .01 level of significance is there a difference in the age (question 15) of seniors (code 4, question 2) based upon their investment intentions (question 11)?

12.22 At the .01 level of significance is there a difference in the grade-point index (question 5) of unemployed students (code 3, question 9) based upon their investment intentions (question 11)?

12.23 Use your own sample data (Problem 3.12) to select problems from among Problems 12.11 through 12.22.

12.24 For your own class (Problem 3.13) select problems from among Problems 12.11 through 12.22.

REFERENCES

1. DYCKMAN, T. R., AND L. J. THOMAS, *Fundamental Statistics for Business and Economics* (Englewood Cliffs, N.J.: Prentice-Hall, 1977).
2. HAYS, W. L., *Statistics for the Social Sciences* (New York: Holt, Rinehart and Winston, 1972).
3. HICKS, C. R., *Fundamental Concepts in the Design of Experiments,* 2nd ed. (New York: Holt, Rinehart and Winston, 1973).
4. NETER, J., AND W. WASSERMAN, *Applied Linear Statistical Models* (Homewood Ill.: Richard D. Irwin, 1974).
5. NIE, N., C. HULL, J. JENKINS, K. STEINBRENNER, AND D. BENT, *Statistical Package for the Social Sciences,* 2nd ed. (New York: McGraw-Hill, 1975).
6. SNEDECOR, G. W., AND W. G. COCHRAN, *Statistical Methods,* 6th ed. (Ames, Iowa: Iowa State University Press, 1967).
7. SOKAL, R. R., AND F. J. ROHLF, *Biometry* (San Francisco: W. H. Freeman, 1969).
8. WINER, B. J., *Statistical Principles in Experimental Design,* 2nd ed. (New York: McGraw-Hill, 1971).

CHAPTER 13

NONPARAMETRIC METHODS OF HYPOTHESIS TESTING

13.1 INTRODUCTION: CLASSICAL VERSUS NONPARAMETRIC PROCEDURES

In the previous four chapters we have considered a variety of hypothesis testing situations, some of which employed classical or parametric methods of hypothesis testing while others required nonparametric methods of hypothesis testing.

13.1.1 Classical Procedures

Classical or **parametric procedures** have three distinguishing characteristics. First, the classical procedures require that a "fairly sophisticated" level of measurement be attained on the collected data. In the literature such levels of measurement are known as *interval* scaling or *ratio* scaling (see References 5 or 13 for a detailed discussion of measurement and scaling concepts). That is, the level of measurement attained must be of "sufficient strength" so that the arithmetic operations of addition, subtraction, multiplication, and division on the collected data are meaningful. This, of course, implies that the data must be continuous and quantitative—not qualitative (see Section 2.5). Hence for classical procedures to be applied, the data cannot be attained by merely classifying the observations into various distinct categories (*nominal* scaling) such as "yes," "no," and "not sure"; nor can the data be attained by merely ranking the observations (*ordinal* scaling) based on some scheme such as highest to lowest.

The second distinguishing feature possessed by the classical procedures is that they involve the testing of hypothesized parameters. As examples, for situations containing but one sample, the t test is concerned with a specific value for the population mean, while for situations containing two or more samples, the t test and F test are concerned with testing for differences between the true population means.

The third characteristic of the classical procedures is that they require one to make some very stringent assumptions and are valid only if these assumptions hold. Among these assumptions are

1. That the sample data be randomly drawn from a population which is normally distributed
2. That the observations be independent of each other
3. For situations concerning central tendency for which two or more samples have been drawn, that they be drawn from normal populations having equal variances (so that any differences between the populations will be in central tendency)

The sensitivity of classical procedures to violations in the assumptions has been considered in the statistical literature (References 2 and 12). Some test procedures are said to be "robust" because they are relatively insensitive to slight violations in the assumptions. However, with gross violations in the assumptions both the true level of significance and the power of a test may be quite different than what otherwise would be expected.

13.1.2 Nonparametric Procedures

What then should the statistician do when the classical procedures are not applicable? In these situations an appropriate nonparametric test could be selected.

Nonparametric test procedures (Reference 3 and 4) **may be broadly thought of as either (1) those whose test statistic does not depend upon the form of the underlying population distribution from which the sample data were drawn; (2) those which are not concerned with the parameters of a population; or (3) those for which the data are of "insufficient strength" to warrant meaningful arithmetic operations.**

It may now be recalled that the one, two, and c sample tests for proportions as well as the χ^2 tests for independence and for goodness of fit, all of which were discussed in Chapters 9 and 10, appear to fit such a broad definition. Indeed, when testing a hypothesis about a proportion we have previously employed what is called (in the nonparametric literature) the binomial test. In the case study involving the manufacturer's quality control of tires, for example, the binomial test may be broadly categorized as nonparametric for two reasons. First, each item (tire) was "grossly measured" (as defective or nondefective). Second, the test statistic did not depend upon the underlying population from which the sample data were drawn. Regardless of whether the original phenomenon of interest—tread life of a tire (in miles)—actually follows a normal distribution or any other shaped distribution, the binomial test statistic is obtained from the classification of each tire as defective or non-defective based upon some preestablished mileage criterion or policy utilized by the company. This binomial test statistic is independent of the form of the particular underlying distribution.

On the other hand, aside from the fact that the data attained are qualitative (nominal scaling), the χ^2 test for independence may be broadly classified as nonparametric for a different reason—it is not concerned with parameters. We may recall, for example, from Section 10.7 that the dean and the researcher were concerned with ascertaining whether or not one's major area of study is independent of one's intention of pursuing graduate education. Other kinds of nonparametric procedures that are in this category are tests for goodness of fit (see Section 10.8), tests for randomness, tests for trend, tests for cyclical effects, and tests for symmetry. Two such nonparametric procedures—the Wald–Wolfowitz one-sample-runs test for randomness and the Cox–Stuart unweighted sign test for trend—will be described in this chapter along with a few other highly useful nonparametric techniques based on ranks (ordinal scaling) or similar "scoring" methods. Prior to this discussion on the selected nonparametric procedures, however, it is important to consider those situations in which nonparametric techniques may be most advantageously used as well as to point out some of their potential shortcomings.

13.2 ADVANTAGES AND DISADVANTAGES OF USING NONPARAMETRIC METHODS

There are numerous advantages to using nonparametric methods. Six such advantages are summarized below.

1. Nonparametric methods may be used on all types of data—qualitative data (nominal scaling), data in rank form (ordinal scaling), as well as data that have been measured more precisely (interval or ratio scaling).
2. Nonparametric methods are generally easy to apply and quick to compute when the sample sizes are small. Thus they are often used for pilot or preliminary studies and/or in situations in which answers are rapidly desired.
3. Nonparametric methods make fewer, less stringent assumptions (which are more easily met) than do the classical procedures. Hence they enjoy wider applicability and yield a more general, broad-based set of conclusions.
4. Nonparametric methods permit the solution of problems that do not involve the testing of population parameters.
5. Nonparametric methods may be more economical than classical pro-

cedures since the researcher may increase power and yet save money, time, and labor by collecting larger samples of data which are more grossly measured (that is, qualitative data or data in rank form) and by solving the problem faster.

6. Depending on the particular procedure selected, nonparametric methods may be as (or almost as) powerful as the classical procedure when the assumptions of the latter are met and may be quite a bit more powerful when the assumptions of the classical procedure are not met.

It should now be apparent that nonparametric methods may be advantageously employed in a variety of situations. On the other hand, nonparametric procedures possess some shortcomings which must be mentioned.

1. It is disadvantageous to use nonparametric methods when all the assumptions of the classical procedures can be met and the data are measured in either an interval or ratio scale. Unless classical procedures are employed in these instances, the researcher is not taking full advantage of the data. Information is lost when we convert such collected data (from an interval or ratio scale) to either ranks (ordinal scale) or categories (nominal scale). In particular, in such circumstances, some very quick and simple nonparametric tests have much less power than the classical procedures and should usually be avoided.
2. As the sample sizes get larger, data manipulations required for nonparametric procedures are sometimes laborious unless packaged computer programs (see Section 13.9) are available.
3. There are several types of statistical problems in experimental design for which nonparametric procedures have not as yet been devised. Such topics, however, are beyond the scope of this text.

Now that we have discussed some of the advantages and disadvantages of nonparametric methods, it is apparent that such procedures do not, in the words of Jean Dickinson Gibbons (see page 26, Reference 4), "provide a panacea for all problems of quantitative analysis. . . . Nevertheless, they comprise an important and useful body of statistical techniques." In the following sections, we will consider the development of six simple but important nonparametric techniques that may prove highly useful under differing circumstances.

13.3 WALD–WOLFOWITZ ONE-SAMPLE-RUNS TEST FOR RANDOMNESS

In statistical inference it is usually assumed that the collected data constitute a random sample. Such an assumption, however, may be tested by the employment of a nonparametric procedure which may be called the Wald–Wolfowitz one-sample-runs test for randomness. Moreover, the test is also useful for determining whether any sequence of collected data may be considered random—no matter how it was generated—or whether there is some underlying nonrandom pattern in the data. Such a test then may be helpful in determining whether or not changes in the daily closing prices of a particular stock are occurring randomly over time; whether or not the numbers selected in a state lottery follow a random sequence; and whether or not the finished items emerging from the production process contain a random sequence of defectives over time.

In all these cases the null hypothesis of randomness may be tested by observing the order or sequence in which the items are obtained. If each item is assigned one of two symbols, such as success and failure, depending on whether the item either possesses a particular property or in what amounts or magnitude the property is possessed, the randomness of the sequence may be investigated. If the sequence is randomly generated, the items would be independent and identically distributed. This means that the value of an item would be independent both of its position in the sequence and of the values of the items which precede and follow it. On the other hand, if an item in the sequence is affected by the items which precede it or succeed it so that the probability of its occurrence varies from one position to another, the process would not be considered random. In such cases either similar items would tend to cluster together (such as when a trend in the data is present) or the similar items would alternatingly mix so that some systematic periodic effect would exist.

To study whether or not an observed sequence is random, we will consider as the test statistic the number of runs present in the data.

A run may be defined as a consecutive series of similar items which are bounded by items of a different type or, at the beginning or ending of the sequence, by none at all.

For instance, suppose the following is the observed occurrence of an experiment for flipping a new coin 20 times:

HHHHHHHHHHTTTTTTTTTT

In the above sequence there are 2 runs—a run of 10 heads followed by a run of 10 tails. With similar items tending to cluster, such a sequence could not be considered random even though, as would theoretically be expected, 10 of the 20 outcomes were heads and 10 were tails.

At the other extreme, suppose the following sequence is obtained when flipping a coin 20 times:

HTHTHTHTHTHTHTHTHTHT

In this sequence there are 20 runs—10 runs of one head each and 10 runs of one tail each. With such an alternating pattern created, this sequence could not be considered random because there are too many runs.

On the other hand if, as shown below, the sequence of responses to the 20 coin tosses is thoroughly mixed, the number of runs will neither be too few nor too many, and the process may then be considered random:

HHTTHHHHTTTTTHTHTTHH

Therefore in testing for randomness, what is essential is the ordering or positioning of the items in the sequence, not the frequency of items of each type.

13.3.1 Procedure

To perform the test of the null hypothesis of randomness, let the total sample size n be decomposed into two parts, n_1 successes and n_2 failures, so that $n_1 + n_2 = n$. The test statistic U, the total number of runs, is then obtained by counting. For a two-tailed test, if U is either too large or too small, we may reject the null hypothesis of randomness in favor of the alternative that the sequence is not random. If both n_1 and n_2 are less than or equal to 20, Appendix E, Table E.8, parts 1 and 2 present the critical values for the test statistic U at the .05 level of significance (two-tailed). If, for a given combination of n_1 and n_2, U is either greater than or equal to the upper critical value or less than or equal to the lower critical value, the null hypothesis of randomness may be rejected at the .05 level. However, if U, the total number of runs, lies between these limits, the null hypothesis of randomness cannot be rejected.

On the other hand, tests for randomness are not always two-tailed. If we are interested in testing for randomness against the specific alternative of a *trend effect*—that there is a tendency for like items to cluster together—a one-tailed test is needed. Here we would reject the null hypothesis only if too few runs occur—if the observed value of U is less than or equal to the critical value presented in Appendix E, Table E.8, part 1 at the .025 level of significance. At the other extreme, if we are interested in testing for randomness against a *systematic or periodic effect* we would use a one-tailed test that only rejects if too many runs occur—if the observed value of U is greater than or equal to the critical value given in Appendix E, Table E.8, part 2 at the .025 level of significance.

Regardless of whether the test is one-tailed or two-tailed, however, for a sample size n greater than 40 or when either n_1 or n_2 exceeds 20, the test statistic U is approximately normally distributed. Therefore the following large sample approximation formula may be used to test the hypothesis of randomness:

$$Z \cong \frac{U - \mu_U}{\sigma_U} \tag{13.1}$$

where U = total number of runs

$$\mu_U = \text{mean value of } U;\ \mu_U = \frac{2n_1n_2}{n} + 1$$

$$\sigma_U = \text{standard deviation of } U;\ \sigma_U = \sqrt{\frac{2n_1n_2(2n_1n_2 - n)}{n^2(n-1)}}$$

n_1 = number of successes in sample

n_2 = number of failures in sample

n = sample size; $n = n_1 + n_2$

that is,

$$Z \cong \frac{U - \left(\dfrac{2n_1n_2}{n} + 1\right)}{\sqrt{\dfrac{2n_1n_2(2n_1n_2 - n)}{n^2(n-1)}}} \qquad (13.2)$$

and, based on the level of significance selected, the null hypothesis may be rejected if the computed Z value falls in the appropriate region of rejection depending on whether a two-tailed test or a one-tailed test is used (see Figure 13.1).

13.3.2 Applications

As one example employing the Wald–Wolfowitz one-sample-runs test suppose that the quality control engineer was observing the production process of a new brand of tires at the manufacturing plant. To check whether the process is in control, the quality control engineer examined a sample of 50 consecutive tires emerging from the production line and classified each as defective or nondefective. The resulting sequence follows:

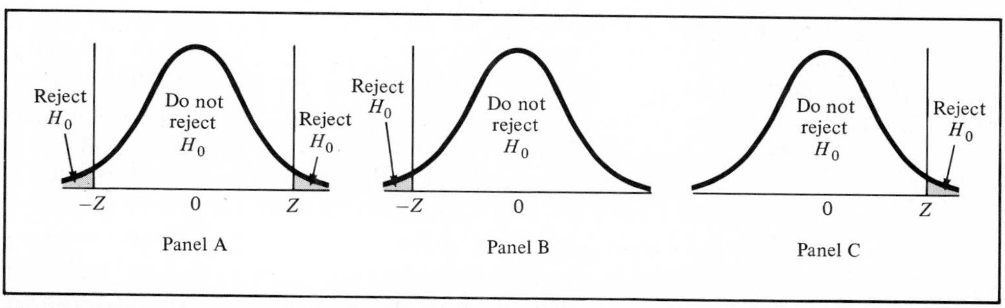

Figure 13.1 Determining the rejection region. Panel A, two-tailed test. Panel B, one-tailed test, trend effect. Panel C, one-tailed test, periodic effect.

Sequence NNNNNNDNNN NNNNNDNNDN NNNNNNNNNN NNNDDNNNNN NNNNNNNNDN
Runs 1 2 3 4 5 6 7 8 9 10 11

The null and alternative hypotheses are

H_0: The process generates defective and nondefective tires randomly

H_1: The process does not generate defective and nondefective tires randomly but in clusters (one-tailed)

If the null hypothesis is not rejected, the production process is said to be in control. If the null hypothesis is rejected, the process is said to be generating defective items in clusters. From the above sequence it is observed that 6 of the 50 tires (12%) in the sample are defective and that the data contain a total of 11 runs, that is, $n = 50$, $n_1 = 6$, $n_2 = 44$, and $U = 11$.

Since the sample size exceeds 40, the large sample approximation formula [Equation (13.2)] is used to test the null hypothesis. If the level of significance is selected as .05, the null hypothesis of randomness would be rejected if $Z < -1.645$ since the test is one-tailed.

$$Z \cong \frac{U - \left(\frac{2n_1n_2}{n} + 1\right)}{\sqrt{\frac{2n_1n_2(2n_1n_2 - n)}{n^2(n-1)}}} = \frac{11 - \left(\frac{(2)(6)(44)}{50} + 1\right)}{\sqrt{\frac{[(2)(6)(44)][(2)(6)(44) - 50]}{(50)^2(49)}}} = \frac{-.56}{\sqrt{2.0602}}$$

$$= -0.39$$

Since $Z = -0.39$ exceeds the critical Z value of -1.645, the null hypothesis of randomness cannot be rejected, and the process is said to be in control in terms of the ways in which the defective items emerge mixed in the sequence. On the other hand, if it is company policy that no more than 8% of the tires being produced can be defective, Equation (9.3) or (9.4) would now be used to test whether the process is in control in terms of the proportion of defective items in the sample (see Section 9.8).

In the above example the data were measured on the strength of a nominal scale. Each tire was merely classified as defective or nondefective. As a second application of the runs test of randomness Table 13.1 presents the unemployment rates (per thousand) of clerical workers in the United States from 1958 through 1974.

A distinguishing feature of the Wald–Wolfowitz one-sample-runs test for randomness is that it may be used not only on data which constitute a nominal scale—where each of the items is classified as success or failure—but also on data measured in the strength of an interval or ratio scale—where each of the items is classified according to its position with respect to the median of the sequence. For example, from Table 13.1 we may wish to test the null hypothesis that the unemployment rates of clerical workers are randomly distributed with respect to the median over time against the al-

Table 13.1
U.S. CLERICAL WORKERS' UNEMPLOYMENT RATES (1958–1974)

Year	Unemployment Rates (per thousand)	Relationship to Median Rate of 4.0 (A = equal or above; B = below)
1958	4.4	A
1959	3.7	B
1960	3.8	B
1961	4.6	A
1962	4.0	A
1963	4.0	A
1964	3.7	B
1965	3.3	B
1966	2.9	B
1967	3.1	B
1968	3.0	B
1969	3.0	B
1970	4.1	A
1971	4.8	A
1972	4.7	A
1973	4.2	A
1974	4.6	A

SOURCE: Data are extracted from Table 65 of *Handbook of Labor Statistics* 1975, reference edition, U.S. Department of Labor, Bureau of Labor Statistics.

ternative that these rates are not randomly distributed with respect to the median over time, that is,

H_0: Unemployment rates of clerical workers are random over time

H_1: Unemployment rates of clerical workers are not random over time (two-tailed)

To perform the runs test we assign the symbol A to each rate that equals or exceeds the median rate, and we assign the symbol B to each rate that is below the median rate. From Table 13.1 the median rate is 4.0. Thus for the 17 rates, 9 are equal to or above the median rate, while 8 are below the median rate. Now Appendix E, Table E.8, parts 1 and 2 present the critical values of the runs test at the .05 level of significance. As demonstrated in Table 13.2, since $n_1 = 9$ and $n_2 = 8$, we would reject the null hypothesis at the .05 level if $U \geq 14$ or if $U \leq 5$ for this two-tailed test. Since the observed number of runs is 5, we may reject the null hypothesis of randomness in favor of the alternative. There is apparently a periodic or cyclical pattern over time to the rates of unemployment for clerical workers. If the null hypothesis were true, the probability of obtaining such a result as this or one even more extreme would be less than .05.

13.4 COX–STUART UNWEIGHTED SIGN TEST FOR TREND

A *trend* is said to exist in a sequence of numbers if the values in the sequence tend to be either increasing (upward or positive trend) or decreasing (downward or negative trend) over the entire sequence. Cox and Stuart developed

Table 13.2
OBTAINING THE LOWER AND UPPER TAIL CRITICAL VALUES U FOR THE RUNS TEST WHERE $n_1 = 9$, $n_2 = 8$, AND $\alpha = .05$

Part 1 Lower Tail (α = .025)

n_1 \ n_2	2	3	4	5	6	7	8	9	10	11	12	13	14	15	16	17	18	19	20
2											2	2	2	2	2	2	2	2	2
3					2	2	2	2	2	2	2	2	2	3	3	3	3	3	3
4				2	2	2	3	3	3	3	3	3	3	3	4	4	4	4	4
5			2	2	3	3	3	3	3	4	4	4	4	4	4	4	5	5	5
6		2	2	3	3	3	3	4	4	4	4	5	5	5	5	5	5	6	6
7		2	2	3	3	3	4	4	5	5	5	5	5	6	6	6	6	6	6
8		2	3	3	3	4	4	5	5	5	6	6	6	6	6	7	7	7	7
9		2	3	3	4	4	[5]	5	5	6	6	6	7	7	7	7	8	8	8
10		2	3	3	4	5	5	5	6	6	7	7	7	7	8	8	8	8	9

Part 2 Upper Tail (α = .025)

n_1 \ n_2	2	3	4	5	6	7	8	9	10	11	12	13	14	15	16	17	18	19	20
2																			
3																			
4				9	9														
5			9	10	10	11	11												
6			9	10	11	12	12	13	13	13	13								
7				11	12	13	13	14	14	14	14	15	15	15					
8				11	12	13	14	14	15	15	16	16	16	16	17	17	17	17	17
9					13	14	[14]	15	16	16	16	17	17	18	18	18	18	18	18
10					13	14	15	16	16	17	17	18	18	18	19	19	19	20	20

SOURCE: Extracted from Appendix E, Table E.8, parts 1 and 2.

a series of tests to determine the existence of trend in a set of data. One of their procedures, the Cox–Stuart unweighted sign test for trend, is presented here.

13.4.1 Procedure

The unweighted sign test for trend is developed by pairing the observations appearing later in the sequence with those observations appearing earlier in the sequence and then examining the direction of differences within each pair. Thus the level of measurement must be at least ordinal. If an upward trend exists in the sequence, the great majority of pairs would have the later number larger than the earlier number. On the other hand, if a downward trend is present the reverse would be true—a great majority of pairs would have the later number smaller than the earlier number. If, however, no trend is present in the data, we would expect, as when flipping a fair coin several times, that approximately half the time the later number in the pair is the larger and half the time the later number in the pair is the smaller.

To obtain the necessary data for the test, the order in which the observations are collected constitutes the sequence of interest, that is, X_1 is the first observation in the sequence, X_2 is the second, . . . , and $X_{n'}$ is the last observation in the sequence. For the unweighted sign test the pairs are formed by splitting the number of observations in the sequence, n', in half. If n' is an even number, the $n'/2$ pairs are formed as:

$$(X_1, X_{(n'/2)+1}), (X_2, X_{(n'/2)+2}), \ldots, (X_i, X_{(n'/2)+i}), \ldots, (X_{n'/2}, X_{n'})$$

For instance, if n' were 10, then the five pairs would be formed as (X_1, X_6), (X_2, X_7), (X_3, X_8), (X_4, X_9), and (X_5, X_{10}). If, however, n' is odd the middle value in the sequence, $X_{n'+1/2}$, is discarded and the $(n' - 1)/2$ pairs are formed as:

$$(X_1, X_{(n'+3)/2}), (X_2, X_{[(n'+3)/2]+1}), \ldots, (X_i, X_{[(n'+3)/2]+i-1}), \ldots, (X_{(n'-1)/2}, X_{n'})$$

For example, if n' were 11, X_6 would be discarded and the five pairs would then be formed as (X_1, X_7), (X_2, X_8), (X_3, X_9), (X_4, X_{10}), and (X_5, X_{11}).

Once the pairs are obtained, we individually assign the symbol + to each pair in which the later number exceeds the earlier number, we assign the symbol − to each pair in which the later number is exceeded by the earlier number, and we eliminate from further consideration any pairs in which the earlier and later numbers are equal (ties). Thus n, the number of untied pairs, is equal to the number of pairs minus the number of tied pairs that are discarded:

n' even	n' odd
$n = \frac{n'}{2}$ − number of tied pairs	$n = \frac{n'-1}{2}$ − number of tied pairs

The Cox–Stuart unweighted sign test statistic V may therefore be considered as the number of + pairs (that is, the number of pairs in which the later number exceeds the earlier number) out of a possible n untied pairs. Under the null hypothesis of no trend there is, in each pair, a .5 probability that the earlier number is larger and a .5 probability that the earlier number is smaller. When testing the null hypothesis, then, the test statistic V is binomially distributed with parameters n and .5. For a two-tailed test the null and alternate hypotheses are

H_0: There is no trend in the data

H_1: There is either a positive (upward) trend or a negative (downward) trend in the data

On the other hand, the researcher may be interested in a specific direction for the trend, and a one-tailed test would be needed. For a test against positive trend the hypotheses are

H_0: There is no positive (upward) trend in the data

H_1: There is a positive (upward) trend in the data

while for a test against negative trend the hypotheses are

H_0: There is no negative (downward) trend in the data

H_1: There is a negative (downward) trend in the data

The exact probability of obtaining a particular value v of the test statistic V or one even more extreme can be computed from the binomial model [Equa-

Table 13.3
OBTAINING THE LOWER TAIL CRITICAL VALUE V FOR THE COX–STUART UNWEIGHTED SIGN TEST FOR TREND WHERE $n = 15$ AND $\alpha = .01$

Number of Untied Pairs	One-Tailed: / Two-Tailed:	$\alpha = .05$ / $\alpha = .10$	$\alpha = .025$ / $\alpha = .05$	$\alpha = .01$ / $\alpha = .02$	$\alpha = .005$ / $\alpha = .01$
n		(Lower, Upper)			
5		0,5	—,—	—,—	—,—
6		0,6	0,6	—,—	—,—
7		0,7	0,7	0,7	—,—
8		1,7	0,8	0,8	0,8
9		1,8	1,8	0,9	0,9
10		1,9	1,9	0,10	0,10
11		2,9	1,10	1,10	0,11
12		2,10	2,10	1,11	1,11
13		3,10	2,11	1,12	1,12
14		3,11	2,12	2,12	1,13
15		3,12	3,12	2,13	2,13
16		4,12	3,13	2,14	2,14

SOURCE: Extracted from Appendix E, Table E.9.

tion (6.7)] in Chapter 6. To save time and labor, however, when n, the number of untied pairs, is less than or equal to 20, Appendix E, Table E.9 may be used to obtain the critical values of the test statistic V for both one- and two-tailed tests at various levels of significance. For a two-tailed test and for a particular level of significance, if the computed value of V equals or exceeds the upper critical value or is less than or equal to the lower critical value, the null hypothesis may be rejected. For one-tailed tests against positive trend the decision rule is to reject the null hypothesis if the observed value of V equals or exceeds the upper critical value, while for one-tailed tests for negative trend the decision rule is to reject if V is less than or equal to the lower critical value. As an example, suppose the number of untied pairs is 15 and we desire a one-tailed test for negative trend at the .01 level of significance. From Table 13.3 we observe that the lower critical value is 2.

For larger samples, however, when the number of untied pairs exceeds 20, the normal approximation to the binomial (Section 6.10) may be used to test for trend regardless of whether the test is one-tailed or two-tailed:

$$Z \cong \frac{V - \mu_V}{\sigma_V} \qquad (13.3)$$

where V = number of untied pairs in which the later number is larger than the earlier number

n = number of untied pairs

μ_V = expected value of V under the null hypothesis; $\mu_V = .5n$

σ_V = standard deviation of V under the null hypothesis; $\sigma_V = \sqrt{.25n}$

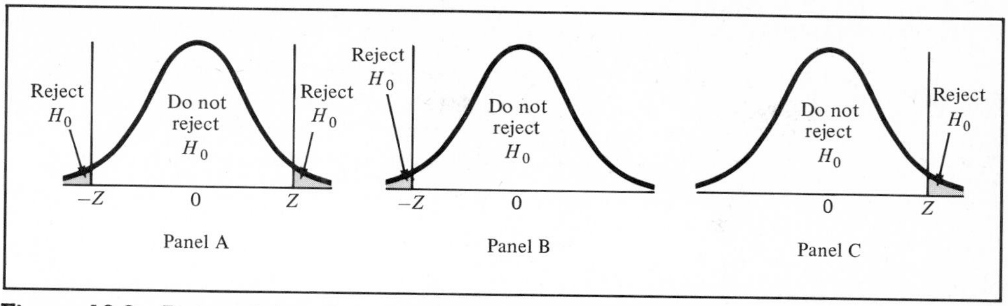

Figure 13.2 Determining the rejection region. Panel A, two-tailed test. Panel B, one-tailed test, negative trend. Panel C, one-tailed test, positive trend.

that is,

$$Z \cong \frac{V - .5n}{\sqrt{.25n}} \tag{13.4}$$

and, based on the level of significance chosen, the null hypothesis of no trend may be rejected if the computed Z value falls in the appropriate rejection region, depending on whether a two-tailed or a one-tailed test is used (see Figure 13.2).

13.4.2 Application

As one application of the Cox–Stuart unweighted sign test we may examine whether there is evidence of a downward trend in the amount of average weekly hours spent by production workers of both durable and nondurable goods over a 50-year period (see Table 13.4). (Other examples concerning trend will be investigated in Chapters 14 and 17.)

The null and alternate hypotheses are

H_0: There is no negative (downward) trend in the data

H_1: There is a negative (downward) trend in the data

Since the number of observations, $n' = 50$, is even, we form the pairs of values starting with the first year (1925) and the twenty-sixth year (1950), etc. For each pair the differences between the later year and the earlier year are compared, and the test statistic V, the number of + differences, is obtained. In this example $n = n'/2 = 25$ and $V = 13$. Since the number of untied pairs exceeds 20, the large sample approximation formula [Equation (13.4)] is used to test the null hypothesis. If the level of significance is

Table 13.4
AVERAGE WEEKLY HOURS OF
U.S. MANUFACTURING PRODUCTION WORKERS
(1925–1974)

Year	Earlier Average Hours	Year	Later Average Hours	Difference (Later − Earlier)
1925	44.5	1950	40.5	−
1926	45.0	1951	40.6	−
1927	45.0	1952	40.7	−
1928	44.4	1953	40.5	−
1929	44.2	1954	39.6	−
1930	42.1	1955	40.7	−
1931	40.5	1956	40.4	−
1932	38.3	1957	39.8	+
1933	38.1	1958	39.2	+
1934	34.6	1959	40.3	+
1935	36.6	1960	39.7	+
1936	39.2	1961	39.8	+
1937	38.6	1962	40.4	+
1938	35.6	1963	40.5	+
1939	37.7	1964	40.7	+
1940	38.1	1965	41.2	+
1941	40.6	1966	41.3	+
1942	43.1	1967	40.6	−
1943	45.0	1968	40.7	−
1944	45.2	1969	40.6	−
1945	43.5	1970	39.8	−
1946	40.3	1971	39.9	−
1947	40.4	1972	40.6	+
1948	40.0	1973	40.7	+
1949	39.1	1974	40.0	+

SOURCE: Data are extracted from Table 78 of *Handbook of Labor Statistics* 1975, reference edition, U.S. Department of Labor, Bureau of Labor Statistics.

selected as .05, the null hypothesis would be rejected if $Z < -1.645$ since the test is one-tailed:

$$Z \cong \frac{V - .5n}{\sqrt{.25n}} = \frac{13 - (.5)(25)}{\sqrt{(.25)(25)}} = \frac{+.50}{\sqrt{6.25}} = +0.20$$

Since $Z = +0.20$ exceeds the critical Z value of -1.645, the null hypothesis of no trend cannot be rejected, and it may be said that there is no evidence of a downward trend in the average number of weekly hours spent by production workers of both durable and nondurable goods over a 50-year period.

13.5 WILCOXON ONE-SAMPLE SIGNED-RANKS TEST

Thus far the nonparametric procedures presented were not concerned with the testing of any particular parameters and as such have no classical counterparts. However, the procedure presented in this section may be used in lieu of the classical one-sample t test (Section 9.6) when it is desired to test a hypothesis regarding a parameter reflecting central tendency. This procedure, known as the Wilcoxon one-sample signed-ranks test, may be chosen over its classical counterpart when the researcher is able to obtain data measured at a higher level than an ordinal scale but does not believe that the assumptions of the classical procedure are sufficiently met. When the assumptions of the t test are violated, the Wilcoxon procedure (which makes fewer and less stringent assumptions than does the t test) is likely to be more powerful in detecting the existence of significant differences than its classical counterpart. Moreover, even under conditions appropriate to the classical t test, the Wilcoxon one-sample signed-ranks test has proven to be almost as powerful.

13.5.1 Procedure

The assumptions necessary for performing the Wilcoxon one-sample signed-ranks test are

1. That the data be randomly and independently drawn
2. That the underlying random phenomenon of interest be continuous
3. That the observed data be measured at a higher level than the ordinal scale
4. That the distribution of differences between the observed data and the hypothesized median be (approximately) symmetric

Such an assumption as symmetry, however, is not as stringent as an assumption of normality since all symmetrical distributions are not normal (see, for example, Figures 4.10 and 4.11) while all normal distributions are also symmetrical.

The Wilcoxon one-sample signed-ranks test deals with both direction and magnitude, that is, the test statistic considers not only whether an observed value is larger or smaller than the hypothesized median but also how much larger or how much smaller. The test of the null hypothesis for a specified population median M_0 may be one-tailed or two-tailed:

Two-Tailed Test	One-Tailed Test	One-Tailed Test
H_0: Median $= M_0$	H_0: Median $\geq M_0$	H_0: Median $\leq M_0$
H_1: Median $\neq M_0$	H_1: Median $< M_0$	H_1: Median $> M_0$

To perform the Wilcoxon one-sample signed-ranks test the following six-step procedure may be used:

1. From a sample of n' items we obtain a set of difference scores D_i between each of the observed values and the hypothesized median

$$D_i = X_i - M_0$$

where $i = 1, 2, \ldots, n'$.

2. We then neglect the $+$ and $-$ signs and obtain a set of n' absolute differences $|D_i|$:

$$|D_i| = |X_i - M_0|$$

where $i = 1, 2, \ldots, n'$.

3. We omit from further analysis any absolute difference of score of zero, thereby yielding a set of n nonzero absolute difference scores where $n \leq n'$.

4. We then assign ranks R_i from 1 to n to each of the $|D_i|$ such that the smallest absolute difference score gets rank 1 and the largest gets rank n. Due to a lack of precision in the measuring process, if two or more $|D_i|$ are equal, they are each assigned the "average rank" of the ranks they otherwise would have individually been assigned had ties in the data not occurred.

5. We now reassign the symbol + or − to each of the n nonzero absolute difference scores $|D_i|$, depending on whether $X_i - M_0$ was positive or negative.

6. The Wilcoxon test statistic W is obtained as the sum of the + ranks.

$$W = \sum_{i=1}^{n} R_i^{(+)} \tag{13.5}$$

Since the sum of the first n integers $(1, 2, \ldots, n)$ is given by $n(n + 1)/2$, the Wilcoxon test statistic W may range from a minimum of 0 (where all the observed differences around the hypothesized median are negative) to a maximum of $n(n + 1)/2$ (where all the observed differences are positive). If the null hypothesis were true, we would expect the test statistic W to take on a value close to its mean, $\mu_W = n(n+ 1)/4$, while if the null hypothesis were false, we would expect the observed value of the test statistic to be close to one of the extremes.

For samples of $n \leq 20$, Appendix E, Table E.10 may be used for obtaining the critical values of the test statistic W for both one- and two-tailed tests at various levels of significance. For a two-tailed test and for a particular level of significance, if the observed value of W equals or exceeds the upper critical value or is equal to or less than the lower critical value, the null hypothesis may be rejected. For a one-tailed test in the positive direction (if we are testing against the alternative that the median is greater than M_0) the decision rule is to reject the null hypothesis if the observed value of W equals or exceeds the upper critical value, while for a one-tailed test in the negative direction (if we are testing against the alternative that the median is less than M_0) the decision rule is to reject the null hypothesis if the observed value of W is less than or equal to the lower critical value. For samples of $n > 20$, however, the test statistic W is approximately normally distributed, and the following large sample approximation formula may be used for testing the null hypothesis:

$$Z \cong \frac{W - \mu_W}{\sigma_W} \tag{13.6}$$

where W = sum of the positive ranks; $W = \sum_{i=1}^{n} R_{i}^{(+)}$

μ_W = mean value of W; $\mu_W = \frac{n(n+1)}{4}$

σ_W = standard deviation of W; $\sigma_W = \sqrt{\frac{n(n+1)(2n+1)}{24}}$

n = number of nonzero absolute difference scores in sample

that is,

$$Z \cong \frac{W - \frac{n(n+1)}{4}}{\sqrt{\frac{n(n+1)(2n+1)}{24}}} \tag{13.7}$$

and, based on the level of significance selected, the null hypothesis may be rejected if the computed Z value falls in the appropriate region of rejection depending on whether a two-tailed or one-tailed test is used (see Figure 13.3).

13.5.2 Applications

As one example using the Wilcoxon one-sample signed-ranks test suppose that a manufacturer of batteries claims that the median capacity of a certain type of battery the company produces is 140 ampere-hours. An independent

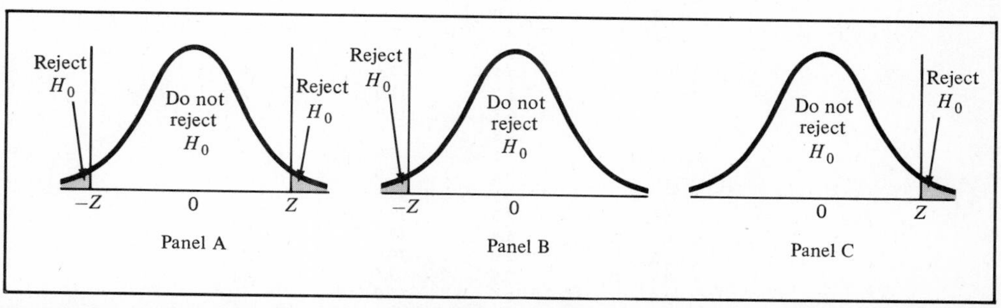

Figure 13.3 Determining the rejection region. Panel A, two-tailed test, median $\neq M_0$. Panel B, one-tailed test, median $< M_0$. Panel C, one-tailed test, median $> M_0$.

consumer protection agency wishes to test the credibility of the manufacturer's claim and measures the capacity of a random sample of $n' = 20$ batteries from a recently produced batch. The results are as follows:

137.0	140.0	138.3	139.0	144.3	139.1	141.7	137.3	133.5	138.2
141.1	139.2	136.5	136.5	135.6	138.0	140.9	140.6	136.3	134.1

Since the consumer protection agency is interested in whether or not the manufacturer's claim is being overstated, the test is one-tailed and the following null and alternate hypotheses are established:

$$H_0: \quad \text{Median} \geq 140.0 \text{ ampere-hours}$$

$$H_1: \quad \text{Median} < 140.0 \text{ ampere-hours}$$

The six-step procedure to perform the test is developed in Table 13.5.

The test statistic W is obtained as the sum of the positive ranks.

$$W = \sum_{i=1}^{n=19} R_i^{(+)} = 16 + 7.5 + 6 + 3.5 + 1 = 34$$

From Table 13.5 we note that only 5 of the 19 nonzero absolute difference scores exceed the claimed median of 140 ampere-hours. Thus to test for significance, we compare the observed value of the test statistic, $W = 34$, to the lower tail critical value presented in Appendix E, Table E.10 for $n = 19$

Table 13.5
SETTING UP THE WILCOXON ONE-SAMPLE SIGNED-RANKS TEST

$D_i = X_i - 140.0$	$\lvert D_i \rvert$	R_i	Sign of D_i
−3.0	3.0	12.0	−
0.0	0.0	. . .	Discard
−1.7	1.7	7.5	−
−1.0	1.0	5.0	−
+4.3	4.3	16.0	+
−0.9	0.9	3.5	−
+1.7	1.7	7.5	+
−2.7	2.7	11.0	−
−6.5	6.5	19.0	−
−1.8	1.8	9.0	−
+1.1	1.1	6.0	+
−0.8	0.8	2.0	−
−3.5	3.5	13.5	−
−3.5	3.5	13.5	−
−4.4	4.4	17.0	−
−2.0	2.0	10.0	−
+0.9	0.9	3.5	+
+0.6	0.6	1.0	+
−3.7	3.7	15.0	−
−5.9	5.9	18.0	−

Table 13.6
OBTAINING LOWER TAIL CRITICAL VALUE W FOR WILCOXON ONE-SAMPLE SIGNED-RANKS TEST WHERE $n = 19$ AND $\alpha = .05$

n	One-Tailed: $\alpha = .05$ / Two-Tailed: $\alpha = .10$	$\alpha = .025$ / $\alpha = .05$	$\alpha = .01$ / $\alpha = .02$	$\alpha = .005$ / $\alpha = .01$
	(Lower, Upper)			
5	0,15	—,—	—,—	—,—
6	2,19	0,21	—,—	—,—
7	3,25	2,26	0,28	—,—
.	.	.	.	.
.	.	.	.	.
.	.	.	.	.
17	41,112	34,119	27,126	23,130
18	47,124	40,131	32,139	27,144
19 →	[53],137	46,144	37,153	32,158
20	60,150	52,158	43,167	37,173

SOURCE: Extracted from Appendix E, Table E.10.

and for an α level selected at .05. As shown in Table 13.6, this critical value is 53. Since $W < 53$, the null hypothesis may be rejected at the .05 level of significance. There is evidence to believe that the manufacturing claim is overstated, and the protection agency should initiate some corrective measure against the manufacturer.

It is interesting to note that the large sample approximation formula [Equation (13.7)] for the test statistic W yields excellent results for samples as small as 8. With the above data, for a sample of $n = 19$,

$$Z \cong \frac{W - \dfrac{n(n+1)}{4}}{\sqrt{\dfrac{n(n+1)(2n+1)}{24}}} = \frac{34 - 95}{\sqrt{617.5}} = \frac{-61}{24.89} = -2.45$$

Since $Z = -2.45$ is less than the critical Z value of -1.645, the null hypothesis would also be rejected. However, since Appendix E, Table E.10 is available for $n \leq 20$, it is both simpler and more accurate to merely look up the critical value in the table and avoid these computations when possible.

In order to demonstrate the power and yet simplicity of the Wilcoxon one-sample signed-ranks test we may use, as a second application, the data on grade-point indexes of accounting majors which were obtained from the survey developed by the researcher and the dean. These survey data, depicted in Figure 2.6, were previously analyzed in Section 9.6 by use of a t test [Equation (9.2)] under more stringent assumptions. We may recall (see Section 9.2) that one of the questions of interest to the researcher and the dean was to determine whether the grade-point indexes of accounting majors had changed over the past 10 years. The average grade-point index had been 2.705.

Table 13.7
SETTING UP THE WILCOXON ONE-SAMPLE SIGNED-RANKS TEST

$X_i - 2.705 = D_i$	$\lvert D_i \rvert$	R_i	Sign of D_i
2.80 − 2.705 = 0.095	0.095	6.0	+
3.17 − 2.705 = 0.465	0.465	22.0	+
2.70 − 2.705 = −0.005	0.005	1.0	−
2.00 − 2.705 = −0.705	0.705	29.0	−
2.90 − 2.705 = 0.195	0.195	9.5	+
3.20 − 2.705 = 0.495	0.495	25.0	+
2.56 − 2.705 = −0.145	0.145	7.0	−
2.28 − 2.705 = −0.425	0.425	21.0	−
2.92 − 2.705 = 0.215	0.215	13.0	+
3.00 − 2.705 = 0.295	0.295	17.0	+
3.86 − 2.705 = 1.155	1.155	35.0	+
2.00 − 2.705 = −0.705	0.705	29.0	−
2.51 − 2.705 = −0.195	0.195	9.5	−
2.62 − 2.705 = −0.085	0.085	5.0	−
2.76 − 2.705 = 0.055	0.055	3.0	+
2.91 − 2.705 = 0.205	0.205	11.5	+
3.20 − 2.705 = 0.495	0.495	25.0	+
2.91 − 2.705 = 0.205	0.205	11.5	+
3.60 − 2.705 = 0.895	0.895	32.0	+
3.75 − 2.705 = 1.045	1.045	34.0	+
2.55 − 2.705 = −0.155	0.155	8.0	−
2.73 − 2.705 = 0.025	0.025	2.0	+
3.12 − 2.705 = 0.415	0.415	20.0	+
2.48 − 2.705 = −0.225	0.225	14.5	−
3.09 − 2.705 = 0.385	0.385	19.0	+
3.54 − 2.705 = 0.835	0.835	31.0	+
3.18 − 2.705 = 0.475	0.475	23.0	+
2.64 − 2.705 = −0.065	0.065	4.0	−
3.62 − 2.705 = 0.915	0.915	33.0	+
3.20 − 2.705 = 0.495	0.495	25.0	+
3.00 − 2.705 = 0.295	0.295	17.0	+
2.10 − 2.705 = −0.605	0.605	27.0	−
3.00 − 2.705 = 0.295	0.295	17.0	+
2.00 − 2.705 = −0.705	0.705	29.0	−
2.93 − 2.705 = 0.225	0.225	14.5	+

To employ the Wilcoxon test we state the following null and alternate hypotheses:

$$H_0: \quad \text{Median} = 2.705$$

$$H_1: \quad \text{Median} \neq 2.705$$

Following the six-step procedure we obtain Table 13.7.

The test statistic $W = \sum_{i=1}^{n=35} R_i^{(+)} = 446$ is the sum of the positive ranks.

Since $n = 35$, we use Equation (13.7) to obtain

$$Z \cong \frac{W - \dfrac{n(n+1)}{4}}{\sqrt{\dfrac{n(n+1)(2n+1)}{24}}} = \frac{446 - 315}{\sqrt{3{,}727.5}} = \frac{131}{61.05} = 2.15$$

For a two-tailed test with an α level of .05 the critical Z values are ± 1.96. Since $Z = 2.15$ exceeds the upper critical Z value of $+1.96$, the null hypothesis may be rejected—as was the case with the t test of Section 9.6—and it may be concluded that the median grade-point index of accounting majors has changed over the past 10 years.

13.6 WILCOXON RANK SUM TEST

In Section 10.2, when the researcher was interested in testing for differences in the means of two mutually independent groups of data, the two-sample t test [Equation (10.2)] was selected. To use the two-sample t test, however, it is necessary to make a set of stringent assumptions (see Section 13.1.1). In particular, it is necessary that the two independent samples be randomly drawn from normal populations having equal variances and that the data be measured in at least the strength of an interval scale. However, it is frequently the case in studies of consumer behavior, marketing research, and experimental psychology that only ordinal type data can be obtained. Since the classical two-sample t test could not be used in such situations, an appropriate nonparametric technique is needed. On the other hand, even if data possessing the properties of an interval or ratio scale are obtained, the researcher may feel that the assumptions of the t test are unrealistic for the set of data. In such a circumstance a very simple but powerful nonparametric procedure known as the Wilcoxon rank sum test may be used. Like the Wilcoxon one-sample test of the previous section, the Wilcoxon rank sum test has proven to be almost as powerful as its classical counterpart under conditions appropriate to the latter and is likely to be more powerful when the assumptions of the t test are not met.

13.6.1 Procedure

When at least ordinal measurement has been achieved, the Wilcoxon rank sum test can be used to test whether two mutually independent sample groups have been drawn from the same population or identical populations, that is, the Wilcoxon rank sum test is designed to detect any kind of difference between the two groups—location, dispersion, and/or shape (as was discussed in Chapter 4). The only assumptions necessary for such a test are

1. Both samples are randomly and independently drawn from their respective populations.
2. The underlying random variable of interest is continuous (to avoid ties).
3. The obtained data constitute at least an ordinal level of measurement both within and between the two samples.

Nevertheless, it is usually of interest to use the Wilcoxon rank sum test specifically as a test for differences in location (that is, the population medians). For such situations it is necessary to also assume that no other differences between the two groups exist except for possible differences in location. When used in this manner we may compare the assumptions of the nonparametric Wilcoxon rank sum test against those of the classical two-sample t test. Although both tests require that the samples be randomly and independently drawn and that the populations have equal variability (dispersion), the t test specifies that the populations be normally distributed and that the data be measured on an interval or ratio scale while the Wilcoxon rank sum test specifies only that the populations be continuously distributed and have the same shape (symmetric or skewed) and that the data need only be measured on an ordinal scale (over both samples). Thus it is clear that the Wilcoxon procedure requires a much less stringent set of assumptions than does its parametric counterpart. On the other hand, it may be reemphasized that even when the assumptions of the classical t test can be met, the Wilcoxon procedure is almost as powerful as its parametric counterpart.

To perform the Wilcoxon rank sum test we must replace the observations in the two samples of size n_1 and n_2 with their combined ranks (unless, of course, the obtained data contained the ranks initially). The ranks are assigned in such manner that rank 1 is given to the smallest of the $n = n_1 + n_2$ combined observations, rank 2 is given to the second smallest, and so on until rank n is given to the largest. If several values are tied, we assign each the average of the ranks that would otherwise have been assigned. Now for convenience we must establish that whenever the two sample sizes are unequal, we will let n_1 represent the *smaller* sized sample and n_2 the *larger* sized sample. The Wilcoxon rank sum test statistic T_{n_1} is merely the summation of the ranks assigned to the n_1 observations in the smaller sample. For equal sized samples, however, either group may be selected for determining T_{n_1}. Since it is well known that for any integer value n the summation over the first n consecutive integers may easily be calculated from $n(n + 1)/2$, the test statistic T_{n_1} plus the summation of the ranks assigned to the n_2 items in the second sample, T_{n_2}, must be equal to this value, that is,

$$T_{n_1} + T_{n_2} = \frac{n(n+1)}{2} \qquad (13.8)$$

and Equation (13.8) can serve as a check on the ranking procedure.

The test of the null hypothesis may be one-tailed or two-tailed:

Two-Tailed Test	One-Tailed Test	One-Tailed Test
H_0: $M_1 = M_2$	H_0: $M_1 \geq M_2$	H_0: $M_1 \leq M_2$
H_1: $M_1 \neq M_2$	H_1: $M_1 < M_2$	H_1: $M_1 > M_2$

where M_1 = population median for the first group having n_1 sample observations

M_2 = population median for the second group having n_2 sample observations

When both samples n_1 and n_2 are ≤ 10, Appendix E, Table E.11 may be used to obtain the critical values of the test statistic T_{n_1} for both one- and two-tailed tests at various levels of significance. For a two-tailed test and for a particular level of significance, if the computed value of T_{n_1} equals or exceeds the upper critical value or is less than or equal to the lower critical value, the null hypothesis may be rejected. For one-tailed tests having the alternative H_1: $M_1 > M_2$, the decision rule is to reject the null hypothesis if the observed value of T_{n_1} equals or exceeds the upper critical value, while for one-tailed tests having the alternative H_1: $M_1 < M_2$ the decision rule is to reject the null hypothesis if the observed value of T_{n_1} is less than or equal to the lower critical value. For large sample sizes the test statistic T_{n_1} is approximately normally distributed. Thus the following large sample approximation formula may be used for testing the null hypothesis when sample sizes are outside the range of Appendix E, Table E.11:

$$Z \cong \frac{T_{n_1} - \mu_{T_{n_1}}}{\sigma_{T_{n_1}}} \qquad (13.9)$$

where T_{n_1} = summation of the ranks assigned to the n_1 observations in the first sample

$\mu_{T_{n_1}}$ = mean value of T_{n_1}; $\mu_{T_{n_1}} = n_1(n + 1)/2$

$\sigma_{T_{n_1}}$ = standard deviation of T_{n_1}; $\sigma_{T_{n_1}} = \sqrt{n_1 n_2 (n + 1)/12}$

and $n_1 \leq n_2$, that is,

$$Z \cong \frac{T_{n_1} - \dfrac{n_1(n + 1)}{2}}{\sqrt{\dfrac{n_1 n_2 (n + 1)}{12}}} \qquad (13.10)$$

and, based on the level of significance selected, the null hypothesis may be rejected if the computed Z value falls in the appropriate region of rejection depending on whether a two-tailed or a one-tailed test is used (see Figure 13.4).

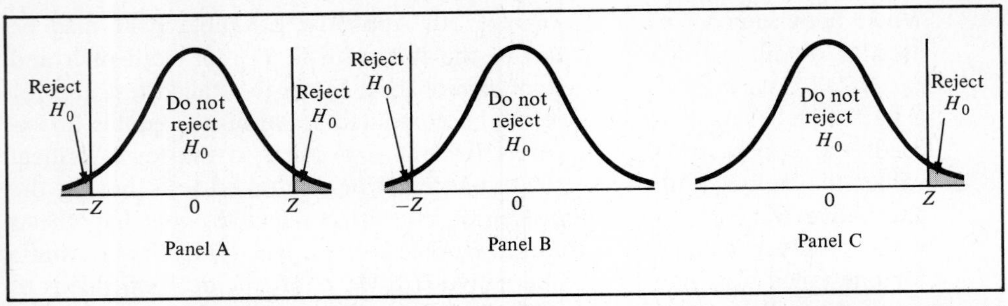

Figure 13.4 Determining the rejection region. Panel A, two-tailed test, $M_1 \neq M_2$. Panel B, one-tailed test, $M_1 < M_2$. Panel C, one-tailed test, $M_1 > M_2$.

13.6.2 Applications

As an example using the Wilcoxon rank sum test suppose that a security analyst wishes to compare the dividend yields of stocks traded on the American Stock Exchange with those traded on the New York Stock Exchange. Random samples of 8 issues from the American Stock Exchange and 10 issues from the New York Stock Exchange are selected with the results presented in Table 13.8.

If the security analyst is specifically concerned with comparing the median dividend yields rather than just any differences whatsoever in the dividend yields, it must be assumed that the distributions of dividend yields in both populations from which the random samples were drawn are identical except possibly for differences in location (the medians).

Since the security analyst is not specifying which of the two groups is likely to possess a greater median dividend yield, the test is two-tailed and the following null and alternate hypotheses are established:

Table 13.8
COMPUTING DIVIDEND YIELDS OF SELECTED ISSUES FROM THE AMERICAN AND NEW YORK STOCK EXCHANGES

American Stock Exchange ($n_1 = 8$)			New York Stock Exchange ($n_2 = 10$)		
(1) Dividends per Share	(2) Price per Share	Dividend Yield (1) ÷ (2)	(1) Dividends per Share	(2) Price per Share	Dividend Yield (1) ÷ (2)
\$ .60	\$11.88	5.1%	\$ .22	\$ 8.00	2.8%
.12	8.75	1.4	.60	8.25	7.3
.24	14.75	1.6	2.00	20.38	9.8
.20	3.50	5.7	1.08	15.50	7.0
.80	8.25	9.7	1.59	16.75	9.5
1.30	14.25	9.1	1.00	18.25	5.5
.35	3.12	11.2	.80	14.25	5.6
.32	3.88	8.2	3.60	33.25	10.8
			.80	17.12	4.7
			.50	7.75	6.5

Table 13.9
FORMING THE COMBINED RANKS

American Stock Exchange Dividend Yield Rankings ($n_1 = 8$)	New York Stock Exchange Dividend Yield Rankings ($n_2 = 10$)
5	3
1	11
2	16
8	10
15	14
13	6
18	7
12	17
	4
	9

SOURCE: Data are extracted from Table 13.8.

H_0: $M_1 = M_2$ (the median dividend yields are equal)

H_1: $M_1 \neq M_2$ (the median dividend yields are different)

To perform the Wilcoxon rank sum test we form the combined ranking of the $n_1 = 8$ dividend yields from the issues on the American Stock Exchange with the $n_2 = 10$ dividend yields from the issues on the New York Stock Exchange as in Table 13.9.

We then obtain the test statistic T_{n_1}, the sum of the ranks assigned to the smaller sample:

$$T_{n_1} = 5 + 1 + 2 + 8 + 15 + 13 + 18 + 12 = 74$$

As a check on the ranking procedure we also obtain T_{n_2} and use Equation (13.8) to show that the sum of the first $n = 18$ integers in the combined ranking is equal to $T_{n_1} + T_{n_2}$:

$$T_{n_1} + T_{n_2} = \frac{n(n+1)}{2} \tag{13.8}$$

$$74 + 97 = \frac{18(19)}{2} = 171$$

Had one or both of the samples been greater than 10, the large sample approximation formula [Equation (13.10)] would have been used for testing. Here, since both n_1 and $n_2 \leq 10$, Appendix E, Table E.11 is employed to obtain the critical values for the T_{n_1} statistic. With $n_1 = 8$ and $n_2 = 10$ it is observed (see Table 13.10) that at the .05 level of significance the lower and upper critical values for the two-tailed test are, respectively, 53 and 99. Since the observed value of the test statistic $T_{n_1} = 74$ falls between these critical values, the null hypothesis cannot be rejected. It may therefore be

Table 13.10
OBTAINING THE LOWER AND UPPER TAIL CRITICAL VALUES T_{n_1} FOR THE WILCOXON RANK SUM TEST WHERE $n_1 = 8$, $n_2 = 10$, AND $\alpha = .05$

	α		n_1						
n_2	One-Tailed	Two-Tailed	4	5	6	7	8 (Lower, Upper)	9	10
9	.025	.05	14,42	22,53	31,65	40,79	51,93	62,109	
	.01	.02	13,43	20,55	28,68	37,82	47,97	59,112	
	.005	.01	11,45	18,57	26,70	35,84	45,99	56,115	
10	.05	.10	17,43	26,54	35,67	45,81	56,96	69,111	82,128
	.025	.05	15,45	23,57	32,70	42,84	**53,99**	65,115	78,132
	.01	.02	13,47	21,59	29,73	39,87	49,103	61,119	74,136
	.005	.01	12,48	19,61	27,75	37,89	47,105	58,122	71,139

SOURCE: Extracted from Appendix E, Table E.11.

said that there is no evidence of any differences between the median dividend yields of stocks traded on the American Stock Exchange versus those traded on the New York Stock Exchange.

As a second application of the Wilcoxon rank sum test we may recall the problem faced by the researcher and the dean in Section 10.2. There a two-sample t test was used to study whether or not there were any significant differences among the grade-point indexes of accounting majors versus non-accounting majors. Here, using less stringent assumptions, we may test:

$$H_0: \quad M_{\text{acc}} = M_{\text{n-acc}}$$

$$H_1: \quad M_{\text{acc}} \neq M_{\text{n-acc}}$$

Using the ordered arrays for the survey data presented in Figure 3.6, the ranks of the grade-point indexes for the samples of 35 accounting majors and 59 non-accounting majors are presented in Table 13.11.

T_{n_1}, the sum of the ranks assigned to the smaller sample, is 1,691.5. As a check on the ranking procedure we obtain $T_{n_2} = 2{,}773.5$, so that from Equation (13.8),

$$T_{n_1} + T_{n_2} = \frac{n(n+1)}{2}$$

where $n = n_1 + n_2 = 94$, and thus,

$$1{,}691.5 + 2{,}773.5 = \frac{(94)(95)}{2}$$

$$4{,}465 = 4{,}465$$

Table 13.11
RANKING THE GRADE-POINT INDEXES

Accounting Majors ($n_1 = 35$)		Non-accounting Majors ($n_2 = 59$)		
4.0	61.0	1.0	33.5	61.0
4.0	61.0	4.0	35.5	65.0
4.0	61.0	4.0	38.5	66.0
7.0	68.0	8.0	38.5	67.0
9.0	69.0	10.5	38.5	74.0
16.0	70.0	10.5	41.0	74.0
20.5	71.0	12.0	42.5	77.0
22.0	74.0	13.0	42.5	78.0
23.0	74.0	14.5	44.0	79.0
25.0	74.0	14.5	45.0	80.0
27.0	85.0	18.0	46.5	81.0
31.0	86.0	18.0	46.5	82.0
32.0	88.0	18.0	48.0	83.0
35.5	89.5	20.5	49.0	84.0
38.5	92.5	24.0	51.0	87.0
51.0		27.0	51.0	89.5
53.5		27.0	57.0	91.0
53.5		29.0	61.0	92.5
55.0		30.0	61.0	94.0
56.0		33.5	61.0	

SOURCE: Figure 3.6.

Since the sample sizes are large, we use Equation (13.10) to test the null hypothesis. Therefore,

$$Z \cong \frac{T_{n_1} - \dfrac{n_1(n+1)}{2}}{\sqrt{\dfrac{n_1 n_2 (n+1)}{12}}} = \frac{1{,}691.5 - \dfrac{(35)(95)}{2}}{\sqrt{\dfrac{(35)(59)(95)}{12}}} = 0.23$$

and, as was the case with the two-sample t test presented in Section 10.2, the null hypothesis cannot be rejected because the critical Z values for a 1% level of significance are ± 2.58.

13.7 THE ABSOLUTE NORMAL SCORES TEST FOR RELATED SAMPLES

In the previous section we considered a nonparametric approach to testing for differences between two independent groups. However, in the social sciences and in marketing research, it is frequently of interest to examine differences among two related groups. For example, in test marketing a product under two different advertising conditions, a sample of test markets can be matched (paired) based on the test market population size and/or other socioeconomic and demographic variables. Moreover, when performing a taste-testing experiment, each subject in the sample could be used as his or her own control so that repeated measurements on the same individual are obtained. We may recall from Section 10.3 that for situations involving either matched (paired) items or repeated measurements of the same item, the parametric t

test for the differences in the related samples could be used. On the other hand, if the researcher does not wish to make the rigorous assumptions of this classical t test, an extremely powerful nonparametric competitor developed by Thompson, Govindarajulu, and Doksum (References 9 and 14) could be used provided the level of measurement attained on the data is higher than an ordinal scale. This procedure is known as the **absolute normal scores test for related samples.** When the assumptions of the t test are violated, the absolute normal scores test is likely to be more powerful than its classical counterpart. Interestingly, even under conditions appropriate to the classical t test, the absolute normal scores test has proven to be at least as powerful.

13.7.1 Procedure

The assumptions necessary for performing the absolute normal scores test for related samples are

1. That the data be randomly and independently drawn
2. That the underlying variable of interest be continuous
3. That the observed data be measured at a higher level than an ordinal scale
4. That the distribution of the population of difference scores between matched (paired) items or individuals, or between repeated measurements, be (approximately) symmetric

Like the Wilcoxon one-sample signed-ranks test, the absolute normal scores test for related samples is concerned with both size and direction. Thus, the normal scores test statistic K considers not only whether an observed difference in two values is larger or smaller than the hypothesized median difference, but also how much larger or how much smaller. The major distinction between the Wilcoxon signed-ranks procedure and the absolute normal scores procedure is that in the former the observed values are converted to a set of ranks, while in the latter the values are converted to a set of absolute normal scores—the observations we would expect if we were sampling from the right portion of a standardized normal distribution (see Figure 13.5).

This aspect of data conversion is what accounts for the increase in power of the normal scores tests over those based merely on ranks. However, we shall observe that while the normal scores tests are easy to perform [that is, the researcher need only use a standardized normal distribution table (Appendix E, Table E.2) both to convert the scores and to make the hypothesis test], nevertheless, the computations are much more tedious than those required for tests based on ranks. Hence, when selecting a particular type of test there is a trade-off. If the researcher were to consider a normal scores test over one based on ranks, he or she must weigh the advantage of potential small increases in statistical power against the disadvantage of increasing the computational effort that would be involved.

The test of the null hypothesis that the population median difference M_D

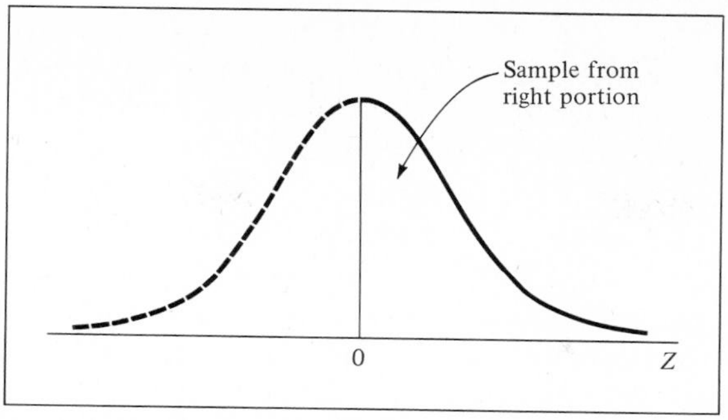

Figure 13.5 Obtaining absolute normal scores.

is zero may be one-tailed or two-tailed:

Two-Tailed Test	One-Tailed Test	One-Tailed Test
H_0: $M_D = 0$	H_0: $M_D \geq 0$	H_0: $M_D \leq 0$
H_1: $M_D \neq 0$	H_1: $M_D < 0$	H_1: $M_D > 0$

As hypothetically depicted in Table 13.12, the following eight-step procedure may be used to develop the absolute normal scores test for related samples:

1. Either for samples containing n' matched (paired) items or individuals, or for a sample of n' items or individuals each with two measurements, we obtain a set of difference scores

$$D_i = X_{1i} - X_{2i}$$

where $i = 1, 2, \ldots, n'$.

2. We then neglect the + and − signs and obtain a set of n' absolute difference scores $|D_i|$:

$$|D_i| = |X_{1i} - X_{2i}|$$

where $i = 1, 2, \ldots, n'$.

3. We omit from further analysis any absolute difference score of zero, thereby yielding a set of n nonzero absolute difference scores where $n \leq n'$.

4. We assign the ranks R_i to each of the $|D_i|$ such that the smallest absolute difference score gets rank 1 while the largest gets rank n. If two or more $|D_i|$ are equal, they are each assigned the average rank of the ranks they otherwise would have been assigned individually had ties not occurred.

5. For each of our n ranked values we determine the proportions $R_i/2(n+1)$. These values represent areas under the normal curve to the right of the mean (and should be rounded to four decimal places to facilitate the use of the normal tables).

6. For each of our n ranked values we then use the table of the standardized normal distribution (Appendix E, Table E.2) to replace the ranked absolute difference scores R_i by z_i, the absolute normal scores (Reference 15) corresponding to areas of $R_i/2(n+1)$ under the normal curve (to the right of the mean) as demonstrated in Table 6.7 of Section 6.9.4.

7. We now reassign the symbols + or − to each of the n nonzero absolute normal scores z_i, depending on whether D_i was positive or negative.

8. The absolute normal scores test statistic K is obtained as the sum of the absolute normal scores corresponding only to the positive difference scores D_i:

$$K = \sum_{i=1}^{n} z_i^{(+)} \tag{13.11}$$

Table 13.12
SETTING UP THE ABSOLUTE NORMAL SCORES TEST FOR RELATED SAMPLES*

(1) Initial n' Observations	(2) Group 1	(3) Group 2	(4) Difference Scores D	(5) Absolute Difference Scores $\lvert D\rvert$	(6) Ranks of Nonzero $\lvert D\rvert$	(7) Areas Under Normal Curve	(8) Absolute Normal Scores Corresponding to Areas	(9) Sign of the Nonzero D	(10) Obtaining K
1	X_{11}	X_{21}	$D_1 = X_{11} - X_{21}$	$\lvert D_1\rvert$	R_1	$R_1/2(n+1)$	z_1	−	—
2	X_{12}	X_{22}	$D_2 = X_{12} - X_{22}$	$\lvert D_2\rvert$	R_2	$R_2/2(n+1)$	z_2	+	z_2
3	X_{13}	X_{23}	$D_3 = X_{13} - X_{23}$	$\lvert D_3\rvert$	R_3	$R_3/2(n+1)$	z_3	+	z_3
.	.	.	.	.	.	.	.	.	.
.	.	.	.	.	.	.	.	.	.
i	X_{1i}	X_{2i}	$D_i = X_{1i} - X_{2i}$	$\lvert D_i\rvert$	R_i	$R_i/2(n+1)$	z_i	+	z_i
.	.	.	.	.	.	.	.	.	.
.	.	.	.	.	.	.	.	.	.
n	X_{1n}	X_{2n}	$D_n = X_{1n} - X_{2n}$	$\lvert D_n\rvert$	R_n	$R_n/2(n+1)$	z_n	+	z_n
n'	$X_{1n'}$	$X_{2n'}$	$D_{n'} = X_{1n'} - X_{2n'}$	$\lvert D_{n'}\rvert$	—	—	—	—	—

* Hypothetical example in which only the first observation has a negative score and only the last observation has a zero difference score as determined in column (4).

To perform the absolute normal scores test for related samples we note that the test statistic K can range from 0 (where all difference scores D_i are negative) to $\sum_{i=1}^{n} z_i$ (where all difference scores D_i are positive). Rather than obtaining tables of critical values for the test statistic K,[1] we shall test the null hypothesis by utilizing the following large-sample approximation formula which has been shown to give fairly good results for related samples even as small as $n = 7$ (Reference 9):

[1] Tables of critical values are found in References 7 and 14. However, these are based on different absolute normal scoring systems than the one described here and such discussion is beyond the scope of this text.

$$Z \cong \frac{K - \mu_K}{\sigma_K} \tag{13.12}$$

where K = sum of the positive normal scores; $K = \sum_{i=1}^{n} z_i^{(+)}$

μ_K = mean value of K; $\mu_K = \dfrac{\sum_{i=1}^{n} z_i}{2}$

σ_K = standard deviation of K; $\sigma_K = \dfrac{\sqrt{\sum_{i=1}^{n} z_i^2}}{2}$

n = number of nonzero absolute difference scores in sample

that is,

$$Z \cong \frac{K - \dfrac{\sum_{i=1}^{n} z_i}{2}}{\dfrac{\sqrt{\sum_{i=1}^{n} z_i^2}}{2}} \tag{13.13}$$

and, based on the level of significance selected, the null hypothesis may be rejected if the computed Z value falls in the appropriate region of rejection depending on whether a two-tailed or one-tailed test is employed (see Figure 13.6).

13.7.2 Applications

As one application of the absolute normal scores test for related samples, let us suppose that a financial analyst was interested in determining whether or not the reported earnings per share (in dollars) of the 500 largest industrial corporations in the United States had significantly increased from one annual period to another. Using the directory compiled by *Fortune* (based on sales revenues), the analyst decides to take a random sample of 20 companies from this population listing, with the results recorded in Table 13.13. Since the direction of the alternative has been specified (that is, the analyst is interested in showing that earnings per share have increased) the test is one-tailed. If the analyst does not wish to make the more rigorous assumptions of

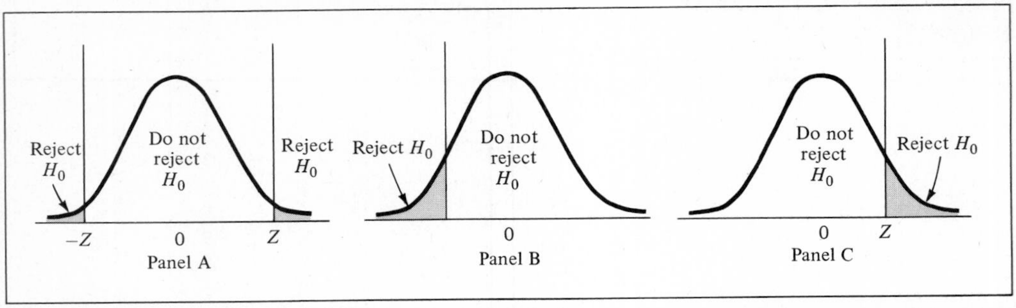

Figure 13.6 Determining the rejection region. Panel A, two-tailed test, $M_D = 0$. Panel B, one-tailed test, $M_D < 0$. Panel C, one-tailed test, $M_D > 0$.

the classical t test for related samples (Section 10.3), the following null and alternate hypotheses may be stated:

$$H_0: \quad M_D \leq 0 \qquad \text{(that is, } M_{1977} \leq M_{1976}\text{)}$$

$$H_1: \quad M_D > 0 \qquad \text{(that is, } M_{1977} > M_{1976}\text{)}$$

The eight-step procedure to perform the absolute normal scores test is developed in Table 13.13.

As mentioned in Step 6, to determine the absolute normal score for a particular item z_i, we first divide its rank R_i by $2(n + 1)$ to obtain an area under the normal curve (to the right of the mean). As demonstrated in Table 6.7 of Section 6.9.4, we then find the particular standardized z_i value corresponding to the area $R_i/2(n + 1)$. For example, from Table 13.13 we note that out of the $n = 20$ companies in our sample, Hewlett-Packard had a rank R_{12} of 16. Dividing this rank by $2(n + 1) = 42$ yields an area under the normal curve (to the right of the mean) of .3810 as depicted in Figure 13.7. As shown in Table 13.14, this area corresponds to a standardized value of 1.18 and the absolute normal score z_{12} is obtained.

It should be noted that usually we will not be able to directly locate the desired area $R_i/2(n + 1)$ in the normal table (Appendix E, Table E.2). This, however, presents no problem. The rule of thumb is to select the standardized value corresponding to an area under the normal curve closest to the desired area $R_i/2(n + 1)$. For example, from Table 13.13, the Aluminum Co. of America had a rank R_1 of 18—yielding a desired area of .4286. As seen from Table 13.15, since this desired area is closer to .4292 than it is to .4279, the absolute normal score $z_1 = 1.47$ is selected.

After obtaining all absolute normal scores in a similar manner, the test statistic K is determined as the sum of the absolute normal scores corresponding only to the positive difference scores D_i [Equation (13.11)]. Thus,

$$K = \sum_{i=1}^{n=20} z_i^{(+)} = 12.19$$

To test the null hypothesis using the large sample approximation formula

Table 13.13
SETTING UP THE ABSOLUTE NORMAL SCORES TEST FOR RELATED SAMPLES

Company	1977 Company Position According to Sales Dollars (Highest = 1) (Lowest = 500)	Earnings Per Share (in dollars) 1977 X_{1i}	1976 X_{2i}	D_i	$\lvert D_i \rvert$	R_i	$\frac{R_i}{2(n+1)}$	z_i	Sign of the Nonzero D_i
Aluminum Co. of America	70	5.58	4.14	1.44	1.44	18	.4286	1.47	+
Amsted Industries	371	6.62	6.38	.24	.24	7	.1667	0.43	+
Chesebrough-Pond's	285	1.86	1.69	.17	.17	5.5	.1310	0.33	+
Chicago Bridge & Iron	337	6.05	5.73	.32	.32	8	.1905	0.50	+
Crown Central Petroleum	342	6.44	6.33	.11	.11	3	.0714	0.18	+
Cummins Engine	196	8.22	7.66	.56	.56	15	.3571	1.07	+
Ford Motor	3	14.16	8.36	5.80	5.80	20	.4762	1.98	+
Globe-Union	461	7.29	3.67	3.62	3.62	19	.4524	1.67	+
Gulf Oil	8	3.86	4.19	− .33	.33	9	.2143	0.57	−
Hanes	446	4.87	4.50	.37	.37	10	.2381	0.64	+
Hercules	148	1.36	2.44	−1.08	1.08	17	.4048	1.31	−
Hewlett-Packard	184	4.27	3.24	1.03	1.03	16	.3810	1.18	+
Koehring	496	3.03	2.52	.51	.51	13	.3095	0.88	+
Koppers	186	2.64	2.67	− .03	.03	1	.0238	0.06	−
MBPXL	244	2.38	2.90	− .52	.52	14	.3333	0.97	−
National Service Industries	379	1.93	1.80	.13	.13	4	.0952	0.24	+
Quaker State Oil Refining	428	1.53	1.46	.07	.07	2	.0476	0.12	+
Reynolds Metals	104	4.61	4.16	.45	.45	11	.2619	0.71	+
Sperry & Hutchinson	303	2.31	2.48	− .17	.17	5.5	.1310	0.33	−
U.S. Industries	190	1.25	0.76	.49	.49	12	.2857	0.79	+

SOURCE: Data are taken from "The Fortune Directory of the 500 Largest U.S. Industrial Corporations," *Fortune,* vol. 97, no. 9, pp. 238–290, May 8, 1978.

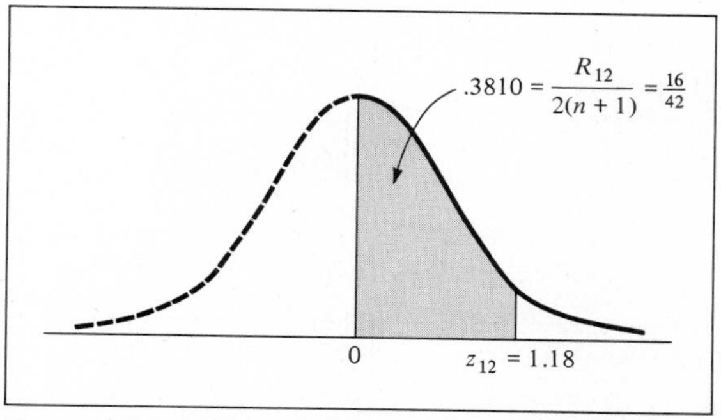

Figure 13.7 Finding the area under the normal curve when converting from ranks to absolute normal scores.

Table 13.14
OBTAINING THE ABSOLUTE NORMAL SCORE
CORRESPONDING TO AN AREA OF .3810

Z	.00	.01	.02	.03	.04	.05	.06	.07	.08	.09
0.0	.0000	.0040	.0080	.0120	.0160	.0199	.0239	.0279	.0319	.0359
0.1	.0398	.0438	.0478	.0517	.0557	.0596	.0636	.0675	.0714	.0753
0.2	.0793	.0832	.0871	.0910	.0948	.0987	.1026	.1064	.1103	.1141
0.3	.1179	.1217	.1255	.1293	.1331	.1368	.1406	.1443	.1480	.1517
0.4	.1554	.1591	.1628	.1664	.1700	.1736	.1772	.1808	.1844	.1879
0.5	.1915	.1950	.1985	.2019	.2054	.2088	.2123	.2157	.2190	.2224
0.6	.2257	.2291	.2324	.2357	.2389	.2422	.2454	.2486	.2518	.2549
0.7	.2580	.2612	.2642	.2673	.2704	.2734	.2764	.2794	.2823	.2852
0.8	.2881	.2910	.2939	.2967	.2995	.3023	.3051	.3078	.3106	.3133
0.9	.3159	.3186	.3212	.3238	.3264	.3289	.3315	.3340	.3365	.3389
1.0	.3413	.3438	.3461	.3485	.3508	.3531	.3554	.3577	.3599	.3621
1.1	.3643	.3665	.3686	.3708	.3729	.3749	.3770	.3790	.3810	.3830
1.2	.3849	.3869	.3888	.3907	.3925	.3944	.3962	.3980	.3997	.4015
1.3	.4032	.4049	.4066	.4082	.4099	.4115	.4131	.4147	.4162	.4177

SOURCE: Data are extracted from Appendix E, Table E.2.

Table 13.15
OBTAINING THE ABSOLUTE NORMAL SCORE
CORRESPONDING TO AN AREA OF .4286

Z	.00	.01	.02	.03	.04	.05	.06	.07	.08	.09
0.0	.0000	.0040	.0080	.0120	.0160	.0199	.0239	.0279	.0319	.0359
0.1	.0398	.0438	.0478	.0517	.0557	.0596	.0636	.0675	.0714	.0753
0.2	.0793	.0832	.0871	.0910	.0948	.0987	.1026	.1064	.1103	.1141
0.3	.1179	.1217	.1255	.1293	.1331	.1368	.1406	.1443	.1480	.1517
0.4	.1554	.1591	.1628	.1664	.1700	.1736	.1772	.1808	.1844	.1879
0.5	.1915	.1950	.1985	.2019	.2054	.2088	.2123	.2157	.2190	.2224
0.6	.2257	.2291	.2324	.2357	.2389	.2422	.2454	.2486	.2518	.2549
0.7	.2580	.2612	.2642	.2673	.2704	.2734	.2764	.2794	.2823	.2852
0.8	.2881	.2910	.2939	.2967	.2995	.3023	.3051	.3078	.3106	.3133
0.9	.3159	.3186	.3212	.3238	.3264	.3289	.3315	.3340	.3365	.3389
1.0	.3413	.3438	.3461	.3485	.3508	.3531	.3554	.3577	.3599	.3621
1.1	.3643	.3665	.3686	.3708	.3729	.3749	.3770	.3790	.3810	.3830
1.2	.3849	.3869	.3888	.3907	.3925	.3944	.3962	.3980	.3997	.4015
1.3	.4032	.4049	.4066	.4082	.4099	.4115	.4131	.4147	.4162	.4177
1.4	.4192	.4207	.4222	.4236	.4251	.4265	.4279	.4292	.4306	.4319
1.5	.4332	.4345	.4357	.4370	.4382	.4394	.4406	.4418	.4429	.4441
1.6	.4452	.4463	.4474	.4484	.4495	.4505	.4515	.4525	.4535	.4545
1.7	.4554	.4564	.4573	.4582	.4591	.4599	.4608	.4616	.4625	.4633
1.8	.4641	.4649	.4656	.4664	.4671	.4678	.4686	.4693	.4699	.4706
1.9	.4713	.4719	.4726	.4732	.4738	.4744	.4750	.4756	.4761	.4767

SOURCE: Data are extracted from Appendix E, Table E.2.

[Equations (13.12) and (13.13)] we first must obtain $\sum_{i=1}^{n} z_i$, the summation of all the absolute normal scores in Table 13.13, as well as $\sum_{i=1}^{n} z_i^2$, the summation of the squares of all the scores.

$$\sum_{i=1}^{n=20} z_i = (1.47) + (0.43) + \cdots + (0.33) + (0.79) = 15.43$$

$$\sum_{i=1}^{n=20} z_i^2 = (1.47)^2 + (0.43)^2 + \cdots + (0.33)^2 + (0.79)^2 = 17.4623$$

Hence,

$$\mu_K = \frac{\sum_{i=1}^{n} z_i}{2} = \frac{15.43}{2} = 7.715$$

and

$$\sigma_K = \frac{\sqrt{\sum_{i=1}^{n} z_i^2}}{2} = \frac{\sqrt{17.4623}}{2} = 2.089$$

and from Equation (13.13),

$$Z \cong \frac{K - \dfrac{\sum_{i=1}^{n} z_i}{2}}{\dfrac{\sqrt{\sum_{i=1}^{n} z_i^2}}{2}} = \frac{12.19 - 7.715}{2.089} = 2.14$$

Using a level of significance .05, our one-tailed test has a critical Z value of $+1.645$. Since $Z = 2.14$ exceeds this critical value of $+1.645$, the null hypothesis may be rejected and the financial analyst may conclude that the reported earnings per share among the 500 largest industrial corporations in the United States has significantly increased from one annual period to another.

As a second application of the absolute normal scores test for related samples, we may recall from Section 10.3 that, in analyzing the survey, the dean was interested in measuring the accuracy of responses to question 5, grade-point index. To accomplish this, the dean utilized the paired difference t test for related samples and compared the reported grade-point indexes and actual grade-point indexes for a sample of 11 juniors majoring in accounting.

Table 13.16
SETTING UP THE ABSOLUTE NORMAL SCORES TEST FOR RELATED SAMPLES

Student	Reported Grade-Point Index X_{1i}	Actual Grade-Point Index X_{2i}	D_i	$\lvert D_i \rvert$	R_i	$\frac{R_i}{2(n+1)}$	z_i	Sign of the Nonzero D_i
1	2.80	2.69	+.11	.11	8	.3636	1.10	+
2	2.00	1.96	+.04	.04	4	.1818	0.47	+
3	2.92	2.80	+.12	.12	9	.4091	1.34	+
4	2.62	2.48	+.14	.14	10	.4545	1.69	+
5	2.76	2.83	−.07	.07	5	.2273	0.60	−
6	2.91	2.92	−.01	.01	1	.0455	0.11	−
7	3.75	3.75	0	0	—	—	—	Discard
8	2.55	2.65	−.10	.10	6.5	.2955	0.83	−
9	2.73	2.70	+.03	.03	3	.1364	0.35	+
10	3.12	3.10	+.02	.02	2	.0909	0.23	+
11	3.00	2.90	+.10	.10	6.5	.2955	0.83	+

SOURCE: Data are taken from Table 10.3.

Using less stringent assumptions, we may now employ the absolute normal scores procedure to test:

H_0: $M_D = 0$ ($M_1 = M_2$—There is no difference in reported and actual GPI)

H_1: $M_D \neq 0$ ($M_1 \neq M_2$—There is a difference in reported and actual GPI)

The eight-step procedure to perform the test is developed in Table 13.16.

From Table 13.16 we note that the seventh student had a difference score of zero and was removed from further analysis. Thus, based on the $n = 10$ remaining nonzero absolute difference scores we obtain the test statistic K, the sum of the absolute normal scores corresponding only to the positive difference scores

$$K = \sum_{i=1}^{n=10} z_i^{(+)} = 6.01$$

Furthermore, we compute

$$\sum_{i=1}^{n=10} z_i = (1.10) + (0.47) + \cdots + (0.23) = 7.55$$

and

$$\sum_{i=1}^{n=10} z_i^2 = (1.10)^2 + (0.47)^2 + \cdots + (0.23)^2 = 8.0079$$

so that

$$\mu_K = \frac{\sum_{i=1}^{n} z_i}{2} = \frac{7.55}{2} = 3.775$$

and

$$\sigma_K = \frac{\sqrt{\sum_{i=1}^{n} z_i^2}}{2} = \frac{\sqrt{8.0079}}{2} = 1.415$$

To test the null hypothesis using the large-sample approximation formula [Equation (13.13)], we obtain

$$Z \cong \frac{K - \frac{\sum_{i=1}^{n} z_i}{2}}{\frac{\sqrt{\sum_{i=1}^{n} z_i^2}}{2}} = \frac{6.01 - 3.775}{1.415} = 1.58$$

For a two-tailed test with an α level of .05 the critical Z values are ± 1.96. Since $Z = 1.58$ falls between these critical Z values, the null hypothesis cannot be rejected (as was the case with the t test of Section 10.3), and the dean may conclude that, for the types of students studied, there is no evidence of a difference between the actual and reported grade-point index.

13.8 KRUSKAL–WALLIS TEST FOR c INDEPENDENT SAMPLES

The Kruskal–Wallis test for c independent samples (where $c > 2$) may be considered an extension of the Wilcoxon rank sum test for two independent samples discussed in Section 13.6. Thus the Kruskal–Wallis test enjoys the same power properties relative to the analysis of variance F test (Chapter 12) as does the Wilcoxon rank sum test relative to the t test for two independent samples (Section 10.2). The Kruskal–Wallis procedure has proven to be almost as powerful as the F test under conditions appropriate to the latter and even more powerful than the classical procedure when its assumptions are violated (see Section 12.6).

13.8.1 Procedure

When at least ordinal measurement has been obtained, the Kruskal–Wallis test can be used to test whether c independent sample groups have been drawn from the same population or identical populations. Thus the Kruskal–Wallis test is designed to detect any type of differences between the c groups —location, dispersion, and/or shape. As with the Wilcoxon rank sum test the only assumptions necessary for such a test are

1. The c samples are randomly and independently drawn from their respective populations.
2. The underlying random phenomenon of interest is continuous (to avoid ties).
3. The observed data constitute at least an ordinal scale of measurement, both within and between the c samples.

However, as with the Wilcoxon rank sum test, it is usually of interest to use the Kruskal–Wallis test specifically as a test for differences in location (that is, differences in the c population medians). For such situations it is necessary to also assume that no other differences between the c groups exist, except for possible differences in location. Even with the *equality of variance* assumption used, the Kruskal–Wallis test still makes less stringent assumptions than does the F test. To employ the Kruskal–Wallis procedure, the measurements need only be ordinal over all sample groups, and the common population distributions need only be continuous—their common shapes are irrelevant. On the other hand, to utilize the classical F test the level of measurement must be more sophisticated, and we must assume that the c samples are coming from underlying normal populations.

The Kruskal–Wallis test is very simple to use. To perform the test we must first (if necessary) replace the observations in the c samples with their combined ranks such that rank 1 is given to the smallest of the combined observations and rank n to the largest of the combined observations (where $n = n_1 + n_2 + \cdots + n_c$). If any values are tied, they are assigned the average of the ranks that would otherwise have been assigned if ties had not been present in the data. Although the derivation is beyond the scope of this book (see Reference 3), the Kruskal–Wallis test statistic H may be given by

$$H = \left[\frac{12}{n(n+1)} \sum_{i=1}^{c} \frac{T_{n_i}^2}{n_i}\right] - 3(n+1) \qquad (13.14)$$

where n = total number of observations over the combined samples; $n = n_1 + n_2 + \cdots + n_c$

n_i = number of observations in the ith sample; $i = 1, 2, \ldots, c$

$T_{n_i}^2$ = square of the summation of the ranks assigned to the ith sample

As the sample sizes in each group get large (>5), the test statistic H may be approximated by the χ^2 distribution with $c - 1$ degrees of freedom. Thus for any selected level of significance α, the decision rule would be to reject the null hypothesis if the computed value of H equals or exceeds the critical χ^2 value and not to reject the null hypothesis if H is less than the critical χ^2 value (see Figure 13.8). We may recall from Section 10.5 that the critical χ^2 values are given in Appendix E, Table E.4.

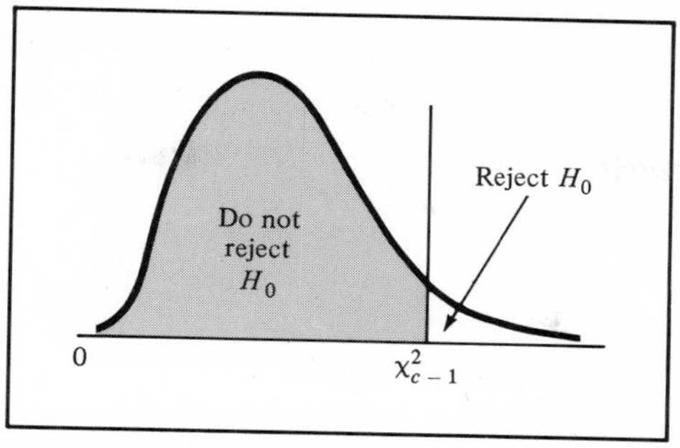

Figure 13.8 Determining the rejection region.

13.8.2 Applications

As an application of the Kruskal–Wallis procedure suppose that the personnel manager of a large insurance company wished to evaluate the effectiveness of four different sales-training programs that have been designed for new employees. To evaluate this, 30 recently hired college graduates were randomly assigned to the four programs so that there were seven subjects in programs A and B and eight subjects in programs C and D. At the end of the month-long training period a standard exam was administered to the 30 subjects and the scores are given in Table 13.17.

Table 13.17
EXAMINATION RESULTS OF SUBJECTS ASSIGNED TO FOUR PROGRAMS

A	B	C	D
66	72	61	63
74	51	60	61
82	59	57	76
75	62	60	84
73	74	81	58
97	64	55	65
87	78	70	69
		71	80

The null hypothesis to be tested is that the median scores for the four groups are equal; the alternative is that at least two of the groups possess median scores which differ. Thus,

$$H_0: \quad M_1 = M_2 = M_3 = M_4$$

$$H_1: \quad \text{Not all } M_i\text{'s are equal (where } i = 1, 2, 3, 4)$$

Converting the 30 test scores to ranks, we obtain Table 13.18.

Table 13.18
CONVERTING DATA TO RANKS

A	B	C	D
14.0	18.0	8.5	11.0
20.5	1.0	6.5	8.5
27.0	5.0	3.0	23.0
22.0	10.0	6.5	28.0
19.0	20.5	26.0	4.0
30.0	12.0	2.0	13.0
29.0	24.0	16.0	15.0
		17.0	25.0

SOURCE: Data are taken from Table 13.17.

Rank Sums: $T_{n_1} = 161.5$ $\quad T_{n_2} = 90.5$ $\quad T_{n_3} = 85.5$ $\quad T_{n_4} = 127.5$

We note that in the combined ranking, the second subject in program B had the lowest score, 51, and received a rank of 1; the sixth subject in program C had the second lowest score, 55, and received a rank of 2; and so on. We also note, for example, that in program C the second and fourth subjects both had scores of 60 and received a rank of 6.5 since they were tied for the sixth and seventh lowest scores. After all the ranks were assigned we then obtained the summation of the ranks for each group. As a check on the rankings we have:

$$T_{n_1} + T_{n_2} + T_{n_3} + T_{n_4} = \frac{n(n+1)}{2}$$

$$161.5 + 90.5 + 85.5 + 127.5 = \frac{(30)(31)}{2}$$

$$465 = 465$$

Now Equation (13.14) is employed to test the null hypothesis:

$$H = \left[\frac{12}{n(n+1)} \sum_{i=1}^{c} \frac{T_{n_i}^2}{n_i}\right] - 3(n+1) \tag{13.14}$$

$$= \left\{\frac{12}{(30)(31)}\left[\frac{(161.5)^2}{7} + \frac{(90.5)^2}{7} + \frac{(85.5)^2}{8} + \frac{(127.5)^2}{8}\right]\right\} - 3(31)$$

$$= \left(\frac{12}{930}\right)\left[\frac{26{,}082.25}{7} + \frac{8{,}190.25}{7} + \frac{7{,}310.25}{8} + \frac{16{,}256.25}{8}\right] - 3(31)$$

$$= \left(\frac{12}{930}\right)[3{,}726.04 + 1{,}170.04 + 913.78 + 2{,}032.03] - 93$$

$$= \left(\frac{12}{930}\right)[7{,}841.89] - 93 = 101.19 - 93 = 8.19$$

Table 13.19

OBTAINING THE APPROXIMATE χ^2 CRITICAL VALUE FOR THE KRUSKAL–WALLIS TEST AT THE 5% LEVEL OF SIGNIFICANCE WITH 3 DEGREES OF FREEDOM

Degrees of Freedom	Lower Tail Areas (Percentiles)									
	.005	.01	.025	.05	.10	.25	.75	.90	.95	.975
1	-	-	0.001	0.004	0.016	0.102	1.323	2.706	3.841	5.024
2	0.010	0.020	0.051	0.103	0.211	0.575	2.773	4.605	5.991	7.378
3	0.072	0.115	0.216	0.352	0.584	1.213	4.108	6.251	7.815	9.348
4	0.207	0.297	0.484	0.711	1.064	1.923	5.385	7.779	9.488	11.143
5	0.412	0.554	0.831	1.145	1.610	2.675	6.626	9.236	11.071	12.833
6	0.676	0.872	1.237	1.635	2.204	3.455	7.841	10.645	12.592	14.449
7	0.989	1.239	1.690	2.167	2.833	4.255	9.037	12.017	14.067	16.013
8	1.344	1.646	2.180	2.733	3.490	5.071	10.219	13.362	15.507	17.535
9	1.735	2.088	2.700	3.325	4.168	5.899	11.389	14.684	16.919	19.023
10	2.156	2.558	3.247	3.940	4.865	6.737	12.549	15.987	18.307	20.483
11	2.603	3.053	3.816	4.575	5.578	7.584	13.701	17.275	19.675	21.920
12	3.074	3.571	4.404	5.226	6.304	8.438	14.845	18.549	21.026	23.337
13	3.565	4.107	5.009	5.892	7.042	9.299	15.984	19.812	22.362	24.736

SOURCE: Extracted from Appendix E, Table E.4.

Using Appendix E, Table E.4, the critical χ^2 value having $c - 1 = 3$ degrees of freedom and corresponding to a .05 level of significance is 7.815 (see Table 13.19). Since the computed value of the test statistic H exceeds this critical value, we may reject the null hypothesis and conclude that not all the training programs were the same with respect to median test performance. (That is, if the null hypothesis were really true, the probability of obtaining such a result or one even more extreme is less than .05.)

As a second application of the Kruskal–Wallis test we may recall the problem of interest to the researcher and the dean in Section 12.5. The analysis of variance (F test) had been performed to study whether or not there were any differences among the grade-point indexes of students based upon their attitudes toward standards of admission. Now, using less stringent assumptions, we may test:

$$H_0: \quad M_{\text{no}} = M_8 = M_{10} = M_{12}$$

$$H_1: \quad \text{Not all } M_i\text{'s are equal (where } i = 1, 2, 3, 4)$$

Using the survey data from Figure 2.6, the ranks of the grade-point indexes based upon attitudes toward admission are presented in Table 13.20.

As a check on the rankings we have

$$T_{n_1} + T_{n_2} + T_{n_3} + T_{n_4} = \frac{n(n+1)}{2}$$

$$306 + 273 + 2{,}156 + 1{,}730 = \frac{(94)(95)}{2}$$

$$4{,}465 = 4{,}465$$

Table 13.20
RANKS OF GRADE-POINT INDEXES
BASED ON ATTITUDES TOWARD ADMISSION

None		8th Grade		10th Grade		12th Grade	
Grade-Point Index	*Rank*	*Grade-Point Index*	*Rank*	*Grade-Point Index*	*Rank*	*Grade-Point Index*	*Rank*
2.00	4.0	2.55	22.0	2.80	38.5	2.00	4.0
2.30	10.5	3.20	74.0	3.17	70.0	2.76	35.5
3.00	61.0	2.00	4.0	2.70	31.0	2.91	53.5
3.75	89.5	2.36	12.0	2.90	51.0	3.20	74.0
3.25	78.0	2.45	14.5	3.20	74.0	3.75	89.5
2.84	45.0	2.80	38.5	2.56	23.0	3.09	68.0
2.50	18.0	2.69	30.0	2.28	9.0	3.54	85.0
$n_1 = 7$		2.86	48.0	2.92	55.0	3.18	71.0
$T_{n_1} = 306$		2.66	29.0	3.00	61.0	3.00	61.0
		1.83	1.0	3.86	92.5	2.93	56.0
		$n_2 = 10$		2.51	20.5	2.00	4.0
		$T_{n_2} = 273$		2.62	25.0	3.51	83.0
				2.91	53.5	3.32	79.0
				3.60	86.0	2.80	38.5
				2.73	32.0	2.90	51.0
				3.12	69.0	3.00	61.0
				2.48	16.0	2.75	33.5
				2.64	27.0	2.60	24.0
				3.62	88.0	3.34	80.0
				2.10	7.0	2.50	18.0
				3.00	61.0	3.20	74.0
				3.43	82.0	3.00	61.0
				2.40	13.0	3.01	65.0
				3.82	91.0	2.25	8.0
				2.82	42.5	2.80	38.5
				3.00	61.0	3.07	66.0
				2.64	27.0	2.81	41.0
				2.45	14.5	3.08	67.0
				2.82	42.5	3.53	84.0
				3.86	92.5	2.75	33.5
				2.50	18.0	3.61	87.0
				2.30	10.5	2.76	35.5
				3.20	74.0	$n_4 = 32$	
				2.64	27.0	$T_{n_4} = 1{,}730$	
				3.21	77.0		
				2.94	57.0		
				3.39	81.0		
				2.85	46.5		
				2.89	49.0		
				2.51	20.5		
				2.83	44.0		
				3.89	94.0		
				2.85	46.5		
				2.90	51.0		
				2.00	4.0		
				$n_3 = 45$			
				$T_{n_3} = 2{,}156$			

SOURCE: Data are taken from Figure 2.6.

Using Equation (13.14) to test the null hypothesis we obtain

$$H = \left[\frac{12}{n(n+1)} \sum_{i=1}^{c} \frac{T_{n_i}^2}{n_i}\right] - 3(n+1) \tag{13.14}$$

$$= \left\{\frac{12}{(94)(95)}\left[\frac{(306)^2}{7} + \frac{(273)^2}{10} + \frac{(2{,}156)^2}{45} + \frac{(1{,}730)^2}{32}\right]\right\} - 3(95)$$

$$= 292.48 - 285 = 7.48$$

and, as was the case with the analysis of variance F test presented in Section 12.5, the null hypothesis cannot be rejected at the $\alpha = .05$ level of significance since $H = 7.48 < 7.815$ (see Table 13.19).

13.9 COMPUTERS AND NONPARAMETRIC PROCEDURES

We may recall from Section 13.2 that a possible disadvantage to using nonparametric procedures arises because the numerical manipulations sometimes become laborious as the sample sizes get larger. Fortunately, in such situations, various packaged computer programs can be utilized to ease the computational complexity. For instance, the SPSS computer package (Reference 10) contains a subprogram NPAR TESTS which includes, among others, all the procedures discussed in this chapter except the Cox–Stuart test and the absolute normal scores test. As an example, the application of the NPAR TESTS subprogram is illustrated in Figure 13.9 for the Kruskal–Wallis test on the data presented in Table 13.20.

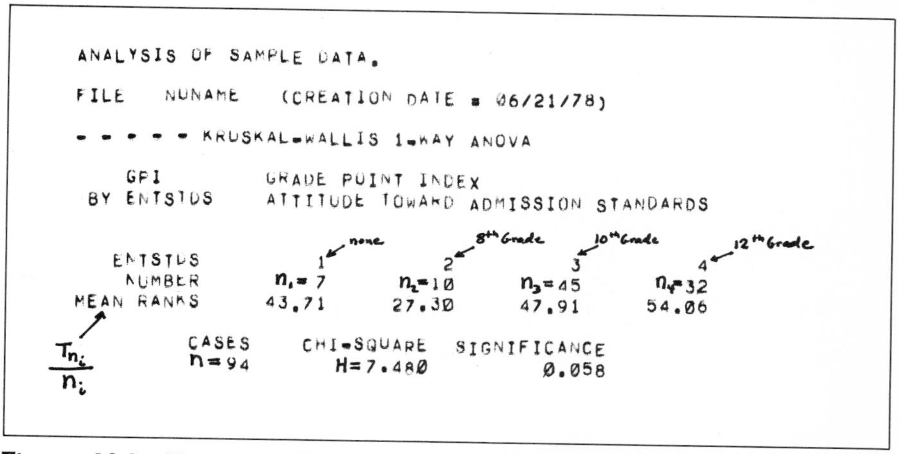

Figure 13.9 Using SPSS subprogram NPAR TESTS to perform Kruskal-Wallis test.

SOURCE: Data are taken from Table 13.20.

13.10 SUMMARY AND OVERVIEW OF NONPARAMETRIC TEST PROCEDURES

In this chapter we have provided but a brief introduction to the very broad subject of nonparametric methods of hypothesis testing. The need for such methods arose for the following three reasons:

1. The measurements attained on behavioral science data were often only qualitative (nominal scale) or in ranks (ordinal scale), and procedures were required to make appropriate tests.

2. The researcher often had no knowledge of the form of the population from which the sample data were drawn and believed that the assumptions for the classical procedure were unrealistic for a given endeavor.

3. The researcher was not always interested in testing hypotheses about particular population parameters and required methods for treating problems regarding randomness, trend, cyclical effects, symmetry, and goodness of fit.

Thus for these reasons a set of nonparametric procedures were devised. Six such nonparametric methods were presented here. Such methods may be broadly described as those which: (1) make fewer and less stringent assumptions than do their parametric counterparts, (2) are generally easier to apply and quicker to compute than their classical counterparts, and (3) yield a more general set of conclusions than do their parametric counterparts. In addition, nonparametric methods are applicable to very small sample sizes (from which exact probabilities may be computed) and also lend themselves readily to preliminary or pilot studies from which more detailed and sophisticated analyses can later be undertaken. Finally, for those situations in which the researcher has a choice, it has been found that some of the nonparametric techniques are either as powerful or almost as powerful in their ability to detect statistically significant differences (when they do exist) as their classical counterparts under conditions appropriate to the latter, and may actually be much more powerful than the classical procedures when their assumptions do not hold. On the other hand, especially with larger sample sizes, when the conditions of the classical methods are met it would be considered wasteful of information to merely convert sophisticated levels of measurement into normal scores or ranks (ordinal scale) or into categorized groupings (nominal scale) just so that a nonparametric procedure could be employed. Thus, in such circumstances as these, the classical procedures should be utilized. For a more detailed study of the subject of nonparametric methods and its applications, the reader may wish to consult References 2, 3, 4, 6, 8, 9, 11, and 13.

PROBLEMS

13.1 Flip a coin 50 times recording the sequence of heads (H) and tails (T). Can the resulting sequence be considered random? ($\alpha = .05$)

13.2 Roll a die 30 times recording the sequence of odd faces (1,3,5) and even faces (2,4,6). Can the resulting sequence be considered random? ($\alpha = .05$)

13.3 Starting at the beginning of the table of random numbers (Appendix E, Table E.1) record the sequence of high digits (5,6,7,8,9) and low digits (0,1,2,3,4) for the first 100 digits observed. Can this resulting sequence of high and low digits be considered random? ($\alpha = .05$)

13.4 Select a page from your local telephone directory and, examining the last digit of each telephone number only, record the sequence of high (5,6,7,8,9) and low (0,1,2,3,4) digits. Can the resulting sequence be considered random? ($\alpha = .05$)

13.5 Sampling with replacement from a deck of cards, record the sequence of red (r = diamonds or hearts) and black (b = clubs or spades) cards for a sample of 20 selections. Be sure to thoroughly shuffle the deck before each selection is made. Can the resulting sequence be considered random? ($\alpha = .05$)

13.6 Using the data in Figure 2.6 corresponding to question 4 of the survey prepared by the researcher and the dean (Figure 2.2), examine the responses of the 94 students recording the sequence of accounting (A) and non-accounting (N) majors. Can this resulting sequence be considered random? ($\alpha = .05$)

* 13.7 A psychologist wishes to construct a true–false examination containing 30 questions so that the sequence of correct responses is random. The following sequence is obtained:

FTTTFTFFFTTFTFFFTTTTFFTTTFFTFF

Can this sequence be considered random? ($\alpha = .05$)

13.8 A machine being used for packaging cereals has been set so that on the average 13 ounces of cereal will be packaged per box. The quality control engineer wishes to test the machine setting and selects a sample of 30 consecutive cereal packages filled during the production process. Their weights are recorded below in row sequence (from left to right):

13.2	13.3	13.1	13.7	13.3	13.0	13.1	12.3	12.6	12.5
13.0	13.2	13.4	13.6	13.7	13.4	13.3	12.9	12.8	12.6
12.3	12.4	13.5	13.4	13.2	13.5	13.6	13.1	13.3	13.1

Do these data indicate a lack of randomness in the sequence of underfills and overfills, or can the production process be considered as "in control"? ($\alpha = .05$)

13.9 The number of cars paying a toll at a particular bridge are observed hourly for a 24-hour period on a Wednesday in June. The results are

Time	Number of Cars	Time	Number of Cars
12– 1 AM	196	12– 1 PM	452
1– 2 AM	123	1– 2 PM	389
2– 3 AM	91	2– 3 PM	461
3– 4 AM	75	3– 4 PM	650
4– 5 AM	89	4– 5 PM	706
5– 6 AM	128	5– 6 PM	893
6– 7 AM	264	6– 7 PM	754
7– 8 AM	721	7– 8 PM	624
8– 9 AM	687	8– 9 PM	406
9–10 AM	671	9–10 PM	311
10–11 AM	629	10–11 PM	279
11–12 noon	518	11–12 midnight	267

Can this pattern be considered random in its fluctuation above and below its median? ($\alpha = .05$)

13.10 During the period from 1960 to 1975 there was an increase in the federal budget outlays for veterans' benefits and services. During this period, however, total

Table 13.21
PERCENTAGE OF TOTAL FEDERAL OUTLAYS TO VETERANS' BENEFITS AND SERVICES

Year	Percentage
1960	6.0
1961	5.8
1962	5.3
1963	5.0
1964	4.8
1965	4.8
1966	4.4
1967	4.4
1968	3.8
1969	4.1
1970	4.4
1971	4.6
1972	4.6
1973	4.9
1974	5.0
1975	4.9

SOURCE: Data are extracted from Table 505 of *Statistical Abstract of the United States* 1975, U.S. Department of Commerce, Bureau of the Census.

federal outlays for all functions also increased. The data in Table 13.21 present the percentage of total federal outlays for veterans' benefits and services during the 16-year period from 1960 to 1975.

Is there any evidence of trend over the 16-year period? ($\alpha = .05$)

* 13.11 The data in Table 13.22 represent the average number of acres per farm in the United States from 1950 to 1975.

Is there any evidence of an increasing trend over the 26-year period? ($\alpha = .05$)

Table 13.22
AVERAGE NUMBER OF ACRES PER FARM

Year	Average Number of Acres per Farm	Year	Average Number of Acres per Farm
1950	213	1963	322
1951	222	1964	332
1952	232	1965	340
1953	242	1966	348
1954	251	1967	355
1955	258	1968	363
1956	265	1969	369
1957	273	1970	373
1958	280	1971	377
1959	288	1972	381
1960	297	1973	383
1961	305	1974	384
1962	314	1975 (Prelim.)	385

SOURCE: Data are extracted from Table 1027 of *Statistical Abstract of the United States* 1975, U.S. Department of Commerce, Bureau of the Census.

13.12 In Problem 13.8 a machine used for packaging cereals has been set so that on the average 13 ounces of cereal will be packaged per box. Using the data of Problem 13.8, the weights from a sample of 30 cereal packages are obtained. Is the claim that the (median) weight per box is 13 ounces valid? ($\alpha = .05$)

Note: To test this claim we must assume that the observed sequence is random. Therefore let us assume then that the data were collected in column sequence (from top to bottom) rather than in row sequence (from left to right).

* 13.13 A task on an assembly line has, in the past, required 30 seconds to complete. An industrial engineer has developed a new method for performing the task which she believes will speed up the process. A random sample of 15 trials is obtained for a worker trained under the new method as shown below:

27.2	31.1	29.0	26.7	28.1	27.3	29.6	30.5	30.0	30.2	25.9	31.3
28.8	27.4	27.0									

Is there evidence to suggest that the median time under the new method is significantly less than 30 seconds? What would you recommend to the management? ($\alpha = .05$)

13.14 A cigarette manufacturer claims that the tar content of a new brand of cigarettes is 17 milligrams. A random sample of 24 cigarettes is selected and the tar content measured. The results are shown below in milligrams:

16.9	16.6	17.3	17.5	17.0	17.2	16.1	16.4	17.3	15.9	17.7	18.3
15.6	16.8	17.1	17.2	16.4	18.1	17.4	16.7	16.9	16.0	16.5	17.8

Is the claim that the (median) tar content of this new brand is 17 milligrams valid? ($\alpha = .01$)

13.15 An actuary of a particular insurance company wants to examine the records of larceny claims filed by persons insured under a household goods policy. In the past the median claim was for \$85. A random sample of 18 claims is taken and the results are shown below:

\$140	\$92	\$35	\$202	\$80	\$87	\$80	\$100	\$47	\$25	\$160	\$68
\$50	\$65	\$310	\$90	\$75	\$120						

Has the median claim significantly increased? ($\alpha = .05$)

13.16 A pharmaceutical manufacturer is producing a new vaccine for a particular strain of influenza. In testing the effects of this vaccine it is desired to study the change in body temperature taken immediately before and 1 hour after the innoculation. The results are shown below for a sample of 10 male patients whose age is 60 and over:

Temperature before (F°)	98.8	98.6	97.5	98.0	98.7	98.4	98.7	98.6	98.3	98.6
Temperature after (F°)	99.8	98.8	98.4	99.9	99.3	98.4	98.6	101.2	99.0	99.1
Change in temeprature (F°)	1.0	0.2	0.9	1.9	0.6	0.0	−0.1	2.6	0.7	0.5

For males aged 60 and over is there any evidence of a significant increase in the body temperature 1 hour after the vaccination? ($\alpha = .05$) **Hint:** Test that the median increase is 0.0°F. Would you generalize your results to the entire population? Discuss.

13.17 Using the data in Figure 2.6 corresponding to questions 2 and 6 from the survey (Figure 2.2), test the claim that the median high-school average of sophomore students was 85. ($\alpha = .01$)

13.18 Using the data in Figure 2.6 corresponding to questions 3 and 5 from the survey (Figure 2.2), test the claim that the median grade-point index of students planning to attend graduate school is 3.20. ($\alpha = .01$)

13.19 Using the data in Figure 2.6 corresponding to questions 2 and 7 from the survey (Figure 2.2), test the claim that the median anticipated starting salaries of seniors is $15,000. ($\alpha = .01$)

13.20 Using the data from Problem 4.9 test the claim that the median amount of sales tax receipts collected quarterly from retail outlets in a particular community is $15,000. ($\alpha = .05$)

13.21 Using the data from Problem 4.10 test the claim that the median amount owed to the company by individuals purchasing books by mail is $20. ($\alpha = .05$)

13.22 A consumer protection agency wishes to perform a "life test" to investigate the differences in the median time to failure of electronic calculators supplied with AAA alkaline batteries operated under normal temperature versus high temperature conditions. A total of 16 electronic calculators of the same type are randomly assigned to the two groups so that 8 are examined under normal temperature and 8 are examined under high temperature. All the calculators are started simultaneously, and the data below are the times to failure (in hours):

Normal Temperature	9.3	12.7	12.9	14.9	16.1	16.3	17.9	23.1
High Temperature	6.0	8.2	8.8	11.9	12.3	14.7	14.8	15.6

Do calculators operating under high temperature have a significantly shorter life than those operating under normal temperatures? ($\alpha = .05$)

* 13.23 A women's organization wishes to know whether starting salaries of men and women entering business differ. A random sample of 10 male seniors with B+ indexes and a random sample of 9 female seniors with B+ indexes are selected, and each person is permitted a total of three interviews with companies in marketing, public relations, or advertising. The data below are the reported "best offers" made to the individuals:

Male		Female	
$14,000	$15.200	$14,200	$10,000
$11,800	$14,500	$10,250	$12,750
$17,000	$10,500	$13,100	$11,250
$13,000	$12,500	$ 9,000	$12,000
$14,000	$15,000	$11,500	

Is there any evidence to support the claim that women are not as well paid as men when their qualifications are similar? ($\alpha = .05$)

13.24 An industrial psychologist wishes to study the effects of motivation on sales in a particular firm. Of 24 new sales persons being trained, 12 are to be paid at an hourly rate, while 12 are to be paid on a commission basis. The 24 persons were randomly assigned to the two groups. The data below represent the sales volume achieved during the first month on the job.

Hourly Rate		Commission	
256	212	224	261
239	216	254	228
222	236	273	234
207	219	285	225
228	225	237	232
241	230	277	245

Is there any evidence to support the claim that wage incentives (through commission) yield significantly greater sales volume? ($\alpha = .01$)

13.25 A statistics professor taught two special sections of a basic course in which the 10 students in each section were considered outstanding. She used a "traditional" method of instruction (T) in one section and an "experimental" method of instruction (E) in the other. At the end of the semester she ranked the students based on their performance from 1 (worst) to 20 (best).

T	1	2	3	5	9	10	12	13	14	15
E	4	6	7	8	11	16	17	18	19	20

For this instructor was there any difference in performance based on the two methods? ($\alpha = .05$)

13.26 Using the data in Figure 2.6 corresponding to questions 3 and 5 from the survey (Figure 2.2), test whether or not there are any differences in the median grade-point indexes of students planning to attend graduate school and those that are not. ($\alpha = .05$)

13.27 Using the data in Figure 2.6 corresponding to questions 4, 21, and 5 from the survey (Figure 2.2), test whether or not there are any differences in the median grade-point indexes of accounting students favoring twelfth-year admission standards versus the non-accounting students favoring twelfth-year admission standards. ($\alpha = .05$)

13.28 Using the data in Figure 2.6 corresponding to questions 2 and 6 from the survey (Figure 2.2), test whether or not there are any differences in the median high-school averages of sophomores versus juniors at this business school. ($\alpha = .01$)

13.29 Using the data in Figure 2.6 corresponding to questions 1, 3, and 5 from the survey (Figure 2.2), test whether or not there are any differences in the median grade-point indexes of male students who are planning to attend graduate school versus female students who are planning to attend graduate school. (α = .05)

13.30 Using the data from Problem 3.18 test whether or not there are any differences in the median batting averages of baseball players in the National versus American leagues. (α = .01)

13.31 Using the data from Problem 3.20 test whether or not there are any differences in the median closing prices of stocks traded on the New York Stock Exchange versus the American Stock Exchange. (α = .01)

13.32 Using the data from Problem 4.7 test whether or not there is a significant difference in the median ages of United States versus Western European automobile salesmen. (α = .01)

13.33 Using the data from Problem 4.8 test whether one-bedroom unfurnished apartments renting in Manhattan are significantly higher than those renting in Brooklyn. (α = .05)

13.34 Using the data from Problem 4.17 test whether or not there are any differences in the median expected starting salaries of juniors versus seniors. (α = .05)

13.35 Using the data from Problem 4.23 test whether or not there are any differences in the median grade-point indexes of juniors versus seniors. (α = .05)

13.36 Using the data from Problem 4.29 test whether students favoring more stringent retention standards have a significantly higher median grade-point index than those who do not favor more rigorous retention standards. (α = .01)

* 13.37 Refer to Problem 13.16.
(a) Use the absolute normal scores test on the data. (α = .05)
(b) Compare your results with those obtained from Problem 13.16.

13.38 Refer to Problem 10.6. Use the absolute normal scores test to determine (at the .01 level of significance) whether the unknown brand sells for a lower price.

13.39 Refer to Problem 10.7. Use the absolute normal scores test to determine (at the .01 level of significance) whether the new system uses less processing time.

13.40 Refer to Problem 10.8. Use the absolute normal scores test to determine (at the .05 level of significance) whether stock market prices have significantly increased over the 2-day period.

13.41 Refer to Problem 10.9. At the .01 level of significance, is there a difference in earnings per share between the 2 years?

13.42 Refer to Problem 10.10. At the .05 level of significance, can the research director conclude that the sales campaign has increased the sales of nonsaleable items?

13.43 A batch of 20 AAA carbon zinc batteries are randomly assigned to four groups (so that there are five batteries per group). Each group of batteries is then subjected to a particular temperature level—low, normal, high, and very high. The batteries are simultaneously tested under these temperatures, and the times to failure (in hours) are recorded below:

Low	Normal	High	Very High
8.0	7.6	6.0	5.1
8.1	8.2	6.3	5.6
9.2	9.8	7.1	5.9
9.4	10.9	7.7	6.7
11.7	12.3	8.9	7.8

Do the four temperature levels yield the same median battery lives? ($\alpha = .05$)

13.44 An industrial psychologist desires to test whether the reaction times of assembly line workers are equivalent under three different learning methods. From a group of 25 new employees, nine are randomly assigned to method A, eight workers are assigned to method B, and eight workers are randomly assigned to method C. The data below represent the rankings from 1 (fastest) to 25 (slowest) of the reaction times to complete a task given by the industrial psychologist after the learning period.

	Method	
A	*B*	*C*
2	1	5
3	6	7
4	8	11
9	15	12
10	16	13
14	17	18
19	21	24
20	22	25
23		

Are there any differences in the reaction times for these learning methods? ($\alpha = .01$)

* 13.45 In a test of the effectiveness of advertising on product perception 30 students were randomly assigned to five treatment groups (so that there are six students per group). Each group of students then received a particular advertisement regarding a ball point pen. Treatments A and B tended to undersell the pen's characteristics; treatments C and D tended to oversell the pen's characteristics; and treatment E attempted to correctly state the pen's characteristics. After reading the advertisement and developing a sense of "product expectancy," the students unknowingly all received the same type of pen to evaluate. The students were permitted to test out their pen and the plausibility of the advertising copy. The students were then asked to rate the pen from 1 to 7 on a "product characteristic" scale, and the combined scores of three ratings—appearance, durability, and writing performance—are given as follows:

Treatment				
A	B	C	D	E
15	16	8	5	12
18	17	6	7	13
17	21	10	13	10
19	16	13	11	12
19	19	11	9	13
20	17	12	10	14

Is product perception different for various levels of expectations? ($\alpha = .05$)

13.46 Using the data in Figure 2.6 corresponding to questions 5 and 9 from the survey (Figure 2.2), test whether there are any differences among grade-point indexes of students working full-time, part-time, or not at all. ($\alpha = .01$)

13.47 Using the data in Figure 2.6 corresponding to questions 5 and 23 from the survey (Figure 2.2), test whether there are any differences among grade-point indexes of students who are not affiliated with any clubs at the college versus those who are affiliated with one group versus those who are affiliated with two or more groups. ($\alpha = .05$)

13.48 Using the data in Figure 2.6 (corresponding to questions 1, 16, and 19 from the survey (Figure 2.2), test whether there are any differences in the heights of male students who do not smoke versus those that smoke less than one pack daily versus those that smoke at least one pack daily. ($\alpha = .01$)

13.49 Use an appropriate nonparametric technique to solve Problem 12.3. Are there any differences in your present results from that of the analysis of variance F test? Discuss.

13.50 Use an appropriate nonparametric technique to solve Problem 12.5. Are there any differences in your present results from that of the analysis of variance F test? Discuss.

13.51 Use an appropriate nonparametric technique to solve Problem 12.8. What differences are there in the assumptions for your selected nonparametric test and those of the analysis of variance F test? Discuss.

13.52 Use an appropriate nonparametric technique to solve Problem 12.10. What differences are there in the assumptions for your selected nonparametric test and those of the analysis of variance F test? Discuss.

13.53 You are having cocktails with a client and during the course of your conversation she mentions that someone has suggested that nonparametric techniques might be helpful for her current project. As her statistical confidant, she asks you "what are nonparametric techniques and when or why may they be helpful?" You sip your drink and you reply. . . .

REFERENCES

1. Bierman, H., C. Bonini, and W. Hausman, *Quantitative Analysis for Business Decisions,* 4th ed., (Homewood, Ill.: Richard D. Irwin, 1973).
2. Bradley, J. V., *Distribution-Free Statistical Tests* (Englewood Cliffs, N.J.: Prentice-Hall, 1968).
3. Conover, W. J., *Practical Nonparametric Statistics* (New York: Wiley, 1971).
4. Gibbons, J. D., *Nonparametric Methods for Quantitative Analysis* (New York: Holt, Rinehart and Winston, 1976). Copyright 1976 by Holt, Rinehart and Winston, reprinted by permission of Holt, Rinehart and Winston.
5. Green, P., and D. Tull, *Research for Marketing Decisions,* 4th ed. (Englewood Cliffs, N.J.: Prentice-Hall, 1978).
6. Hollander, M., and D. A. Wolfe, *Nonparametric Statistical Methods* (New York: Wiley, 1973).
7. Klotz, J., "Small Sample Power and Efficiency for the One Sample Wilcoxon and Normal Scores Tests," *Annals of Mathematical Statistics,* vol. 34, no. 2, pp. 624–632, 1963.
8. Lehmann, E. L., *Nonparametrics: Statistical Methods Based on Ranks* (San Francisco: Holden-Day, 1975).
9. Marascuilo, L., and M. McSweeney, *Nonparametric and Distribution-Free Methods for the Social Sciences* (Belmont, Calif.: Wadsworth, 1977).
10. Nie, N., C. Hull, J. Jenkins, K. Steinbrenner, and D. Bent, SPSS Batch Release 7 (Update Manual, March 1977) to *Statistical Package for the Social Sciences,* 2nd ed. (New York: McGraw-Hill, 1975).
11. Noether, G., *Introduction to Statistics: A Nonparametric Approach,* 2nd ed. (Boston: Houghton Mifflin, 1976).
12. Scheffé, H., *The Analysis of Variance* (New York: Wiley, 1959).
13. Siegel, S., *Nonparametric Statistics for the Behavioral Sciences* (New York: McGraw-Hill, 1956).
14. Thompson, R., Z. Govindarajulu, and K. Doksum, "Distribution and Power of the Absolute Normal Scores Test," *Journal of the American Statistical Association,* vol. 62, no. 319, pp. 966–975, 1967.
15. van der Waerden, B. L., "The Computation of the χ-Distribution," *Proceedings of the Third Berkeley Symposium on Mathematical Statistics and Probability* (Berkeley, Calif.: University of California Press, 1956), pp. 207–208.

CHAPTER 14

SIMPLE LINEAR REGRESSION AND CORRELATION

14.1 INTRODUCTION

In our discussion of statistical methods in previous chapters we have primarily focused upon a single variable of interest, such as mileage life of tires, amount of credit card purchases, or grade-point index of accounting majors. In discussing these variables we examined various measures of description (see Chapter 4) and applied different techniques of statistical inference, such as confidence intervals and tests of hypothesis, to make estimates and draw conclusions about them (see Chapters 8 to 10, 12, and 13). In this and the following chapter we will concern ourselves with problems involving two or more variables as a means of viewing the relationships that exist between them. Two techniques will be discussed: regression and correlation.

Regression analysis is utilized for the purpose of prediction. For the case of two variables, a model is developed that uses the independent variable X to obtain a better prediction of the other variable, the dependent variable Y. For example, the dean and the researcher would like to develop a statistical model that would use high school average as a predictor of college performance (as measured by the grade-point index). The dependent variable Y, the one to be predicted, would be grade-point index, while the variable used to obtain a better prediction (the independent variable X) is high school average.

Correlation analysis, in contrast to regression, is used to measure the strength of the association between variables. For example, in the dean's survey we could determine the correlation between the heights and weights of students. In this instance the objective is not to use one variable to predict another, but only to measure the strength of the association or "covariation" between two variables.

14.2 THE SCATTER DIAGRAM

Methods of regression and correlation analysis will be applied to two problems in this chapter. The first problem involves a company that constructs new single-family homes. The cost accountant for the company would like to estimate the construction cost of new single-family homes in the coming year so that a selling price can be assigned to each house. The construction cost of all single-family homes built by the company in the past year is available from the company records. Rather than simply using last year's costs as an estimate of costs in the coming year, the accountant believes that construction cost is strongly related to the size of the property. A random sample of 12 houses constructed in the last year is selected and the information collected is given in Table 14.1.

In this example the dependent variable Y would be the construction cost (in thousands of dollars), since this is the variable to be predicted. The independent variable X is property size (in thousands of square feet), since this variable is used to predict the construction cost. Now that we have obtained the data shown in Table 14.1, they can be presented in a form that is more visually interpretable. In Chapter 3, when information concerning the grade-point index of students was collected, various graphs (such as histograms, polygons, and ogives) were developed for data presentation. In a regression

Table 14.1
CONSTRUCTION COST AND PROPERTY SIZE FOR A RANDOM SAMPLE OF 12 ONE-FAMILY HOMES

Observation	Property Size (thousands of square feet)	Construction Cost (thousands of dollars)
1	5	31.6
2	7	32.4
3	10	41.7
4	10	50.2
5	12	46.2
6	20	58.5
7	22	59.3
8	15	48.4
9	30	63.7
10	40	85.3
11	12	53.4
12	15	54.5

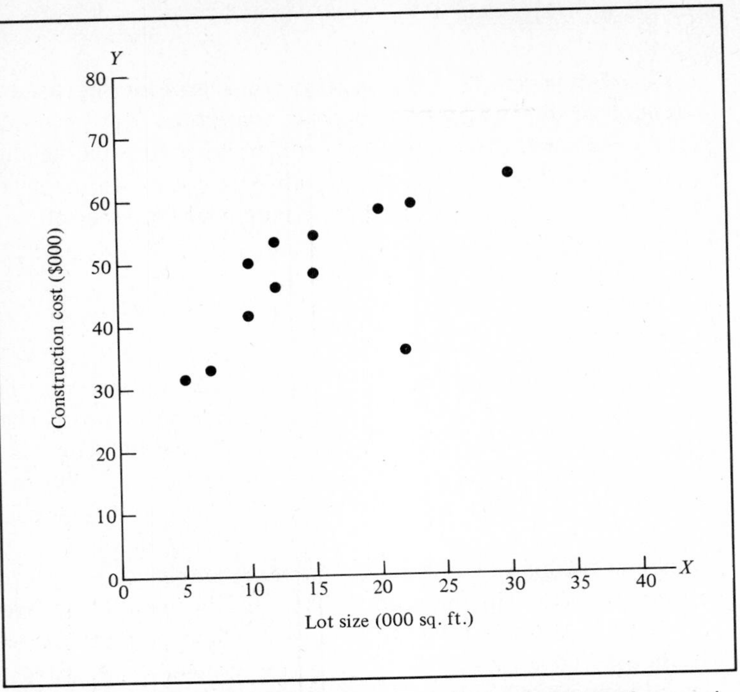

Figure 14.1 Scatter diagram of lot size vs. construction cost.
SOURCE: Data are taken from Table 14.1.

analysis (involving one independent and one dependent variable) the individual values are plotted on a two-dimensional graph called a **scatter diagram.** Each value is plotted at its particular X and Y coordinates. The scatter diagram for the construction cost problem of Table 14.1 is shown in Figure 14.1.

A brief examination of Figure 14.1 indicates a clearly increasing relationship between lot size X and construction cost Y. As the lot size increases, the construction cost also increases. The exact mathematical form of the model expressing the relationship, as well as methods for estimating the construction costs for a given property size, will be examined in subsequent sections of this chapter.

The second problem that we shall consider refers to the previously mentioned desire of the dean to evaluate the relationship between high school average and grade-point index. Since the grade-point index may fluctuate until the student has accumulated many credits, only the seniors in the sample will be evaluated. Referring to the data of Figure 2.6, the high school averages (question 6) and the corresponding grade-point indexes (question 5) of the 15 seniors (question 2, code 4) were extracted, and the results are presented in Table 14.2.

The scatter diagram for this problem can be plotted with grade-point index (dependent variable) on the Y axis and high school average (independent variable) on the X axis, as illustrated in Figure 14.2.

Table 14.2
GRADE-POINT INDEX AND HIGH SCHOOL AVERAGE OF THE 15 SENIORS OBTAINED FROM A RANDOM SAMPLE OF 94 STUDENTS

Student	High School Average (X)	Grade-Point Index (Y)
1	87	2.30
2	88	2.80
3	80	2.90
4	83	3.00
5	80	2.82
6	98	3.86
7	78	2.60
8	85	3.34
9	80	2.50
10	92	3.00
11	76	3.20
12	81	3.20
13	82	2.64
14	89	3.21
15	78	2.66

SOURCE: Data are taken from Figure 2.6.

A quick scan of Figure 14.2 appears to indicate that students with low high school averages achieve both low and high grade-point indexes while students with high averages also achieve both low and high grade-point indexes. The questions that will be examined in subsequent sections of this chapter, then, are whether there is a relationship between these two variables, and if so, how

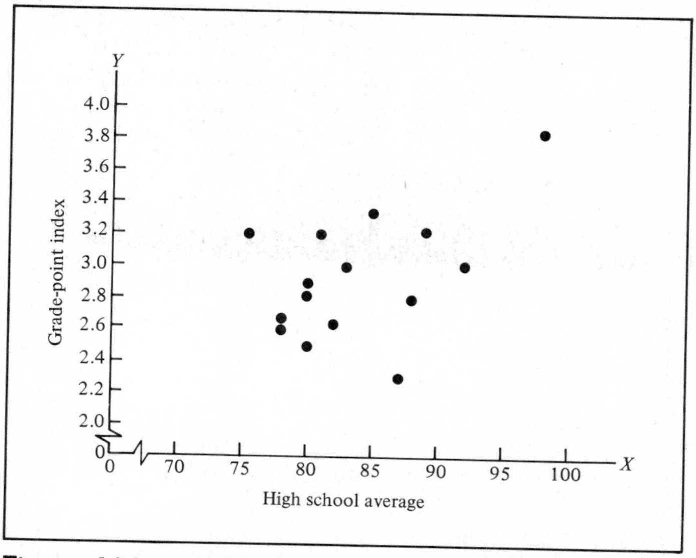

Figure 14.2 Scatter diagram of high school average versus grade-point index of 15 seniors.
SOURCE: Data are taken from Table 14.2.

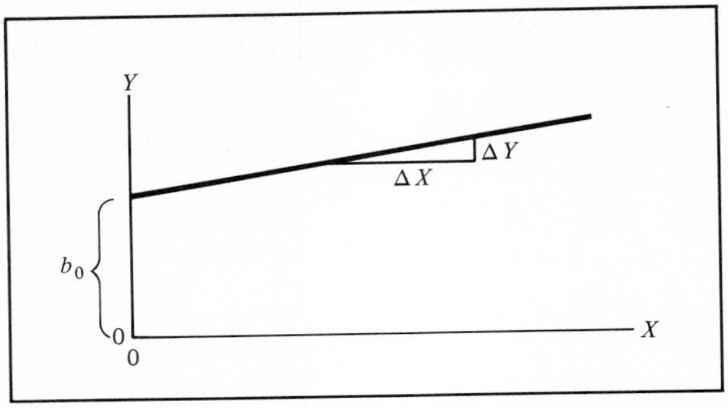

Figure 14.3 A positive straight-line relationship.

the existence of a relationship can provide a better prediction of the dependent variable Y.

14.3 TYPES OF REGRESSION MODELS

In the scatter diagrams plotted in Figures 14.1 and 14.2 a rough idea of the type of relationship that exists between the variables can be observed. The nature of the relationship can take many forms, ranging from simple mathematical functions to extremely complicated ones. The simplest relationship consists of a straight line or **linear** relationship. An example of this relationship is shown in Figure 14.3.

The straight-line (linear) model can be represented as

$$Y_i = \beta_0 + \beta_1 X_i + \varepsilon_i \qquad (14.1)$$

where β_0 = true Y intercept for the population

β_1 = true slope for the population

ε_i = random error in Y for observation i

In this model, the slope of the line β_1 represents the unit change in Y, ΔY, per unit change in X, ΔX, that is, it represents the amount that Y changes (either positively or negatively) for a particular unit change in X. On the other hand, the Y intercept β_0 represents a constant factor that is included in the equation. It represents the value of Y when X equals zero. Moreover, the last component of the model, ε_i, represents the random error in Y for each observation i that occurs. This term is included because the statistical model is only an approximation of the exact relationship between the two variables.

The proper mathematical model to be selected is influenced by the distribution of the X and Y values on the scatter diagram. This can be seen readily from an examination of panels A through F in Figure 14.4. Clearly, from

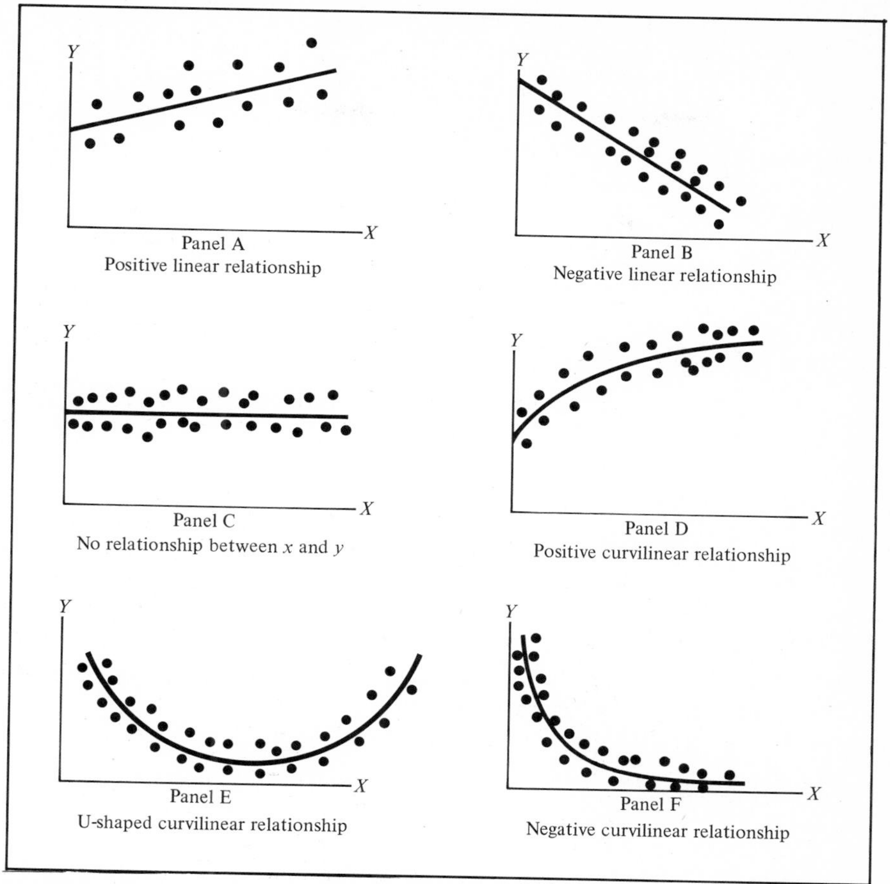

Figure 14.4 Examples of types of relationships found in scatter diagrams.

panel A in Figure 14.4 it can be seen that the values of Y are generally increasing linearly as X increases. This panel is similar to Figure 14.1, which illustrates the relationship between construction cost and lot size. Panel B is an example of a negative linear relationship. As X increases, we note that the value of Y is decreasing. An example of this type of relationship might be the price of a particular product and the amount of sales. Panel C shows a set of data in which there is very little or no relationship between X and Y. High and low values of Y appear at each value of X. We may note that the relationship between high school average and grade-point index, as indicated in Figure 14.2, is similar to panel C. The data in panel D show a positive curvilinear relationship between X and Y. The values of Y are increasing as X increases, but this increase tapers off beyond certain values of X. An example of this positive curvilinear relationship might be the age and maintenance cost of a machine. As a machine gets older, the maintenance cost may rise rapidly at first, but then level off beyond a certain number of years. Panel E shows a parabolic or U-shaped relationship between X and Y. As X increases, at first Y decreases; but as X continues to increase, Y not only stops decreasing but actually increases above its minimum value. An example of this type of relationship could be the number of errors per hour at a task and the number of hours worked. The number of errors per hour would decrease as the individual becomes more proficient at the task, but then would increase beyond a certain point due to factors such as fatigue and boredom. Finally, panel F indicates an exponential or negative curvilinear relationship between X and Y. In this case Y decreases very rapidly as X first increases, but then decreases much less rapidly as X increases further. An example of this exponential relationship could be the resale value of a particular type of automobile and its age. In the first year the resale value drops drastically from its original price; however, the resale value then decreases much less rapidly in subsequent years.

In this section we have briefly examined a variety of different models that could be used to represent the relationship between two variables. Although scatter diagrams can be extremely helpful in determining the mathematical form of the relationship, more sophisticated statistical procedures are available (References 1, 3, and 8) to determine the most appropriate model for a set of variables. In subsequent sections of this chapter we shall primarily focus on building statistical models for fitting *linear* relationships between variables.

14.4 DETERMINING THE SIMPLE LINEAR REGRESSION EQUATION

If we were to refer to the scatter diagrams in the previous sections, it can be seen that various linear (straight-line) relationships can be fit to the data. In particular, in Figure 14.1 we noticed that construction cost appeared to increase linearly as a function of property size. The question that must be addressed in regression analysis involves the determination of the particular straight-line model that is the best fit to these data.

14.4.1 The Least-Squares Method

In our discussions in the previous section we have hypothesized a statistical model to represent the relationship between two variables in a population. In both the construction cost example and the dean's survey we have obtained data from only a random sample of the population. If certain assumptions are valid (see Section 14.10), the sample Y intercept (b_0) and the sample slope (b_1) can be used as estimates of the respective population parameters (β_0 and β_1). Thus, the sample regression equation representing the straight-line regression model would be

$$\hat{Y}_i = b_0 + b_1 X_i \qquad (14.1a)$$

where $\hat{Y}_i$ = predicted value of Y for observation i.

The prediction of Y using this equation involves the determination of two coefficients: b_0, the Y intercept, and b_1, the slope. Once b_0 and b_1 are obtained, the straight line is known and can be plotted on the scatter diagram. We could then make a visual comparison of how well our particular statistical model (a straight line) fits the original data, that is, we can see whether the original data lie close to the fitted line or deviate greatly from the fitted line. Simple linear regression analysis is concerned with finding the straight line that "fits" the data best. The best fit means that we wish to find the straight line for which the difference between the actual value of Y, Y_i, and the value that would be predicted from the fitted line of regression, $\hat{Y}_i$, is as small as possible. Because these differences will be both positive and negative for different observations, mathematically we minimize

$$\sum_{i=1}^{n} (Y_i - \hat{Y}_i)^2$$

where Y_i = actual value of Y for observation i

$\hat{Y}_i$ = predicted value of Y for observation i

Since $\hat{Y}_i = b_0 + b_1X_i$, we are minimizing

$$\sum_{i=1}^{n} [Y_i - (b_0 + b_1X_i)]^2$$

which has two unknowns, b_0 and b_1. A mathematical technique which determines the values of b_0 and b_1 that best fit the observed data is known as the **least-squares method.** Any values for b_0 and b_1 other than those determined by the least-squares method would result in a greater sum of squared differences between the actual value of Y and the predicted value of Y. In using the least-squares method, we obtain the following two equations, called the normal equations:

$$\sum_{i=1}^{n} Y_i = nb_0 + b_1 \sum_{i=1}^{n} X_i \qquad (14.2a)$$

$$\sum_{i=1}^{n} X_iY_i = b_0 \sum_{i=1}^{n} X_i + b_1 \sum_{i=1}^{n} X_i^2 \qquad (14.2b)$$

Since there are two equations with two unknowns, we can solve these equations simultaneously for b_1 and b_0 as follows:

$$b_1 = \frac{n\sum_{i=1}^{n} X_iY_i - \left(\sum_{i=1}^{n} X_i\right)\left(\sum_{i=1}^{n} Y_i\right)}{n\sum_{i=1}^{n} X_i^2 - \left(\sum_{i=1}^{n} X_i\right)^2} \tag{14.3}$$

$$b_0 = \overline{Y} - b_1\overline{X} \tag{14.4}$$

where $\overline{Y} = \dfrac{\sum_{i=1}^{n} Y_i}{n}$ and $\overline{X} = \dfrac{\sum_{i=1}^{n} X_i}{n}$

or, alternatively,

$$b_1 = \frac{\sum_{i=1}^{n} X_iY_i - \dfrac{\left(\sum_{i=1}^{n} X_i\right)\left(\sum_{i=1}^{n} Y_i\right)}{n}}{\sum_{i=1}^{n} X_i^2 - \dfrac{\left(\sum_{i=1}^{n} X_i\right)^2}{n}} \tag{14.3a}$$

and

$$b_0 = \frac{\sum_{i=1}^{n} Y_i}{n} - b_1 \frac{\sum_{i=1}^{n} X_i}{n} \tag{14.4a}$$

Examining Equations (14.3) and (14.4), we see that there are five quantities that must be calculated in order to determine b_0 and b_1. These are n, the sample size; $\sum_{i=1}^{n} X_i$, the sum of the X values; $\sum_{i=1}^{n} Y_i$, the sum of the Y values; $\sum_{i=1}^{n} X_i^2$, the sum of the squared X values, and $\sum_{i=1}^{n} X_iY_i$, the sum of the cross product of X and Y. In our construction cost problem (Table 14.1) the size of the property was to be used to predict the construction cost of the single-family house. The computations of the various sums needed (including

Table 14.3
COMPUTATIONS FOR THE CONSTRUCTION COST PROBLEM

Observation	Property Size (thousands of square feet) X	Construction Cost (thousands of dollars) Y	XY	X^2	Y^2
1	5	31.6	158.0	25	998.56
2	7	32.4	226.8	49	1,049.76
3	10	41.7	417.0	100	1,738.89
4	10	50.2	502.0	100	2,520.04
5	12	46.2	554.4	144	2,134.44
6	20	58.5	1,170.0	400	3,422.25
7	22	59.3	1,304.6	484	3,516.49
8	15	48.4	726.0	225	2,342.56
9	30	63.7	1,911.0	900	4,057.69
10	40	85.3	3,412.0	1,600	7,276.09
11	12	53.4	640.8	144	2,851.56
12	15	54.5	817.5	225	2,970.25
	198	625.2	11,840.1	4,396	34,878.58

$\sum_{i=1}^{n} Y_i^2$, the sum of the squared Y values which will be used in Section 14.5) are presented in Table 14.3. Using Equations (14.3) and (14.4) we can compute the values of b_0 and b_1.

$$b_1 = \frac{n\sum_{i=1}^{n} X_iY_i - \left(\sum_{i=1}^{n} X_i\right)\left(\sum_{i=1}^{n} Y_i\right)}{n\sum_{i=1}^{n} X_i^2 - \left(\sum_{i=1}^{n} X_i\right)^2} = \frac{12(11{,}840.1) - (198)(625.2)}{12(4{,}396) - (198)^2}$$

$$= \frac{142{,}081.2 - 123{,}789.6}{52{,}752 - 39{,}204} = \frac{18{,}291.6}{13{,}548} = +1.35$$

$$b_0 = \overline{Y} - b_1\overline{X}$$

$$\overline{Y} = \frac{\sum_{i=1}^{n} Y_i}{n} = \frac{625.2}{12} = 52.1$$

$$\overline{X} = \frac{\sum_{i=1}^{n} X_i}{n} = \frac{198}{12} = 16.5$$

$$b_0 = 52.1 - (+1.35)(16.5) = 52.1 - 22.275 = 29.825$$

Thus the equation for the best straight line for these data is

$$\hat{Y}_i = 29.825 + 1.35X_i \qquad (14.5)$$

The slope b_1 was computed as $+1.35$. This value means that for each increase of one unit in X, the value of Y increases by 1.35 units. In this construction cost problem, for each increase in property size of 1,000 square feet, the construction cost will increase by \$1,350 (1.35 thousands of dollars). This slope then can be viewed as representing the variable portion of the cost of construction—varying with the size of the property. The Y intercept b_0 was computed to be 29.825. The Y intercept represents the value of Y (in this problem \$29,825) when X equals zero. This Y intercept can be viewed as representing the fixed portion of the cost of construction—the cost that does not vary with the size of the property.

The regression equation that has been fit to the data can now be used for predicting the Y value for a given value of X. For example, the cost accountant would like to predict the average construction cost of houses that are to be built on property of 15,000 square feet. The predicted cost $\hat{Y}_i$ can be determined by substituting $X = 15$ into Equation (14.5) since X represents thousands of square feet:

$$\hat{Y}_i = 29.825 + 1.35(15) = 29.825 + 20.25 = 50.075$$

Since Y represents thousands of dollars, the predicted average cost is \$50,075 for houses built on a property of 15,000 square feet.

Now that we have fitted a regression line and predicted construction cost, let us turn to our second problem, that of examining the relationship between high school average X and college grade-point index Y for our sample of 15 seniors. The computations of the various sums needed to solve this problem are shown in Table 14.4.

Therefore we can compute the values of b_0 and b_1 in this problem using Equations (14.3) and (14.4) as follows:

Table 14.4
COMPUTATIONS FOR THE RELATIONSHIP BETWEEN HIGH SCHOOL AVERAGE AND GRADE-POINT INDEX

Student	High School Average X	Grade-Point Index Y	XY	X^2	Y^2
1	87	2.30	200.1	7,569	5.29
2	88	2.80	246.4	7,744	7.84
3	80	2.90	232.0	6,400	8.41
4	83	3.00	249.0	6,889	9.00
5	80	2.82	225.6	6,400	7.9524
6	98	3.86	378.28	9,604	14.8996
7	78	2.60	202.8	6,084	6.76
8	85	3.34	283.9	7,225	11.1556
9	80	2.50	200.0	6,400	6.25
10	92	3.00	276.0	8,464	9.00
11	76	3.20	243.2	5,776	10.24
12	81	3.20	259.2	6,561	10.24
13	82	2.64	216.48	6,724	6.9696
14	89	3.21	285.69	7,921	10.3041
15	78	2.66	207.48	6,084	7.0756
	1,257	44.03	3,706.13	105,845	131.3869

$$b_1 = \frac{n\sum_{i=1}^{n} X_iY_i - \left(\sum_{i=1}^{n} X_i\right)\left(\sum_{i=1}^{n} Y_i\right)}{n\sum_{i=1}^{n} X_i^2 - \left(\sum_{i=1}^{n} X_i\right)^2} = \frac{15(3{,}706.13) - (1{,}257)(44.03)}{15(105{,}845) - (1{,}257)^2}$$

$$= \frac{55{,}591.95 - 55{,}345.71}{1{,}587{,}675 - 1{,}580{,}049} = \frac{246.24}{7{,}626} = +.0323$$

$$\overline{Y} = \frac{\sum_{i=1}^{n} Y_i}{n} = \frac{44.03}{15} = 2.9353$$

$$\overline{X} = \frac{\sum_{i=1}^{n} X_i}{n} = \frac{1{,}257}{15} = 83.8$$

$$b_0 = \overline{Y} - b_1\overline{X} = 2.9353 - (.0323)(83.8) = .2286$$

The equation for the best straight line for these data then is

$$\hat{Y}_i = .2286 + .0323X_i \tag{14.6}$$

The slope in this problem, +.0323, means that for each increase of one point in high school average, the college grade-point index increases by .0323 unit. The Y intercept, equal to .2286, represents a constant portion of the grade-point index that does not vary with the high school average.

Now that Equation (14.6) has been obtained, it can be used to predict the average grade-point index of seniors with a given high school average. If the dean wished to predict the grade-point index for seniors whose high school average was 85, the value of 85 would be substituted into Equation (14.6):

$$\hat{Y}_i = .2286 + .0323(85) = .2286 + 2.7455 = 2.9741$$

Therefore the predicted grade-point index for seniors who had a high school average of 85 would be 2.9741.

14.4.2 Predictions in Regression Analysis: Interpolation versus Extrapolation

When using regression analysis for prediction purposes, it is important that we only consider the relevant range of the independent variable in making our predictions. This relevant range encompasses all values from the smallest to the largest X used in developing the regression equation. Hence, when predicting Y for a given value of X, we may *interpolate* within this relevant range of the X values, but we may not *extrapolate* beyond the range of X values. For example, when we used property size to predict the construction cost for single-family homes, we recall from Table 14.3 that the property sizes varied from 5,000 to 40,000 square feet. Therefore, predictions of construction cost should only be made for homes with property sizes between 5,000 and 40,000 square feet. Any prediction of construction cost outside this range of property size presumes that the fitted relationship holds outside the range of 5,000 to 40,000 square feet. Similarly, for the dean's study, we note from Table 14.4 that the regression equation was developed based on high school averages ranging from 76 to 98 and thus any high school averages within that range could be used to predict grade-point index.

14.5 STANDARD ERROR OF THE ESTIMATE

In the previous section we have used the least-squares method to develop an equation to predict the construction cost based upon property size and to predict the grade-point index based upon high school average. Although the least-squares method results in the line that fits the data with the minimum amount of variation, the regression equation is not a perfect predictor, especially when samples are taken from a population, unless all the observed data points fall on the predicted regression line. Just as we cannot expect all data values to be located exactly at their arithmetic mean, in the same way we cannot expect all data points to fall exactly on the regression line. Thus the regression line serves only as an approximate predictor of a Y value for a given value of X. Therefore we need to develop a statistic that measures this variability of the actual Y values, Y_i, from the predicted Y values, $\hat{Y}_i$, in the same way that we developed (see Chapter 4) a measure of the variability of each observation around its mean. The measure of variability around

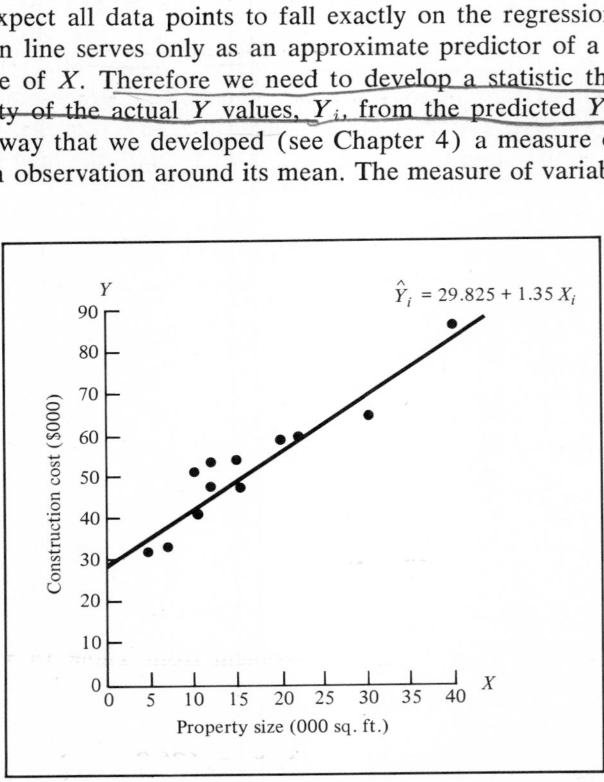

Figure 14.5 Scatter diagram and line of regression for construction cost problem.

the line of regression is called the **standard error of the estimate.** The variability around the line of regression is illustrated in Figure 14.5 for the construction cost problem.

We can see from Figure 14.5 that, although the predicted line of regression falls near many of the actual values of Y, there are values above the line of regression as well as below the line of regression so that

$$\sum_{i=1}^{n} (Y_i - \hat{Y}_i) = 0$$

The standard error of the estimate, given by the symbol S_{YX}, is defined as

$$S_{YX} = \sqrt{\frac{\sum_{i=1}^{n} (Y_i - \hat{Y}_i)^2}{n-2}} \qquad (14.7a)$$

where Y_i = actual value of Y for a given X_i

$\hat{Y}_i$ = predicted value of Y for a given X_i

The computation of the standard error of the estimate using Equation (14.7a) would require the calculation of the predicted value of Y for each X value in the sample. The calculation can be simplified, however, because of the following identity:

$$\sum_{i=1}^{n} (Y_i - \hat{Y}_i)^2 = \sum_{i=1}^{n} Y_i^2 - b_0 \sum_{i=1}^{n} Y_i - b_1 \sum_{i=1}^{n} X_i Y_i$$

The standard error of the estimate S_{YX} can thus be obtained using the following computational formula:

$$S_{YX} = \sqrt{\frac{\sum_{i=1}^{n} Y_i^2 - b_0 \sum_{i=1}^{n} Y_i - b_1 \sum_{i=1}^{n} X_i Y_i}{n-2}} \qquad (14.7b)$$

For the construction cost problem from Table 14.3 we have determined that:

$$\sum_{i=1}^{n} Y_i^2 = 34{,}878.58 \qquad \sum_{i=1}^{n} Y_i = 625.2 \qquad \sum_{i=1}^{n} X_i Y_i = 11{,}840.1$$

$$b_0 = 29.825 \qquad b_1 = +1.35$$

Therefore, using Equation (14.7*b*), the standard error of the estimate S_{YX} can be computed as

$$S_{YX} = \sqrt{\frac{\sum_{i=1}^{n} Y_i^2 - b_0 \sum_{i=1}^{n} Y_i - b_1 \sum_{i=1}^{n} X_i Y_i}{n - 2}}$$

$$= \sqrt{\frac{34{,}878.58 - (29.825)(625.2) - (+1.35)(11{,}840.1)}{12 - 2}}$$

$$= \sqrt{\frac{34{,}878.58 - 18{,}646.59 - 15{,}984.135}{10}} = \sqrt{\frac{247.855}{10}} = \sqrt{24.7855}$$

$S_{YX} = 4.9785$ (in thousands of dollars) or \$4,978.50

This standard error of the estimate, equal to \$4,978.50, represents a measure of the variation around the fitted line of regression. It is measured in units of the dependent variable Y. The interpretation of the standard error of the estimate, then, is analogous to that of the standard deviation. Just as the standard deviation measured variability around the arithmetic mean, the standard error of the estimate measures variability around the fitted line of regression. Moreover, as we shall see in Sections 14.8 and 14.9, the standard error of the estimate can be used in making inferences about a predicted value of Y and in determining whether a statistically significant relationship exists between the two variables.

Turning to the problem involving the relationship between high school average and grade-point index, we determined the following from Table 14.4:

$$\sum_{i=1}^{n} Y_i^2 = 131.3869 \qquad \sum_{i=1}^{n} Y_i = 44.03 \qquad \sum_{i=1}^{n} X_i Y_i = 3{,}706.13$$

$$b_0 = .2286 \qquad b_1 = +.0323$$

Using Equation (14.7*b*) the standard error of the estimate would be computed as

$$S_{YX} = \sqrt{\frac{\sum_{i=1}^{n} Y_i^2 - b_0 \sum_{i=1}^{n} Y_i - b_1 \sum_{i=1}^{n} X_i Y_i}{n - 2}}$$

$$= \sqrt{\frac{131.3869 - (.2286)(44.03) - (+.0323)(3{,}706.13)}{15 - 2}}$$

$$= \sqrt{\frac{1.613652}{13}} = 0.3523$$

14.6 MEASURES OF VARIATION IN REGRESSION AND CORRELATION

In order to examine how well the independent variable predicts the dependent variable in our statistical model, we need to develop several measures of

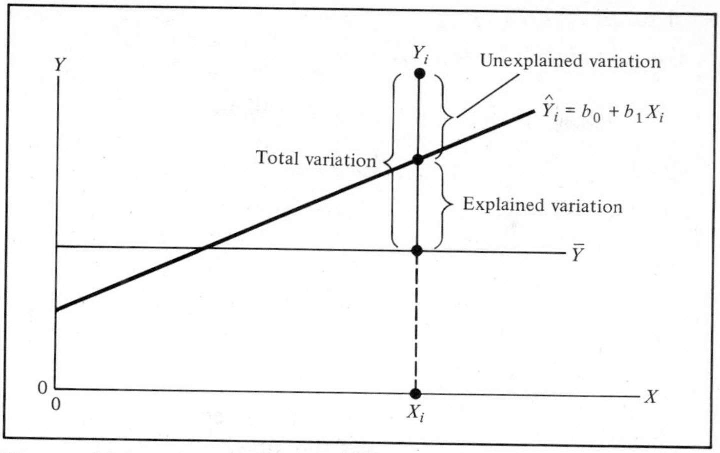

Figure 14.6 Measures of variation in regression.

variation. The first measure, the **total variation**, is a measure of variation of the Y values around their mean, $\bar{Y}$. As we have seen previously in Section 12.2, the total variation can be subdivided into two components. In a regression problem the total variation in Y, the dependent variable, can be subdivided into **explained variation,** that which is attributable to the relationship between X and Y, and **unexplained variation,** that which is attributable to factors other than the relationship between X and Y. These different measures of variation can be seen in Figure 14.6.

The explained variation represents the difference between $\bar{Y}$ (the average value of Y) and $\hat{Y}_i$ (the value of Y that would be predicted from the regression relationship). The unexplained variation represents that part of the variation in Y that is not explained by the regression and is based upon the difference between Y_i (the actual value of Y) and $\hat{Y}_i$ (the predicted value of Y for a given X). These measures of variation can be represented as follows:

$$\text{Total Variation} = \text{Explained Variation} + \text{Unexplained Variation} \tag{14.8}$$

where

$$\text{Total Variation} = \sum_{i=1}^{n} (Y_i - \bar{Y})^2 = \sum_{i=1}^{n} Y_i^2 - \frac{\left(\sum_{i=1}^{n} Y_i\right)^2}{n} \tag{14.9}$$

Unexplained Variation =

$$\sum_{i=1}^{n} (Y_i - \hat{Y}_i)^2 = \sum_{i=1}^{n} Y_i^2 - b_0 \sum_{i=1}^{n} Y_i - b_1 \sum_{i=1}^{n} X_i Y_i \qquad (14.10)$$

$$\text{Explained Variation} = \sum_{i=1}^{n} (\hat{Y}_i - \overline{Y})^2$$

$$= \text{Total Variation} - \text{Unexplained Variation}$$

$$= b_0 \sum_{i=1}^{n} Y_i + b_1 \sum_{i=1}^{n} X_i Y_i - \frac{\left(\sum_{i=1}^{n} Y_i\right)^2}{n} \qquad (14.11)$$

Examining the unexplained variation [Equation (14.10)], we may recall that $\sum_{i=1}^{n} (Y_i - \hat{Y}_i)^2$ was the numerator under the square root in the computation of the standard error of the estimate [see Equation (14.7*a*)]. Therefore, in the process of computing the standard error of the estimate, we have already computed the following unexplained variation:

$$\text{Unexplained Variation} = \sum_{i=1}^{n} Y_i^2 - b_0 \sum_{i=1}^{n} Y_i - b_1 \sum_{i=1}^{n} X_i Y_i$$

$$= 34{,}878.58 - (29.825)(625.2) - (+1.35)(11{,}840.1) = 247.855$$

Moreover,

$$\text{Total Variation} = \sum_{i=1}^{n} Y_i^2 - \frac{\left(\sum_{i=1}^{n} Y_i\right)^2}{n}$$

$$= 34{,}878.58 - \frac{(625.2)^2}{12} = 34{,}878.58 - 32{,}572.92$$

$$= 2{,}305.66$$

so that

$$\text{Explained Variation} = \text{Total Variation} - \text{Unexplained Variation}$$

$$= 2{,}305.66 - 247.855 = 2{,}057.805$$

Now the **coefficient of determination** r^2 can be defined as

$$r^2 = \frac{\text{Explained Variation}}{\text{Total Variation}} \qquad (14.12)$$

that is, the coefficient of determination measures the proportion of variation that is explained by the independent variable for the regression model. For the construction cost problem,

$$r^2 = \frac{2{,}057.805}{2{,}305.66} = .8925$$

This can be interpreted to mean that 89.25% of the variation in construction cost from house to house can be explained by the variability in the property size of houses. This is an example in which there is a very strong linear relationship between two variables since the use of a regression model has reduced the variability in predicting construction cost by 89.25%. Only 10.75% of the variability in construction cost can be explained by factors other than property size. On the other hand, from the regression model between high school average and grade-point index we obtain the following:

$$\text{Unexplained Variation} = \sum_{i=1}^{n} Y_i^2 - b_0 \sum_{i=1}^{n} Y_i - b_1 \sum_{i=1}^{n} X_i Y_i$$

$$= 131.3869 - (.2286)(44.03) - (+.0323)(3{,}706.13) = 1.613652$$

$$\text{Total Variation} = \sum_{i=1}^{n} Y_i^2 - \frac{\left(\sum_{i=1}^{n} Y_i\right)^2}{n}$$

$$= 131.3869 - \frac{(44.03)^2}{15} = 131.3869 - 129.24272$$

$$= 2.14418$$

$$\text{Explained Variation} = \text{Total Variation} - \text{Unexplained Variation}$$

$$= 2.14418 - 1.613652 = .530528$$

$$r^2 = \frac{\text{Explained Variation}}{\text{Total Variation}} = \frac{.530528}{2.14418} = .2474$$

We have seen in the above computations that the coefficient of determination has been computed as .2474. This result means that only 24.74% of the variation in grade-point index can be explained by a variation in the high school average of the students. This value of r^2 seems to indicate a moderate

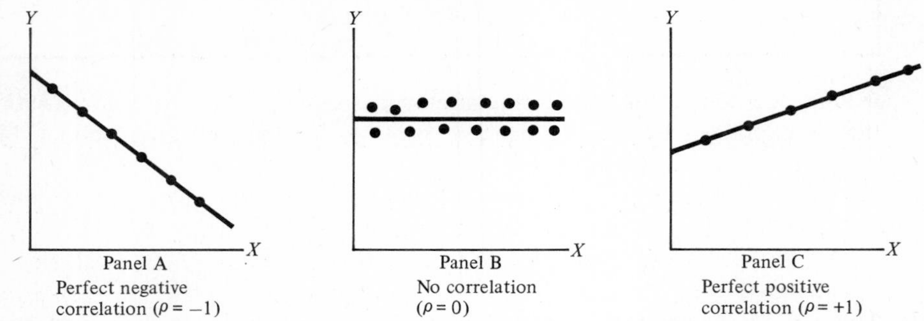

Figure 14.7 Types of association between variables.

linear relationship between the two variables since more than 75% of the variation in grade-point index is explained by factors other than high school average.

14.7 CORRELATION—MEASURING THE STRENGTH OF THE ASSOCIATION

In our discussion of the relationship betwen two variables thus far in this chapter we have been concerned with the prediction of the dependent variable Y based upon the independent variable X. On the other hand, as we have previously mentioned, correlation measures the degree of association between two variables. Figure 14.7 illustrates three different types of association between variables: perfect negative correlation, perfect positive correlation, and no correlation.

The strength of a relationship between two variables is usually measured by the coefficient of correlation ρ, whose values range from -1 for perfect negative correlation up to $+1$ for perfect positive correlation. Panel A of Figure 14.7 illustrates a perfect negative linear relationship between X and Y. Thus there is a perfect one-to-one relationship between X and Y so that Y will decrease in a perfectly predictable manner as X increases. Panel B of Figure 14.7 illustrates an example in which there is no relationship between X and Y. As X increases there is no change in Y so that there is no association between the value of X and the value of Y. On the other hand, panel C of Figure 14.7 depicts a perfect positive correlation between X and Y. In this case Y increases in a perfectly predictable manner as X increases.

For regression-oriented problems, the sample coefficient of correlation (r) may easily be obtained from Equation (14.12) as follows:

$$r^2 = \frac{\text{Explained Variation}}{\text{Total Variation}} \tag{14.12}$$

so that

$$r = \sqrt{r^2} \tag{14.13a}$$

In simple linear regression r takes the sign of b_1. If b_1 is positive, r is positive. If b_1 is negative, r is negative. If b_1 is zero, r is zero.

In the construction cost problem, since $r^2 = .8925$ and the slope b_1 is positive, the coefficient of correlation is computed as $+.945$. The closeness of the correlation coefficient to $+1.0$ implies a very strong association between construction cost and property size. On the other hand, in the student survey, since $r^2 = .2474$ and the slope b_1 is positive, the coefficient of correlation is computed as $+.497$. This correlation seems to indicate a moderate association between high school average and the grade-point index of seniors.

In our discussion of correlation in the previous two problems we have computed and interpreted the correlation coefficient in terms of its regression viewpoint. As we mentioned at the beginning of this chapter, regression and correlation are two separate techniques, with regression being concerned with prediction and correlation with association. In many applications, particularly in the social sciences, the researcher is only concerned with measuring association and covariation between variables, not in using one variable to predict another. This can be illustrated by reference to the survey conducted by the dean and the researcher. If, for example, they were interested in studying the association between the height (question 16) and weight (question 17) of females in our population, the problem would be one of correlation not regression. The objective of the correlation analysis would be to measure the strength of the association between height and weight—not to use one variable to predict values of the other.

Equation (14.12) gives the sample correlation coefficient in terms of the explained variation. This formula is particularly appropriate when correlation analysis is being done in conjunction with regression. However, if only correlation analysis is being performed on a set of data, the sample correlation coefficient r can be computed directly using the following formula:

$$r = \frac{\sum_{i=1}^{n} (X_i - \bar{X})(Y_i - \bar{Y})}{\sqrt{\sum_{i=1}^{n} (X_i - \bar{X})^2}\sqrt{\sum_{i=1}^{n} (Y_i - \bar{Y})^2}} \tag{14.13b}$$

or, alternatively,

$$r = \frac{\sum_{i=1}^{n} X_i Y_i - \frac{\left(\sum_{i=1}^{n} X_i\right)\left(\sum_{i=1}^{n} Y_i\right)}{n}}{\sqrt{\sum_{i=1}^{n} X_i^2 - \frac{\left(\sum_{i=1}^{n} X_i\right)^2}{n}}\sqrt{\sum_{i=1}^{n} Y_i^2 - \frac{\left(\sum_{i=1}^{n} Y_i\right)^2}{n}}} \qquad (14.13c)$$

The heights and weights of the 43 females in the survey are shown in Table 14.5.

For the data of Table 14.5 we can compute the following values:

$$\sum_{i=1}^{n} X_i = 2{,}744 \qquad \sum_{i=1}^{n} X_i^2 = 175{,}502 \qquad \sum_{i=1}^{n} Y_i = 5{,}261$$

$$n = 43 \qquad \sum_{i=1}^{n} Y_i^2 = 655{,}227 \qquad \sum_{i=1}^{n} X_i Y_i = 336{,}642$$

Using Equation (14.13*c*), we obtain

$$r = \frac{\sum_{i=1}^{n} X_i Y_i - \frac{\left(\sum_{i=1}^{n} X_i\right)\left(\sum_{i=1}^{n} Y_i\right)}{n}}{\sqrt{\sum_{i=1}^{n} X_i^2 - \frac{\left(\sum_{i=1}^{n} X_i\right)^2}{n}}\sqrt{\sum_{i=1}^{n} Y_i^2 - \frac{\left(\sum_{i=1}^{n} Y_i\right)^2}{n}}}$$

$$= \frac{336{,}642 - \frac{(2{,}744)(5{,}261)}{43}}{\sqrt{175{,}502 - \frac{(2{,}744)^2}{43}}\sqrt{655{,}227 - \frac{(5{,}261)^2}{43}}}$$

$$= \frac{336{,}642 - 335{,}725.2}{\sqrt{175{,}502 - 175{,}105.48}\ \sqrt{655{,}227 - 643{,}677.23}}$$

$$= \frac{+916.8}{\sqrt{396.52}\ \sqrt{11{,}549.77}} = \frac{+916.8}{(19.91281)(107.46985)} = +.4284$$

The coefficient of correlation between the height and weight of females in the sample indicates a moderate positive association, that is, increased height is somewhat associated with increased weight. In Section 14.9 we shall use these sample results to determine whether there is any evidence of association between these variables in the population.

Table 14.5
HEIGHT AND WEIGHT OF THE 43 FEMALES OBTAINED FROM A RANDOM SAMPLE OF 94 STUDENTS

Student	Height (inches)	Weight (pounds)
1	60	115
2	62	98
3	61	115
4	69	125
5	67	131
6	63	162
7	69	140
8	65	103
9	61	95
10	60	125
11	68	132
12	65	120
13	61	145
14	65	127
15	66	114
16	60	115
17	61	110
18	64	115
19	63	160
20	63	125
21	61	110
22	67	135
23	61	105
24	61	105
25	69	118
26	60	105
27	65	140
28	70	145
29	65	123
30	66	121
31	62	120
32	64	105
33	60	110
34	63	117
35	71	135
36	64	108
37	63	115
38	62	115
39	67	144
40	65	142
41	60	93
42	63	133
43	62	145

SOURCE: Data are taken from Figure 2.6.

14.8 CONFIDENCE INTERVAL ESTIMATES FOR PREDICTING μ_{YX}

In our discussion of regression and correlation analysis in the previous sections of this chapter we have been concerned with the use of these methods solely for the purpose of description. The least-squares method has been utilized to determine the regression coefficients and to predict the value of Y from a given value of X. In addition, the standard error of the estimate has been discussed along with the coefficients of correlation and determination. Nevertheless, since we are usually sampling from large populations, we are often concerned with making inferences about the relationship between the variables in the entire population based upon our sample results. In this section, then, we will discuss methods of making predictive inferences about the mean of Y, while in the following section we will discuss techniques that are used to test hypotheses about the existence of relationships in the population.

We may recall that in Section 14.4 the fitted regression equation was used to make predictions about the value of Y for a given X. In the construction cost problem, for example, we predicted that the average cost of building houses with property sizes of 15,000 square feet would be \$50,075, while in the student survey problem we predicted a grade-point index of 2.9741 for students who had a high school average of 85. These estimates, however, are merely point estimates of the true average value since they are based upon sample results. In Chapter 8 we developed the concept of the confidence interval as an estimate of the true population value. In a similar fashion, a confidence interval estimate can now be developed to make inferences about the predicted value of Y.

Confidence Interval Estimate of the Mean of $Y(\mu_{YX})$ for a Given X

$$\hat{Y}_i \pm t_{n-2} S_{YX} \sqrt{\frac{1}{n} + \frac{(X_i - \bar{X})^2}{\sum_{i=1}^{n} X_i^2 - \frac{\left(\sum_{i=1}^{n} X_i\right)^2}{n}}} \tag{14.14}$$

where $\hat{Y}_i$ = predicted value of Y; $\hat{Y}_i = b_0 + b_1 X_i$

S_{YX} = standard error of the estimate

n = sample size

X_i = given value of X

An examination of Equation (14.14) indicates that the width of the confidence interval is dependent on several factors. For a given level of confidence, increased variation around the line of regression, as measured by the standard error of the estimate, results in a wider interval. However, as would be expected, increased sample size reduces the width of the interval. Moreover, the width of the interval also varies at different values of X. When predicting Y for values of X close to $\bar{X}$, the interval is much narrower than for

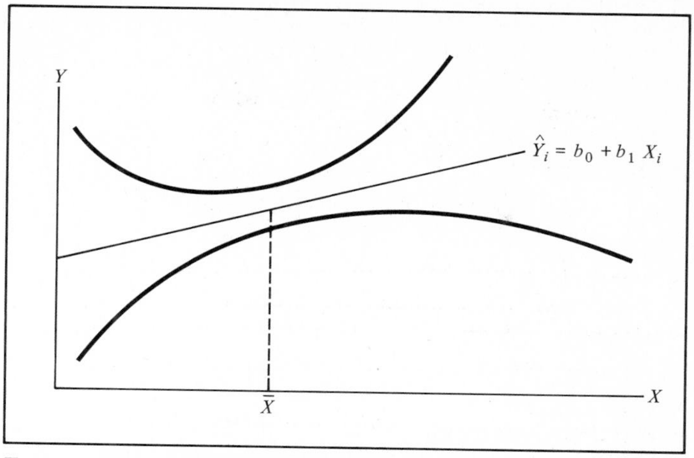

Figure 14.8 Interval estimates of μ_{YX} for different values of X.

predictions for X values more distant from the mean. This effect can be seen from the square-root portion of Equation (14.14) and from Figure 14.8.

The interval estimate of the true mean of Y varies *hyperbolically* as a function of the closeness of the given X to $\bar{X}$. When predictions are to be made for X values which are distant from the average value of X, the much wider interval is the trade-off for predicting at such values of X. Thus, as depicted in Figure 14.8, we observe a *confidence band effect* for the predictions.

Now that Equation (14.14) has been presented, confidence interval estimates can be developed for the two problems that we have been discussing. For example, in the construction cost problem, if we desire a 95% confidence interval estimate of the true average cost for houses with property of 15,000 square feet, we compute the following:

$$\hat{Y}_i = 29.825 + 1.35X_i$$

and for $X_i = 15$, $\hat{Y}_i = 50.075$. Also, $\bar{X} = 16.5$, $S_{YX} = 4.9785$,

$\sum_{i=1}^{n} X_i = 198$, $\sum_{i=1}^{n} X_i^2 = 4{,}396$; from Appendix E, Table E.3, $t_{10} = 2.228$.

Thus,

$$\hat{Y}_i \pm t_{n-2} S_{YX} \sqrt{\frac{1}{n} + \frac{(X_i - \bar{X})^2}{\sum_{i=1}^{n} X_i^2 - \frac{\left(\sum_{i=1}^{n} X_i\right)^2}{n}}}$$

and

$$50.075 \pm (2.228)(4.9785)\sqrt{\frac{1}{12} + \frac{(15 - 16.5)^2}{4{,}396 - \frac{(198)^2}{12}}}$$

$$= 50.075 \pm (11.0921)\sqrt{\frac{1}{12} + \frac{(-1.5)^2}{4{,}396 - 3{,}267}}$$

$$= 50.075 \pm (11.0921)\sqrt{.0833 + \frac{2.25}{1{,}129}}$$

$$= 50.075 \pm (11.0921)\sqrt{.0853}$$

$$= 50.075 \pm (11.0921)(.2921) = 50.075 \pm 3.24$$

$$46.835 \leq \mu_{YX} \leq 53.315$$

so that $\$46{,}835 \leq \mu_{YX} \leq \$53{,}315$

Therefore our estimate is that the true average construction cost is between \$46,835 and \$53,315 for houses with property size of 15,000 square feet. We have 95% confidence that this interval correctly estimates the true construction cost for this type of house.

Turning to the student survey example, a 95% confidence interval estimate of the average predicted grade-point index can be developed for students with high school averages of 85. From Sections 14.4 and 14.5 we have $\hat{Y}_i = .2286 + .0323X_i$, and for $X_i = 85$, $\hat{Y}_i = 2.9741$. Also, $\bar{X} = 83.8$, $S_{YX} = .3523$, $\sum_{i=1}^{n} X_i = 1{,}257$, $\sum_{i=1}^{n} X_i^2 = 105{,}845$, and from Appendix E. Table E.3, $t_{13} = 2.16$. Thus,

$$\hat{Y}_i \pm t_{n-2}S_{YX}\sqrt{\frac{1}{n} + \frac{(X_i - \bar{X})^2}{\sum_{i=1}^{n} X_i^2 - \frac{\left(\sum_{i=1}^{n} X_i\right)^2}{n}}}$$

and

$$2.9741 \pm (2.16)(.3523)\sqrt{\frac{1}{15} + \frac{(85 - 83.8)^2}{105{,}845 - \frac{(1{,}257)^2}{15}}}$$

$$= 2.9741 \pm .761\sqrt{.066 + \frac{1.44}{105{,}845 - 105{,}336.6}}$$

$$= 2.9741 \pm .761\sqrt{.0695} = 2.9741 \pm .2006$$

$$2.7735 \leq \mu_{YX} \leq 3.1747$$

Therefore we estimate that the true average grade-point index of students whose high school average was 85 is between 2.7735 and 3.1747. We have

95% confidence that this interval correctly estimates the true grade-point index under these circumstances.

14.9 INFERENCES ABOUT THE POPULATION PARAMETERS IN REGRESSION AND CORRELATION

In the previous section we used statistical inference to develop a confidence interval estimate for μ_{YX}, the true mean value of Y. In this section, statistical inference will be used to draw conclusions about the population slope β_1 and the population correlation coefficient ρ.

We can determine whether a significant relationship between the variables X and Y exists by testing whether β_1 (the true slope) is equal to zero. If this hypothesis is rejected, one could conclude that there is evidence of a linear relationship. Thus the null and alternative hypotheses could be stated as follows:

$$H_0: \quad \beta_1 = 0 \qquad \text{(no relationship)}$$

$$H_1: \quad \beta_1 \neq 0 \qquad \text{(there is a relationship)}$$

and the test statistic for this is given by

$$t_{n-2} = \frac{b_1}{S_{b_1}} \tag{14.15}$$

where $S_{b_1} = \dfrac{S_{YX}}{\sqrt{\displaystyle\sum_{i=1}^{n} X_i^2 - \frac{\left(\sum_{i=1}^{n} X_i\right)^2}{n}}}$

Turning to our construction cost problem, let us now test whether the sample results enable us to conclude that a significant relationship between construction cost and property size exists at the .01 level of significance. The results from Sections 14.4 and 14.5 gave the following information:

$b_1 = +1.35$, $n = 12$, $S_{YX} = 4.9785$, $\sum_{i=1}^{n} X_i = 198$, and $\sum_{i=1}^{n} X_i^2 = 4{,}396$.

Therefore, to test the existence of a relationship at the .01 level of significance we have (see Figure 14.9):

$$S_{b_1} = \frac{S_{YX}}{\sqrt{\displaystyle\sum_{i=1}^{n} X_i^2 - \frac{\left(\sum_{i=1}^{n} X_i\right)^2}{n}}} = \frac{4.9785}{\sqrt{4{,}396 - \frac{(198)^2}{12}}} = \frac{4.9785}{\sqrt{1{,}129}} = .1482$$

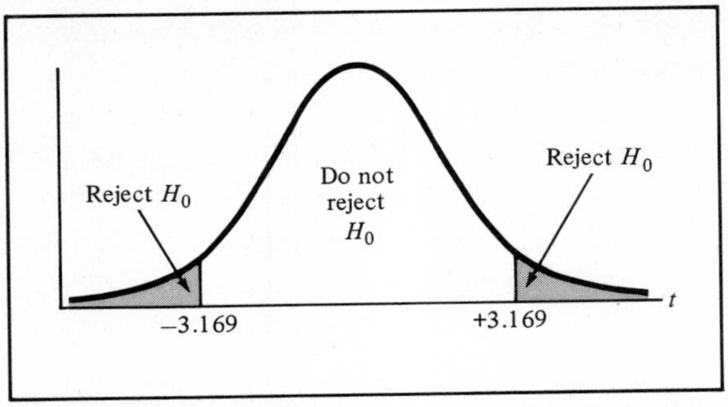

Figure 14.9 Testing a hypothesis about the population slope at the 1% level of significance with 10 degrees of freedom.

and

$$t_{n-2} = \frac{b_1}{S_{b_1}}$$

$$t_{10} = \frac{1.35}{.1482} = 9.112$$

Now since $9.112 > 3.169$, reject H_0.

Hence the conclusion reached is to reject the null hypothesis and state that there is a significant relationship between construction cost and property size.

A second, equivalent method for testing the existence of a linear relationship between the variables is to set up a confidence interval estimate of β_1 and to determine whether the hypothesized value ($\beta_1 = 0$) is included in the interval. The confidence interval estimate of β_1 would be obtained by using the following formula:

$$b_1 \pm t_{n-2} S_{b_1} \qquad (14.16)$$

If a 99% confidence interval estimate was desired, we would have $b_1 = +1.35$, $t_{10} = 3.169$, and $S_{b_1} = 0.1482$. Thus,

$$b_1 \pm t_{n-2} S_{b_1} = +1.35 \pm (3.169)(.1482) = +1.35 \pm .469$$

$$+.881 \le \beta_1 \le +1.819$$

From Equation (14.16) the true slope is estimated with 99% confidence to be between +.881 and +1.819. Since these values are clearly above zero, we can conclude that there is a significant positive relationship between construction cost and property size. On the other hand, had the interval included zero, no relationship would have been determined.

Hypothesis testing and confidence intervals can also be applied to our second problem, the dean's study, to determine whether one can conclude, based upon the sample results, that there is a relationship between grade-point index of seniors and high-school average. For these data, from Sections 14.4 and 14.5, we have $b_1 = +.0323$, $n = 15$, $S_{YX} = .3523$, $\sum_{i=1}^{n} X_i = 1{,}257$, and $\sum_{i=1}^{n} X_i^2 = 105{,}845$. To test the existence of a relationship at the .01 level of significance, we have (see Figure 14.10)

$$S_{b_1} = \frac{S_{YX}}{\sqrt{\sum_{i=1}^{n} X_i^2 - \frac{\left(\sum_{i=1}^{n} X_i\right)^2}{n}}}$$

$$= \frac{.3523}{\sqrt{105{,}845 - \frac{(1{,}257)^2}{15}}} = \frac{.3523}{508.4} = .0156$$

and we obtain

$$t_{n-2} = \frac{b_1}{S_{b_1}}$$

$$t_{13} = \frac{+.0323}{.0156} = 2.07$$

But since $2.07 < 3.012$, do not reject H_0.

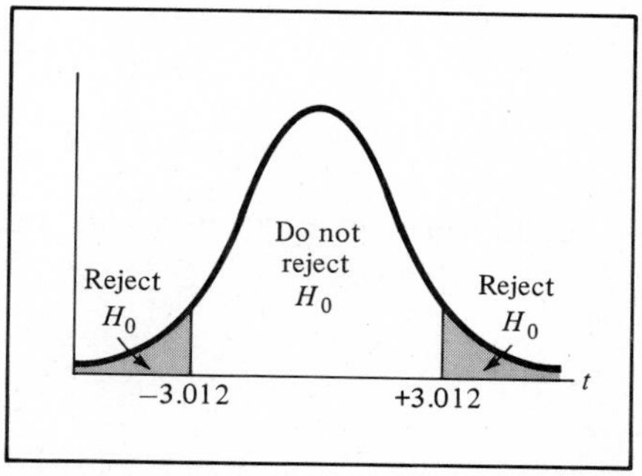

Figure 14.10 Testing a hypothesis about the population slope at the 1% level of significance with 13 degrees of freedom.

Therefore since the null hypothesis cannot be rejected, the conclusion can be drawn that there is no evidence of a relationship between grade-point index of seniors and high school average.

A similar conclusion can be reached in an alternate manner by developing a 99% confidence interval estimate of the true population slope. Using Equation (14.16), we have

$$b_1 \pm t_{n-2} S_{b_1}$$

When $b_1 = +.0323$, $t_{13} = 3.012$, and $S_{b_1} = .0156$, we have

$$b_1 \pm t_{n-2} S_{b_1} = +.0323 \pm (3.012)(.0156) = +.0323 \pm .047$$

$$-.0147 \leq \beta_1 \leq .0793$$

Since this interval includes a slope of zero, we can conclude that there is no evidence of a relationship between the grade-point index of seniors and their high school average.

A third method for examining the existence of a linear relationship between two variables involves the sample correlation coefficient r. The existence of a relationship between X and Y, which was tested using Equation (14.15), could be tested in terms of the correlation coefficient with equivalent results. Testing for the existence of a linear relationship between two variables is the same as determining whether there is any significant correlation between them. The population correlation coefficient ρ is hypothesized as equal to zero. Thus the null and alternative hypotheses would be

H_0: $\rho = 0$ (no correlation)

H_1: $\rho \neq 0$ (there is correlation)

The test statistic for determining the existence of correlation is given by Equation (14.17):

$$t_{n-2} = \frac{r}{\sqrt{\dfrac{1 - r^2}{n - 2}}} \qquad (14.17)$$

In order to demonstrate that this statistic produces the same result as the test for the existence of a slope [Equation (14.15)], we will use the data from the construction cost problem. For these data $r = +.945$, $r^2 = .8925$, and n = 12 and we have

$$t_{n-2} = \frac{r}{\sqrt{\dfrac{1 - r^2}{n - 2}}}$$

$$t_{10} = \frac{+.945}{\sqrt{\dfrac{1 - .8925}{12 - 2}}} = +9.112$$

We may note that this t value is exactly the same as that obtained by using Equation (14.15). Therefore, in a linear regression problem Equations (14.15) and (14.17) give equivalent alternative ways of determining the existence of a relationship between two variables. However, if the sole purpose of a particular study is to determine the existence of correlation, then Equation (14.17) is more appropriate. For instance, in Section 14.7 we studied the association of the heights and weights of females in the sample of students. Had we wanted to determine the significance of the correlation between height and weight we could have used Equation (14.17) as follows:

H_0: $\rho = 0$ (no correlation)

H_1: $\rho \neq 0$ (there is correlation)

If a level of significance of .05 was selected, we would have (see Figure 14.11)

$$t_{n-2} = \frac{r}{\sqrt{\dfrac{1 - r^2}{n - 2}}}$$

$$t_{41} = \frac{+.497}{\sqrt{\dfrac{1 - (.497)^2}{43 - 2}}} = \frac{+.497}{\sqrt{\dfrac{1 - .247}{41}}} = \frac{+.497}{.1355} = 3.668$$

But, $3.668 > 2.02$, so reject H_0.

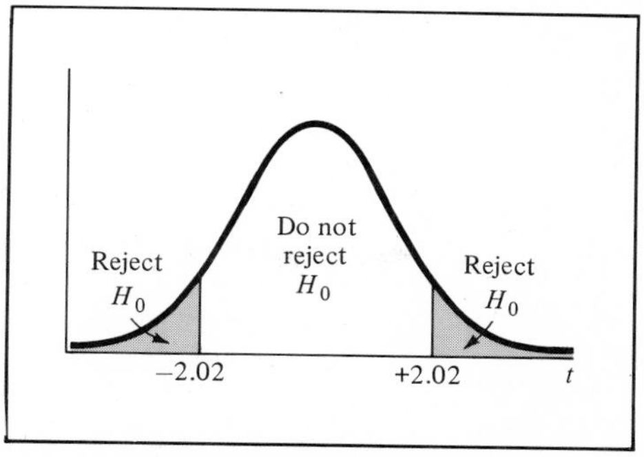

Figure 14.11 Testing for the existence of correlation at the 5% level of significance with 41 degrees of freedom.

Since the null hypothesis has been rejected, the researcher would conclude that there is a significant association between the heights and weights of the female students.

When inferences concerning the population slope were discussed, confidence intervals and tests of hypothesis were used interchangeably. However, when examining the correlation coefficient, the development of a confidence interval becomes more complicated because the shape of the sampling distribution of the statistic r varies for different values of the true correlation coefficient. Methods for developing a confidence interval estimate for the correlation coefficient are presented in References 1, 2, 8, and 9.

14.10 ASSUMPTIONS OF REGRESSION AND CORRELATION

In our investigations into hypothesis testing and the analysis of variance we have noted that the appropriate application of a particular statistical procedure is dependent on how well a set of assumptions for that procedure are met. The assumptions necessary for regression and correlation analysis are analogous to those of the analysis of variance since they fall under the general heading of "linear models" (Reference 8). Although there are some differences in the assumptions made by the regression model and by correlation (see References 3 and 8), this topic is beyond the scope of this text and we will consider only the former.

The three major assumptions of regression are

1. Normality
2. Homoscedasticity
3. Independence of error

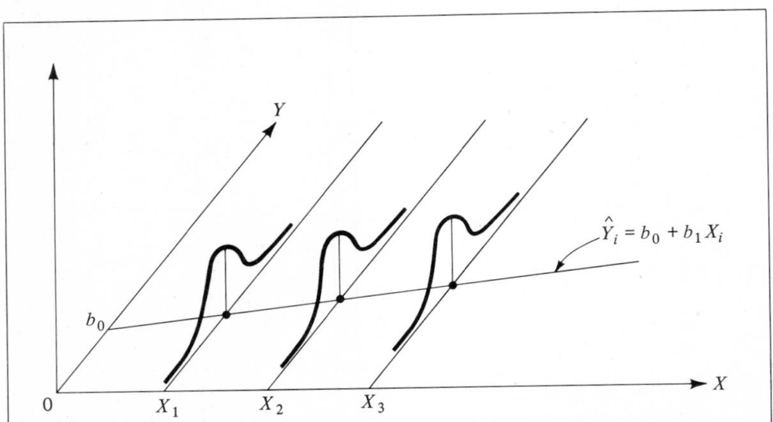

Figure 14.12 Assumptions of regression.

The first assumption, **normality,** requires that the value of Y be normally distributed at each value of X (see Figure 14.12). Like the t test and the analysis of variance F test, regression analysis is "robust" against departures from the normality assumption, that is, as long as the distribution of Y values around each X is not extremely different from a normal distribution, inferences about the line of regression and the regression coefficients will not be seriously affected.

The second assumption, **homoscedasticity,** requires that the variation around the line of regression be constant for all values of X. This means then that Y varies the same amount when X is a low value as when X is a high value (see Figure 14.12). The homoscedasticity assumption is im-

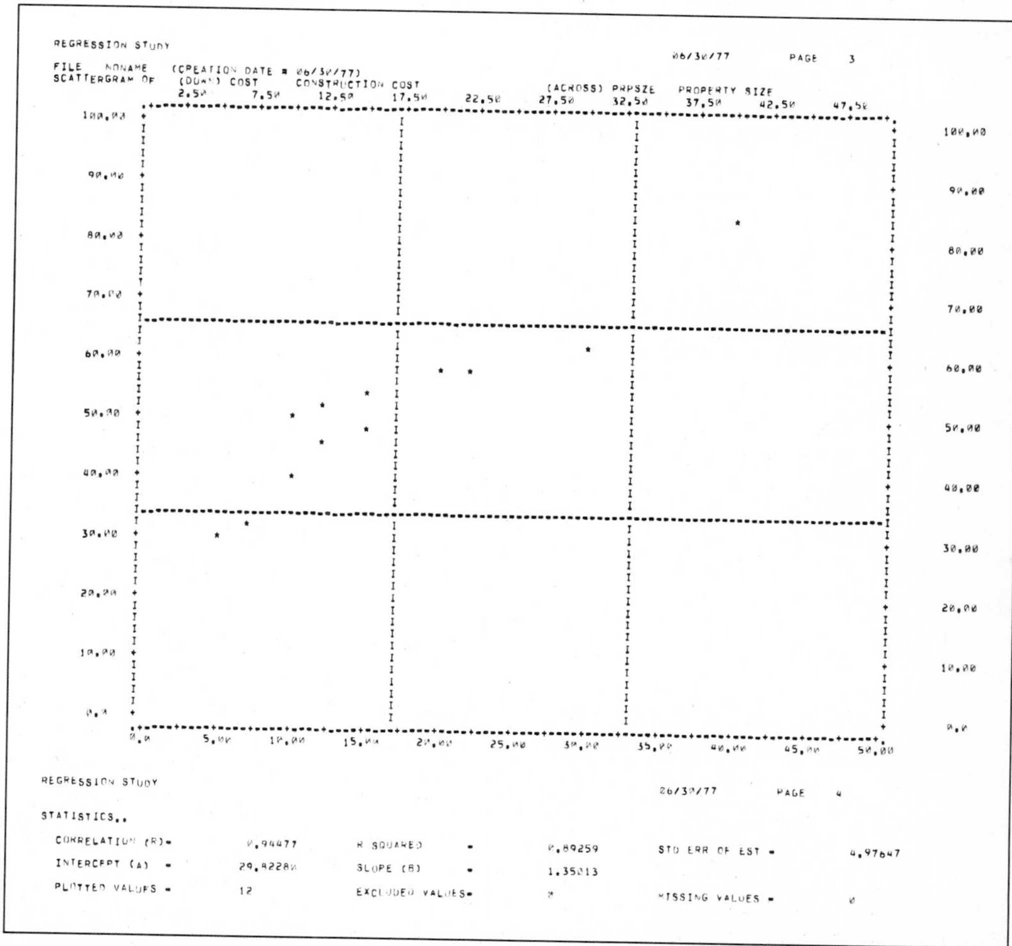

Figure 14.13 Partial output for the data of Table 14.1 from SPSS subprogram SCATTERGRAM.

portant for using the least-squares method of determining the regression coefficients. If there are serious departures from this assumption, either data transformations or weighted least-squares methods (References 3 and 8) can be applied.

The third assumption, **independence of error,** requires that the error ("residual" difference between an observed and predicted value of Y) should be independent for each value of X. This assumption often refers to data that are collected over a period of time. For example, in economic data the values for a particular time period are correlated with the values of the previous time period. These types of models fall under the general heading of time series and will be considered in Chapter 17.

14.11 USE OF COMPUTER PACKAGES IN SIMPLE LINEAR REGRESSION AND CORRELATION

In previous chapters, we have briefly considered how computers could be used to ease computational complexity. The role of the computer becomes even more important when applied to regression and correlation analysis and in particular to problems in multiple regression which will be discussed in Chapter 15. Nevertheless, for simple linear regression problems the computer is important, especially when large sample sizes or unwieldy data values are involved. Hence, in order to illustrate the use of the computer in regression and correlation analysis the SCATTERGRAM subprogram of SPSS was applied to the data of Table 14.1 (see Figure 14.13) so that we may see what kind of information can be obtained.

PROBLEMS

* 14.1 The marketing manager of a large supermarket chain would like to determine the effect of shelf space on the sales of pet food. A random sample of 12 equal-sized stores was selected and the results were presented in Table 14.6.

Table 14.6
RELATING SALES WITH SHELF SPACE

Store	Shelf Space X (feet)	Weekly Sales Y (hundreds of dollars)
1	5	1.6
2	5	2.2
3	5	1.4
4	10	1.9
5	10	2.4
6	10	2.6
7	15	2.3
8	15	2.7
9	15	2.8
10	20	2.6
11	20	2.9
12	20	3.1

(a) Set up a scatter diagram.

(b) Assuming a linear relationship, use the least-squares method to compute the regression coefficients b_0 and b_1.

(c) Interpret the meaning of the slope b_1 in this problem.

(d) Predict the weekly sales (in hundreds of dollars) of pet food for a store with 8 feet of shelf space for pet food.

(e) Compute the standard error of the estimate.

(f) Compute the coefficient of determination r^2 and interpret its meaning in this problem.

(g) Compute the coefficient of correlation r.

(h) Set up a 90% confidence interval estimate of the average weekly sales for a store that has 8 feet of shelf space for pet food.

(i) At the .10 level of significance is there a linear relationship between shelf space and sales?

14.2 A personnel manager for a large corporation feels that there may be a relationship between absenteeism and age, and would like to use the age of a worker to predict the number of days absent during a calendar year. A random sample of 10 workers was selected with the results given in Table 14.7.

(a) Set up a scatter diagram.

(b) Assuming a linear relationship, use the least-squares method to find the regression coefficients b_0 and b_1.

(c) Interpret the meaning of the slope b_1 in this problem.

(d) How many days (on the average) would you predict that a 40-year-old worker would be absent?

(e) Compute the standard error of the estimate.

(f) Compute the coefficient of determination r^2 and interpret its meaning in this problem.

(g) Compute the coefficient of correlation r.

(h) Set up a 99% confidence interval estimate of the average number of days absent for a 40-year-old worker.

(i) At the .01 level of significance is there a linear relationship between age and absenteeism?

Table 14.7
RELATING ABSENTEEISM WITH AGE

Worker	Age X (years)	Days Absent Y
1	27	15
2	61	6
3	37	10
4	23	18
5	46	9
6	58	7
7	29	14
8	36	11
9	64	5
10	40	8

Table 14.8
RELATING DELIVERY TIME WITH OPTIONS ORDERED

Car	Number of Options Ordered X	Delivery Time Y (days)
1	3	25
2	4	32
3	4	26
4	7	38
5	7	34
6	8	41
7	9	39
8	11	46
9	12	44
10	12	51
11	14	53
12	16	58
13	17	61
14	20	64
15	23	66
16	25	70

* 14.3 A statistician for a large American automobile manufacturer would like to develop a statistical model for predicting delivery time (the days between the ordering of the car and the actual delivery of the car) of custom-ordered new automobiles. The statistician believes that there is a **linear** relationship between the number of options ordered on the car and delivery time. A random sample of 16 is selected with the results given in Table 14.8.

(a) Set up a scatter diagram.

(b) Use the least-squares method to find the regression coefficients b_0 and b_1.

(c) Interpret the meaning of the Y intercept b_0 and the slope b_1 in this problem.

(d) If a car was ordered that had 16 options, how many days would you predict it would take to be delivered?

(e) Compute the standard error of the estimate.

(f) Compute the coefficient of determination r^2 and interpret its meaning in this problem.

(g) Compute the coefficient of correlation r.

(h) Set up a 95% confidence interval estimate of the average delivery time for a car ordered with 16 options.

(i) At the .05 level of significance is there a linear relationship between number of options and delivery time?

(j) Set up a 95% confidence interval estimate of the true slope.

14.4 At the beginning of a statistics course a diagnostic mathematics quiz is administered to all students. We wish to determine whether there is a relationship between quiz score and final exam score. A sample of 13 students is selected with the results given in Table 14.9.

(a) Set up a scatter diagram.

(b) Assuming a linear relationship, use the least-squares method to find the regression coefficients b_0 and b_1.

Table 14.9
RELATING FINAL EXAM RESULTS TO QUIZ RESULTS

Student	Mathematics Quiz Score X	Final Exam Score Y
1	16	92
2	20	100
3	16	77
4	17	84
5	19	82
6	12	89
7	14	73
8	11	70
9	15	63
10	15	68
11	15	49
12	17	87
13	13	36

(c) What would you predict the final exam score to be for a student whose mathematics quiz score was 15?

(d) Compute the standard error of the estimate.

(e) Compute the coefficient of determination r^2 and interpret its meaning in this problem.

(f) Compute the coefficient of correlation r.

(g) Set up a 90% confidence interval estimate of the average final exam score for students that score 15 on the mathematics quiz.

(h) At the .10 level of significance is there a linear relationship between mathematics quiz and final exam scores?

14.5 An official of a local racetrack would like to forecast the amount of money bet (in millions of dollars) based on attendance. A random sample of 10 days is selected with the results given in Table 14.10. **Hint:** Determine which are the independent and dependent variables.

(a) Set up a scatter diagram.

(b) Assuming a linear relationship, use the least-squares method to find the regression coefficients b_0 and b_1.

(c) Interpret the meaning of the slope b_1 in this problem.

(d) Predict the amount bet for a day on which attendance is 20,000.

Table 14.10
RELATING BETTING WITH ATTENDANCE

Day	Attendance (thousands)	Amount Bet (millions of dollars)
1	14.5	.70
2	21.2	.83
3	11.6	.62
4	31.7	1.10
5	46.8	1.27
6	31.4	1.02
7	40.0	1.15
8	21.0	.80
9	16.3	.71
10	32.1	1.04

(e) Compute the standard error of the estimate.

(f) Compute the coefficient of determination r^2 and interpret its meaning in this problem.

(g) Compute the coefficient of correlation r.

(h) Set up a 99% confidence interval estimate of the average amount of money bet when attendance is 20,000.

(i) At the .01 level of significance is there a linear relationship between the amount of money bet and attendance?

(j) Set up a 99% confidence interval estimate of the true slope.

(k) Discuss why you should not predict the amount bet on a day on which the attendance exceeded 46,800 or was below 11,600.

14.6 The controller of a large department store chain would like to predict the account balance at the end of a billing period based upon the number of transactions made during the billing period. A random sample of 12 accounts was selected with the results given in Table 14.11. **Hint:** Determine which are the independent and dependent variables.

Table 14.11
RELATING ACCOUNT BALANCE WITH TRANSACTIONS

Account	Number of Transactions	Account Balance (dollars)
1	1	$ 15
2	2	36
3	3	40
4	3	63
5	4	69
6	5	78
7	6	84
8	7	100
9	10	175
10	10	120
11	12	150
12	15	198

(a) Set up a scatter diagram.

(b) Assuming a linear relationship, use the least-squares method to find the regression coefficients b_0 and b_1.

(c) Interpret the meaning of the slope b_1 in the problem.

(d) Predict the account balance for an account which has had five transactions in the last billing period.

(e) Compute the standard error of the estimate.

(f) Compute the coefficient of determination r^2 and interpret its meaning in this problem.

(g) Compute the coefficient of correlation r.

(h) Set up a 95% confidence interval estimate of the average account balance for an account in which there have been five transactions in the last billing period.

(i) At the .05 level of significance is there a linear relationship between the number of transactions and account balance?

(j) Set up a 95% confidence interval estimate of the true slope.

Table 14.12
RELATING YIELD WITH LEVEL OF FERTILIZER USED

Plot	Amount of Fertilizer X (pounds per 100 square feet)	Yield Y (pounds)
1	0	6
2	0	8
3	10	11
4	10	14
5	20	18
6	20	23
7	30	25
8	30	28
9	40	30
10	40	34

* 14.7 An agronomist would like to determine the effect of a natural organic fertilizer on the yield of tomatoes. Five differing amounts of fertilizer are to be used on 10 equivalent plots of land: 0, 10, 20, 30, and 40 pounds per 100 square feet. The levels of fertilizer are randomly assigned to the plots of land with the results given in Table 14.12.

(a) Set up a scatter diagram.

(b) Assuming a linear relationship, use the least-squares method to find the regression coefficients b_0 and b_1.

(c) Interpret the meaning of the Y intercept b_0 and the slope b_1 in this problem.

(d) Predict the yield of tomatoes for a plot that has been given 15 pounds per 100 square feet of natural organic fertilizer.

(e) Compute the standard error of the estimate.

(f) Compute the coefficient of determination r^2 and interpret its meaning in this problem.

(g) Compute the coefficient of correlation r.

(h) Set up a 90% confidence interval estimate of the average yield for tomatoes that have been fertilized with 15 pounds per 100 square feet of natural organic fertilizer.

(i) At the .10 level of significance is there a linear relationship between the amount of fertilizer used and the yield of tomatoes?

(j) Set up a 90% confidence interval estimate of the true slope.

14.8 The owner of a large chain of ice-cream stores would like to study the effect of atmospheric temperature on sales during the summer season. A random sample of 14 days is selected with the results given in Table 14.13.

(a) Set up a scatter diagram.

(b) Assuming a linear relationship, use the least-squares method to compute the regression coefficients b_0 and b_1.

(c) Interpret the meaning of the slope b_1 in this problem.

(d) Predict the sales per store for a day in which the temperature is 83°F.

(e) Compute the standard error of the estimate.

(f) Compute the coefficient of determination r^2 and interpret its meaning in this problem.

(g) Compute the coefficient of correlation r.

Table 14.13
RELATING SALES TO TEMPERATURE

Day	Temperature (°F)	Sales per Store (thousands of dollars)
1	63	1.52
2	70	1.68
3	73	1.80
4	75	2.05
5	80	2.36
6	82	2.25
7	85	2.68
8	88	2.90
9	90	3.14
10	91	3.06
11	92	3.24
12	75	1.92
13	98	3.40
14	100	3.28

(h) Set up a 95% confidence interval estimate of the average sales per store for a day in which the temperature is 83°F.

(i) At the .05 level of significance is there a linear relationship between temperature and sales?

(j) Set up a 95% confidence interval estimate of the true slope.

(k) Discuss how different your results might be if the model had been based upon temperature measured according to the Celsius (°C) scale.

14.9 An auditor for a local county government would like to study the relationship between county taxes and the age of single-family homes in the county. A random sample of 19 single-family homes is selected with the results given in Table 14.14.

(a) Set up a scatter diagram.

(b) Assuming a linear relationship, use the least-squares method to compute the regression coefficients b_0 and b_1.

Table 14.14
RELATING TAXES WITH AGE OF HOMES

Age X (years)	County Taxes Y (dollars)
1	925
2	870
4	809
4	720
5	694
8	630
10	626
10	562
12	546
15	523
20	480
22	486
25	462
25	441
30	426
35	368
40	350
50	348
50	322

(c) Predict the county taxes for a single-family house that is 16 years old.

(d) Compute the standard error of the estimate.

(e) Compute the coefficient of determination r^2 and interpret its meaning in this problem.

(f) Compute the coefficient of correlation r.

(g) Set up a 95% confidence interval estimate of the average county taxes for a house that is 16 years old.

(h) At the .05 level of significance is there a linear relationship between the age of a house and county taxes?

(i) Discuss why the auditor should not try to predict the county taxes for a 100-year-old house based upon this model.

14.10 An industrial psychologist would like to predict the number of typing errors based upon the amount of alcohol consumed by a group of typists. 15 typists were randomly assigned to one of three consumption levels (0 ounces, 1 ounce, or 2 ounces). Each typist was instructed to type a standard page of print. The number of errors made was recorded with the results given in Table 14.15. **Hint:** First determine the dependent and independent variables.

(a) Assuming a linear relationship, use the least-squares method to find the regression coefficients b_0 and b_1.

(b) Interpret the meaning of the Y intercept b_0 and the slope b_1 in this problem.

(c) Use the regression equation developed in (a) to predict the average number of errors for typists who have consumed 1.5 ounces of alcohol.

(d) Compute the standard error of the estimate.

(e) Compute the coefficient of determination r^2 and interpret its meaning in this problem.

(f) Compute the coefficient of correlation r.

(g) Set up a 95% confidence interval estimate of the average number of errors made by typists who have consumed 1.5 ounces of alcohol.

(h) At the .05 level of significance is there a significant linear relationship between these two variables?

(i) Set up a 95% confidence interval of the true slope.

Table 14.15
RELATING TYPING ERRORS TO ALCOHOLIC CONSUMPTION

Typist	Alcoholic Consumption (ounces)	Number of Errors
1	0	2
2	0	5
3	0	3
4	0	6
5	0	4
6	1	7
7	1	5
8	1	6
9	1	3
10	1	9
11	2	10
12	2	6
13	2	10
14	2	12
15	2	12

* 14.11 The director of a management training program for a large company would like to study the relationship between the score on an aptitude test given at the beginning of the program and the final rating given by a committee of executives upon completion of the program. A random sample of eight trainees revealed the results given in Table 14.16.

Table 14.16
RELATING RATINGS TO TEST SCORES

Trainee	Test Score	Final Rating
1	63	81
2	49	83
3	72	91
4	66	78
5	58	84
6	78	90
7	61	74
8	59	80

(a) Compute the coefficient of correlation r between test score and final rating.

(b) At the .05 level of significance is there a significant linear relationship between test score and final rating?

14.12 An economist for a large automobile company wanted to measure the relationship between family income and the purchase price of new cars. A random sample of 10 persons purchasing new cars is selected with the results given in Table 14.17.

(a) Compute the coefficient of correlation r between family income and purchase price.

(b) At the .01 level of significance is there a significant linear relationship between family income and purchase price?

14.13 A statistician for a consumer organization wanted to determine the relationship between the price of a transistor radio battery and the number of

Table 14.17
RELATING PURCHASE PRICE TO INCOME

Purchaser	Family Income (thousands of dollars)	Purchase Price (thousands of dollars)
1	10.2	3.6
2	14.4	4.1
3	16.3	3.9
4	20.0	5.2
5	24.3	5.1
6	11.6	3.9
7	32.8	7.8
8	9.4	3.4
9	26.7	9.1
10	18.3	5.0

hours the battery lasted. A sample of 16 batteries was purchased with the results given in Table 14.18.

Table 14.18
RELATING PRODUCT LIFE TO PRICE

Battery	Price (dollars)	Life (hours)
1	.24	5.4
2	.32	4.8
3	.49	6.3
4	.49	7.2
5	.39	6.3
6	.69	7.4
7	.69	6.8
8	.89	10.2
9	1.19	13.1
10	.79	9.2
11	.35	6.0
12	.80	8.3
13	1.09	10.2
14	.72	7.6
15	.45	6.3
16	.30	4.7

(a) Compute the coefficient of correlation r between the price and the life of transistor radio batteries.

(b) At the .05 level of significance is there a significant linear relationship between the price and the life of transistor radio batteries?

The following problems refer to the sample data obtained from the questionnaire of Figure 2.2 and presented in Figure 2.6.

14.14 For the 15 seniors (code 4, question 2) in the sample we would like to predict the expected starting salary (question 7) based upon the college grade-point index (question 5). For these data,

(a) Assuming a linear relationship, use the least-squares method to compute the regression coefficients.

(b) Interpret the meaning of the slope b_1 in this problem.

(c) Predict the expected annual starting salary for a senior with a grade-point index of 3.0.

(d) Compute the standard error of the estimate.

(e) Compute the coefficient of determination r^2 and interpret its meaning in this problem.

(f) Compute the coefficient of correlation r.

(g) Set up a 90% confidence interval estimate of the average expected starting salary of a senior with a grade-point index of 3.0.

(h) At the .10 level of significance is there a linear relationship between the grade-point index and the expected starting salary?

14.15 For the juniors (code 3, question 2) in the sample we would like to predict the expected starting salary (question 7) based upon the college grade-point index (question 5). For these data,

(a) Assuming a linear relationship, use the least-squares method to compute the regression coefficients.

(b) Predict the expected annual starting salary for a junior with a grade-point index of 3.0.

(c) Compute the coefficient of determination r^2 and interpret its meaning in this problem.

(d) Compute the standard error of the estimate.

(e) Compute the coefficient of correlation r.

(f) Set up a 95% confidence interval estimate of the average expected starting salary of a junior with a grade-point index of 3.0.

(g) At the .05 level of significance is there a linear relationship between grade-point index and expected starting salary?

14.16 For the freshmen and sophomores (codes 1 and 2, question 2), we would like to predict maximum yearly tuition (question 8) based upon high school average (question 6). For these data,

(a) Assuming a linear relationship, use the least-squares method to compute the regression coefficients b_0 and b_1.

(b) Interpret the meaning of the slope b_1 in this problem.

(c) Predict the maximum yearly tuition for a freshman or sophomore student with a high school average of 85.

(d) Compute the coefficient of determination r^2 and interpret its meaning in this problem.

(e) Compute the coefficient of correlation r.

(f) Set up a 95% confidence interval estimate of the average maximum yearly tuition for a freshman or sophomore student with a high school average of 85.

(g) At the .05 level of significance is there a linear relationship between high-school average and maximum yearly tuition?

14.17 For the 51 males (code 1, question 1) in the sample, compute the coefficient of correlation between height (question 16) and weight (question 17) and determine whether there is a linear relationship between these variables at the .05 level of significance.

14.18 For the 35 accounting majors (code 2, question 4) compute the coefficient of correlation between the expected starting salary (question 7) and maximum yearly tuition (question 8) and determine whether there is any linear relationship between these variables at the .01 level of significance.

14.19 Use your own sample data (Problem 3.12) to select problems from among Problems 14.14 through 14.18.

14.20 For your own class (Problem 3.13) select problems from among Problems 14.14 through 14.18.

REFERENCES

1. BROWNLEE, K. A., *Statistical Theory and Methodology in Science and Engineering,* 2nd ed. (New York: Wiley, 1965).
2. COHEN, J., AND P. COHEN, *Applied Multiple Regression/Correlation Analysis for the Behavioral Sciences* (Hillsdale, N.J.: Lawrence Erlbaum, 1975).
3. DRAPER, N. R., AND H. SMITH, *Applied Regression Analysis* (New York: Wiley, 1966).
4. EZEKIEL, M., AND K. A. FOX, *Methods of Correlation and Regression Analysis,* 3rd ed. (New York: Wiley, 1959).
5. KLEINBAUM, D. G., AND L. L. KUPPER, *Applied Regression Analysis and Other Multivariate Methods* (North Scituate, Mass.: Duxbury Press, 1978).
6. LI, J. C. R., *Statistical Inference,* vols. I and II (Ann Arbor, Mich.: Edward Bros., 1964).
7. MOSTELLER, F., AND J. W. TUKEY, *Data Analysis and Regression: A Second Course in Statistics* (Reading, Mass.: Addison-Wesley, 1977).
8. NETER, J., AND W. WASSERMAN, *Applied Linear Statistical Models* (Homewood, Ill.: Richard D. Irwin, 1974).
9. NIE, N. H., C. H. HULL, J. G. JENKINS, K. STEINBRENNER, AND D. H. BENT, *SPSS Statistical Package for the Social Sciences,* 2nd ed. (New York: McGraw-Hill, 1975).
10. SNEDECOR, G. W., AND W. G. COCHRAN, *Statistical Methods,* 6th ed. (Ames, Iowa: Iowa State University Press, 1967).

CHAPTER 15

MULTIPLE REGRESSION ANALYSIS

15.1 INTRODUCTION

In our discussion of regression and correlation in the previous chapter we primarily focused upon a linear relationship between a single independent variable and a dependent variable. In this chapter we shall extend this discussion in order to explore the basic principles of multiple regression. Several independent variables will be used to predict the value of a dependent variable.

In the construction cost problem of Chapter 14 we may recall that property size was used as an independent variable to predict the construction cost for single-family houses. Suppose that this home construction company also wanted to study the effect of average daily atmospheric temperature and the amount of attic insulation upon the monthly consumption of home heating oil. In order to study the relationship between these variables, records were obtained from 15 homes in a similar housing development at various times during the past year. These data are presented in Table 15.1.

With two independent variables in the multiple regression problem, a scatter diagram of the points can be plotted on a three-dimensional graph as shown in Figure 15.1.

For a particular investigation, when there are several independent variables

Table 15.1
MONTHLY CONSUMPTION OF HEATING OIL, DAILY ATMOSPHERIC TEMPERATURE, AND AMOUNT OF ATTIC INSULATION FOR A RANDOM SAMPLE OF 15 SINGLE-FAMILY HOMES

Observation	Monthly Consumption Heating Oil (gallons) Y	Average Daily Atmospheric Temperature (°F) X_1	Amount of Attic Insulation (inches) X_2
1	275.3	40	3
2	363.8	27	3
3	164.3	40	10
4	40.8	73	6
5	94.3	64	6
6	230.9	34	6
7	366.7	9	6
8	300.6	8	10
9	237.8	23	10
10	121.4	63	3
11	31.4	65	10
12	203.5	41	6
13	441.1	21	3
14	323.0	38	3
15	52.5	58	10

present, the simple linear regression model of the previous chapter [Equation (14.1)] can be extended by assuming a linear relationship between each independent variable and the dependent variable. For example, as in our current problem with two independent variables, the multiple linear regression model is expressed as

$$Y_i = \beta_0 + \beta_1 X_{1i} + \beta_2 X_{2i} + \varepsilon_i \tag{15.1}$$

where β_0 = Y intercept
β_1 = slope of Y with variable X_1 holding variable X_2 constant
β_2 = slope of Y with variable X_2 holding variable X_1 constant
ε_i = random error in Y for observation i

This multiple linear regression model can be compared to the simple linear regression model [Equation (14.1)] expressed as

$$Y_i = \beta_0 + \beta_1 X_i + \varepsilon_i$$

In the case of the simple linear regression model we should note that the slope β_1 represents the unit change in Y per unit change in X and does not take into account any other variables besides the single independent variable that is included in the model. On the other hand, in the multiple linear regression model [Equation (15.1)], the slope β_1 represents the unit change in

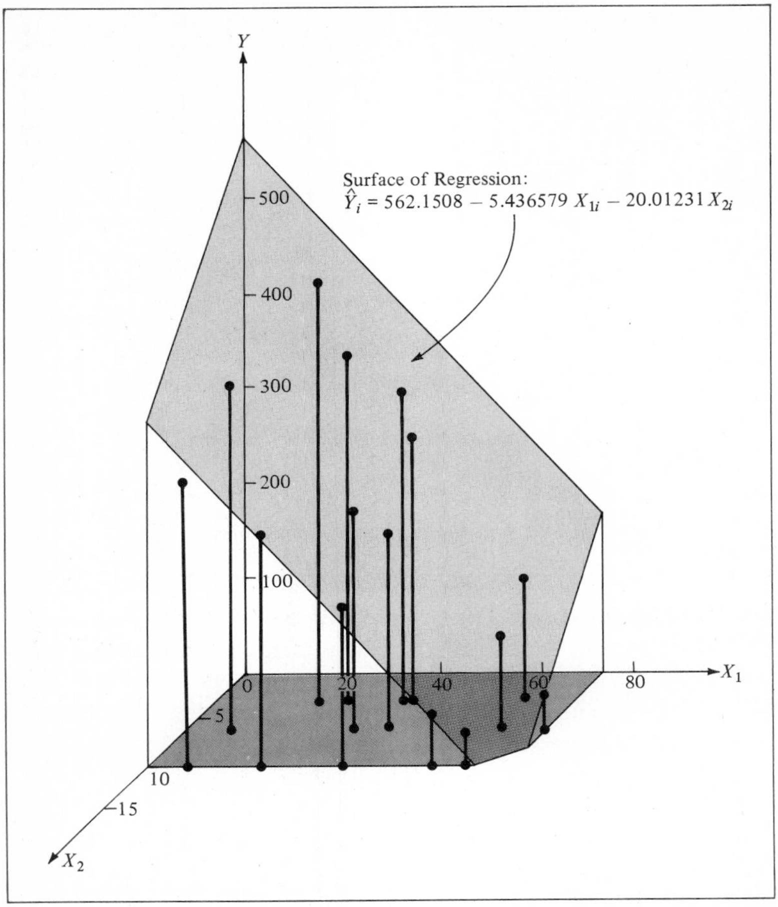

Figure 15.1 Scatter diagram of average daily atmospheric temperature X_1, amount of attic insulation X_2, and monthly consumption of heating oil Y with indicated regression plane fitted by least-squares method.

Y per unit change in X_1, taking into account the effect of X_2 and is referred to as a **net regression coefficient.**

As in the case of simple linear regression, when sample data are analyzed, the sample regression coefficients (b_0, b_1, and b_2) are used as estimates of the true parameters (β_0, β_1, and β_2). Thus the regression equation for the multiple linear regression model with two independent variables would be

$$\hat{Y}_i = b_0 + b_1X_{1i} + b_2X_{2i} \qquad (15.1a)$$

15.2 FINDING THE REGRESSION COEFFICIENTS

If the least-squares method (see Section 14.4.1) is utilized to compute the sample regression coefficients (b_0, b_1, and b_2), we would have the following three normal equations:

I. $$\sum_{i=1}^{n} Y_i = nb_0 + b_1 \sum_{i=1}^{n} X_{1i} + b_2 \sum_{i=1}^{n} X_{2i} \qquad (15.2a)$$

II. $$\sum_{i=1}^{n} X_{1i}Y_i = b_0 \sum_{i=1}^{n} X_{1i} + b_1 \sum_{i=1}^{n} X_{1i}^2 + b_2 \sum_{i=1}^{n} X_{1i}X_{2i} \qquad (15.2b)$$

III. $$\sum_{i=1}^{n} X_{2i}Y_i = b_0 \sum_{i=1}^{n} X_{2i} + b_1 \sum_{i=1}^{n} X_{1i}X_{2i} + b_2 \sum_{i=1}^{n} X_{2i}^2 \qquad (15.2c)$$

For the data of Table 15.1, the values of the various sums needed to solve Equations (15.2a), (15.2b), and (15.2c) are given below:

$$\sum_{i=1}^{n} Y_i = 3{,}247.4 \qquad \sum_{i=1}^{n} X_{1i}Y_i = 98{,}060.1 \qquad \sum_{i=1}^{n} X_{1i} = 604$$

$$\sum_{i=1}^{n} X_{2i} = 95 \qquad \sum_{i=1}^{n} X_{2i}Y_i = 18{,}057 \qquad \sum_{i=1}^{n} X_{1i}^2 = 30{,}308$$

$$\sum_{i=1}^{n} X_{2i}^2 = 725 \qquad \sum_{i=1}^{n} X_{1i}X_{2i} = 3{,}833 \qquad n = 15$$

$$\sum_{i=1}^{n} Y_i^2 = 939{,}175.68$$

Therefore, for these data the three normal equations are:

I. $3{,}247.4 = 15b_0 + 604b_1 + 95b_2$
II. $98{,}060.1 = 604b_0 + 30{,}308b_1 + 3{,}833b_2$
III. $18{,}057.0 = 95b_0 + 3{,}833b_1 + 725b_2$

The values of the three sample regression coefficients (b_0, b_1, and b_2) can be obtained by using various methods for the solution of simultaneous equations (see Reference 5), through matrix algebra (see References 4, 6, 7, 8, and 10), or by using an available computer package (see References 3, 9, and 11). Figure 15.2 presents the output from the SPSS subprogram REGRESSION for the data of Table 15.1.

From Figure 15.2, we observe that the computed values of the regression coefficients in this problem are

$$b_0 = 562.1508 \qquad b_1 = -5.436579 \qquad b_2 = -20.01231$$

```
                 MULTIPLE REGRESSION STUDY OF HOME HEATING OIL USAGE

                 VARIABLE               MEAN        STANDARD DEV       CASES

              Y ←HTNGOIL            216.4933           120.8722          15
              X1←TEMPF               40.2667            20.6794          15
              X2←INSU                 6.3333             2.9681          15

                                  CORRELATION COEFFICIENTS

                                    HTNGOIL     TEMPF      INSU

                       HTNGOIL      1.00000   -0.86974   -0.46508
                       TEMPF       -0.86974    1.00000    0.00892
                       INSU        -0.46508    0.00892    1.00000

DEPENDENT VARIABLE..         HTNGOIL      NUMBER OF GALLONS OF HEATING OIL USED

VARIABLE(S) ENTERED          INSU         NUMBER OF INCHES OF ATTIC INSULATION
                             TEMPF        AVERAGE DAILY TEMPERATURE DEGREES FHT


MULTIPLE R                 0.98265  ← r_Y.12
R SQUARE                   0.96561  ← r²_Y.12

STANDARD ERROR            26.01377  ← S_YX


ANALYSIS OF VARIANCE        DF     SUM OF SQUARES        MEAN SQUARE          F
REGRESSION                  2.       228014.45924       114007.22962   168.47123
RESIDUAL                   12.         8120.59574          676.71631

      ---------------- VARIABLES IN THE EQUATION ----------------

      VARIABLE               B                     STD ERROR B          F

      INSU           -20.01231 ← b2              Sb2 = 2.34250       72.985
      TEMPF          -5.436579 ← b1              Sb1 = 0.33622      261.466
      (CONSTANT)      562.1508 ← b0
```

Figure 15.2 Partial output for the data of Table 15.1 from SPSS subprogram REGRESSION.

Therefore the multiple regression equation can be expressed as:

$$\hat{Y}_i = 562.1508 - 5.436579X_{1i} - 20.01231X_{2i}$$

where $\hat{Y}_i$ = predicted amount of home heating oil consumed per month (gallons) for observation i

X_{1i} = average daily atmospheric temperature (°F) for observation i

X_{2i} = amount of attic insulation (inches) for observation i

The interpretation of the regression coefficients is analogous to that of the simple linear regression model. The Y intercept b_0, computed as 562.1508, represents the number of gallons of home heating oil that would be consumed in a month in which the average daily atmospheric temperature was 0°F for a home that was not insulated (that is, a house with 0 inches of attic insulation). The slope of average daily atmospheric temperature with consumption of heating oil (b_1, computed as -5.436579) can be interpreted to mean that for a home with a *given* number of inches of attic insulation, the monthly consumption of heating oil will *decrease* by 5.436579 gallons per month for each 1°F increase in average daily atmospheric temperature. Furthermore, the slope of amount of attic insulation with consumption of heating oil (b_2, computed as -20.01231) can be interpreted to mean that for a month with a *given* average daily atmospheric temperature, the monthly consumption of heating oil will *decrease* by 20.01231 gallons for each additional inch of attic insulation.

15.3 PREDICTION OF THE DEPENDENT VARIABLE Y FOR GIVEN VALUES OF THE INDEPENDENT VARIABLES

Now that the multiple-regression model has been fitted to these data, various procedures, analogous to those discussed for simple linear regression, could be developed. In this section we shall use the multiple regression equation to predict the monthly consumption of heating oil.

Suppose that we wanted to predict the number of gallons of heating oil consumed in a house that had 6 inches of attic insulation during a month in which the average daily atmospheric temperature was 30°F. Using our multiple regression equation

$$\hat{Y}_i = 562.1508 - 5.436579X_{1i} - 20.01231X_{2i}$$

with $X_{1i} = 30$ and $X_{2i} = 6$, we have

$$\hat{Y}_i = 562.1508 - 5.436579(30) - 20.01231(6)$$

and thus

$$\hat{Y}_i = 278.97957$$

Therefore, we would predict that approximately 278.98 gallons of heating oil would be used by that house during that particular month.

15.4 TESTING FOR THE SIGNIFICANCE OF THE RELATIONSHIP BETWEEN THE DEPENDENT VARIABLE AND THE INDEPENDENT VARIABLES

Once a regression model has been fitted to a set of data, we can determine whether there is a significant relationship between the dependent variable and the set of independent variables. Since there is more than one independent variable, the null and alternative hypotheses can be set up as follows:

H_0: $\beta_1 = \beta_2 = 0$ (There is no linear relationship between the dependent variable and the independent variables)

H_1: $\beta_1 \neq \beta_2 \neq 0$ (At least one regression coefficient is not equal to zero)

This null hypothesis may be tested by utilizing an F test, as indicated in Table 15.2.

We may recall from Sections 10.10 and 12.3 that the F test is used when testing the ratio of two variances. When testing for the significance of the regression coefficients, the measure of random error is called the **error variance** so that the F test is the ratio of the variance due to the regression divided by the error variance as shown in Equation (15.3).

$$F_{p,n-p-1} = \frac{S_{\text{reg}}^2}{S_{YX}^2} \quad (15.3)$$

Interestingly, the error variance is also the square of the standard error of the estimate (S_{YX}) which was described in Section 14.5.

For the data of the heating oil consumption problem, the ANOVA table (also displayed as part of Figure 15.2) is presented in Table 15.3.

If a level of significance of .05 is chosen, from Appendix E, Table E.5, we determine that the critical value on the F distribution (with 2 and 12 degrees of freedom) is 3.89, as depicted in Figure 15.3. From Equation (15.3), since $F_{2,12} = S_{\text{reg}}^2/S_{YX}^2 = 168.47123 > 3.89$, we can reject H_0 and conclude that *at least* one of the independent variables (temperature and/or insulation) is related to heating oil consumption.

Table 15.2
ANALYSIS OF VARIANCE TABLE FOR TESTING THE SIGNIFICANCE OF A SET OF REGRESSION COEFFICIENTS IN MULTIPLE REGRESSION CONTAINING P INDEPENDENT VARIABLES

Source	DF	Sums of Squares (SS)	Variance	F
Regression	p	$b_0 \sum_{i=1}^{n} Y_i + b_1 \sum_{i=1}^{n} X_{1i}Y_i + b_2 \sum_{i=1}^{n} X_{2i}Y_i - \frac{\left(\sum_{i=1}^{n} Y_i\right)^2}{n}$	$S_{\text{reg}}^2 = \frac{SS_{\text{reg}}}{p}$	$F = \frac{S_{\text{reg}}^2}{S_{YX}^2}$
Error	$n - p - 1$	$\sum_{i=1}^{n} Y_i^2 - b_0 \sum_{i=1}^{n} Y_i - b_1 \sum_{i=1}^{n} X_{1i}Y_i - b_2 \sum_{i=1}^{n} X_{2i}Y_i$	$S_{YX}^2 = \frac{SS_{\text{error}}}{n-p-1}$	
Total	$n - 1$	$\sum_{i=1}^{n} Y_i^2 - \frac{\left(\sum_{i=1}^{n} Y_i\right)^2}{n}$		

Table 15.3
ANALYSIS OF VARIANCE TABLE FOR TESTING THE SIGNIFICANCE OF THE SET OF REGRESSION COEFFICIENTS FOR THE HEATING OIL CONSUMPTION PROBLEM

Source	DF	Sums of Squares (SS)	Variance	F
Regression	2	$(562.1508)(3{,}247.4) + (-5.436579)(98{,}060.1) + (-20.01231)(18{,}057) - \frac{(3{,}247.4)^2}{15} = 228{,}014.45924$	$\frac{228{,}014.45924}{2} = 114{,}007.22962$	$\frac{114{,}007.22962}{676.71631} = 168.47123$
Error	$15 - 2 - 1 = 12$	$939{,}175.68 - (562.1508)(3{,}247.4) - (-5.436579)(98{,}060.1) - (-20.01231)(18{,}057) = 8{,}120.59574$	$\frac{8{,}120.59574}{12} = 676.71631$	
Total	$15 - 1 = 14$	$939{,}175.68 - \frac{(3{,}247.4)^2}{15} = 236{,}135.05498$		

SOURCE: Format of Table 15.2.

15.5 MEASURING ASSOCIATION IN THE MULTIPLE-REGRESSION MODEL

We may recall from Section 14.6 that once a regression model has been developed, the coefficient of determination r^2 could be computed. In multiple regression, since there are at least two independent variables, the coefficient of multiple determination represents the proportion of the variation in Y that is explained by the set of independent variables selected. In our example, containing two independent variables, the coefficient of multiple determination ($r_{Y.12}^2$) is given by

$$r_{Y.12}^2 = \frac{SS_{reg}}{SS_{tot}} \tag{15.4}$$

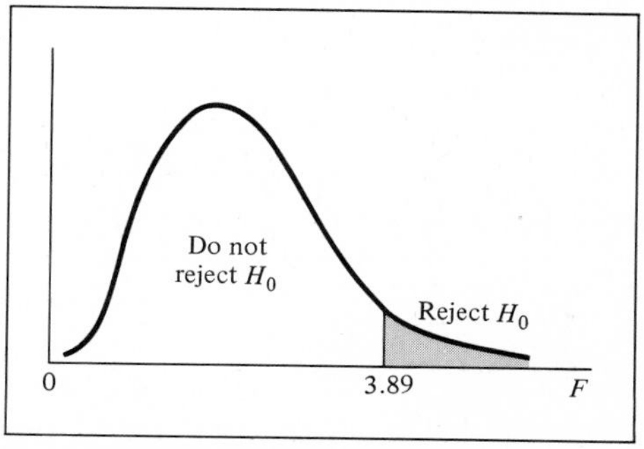

Figure 15.3 Testing for the significance of a set of regression coefficients at the .05 level of significance with 2 and 12 degrees of freedom.

where $$SS_{reg} = b_0 \sum_{i=1}^{n} Y_i + b_1 \sum_{i=1}^{n} X_{1i}Y_i + b_2 \sum_{i=1}^{n} X_{2i}Y_i - \frac{\left(\sum_{i=1}^{n} Y_i\right)^2}{n}$$

$$SS_{tot} = \sum_{i=1}^{n} Y_i^2 - \frac{\left(\sum_{i=1}^{n} Y_i\right)^2}{n}$$

In the home heating oil consumption problem, we have already computed $SS_{reg} = 228{,}014.45924$ and $SS_{tot} = 236{,}135.05498$. Thus, as is displayed in the computer output of Figure 15.2,

$$r_{Y.12}^2 = \frac{SS_{reg}}{SS_{tot}} = \frac{228{,}014.45924}{236{,}135.05498} = .96561$$

This coefficient of multiple determination, computed as .96561, can be interpreted to mean that 96.561% of the variation in the monthly consumption of home heating oil can be explained by the variation in the average daily atmospheric temperature and the variation in the amount of attic insulation.

In order to further study the relationship among the variables, it is often useful to examine the correlation between each pair of variables included in the model. Referring to Figure 15.2, we can observe that a correlation "matrix" or table is obtained that indicates the coefficient of correlation between each pair of variables. This matrix is displayed in Table 15.4.

From Table 15.4, we observe that the correlation between the amount of heating oil consumed and temperature is $-.86974$, indicating a strong negative association between the variables. We may also observe that the correlation between the amount of heating oil consumed and attic insulation is $-.46508$, indicating a moderate negative correlation between these variables. Furthermore, we also note that there is virtually no correlation (.00892) between the two independent variables, temperature and attic insulation. Finally, we may note that the correlation coefficients along the main diagonal of the matrix (r_{YY}, r_{11}, r_{22}) are each 1.0 since there will be perfect correlation between a variable and itself.

Table 15.4
CORRELATION MATRIX FOR THE HEATING OIL CONSUMPTION PROBLEM

	Y (Heating Oil)	X_1 (Temperature)	X_2 (Attic Insulation)
Y (heating oil)	$r_{YY} = 1.0$	$r_{Y1} = -.86974$	$r_{Y2} = -.46508$
X_1 (temperature)	$r_{Y1} = -.86974$	$r_{11} = 1.0$	$r_{12} = .00892$
X_2 (attic insulation)	$r_{Y2} = -.46508$	$r_{12} = .00892$	$r_{22} = 1.0$

15.6 EVALUATING THE CONTRIBUTION OF EACH INDEPENDENT VARIABLE TO A MULTIPLE-REGRESSION MODEL

In developing a multiple-regression model, the objective is to utilize only those independent variables that are useful in predicting the value of a dependent variable. If an independent variable is not helpful in making this prediction, then it could be deleted from the multiple-regression model and a model with fewer independent variables could be utilized in its place.

One method for determining the contribution of an independent variable is called **the partial F test criterion** (see Reference 4). It involves determining the contribution to the regression sum of squares made by each independent variable after all the other independent variables have been included in a model. The new independent variable would only be included if it significantly improved the model. To apply the partial F test criterion in our heating oil consumption problem containing two independent variables, we need to evaluate the contribution of the variable attic insulation (X_2) once average daily atmospheric temperature (X_1) has been included in the model and, conversely, we also must evaluate the contribution of the variable average daily atmospheric temperature (X_1) once attic insulation (X_2) has been included in the model.

15.6.1 Determining the Contribution of an Independent Variable by Comparing Different Regression Models

The contribution of each independent variable to be included in the model can be determined in two ways, depending upon the information available from the computer output or the solution of the normal equations. In this section we shall evaluate the contribution of each independent variable by taking into account the regression sum of squares of a model that includes all independent variables except the one of interest, SS_{reg} (slopes of all variables except k). Thus, in general, to determine the contribution of variable k given that all other variables are already included, we would have:

$$SS_{reg}(b_k|\text{slopes of all variables except } k) = SS_{reg}\ (\text{slopes of all variables including } k) - SS_{reg}\ (\text{slopes of all variables except } k) \qquad (15.5a)$$

If, as in the heating oil consumption problem, there are two independent variables, the contribution of each variable can be determined from Equations (15.5b) and (15.5c).

Contribution of Variable X_1 Given X_2 Has Been Included:

$$SS_{reg}(b_1|b_2) = SS_{reg}(b_1 \text{ and } b_2) - SS_{reg}(b_2) \qquad (15.5b)$$

Contribution of Variable X_2 Given X_1 Has Been Included:

$$SS_{reg}(b_2|b_1) = SS_{reg}(b_1 \text{ and } b_2) - SS_{reg}(b_1) \qquad (15.5c)$$

The term $SS_{reg}(b_2)$ represents the regression sum of squares for a model that includes only the independent variable X_2 (amount of attic insulation) while the term $SS_{reg}(b_1)$ represents the regression sum of squares for a model that includes only the independent variable X_1 (average daily atmospheric temperature). Computer output obtained from the SPSS subprogram REGRESSION for these two models is presented in Figures 15.4 and 15.5.

We can observe from Figure 15.4 that

$$SS_{reg}(b_2) = 51{,}076.41665$$

and, therefore, from Equation (15.5*b*)

$$SS_{reg}(b_1|b_2) = SS_{reg}(b_1 \text{ and } b_2) - SS_{reg}(b_2)$$

```
DEPENDENT VARIABLE..      HTNGOIL <-Y  NUMBER OF GALLONS OF HEATING OIL USED

VARIABLE(S) ENTERED       INSU    <-X2 NUMBER OF INCHES OF ATTIC INSULATION

MULTIPLE R              0.46508   <- r
R SQUARE                0.21030   <- r2

STANDARD ERROR        119.31169   <- SYX


ANALYSIS OF VARIANCE      DF      SUM OF SQUARES        MEAN SQUARE        F
REGRESSION                1.        51076.41665         51076.41665     3.58802
RESIDUAL                 13.       185058.63833         14235.27987


    ----------------- VARIABLES IN THE EQUATION -----------------

    VARIABLE               B                    STD ERROR B         F

    INSU          -20.35026  <-b2            Sb2= 10.74343        3.588
    (CONSTANT)    345.3783   <-b0
```

Figure 15.4 Partial output of simple linear regression model of amount of heating oil consumed and amount of attic insulation (obtained from SPSS program REGRESSION).

```
DEPENDENT VARIABLE..      HTNGOIL ←Y  NUMBER OF GALLONS OF HEATING OIL USED
VARIABLE(S) ENTERED       TEMPF  ←X1  AVERAGE DAILY TEMPERATURE DEGREES FHT

MULTIPLE R             0.86974   ← r
R SQUARE               0.75645   ← r²

STANDARD ERROR        66.51243   ← S_YX


ANALYSIS OF VARIANCE      DF        SUM OF SQUARES          MEAN SQUARE        F
REGRESSION                1.          178624.30414         178624.30414    40.37708
RESIDUAL                 13.           57510.75084           4423.90391


---------------------- VARIABLES IN THE EQUATION ----------------------

VARIABLE                  B                        STD ERROR B            F

TEMPF              -5.462206  ← b1              S_b1 = 0.85961        40.377
(CONSTANT)        436.4381    ← b0
```

Figure 15.5 Partial output of simple linear regression model of amount of heating oil consumed and average daily atmospheric temperature (obtained from SPSS subprogram REGRESSION).

we have

$$\mathrm{SS}_{\mathrm{reg}}(b_1|b_2) = 228{,}014.45924 - 51{,}076.41665 = 176{,}938.04259$$

In order to determine whether X_1 significantly improves the model after X_2 has been included, we can now subdivide the regression sum of squares into two component parts as shown in Table 15.5.

The null and alternative hypotheses to test for the contribution of X_1 to the model would be:

Table 15.5
ANALYSIS OF VARIANCE TABLE DIVIDING THE REGRESSION SUM OF SQUARES INTO COMPONENTS TO DETERMINE THE CONTRIBUTION OF VARIABLE X_1

Source	DF	Sums of Squares (SS)	Variance	F
Regression	2	228,014.45924	114,007.22962	
$\{b_2\}$	{1}	{51,076.41665}	51,076.41665	
$\{b_1\|b_2\}$	{1}	{176,938.04259}	176,938.04259	261.466
Error	12	8,120.59374	$S_{YX}^2 = 676.71631$	
Total	14	236,135.05498		

H_0: Variable X_1 does not significantly improve the model once variable X_2 has been included

H_1: Variable X_1 significantly improves the model once variable X_2 has been included

The partial F test criterion is expressed by

$$F_{1,n-p-1} = \frac{SS_{reg}(b_k|\text{slopes of all variables except } k)}{S_{YX}^2} \qquad (15.6)$$

Thus, from Table 15.5 we have

$$F_{1,12} = \frac{176{,}938.04259}{676.71631} = 261.466$$

If a level of significance of .05 is selected, from Appendix E, Table E.5, we can observe that the critical value is 4.75 (see Figure 15.6). Since the computed F value exceeds this critical F value ($261.466 > 4.75$), our decision would be to reject H_0 and conclude that the addition of variable X_1 (average daily atmospheric temperature) significantly improves a multiple-regression model that already contains variable X_2 (attic insulation).

In order to evaluate the contribution of variable X_2 (attic insulation) to a model in which variable X_1 has been included, we need to compute

$$SS_{reg}(b_2|b_1) = SS_{reg}(b_1 \text{ and } b_2) - SS_{reg}(b_1) \qquad (15.5c)$$

From Figure 15.5 we determine that

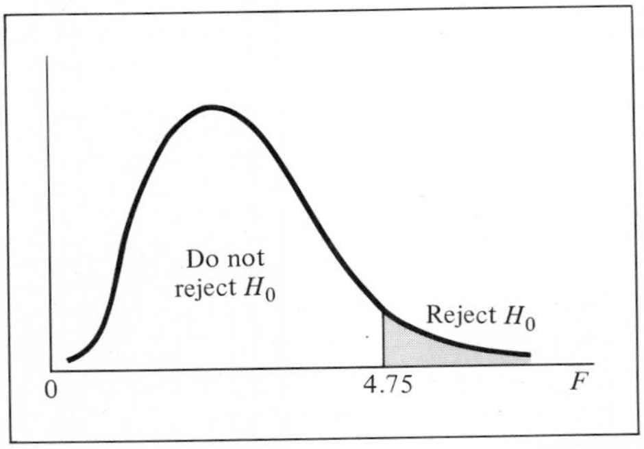

Figure 15.6 Testing for the contribution of a regression coefficient to a multiple-regression model at the .05 level of significance with 1 and 12 degrees of freedom.

$$SS_{reg}(b_1) = 178{,}624.30414$$

Therefore,

$$SS_{reg}(b_2|b_1) = 228{,}014.45924 - 178{,}624.30414 = 49{,}390.15510$$

Thus, in order to determine whether X_2 significantly improves a model after X_1 has been included, the regression sum of squares can be subdivided into two component parts as shown in Table 15.6.

The null and alternative hypotheses to test for the contribution of X_2 to the model would be:

H_0: Variable X_2 does not significantly improve the model once variable X_1 has been included

H_1: Variable X_2 significantly improves the model once variable X_1 has been included

Using Equation (15.6) we obtain

$$F_{1,12} = \frac{49{,}390.15510}{676.71631} = 72.985$$

as indicated in Table 15.6. Since there are 1 and 12 degrees of freedom, respectively, if a .05 level of significance is selected, we again observe from Figure 15.6 that the critical value of F is 4.75. Since the computed F value exceeds this critical value ($72.985 > 4.75$), our decision is to reject H_0 and conclude that the addition of variable X_2 (attic insulation) significantly improves the multiple-regression model already containing X_1 (average daily atmospheric temperature).

Thus, by testing for the contribution of each independent variable after the other had been included in the model, we determined that each of the two independent variables contributed by significantly improving the model. Therefore, our multiple-regression model should include both average

Table 15.6
ANALYSIS OF VARIANCE TABLE SUBDIVIDING THE REGRESSION SUM OF SQUARES INTO COMPONENTS TO DETERMINE THE CONTRIBUTION OF VARIABLE X_2

Source	DF	Sums of Squares (SS)	Variance	F
Regression	2	228,014.45924	114,007.22962	
b_1	1	178,624.30414	178,624.30414	
$b_2 \mid b_1$	1	49,390.15510	49,390.15510	72.985
Error	12	8,120.59574	S_{YX}^2 = 676.71631	
Total	14	236,135.05498		

daily atmospheric temperature X_1 and the amount of attic insulation X_2 in predicting the monthly consumption of home heating oil.

15.6.2 Determining the Contribution of an Independent Variable Based on the Standard Error of Its Regression Coefficient

A second approach to evaluating the contribution made by an independent variable is based upon the standard error of its regression coefficient. If the standard errors of the regression coefficients S_{b_k} (see Sections 14.9 and 15.7) are available from the output of a computer package (see Figure 15.2), the contribution of a particular variable to the regression sum of squares can be determined in the following manner:

$$SS_{\text{reg}}(b_k|\text{slopes of all variables except } k) = \frac{b_k^2 S_{YX}^2}{S_{b_k}^2} \qquad (15.7)$$

We can observe from Figure 15.2 that the standard errors of the regression coefficients for each independent variable (S_{b_1} and S_{b_2}) are available as part of the SPSS output. Thus if we wish to determine the contribution of variable X_2 after X_1 has been included, from Equation (15.7) we have

$$SS_{\text{reg}}(b_2|b_1) = \frac{b_2^2 S_{YX}^2}{S_{b_2}^2} = \frac{(-20.01231)^2(676.71631)}{(2.3425)^2} = 49{,}390.1551$$

However, as indicated in Table 15.6, we can observe that this result is the same as that obtained by using Equation (15.5*c*).

15.7 INFERENCES CONCERNING THE POPULATION REGRESSION COEFFICIENTS

In Section 14.9 we may recall that tests of hypotheses were performed on the regression coefficients in a simple linear regression problem in order to determine the significance of the relationship between X and Y. In addition, confidence intervals were used to estimate the population values of these regression coefficients. In this section, these procedures will be extended to situations involving multiple regression.

15.7.1 Tests of Hypothesis

To test a hypothesis regarding a regression coefficient, we used Equation (14.15):

$$t_{n-2} = \frac{b_1}{S_{b_1}} \qquad (14.15)$$

However, this equation can be generalized for multiple regression as follows:

$$t_{n-p-1} = \frac{b_k}{S_{b_k}} \tag{15.8}$$

where p = number of independent variables in the regression equation
S_{b_k} = standard error of the regression coefficient b_k

Although the formulas for the standard errors of the regression coefficients become unwieldy with a large number of variables [and are expressed better by using a matrix approach (see References 4, 6, 7, 8 and 10)], they can be written for the case of two independent variables as follows:

$$S_{b_0} = S_{YX}\sqrt{\frac{\sum_{i=1}^{n} X_{1i}^2 \sum_{i=1}^{n} X_{2i}^2 - \left(\sum_{i=1}^{n} X_{1i}X_{2i}\right)^2}{n\sum_{i=1}^{n} X_{1i}^2 \sum_{i=1}^{n} X_{2i}^2 + 2\left(\sum_{i=1}^{n} X_{1i}\right)\left(\sum_{i=1}^{n} X_{2i}\right)\left(\sum_{i=1}^{n} X_{1i}X_{2i}\right) - n\left(\sum_{i=1}^{n} X_{1i}X_{2i}\right)^2 - \left(\sum_{i=1}^{n} X_{1i}\right)^2 \sum_{i=1}^{n} X_{2i}^2 - \left(\sum_{i=1}^{n} X_{2i}\right)^2 \sum_{i=1}^{n} X_{1i}^2}} \tag{15.9}$$

$$S_{b_1} = S_{YX}\sqrt{\frac{n\sum_{i=1}^{n} X_{2i}^2 - \left(\sum_{i=1}^{n} X_{2i}\right)^2}{n\sum_{i=1}^{n} X_{1i}^2 \sum_{i=1}^{n} X_{2i}^2 + 2\left(\sum_{i=1}^{n} X_{1i}\right)\left(\sum_{i=1}^{n} X_{2i}\right)\left(\sum_{i=1}^{n} X_{1i}X_{2i}\right) - n\left(\sum_{i=1}^{n} X_{1i}X_{2i}\right)^2 - \left(\sum_{i=1}^{n} X_{1i}\right)^2 \sum_{i=1}^{n} X_{2i}^2 - \left(\sum_{i=1}^{n} X_{2i}\right)^2 \sum_{i=1}^{n} X_{1i}^2}} \tag{15.10}$$

$$S_{b_2} = S_{YX}\sqrt{\frac{n\sum_{i=1}^{n} X_{1i}^2 - \left(\sum_{i=1}^{n} X_{1i}\right)^2}{n\sum_{i=1}^{n} X_{1i}^2 \sum_{i=1}^{n} X_{2i}^2 + 2\left(\sum_{i=1}^{n} X_{1i}\right)\left(\sum_{i=1}^{n} X_{2i}\right)\left(\sum_{i=1}^{n} X_{1i}X_{2i}\right) - n\left(\sum_{i=1}^{n} X_{1i}X_{2i}\right)^2 - \left(\sum_{i=1}^{n} X_{1i}\right)^2 \sum_{i=1}^{n} X_{2i}^2 - \left(\sum_{i=1}^{n} X_{2i}\right)^2 \sum_{i=1}^{n} X_{1i}^2}} \tag{15.11}$$

From Equations (15.9), (15.10), and (15.11) we note that the denominators are all equal. Moreover, all the terms comprising these formulas are readily obtainable from the original data (see Table 15.1).

When a computer package is available the values of the standard errors of the regression coefficients are usually part of the multiple regression output (see Figure 15.2 for the SPSS output and Figure 15.13 for the BMDP output). For example, if we wish to determine whether variable X_2 (amount of attic insulation) has a significant effect on the monthly consumption of home heating oil, taking into account the average daily atmospheric temperature, the null and alternative hypotheses would be:

$$H_0\colon \quad \beta_2 = 0 \qquad \text{(there is no relationship)}$$

$$H_1\colon \quad \beta_2 \neq 0 \qquad \text{(there is a relationship)}$$

From Equations (15.8) and (15.11) we have

$$t_{n-p-1} = \frac{b_2}{S_{b_2}}$$

and from the data of this problem,

$$b_2 = -20.01231$$

$$S_{b_2} = (26.01377)\sqrt{\frac{15(30{,}308) - (604)^2}{15(30{,}308)(725) + 2(604)(95)(3{,}833) - 15(3{,}833)^2 - (604)^2(725) - (95)^2(30{,}308)}}$$

$$= (26.01377)(.0900486) = 2.3425$$

so that

$$t_{12} = \frac{-20.01231}{2.3425} = -8.543$$

If a level of significance of .05 is selected, from Appendix E, Table E.3, we can observe that for 12 degrees of freedom the critical values of t are -2.1788 and $+2.1788$ (see Figure 15.7).

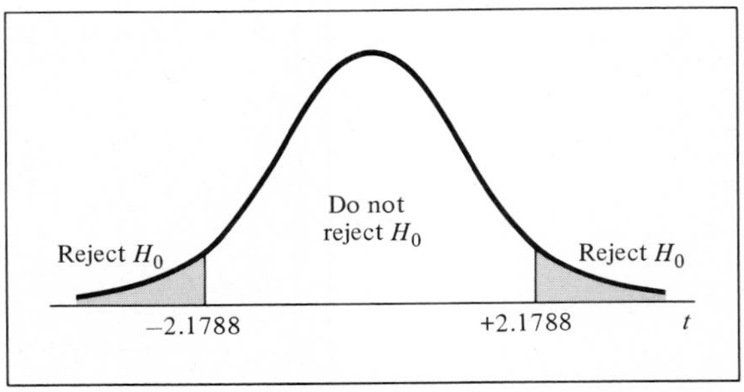

Figure 15.7 Testing for the significance of a regression coefficient at the .05 level of significance with 12 degrees of freedom.

Since we have $t_{12} = -8.543 < -2.1788$, we reject H_0 and conclude that there is a significant relationship between variable X_2 (amount of attic insulation) and the monthly consumption of heating oil, taking into account the average daily atmospheric temperature X_1.

In order to focus upon the interpretation of this conclusion, we should note that there is a relationship between the value of the t test statistic obtained from Equation (15.8) and the partial F test statistic [Equation (15.6)] used to determine the contribution of X_2 to the multiple-regression model. While the t value was computed to be -8.543, the corresponding computed value of F was 72.985—which happens to be the square of -8.543. This points up the following relationship between t and F.[1]

$$t_a^2 = F_{1,a} \qquad (15.12)$$

where a = number of degrees of freedom.

Thus, the test of significance for a particular regression coefficient (in this case b_2) is actually a test for the significance of adding a particular variable into a regression model given that the other variables have been included. Hence the t test for the regression coefficient is equivalent to testing for the contribution of each independent variable as discussed in Section 15.6.

15.7.2 Confidence Interval Estimation

Rather than attempting to determine the significance of a regression coefficient, we may be more concerned with estimating the true population value of a regression coefficient. In multiple regression a confidence interval estimate can be obtained from

$$b_k \pm t_{n-p-1} S_{b_k} \qquad (15.13)$$

For example, if we wish to obtain a 95% confidence interval estimate of the true slope β_1 (that is, the effect of average daily temperature X_1 on monthly consumption of heating oil Y, holding constant the effect of attic insulation X_2), we would have, from Equations (15.10) and (15.13) and Figure 15.2,

$$b_1 \pm t_{12} S_{b_1}$$

Since the critical value of t at the 95% confidence level with 12 degrees of freedom is 2.1788 (see Appendix E, Table E.3), we have

[1] The relationship between t and F indicated in Equation (15.12) holds where t is a two-tailed test.

$$-5.436579 \pm (2.1788)(.33622)$$
$$-5.436579 \pm .732556$$
$$-6.169135 \leq \beta_1 \leq -4.704023$$

Thus, taking into account the effect of attic insulation, we estimate that the true effect of average daily atmospheric temperature is to reduce monthly consumption of heating oil by between approximately 4.7 and 6.17 gallons for each 1°F increase in temperature. Furthermore, we have 95% confidence that this interval correctly estimates the true relationship between these variables. Of course, from a hypothesis-testing viewpoint, since this confidence interval does not include zero, the regression coefficient β_1 would be considered to have a significant effect.

It should be noted here that when confidence interval estimates are desired for a set of regression coefficients, it is more appropriate to *simultaneously* estimate all the parameters since the various regression parameters may *covary* among themselves. This topic of joint confidence interval estimation, however, is beyond the scope of this text and is discussed in greater depth in References 6 and 8.

15.8 CONFIDENCE INTERVAL ESTIMATES FOR PREDICTING μ_{YX}

In Section 15.3, we used the multiple regression equation to obtain a prediction of the monthly consumption of heating oil for a house that has 6 inches of attic insulation in a month in which the average daily temperature was 30°F. A confidence interval estimate of the true mean value of Y, μ_{YX}, can be obtained by extending the procedures discussed in Section 14.8 to the multiple-regression model. However, as indicated in the previous section in our discussion of the standard error of the regression coefficients, the formulas used for predicting the true mean value of Y also become unwieldy when there are several independent variables included in a multiple-regression model and usually are expressed in terms of matrix notation (see Reference 8). The prediction of the true mean μ_{YX} depends not only on the standard error of the estimate S_{YX} and the standard error of the regression coefficients S_{b_k}, but also upon the *covariance* of the regression coefficients (that is, the way in which the coefficients vary together). We should realize, however, that computer programs may provide us with a matrix of the standard errors and covariances of the regression coefficients as part of the output. Thus, for practical purposes, it merely becomes a matter of understanding the concepts and extracting the correct terms from such computer output.

If we focus on the case of two independent variables, six standard errors and covariances of the regression coefficients must be obtained in order to predict μ_{YX}:

1. S_{b_0} = standard error of b_0
2. S_{b_1} = standard error of b_1

3. S_{b_2} = standard error of b_2
4. $S_{b_0b_1}$ = covariance of b_0 and b_1
5. $S_{b_0b_2}$ = covariance of b_0 and b_2
6. $S_{b_1b_2}$ = covariance of b_1 and b_2

The standard errors of the regression coefficients have been discussed in Section 15.7 and are given by Equations (15.9), (15.10), and (15.11). The equations for each of the three covariances of the regression coefficients may be expressed as follows:

$$S_{b_0b_1} = S_{YX}^2 \left\{ \frac{\sum_{i=1}^{n} X_{2i} \sum_{i=1}^{n} X_{1i}X_{2i} - \left(\sum_{i=1}^{n} X_{1i}\right)\left(\sum_{i=1}^{n} X_{2i}^2\right)}{n \sum_{i=1}^{n} X_{1i}^2 \sum_{i=1}^{n} X_{2i}^2 + 2\left(\sum_{i=1}^{n} X_{1i}\right)\left(\sum_{i=1}^{n} X_{2i}\right)\left(\sum_{i=1}^{n} X_{1i}X_{2i}\right) - n\left(\sum_{i=1}^{n} X_{1i}X_{2i}\right)^2 - \left(\sum_{i=1}^{n} X_{1i}\right)^2 \sum_{i=1}^{n} X_{2i}^2 - \left(\sum_{i=1}^{n} X_{2i}\right)^2 \sum_{i=1}^{n} X_{1i}^2} \right\} \quad (15.14)$$

$$S_{b_0b_2} = S_{YX}^2 \left\{ \frac{\sum_{i=1}^{n} X_{1i} \sum_{i=1}^{n} X_{1i}X_{2i} - \left(\sum_{i=1}^{n} X_{2i}\right)\left(\sum_{i=1}^{n} X_{1i}^2\right)}{n \sum_{i=1}^{n} X_{1i}^2 \sum_{i=1}^{n} X_{2i}^2 + 2\left(\sum_{i=1}^{n} X_{1i}\right)\left(\sum_{i=1}^{n} X_{2i}\right)\left(\sum_{i=1}^{n} X_{1i}X_{2i}\right) - n\left(\sum_{i=1}^{n} X_{1i}X_{2i}\right)^2 - \left(\sum_{i=1}^{n} X_{1i}\right)^2 \sum_{i=1}^{n} X_{2i}^2 - \left(\sum_{i=1}^{n} X_{2i}\right)^2 \sum_{i=1}^{n} X_{1i}^2} \right\} \quad (15.15)$$

$$S_{b_1b_2} = S_{YX}^2 \left\{ \frac{\sum_{i=1}^{n} X_{1i} \sum_{i=1}^{n} X_{2i} - n \sum_{i=1}^{n} X_{1i}X_{2i}}{n \sum_{i=1}^{n} X_{1i}^2 \sum_{i=1}^{n} X_{2i}^2 + 2\left(\sum_{i=1}^{n} X_{1i}\right)\left(\sum_{i=1}^{n} X_{2i}\right)\left(\sum_{i=1}^{n} X_{1i}X_{2i}\right) - n\left(\sum_{i=1}^{n} X_{1i}X_{2i}\right)^2 - \left(\sum_{i=1}^{n} X_{1i}\right)^2 \sum_{i=1}^{n} X_{2i}^2 - \left(\sum_{i=1}^{n} X_{2i}\right)^2 \sum_{i=1}^{n} X_{1i}^2} \right\} \quad (15.16)$$

Again, we note that the denominators in these three formulas are the same as those previously given for the standard error of the regression coefficients [Equations (15.9), (15.10), and (15.11)] and the numerous terms are readily obtainable from the original data (see Table 15.1).

For our heating oil consumption problem, from the computer output in Figure 15.2 we can extract:

$$S_{b_1} = .33622 \qquad S_{b_2} = 2.3425$$

Moreover, from Equations (15.9), (15.14), (15.15), and (15.16) we can compute:

$$S_{b_0}{}^2 = (676.71631)\left(\frac{(30{,}308)(725) - (3{,}833)^2}{15(30{,}308)(725) + 2(604)(95)(3{,}833) - 15(3{,}833)^2 - (604)^2(725) - (95)^2(30{,}308)}\right)$$

$$= (676.71631)\left(\frac{7{,}281{,}411}{11{,}074{,}945}\right)$$

$$= (676.71631)(.657467) = 444.91865$$

and

$$S_{b_0b_1} = (676.71631)\left(\frac{95(3{,}833) - (604)(725)}{11{,}074{,}945}\right)$$

$$= (676.71631)\left(\frac{-73{,}765}{11{,}074{,}945}\right)$$

$$= (676.71631)(-.0066605) = -4.50729$$

and

$$S_{b_0b_2} = (676.71631)\left(\frac{604(3{,}833) - (95)(30{,}308)}{11{,}074{,}945}\right)$$

$$= (676.71631)\left(\frac{-564{,}128}{11{,}074{,}945}\right)$$

$$= (676.71631)(-.0509373) = -34.470114$$

and

$$S_{b_1b_2} = (676.71631)\left(\frac{604(95) - 15(3{,}833)}{11{,}074{,}945}\right)$$

$$= (676.71631)\left(\frac{-115}{11{,}074{,}945}\right)$$

$$= (676.71631)(.0000104) = -0.007027$$

Thus the confidence interval estimate of the true mean predicted value of Y can be obtained from:

$$\hat{Y}_i \pm t_{n-p-1} S_{\hat{Y}_i} \qquad (15.17)$$

where $S_{\hat{Y}_i}{}^2 = S_{b_0}{}^2 + X_{1i}{}^2 S_{b_1}{}^2 + X_{2i}{}^2 S_{b_2}{}^2 + 2X_{1i}S_{b_0b_1} + 2X_{2i}S_{b_0b_2} + 2X_{1i}X_{2i}S_{b_1b_2}$

Since we were predicting a monthly consumption of heating oil when the average daily atmospheric temperature was 30°F for a home with 6 inches of insulation, from Section 15.3 we have $\hat{Y}_i = 278.97957$, $X_{1i} = 30$, $X_{2i} = 6$ and thus,

$$\begin{aligned} S_{\hat{Y}_i}{}^2 &= 444.91865 + (30)^2(.33622)^2 + (6)^2(2.3425)^2 \\ &\quad + 2(30)(-4.50729) + 2(6)(-34.470114) \\ &\quad + 2(30)(6)(-.007027) \\ &= 444.91865 + 101.7395 + 197.54303 + (-270.4374) \\ &\quad + (-413.64137) + (-2.52972) \\ &= 57.5927 \\ S_{\hat{Y}_i} &= 7.589 \end{aligned}$$

If a 95% confidence interval estimate is desired, the critical value of t with 12 degrees of freedom will be 2.1788 (see Appendix E, Table E.3) and, therefore, from Equation (15.17) we have

$$\begin{gathered} \hat{Y}_i \pm t_{n-p-1} S_{\hat{Y}_i} \\ 278.97957 \pm (2.1788)(7.589) \\ 278.97957 \pm 16.53491 \\ 262.44466 \leq \mu_{YX} \leq 295.51448 \end{gathered}$$

Thus, we estimate that when the average daily atmospheric temperature is 30°F for a house with 6 inches of attic insulation, the average monthly consumption of home heating oil is between approximately 262.44 and 295.51 gallons. We have 95% confidence that this correctly estimates the true average monthly consumption under these circumstances.

15.9 COEFFICIENT OF PARTIAL DETERMINATION

In Section 15.5 we discussed the coefficient of multiple determination ($r_{Y.12}{}^2$) which measured the proportion of the variation in Y that was explained by variation in the two independent variables. Now that we have examined ways in which the contribution of each independent variable to the multiple-regression model can be evaluated, we can also compute the coefficients of partial determination ($r_{Y1.2}{}^2$ and $r_{Y2.1}{}^2$). These coefficients measure the proportion of the variation in the dependent variable that is explained by each independent variable while controlling for, or holding constant, the other independent variable(s). Thus, in a multiple-regression model with two independent variables we have

$$r_{Y1.2}{}^2 = \frac{SS_{reg}(b_1|b_2)}{SS_{tot} - SS_{reg}(b_1 \text{ and } b_2) + SS_{reg}(b_1|b_2)} \qquad (15.18a)$$

and also

$$r_{Y2.1}^2 = \frac{\text{SS}_{\text{reg}}(b_2|b_1)}{\text{SS}_{\text{tot}} - \text{SS}_{\text{reg}}(b_1 \text{ and } b_2) + \text{SS}_{\text{reg}}(b_2|b_1)} \tag{15.18b}$$

where $\text{SS}_{\text{reg}}(b_1|b_2)$ = sum of squares of the contribution of variable X_1 to the regression model given that variable X_2 has been included in the model

SS_{tot} = total sum of squares for Y

$\text{SS}_{\text{reg}}(b_1 \text{ and } b_2)$ = regression sum of squares when variables X_1 and X_2 are both included in the multiple-regression model

$\text{SS}_{\text{reg}}(b_2|b_1)$ = sum of squares of the contribution of variable X_2 to the regression model given that variable X_1 has been included in the model

while in a multiple-regression model containing several (p) independent variables, we have

$$r^2_{Yk.(\text{all variables except } k)} = \frac{\text{SS}_{\text{reg}}(b_k|\text{slopes of all variables except } k)}{\text{SS}_{\text{tot}} - \text{SS}_{\text{reg}}(\text{slopes of all variables including } k) + \text{SS}_{\text{reg}}(b_k|\text{slopes of all variables except } k)} \tag{15.19}$$

For our heating oil consumption problem we can compute

$$r_{Y1.2}^2 = \frac{176{,}938.04259}{236{,}135.05498 - 228{,}014.45924 + 176{,}938.04259}$$
$$= 0.9561$$

and

$$r_{Y2.1}^2 = \frac{49{,}390.15510}{236{,}135.05498 - 228{,}014.45924 + 49{,}390.15510}$$
$$= 0.8588$$

The coefficient of partial determination of variable Y with X_1 while holding X_2 constant ($r_{Y1.2}^2$) can be interpreted to mean that for a fixed (constant) amount of attic insulation, 95.61% of the variation in the monthly consumption of heating oil can be explained by the variation in the average daily atmospheric temperature from month to month. Moreover, the coefficient of partial determination of variable Y with X_2 while holding X_1 constant ($r_{Y2.1}^2$) can be interpreted to mean that for a given (constant) average daily atmospheric temperature, 85.88% of the variation in the monthly consumption of heating oil can be explained by variation in the amount of attic insulation.

15.10 CURVILINEAR REGRESSION

In our discussion of simple regression (Chapter 14) and multiple regression, we have thus far assumed that the relationship between Y and each independent variable is linear. However, we may recall that several different types of relationships between variables were introduced in Section 14.3. One of the more common nonlinear relationships that was illustrated was a curvilinear polynomial relationship between two variables (see Figure 14.4, panels D to F) in which Y increases (or decreases) at a *changing rate* for various values of X. This model of a polynomial relationship between X and Y can be expressed as

$$Y_i = \beta_0 + \beta_1 X_{1i} + \beta_{11} X_{1i}^2 + \varepsilon_i \qquad (15.20)$$

where β_0 = Y intercept
β_1 = *linear* effect on Y
β_{11} = *curvilinear* effect on Y
ε_i = random error in Y_i for observation i

This regression model is similar to the multiple-regression model with two independent variables [see Equation (15.1)] except that the second "independent" variable in this instance is merely the square of the first independent variable.

As in the case of multiple linear regression, when sample data are analyzed, the sample regression coefficients (b_0, b_1, and b_{11}) are used as estimates of the true parameters (β_0, β_1, and β_{11}). Thus, the regression equation for the curvilinear polynomial model having one independent variable (X_1) and a dependent variable (Y) is

$$\hat{Y}_i = b_0 + b_1 X_{1i} + b_{11} X_{1i}^2 \qquad (15.20a)$$

15.10.1 Finding the Regression Coefficients and Predicting Y

In order to illustrate the curvilinear regression model, let us suppose that the personnel director of a large corporation wished to study the effect of length

Table 15.7
WEEKLY SALARY (DOLLARS) AND QUARTERS (OF YEARS) OF EMPLOYMENT FOR A RANDOM SAMPLE OF 19 MALE EMPLOYEES

Weekly Salary, Y	Quarters (of Years) of Employment, X_1
$163.47	2
181.56	3
204.27	3
181.56	5
188.26	6
242.12	7
193.22	10
205.55	11
264.83	17
242.12	19
204.16	25
361.43	46
414.09	62
467.74	70
377.93	74
397.93	81
278.05	102
517.88	115
450.18	167

of employment on the weekly salary of male employees. A random sample of 19 male employees was selected at the beginning of the year. The weekly salary (in dollars) and the number of quarters (of years) of employment for each employee selected are presented in Table 15.7.

In order to evaluate the proper model expressing the relationship between length of employment and weekly salary, a scatter diagram is plotted as in Figure 15.8.

An examination of Figure 15.8 indicates that the increase in salary continues to level off for an increasing length of employment. Therefore, it is appropriate to use a curvilinear model to estimate weekly salary based upon length of employment.

If the least-squares method (see Sections 14.4.1 and 15.2) is utilized to compute the sample regression coefficients (b_0, b_1, and b_{11}), we would have the following three normal equations:

$$\text{I.} \quad \sum_{i=1}^{n} Y_i = nb_0 + b_1 \sum_{i=1}^{n} X_{1i} + b_{11} \sum_{i=1}^{n} X_{1i}^2 \qquad (15.21a)$$

$$\text{II.} \quad \sum_{i=1}^{n} X_{1i}Y_i = b_0 \sum_{i=1}^{n} X_{1i} + b_1 \sum_{i=1}^{n} X_{1i}^2 + b_{11} \sum_{i=1}^{n} X_{1i}^3 \qquad (15.21b)$$

$$\text{III.} \quad \sum_{i=1}^{n} X_{1i}^2Y_i = b_0 \sum_{i=1}^{n} X_{1i}^2 + b_1 \sum_{i=1}^{n} X_{1i}^3 + b_{11} \sum_{i=1}^{n} X_{1i}^4 \qquad (15.21c)$$

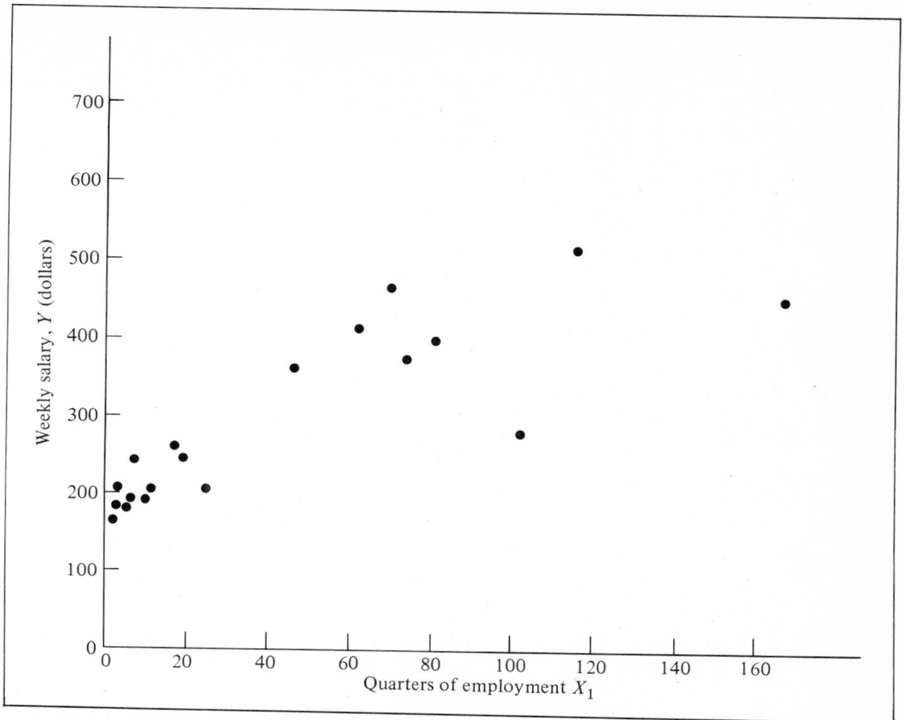

Figure 15.8 Scatter diagram of quarters of employment X_1 and weekly salary Y.

For the data of Table 15.7, the values of the various sums needed to solve Equations (15.21*a*), (15.21*b*), and (15.21*c*) are given below:

$$\sum_{i=1}^{n} Y_i = 5{,}536.35 \qquad \sum_{i=1}^{n} X_{1i}Y_i = 321{,}953.94 \qquad \sum_{i=1}^{n} X_{1i}^3 = 8{,}885{,}349$$

$$\sum_{i=1}^{n} X_{1i}^2 = 76{,}043 \qquad \sum_{i=1}^{n} X_{1i}^4 = 1{,}177{,}870{,}900 \qquad \sum_{i=1}^{n} Y_i^2 = 1{,}846{,}699.5$$

$$\sum_{i=1}^{n} X_{1i}^2 Y_i = 31{,}988{,}744 \qquad \sum_{i=1}^{n} X_{1i} = 825 \qquad n = 19$$

The three normal equations would be:

$$\begin{array}{lrcl} \text{I.} & 5{,}536.35 &=& 19b_0 + 825b_1 + 76{,}043b_{11} \\ \text{II.} & 321{,}953.94 &=& 825b_0 + 76{,}043b_1 + 8{,}885{,}349b_{11} \\ \text{III.} & 31{,}988{,}744 &=& 76{,}043b_0 + 8{,}885{,}349b_1 + 1{,}177{,}870{,}900b_{11} \end{array}$$

```
DEPENDENT VARIABLE..      SAL   ← Y    WEEKLY SALARY AS OF JAN.

VARIABLE(S) ENTERED       EMP   ← X1   QUARTERS OF EMPLOYMENT AS OF JAN.

                          EMPSQ ← X1^2


MULTIPLE R            0.88981 ← R
R SQUARE              0.79177 ← R^2

STANDARD ERROR       54.90074 ← SYX



ANALYSIS OF VARIANCE      DF      SUM OF SQUARES      MEAN SQUARE       F
REGRESSION                2.      183371.63025        91685.81513    30.41905
RESIDUAL                 16.       48225.46765         3014.09173


---------------- VARIABLES IN THE EQUATION ----------------

VARIABLE              B                       STD ERROR B          F

EMP              4.037787          ← b1       Sb1 = 0.81555      24.513
EMPSQ          -0.1461349D-01      ← b11      Sb11 = 0.00553      6.973
(CONSTANT)     173.4962            ← b0
```

Figure 15.9 Partial output for the data of Table 15.7 from SPSS subprogram REGRESSION.

As in the case of multiple linear regression, the values of the three sample regression coefficients (b_0, b_1, and b_{11}) can be obtained by using various methods for the solution of simultaneous equations, through a matrix algebra approach, or by using an available computer package. Figure 15.9 presents the partial output from the SPSS subprogram REGRESSION for the data of Table 15.7.

From Figure 15.9, we observe that the computed values of the regression coefficients in this problem are

$$b_0 = 173.4962 \qquad b_1 = 4.037787 \qquad b_{11} = -.01461349$$

Therefore, the curvilinear regression equation can be expressed as:

$$\hat{Y}_i = 173.4962 + 4.037787X_{1i} - .01461349X_{1i}^2$$

where $\hat{Y}_i$ = predicted weekly salary (dollars) of male employee i
X_{1i} = number of quarters (of years) of employment for male employee i

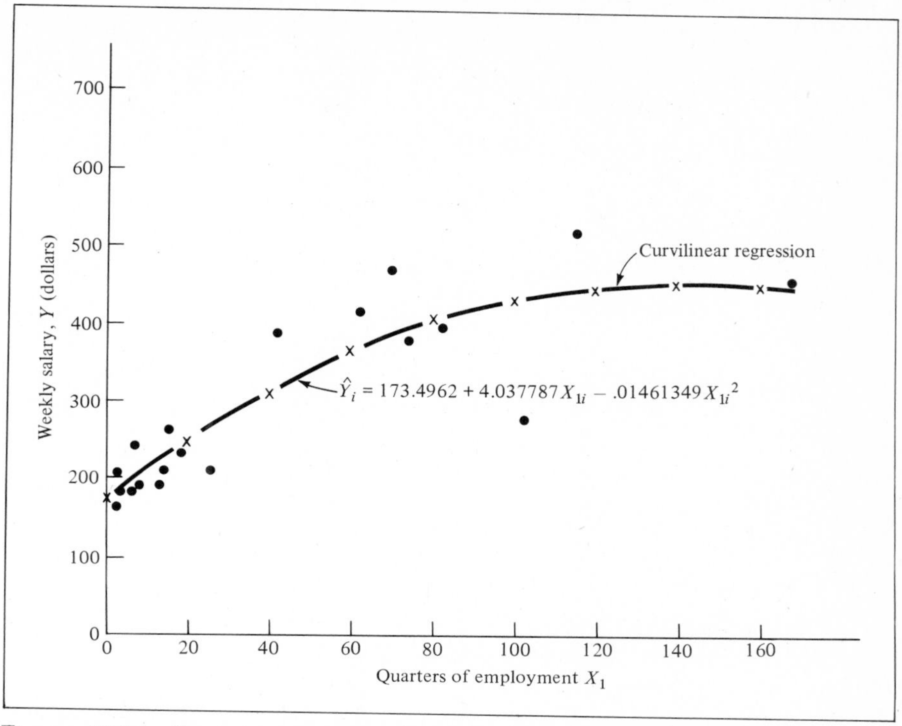

Figure 15.10 Scatter diagram expressing the curvilinear relationship between length of employment X_1 and weekly salary Y.

As depicted in Figure 15.10, this curvilinear regression equation is plotted on the scatter diagram to indicate how well the selected regression model fits the original data. Moreover, Figure 15.10 helps us to interpret the meaning of our computed regression coefficients. From our curvilinear regression equation and Figure 15.10, the Y intercept (b_0, computed as 173.4962) can be interpreted to mean that the predicted weekly salary of a new employee (without any length of employment) is approximately \$173.50. Moreover, to interpret the coefficients b_1 and b_{11}, we see from Figure 15.10 that weekly salary rises with longer periods of employment; nevertheless, we also observe that these increases in salary level off or become reduced with increasing length of employment. This can also be seen by predicting the average weekly salary for male employees who have been employed by this corportion for 5, 10, 15, 20, or 25 years (that is, for 20, 40, 60, 80, or 100 quarters). Using our curvilinear regression equation:

$$\hat{Y}_i = 173.4962 + 4.037787X_{1i} - .01461349X_{1i}^2$$

for $X_{1i} = 20$ (quarters of employment) we have

$$\hat{Y}_i = 173.4962 + 4.037787(20) - .01461349(20)^2 = \$248.41$$

for $X_{1i} = 40$ (quarters of employment) we have

$$\hat{Y}_i = 173.4962 + 4.037787(40) - .01461349(40)^2 = \$311.63$$

for $X_{1i} = 60$ (quarters of employment) we have

$$\hat{Y}_i = 173.4962 + 4.037787(60) - .01461349(60)^2 = \$363.15$$

for $X_{1i} = 80$ (quarters of employment) we have

$$\hat{Y}_i = 173.4962 + 4.037787(80) - .01461349(80)^2 = \$402.99$$

while for $X_{1i} = 100$ (quarters of employment) we have

$$\hat{Y}_i = 173.4962 + 4.037787(100) - .01461349(100)^2 = \$431.14$$

Thus, we can see that a male employee with 5 years' experience is expected to earn \$74.91 per week more than a new employee (\$248.41 − \$173.50). However, a male employee with 10 years' experience is expected to earn only \$63.22 per week more than an employee with 5 years' experience (\$311.63 − \$248.41). As the number of years of experience of employees increases, we note that their expected weekly salary continues to increase at an even lower rate. An employee with 15 years' experience can be expected to make \$51.62 more than one with 10 years' experience; an employee with 20 years' experience can be expected to make \$39.84 per week more than one with 15 years' experience; and an employee with 25 years' experience can be expected to make \$28.15 per week more than one with 20 years' experience.

15.10.2 Testing for the Significance of the Curvilinear Model

Now that the curvilinear model has been fitted to the data we can determine whether there is a significant curvilinear relationship between weekly salary Y and the number of quarters of employment X_1. In a manner similar to multiple regression (see Section 15.4), the null and alternative hypotheses can be set up as follows:

H_0: $\beta_1 = \beta_{11} = 0$ (there is no curvilinear relationship between X_1 and Y)

H_1: $\beta_1 \neq \beta_{11} \neq 0$ (at least one regression coefficient is not equal to zero)

The null hypothesis can be tested by utilizing an F test [Equation (15.3)] as indicated in Table 15.8.

Table 15.8
ANALYSIS OF VARIANCE TABLE FOR TESTING THE SIGNIFICANCE OF A CURVILINEAR POLYNOMIAL RELATIONSHIP

Source	DF	Sums of Squares (SS)	Variance	F
Regression	2	$b_0 \sum_{i=1}^{n} Y_i + b_1 \sum_{i=1}^{n} X_{1i}Y_i + b_{11} \sum_{i=1}^{n} X_{1i}^2 Y_i - \frac{\left(\sum_{i=1}^{n} Y_i\right)^2}{n}$	$S_{reg}^2 = \frac{SS_{reg}}{2}$	$\frac{S_{reg}^2}{S_{YX}^2}$
Error	$n-3$	$\sum_{i=1}^{n} Y_i^2 - b_0 \sum_{i=1}^{n} Y_i - b_1 \sum_{i=1}^{n} X_{1i}Y_i - b_{11} \sum_{i=1}^{n} X_{1i}^2 Y_i$	$S_{YX}^2 = \frac{SS_{error}}{n-3}$	
Total	$n-1$	$\sum_{i=1}^{n} Y_i^2 - \frac{\left(\sum_{i=1}^{n} Y_i\right)^2}{n}$		

For the data of Table 15.7 the ANOVA table (also displayed as part of the computer output in Figure 15.9) is presented in Table 15.9.

If a level of significance of .05 is chosen, from Appendix E, Table E.5, we find that for 2 and 16 degrees of freedom the critical value on the F distribution is 3.63 (see Figure 15.11). Utilizing Equation (15.3), since $F_{2,16} = S_{reg}^2/S_{YX}^2 = 30.419 > 3.63$, we can reject the null hypothesis (H_0) and conclude that there is a significant curvilinear relationship between weekly salary Y and length of employment X_1.

15.10.3 Index of Determination

In the multiple-regression model we computed the coefficient of multiple determination $r_{Y1.2}^2$ (see Section 15.5) to represent the proportion of variation in Y that is explained by variation in the independent variables. In curvilinear regression analysis, this coefficient, called the **index of determination** (R^2), can be computed from

Table 15.9
ANALYSIS OF VARIANCE TABLE FOR TESTING THE SIGNIFICANCE OF THE CURVILINEAR RELATIONSHIP IN THE SALARY–LENGTH OF EMPLOYMENT PROBLEM

Source	DF	Sums of Squares (SS)	Variance	F
Regression	2	$(173.4962)(5{,}536.35) + (4.037787)(321{,}953.94) + (-.01461349)(31{,}988{,}744) - \frac{(5{,}536.35)^2}{19} = 183{,}371.63025$	$\frac{183{,}371.63025}{2} = 91{,}685.81513$	$\frac{91{,}685.81513}{3{,}014.09173} = 30.419$
Error	$19-3 = 16$	$(1{,}848{,}699.5) - (173.4962)(5{,}536.35) - (4.037787)(321{,}953.94) - (-.01461349)(31{,}988{,}744) = 48{,}255.46765$	$\frac{48{,}255.46765}{16} = 3{,}014.09173$	
Total	$19-1 = 18$	$1{,}848{,}699.5 - \frac{(5{,}536.35)^2}{19} = 231{,}597.0979$		

SOURCE: Format of Table 15.8.

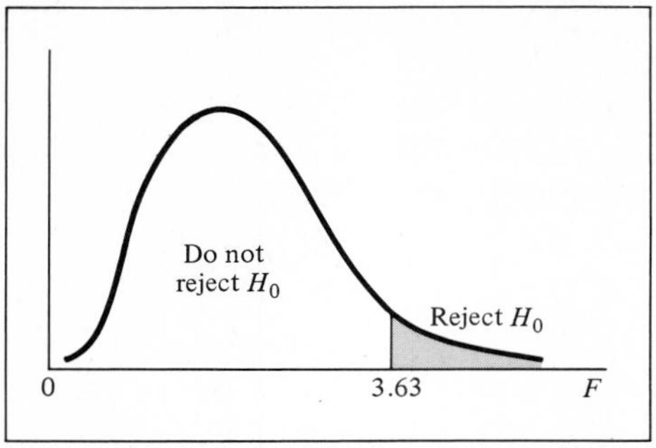

Figure 15.11 Testing for the existence of a curvilinear relationship at the .05 level of significance with 2 and 16 degrees of freedom.

$$R^2 = \frac{SS_{reg}}{SS_{tot}} \qquad (15.22)$$

where

$$SS_{reg} = b_0 \sum_{i=1}^{n} Y_i + b_1 \sum_{i=1}^{n} X_{1i}Y_i + b_{11} \sum_{i=1}^{n} X_{1i}^2 Y_i - \frac{\left(\sum_{i=1}^{n} Y_i\right)^2}{n}$$

$$SS_{tot} = \sum_{i=1}^{n} Y_i^2 - \frac{\left(\sum_{i=1}^{n} Y_i\right)^2}{n}$$

In our weekly salary–length of employment problem, we have already computed

$$SS_{reg} = 183{,}371.63025 \qquad \text{and} \qquad SS_{tot} = 231{,}597.0979$$

Thus, as displayed in Figure 15.9,

$$R^2 = \frac{SS_{reg}}{SS_{tot}} = \frac{183{,}371.63025}{231{,}597.0979} = .79177$$

This index of determination, computed as .79177, can be interpreted to mean that 79.177% of the variation in weekly salary can be explained by the curvilinear relationship between weekly salary Y and length of employment X_1.

15.11 COMPARING THE CURVILINEAR MODEL TO THE LINEAR MODEL

In using a regression model to examine a relationship between two variables, we would like to fit not only the most accurate model but also the simplest model expressing that relationship. Therefore, it becomes important to examine whether there is a significant difference between the curvilinear model

$$Y_i = \beta_0 + \beta_1 X_{1i} + \beta_{11} X_{1i}^2 + \varepsilon_i$$

and the linear model

$$Y_i = \beta_0 + \beta_1 X_i + \varepsilon_i$$

When we evaluated the multiple-regression model in Section 15.6.2, we determined the contribution of each independent variable from Equation (15.7):

$$SS_{\text{reg}}(b_k|\text{slopes of all variables except } k) = \frac{b_k^2 S_{YX}^2}{S_{b_k}^2} \quad (15.7)$$

Since the standard errors of each regression coefficient are available from the computer output of Figure 15.9, we can determine the contribution of the curvilinear effect to the model by using Equation (15.7) as follows:

$$SS_{\text{reg}}(b_{11}|b_1) = \frac{b_{11}^2 S_{YX}^2}{S_{b_{11}}^2}$$

and for our data

$$SS_{\text{reg}}(b_{11}|b_1) = \frac{(-.01461349)^2(3{,}014.09173)}{(.00553)^2} = 21{,}018.06029$$

Thus the regression sum of squares can be subdivided into two component parts: a *linear* effect and a *curvilinear* effect as shown in Table 15.10.

The null and alternative hypotheses to test for the contribution of the curvilinear effect to the regression model are:

H_0: Including the curvilinear effect does not significantly improve the model ($\beta_{11} = 0$)
H_1: Including the curvilinear effect significantly improves the model ($\beta_{11} \neq 0$)

If a level of significance of .05 is selected, from Appendix E, Table E.5, we find that for 1 and 16 degrees of freedom the critical value is 4.49 (see Figure

Table 15.10
ANALYSIS OF VARIANCE TABLE DIVIDING THE REGRESSION SUM OF SQUARES INTO COMPONENTS TO DETERMINE THE LINEAR AND CURVILINEAR EFFECTS

Source	DF	Sums of Squares (SS)	Variance	F
Regression	2	183,371.63025	91,685.81513	
b_1 (linear)	1	162,353.56996	162,353.56996	
$b_{11}\|b_1$ (curvilinear)	1	21,018.06029	21,018.06029	6.973
Error	16	48,225.46765	S_{YX}^2 = 3,014.09173	
Total	18	231,597.0979		

15.12). Using Equation (15.7), since $F_{1,16} = 6.973 > 4.49$, our decision would be to reject H_0 and conclude that the curvilinear model is significantly better than the linear model in representing the relationship between weekly salary and length of employment.

15.12 THE PROBLEM OF MULTICOLLINEARITY

One important problem in the application of multiple regression analysis involves the possible *multicollinearity* of the independent variables. This condition refers to situations in which some of the independent variables are highly correlated with each other. In such cases the values of the regression coefficients for the correlated variables may fluctuate drastically, depending on which variables are included in the selected model. If we refer to the heating oil consumption example, we can observe from Figures 15.2 and 15.5 that since the two independent variables were uncorrelated (r = .00892 as depicted in Table 15.4), the slope for the variable average daily atmospheric

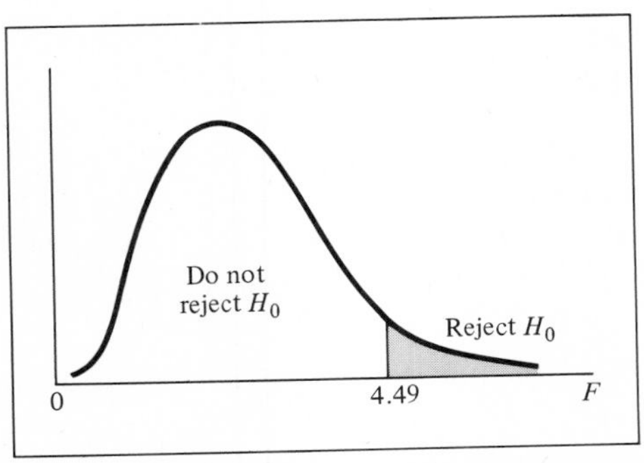

Figure 15.12 Testing for the contribution of the curvilinear effect to a regression model at the .05 level of significance with 1 and 16 degrees of freedom.

temperature is relatively unaffected by whether the variable X_2 (amount of attic insulation) is included in the model. This problem of multicollinearity is discussed in greater depth in References 2 and 8.

15.13 MULTIPLE REGRESSION: AN OVERVIEW AND THE ROLE OF THE COMPUTER

In this chapter we have extended the simple linear regression model of Chapter 14. We have considered situations (multiple linear regression) which involve two independent variables as well as situations (curvilinear regression) which involve a nonlinear relationship between a single independent variable and a dependent variable. The procedures that we have discussed when analyzing the multiple-regression model can clearly be extended to models that have several independent variables, each of which may vary differently with respect to the dependent variable. For these more complex cases, the matrix algebra approach (see References 4, 6, 7, 8, and 10) offers a more concise means of expression. In addition, it indeed becomes quite clear that for multiple regression problems the role of the computer is crucial to data analysis. Even for multiple regression problems having only two independent variables the computations involved in the solution of simultaneous equations can be extremely time-consuming. Various computer packages (such as SPSS and BMDP), each with their advantages and disadvantages (see Reference 1), aid the researcher in analyzing the results of multiple regression problems.

In order to observe some of the similarities and differences between these two packages, the data for the heating oil consumption problem (Table 15.1) that was analyzed by using the SPSS subprogram REGRESSION (Figure 15.2) was also analyzed by using the BMDP program P2R, Stepwise Regression. The output of this package is presented in Figure 15.13.

VARIABLES TO BE USED

1 HTNGOIL	2 TEMPF	3 INSU
Y	X_1	X_2

VARIABLE NO.	NAME	MEAN	STANDARD DEVIATION	COEFFICIENT OF VARIATION	SKEWNESS	SMALLEST VALUE	LARGEST VALUE
1	HTNGOIL	216.4932	129.8721	0.5999	0.0284	31.4000	441.0999
2	TEMPF	40.2666	20.6794	0.5136	0.0139	8.0000	73.0000
3	INSU	6.3333	2.9681	0.4686	0.1558	3.0000	10.0000

CORRELATION MATRIX

		HTNGOIL 1	TEMPF 2	INSU 3
HTNGOIL	1	1.0000		
TEMPF	2	-0.8697	1.0000	
INSU	3	-0.4651	0.0089	1.0000

Figure 15.13 Partial output for the data of Table 15.1 from the BMDP program P2R.

MULTIPLE R 0.9827 ← $r_{Y.12}$
MULTIPLE R-SQUARE 0.9656 ← $r^2_{Y.12}$

STD. ERROR OF EST. 26.0132 ← S_{YX}

ANALYSIS OF VARIANCE

	SUM OF SQUARES	DF	MEAN SQUARE	F RATIO
REGRESSION	228014.38	2	114007.2	168.48
RESIDUAL	8120.2305	12	676.6858	

VARIABLES IN EQUATION

VARIABLE		COEFFICIENT	STD. ERROR OF COEFF
(Y-INTERCEPT		b_0= 562.150)	
TEMPF	2	b_1= -5.437	S_{b_1}=0.336
INSU	3	b_2= -20.012	S_{b_2}=2.342

LIST OF PREDICTED VALUES, RESIDUALS, AND VARIABLES

VARIABLES

CASE NO.	LABEL	X_1 2 TEMPF	X_2 3 INSU	Y 1 HTNGOIL	$\hat{Y}$ PREDICTED	$Y-\hat{Y}$ RESIDUAL
1		40.0000	3.0000	275.2998	284.6501	-9.3503
2		27.0000	3.0000	363.7998	355.3257	8.4741
3		40.0000	10.0000	164.3000	144.5642	19.7358
4		73.0000	6.0000	40.8000	45.2063	-4.4063
5		64.0000	6.0000	94.3000	94.1355	0.1645
6		34.0000	6.0000	230.9000	257.2329	-26.3329*
7		9.0000	6.0000	366.7000	393.1472	-26.4473*
8		8.0000	10.0000	300.5999	318.5347	-17.9348
9		23.0000	10.0000	237.8000	236.9861	0.8139
10		63.0000	3.0000	121.4000	159.6089	-38.2089*
11		65.0000	10.0000	31.4000	8.6497	22.7503
12		41.0000	6.0000	203.5000	219.1768	-15.6768
13		21.0000	3.0000	441.0999	387.9451	53.1548**
14		38.0000	3.0000	323.0000	295.5234	27.4766*
15		58.0000	10.0000	52.5000	46.7058	5.7942

EACH ASTERISK REPRESENTS ONE STANDARD DEVIATION

STEPWISE REGRESSION COEFFICIENTS / PARTIAL CORRELATIONS

VARIABLES STEP	0 Y-INTCPT	2 TEMPF	3 INSU	2 TEMPF	3 INSU
0	216.4932*	-5.4622	-20.3503	-0.8697	-0.4651
1 (see Figure 15.5)→	436.4380*	-5.4622*	-20.0123	-0.8697*	-0.9267
2	562.1504*	-5.4366*	-20.0123*	-0.9778*	-0.9267*
	b_0	b_1	b_2	$r_{Y1\cdot 2}$	$r_{Y2\cdot 1}$

NOTE-
1) REGRESSION COEFFICIENTS FOR VARIABLES IN THE EQUATION ARE INDICATED BY AN ASTERISK
2) THE REMAINING COEFFICIENTS ARE THOSE WHICH WOULD BE OBTAINED IF THAT VARIABLE WERE TO ENTER IN THE NEXT STEP

Figure 15.13 (continued)

PROBLEMS

* 15.1 Referring to Problem 14.2, the personnel manager feels that there may be a relationship between a second independent variable, salary, and absenteeism. The annual salary (in thousands of dollars) of the workers in the sample was:

Worker	Annual Salary (thousands of dollars)
1	8.5
2	16.4
3	14.3
4	8.0
5	14.7
6	14.0
7	11.2
8	15.3
9	16.9
10	17.2

(a) Assuming that each independent variable (age and annual salary) is related linearly to absenteeism, use the least-squares method to find the multiple regression coefficients (b_0, b_1, b_2).

(b) Interpret the meaning of the slopes in this problem.

(c) Predict the number of days absent for a 40-year-old worker who earns $13,000 per year.

(d) Determine whether there is a significant relationship between absenteeism and the two independent variables (age and salary) at the .05 level of significance.

(e) Compute the coefficient of multiple determination $r_{Y.12}^2$ and interpret its meaning in this problem.

(f) At the .05 level of significance, determine whether each independent variable makes a contribution to the regression model. Based upon these results, indicate the regression model that should be used in this problem.

(g) Compute the standard error of the estimate.

(h) Set up a 95% confidence interval estimate of the true population slope for the relationship of absenteeism to age.

(i) Set up a 95% confidence interval estimate of the average number of days absent for a 40-year-old worker who earns $13,000 per year.

(j) Compute the coefficients of partial determination $r_{Y1.2}^2$ and $r_{Y2.1}^2$ and interpret their meaning in this problem.

(**Note:** This problem can be solved using matrix algebra or simultaneous equation methods, or by using an available computer package.)

15.2 Referring to Problem 14.3, the statistician for the automobile manufacturer believes that a second independent variable, shipping mileage, may be related to delivery time. The shipping mileage (in hundreds of miles) of the 16 cars in the sample was:

Car	Shipping Mileage (hundreds of miles)
1	7.5
2	13.3
3	4.7
4	14.6
5	8.4
6	12.6
7	6.2
8	16.4
9	9.7
10	17.2
11	10.6
12	11.3
13	9.0
14	12.3
15	8.2
16	11.5

(a) Assuming that each independent variable (number of options ordered and shipping mileage) is related linearly to delivery time, use the least-squares method to find the multiple regression coefficients (b_0, b_1, b_2).

(b) Interpret the meaning of the slopes in this problem.

(c) If a car was ordered with 10 options and had to be shipped 800 miles, what would you predict the delivery time to be?

(d) Determine whether there is a significant relationship between delivery time and the two independent variables, number of options and shipping mileage, at the .05 level of significance.

(e) Compute the coefficient of multiple determination $r_{Y.12}^2$ and determine its meaning in this problem.

(f) At the .05 level of significance, determine whether each independent variable makes a contribution to the regression model. Based upon these results, indicate the regression model that should be utilized in this problem.

(g) Compute the standard error of the estimate.

(h) Set up a 95% confidence interval estimate of the average delivery time for a car that was ordered with 10 options and had to be shipped 800 miles.

(i) Set up a 95% confidence interval estimate of the true population slope between delivery time and shipping mileage.

(j) Compute the coefficients of partial determination $r_{Y1.2}^2$ and $r_{Y2.1}^2$ and interpret their meaning in this problem.

(**Note:** This problem can be solved using matrix algebra or simultaneous equation methods, or by using an available computer package.)

* 15.3 Referring to Problem 14.9, the auditor also believes that a second independent variable, number of rooms in the house, may be related to the amount of county taxes. The number of rooms for the 19 single-family homes in the sample were as follows:

Home	Number of Rooms
1	9
2	10
3	10
4	7
5	9
6	8
7	7
8	6
9	9
10	8
11	7
12	8
13	8
14	6
15	9
16	6
17	7
18	8
19	7

(a) Assuming that each independent variable (age and number of rooms) is related linearly to county taxes, use the least-squares method to find the multiple-regression coefficients (b_0, b_1, b_2).

(b) Interpret the meaning of the slopes in this problem.

(c) Predict the county taxes for a house that has eight rooms and is 10 years old.

(d) Determine whether there is a significant relationship between county taxes and the two independent variables (age and number of rooms) at the .05 level of significance.

(e) Compute the coefficient of multiple determination $r_{Y.12}^2$ and interpret its meaning in this problem.

(f) At the .05 level of significance, determine whether each independent variable makes a contribution to the regression model. Based upon these results, indicate the regression model that should be utilized in this problem.

(g) Compute the standard error of the estimate.

(h) Set up a 95% confidence interval estimate of the true population slope between county taxes and age of a home.

(i) Set up a 95% confidence interval estimate of the average county taxes for a home that is 24 years old and has eight rooms.

(j) Compute the coefficients of partial determination $r_{Y1.2}^2$ and $r_{Y2.1}^2$ and interpret their meaning in this problem.

(**Note:** This problem can be solved using matrix algebra or simultaneous equation methods, or by using an available computer package.)

15.4 The personnel department of a large industrial corporation would like to study the relationship between the weekly salary, and length of employment and age of its managerial employees. A random sample of 16 managerial employees is selected with the results shown in Table 15.11.

(a) Assuming that we wish to predict the weekly salary based upon the length of employment and the age of an employee, and that these variables are related linearly, use the least-squares method to find the regression coefficients (b_0, b_1, b_2).

(b) Interpret the meaning of the slopes in this problem.

(c) Predict the weekly salary for a managerial employee who has been employed for 15 *years* and is 47 years old.

(d) Determine whether there is a significant relationship between weekly salary and the two independent variables (length of employment and age) at the .05 level of significance.

(e) Compute the coefficient of multiple determination $r_{Y.12}^2$ and interpret its meaning in this problem.

(f) At the .05 level of significance, determine whether each independent variable makes a significant contribution to the regression model. Based upon these results, indicate the regression model that should be utilized in this problem.

(g) Compute the standard error of the estimate.

Table 15.11
MANAGERIAL EMPLOYEES

Employee	Weekly Salary	Length of Employment (months)	Age (years)
1	$519	330	46
2	636	569	65
3	558	375	57
4	408	113	47
5	442	215	41
6	502	343	59
7	408	252	45
8	490	348	57
9	432	352	55
10	399	256	61
11	336	87	28
12	544	337	51
13	316	42	28
14	419	129	37
15	420	216	46
16	482	327	56

(h) Set up 95% confidence interval estimates for the true population slopes between weekly salary and length of employment and weekly salary and age.

(i) Set up a 95% confidence interval estimate for the average weekly salary of a managerial employee who has been employed for 10 *years* and is 47 years old.

(j) Compute the coefficients of partial determination $r_{Y1.2}^2$ and $r_{Y2.1}^2$ and interpret their meaning in this problem.

(**Note:** This problem can be solved using matrix algebra or simultaneous equation methods, or by using an available computer package.)

15.5 Referring to Problem 15.4, the personnel department would like to perform a similar analysis for technical employees. A random sample of 33 technical employees was selected with the results shown in Table 15.12.

Perform the same analysis for the technical employees as was performed for the managerial employees in Problem 15.4 and compare the results.

15.6 The personnel department described in Problem 15.4 and 15.5 also wished to analyze the results for clerical employees. A random sample of 16 clerical employees was selected with the results shown in Table 15.13.

Perform the same analysis for the clerical employees as was performed for the

Table 15.12
TECHNICAL EMPLOYEES

Employee	Weekly Salary	Length of Employment (months)	Age (years)
1	$269	69	47
2	219	46	40
3	318	125	39
4	288	20	45
5	323	173	56
6	241	37	25
7	384	237	48
8	266	52	28
9	266	67	46
10	209	124	30
11	189	12	20
12	425	313	46
13	357	291	47
14	243	34	23
15	382	275	48
16	288	111	56
17	223	14	27
18	296	89	29
19	309	188	58
20	291	44	34
21	247	21	24
22	291	35	26
23	271	46	21
24	257	43	25
25	291	27	22
26	257	19	24
27	371	229	58
28	341	276	58
29	309	330	52
30	377	331	60
31	317	72	41
32	238	85	27
33	249	84	47

Table 15.13
CLERICAL EMPLOYEES

Employee	Weekly Salary	Length of Employment (months)	Age (years)
1	$189	25	21
2	412	220	39
3	254	31	25
4	370	300	55
5	398	311	50
6	183	6	32
7	234	18	44
8	270	89	46
9	220	76	40
10	254	53	47
11	221	17	53
12	402	354	58
13	259	64	42
14	240	88	34
15	213	11	21
16	453	407	53

managerial and technical employees in Problems 15.4 and 15.5 and compare the results.

15.7 The *Fortune* 500 consists of the 500 largest industrial corporations ranked by sales. We would like to be able to predict the net income of a corporation based upon sales and assets. A random sample of 27 corporations was selected from the *Fortune* 500 for the year 1977 with the results shown in Table 15.14.

Table 15.14
PREDICTING NET INCOME

Corporation	Net Income (millions)	Sales (millions)	Assets (millions)
General Electric	$1,088.20	$17,518.60	$13,696.8
United Technologies	195.98	5,550.67	2,979.3
Cities Service	210.20	4,388.20	3,739.6
Gulf & Western Ind.	150.33	3,643.00	4,159.1
General Mills	117.03	2,909.40	1,447.3
Dresser Industries	185.10	2,538.80	2,173.3
Singer	94.20	2,294.30	1,461.9
Iowa Beef Processors	29.97	2,023.77	277.5
Interco	76.76	1,566.40	778.0
SCM	37.41	1,377.64	767.9
Cummins Engine	67.02	1,263.81	818.4
Oscar Mayer	35.02	1,188.43	377.7
Crown Cork & Seal	53.80	1,049.14	631.1
Libbey-Owens-Ford	58.94	978.69	652.3
Akzona	7.52	808.85	674.9
A. O. Smith	18.68	727.28	355.6
Joy Manufacturing	48.22	677.69	551.2
General Cable	20.38	620.51	496.6
Adolph Coors	67.70	593.12	691.6
Cabot	39.90	574.84	579.1
Macmillan	19.40	512.73	525.0
Bell & Howell	11.80	491.57	366.8
Eagle-Picher Ind.	26.00	474.04	282.7
Ward Foods	7.74	443.67	125.2
Hanes	20.82	414.16	229.0
Inland Container	21.92	396.73	308.3
Koehring	10.53	361.61	277.2

SOURCE: *Fortune* Magazine, May 8, 1978.

(a) Assuming that each independent variable (sales and assets) is related linearly to net income, use the least-squares method to find the regression coefficients (b_0, b_1, b_2).

(b) Interpret the meaning of the slopes in this problem.

(c) Predict the net income for a corporation that has sales of $1 billion and has assets of $1 billion.

(d) Determine whether there is a significant linear relationship between net income and the two independent variables (sales and assets) at the .05 level of significance.

(e) Compute the coefficient of multiple determination $r_{Y.12}^2$ and interpret its meaning in this problem.

(f) At the .05 level of significance determine whether each independent variable makes a contribution to the regression model. Based upon these results indicate the regression model that should be used in this problem.

(g) Compute the standard error of the estimate.

Table 15.15
PREDICTING NET INCOME

Corporation	Net Income (millions)	Sales (millions)	Assets (millions)
Royal Crown Companies	$18.67	$349.62	$168.6
Wyman-Gordon	20.97	291.50	195.9
Rorer Group	18.14	243.80	169.2
Prentice-Hall	23.72	230.64	177.9
Capital Industries-EMI	16.16	209.77	137.1
King-Seeley Thermos	19.16	209.31	134.6
Pettibone	8.79	206.68	180.7
Cox Broadcasting	25.46	186.43	284.5
Storage Technology	11.43	162.27	159.7
First Mississippi	14.44	161.49	210.7
Overhead Door	9.64	157.63	79.9
American Sterilizer	2.88	155.71	104.5
Worthington Industries	7.48	148.60	68.1
Keller Industries	5.29	147.12	87.6
G. F. Business Equip.	0.78	145.11	87.8
Justin Ind.	6.21	144.40	97.0
Ideal Toy	5.15	137.63	78.5
Diversey	4.13	130.56	72.3
Compugraphic	9.61	129.95	86.2
Florida Steel	1.79	127.32	88.6
Robintech	−1.46	126.30	101.7
Filmways	2.70	125.33	68.0
Twin Disc	7.41	115.88	97.2
Mansfield Tire and Rubber	−4.88	112.49	69.6
Bayly	2.71	108.73	44.5

SOURCE: *Fortune* Magazine, June 19, 1978.

(h) Set up a 95% confidence interval estimate of the true population slope for the relationship between sales and net income.

(i) Set up a 95% confidence interval estimate of the average net income for a corporation which has sales of $1 billion and has assets of $1 billion.

(j) Compute the coefficients of partial determination $r_{Y1.2}^2$ and $r_{Y2.1}^2$ and interpret their meaning in this problem.

(**Note:** This problem can be solved using matrix algebra or simultaneous equation methods, or by using an available computer package.)

15.8 The *Fortune* Second 500 consists of the industrial corporations that are ranked 501st to 1,000th in sales. As in Problem 15.7 for the *Fortune* 500, we would like to predict the net income of a corporation based upon sales and assets. A random sample of 25 corporations was selected from the *Fortune* Second 500 for the year 1977 with the results shown in Table 15.15.

Perform the same kind of analysis for the *Fortune* Second 500 as was performed for the *Fortune* 500 in Problem 15.7 and compare the results. In parts (c) and (i), however, make your predictions for a corporation that has sales of $200 million and has assets of $200 million.

* 15.9 Referring to the company discussed in Section 15.11, the personnel director also selected a random sample of 13 female employees with the following results:

Employee	Weekly Salary	Quarters of Employment
1	$163.47	1
2	189.51	9
3	242.12	10
4	239.84	13
5	276.73	15
6	289.54	22
7	260.74	30
8	215.79	33
9	297.28	39
10	281.01	49
11	380.33	92
12	352.96	136
13	303.09	144

(a) Set up a scatter diagram of quarters of employment and weekly salary.

(b) Assuming a curvilinear relationship between length of employment and weekly salary, use the least-squares method to find the regression coefficients (b_0, b_1, b_{11}).

(c) Predict the weekly salary for a female who has been employed for 5 years.

(d) Determine whether there is a significant curvilinear relationship between weekly salary and length of employment at the .05 level of significance.

(e) Compute the index of determination R^2 and interpret its meaning in this problem.

(f) At the .05 level of significance, determine whether the curvilinear model is superior to the linear regression model.

(**Note:** This problem can be solved using matrix algebra or simultaneous equation methods, or by using an available computer package.)

15.10 Referring to Problem 14.7, suppose that the agronomist wished to determine whether there was a curvilinear relationship between the amount of a natural organic fertilizer utilized and the yield of tomatoes when either 0, 20, 40, 60, 80, or 100 pounds per hundred square feet of fertilizer was applied. These six different amounts of fertilizer are randomly assigned to plots of land with the following results:

Plot	X Amount of Fertilizer (lb/hundred sq ft)	Y Yield (lb)
1	0	6
2	0	9
3	20	19
4	20	24
5	40	32
6	40	38
7	60	46
8	60	50
9	80	48
10	80	54
11	100	52
12	100	58

(a) Set up a scatter diagram of yield Y and amount of fertilizer X.

(b) Assuming a curvilinear relationship between the amount of fertilizer used and tomato yield, use the least-squares method to find the regression coefficients (b_0, b_1, b_{11}).

(c) Predict the yield of tomatoes (in pounds) for a plot that has been fertilized with 70 pounds per hundred square feet of natural organic fertilizer.

(d) Determine whether there is a significant curvilinear relationship between the amount of fertilizer used and tomato yield at the .05 level of significance.

(e) Compute the index of determination R^2 and interpret its meaning in this problem.

(f) At the .05 level of significance, determine whether the curvilinear model is superior to the linear regression model.

(**Note:** This problem can be solved using matrix algebra or simultaneous equation methods, or by using an available computer package.)

15.11 Referring to Problem 14.10, suppose that the industrial psychologist assigned the 15 typists to one of five different alcoholic consumption levels (0, 1, 2, 3, 4 ounces) with the following results:

Typist	X Alcoholic Consumption (ounces)	Y Number of Errors
1	0	2
2	0	6
3	0	3
4	1	7
5	1	5
6	1	9
7	2	12
8	2	7
9	2	9
10	3	13
11	3	18
12	3	16
13	4	24
14	4	30
15	4	22

(a) Set up a scatter diagram between alcoholic consumption X and the number of errors Y.

(b) Assuming a curvilinear relationship between alcoholic consumption and the number of errors, use the least-squares method to find the regression coefficients (b_0, b_1, b_{11}).

(c) Predict the number of errors made by the typist who has consumed 2.5 ounces of alcohol.

(d) Determine whether there is a significant curvilinear relationship between alcoholic consumption and the number of errors made at the .05 level of significance.

(e) Compute the index of determination R^2 and interpret its meaning in this problem.

(f) At the .05 level of significance, determine whether the curvilinear model is superior to the linear regression model.

(**Note:** This problem can be solved using matrix algebra or simultaneous equation methods, or by using an available computer package.)

15.12 Referring to Problem 14.5, if a curvilinear relationship is assumed between the attendance and amount bet:

(a) Use the least-squares method to find the regression coefficients (b_0, b_1, b_{11}).

(b) Predict the amount bet for a day in which attendance is 30,000.

(c) Determine whether there is a significant curvilinear relationship between attendance and amount bet at the .05 level of significance.

(d) Compute the index of determination R^2 and interpret its meaning here.

(e) At the .05 level of significance, determine whether the curvilinear model is superior to the linear regression model.

(**Note:** This problem can be solved using matrix algebra or simultaneous equation methods, or by using an available computer package.)

15.13 For the sample of 94 students selected by the researcher and the dean (Figure 2.6), use the results of questions 13 (number of pairs of jeans owned), 15 (age), and 16 (height) to perform a multiple regression analysis to predict the jeans owned by a student based upon his or her age and height.

(a) Assuming that each of the independent variables (age and height) is related linearly to jeans owned, use the least-squares method to find the regression coefficients (b_0, b_1, and b_2).

(b) Interpret the meaning of the slopes in this problem.

(c) Predict the number of pairs of jeans owned by a student who is 20 years old and is 5 feet 8 inches tall.

(d) Determine whether there is a significant linear relationship between jeans owned and the two independent variables (age and height) at the .05 level of significance.

(e) Compute the coefficient of multiple determination $r_{Y.12}^2$ and interpret its meaning in this problem.

(f) At the .05 level of significance, determine whether each independent variable makes a contribution to the regression model. Based upon these results, indicate the regression model that should be used in this problem.

(g) Compute the standard error of the estimate.

(h) Set up a 95% confidence interval estimate of the true population slope for the relationship between jeans owned and age.

(i) Set up a 95% confidence interval estimate of the average number of pairs of jeans owned by a student who is 20 years old and is 5 feet 8 inches tall.

(j) Compute the coefficients of partial determination $r_{Y1.2}^2$ and $r_{Y2.1}^2$ and interpret their meaning in this problem.

(**Note:** This problem can be solved using matrix algebra or simultaneous equation methods, or by using an available computer package.)

15.14 For the sample of 94 students selected by the researcher and the dean (Figure 2.6), use the results of question 8 (tuition), question 5 (grade-point index), question 7 (expected annual starting salary) to perform a multiple regression analysis to predict the maximum tuition a student can pay based upon his or her grade-point index and expected annual starting salary.

(a) Assuming that each of the independent variables (grade-point index and expected annual starting salary) is related linearly to maximum tuition, use the least-squares method to find the regression coefficients (b_0, b_1, b_2).

(b) Interpret the meaning of the slopes in this problem.

(c) For a student with a grade-point index of 3.0 who expects an annual starting salary of $13,000, predict the maximum tuition that he or she would be willing to pay.

(d) Determine whether there is a significant linear relationship between maximum tuition and the two independent variables (grade-point index and expected annual starting salary) at the .05 level of significance.

(e) Compute the coefficient of multiple determination $r_{Y.12}^2$ and interpret its meaning in this problem.

(f) At the .05 level of significance, determine whether each independent variable makes a contribution to the regression model. Based upon these results indicate the regression model that should be used in this problem.

(g) Compute the standard error of the estimate.

(h) Set up a 95% confidence interval estimate of the true population slope for the relationship between maximum tuition and expected annual starting salary.

(i) Set up a 95% confidence interval estimate of the average maximum tuition that a student would be willing to pay who has a grade-point index of 3.0 and expects a starting salary of $13,000.

(j) Compute the coefficients of partial determination $r_{Y1.2}^2$ and $r_{Y2.1}^2$ and interpret their meaning in this problem.

(**Note**: This problem can be solved using matrix algebra or simultaneous equation methods, or by using an available computer package.)

REFERENCES

1. BERK, K. N., I. S. FRANCIS, AND M. E. MULLER, "A Review of the Manuals for BMDP and SPSS" (and comments and rejoinder), *J. Amer. Statistical Assoc.*, 73 (March 1978), 65–98.
2. COHEN, J., AND P. COHEN, *Applied Multiple Regression/Correlation Analysis for the Behavioral Sciences* (Hillsdale, N.J.: Lawrence Erlbaum, 1975).
3. DIXON, W. J., AND M. B. BROWN (eds.), *BMDP Biomedical Computer Programs P-Series 1977* (Berkeley: University of California Press, 1977).
4. DRAPER, N. R., AND H. SMITH, *Applied Regression Analysis* (New York: Wiley, 1966).
5. EZEKIEL, J., AND R. A. FOX, *Methods of Correlation and Regression Analysis*, 3rd ed. (New York: Wiley, 1959).
6. KLEINBAUM, D. G., AND L. L. KUPPER, *Applied Regression Analysis and Other Multivariable Methods* (North Scituate, Mass.: Duxbury Press, 1978).
7. MILLER, R. B., AND D. W. WICHERN, *Intermediate Business Statistics: Analysis of Variance, Regression and Time Series* (New York: Holt, Rinehart and Winston, 1977).
8. NETER, J., AND W. WASSERMAN, *Applied Linear Statistical Models* (Homewood, Ill.: Richard D. Irwin, 1974).
9. NIE, N. H., C. H. HULL, J. G. JENKINS, K. STEINBRENNER, AND D. H. BENT, *SPSS Statistical Package for the Social Sciences*, 2nd ed. (New York: McGraw-Hill, 1975).
10. OTT, L., *An Introduction to Statistical Methods and Data Analysis* (North Scituate, Mass.: Duxbury Press, 1977).
11. STATPACK: Statistical Package, 2nd ed., developed by International Business Machines Corporation in February 1970.

CHAPTER 16

INDEX NUMBERS

16.1 INTRODUCTION

In the previous two chapters we discussed the topic of regression analysis as a tool for model building and prediction. In these respects, regression analysis provides a useful guide to managerial decision making. In this and the following chapter we shall develop an understanding of the concepts behind index number construction, time-series analysis, as well as consider other business forecasting methods so that more timely and pertinent information will be available for managerial decision-making purposes.

Over the years, index numbers have become increasingly important to management as indicators of changing economic or business activity. In fact, the use of index numbers has become the most widely accepted procedure of measuring changes in business conditions. But actually, what are index numbers? Generally speaking, **index numbers constructed at a particular point in time measure the size or magnitude of some item at that particular point in time as a percentage of some base or reference object in the past.** Specifically, many kinds of index numbers can be constructed. As examples, numerous price indexes, quantity indexes, value indexes, quality indexes, and sociological indexes (References 1 and 9) have been devised. For simplicity,

Type of Index		Application
Quantity	⟶	Index of industrial production
Value	⟶	College grade-point index
Quality	⟶	Temperature/humidity ratio
Sociological	⟶	IQ

Figure 16.1 Types of indexes

and for business forecasting purposes, however, we shall consider the construction of price indexes in this chapter. Applications of other types of indexes are summarized in Figure 16.1.

16.2 THE PRICE INDEX

Price indexes reflect the percentage change in the price of some commodity (or group of commodities) in a given period of time over the price paid for that commodity (or group of commodities) at a particular point of time in the past. Price indexes, however, are not merely computed once, but rather are obtained over numerous consecutive time periods in order to indicate changing economic or business activity.

16.2.1 Selecting the Base Period for a Price Index

The base period or reference point then is the year or time period in the past against which all these comparisons are made. In selecting the base period for a particular index, two rules should be observed. First, the period selected should, as much as possible, be one of economic normalcy or stability rather than one at or near the *peak* of an expanding economy or the *trough* of a recession or declining economy. Second, the base period should be recent so that comparisons will not be unduly affected by changing technology, changing product quality, and/or changing consumer attitudes, interests, tastes, and habits.

16.2.2 Forming a Price Index for a Particular Commodity

As an example of a price index for a particular commodity, Table 16.1 depicts the average annual prices received by fishermen (in cents per pound) for flounder caught between the years of 1960 and 1975. Using 1960 as the base period or reference point, we may construct a price index for flounder by simply forming the ratio of the price paid to fishermen in any given year to the price paid in the base year, and then multiply the result by 100 to express the index as a percentage. Thus if we let the symbol $I_i^{(t)}$ represent the price index for the ith commodity in time period (year) t, we have

Table 16.1
PRICES PAID TO FISHERMEN FOR FLOUNDER
(1960 = 100.0)
(Annual Average Price in Cents per Pound)

Year	Price	Price Index	Year	Price	Price Index
1960	12.2	100.0	1968	11.4	93.4
1961	10.6	86.9	1969	13.7	112.3
1962	9.7	79.5	1970	15.3	125.4
1963	8.4	68.9	1971	16.7	136.9
1964	8.0	65.6	1972	21.4	175.4
1965	9.5	77.9	1973	23.7	194.3
1966	12.7	104.1	1974	26.5	217.2
1967	11.5	94.3	1975	34.7	284.4

SOURCE: Data are taken from *Statistical Abstract of the United States*, 1976, 1973, and *The Statistical History of the United States from Colonial Times to the Present.*

$$I_i^{(t)} = \frac{P_i^{(t)}}{P_i^{(0)}} \times 100 \qquad (16.1)$$

where $P_i^{(t)}$ = price paid for the *i*th commodity in time period t

$P_i^{(0)}$ = price paid for the *i*th commodity in time period 0—the base period

From Table 16.1 and using Equation (16.1), the price index for the year 1961 is

$$I^{(1)} = \frac{P^{(1)}}{P^{(0)}} \times 100 = \frac{10.6}{12.2} \times 100 = 86.9$$

where the base year is selected as 1960. Obviously, from Equation (16.1) the price index for the base period must be 100.0.

16.2.3 Forming a Price Index for a Group of Commodities

While a price index for any individual commodity [as described by Equation (16.1)] may be of interest, it is not usually considered important for most decision-making purposes. What is important is an index constituting a group of commodities—taken together—which may affect the quality of life enjoyed by a large number of consumers. Basically, four such types of price indexes concerning a group of commodities may be considered: simple aggregate price indexes, simple arithmetic mean of price relatives, weighted aggregate price indexes; and weighted arithmetic mean of price relatives.[1]

[1] In this context a "simple" index is one which is unweighted or equally weighted.

16.3 SIMPLE AGGREGATE PRICE INDEX

When developing a price index for a group of commodities, the easiest index to construct is the simple aggregate price index given by

$$I_{SA}^{(t)} = \frac{\sum_{i=1}^{n} P_i^{(t)}}{\sum_{i=1}^{n} P_i^{(0)}} \times 100 \tag{16.2}$$

where the superscript (t) in the symbol $I_{SA}^{(t)}$ represents the value of the simple aggregate index in time period t, and

$\sum_{i=1}^{n} P_i^{(t)}$ = sum of the prices paid for each of the n commodities in time period t

$\sum_{i=1}^{n} P_i^{(0)}$ = sum of the prices paid for the same commodities in time period 0—the base period

As an example, Table 16.2 depicts the average prices received by fishermen from processors for various quantities of selected types of fish caught every 5 years from 1960 to 1975. Using 1960 as the base period, the simple aggregate price index is obtained as shown in Table 16.3.

Table 16.2
PRICES PAID TO FISHERMEN FOR SELECTED TYPES OF FISH

Fish	Quantities* and Prices†	Year 1960	1965	1970	1975‡
Cod	Q	40	36	53	56
	P	6.5	8.0	11.2	23.3
Flounder	Q	127	180	169	156
	P	12.2	9.5	15.3	34.7
Haddock	Q	119	134	27	16
	P	7.9	8.9	22.7	26.8
Ocean perch	Q	141	84	55	32
	P	4.0	4.1	4.9	10.3
Tuna	Q	298	319	393	391
	P	15.7	15.7	26.2	37.4

SOURCE: Data are taken from *Statistical Abstract of the United States*, 1976. (Extracted from data provided by U.S. National Oceanic and Atmospheric Administration.)

* *Q*—quantity caught in millions of pounds.
† *P*—price received in cents per pound.
‡ Preliminary estimates.

Table 16.3
CONSTRUCTING A SIMPLE AGGREGATE PRICE INDEX
(1960 = 100.0)

	Year			
	1960	1965	1970	1975
Fish	$P_i^{(0)}$	$P_i^{(1)}$	$P_i^{(2)}$	$P_i^{(3)}$
Cod	6.5	8.0	11.2	23.3
Flounder	12.2	9.5	15.3	34.7
Haddock	7.9	8.9	22.7	26.8
Ocean perch	4.0	4.1	4.9	10.3
Tuna	15.7	15.7	26.2	37.4
Totals	46.3	46.2	80.3	132.5
Index	$\frac{46.3}{46.3} \times 100$ $I_{SA}^{(0)} = 100.0$	$\frac{46.2}{46.3} \times 100$ $I_{SA}^{(1)} = 99.8$	$\frac{80.3}{46.3} \times 100$ $I_{SA}^{(2)} = 173.4$	$\frac{132.5}{46.3} \times 100$ $I_{SA}^{(3)} = 286.2$

SOURCE: Data are taken from Table 16.2.

Such an index, however, possesses two distinct shortcomings. First, the index considers each commodity in the group as equally important and thereby permits the most expensive commodities per unit to be overly influential. Second, any change in the unit of measurement of any commodity (for example, tuna could be priced in cents per kilogram while the other fish continue to be priced in cents per pound) alters the value of the index. This latter drawback is overcome by constructing a simple arithmetic mean of price relatives for the group of commodities considered.

16.4 SIMPLE ARITHMETIC MEAN OF PRICE RELATIVES

A price relative for any commodity in any given period of time may be defined as the ratio of the price for that commodity at that given point in time to its price in some base period with the ensuing result multiplied by 100 and expressed as a percentage. To construct the simple arithmetic mean of price relatives, we first form the ratios of the prices of each commodity in time period t to the respective prices in the base period. Once these price ratios for each commodity are obtained, we merely sum up the result and divide this total by the number of commodities comprising the index. The average is then multiplied by 100 to be expressed as a percentage, that is,

$$I_{SM}^{(t)} = \frac{\sum_{i=1}^{n} P_i^{(t)}/P_i^{(0)}}{n} \times 100 \tag{16.3}$$

Table 16.4
CONSTRUCTING A SIMPLE MEAN OF PRICE RELATIVES
(1960 = 100.0)

	Year							
	1960		1965		1970		1975	
Fish	$P_i^{(0)}$	$\left(\frac{P_i^{(0)}}{P_i^{(0)}}\right)$	$P_i^{(1)}$	$\left(\frac{P_i^{(1)}}{P_i^{(0)}}\right)$	$P_i^{(2)}$	$\left(\frac{P_i^{(2)}}{P_i^{(0)}}\right)$	$P_i^{(3)}$	$\left(\frac{P_i^{(3)}}{P_i^{(0)}}\right)$
Cod	6.5	1.000	8.0	1.231	11.2	1.723	23.3	3.585
Flounder	12.2	1.000	9.5	0.779	15.3	1.254	34.7	2.844
Haddock	7.9	1.000	8.9	1.127	22.7	2.873	26.8	3.392
Ocean perch	4.0	1.000	4.1	1.025	4.9	1.225	10.3	2.575
Tuna	15.7	1.000	15.7	1.000	26.2	1.669	37.4	2.382
Totals		5.000		5.162		8.744		14.778
Index		$\frac{5.000}{5} \times 100$ $I_{SM}^{(0)} = 100.0$		$\frac{5.162}{5} \times 100$ $I_{SM}^{(1)} = 103.2$		$\frac{8.744}{5} \times 100$ $I_{SM}^{(2)} = 174.9$		$\frac{14.778}{5} \times 100$ $I_{SM}^{(3)} = 295.6$

SOURCE: Data are taken from Table 16.2.

The construction of $I_{SM}^{(t)}$, the arithmetic mean of price relatives, is demonstrated in Table 16.4, using the price data for the selected types of fish.

From Tables 16.3 and 16.4 we note that the computed index of the simple arithmetic mean of price relatives $I_{SM}^{(t)}$ differs from the computed simple aggregate price index $I_{SA}^{(t)}$. If the results from the two indexes are different, which type of index—one based on aggregates or one based on mean price relatives—is more appropriate? Which type, if either, should we choose? An aggregate price index represents the changes in prices, over time, for an entire group of commodities, whereas an index of the mean of price relatives reflects the average of the changes in prices, over time, of each commodity considered in the index. The major advantage of the simple aggregate price index over the mean of price relatives is that the former is easier to compute. On the other hand, while computationally more cumbersome, the simple arithmetic mean of price relatives is the more useful because, as shown in Figure 16.2, the relative price changes in each commodity are observed as well as the composite price change over all commodities considered. Even more importantly, however, the mean of price relatives is not influenced by the measuring units utilized—a major drawback of the simple aggregate price index.

Nevertheless, both the simple aggregate price index and the simple mean of price relatives suffer from the same shortcoming, that is, the simple (or unweighted) indexes fail to consider the importance of individual commodities comprising the indexes, and, therefore, they are not really meaningful as a measure of how price changes affect large groups of consumers. Unlike the simple aggregate price index which suffers from an overinfluence of large priced commodities, the simple mean of price relatives suffers from an overinfluence of large percentage increases in prices compared to the base period.

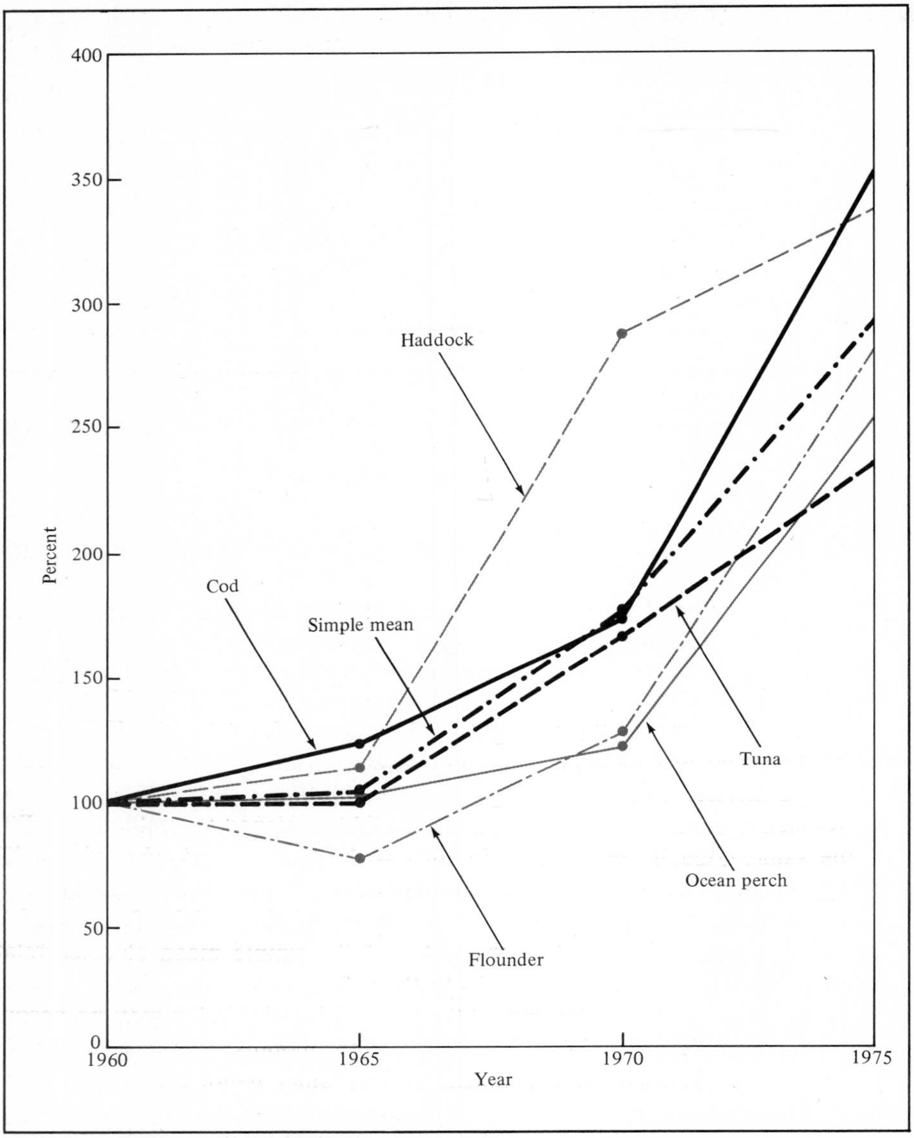

Figure 16.2 Simple mean of price relatives and price relatives for each of five selected types of fish sold to processors every 5 years, 1960–1975, (1960 = 100.0)

SOURCE: Data are taken from Table 16.2.

16.5 WEIGHTED AGGREGATE PRICE INDEX AND WEIGHTED MEAN OF PRICE RELATIVES

To overcome these difficulties we may use the weighted aggregate price index as defined by

$$I_{\mathrm{WA}}{}^{(t)} = \frac{\sum_{i=1}^{n} P_i{}^{(t)} W_i}{\sum_{i=1}^{n} P_i{}^{(0)} W_i} \times 100 \tag{16.4}$$

or the index of the weighted arithmetic mean of price relatives given by

$$I_{\mathrm{WM}}{}^{(t)} = \frac{\sum_{i=1}^{n} \left(P_i{}^{(t)}/P_i{}^{(0)}\right) W_i}{\sum_{i=1}^{n} W_i} \times 100 \tag{16.5}$$

where, in both formulas, W_i represents the "weight of importance" attached to the ith commodity ($i = 1, 2, \ldots, n$) in the groups considered.

Now how should these weights of importance be determined? Subjectively, as discussed in Chapters 5 and 11, each individual or decision maker could assign his/her own set of weights to each commodity and obtain a personal, subjective price index. Clearly, if the weights assigned to each commodity are the same (that is, $W_i = 1/n$ for all $i = 1, 2, \ldots, n$), then the weighted aggregate price index would be identical to the simple aggregate price index ($I_{\mathrm{WA}}{}^{(t)} = I_{\mathrm{SA}}{}^{(t)}$) and also the index of the weighted mean of price relatives would be equal to the index of the simple mean of price relatives ($I_{\mathrm{WM}}{}^{(t)} = I_{\mathrm{SM}}{}^{(t)}$). More objectively then, for an aggregate index the importance of the individual commodities comprising the index may be accounted for by choosing as weights the *quantities* or amounts of each commodity that are produced, used, or consumed. On the other hand, for indexes of mean price relatives, the importance of the individual commodities is reflected in the *total value* (that is, the price per unit multiplied by the quantity of units produced, used, or consumed), and it is these total values per commodity that are selected as the weights.

Now it seems reasonable to hold the weights constant when constructing the price index over time so that changes attributable to price movements may be isolated. Obviously, if both prices and quantities are varying, it would not be possible to isolate the fluctuations in price. But for which periods of time should these weights be chosen? Two different, commonly used approaches to this problem were taken by Laspeyres and by Paasche.

16.5.1 The Laspeyres Indexes

The Laspeyres Aggregate Price Index uses quantity weights from the base period so that $W_i = Q_i^{(0)}$ and Equation (16.4) can be rewritten as

$$I_{LA}^{(t)} = \frac{\sum_{i=1}^{n} P_i^{(t)} Q_i^{(0)}}{\sum_{i=1}^{n} P_i^{(0)} Q_i^{(0)}} \times 100 \qquad (16.6)$$

Correspondingly, the Laspeyres Index of Mean Price relatives uses value weights of the form $W_i = P_i^{(0)} Q_i^{(0)}$ so that Equation (16.5) may be given by

$$I_{LM}^{(t)} = \frac{\sum_{i=1}^{n} (P_i^{(t)}/P_i^{(0)}) P_i^{(0)} Q_i^{(0)}}{\sum_{i=1}^{n} P_i^{(0)} Q_i^{(0)}} \times 100 \qquad (16.7)$$

The Laspeyres indexes measure the percentage change in prices that would occur in any given period had we bought the same commodities in the same quantities that we actually had bought in a selected reference point or base

Table 16.5
CONSTRUCTING A LASPEYRES AGGREGATE PRICE INDEX
(1960 = 100.0)

	Quantities in Base Year	Year					
		1960		1970		1975	
Fish	$Q_i^{(0)}$	$P_i^{(0)}$	$P_i^{(0)}Q_i^{(0)}$	$P_i^{(2)}$	$P_i^{(2)}Q_i^{(0)}$	$P_i^{(3)}$	$P_i^{(3)}Q_i^{(0)}$
Cod	40	6.5	260.0	11.2	448.0	23.3	932.0
Flounder	127	12.2	1,549.4	15.3	1,943.1	34.7	4,406.9
Haddock	119	7.9	940.1	22.7	2,701.3	26.8	3,189.2
Ocean perch	141	4.0	564.0	4.9	690.9	10.3	1,452.3
Tuna	298	15.7	4,678.6	26.2	7,807.6	37.4	11,145.2
Totals		$\sum_{i=1}^{n} P_i^{(0)}Q_i^{(0)} = 7.992.1$		$\sum_{i=1}^{n} P_i^{(2)}Q_i^{(0)} = 13,590.9$		$\sum_{i=1}^{n} P_iQ_i = 21,125.6$	
Index		$\frac{7,992.1}{7,992.1} \times 100$ $I_{LA}^{(0)} = 100.0$		$\frac{13,590.9}{7,992.1} \times 100$ $I_{LA}^{(2)} = 170.1$		$\frac{21,125.6}{7,992.1} \times 100$ $I_{LA}^{(3)} = 264.3$	

SOURCE: Data are taken from Table 16.2.

Table 16.6
CONSTRUCTING A LASPEYRES INDEX OF MEAN PRICE RELATIVES (1960 = 100.0)

Fish	Base Year (1960) Values			Year: 1970			Year: 1975		
	$P_i^{(0)}$	$Q_i^{(0)}$	$P_i^{(0)}Q_i^{(0)}$	$P_i^{(2)}$	$(P_i^{(2)}/P_i^{(0)})$	$(P_i^{(2)}/P_i^{(0)})(P_i^{(0)}Q_i^{(0)})$	$P_i^{(3)}$	$(P_i^{(3)}/P_i^{(0)})$	$(P_i^{(3)}/P_i^{(0)})(P_i^{(0)}Q_i^{(0)})$
Cod	6.5	40	260.0	11.2	1.723	447.980	23.3	3.585	932.100
Flounder	12.2	127	1,549.4	15.3	1.254	1,942.948	34.7	2.844	4,406.494
Haddock	7.9	119	940.1	22.7	2.873	2,700.907	26.8	3.392	3,188.819
Ocean perch	4.0	141	564.0	4.9	1.225	690.900	10.3	2.575	1,452.300
Tuna	15.7	298	4,678.6	26.2	1.669	7,808.583	37.4	2.382	11,144.425
Totals			7,992.1			13,591.318			21,124.138
Index	$\frac{7{,}992.1}{7{,}992.1} \times 100$ $I_{LM}^{(0)} = 100.0$			$\frac{13{,}591.318}{7{,}992.1} \times 100$ $I_{LM}^{(2)} = 170.1$			$\frac{21{,}124.138}{7{,}992.1} \times 100$ $I_{LM}^{(3)} = 264.3$		

SOURCE: Data are taken from Table 16.2.

period. Tables 16.5 and 16.6 demonstrate how the respective aggregate price index and the index of mean price relatives are constructed according to Laspeyres' method. From Tables 16.5 and 16.6 we may note that (aside from "rounding" errors) identical results have been achieved; that is, the index of mean price relatives weighted by base year values, $P_i^{(0)}Q_i^{(0)}$, is algebraically identical to the aggregate price index weighted by base year quantities, $Q_i^{(0)}$.

One major advantage to the Laspeyres method is that it is always possible to make price comparisons, not only from each time period to the base period, but also from period to period as well. Unfortunately, however, fixing quantities in the base period yields a static consumption pattern which becomes more and more unrealistic over time. Hence business researchers would argue that utilizing a consumption pattern based on present attitudes, interests, tastes, and habits would seem to reflect a more realistic set of weights. This is achieved by the methods of Paasche.

16.5.2 The Paasche Indexes

The Paasche Aggregate Price Index uses quantity weights from the given period of interest so that $W_i = Q_i^{(t)}$ and Equation (16.4) takes the form

$$I_{PA}^{(t)} = \frac{\sum_{i=1}^{n} P_i^{(t)}Q_i^{(t)}}{\sum_{i=1}^{n} P_i^{(0)}Q_i^{(t)}} \times 100 \qquad (16.8)$$

while the Paasche Index of Mean Price Relatives uses total value weights of the form $W_i = P_i^{(0)}Q_i^{(t)}$, and Equation (16.5) becomes

$$I_{\text{PM}}^{(t)} = \frac{\sum_{i=1}^{n} (P_i^{(t)}/P_i^{(0)})P_i^{(0)}Q_i^{(t)}}{\sum_{i=1}^{n} P_i^{(0)}Q_i^{(t)}} \times 100 \qquad (16.9)$$

The Paasche indexes measure the percentage changes in prices that would occur in any given period had we bought in a reference or base period the same commodities in the same quantities as we did now in the given period. From Equations (16.8) and (16.9) we note that the index of mean price relatives, whose value weights contain current year quantities, $P_i^{(0)}Q_i^{(t)}$, is algebraically equivalent to the aggregate price index weighted by current year quantities, $Q_i^{(t)}$. Since the ultimate result is the same by either method and since the aggregate index is the simpler to compute, Table 16.7 depicts the construction of the Paasche Aggregate Price Index.

Users of the Paasche indexes argue that since production and purchasing conditions change over time, the quantities utilized as weights in any one reference point or base period of time will not be a good measure of the relative importance of their respective commodities for all other time periods. Thus, they believe, a set of weights that change every period must be utilized. One major problem, though, is that to obtain a detailed and adequate list of weights in every period would be very laborious and costly. Moreover, a second major problem is that an alteration of the weights each period prevents us from obtaining true measures of period-to-period price changes; only comparisons from each period to a base period can appropriately be made.

Table 16.7
CONSTRUCTING A PAASCHE AGGREGATE PRICE INDEX
(1960 = 100.0)

	Year										
	1960			1970				1975			
Fish	$Q_i^{(0)}$	$P_i^{(0)}$	$P_i^{(0)}Q_i^{(0)}$	$Q_i^{(2)}$	$P_i^{(2)}$	$P_i^{(0)}Q_i^{(2)}$	$P_i^{(2)}Q_i^{(2)}$	$Q_i^{(3)}$	$P_i^{(3)}$	$P_i^{(0)}Q_i^{(3)}$	$P_i^{(3)}Q_i^{(3)}$
Cod	40	6.5	260.0	53	11.2	344.5	593.6	56	23.3	364.0	1,304.8
Flounder	127	12.2	1,549.4	169	15.3	2,061.8	2,585.7	156	34.7	1,903.2	5,413.2
Haddock	119	7.9	940.1	27	22.7	213.3	612.9	16	26.8	126.4	428.8
Ocean perch	141	4.0	564.0	55	4.9	220.0	269.5	32	10.3	128.0	329.6
Tuna	298	15.7	4,678.6	393	26.2	6,170.1	10,296.6	391	37.4	6,138.7	14,623.4
Totals			7,992.1			9,009.7	14,358.3			8,660.3	22,099.8
Index	$\frac{7,992.1}{7,992.1} \times 100$ $I_{\text{PA}}^{(0)} = 100.0$			$\frac{14,358.3}{9,009.7} \times 100$ $I_{\text{PA}}^{(2)} = 159.4$				$\frac{22,099.8}{8,660.3} \times 100$ $I_{\text{PA}}^{(3)} = 255.2$			

SOURCE: Data are taken from Table 16.2.

Differences between the Laspeyres and Paasche indexes can be expected to increase over time from the base period to each new (current) period. In general, if consumers alter their purchasing patterns in response to relative price changes for certain commodities, the Laspeyres indexes will tend to exceed the Paasche indexes. In the former, an upward bias in the movement of price changes is created because of the time period used for selecting the weights; in the latter, a downward bias in the movement of price changes is created.

16.5.3 The Marshall–Edgeworth Indexes

To overcome these dilemmas, several compromise solutions have been offered in the form of constructed price indexes. As one compromise solution, the Marshall–Edgeworth indexes utilize weights based upon both a base period and a given time period. The quantity weights $W_i = Q_i^{(0)} + Q_i^{(t)}$ and total value weights $W_i = P_i^{(0)}(Q_i^{(0)} + Q_i^{(t)})$ yield, respectively, the Marshall–Edgeworth Aggregate Price Index

$$I_{\text{M-EA}}^{(t)} = \frac{\sum_{i=1}^{n} [P_i^{(t)}(Q_i^{(0)} + Q_i^{(t)})]}{\sum_{i=1}^{n} [P_i^{(0)}(Q_i^{(0)} + Q_i^{(t)})]} \times 100 \qquad (16.10)$$

and the Marshall–Edgeworth Index of Mean Price Relatives

$$I_{\text{M-EM}}^{(t)} = \frac{\sum_{i=1}^{n}\{(P_i^{(t)}/P_i^{(0)})[P_i^{(0)}(Q_i^{(0)} + Q_i^{(t)})]\}}{\sum_{i=1}^{n} [P_i^{(0)}(Q_i^{(0)} + Q_i^{(t)})]} \times 100 \qquad (16.11)$$

Because of the algebraic equivalence of Equations (16.10) and (16.11), the results of both indexes would be identical. Therefore, Table 16.8 presents only the construction of the Marshall–Edgeworth Aggregate Price Index.

The Marshall–Edgeworth price indexes are not biased with respect to price movements. However, as with the Paasche price indexes, the Marshall–Edgeworth indexes also require a continuously adjusted set of weights, and, unfortunately, this laborious and costly necessity prohibits us from making any period-to-period comparisons of price movements.

Table 16.8
CONSTRUCTING THE MARSHALL–EDGEWORTH AGGREGATE PRICE INDEX
(1960 = 100.0)

Fish	1960 $P_i^{(0)}$	1960 $Q_i^{(0)}$	1970 $P_i^{(2)}$	1970 $Q_i^{(2)}$	1970 $P_i^{(0)}(Q_i^{(0)}+Q_i^{(2)})$	1970 $P_i^{(2)}(Q_i^{(0)}+Q_i^{(2)})$	1975 $P_i^{(3)}$	1975 $Q_i^{(3)}$	1975 $P_i^{(0)}(Q_i^{(0)}+Q_i^{(3)})$	1975 $P_i^{(3)}(Q_i^{(0)}+Q_i^{(3)})$
Cod	6.5	40	11.2	53	604.5	1,041.6	23.3	56	624.0	2,236.8
Flounder	12.2	127	15.3	169	3,611.2	4,528.8	34.7	156	3,452.6	9,820.1
Haddock	7.9	119	22.7	27	1,153.4	3,314.2	26.8	16	1,066.5	3,618.0
Ocean perch	4.0	141	4.9	55	784.0	960.4	10.3	32	692.0	1,781.9
Tuna	15.7	298	26.2	393	10,848.7	18,104.2	37.4	391	10,817.3	25,768.6
Totals					17,001.8	27,949.2			16,652.4	43,225.4
Index	$I_{\text{M-EA}}^{(0)} = 100.0$		$I_{\text{M-EA}}^{(2)} = \frac{27{,}949.2}{17{,}001.8} \times 100 = 164.4$				$I_{\text{M-EA}}^{(3)} = \frac{43{,}225.4}{16{,}652.4} \times 100 = 259.6$			

SOURCE: Data are taken from Table 16.2.

16.5.4 The Fisher Ideal Indexes

Another compromise solution was devised by Irving Fisher (Reference 8). In an effort to cancel out the price biases inherent in the Laspeyres and Paasche indexes, Fisher constructed an "ideal" index as the **geometric average** of the two aforementioned indexes,[2] that is, the Fisher Ideal Aggregate Price Index is given as

$$I_{\text{FIA}}^{(t)} = \sqrt{(I_{\text{LA}}^{(t)})(I_{\text{PA}}^{(t)})} = \sqrt{\left(\frac{\sum_{i=1}^{n} P_i^{(t)}Q_i^{(0)}}{\sum_{i=1}^{n} P_i^{(0)}Q_i^{(0)}}\right)\left(\frac{\sum_{i=1}^{n} P_i^{(t)}Q_i^{(t)}}{\sum_{i=1}^{n} P_i^{(0)}Q_i^{(t)}}\right)} \times 100 \qquad (16.12)$$

while the Fisher Ideal Index of Mean Price Relatives is given by

$$I_{\text{FIM}}^{(t)} = \sqrt{(I_{\text{LM}}^{(t)})(I_{\text{PM}}^{(t)})} = \sqrt{\left(\frac{\sum_{i=1}^{n}\left(\frac{P_i^{(t)}}{P_i^{(0)}}\right)P_i^{(0)}Q_i^{(0)}}{\sum_{i=1}^{n} P_i^{(0)}Q_i^{(0)}}\right)\left(\frac{\sum_{i=1}^{n}\left(\frac{P_i^{(t)}}{P_i^{(0)}}\right)P_i^{(0)}Q_i^{(t)}}{\sum_{i=1}^{n} P_i^{(0)}Q_i^{(t)}}\right)} \times 100 \qquad (16.13)$$

[2] The **arithmetic** mean or average was defined in Chapter 4 as the summation of the values divided by n, the number of terms in the summation. On the other hand, the **geometric** mean is defined as the nth root of the product of the values of the n terms. For the aggregate price index we have but two terms, $I_{\text{LA}}^{(t)}$ and $I_{\text{PA}}^{(t)}$, and, therefore, the second root (that is, the square root) of this product yields the geometric mean—the Fisher Ideal Aggregate Price Index. For the index of mean price relatives the square root of the product $I_{\text{LM}}^{(t)}$ with $I_{\text{PM}}^{(t)}$ gives the Fisher index.

Table 16.9
CONSTRUCTING THE FISHER IDEAL AGGREGATE PRICE INDEX (1960 = 100.0)

Year					
1960		1970		1975	
$I_{LA}^{(0)}$	$I_{PA}^{(0)}$	$I_{LA}^{(2)}$	$I_{PA}^{(2)}$	$I_{LA}^{(3)}$	$I_{PA}^{(3)}$
100.0	100.0	170.1	159.4	264.3	255.2
$I_{FIA}^{(0)} = \sqrt{(100.0) \times (100.0)} = 100.0$		$I_{FIA}^{(2)} = \sqrt{(170.1) \times (159.4)} = 164.7$		$I_{FIA}^{(3)} = \sqrt{(264.3) \times (255.2)} = 259.7$	

SOURCE: Data are taken from Tables 16.2, 16.5, and 16.7.

Again, from Equations (16.12) and (16.13) we may note the algebraic equivalence of the two indexes. Thus the construction of the simpler one—the aggregate price index—is depicted in Table 16.9.

Fisher considered the indexes to be ideal because they conform to certain tests of *consistent behavior* with respect to both time and the value of the commodities (Reference 8). Unfortunately, though, as with the Marshall–Edgeworth indexes, the ideal indexes of Fisher still require a continuously adjusted set of weights which is both laborious and costly to construct and also prohibits a direct period-to-period comparison of price movements.

16.5.5 The "Fixed" Weight Indexes

Thus a compromise solution which has become widely accepted in practice is one based on a "fixed" set of weights—established at a particular point in time or developed as an average over several periods of time. Using the fixed quantity weights $W_i = Q_i^{(f)}$ and the fixed total value weights $W_i = P_i^{(0)}Q_i^{(f)}$, respectively, we obtain the Fixed Weight Aggregate Price Index as

$$I_{FWA}^{(t)} = \frac{\sum_{i=1}^{n} P_i^{(t)}Q_i^{(f)}}{\sum_{i=1}^{n} P_i^{(0)}Q_i^{(f)}} \times 100 \qquad (16.14)$$

and the index of Fixed Weight Mean Price Relatives

$$I_{FWM}^{(t)} = \frac{\sum_{i=1}^{n} (P_i^{(t)}/P_i^{(0)})P_i^{(0)}Q_i^{(f)}}{\sum_{i=1}^{n} P_i^{(0)}Q_i^{(f)}} \times 100 \qquad (16.15)$$

Table 16.10
CONSTRUCTING A FIXED WEIGHT AGGREGATE PRICE INDEX (1960 = 100.0)

		Year					
		1960		1970		1975	
Fish	1970 Fixed Weight $Q_i^{(2)}$	$P_i^{(0)}$	$P_i^{(0)}Q_i^{(2)}$	$P_i^{(2)}$	$P_i^{(2)}Q_i^{(2)}$	$P_i^{(3)}$	$P_i^{(3)}Q_i^{(2)}$
Cod	53	6.5	344.5	11.2	593.6	23.3	1,234.9
Flounder	169	12.2	2,061.8	15.3	2,585.7	34.7	5,864.3
Haddock	27	7.9	213.3	22.7	612.9	26.8	723.6
Ocean perch	55	4.0	220.0	4.9	269.5	10.3	566.5
Tuna	393	15.7	6,170.1	26.2	10,296.6	37.4	14,698.2
Totals			9,009.7		14,358.3		23,087.5
Index		$\frac{9{,}009.7 \times 100}{9{,}009.7}$ $I_{FWA}^{(0)} = 100.0$		$\frac{14{,}358.3}{9{,}009.7} \times 100$ $I_{FWA}^{(2)} = 159.4$		$\frac{23{,}087.5}{9{,}009.7} \times 100$ $I_{FWA}^{(3)} = 256.3$	

SOURCE: Data are taken from Table 16.2.

From these formulas we again note the algebraic equivalence between the aggregate price index and the index of mean price relatives. Hence the construction of the simpler aggregate index is demonstrated in Table 16.10. The fixed quantity weights used are the quantities of the selected types of fish caught, in millions of pounds, in the year 1970.

The major advantages of these more general fixed weight indexes are that they avoid the price biases inherent in the Laspeyres and Paasche indexes, and they permit a direct period-to-period comparison of price movements in addition to comparisons from each period to the base.

16.6 SOME WELL-KNOWN PRICE INDEXES AND THEIR USES

16.6.1 Consumer Price Index

Perhaps the most important and certainly the most familiar of the price indexes developed by the federal government is the Consumer Price Index. The index, which is computed and published monthly by the Bureau of Labor Statistics since 1940, had (since 1964) consisted of some 400 goods and services purchased by urban wage earners and clerical workers. In recent years, however, it was widely believed that the Consumer Price Index needed to be updated in order to better reflect changes in urban consumer purchasing habits. Thus, in 1978, the Bureau of Labor Statistics introduced two versions of the index—a *new* **Consumer Price Index for All Urban Consumers** as well as an updated, *revised* **Consumer Price Index for Urban Wage Earners and Clerical Workers.** The *new* Consumer Price Index for All Urban Consumers includes, in addition to wage earners and clerical workers, groups which historically had been excluded from coverage—professional, managerial, and technical workers; the self-employed; short-term workers; the unemployed; retirees; and others not in the labor force. Thus, this *new,* broad index covers approximately 80 percent of the total civilian noninstitutional population, roughly twice the size of the population covered by the *revised* Consumer Price Index for Urban Wage Earners and Clerical Workers.[3] Nevertheless, the *revised* index was still needed because organized labor believed that it better reflected the purchasing habits of the blue-collar family than did the new index.

The major categories comprising both versions of the index are:

Food and beverages
Housing
Apparel and upkeep
Transportation
Medical care
Entertainment

The Consumer Price Index is, in its most basic form, an index of fixed weight mean price relatives as shown in Equation (16.15). The price base for the items was established in one period of time while the value weights were obtained from a survey on consumer expenditures in another period of time. This massive undertaking involved both a probability sample and a judgment sample (as discussed in Chapter 2). Complete details as to the history, development, and construction of this index are given in Reference 4 while information pertaining to the revision is presented in Reference 5.

The Consumer Price Index was originally devised by the Bureau of Labor Statistics in 1919 so that the effects of inflationary tendencies resulting from World War I could be accounted for in wage negotiations in various industries. Subsequently the index has been tied into numerous pension plans and escalator clauses in union contracts which affect the lives of millions of workers in the United States today. Hence the index has gained acceptance by the public as a cost-of-living indicator.[4] As an example, Table 16.11 presents the cost-of-living adjustment negotiated by Chrysler Corporation and the United Automobile Workers of America for the contractual period, September 1964 to September 1967. The agreement provided for cost-of-living reviews each December, March, June, and September based on the respective published Consumer Price Indexes for the months of October, January, April, and July.

Another important use of the Consumer Price Index results from the fact that economists, forecasters, and business decision makers are concerned with economic models which are representative of the complex workings of our

[3] Neither index, however, includes persons in the military services or in institutions, or persons living outside urban areas such as farm families.

[4] Although the Consumer Price Index is often thought of as a cost-of-living index it is, in fact, a price index. It denotes changes in prices for a fixed group of goods and services but does not reflect changes in purchasing patterns resulting from the changes in prices. Moreover, a complete cost-of-living index would also take into account income and social security taxes which (unlike sales taxes) are excluded from the Consumer Price Index because they do not directly relate to prices of specific goods and services.

Table 16.11
COST-OF-LIVING ADJUSTMENT AT CHRYSLER CORPORATION FOR 3-YEAR PERIOD (SEPTEMBER 1964 TO SEPTEMBER 1967)

Consumer Price Index (1957–1959 = 100)	Hourly Cost-of-Living Allowance
106.4 or less	None
106.5 to 106.8	1 cent
106.9 to 107.2	2 cents
107.3 to 107.6	3 cents
107.7 to 108.0	4 cents
108.1 to 108.4	5 cents
108.5 to 108.8	6 cents
108.9 to 109.2	7 cents
109.3 to 109.6	8 cents
109.7 to 110.0	9 cents
110.1 to 110.4	10 cents
110.5 to 110.8	11 cents
110.9 to 111.2	12 cents
111.3 to 111.6	13 cents
111.7 to 112.0	14 cents
112.1 to 112.4	15 cents
112.5 to 112.8	16 cents
112.9 to 113.2	17 cents
113.3 to 113.6	18 cents
113.7 to 114.0	19 cents
114.1 to 114.4	20 cents
and so forth with a 1-cent adjustment for each 0.4-point change in the index	

SOURCE: Data are taken from B.L.S. Bulletin No. 1515, *Wage Chronology—Chrysler Corporation 1939–66,* (U.S. Department of Labor, June 1967).

economy. Since such models, when used for predictive purposes, deal with *real* wages, the Consumer Price Index is often used to adjust *nominal* wages (in current dollars) to *real* wages (in constant base-year dollars) by adjusting for changes in the cost of living. Using the Consumer Price Index as a "deflator," we have

$$\text{Real Wages (in Constant Dollars)} = \frac{\begin{array}{c}\text{Nominal Wages}\\ \text{(in Current Dollars)}\end{array}}{\begin{array}{c}\text{Consumer Price Index}\\ \text{(Current)}\end{array}} \times 100 \qquad (16.16)$$

This computation transforms current price levels into constant price levels which are more comparable to those of a prior period. To illustrate this, suppose, for example, your 1976 salary had been \$18,000 while your 1967 salary was \$10,000. Now it may be noted that while your *nominal* income would have increased by 80.0% $\left[\text{that is, } \left(\frac{\$18{,}000 - \$10{,}000}{\$10{,}000}\right) \times 100\right]$

over the 10-year period, the purchasing power of your *real* income certainly would not have. Using the Consumer Price Index as a deflator of *nominal* income [Equation (16.16)] we have

$$\begin{array}{c}\text{Real Wages}\\ \text{(in constant 1976 Dollars)}\end{array} = \frac{18{,}000}{170.5} \times 100 = \$10{,}557.19$$

and

$$\begin{array}{c}\text{Real Wages}\\ \text{(in constant 1967 Dollars)}\end{array} = \frac{\$10{,}000}{100.0} \times 100 = \$10{,}000.00$$

Thus it is clear that your *real* income would have increased not by 80.0% but only by approximately 5.6% $\left[\text{that is, } \left(\frac{\$10{,}557.19 - \$10{,}000}{\$10{,}000}\right) \times 100\right]$ since consumer prices increased (from 100.0 to 170.5 as measured by the Consumer Price Index) by 70.5% in the same period of time.

16.6.2 Wholesale Price Index

The Wholesale Price Index is the oldest of the continuous price indexes in the United States and dates back to the turn of this century. Published monthly by the Bureau of Labor Statistics, the Wholesale Price Index is comprised of roughly 2,500 commodities at several stages of the production process and attempts to show price movements in primary markets (that is, the first commercial transaction for a commodity). The major groupings comprising the index are

- Farm products
- Processed foods and feeds
- Industrial commodities
  - Chemicals and allied products
  - Fuels, related products, and power
  - Furniture and household durables
  - Hides, skins, and leather products
  - Lumber and wood products
  - Machinery and equipment
  - Metals and metal products
  - Nonmetallic mineral products
  - Pulp, paper, and allied products
  - Rubber and plastics products
  - Textile products and apparel
  - Transportation equipment

Like the Consumer Price Index, the Wholesale Price Index is also essentially an index of fixed weight mean price relatives. Details as to its history, development, and compilation are given in Reference 2.

The Wholesale Price Index has two interesting uses. Similar to the Consumer Price Index, it is used as a deflator, that is, for price series measured at the primary or wholesale levels, the Wholesale Price Index is utilized to adjust *nominal* price levels into *real* price levels. In addition, and perhaps even more importantly, the magnitude and direction of future changes in consumer prices may be predicted by the Wholesale Price Index. Thus we note that index numbers are useful not only in helping to understand the past and present functions of the economy, but they may also be useful to economists, forecasters, and business decision makers in their predictions of future economic activity.

16.7 INDEX NUMBER ADJUSTMENTS

Index numbers are measures that may be readily adjusted by a decision maker who wishes to examine certain kinds of comparisons. Two such adjustments are (1) shifting the base of the index number; and (2) splicing together two series of index numbers.

16.7.1 Shifting the Base

In studying index numbers the decision maker is interested in comparing the current value of the index to some base period or reference point. Nevertheless, it becomes difficult to relate to price comparisons with reference points too far in the distant past. Under such circumstances, shifting the base period is desirable. Furthermore, a decision maker is frequently involved in comparing two series of index numbers—each with differing reference points. In these cases, too, it is feasible to shift the base period of one of the index number series so that it matches that of the other series.

To shift the base we merely divide each index number in the series by the value of the index number in the newly desired base period. Each result is then multiplied by 100 to yield the new set of index numbers with the shifted base. Hence, symbolically, we have

$$I_{\text{new}}^{(t)} = \frac{I_{\text{old}}^{(t)}}{I_{\text{desired base}}^{(0)}} \times 100 \qquad (16.17)$$

where $I_{\text{new}}^{(t)}$ = index number in time period t under the shifted base

$I_{\text{old}}^{(t)}$ = index number in time period t under the old base

$I_{\text{desired base}}^{(0)}$ = value of the index number under the old base which is to be established as the new base

Table 16.12
SHIFTING THE BASE OF A PRICE INDEX: PRICES PAID TO FISHERMEN FOR FLOUNDER

Year	Price Index with 1960 Base	Price Index with 1967 Base
1960	100.0	(100.0/94.3) × 100.0 = 106.0
1961	86.9	(86.9/94.3) × 100.0 = 92.1
1962	79.5	(79.5/94.3) × 100.0 = 84.3
1963	68.9	(68.9/94.3) × 100.0 = 73.1
1964	65.6	(65.6/94.3) × 100.0 = 69.6
1965	77.9	(77.9/94.3) × 100.0 = 82.6
1966	104.1	(104.1/94.3) × 100.0 = 110.4
1967	94.3	(94.3/94.3) × 100.0 = 100.0
1968	93.4	(93.4/94.3) × 100.0 = 99.0
1969	112.3	(112.3/94.3) × 100.0 = 119.1
1970	125.4	(125.4/94.3) × 100.0 = 133.0
1971	136.9	(136.9/94.3) × 100.0 = 145.2
1972	175.4	(175.4/94.3) × 100.0 = 186.0
1973	194.3	(194.3/94.3) × 100.0 = 206.0
1974	217.2	(217.2/94.3) × 100.0 = 230.3
1975	284.4	(284.4/94.3) × 100.0 = 301.6

SOURCE: Data are taken from Table 16.1.

In other words, we create a series of *new* index numbers relative to the *old* value of the desired (shifted) base period. Hence this new point of reference must have a value of 100.0.

To demonstrate the shifting of the base, suppose we return to the simple price index we had constructed in Table 16.1. We may recall that this price index represented the average annual prices paid to fishermen for flounder caught from 1960 through 1975. The base period that had been selected was 1960. If we now desire to shift the base to the year 1967 so that the results are more comparable to some of the more well-known government price indexes, Equation (16.17) may be used as indicated in Table 16.12.

Thus, for example, from Table 16.12 the computation of the price index for the year 1975 using the shifted (1967) base period is

$$I_{\text{new}}^{(1975)} = \frac{I_{\text{old}}^{(1975)}}{I_{\text{desired base}}^{(1967)}} \times 100 \qquad (16.17)$$

$$= \frac{284.4}{94.3} \times 100 = 301.6$$

The results indicate that prices paid to fishermen for flounder more than tripled (in terms of 1975 dollars) from 1967 to 1975.

16.7.2 Splicing Two Series of Index Numbers

A second important adjustment that may be made is the splicing together of two series of index numbers. To illustrate this concept, suppose we have obtained a series of index numbers representing the prices paid by city wage earners and clerical workers for goods and services purchased over many years. Suppose further, that a new index is now desired—comprised of either a different set of goods and services or the same set but in different quantities—reflecting current consumption patterns. Rather than spending the time and money to develop a brand new index all the way back to the beginning periods, we may develop instead a new index for a desired base period and then "splice" the two index series together from this desired base period. This process is essentially what is accomplished when the Bureau of Labor Statistics revises its Consumer Price Index (Reference 5). Since the Consumer Price Index is basically a fixed weight mean of price relatives [Equation (16.15)], having value weights based on one period of time and a price base established at another, it is permissible to use a splicing adjustment.[5] As the Consumer Price Index dates back to the year 1919, it is clearly evident that adjustment through splicing is preferable to total reconstruction from scratch.

To splice, we must first establish the new index series as having a value of 100.0 in its desired base period and match this against the value corresponding to the old index series for the same period. Once this is done we may reconstruct the new index series back in time. Moreover, as new data become available so that the new index series can be computed, we may then take the old index series forward in time. This type of "carry back" and "carry forward" is usually accomplished only in the initial periods following a splicing.[6] Once the decision maker becomes familiar with the new series, the old series is discontinued.

To demonstrate the splicing, the carry back, and the carry forward, Table 16.13 presents the Consumer Price Index for the years 1960 through 1967

Table 16.13
CONSUMER PRICE INDEX
FOR ALL COMMODITIES:
1960–1967
(1957–1959 = 100.0)

Year	Index
1960	103.1
1961	104.2
1962	105.4
1963	106.7
1964	108.1
1965	109.9
1966	113.1
1967	116.3

SOURCE: Data are taken from *Statistical Abstract of the United States,* 1968.

[5] Splicing can be accomplished provided that the older index was constructed either as a weighted aggregate having base period quantity weights or fixed period quantity weights or as a weighted mean of price relatives having base period value weights or fixed period value weights.

[6] The Bureau of Labor Statistics has, however, historically carried back the Consumer Price Index to the year 1800 (Reference 14).

Table 16.14
CONSUMER PRICE INDEX FOR ALL COMMODITIES: 1967–1976 (1967 = 100.0)

Year	Index
1967	100.0
1968	104.2
1969	109.8
1970	116.3
1971	121.3
1972	125.3
1973	133.1
1974	147.7
1975	161.2
1976	170.5

SOURCE: Data are taken from *Statistical Abstract of the United States,* 1977.

using as its price base period the average prices in the years 1957 through 1959. In addition, Table 16.14 depicts the Consumer Price Index for the years 1967 through 1976 using as its price base period the year 1967. When the new series was established in 1967, it also could have been carried back in time as illustrated in Table 16.15.

We note that the computations are similar to those used for shifting a base period. For example, to carry the new series back to 1965, we may use Equation (16.17) to show that

$$I_{\text{new}}^{(1965)} = \frac{I_{\text{old}}^{(1965)}}{I_{\text{desired base}}^{(1967)}} \times 100 \qquad (16.17)$$

$$= \left(\frac{109.9}{116.3}\right) \times 100 = 94.5$$

On the other hand, to carry the old series forward to time period t we must first rewrite Equation (16.17) in terms of $I_{\text{old}}^{(t)}$, that is,

$$I_{\text{new}}^{(t)} = \frac{I_{\text{old}}^{(t)}}{I_{\text{desired base}}^{(0)}} \times 100 \qquad (16.17)$$

becomes

$$\frac{I_{\text{new}}^{(t)} \times I_{\text{desired base}}^{(0)}}{100} = I_{\text{old}}^{(t)} \qquad (16.18)$$

Table 16.15
SPLICING TWO CONSUMER PRICE INDEX SERIES

Year	Old Series (1957–1959 = 100.0)	New Series (1967 = 100.0)
1960	103.1	$\left(\frac{103.1}{116.3}\right) \times 100 = 88.7$
1961	104.2	$\left(\frac{104.2}{116.3}\right) \times 100 = 89.6$
1962	105.4	$\left(\frac{105.4}{116.3}\right) \times 100 = 90.6$
1963	106.7	$\left(\frac{106.7}{116.3}\right) \times 100 = 91.7$
1964	108.1	$\left(\frac{108.1}{116.3}\right) \times 100 = 92.9$ ↑ CARRY BACK
1965	109.9	$\left(\frac{109.9}{116.3}\right) \times 100 = 94.5$
1966	113.1	$\left(\frac{113.1}{116.3}\right) \times 100 = 97.2$
1967	116.3	100.0
1968 CARRY FORWARD ↓	$\frac{(104.2) \times (116.3)}{100} = 121.2$	104.2
1969	$\frac{(109.8) \times (116.3)}{100} = 127.7$	109.8
1970	$\frac{(116.3) \times (116.3)}{100} = 135.3$	116.3
1971	$\frac{(121.3) \times (116.3)}{100} = 141.1$	121.3
1972	$\frac{(125.3) \times (116.3)}{100} = 145.7$	125.3
1973	$\frac{(133.1) \times (116.3)}{100} = 154.8$	133.1
1974	$\frac{(147.7) \times (116.3)}{100} = 171.8$	147.7
1975	$\frac{(161.2) \times (116.3)}{100} = 187.5$	161.2
1976	$\frac{(170.5) \times (116.3)}{100} = 198.3$	170.5

SOURCE: Data are taken from Tables 16.12 and 16.13.

Using Equation (16.18) we may illustrate this process by carrying the old index series forward to the year 1976 as follows:

$$\frac{I_{\text{new}}^{(1976)} \times I_{\text{desired base}}^{(1967)}}{100} = I_{\text{old}}^{(1976)}$$

$$\frac{(170.5) \times (116.3)}{100} = 198.3 = I_{\text{old}}^{(1976)}$$

Had the old index series been maintained, its value in terms of 1976 (actual or current) prices would have been 198.3.

16.8 INDEX NUMBERS: AN OVERVIEW

In the previous sections of this chapter we have noted that index numbers are useful to economists, forecasters, and business decision makers who study the magnitude and direction of price movements in our economy. Index numbers then are barometers of business change. Nevertheless, index numbers are useful not only in studying the past and present workings of our economy, but they are also important in forecasting future economic activity. Index numbers then are often used in time-series analysis—the historical study of long-term trends, seasonal variations, and business cycle developments (see Section 17.2)—so that business leaders may keep pace with changing economic and business conditions and have better information available for decision-making purposes. Time-series analysis and business forecasting are the topics of our next chapter.

PROBLEMS

16.1 Table 16.1 presented the prices paid by processors to fishermen for flounder caught in 1960–1975. Table 16.16 depicts the Wholesale Price Index for the same period of time.

(a) Determine the **nominal** growth rate for prices paid to fishermen over the 15-year period from 1960 to 1975.

(b) Deflate the 1960 and 1975 figures by the Wholesale Price Index for those years.

(c) Determine the **real** growth rate for prices paid to fishermen over the 15-year period.

(d) Discuss the differences in your results between (a) and (c).

16.2 Table 16.17 shows prices paid by processors to fishermen for shell fish.

(a) Using 1960 as the price base year construct the following weighted aggregate indexes of prices paid for shell fish:

1. Laspeyres
2. Paasche

Table 16.16
WHOLESALE PRICE INDEX
(1967 = 100.0)

Year	Index Value
1960	94.9
1961	94.5
1962	94.8
1963	94.5
1964	94.7
1965	96.6
1966	99.8
1967	100.0
1968	102.5
1969	106.5
1970	110.4
1971	113.9
1972	119.1
1973	134.7
1974	160.1
1975	174.9

SOURCE: Data are taken from *Survey of Current Business* and *Business Statistics Supplement,* 1975.

Table 16.17
PRICES PAID FOR SHELL FISH

Shellfish	Quantity* and Price†	1960	1965	1970	1973	1974	Preliminary 1975
Clams	*Q*	50	71	99	108	120	111
	P	39.7	49.1	47.5	78.5	76.4	88.2
Crabs	*Q*	222	335	277	297	329	301
	P	5.5	7.6	6.6	12.1	12.3	15.7
Lobsters	*Q*	31	30	34	29	28	29
	P	45.7	75.2	94.7	136.5	141.0	161.5
Oysters	*Q*	60	55	54	52	45	33
	P	68.2	78.6	61.1	66.0	68.1	80.7
Shrimp	*Q*	249	244	367	380	370	344
	P	44.4	58.0	75.4	151.9	116.1	125.6

SOURCE: Data are taken from *Statistical Abstract of the United States,* Tables 1178 and 1180, 1976. Data compiled by U.S. National Oceanic and Atmospheric Administration.

* *Q*—quantity in millions of pounds caught in U.S. waters.
† *P*—price paid in cents per pound.

Table 16.18
PRINCIPAL CROP PRICES

Crop	Quantity and Price	1960	1965	1970	Preliminary 1975
Corn for grain	*Q* (millions of bushels)	3,907	4,103	4,152	5,767
	P ($/bushel)	1.00	1.16	1.33	2.49
Wheat	*Q* (millions of bushels)	1,355	1,316	1,352	2,134
	P ($/bushel)	1.74	1.35	1.33	3.49
Oats	*Q* (millions of bushels)	1,153	930	917	657
	P ($/bushel)	.60	.62	.62	1.46
Sugar beets	*Q* (millions of short tons)	16	21	26	29
	P ($/ton)	11.58	11.90	14.82	27.40
Rice rough	*Q* (millions of hundredweights)	55	76	84	128
	P ($/hundredweight)	4.55	4.93	5.17	8.74
Sorghums for grain	*Q* (millions of bushels)	620	673	684	758
	P ($/bushel)	.84	.98	1.14	2.37
Cotton	*Q* (millions of bales, 500 pounds)	14	15	10	8
	P ($/pound)	.3019	.2937	.2293	.4880
Hay	*Q* (millions of short tons)	118	126	127	133
	P ($/ton)	21.70	23.20	26.10	51.90
Soybeans for beans	*Q* (millions of bushels)	555	846	1,127	1,521
	P ($/bushel)	2.13	2.54	2.85	4.63
Irish potatoes	*Q* (millions of hundredweights)	257	291	326	316
	P ($/hundredweight)	2.00	2.53	2.21	4.81
Tobacco	*Q* (millions of pounds)	1,944	1,855	1,906	2,184
	P ($/pound)	.61	.65	.73	1.02

SOURCE: Data are taken from *Statistical Abstract of the United States,* Table 1119, 1976. Data compiled by U.S. Department of Agriculture.

3. Marshall–Edgeworth
4. Fisher Ideal
5. Fixed Weights (using 1970 as the weight period)

(b) Shift the price base from 1960 to 1970 for all five indexes.

16.3 Table 16.18 depicts the prices received by farmers for principal crops.

(a) Using 1960 as the price base year construct the following weighted aggregate indexes of prices received by farmers for principal crops:

1. Laspeyres
2. Paasche
3. Marshall–Edgeworth
4. Fisher Ideal
5. Fixed Weights (using 1970 as the weight period)

(b) Shift the price base from 1960 to 1965 for all five indexes.

16.4 Table 16.19 depicts foreign travel from the continental United States.

(a) Compute $P = PQ/Q$ for each destination and for each year to get the expenditures per person.

(b) Using 1970 as the price base year develop the following weighted aggregate indexes of price expenditures for foreign travel:

1. Laspeyres
2. Paasche
3. Marshall–Edgeworth
4. Fisher Ideal
5. Fixed Weights (using 1972 as the weight period)

(c) Shift the price base from 1970 to 1972 for all five indexes.

Table 16.19
FOREIGN TRAVEL FROM UNITED STATES

International Destination	Number of Persons† and Expenditures‡	Year 1970	1971	1972	1973	1974	Preliminary 1975
Europe and Mediterranean	Q	2,898	3,202	3,843	3,915	3,325	3,185
	PQ	1,425	1,540	1,853	1,993	1,802	1,918
Caribbean and Central America	Q	1,663	1,736	1,992	2,032	2,147	2,065
	PQ	390	408	504	570	685	787
South America	Q	249	254	338	383	423	447
	PQ	90	92	113	132	209	242
Other*	Q	450	475	617	603	572	657
	PQ	279	295	400	409	450	527

SOURCE: Data are taken from *Statistical Abstract of the United States*, Table 369, 1976. Compiled by U.S. Bureau of Economic Analysis.

* Excludes Mexico, Canada, Hawaii, Alaska, Puerto Rico, and Virgin Islands; military and other government travel; and cruises.

† Q—number of persons traveling, in thousands.

‡ PQ—expenditures abroad, in current millions of dollars.

Table 16.20
SELECTED TIMBER SPECIES

Timber	Quantities* and Prices†	Year					
		1969	1970	1971	1972	1973	Preliminary 1974
Soft woods							
Douglas fir	Q	8,059	7,727	8,211	8,459	8,686	7,901
	P	82,200	41,900	49,100	71,700	138,100	202,400
Southern pine	Q	7,181	7,063	7,736	7,884	7,895	6,921
	P	51,700	44,100	52,200	65,600	93,400	76,200
Ponderosa pine	Q	3,684	3,429	3,780	4,001	4,030	3,580
	P	71,000	32,100	37,600	65,800	92,300	100,600
Hemlock	Q	1,902	1,980	2,367	2,692	2,711	2,105
	P	45,100	20,500	20,600	49,000	99,200	110,800
Hard woods							
Oak	Q	3,410	3,250	3,177	3,121	3,227	3,160
	P	28,200	26,600	21,200	26,600	43,600	54,700
Maple	Q	745	742	735	624	623	574
	P	41,100	34,400	37,800	59,400	71,400	79,500

SOURCE: Data are taken from *Statistical Abstract of the United States*, Tables 1157 and 1160, 1976. Data compiled by U.S. Forest Service and Mackay-Shields Economics, Inc.

* *Q*—quantities in millions of board feet.

† *P*—prices in dollars per million board feet.

16.5 Table 16.20 depicts the prices paid for selected timber from 1969 through 1974.

(a) Using 1969 as the price base year develop the following weighted aggregate indexes of prices paid for selected timber species:

1. Laspeyres
2. Paasche
3. Marshall–Edgeworth
4. Fisher Ideal
5. Fixed Weights (using 1972 as the weight period)

(b) Shift the price base from 1969 to 1972 for all five indexes.

(c) Table 16.21 depicts the Wholesale Price Index for lumber and wood products from 1969 through 1974.

(1) Shift the base of this index to 1972.

(2) Compare this index with your results from (b). Discuss.

Table 16.21
WHOLESALE PRICE INDEX FOR LUMBER AND WOOD PRODUCTS (1967 = 100.0)

	Year					
	1969	1970	1971	1972	1973	1974
Index (1967 = 100.0)	125.3	113.6	127.0	144.3	177.2	183.6

SOURCES: Data are taken from *Survey of Current Business* and *Business Statistics Supplement*, 1975.

Table 16.22
PAID CIVILIAN EMPLOYMENT IN FULL-TIME POSITIONS IN FEDERAL GOVERNMENT

Pay System	Number† and Pay‡	Year					
		1960	1965	1970	1973	1974	1975
General schedule	*Q*	976	1,112	1,287	1,302	1,322	1,349
	P	5,705	7,707	11,065	13,204	13,704	14,483
Wage system	*Q*	667	621	674	547	536	528
	P	4,935	5,887	6,976	9,194	10,071	11,197
Postal pay system	*Q*	483	534	673	550	553	566
	P	4,854	6,219	8,120	11,652	11,937	13,329
Other*	*Q*	114	131	172	139	136	139
	P	5,344	7,032	8,741	12,297	12,985	13,951

SOURCE: Data are taken from *Statistical Abstract of the United States,* Table 411, 1976. Data compiled by U.S. Civil Service Commission.

* Excludes Congress and Federal Court employees and Department of Commerce maritime seamen.
† *Q*—number of employees in thousands.
‡ *P*—average pay per annum in dollars.

16.6 Table 16.22 presents the wages paid to full-time civilian employees having certain types of federal government positions.

(a) Using 1960 as the wage pay base year construct the following weighted aggregate indexes of wages paid to full-time civilian federal government employees:

1. Laspeyres
2. Paasche
3. Marshall–Edgeworth
4. Fisher Ideal
5. Fixed Weights (using 1970 as the weight period)

(b) Shift the wage pay base year from 1960 to 1970 for all five indexes.

(c) Table 16.23 depicts the Consumer Price Index for selected periods from 1960 through 1975.

(1) Determine the percentage growth rate in **nominal** wages paid in the postal pay system over the 15-year period 1960 to 1975.

(2) Deflate the 1960 and 1975 figures by the Consumer Price Index for those years.

(3) Determine the percentage growth rate in **real** wages paid in the postal pay system over the 15-year period 1960 to 1975.

Table 16.23
CONSUMER PRICE INDEX (1967 = 100.0)

	Year					
	1960	1965	1970	1973	1974	1975
Index (1967 = 100.0)	88.7	94.5	116.3	133.1	147.7	161.2

SOURCE: Data are taken from *Statistical Abstract of the United States,* 1977.

Table 16.24
AVERAGE WEEKLY EARNINGS

Private Group	Number† and Earnings‡	1950	1955	1960	1965	1970	1975
Manufacturing	*Q*	15,241	16,882	16,796	18,062	19,349	18,347
	P	58.32	75.70	89.72	107.53	133.73	189.51
Contract	*Q*	2,333	2,802	2,885	3,186	3,536	3,457
construction	*P*	69.68	90.90	113.04	138.38	195.45	265.35
Wholesale and	*Q*	9,386	10,535	11,391	12,716	15,040	16,947
retail trade*	*P*	39.71	48.75	57.76	66.61	82.47	108.22

SOURCE: Data are taken from *Economic Report of the President,* Tables B-32 and B-34, 1977. Data compiled by the Bureau of Labor Statistics.

* Includes eating and drinking places.
† *Q*—number of wage and salary workers in nonagricultural establishments in thousands.
‡ *P*—average weekly earnings in current dollars.

(4) Discuss how purchasing power for employees under the postal pay system has changed from 1960 to 1975.

* 16.7 Table 16.24 depicts the average weekly earnings of workers in selected private nonagricultural groups for 5-year periods from 1950 through 1975.

(a) Create a simple aggregate index of weekly earnings with a 1960 pay base.

(b) Create a simple arithmetic mean of price relatives of weekly earnings with a 1960 pay base.

(c) Using 1960 as the pay base year construct the following weighted aggregate indexes of weekly earnings:

1. Laspeyres
2. Paasche
3. Marshall–Edgeworth
4. Fisher Ideal
5. Fixed Weights (using 1970 as the weight period)

(d) Shift the pay base year from 1960 to 1970 for all seven indexes.

(e) Table 16.25 presents the Consumer Price Index for 5-year periods from 1950 through 1975.

(1) Determine the percentage growth rate in **nominal** earnings for private contract construction workers over the 25-year period 1950 to 1975.

(2) Deflate the 1950 and 1975 figures by the Consumer Price Index for those years.

Table 16.25
CONSUMER PRICE INDEX (1967 = 100.0)

	1950	1955	1960	1965	1970	1975
Index (1967 = 100.0)	72.1	80.2	88.7	94.5	116.3	161.2

SOURCE: Data are taken from *Statistical Abstract of the United States,* 1977.

(3) Determine the percentage growth rate in **real** earnings for private contract construction workers over the 25-year period 1950 to 1975.

(4) Discuss how purchasing power for private contract construction workers has changed from 1950 to 1975.

16.8 Figure 16.3 depicts the minimum wage rates in the United States over 10-year periods from 1938 through 1968. For the year 1978 Congress approved an increase in the minimum wage to $2.65 per hour. Table 16.26 presents the Consumer Price Index for the same 10-year periods.

(a) Determine the **nominal** growth rate in the minimum wage over the decade from 1968 to (what was approved for) 1978.

(b) Deflate the 1968 and 1978 figures by the respective actual and predicted Consumer Price Indexes for these years.

(c) Determine the **real** growth rate in the minimum wage over the decade from 1968 to (what was approved for) 1978.

(d) Discuss how the purchasing power of persons paid at the minimum wage rate has changed over the decade from 1968 to 1978.

(e) If you feel strongly about your findings, write to your Representative in Washington, D.C.

(f) Define the following terms:

1. Minimum wage
2. Nominal growth rate
3. Real growth rate
4. Price deflator
5. Purchasing power

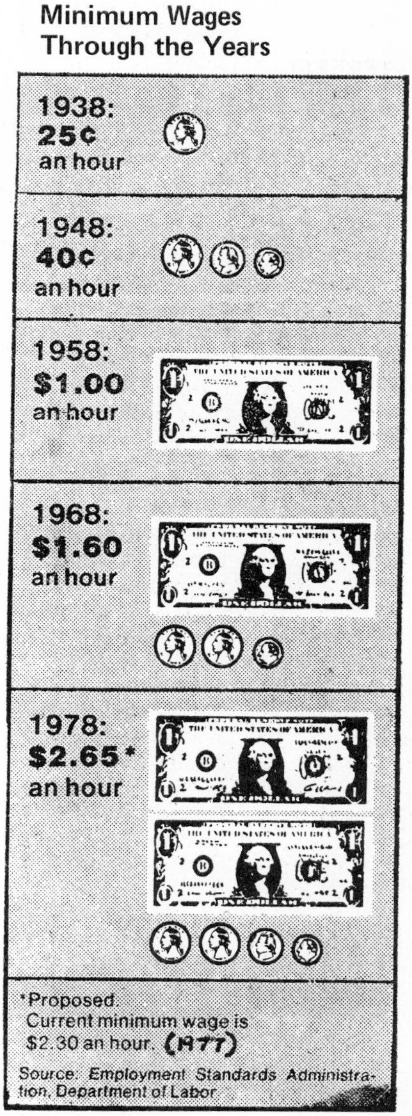

Minimum Wages
Through the Years
1938:
25¢
an hour
1948:
40¢
an hour
1958:
$1.00
an hour
1968:
$1.60
an hour
1978:
$2.65*
an hour
*Proposed.
Current minimum wage is
$2.30 an hour. (1977)
Source: Employment Standards Administration, Department of Labor

Table 16.26
CONSUMER PRICE INDEX
(1967 = 100.0)

	Year				
	1938	1948	1958	1968	(Predicted) 1978
Index (1967 = 100)	42.2	72.1	86.6	104.2	193.7

SOURCE: Data are taken from *Statistical Abstract of the United States,* 1977.

16.9 Go to your college library and, using *The Statistical History of the United States from Colonial Times to the Present,* write a report that briefly describes the composition, characteristics, and uses of the Indexes of Prices Received and Paid by Farmers and the Parity Ratio.

16.10 Go to your college library and, using *Business Statistics, the Biennial Supplement to the Survey of Current Business, 1977,* write a report that briefly describes the composition, characteristics, and uses of the Federal Reserve Board's Index of Industrial Production.

REFERENCES

1. ARKIN, H., AND R. COLTON, *Statistical Methods,* 5th ed., (New York: Barnes and Noble College Outline Series, 1970).
2. Bureau of Labor Statistics Bulletin No. 1458, *Handbook of Methods for Surveys and Studies* (U.S. Department of Labor, 1966).
3. Bureau of Labor Statistics Bulletin No. 1515, *Wage Chronology—Chrysler Corporation, 1939–1966* (U.S. Department of Labor, 1967).
4. Bureau of Labor Statistics Bulletin No. 1517, *The Consumer Price Index: History and Techniques* (U.S. Department of Labor, 1966).
5. Bureau of Labor Statistics Report No. 517, *The Consumer Price Index: Concepts and Content over the Years* (U.S. Department of Labor, 1977).
6. *Business Statistics Supplement* (U.S. Department of Commerce, 1975).
7. CROXTON, F., D. COWDEN, AND S. KLEIN, *Applied General Statistics,* 3rd ed. (Englewood Cliffs, N.J.: Prentice-Hall, 1967).
8. FISHER, I., *The Making of Index Numbers* (Boston: Houghton Mifflin, 1927).
9. JEDAMUS, P., R. FRAME, AND R. TAYLOR, *Statistical Analysis for Business Decisions* (New York: McGraw-Hill, 1976).
10. LAPIN, L., *Statistics for Modern Business Decisions* (New York: Harcourt Brace Jovanovich, 1973).
11. MUDGETT, B. D., *Index Numbers* (New York: Wiley, 1951).
12. *Statistical Abstract of the United States* (U.S. Department of Commerce, 1977).
13. *Survey of Current Business* (U.S. Department of Commerce, 1977).
14. WATTENBERG, B. J., ed., *The Statistical History of the United States from Colonial Times to the Present* (New York: Basic Books, 1976).

CHAPTER 17

TIME-SERIES ANALYSIS AND BUSINESS FORECASTING

17.1 THE NEED FOR BUSINESS FORECASTING: INTRODUCTION TO TIME-SERIES ANALYSIS

17.1.1 Introduction

Since economic and business conditions vary over time, business leaders must find ways to keep abreast of the effects that such changes will have on their particular operations. One method which business leaders may use as an aid in controlling present operations and in planning for future needs (by forecasting likely developments in sales, raw materials, labor, etc.) is time-series analysis. **A time series is a set of quantitative data that are obtained at regular periods over time.**

Thus the *daily* closing prices of a particular stock on the New York Stock Exchange constitutes a time series. Other examples of economic or business time series are the *weekly* percentage changes in department store sales; the *monthly* publication of the Consumer Price Index (see Chapter 16); the *quarterly* statements of gross national product (GNP); as well as the *annually* recorded total sales revenues of a particular firm. Time series, however, are not restricted to economic or business data (for example, see Reference 13). As another example, our dean and researcher may wish to study whether there is an indication of persistent "grade inflation" during the past decade at their business school. To accomplish this, they may, on an annual basis, either investigate the percentage of freshmen and sophomore students on the Dean's List (see Problem 17.2), or they may study the percentage of seniors graduating with honors.

17.1.2 Objectives of Time-Series Analysis

The basic assumption underlying time-series analysis is that those factors which have influenced patterns of economic activity in the past and present will continue to do so in more or less the same manner in the future. Thus the major goals of time-series analysis are to isolate these influencing factors for predictive (forecasting) purposes as well as for managerial planning and control.

17.2 COMPONENT FACTORS OF THE CLASSICAL MULTIPLICATIVE TIME-SERIES MODEL

To achieve these goals, many mathematical models have been devised for exploring the fluctuations among the component factors of a time series. Perhaps the most fundamental is the **classical multiplicative model** for data re-

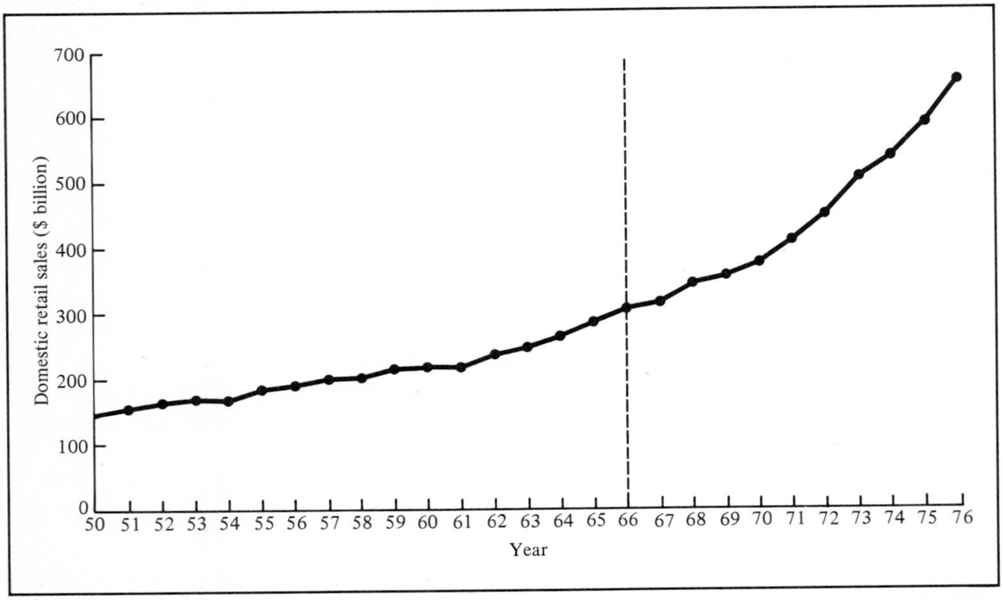

Figure 17.1 Estimated total annual domestic retail sales (in billions of dollars) for all retail stores (1950–1976).

SOURCE: Data are taken from Series S-12, *Survey of Current Business* and *Business Statistics Supplement*, U.S. Department of Commerce.

corded annually or monthly. It is this model that will be considered in this text.

To demonstrate the classical multiplicative time-series model, Figure 17.1 presents the estimated total annual domestic retail sales (in billions of dollars) for all retail establishments in the United States from 1950 through 1976. If we may characterize these time-series data, it is clear that retail sales revenues have shown a tendency to increase (in a curvilinear manner as indicated in Figure 17.3) over this 27-year period. This overall long-term tendency or impression (of upward or downward movements) is known as **trend.**

However, trend is not the only component factor influencing either these particular data or other annual time series. Two other factors—the **cyclical** component and the **irregular** component—are also present in the data. While these two factors will be more easily observed when we "decompose" the classical multiplicative time-series model in Section 17.3.4, it suffices here for us to get some general impression of what these component factors are from the data at hand. As presented in Figure 17.2, the **cyclical** component depicts the up and down swings or movements through the series. Cyclical movements vary in length—usually lasting from 2 to 10 years—and also differ in intensity or amplitude. If we visualize a smooth curve representing trend[1] which passes through the time series (see Figure 17.3), several data points are found to dip far below the trend curve and others are found to be protruding above it. As examples, from Figure 17.3 the observed values for the years 1954, 1958, 1961, 1967, and 1970 are dipping below the trend curve and are

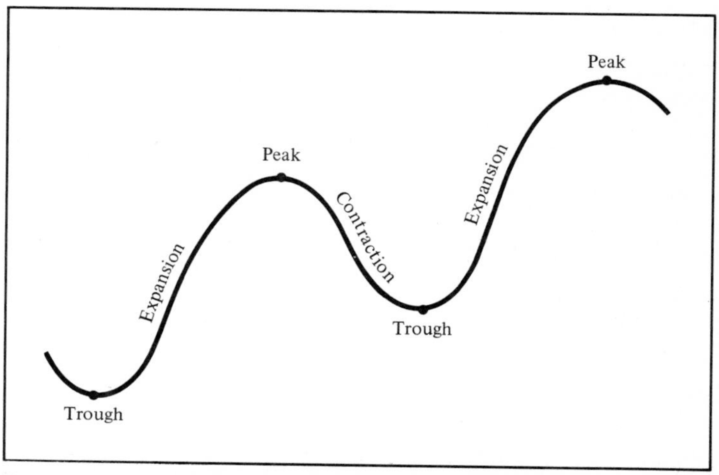

Figure 17.2 The four phases of the business cycle.

[1] The nonlinear trend curve was obtained by a 7-year moving average. The method of moving averages will be discussed in Section 17.4.1.

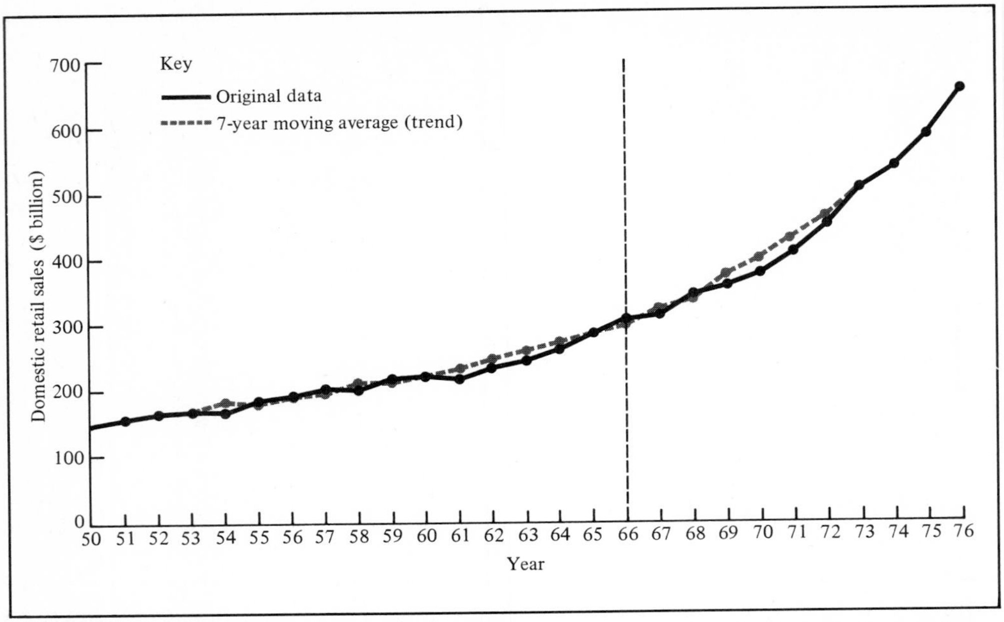

Figure 17.3 7-year moving average fitted to estimated total annual domestic retail sales (1950–1976).

SOURCE: Data are taken from Figure 17.1.

representing the "bottoming out" or "trough" of their respective cycles.[2] At the other extreme, the observed data for the years 1957, 1959, 1966, 1968, and 1973 are protruding above the smoothly fitted trend curve. Therefore they are representing the peaks of their respective business cycles. Any observed data which do not follow the smoothly fitted trend curve modified by the aforementioned cyclical movements are indicative of the irregular or random factors of influence. These random fluctuations will be more easily observed in Figure 17.4.

When data are recorded monthly rather than annually, an additional component factor has an effect on the time series. This fourth factor is called the **seasonal** component. To demonstrate the seasonal effects on a times series, Figure 17.4 presents the estimated monthly total domestic retail sales (in billions of dollars) for all retail outlets in the United States from January 1966 through December 1976. (The annual results for this 11-year period

[2] The time series for retail sales is considered by economists as one which generally "coincides" with the cyclical movements of the overall economy. Thus forecasters have referred to the series of retail sales as a "coinciding indicator" of economic conditions (Reference 2). It should be pointed out that in the past 25 years it is generally recognized by economists that peaks in the cyclical activity of the overall economy occurred in 1957, 1960, 1969, and 1973, while troughs occurred in 1954, 1958, 1961, 1970, and 1975. Thus from our visual impressions of Figure 17.3 retail sales have been a good (coinciding) indicator of cyclical fluctuations in general business activity.

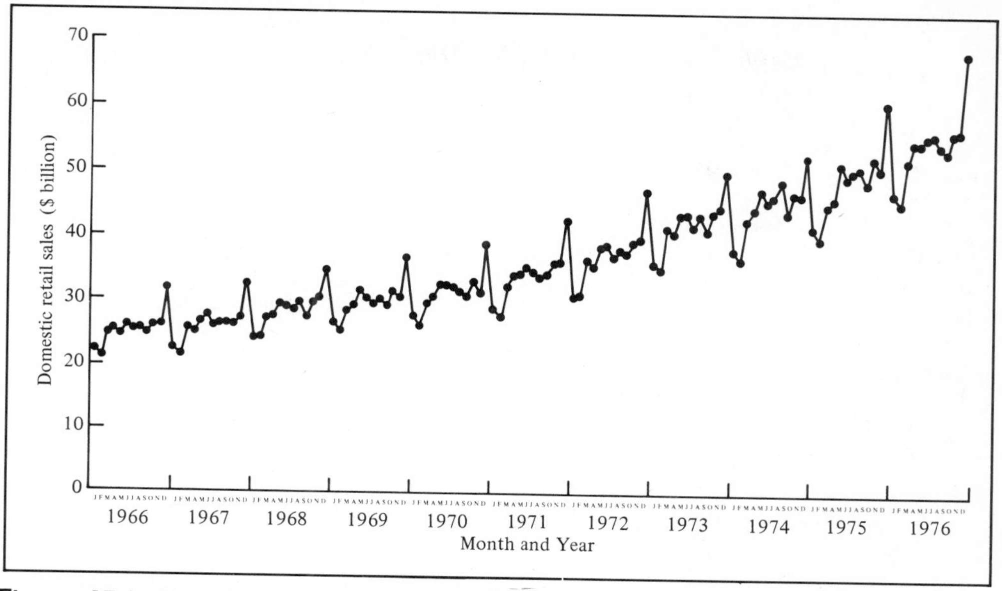

Figure 17.4 Estimated total domestic retail sales (in billions of dollars) for all stores (January 1966 to December 1976).

SOURCE: Data are taken from Series S-12, *Survey of Current Business* and *Business Statistics Supplement,* U.S. Department of Commerce.

can be observed to the right of the vertical dash line in Figures 17.1 and 17.3.) In each of the 11 years the seasonal influences on retail sales revenues are clearly indicated. The peak retail sales volume in each year is seen to occur in December (as the nation prepares for the holiday season) while lulls in sales volume are observed in every January and February.

On the other hand, the irregular or random fluctuations influencing the time series may be observed more or less by comparing the recorded sales revenues for the months of May and June. It is seen from Figure 17.4 that June retail sales revenues exceeded May retail sales revenues in 7 of the 11 years. However, if we were also to compare May and June sales revenues for the years 1950 through 1965 (Reference 3), we would observe that over the 16-year period May revenues were higher on eight occasions and June revenues were higher on eight occasions. In total, then, over the 27 years there is no systematic or observable pattern to the changes in the retail sales revenues for the months of May and June. Thus the obtained results are influenced by irregular or random factors.

In addition, from the monthly time-series data of Figure 17.4, the overall tendency (the curvilinearly increasing trend) is again readily observed. Unfortunately, for this particular set of data we are not able to visualize the cyclical factors which are influencing the time series. However, cyclical effects on monthly time series will be discussed and computed in Section 17.5.5.

Table 17.1
FACTORS INFLUENCING TIME-SERIES DATA

Component	Classification of Component	Definition	Reason for Influence	Duration
Trend	Systematic	Overall or persistent, long-term upward or downward pattern of movements	Due to changes in technology, population, wealth, values	Several years
Seasonal	Systematic	Fairly regular periodic fluctuations which occur within each 12-month period year after year	Due to weather conditions, social customs, religious customs	Within 12 months (for monthly data)
Cyclical	Systematic	Repeating up and down swings or movements through four phases: from peak (prosperity) to contraction (recession) to trough (depression) to expansion (recovery or growth)	Due to interactions of numerous combinations of factors influencing the economy	Usually 2–10 years with differing intensity for a complete cycle
Irregular	Unsystematic	The erratic or "residual" fluctuations in a time series which exist after taking into account the systematic effects—trend, seasonal, and cyclical	Due to random variations in data or due to unforeseen events such as strikes, hurricanes, floods, political assassinations, etc.	Short duration and nonrepeating

From the annual data in Figure 17.1 and from the monthly data in Figure 17.4 we have thus far determined that there are three or four component factors, respectively, which influence an economic or business time series. These are summarized in Table 17.1. Hence the classical multiplicative time-series model states that any observed value in a time series is the *product* of these influencing factors; that is, when the data are obtained annually, an observation recorded in the year *i*, Y_i, may be expressed as

$$Y_i = T_i \cdot C_i \cdot I_i \qquad (17.1)$$

where, in the year *i*,

T_i = value of the trend component

C_i = value of the cyclical component

I_i = value of the irregular component

On the other hand, when the data are obtained monthly, an observation recorded in month *i*, Y_i, may be given as

$$Y_i = T_i \cdot S_i \cdot C_i \cdot I_i \qquad (17.2)$$

where, in the month i, T_i, C_i, and I_i are the values of the trend, cyclical, and irregular components, respectively, and S_i is the value of the seasonal component.

17.3 TIME-SERIES ANALYSIS: ANNUAL DATA

The component factor of a time series most often studied is trend. Primarily we study trend for predictive purposes; that is, we either may wish to study trend directly as an aid in making intermediate and long-range forecasting projections, or we may wish to merely isolate and then eliminate its influencing effects on the time-series model as a guide to short-run (1 year or less) forecasting of general business cycle conditions.[3] As depicted in Figures 17.1 and 17.4, to obtain some visual impression or feeling of the overall long-term movements in a time series we construct a chart in which the observed data (dependent variable) are plotted on the vertical axis and the time periods (independent variable) are plotted on the horizontal axis. If it appears that a straight-line trend could be adequately fitted to the data, the two most widely used methods of trend fitting are the method of least squares (see Section 14.4) and the method of double exponential smoothing (References 1 and 9). However, if the time-series data indicate some long-run downward or upward (see Figure 17.1) curvilinear movement, the two most widely used trend fitting methods are the method of least squares (see Section 15.10) and the method of triple exponential smoothing (Reference 1). In this text, however, we shall focus mainly on the least-squares method for fitting linear trends as a guide to forecasting (see References 4 and 6 for curvilinear trend fitting).

17.3.1 Fitting and Forecasting Linear Trends

We recall from Section 14.4 that essentially, the least-squares method permits us to fit a straight line of the form

$$\hat{Y}_i = b_0 + b_1 X_i \qquad (17.3)$$

such that the values we calculate for the two coefficients—the intercept b_0 and the slope b_1—result in the sum of squared differences between each observed value Y_i in the data and each predicted value $\hat{Y}_i$ along the trend line being minimized, that is,

[3] Before we do this, however, it would not be at all unreasonable if we tested the time-series data for the existence of a statistically significant trend effect. One such procedure which could be used for this purpose is the Cox–Stuart Unweighted Sign test which was discussed in Section 13.4. In this chapter, however, we will be examining the time-series components for descriptive rather than inferential purposes.

$$\sum_{i=1}^{n} (Y_i - \hat{Y}_i)^2 = \text{minimum} \tag{17.4}$$

To obtain such a line, we recall that in linear regression analysis we compute the slope from

$$b_1 = \frac{\sum_{i=1}^{n} X_i Y_i - \frac{\left(\sum_{i=1}^{n} X_i\right)\left(\sum_{i=1}^{n} Y_i\right)}{n}}{\sum_{i=1}^{n} X_i^2 - \frac{\left(\sum_{i=1}^{n} X_i\right)^2}{n}} \tag{17.5}$$

and the intercept from

$$b_0 = \overline{Y} - b_1\overline{X} \tag{17.6}$$

Once this is accomplished and the line $\hat{Y}_i = b_0 + b_1X_i$ is obtained, we may substitute values for X into Equation (17.3) to predict various values for Y. We may note, however, that when using the method of least squares for fitting linear trends in time series, the observed values of the series (Y values) are usually recorded annually over several consecutive years (X values). Therefore, when dealing with annual time-series data, our computational efforts can be simplified if we properly "code" the X values. However, the coding scheme that we choose depends on whether our time-series data have been obtained over an even number of years or an odd number of years.

17.3.2 Fitting a Least-Square Trend for an Even Number of Years

For time-series data observed over an even number of years we choose the first year in our series as the origin and assign that year a code $X = 0$. All successive years are then assigned consecutively increasing integer codes: 1, 2, 3, 4, . . . , so that the last year in the series, the nth year, has code $n - 1$. Thus, for example, if a time series had $n = 6$ years of data, the codes would be 0, 1, 2, 3, 4, 5.

To fit a least-squares trend line to annual data having an even number of years, the time series presented in Table 17.2 and plotted in Figure 17.5 represents the annual payments (in billions of dollars) to life insurance companies both for interest on policy loans and for premium notes over the

Table 17.2
ANNUAL POLICY LOANS AND PREMIUM NOTES OF LIFE INSURANCE COMPANIES (1967–1976)

Year	Payment (billions of dollars)
1967	10.1
1968	11.3
1969	13.8
1970	16.1
1971	17.1
1972	18.0
1973	20.2
1974	22.9
1975	24.5
1976	25.9

SOURCE: Data are taken from Series S-19, *Survey of Current Business* and *Business Statistics Supplement*, U.S. Department of Commerce.

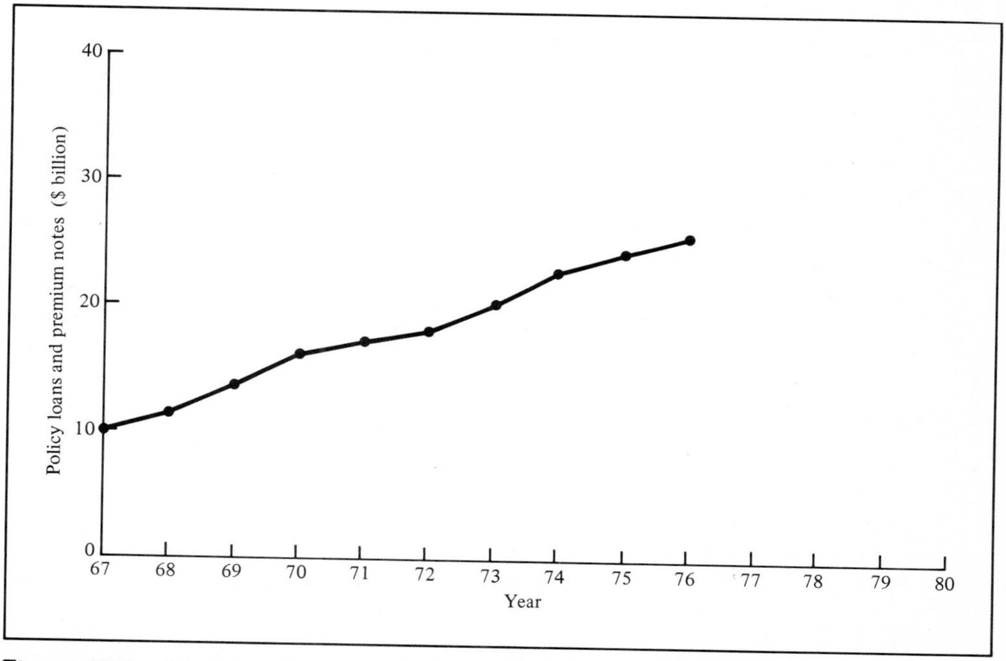

Figure 17.5 Annual policy loans and premium notes of life insurance companies (1967–1976).

SOURCE: Data are taken from Series S-19, *Survey of Current Business* and *Business Statistics Supplement,* U.S. Department of Commerce.

Table 17.3
COMPUTATION OF LEAST-SQUARES TREND LINE FOR 10 YEARS OF DATA

Year	X	Y (billions of dollars)	XY	X^2
1967	0	10.1	0	0
1968	1	11.3	11.3	1
1969	2	13.8	27.6	4
1970	3	16.1	48.3	9
1971	4	17.1	68.4	16
1972	5	18.0	90.0	25
1973	6	20.2	121.2	36
1974	7	22.9	160.3	49
1975	8	24.5	196.0	64
1976	9	25.9	233.1	81
n = 10	45	179.9	956.2	285

SOURCE: Data are taken from Table 17.2.

10-year period of 1967 through 1976. The necessary computations are given in Table 17.3.

Using Equations (17.5) and (17.6) we determine that

$$b_1 = \frac{\sum_{i=1}^{n} X_i Y_i - \frac{\left(\sum_{i=1}^{n} X_i\right)\left(\sum_{i=1}^{n} Y_i\right)}{n}}{\sum_{i=1}^{n} X_i^2 - \frac{\left(\sum_{i=1}^{n} X_i\right)^2}{n}} = \frac{956.2 - \frac{(45)(179.9)}{10}}{285 - \frac{(45)^2}{10}} \cong 1.8$$

and since

$$\bar{Y} = \frac{\sum_{i=1}^{n} Y_i}{n} = \frac{179.9}{10} = 17.99 \qquad \text{and} \qquad \bar{X} = \frac{\sum_{i=1}^{n} X_i}{n} = \frac{45}{10} = 4.5$$

then

$$b_0 = \bar{Y} - b_1\bar{X} = 17.99 - (1.8)(4.5) \cong 9.9$$

Since our first observed value in the time series was obtained for the year 1967, our origin is referenced in the middle of that year. Thus, we have

$$\hat{Y}_i = 9.9 + 1.8X_i \tag{17.7}$$

where origin = 1967 and X units = 1 year.

The intercept $b_0 = 9.9$ is the fitted trend value reflecting the amount of money (in billions of dollars) paid to life insurance companies for interest on loans and for premium notes during the origin or base year, 1967. The slope $b_1 = 1.8$ indicates that such payments are increasing at a rate of 1.8 billions of dollars per year.

To fit the trend line to the observed years of the series we merely substitute the appropriate coded values of X into Equation (17.7). As an example, for the year 1975, where $X = 8$, the predicted (fitted) trend value is given by

$$\hat{Y}_9 = 9.9 + (1.8)(8) = 24.3 \text{ billions of dollars}$$

To use the trend line for forecasting purposes, we may project the fitted line into the future by mathematical extrapolation.[4] For example, to predict the trend in payments for the year 1977, we substitute $X = 10$, the code for the year 1977, into Equation (17.7), and we forecast the trend to be

$$1977 \qquad \hat{Y}_{11} = 9.9 + (1.8)(10) = 27.9 \text{ billions of dollars}$$

Moreover, for the years 1978 through 1980 we forecast the trend in payments to be

$$1978 \qquad \hat{Y}_{12} = 9.9 + (1.8)(11) = 29.7 \text{ billions of dollars}$$

$$1979 \qquad \hat{Y}_{13} = 9.9 + (1.8)(12) = 31.5 \text{ billions of dollars}$$

$$1980 \qquad \hat{Y}_{14} = 9.9 + (1.8)(13) = 33.3 \text{ billions of dollars}$$

The fitted trend line projected to 1980 is plotted in Figure 17.6 along with the original time series.

17.3.3 Fitting a Least-Squares Trend for an Odd Number of Years

For time-series data observed over an odd number of years, the most efficient coding scheme that we can choose to facilitate our computations is the selection of the middle year in the sequence as the designated origin, having a code $X = 0$. All successive years are then assigned consecutively increasing integer codes while all preceding years are assigned consecutively decreasing integer codes. Thus, for example, if a time series had $n = 7$ years of data, the middle (fourth) year would be given a code of $X = 0$, and the coded sequence from the first year to the last would be

$$-3 \quad -2 \quad -1 \quad 0 \quad 1 \quad 2 \quad 3$$

By coding the middle year in the series as $X = 0$ it will always happen that

[4] The method of least squares is an objective mathematical technique for fitting a trend line or curve. Of course, the obtained trend equation may mechanically be projected into the future for predictive purposes. However, we may recall from Section 14.4.2 in our discussion of interpolation versus extrapolation that any such extrapolation beyond the range of the observed data makes assumptions about the continuity of the process that cannot be verified in a probabilistic sense.

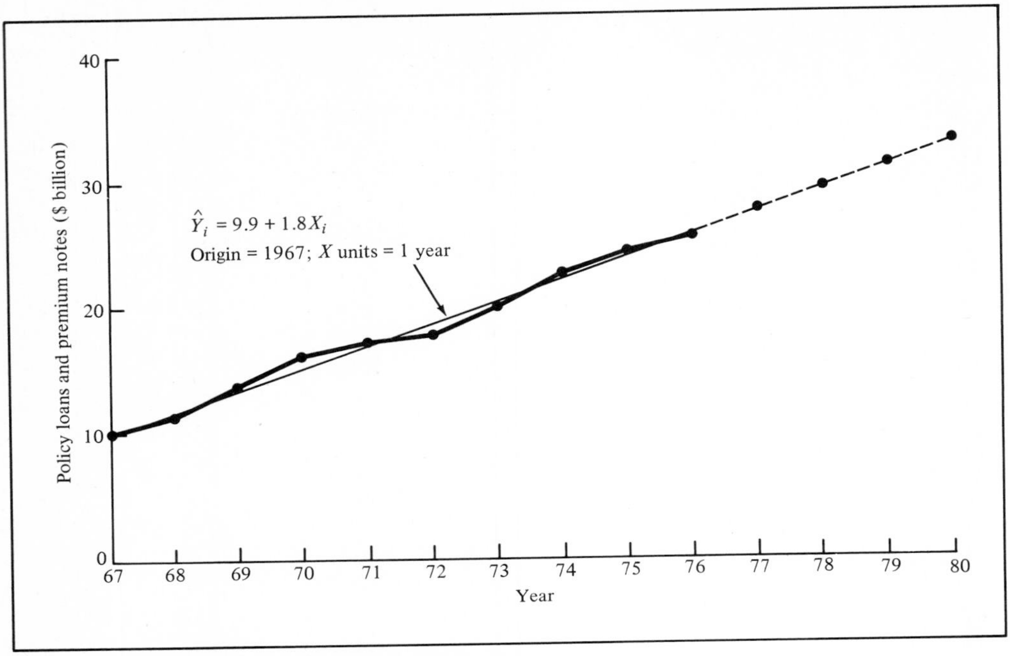

Figure 17.6 Fitting the least-squares trend line.

SOURCE: Data are taken from Tables 17.2 and 17.3.

$\sum_{i=1}^{n} X_i = 0$, and, therefore, the formulas for the slope and intercept [Equations (17.5) and (17.6)] would be altered as follows:

$$b_1 = \frac{\sum_{i=1}^{n} X_i Y_i}{\sum_{i=1}^{n} X_i^2} \tag{17.5a}$$

$$b_0 = \overline{Y} = \frac{\sum_{i=1}^{n} Y_i}{n} \tag{17.6a}$$

and the computational effort is reduced.

To fit a least-squares trend line to annual data having an odd number of years, the time series presented in Table 17.4 and plotted in Figure 17.7

Table 17.4
ANNUAL INDIVIDUAL INCOME TAXES PAID TO FEDERAL GOVERNMENT (1966–1976)

Year	Taxes (billions of dollars)
1966	55.4
1967	61.5
1968	68.7
1969	87.2
1970	90.4
1971	86.2
1972	94.7
1973	103.2
1974	119.0
1975	122.4
1976	131.6

SOURCE: Data are taken from Series S-19, *Survey of Current Business* and *Business Statistics Supplement*, U.S. Department of Commerce.

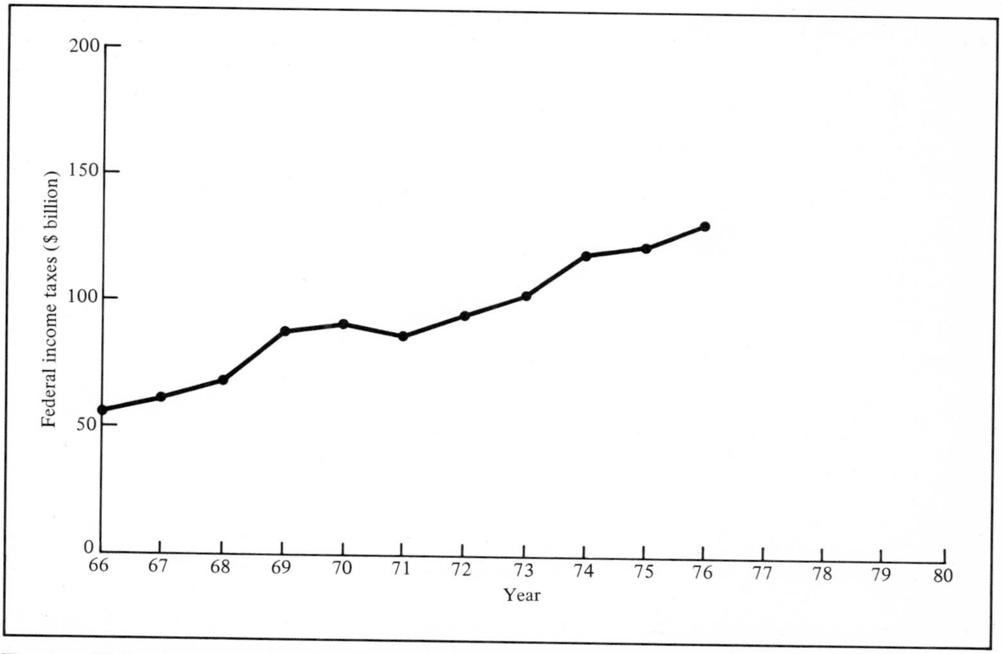

Figure 17.7 Annual income taxes paid to federal government (1966–1976).

SOURCE: Data are taken from Series S-19, *Survey of Current Business* and *Business Statistics Supplement*, U.S. Department of Commerce.

Table 17.5
COMPUTATION OF LEAST-SQUARES TREND FOR 11 YEARS OF DATA

Year	X	Y (billions of dollars)	XY	X^2
1966	−5	55.4	−277.0	25
1967	−4	61.5	−246.0	16
1968	−3	68.7	−206.1	9
1969	−2	87.2	−174.4	4
1970	−1	90.4	− 90.4	1
1971	0	86.2	0	0
1972	1	94.7	94.7	1
1973	2	103.2	206.4	4
1974	3	119.0	357.0	9
1975	4	122.4	489.6	16
1976	5	131.6	658.0	25
$n = 11$	0	1,020.3	811.8	110

SOURCE: Data are taken from Table 17.4.

represents the annual individual income taxes paid to the federal government for the 11-year period 1966 through 1976.

The computations needed for fitting a linear trend to this 11-year series by the method of least squares are shown in Table 17.5.

Using the adjusted formulas for slope and intercept [Equations (17.5*a*) and (17.6*a*)], we compute

$$b_1 = \frac{\sum_{i=1}^{n} X_i Y_i}{\sum_{i=1}^{n} X_i^2} = \frac{811.8}{110} \cong 7.4$$

and

$$b_0 = \overline{Y} = \frac{\sum_{i=1}^{n} Y_i}{n} = \frac{1,020.3}{11} \cong 92.8$$

Since the designated origin was the middle year (1971) of the series, we have

$$\hat{Y}_i = 92.8 + 7.4X_i \tag{17.8}$$

where origin = 1971 and X units = 1 year.

Equation (17.8) may be interpreted as follows: For the designated origin year, 1971, the fitted trend line indicates that 92.8 billions of dollars of income taxes was expected to have been paid by individuals to the federal government. Moreover, the slope $b_1 = 7.4$ indicates that such payments of individual income taxes to the federal government were increasing at the rate of 7.4 billions of dollars per year. Thus to project the trend in individual income tax payments to the year 1980, we substitute $X = 9$, the code for the year

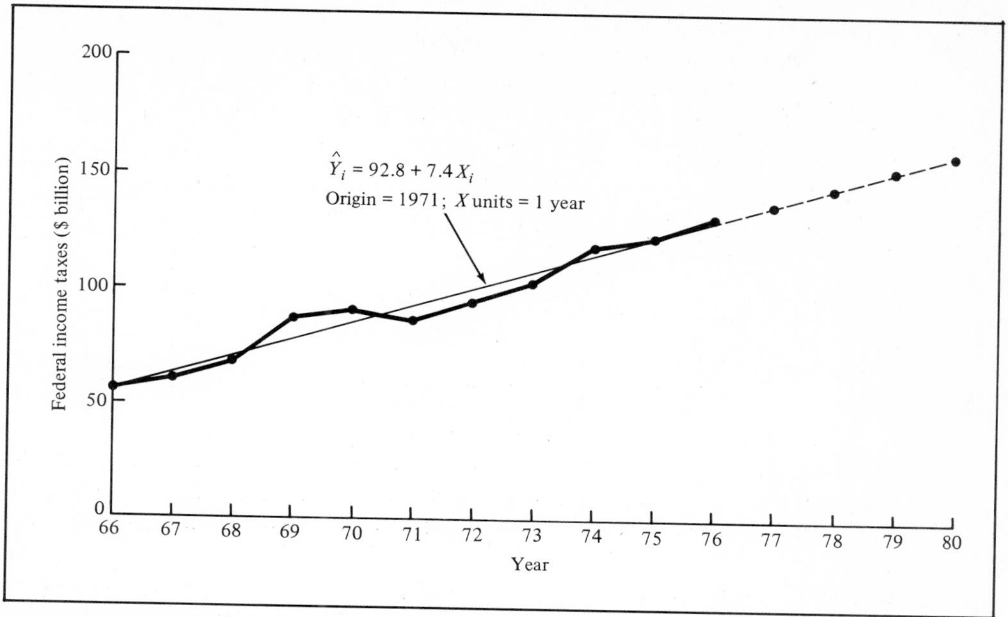

Figure 17.8 Fitting the least-squares trend line.
SOURCE: Data are taken from Tables 17.4 and 17.5.

1980, into Equation (17.8) and we forecast the trend to be

1980 $\quad \hat{Y}_{15} = 92.8 + (7.4)(9) = 159.4$ billions of dollars

The fitted trend line projected to 1980 is plotted in Figure 17.8 along with the original time series.

17.3.4 Isolating and Removing Trend from Annual Data: The Cyclical-Irregular Relatives

In the previous section we studied trend as an aid to intermediate and long-term forecasting. However, as we have already noted, economists and/or business forecasters may also wish to study trend so that its influencing effects may be removed from the classical multiplicative time-series model and thereby provide the framework for short-run forecasting of general business activity. The procedure of isolating and eliminating a component factor from the data is called **decomposing the time series.** Since the method of least squares provides us with "fitted" trend values $\hat{Y}_i$ for each year in the series, we can easily remove the trend component T_i from our classical multiplicative time-series model (because in any given year the trend component T_i is estimated by $\hat{Y}_i$). Thus, from Equation (17.1) the trend component may be removed through division as follows:

$$Y_i = T_i \cdot C_i \cdot I_i \qquad (17.1)$$

so that

$$\frac{Y_i}{\hat{Y}_i} = \frac{T_i \cdot C_i \cdot I_i}{\hat{Y}_i}$$

but since $\hat{Y}_i = T_i$, we have

$$\frac{Y_i}{\hat{Y}_i} = \frac{T_i \cdot C_i \cdot I_i}{T_i} = C_i \cdot I_i$$

The ratios of the observed values to fitted trend values, $Y_i/\hat{Y}_i$, which are computed each year in the series, are called the **cyclical-irregular relatives.** These values, which fluctuate around a base of 1.0, depict both cyclical and irregular activity in the series.

Returning to a previous example (see Table 17.2), the computations of the cyclical-irregular relatives are shown in Table 17.6 for the 10-year time-series data reflecting annual payments to life insurance companies both for interest on policy loans and for premium notes. The fitted trend values [column (4)] are determined by merely substituting the appropriately coded X values [column (2)] into the linear trend model [Equation (17.7)] obtained by the method of least squares. For each year in the series we see that the observed value [column (3)] is then divided by the fitted trend value [column (4)] to

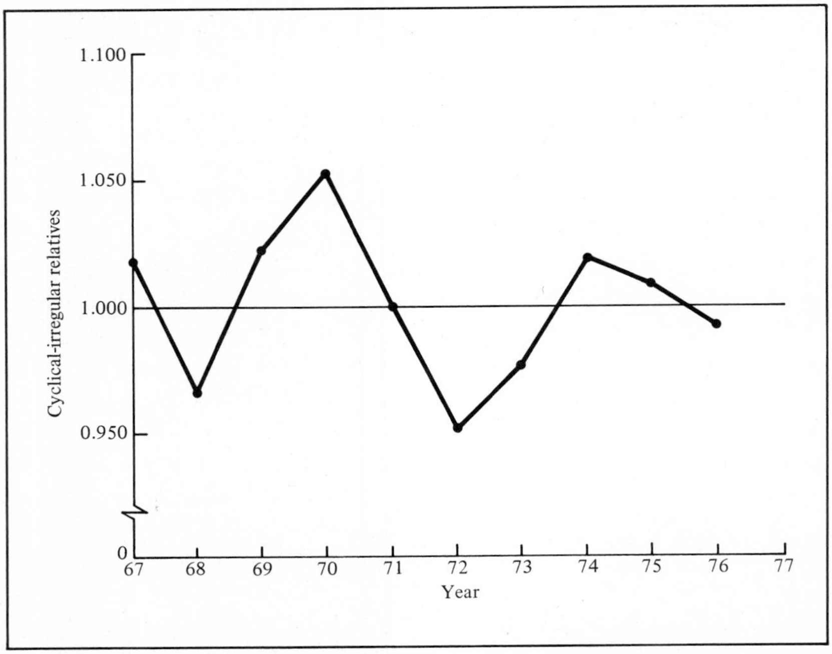

Figure 17.9 Plotting the cyclical-irregular relatives.
SOURCE: Data are taken from Table 17.6.

Table 17.6
OBTAINING THE CYCLICAL-IRREGULAR RELATIVES

(1) Year	(2) X_i	(3) Y_i (billions of dollars)	(4) $\hat{Y}_i = 9.9 + 1.8X_i$ Fitted Trend	(5) $Y_i/\hat{Y}_i$
1967	0	10.1	9.9	1.020
1968	1	11.3	11.7	0.966
1969	2	13.8	13.5	1.022
1970	3	16.1	15.3	1.052
1971	4	17.1	17.1	1.000
1972	5	18.0	18.9	0.952
1973	6	20.2	20.7	0.976
1974	7	22.9	22.5	1.018
1975	8	24.5	24.3	1.008
1976	9	25.9	26.1	0.992

SOURCE: Data are taken from Table 17.2.

yield the cyclical-irregular relative [column (5)]. This series of cyclical-irregular relatives is plotted in Figure 17.9. With annual data, no further "decomposition" of the time series is undertaken.

17.4 OTHER METHODS OF ANNUAL TREND ANALYSIS

Figure 17.10 depicts the number of passenger cars sold annually from automobile plants in the United States over the 17-year period of 1960 through 1976. When examining annual data such as these, our visual impression of the overall long-term tendencies or movements in the series is obscured by wide fluctuations in the cyclical and irregular components. It is then difficult to

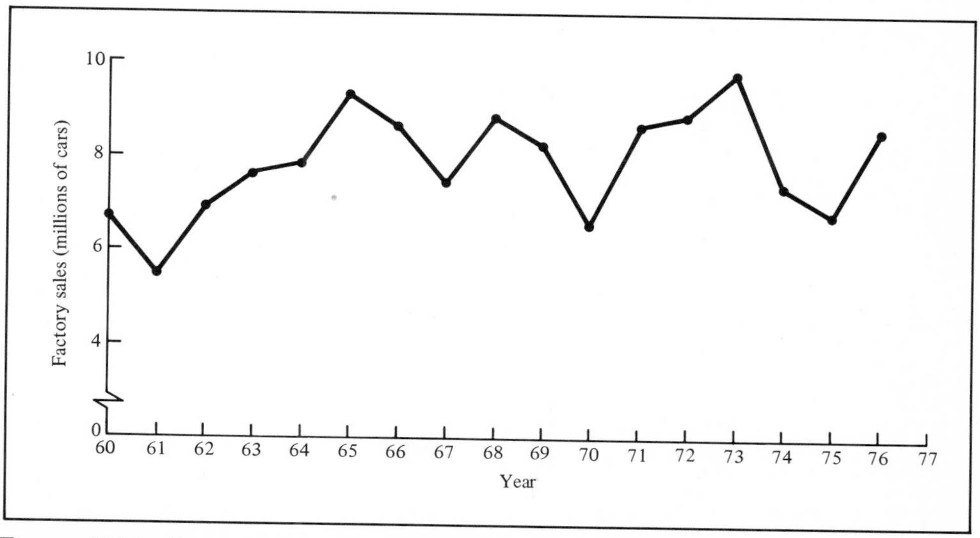

Figure 17.10 Factory sales of passenger cars from U.S. plants (1960–1976).

SOURCE: Data are taken from Series S-40, *Survey of Current Business* and *Business Statistics Supplement*, U.S. Department of Commerce.

judge whether a linear or curvilinear trend will better fit a particular set of data, or, as in Figure 17.10 whether there is really any long-term downward or upward trend effect present in the data at all.

Under such conditions as these, the **method of moving averages** or the **method of exponential smoothing** may be utilized to "smooth" a series and thereby provide us with an impression as to the overall long-term pattern of movement in the data—free from unwanted cyclical or irregular disturbances.

17.4.1 Moving Averages

Unlike the method of least squares, the moving-averages method of studying trend is highly subjective and dependent upon the length of the period selected for constructing the averages. To eliminate the cyclical fluctuations, the period chosen should be an integer value which corresponds to (or is a multiple of) the estimated average length of a cycle in the series.[5] Thus for the retail sales data (Figure 17.1) it was subjectively estimated that the average cyclical length—measured from peak to peak or from trough to trough—was 3.5 years; therefore since 7 is a multiple of 3.5, 7-year moving averages were fitted to the data in order to indicate the overall trend (Figure 17.3).

But what are moving averages and how are they computed? **Moving averages for a chosen period of length l consist of a series of arithmetic means computed over time such that each mean is calculated for a sequence of observed values having that particular length l.**

Table 17.7
3-YEAR AND 7-YEAR MOVING AVERAGES OF FACTORY SALES OF PASSENGER CARS FROM U.S. PLANTS (1960–1976)

(1) Year	(2) Millions of Cars Sold	(3) 3-Year Moving Total	(4) 3-Year Moving Average	(5) 7-Year Moving Total	(6) 7-Year Moving Average
1960	6.7	—	—	—	—
1961	5.5	19.1	6.4	—	—
1962	6.9	20.0	6.7	—	—
1963	7.6	22.3	7.4	52.4	7.5
1964	7.8	24.7	8.2	53.1	7.6
1965	9.3	25.7	8.6	56.4	8.1
1966	8.6	25.3	8.4	57.7	8.2
1967	7.4	24.8	8.3	56.6	8.1
1968	8.8	24.4	8.1	57.4	8.2
1969	8.2	23.5	7.8	56.9	8.1
1970	6.5	23.3	7.8	58.0	8.3
1971	8.6	23.9	8.0	57.9	8.3
1972	8.8	27.1	9.0	55.8	8.0
1973	9.7	25.8	8.6	56.1	8.0
1974	7.3	23.7	7.9	—	—
1975	6.7	22.5	7.5	—	—
1976	8.5	—	—	—	—

SOURCES: Data are taken from Series S-40, *Survey of Current Business* and *Business Statistics Supplement,* U.S. Department of Commerce.

[5] As we shall see in Section 17.5.2, moving averages with an even number of terms are more cumbersome to construct than those with an odd number of terms.

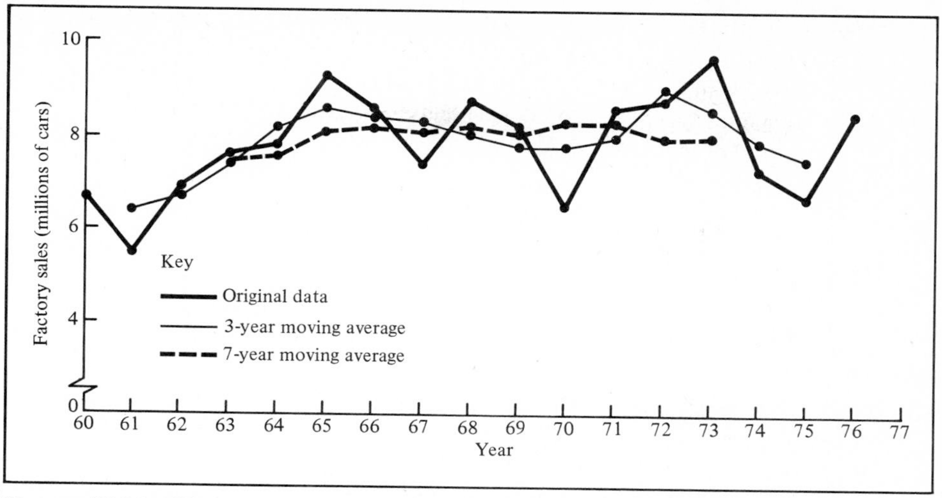

Figure 17.11 Plotting the 3-year and the 7-year moving averages.

SOURCE: Data are taken from Table 17.7.

For example, 3-year moving averages consist of series of means obtained over time by averaging out consecutive sequences containing three observed values. Table 17.7 presents the annual data on passenger car sales along with the computations for 3-year moving averages and 7-year moving averages. Both of these constructed series are plotted in Figure 17.11 with the original data.

In practice, to compute 3-year moving averages, we first obtain a series of 3-year moving totals as indicated in column (3) of Table 17.7 and then divide each of these totals by 3. The results are given in column (4). For example, since our observed time series was first recorded in 1960, the first 3-year moving total consists of the sum of the first three annually recorded values—6.7, 5.5, and 6.9. This moving total, 19.1, is then "centered" so that the recording is made against the year 1961. To obtain the moving total for the year 1962—which consists of the observed annual sales data for the years 1961, 1962, and 1963—we merely add the next observed value in the time series (year 1963) to the previous moving total and then subtract the first (oldest) value in the series. This process continues so that the 3-year moving total for any particular year i in the series represents the sum of the observed value for year i along with the observed values for the year preceding it and the year following it. On the other hand, with 7-year moving totals the result computed and recorded for the year i consists of the observed value in the time series for year i plus the three observed values which precede it and the three observed values which follow it. To "move the seven-year total" from one year to the next, we merely add on to the previous total the next observed value in the time series and remove the oldest value that had appeared in the previous total. This process continues through the series. The 7-year moving averages are then obtained by dividing the series of moving totals by 7.

We note from columns (3) and (4) of Table 17.7 that in obtaining the 3-year moving averages, no result can be computed for the first or last observed value in the time series. Moreover, as seen in columns (5) and (6), when computing 7-year moving averages there are no results for the first three observed values or the last three values. This, of course, occurs because the first 7-year moving total for the data at hand consists of the number of passenger cars sold during the years 1960 through 1966, which is centered at 1963, while the last moving total consists of the number of car sales recorded in 1970 through 1976, which is centered at 1973.

From Figure 17.11 we can clearly see that the 7-year moving averages smooth the series a great deal more than do the 3-year moving averages, since the period is of longer duration. Unfortunately, however, as we previously had noted, the longer the period, the fewer the number of moving average values that can be computed and plotted. Therefore, selecting moving averages with periods of length greater than 7 years is usually undesirable since too many computed data points would be missing at the beginning and end of the series, making it more difficult to obtain an overall impression of the trend through the entire series.

17.4.2 Exponential Smoothing

Exponential smoothing is another technique that may be used to smooth a time series and thereby provide us with an impression as to the overall long-term movements in the data. In addition, the method of exponential smoothing can be utilized for obtaining short-term (one period into the future) forecasts for such time series as depicted in Figure 17.10 for which it is questionable as to what type of long-term trend effect, if any, is present in the data. In this respect, the technique possesses a distinct advantage over the method of moving averages. Essentially, the method of exponential smoothing derives its name because it provides us with an *exponentially weighted* moving average through the time series; that is, throughout the series each smoothing calculation or forecast is dependent upon all previously observed values. This is another advantage over the method of moving averages, which does not take into account all the observed values in this manner. With exponential smoothing, the weights assigned to the observed values decrease over time so that when a calculation is made, the most recently observed value receives the highest weight, the previously observed value receives the second highest weight, and so on, with the initially observed value receiving the lowest weight.

While the magnitude of work involved from this description may seem formidable, the calculations are really quite simple. If we may focus on the smoothing aspects of the technique (rather than the forecasting aspects), the formulas developed for exponentially smoothing a series in any time period

i are based only on three terms—the presently observed value in the time series Y_i, the previously computed exponentially smoothed value ε_{i-1}, and some subjectively assigned weight or smoothing coefficient W. Thus to smooth a series at any time period i we have the following expression:

$$\varepsilon_i = WY_i + (1 - W)\varepsilon_{i-1} \qquad (17.9)$$

where ε_i = value of the exponentially smoothed series being computed in time period i

ε_{i-1} = value of the exponentially smoothed series already computed in time period $i - 1$

Y_i = observed value of the time series in period i

W = subjectively assigned weight or smoothing coefficient[6] (where $0 < W < 1$)

The choice of a smoothing coefficient or weight which we should assign to our time series is quite important since it will affect our results. Unfortunately this selection is rather subjective. However, in regard to smoothing ability, we may observe from Figures 17.11 and 17.12 that a series of l term moving averages is related to an exponentially smoothed series having weight W as follows:

[6] That all i observed values in the time series are included in the computation of the exponentially smoothed value in time period i can be seen by noting that the present smoothed value is calculated using the smoothed value of the previous period, and that value, in turn, was calculated using the smoothed value from its previous period, and so on. Algebraically, this can be stated as follows.

In time period 1,

$$\varepsilon_1 = Y_1$$

In time period 2,

$$\varepsilon_2 = WY_2 + (1 - W)\varepsilon_1 = WY_2 + (1 - W)Y_1$$

In time period 3,

$$\begin{aligned}\varepsilon_3 = WY_3 + (1 - W)\varepsilon_2 &= WY_3 + (1 - W)[WY_2 + (1 - W)Y_1] \\ &= WY_3 + W(1 - W)Y_2 + (1 - W)^2Y_1\end{aligned}$$

In general, in time period i,

$$\begin{aligned}\varepsilon_i = WY_i + (1 - W)\varepsilon_{i-1} = WY_i &+ W(1 - W)Y_{i-1} \\ &+ W(1 - W)^2Y_{i-2} + \cdots + (1 - W)^{(i-1)}Y_1\end{aligned}$$

Thus we see that over time, as the integer value i gets large, the weights assigned to the earlier (older) values in the time series may become negligible.

$$W = \frac{2}{l + 1} \quad (17.10)$$

or

$$l = \frac{2}{W} - 1 \quad (17.11)$$

From Equations (17.10) and (17.11) we note that with respect to smoothing ability, similarities are found between the 3-year series of moving averages (Figure 17.11) and the exponentially smoothed series having weight $W = .50$ (see Figure 17.12). In addition, we see that the series of 7-year moving averages (Figure 17.11) corresponds to the exponentially smoothed series having weight $W = .25$ (see Figure 17.12). By examining how our two smoothing series (one with $W = .25$ and the other with $W = .50$) fit the observed data in Figure 17.12, we may realize that the choice of a particular smoothing coefficient W is dependent upon the purpose of the user. If we desire only to smooth a series by eliminating unwanted cyclical and irregular variations, we should select a small value for W (closer to zero). On the other hand, if our goal is forecasting, we should choose a larger value for W (closer to 1). In the former case, the overall long-term tendencies of the series will be apparent; in the latter case, future short-term directions may be more adequately predicted.

To use the exponentially weighted moving average for purposes of forecasting rather than for smoothing, we merely take the smoothed value in our current period of time (say time period i) as our projected estimate of the observed value of the time series in the following time period, $i + 1$, that is,

$$\hat{Y}_{i+1} = \varepsilon_i \quad (17.12)$$

Thus, for example, to forecast the number of passenger cars to be sold from automobile plants in the United States during the year 1977 we would use the smoothed value for the year 1976 as its estimate. From Table 17.8, for a smoothing coefficient of $W = .50$, that projection is 8.0 million cars. Once the observed data for the year 1977 become available, we can use Equation (17.9) to make a forecast for the year 1978 (by obtaining the smoothed value for 1977) as follows:

$$\varepsilon_{1977} = WY_{1977} + (1 - W)\varepsilon_{1976}$$

Current Smoothed Value = (W)(Current Observed Value)
+ (1 − W)(Previous Smoothed Value)

or, in terms of forecasting,

$$\hat{Y}_{1978} = WY_{1977} + (1 - W)\hat{Y}_{1977}$$

New Forecast = (W)(Current Observed Value)
+ (1 − W)(Current Forecast)

The computations for the two smoothed series (using respective weights of $W = .25$ and $W = .50$) are listed in Table 17.8 and, as previously indicated, are plotted in Figure 17.12 along with the original time series. To demonstrate the computations, let us consider for a moment the exponentially smoothed series having weight $W = .25$. For example, as a starting point we merely use the initial observed value $Y_{1960} = 6.7$ as our first smoothed value ($\varepsilon_{1960} = 6.7$) and as our first forecast value ($\hat{Y}_{1961} = 6.7$). Now using the observed value of the time series for the year 1961 ($Y_{1961} = 5.5$) we may smooth the series for the year 1961 by computing

$$\begin{aligned} \varepsilon_{1961} &= WY_{1961} + (1 - W)\varepsilon_{1960} \\ &= (.25)(5.5) + (.75)(6.7) = 6.4 \text{ million} \end{aligned}$$

Of course, this smoothed value also serves as the forecast value for the following year ($\hat{Y}_{1962} = 6.4$). The process continues in the same manner until

Table 17.8
EXPONENTIALLY SMOOTHED SERIES
OF FACTORY SALES OF PASSENGER CARS
FROM U.S. PLANTS (1960–1976)

Year	Millions of Cars Sold	W = .50	W = .25
1960	6.7	6.7	6.7
1961	5.5	6.1	6.4
1962	6.9	6.5	6.5
1963	7.6	7.0	6.8
1964	7.8	7.4	7.0
1965	9.3	8.4	7.6
1966	8.6	8.5	7.8
1967	7.4	8.0	7.7
1968	8.8	8.4	8.0
1969	8.2	8.3	8.0
1970	6.5	7.4	7.6
1971	8.6	8.0	7.8
1972	8.8	8.4	8.0
1973	9.7	9.0	8.4
1974	7.3	8.2	8.1
1975	6.7	7.4	7.8
1976	8.5	8.0	8.0

SOURCE: Data are taken from Table 17.7.

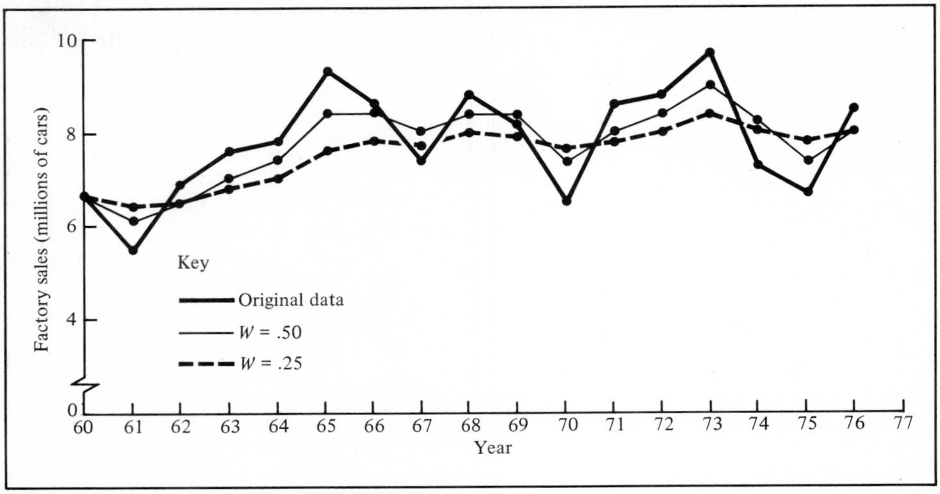

Figure 17.12 Plotting the exponentially smoothed series (W = .50 and W = .25).

SOURCE: Data are taken from Table 17.8.

all the values in the series have been smoothed and the results plotted in Figure 17.12.

The exponentially weighted moving-average methods for smoothing and forecasting have gained wide recognition over the past decade as a guide to managerial planning and control. While a description of the more sophisticated methods (such as double exponential smoothing for linear movements and triple exponential smoothing for curvilinear movements) are beyond the scope of this text, available computer packages such as STATPACK are invaluable for handling the laborious calculations (Reference 11).

17.5 TIME-SERIES ANALYSIS: MONTHLY DATA

Figure 17.13 depicts the monetary value (in billions of dollars) of residential construction contracts issued on a monthly basis from January 1971 through December 1976. For such monthly time series as these the classical multiplicative time-series model includes the **seasonal** component in addition to the trend, cyclical, and irregular components. The model is expressed by Equation (17.2) as

$$Y_i = T_i \cdot S_i \cdot C_i \cdot I_i \tag{17.2}$$

Basically there are two major goals of time-series analysis with monthly data. Either we are interested in *forecasting* some future monthly movements, or we are interested in *decomposing* the time series by isolating and removing the trend, seasonal, and irregular components so that we can concentrate on

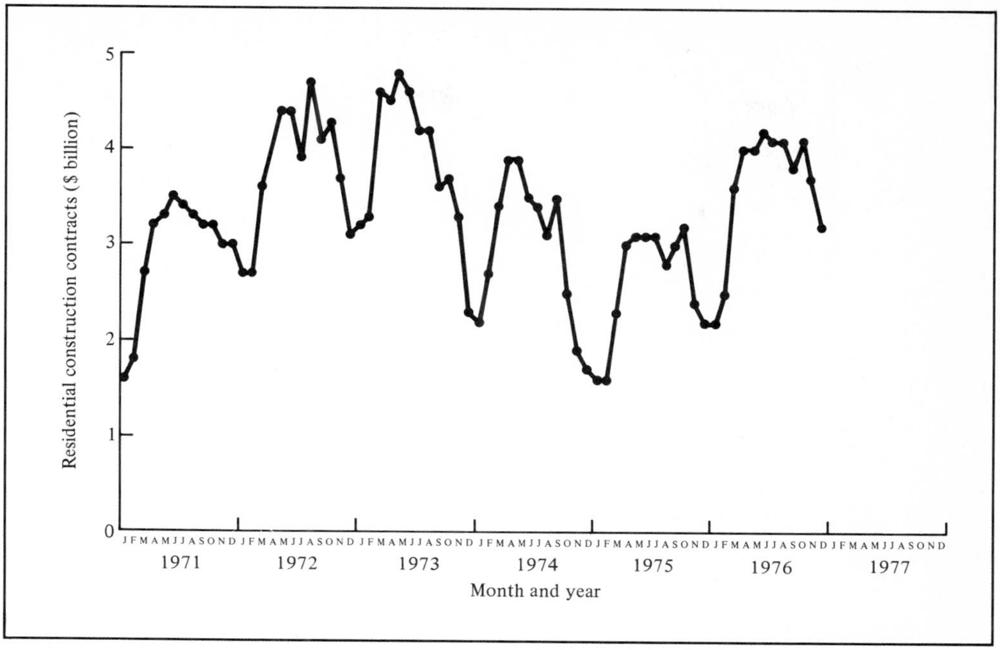

Figure 17.13 Value of monthly residential construction contracts issued (in billions of dollars) (January 1971 to December 1976).

SOURCE: Data are taken from Series S-10, *Survey of Current Business* and *Business Statistics Supplement,* U.S. Department of Commerce.

how a particular series correlates with overall business activity; that is, we may determine whether a particular series can be considered as a *leading, coinciding,* or *lagging* indicator of overall economic activity based upon whether the cyclical component of the series exhibits tendencies to precede, match, or follow, respectively, the cyclical behavior of the overall economy.

17.5.1 Fitting and Forecasting Linear Trends: Converting Annual Series to Monthly Series

When dealing with monthly time series which can be fitted with a linear trend much labor can be saved without too much loss in accuracy if we form the *annual aggregates* from our monthly totals and fit a least-squares trend line to the annual data. The resulting expression, in annual terms, can easily be converted to monthly terms by dividing the intercept by 12 and the slope by 144 and then "shifting the series." To demonstrate this, Table 17.9 presents the calculations necessary for fitting a least-squares trend line to the annual value of residential construction contracts issued over the 6-year period of 1971 through 1976.

Since this series contains an even number of years, the appropriate coding scheme given in Section 17.3.2 is employed, and, using Equations (17.5) and (17.6), we have

Table 17.9
COMPUTATION OF LEAST-SQUARES TREND LINE FOR 6 YEARS OF DATA

Year	X	Y (billions of dollars)	XY	X^2
1971	0	34.7	0	0
1972	1	45.0	45.0	1
1973	2	45.7	91.4	4
1974	3	34.4	103.2	9
1975	4	31.3	125.2	16
1976	5	43.7	218.5	25
$n = 6$	15	234.8	583.3	55

SOURCE: Data are taken from Series S-10, *Survey of Current Business* and *Business Statistics Supplement*, U.S. Department of Commerce.

$$b_1 = \frac{\sum_{i=1}^{n} X_i Y_i - \frac{\left(\sum_{i=1}^{n} X_i\right)\left(\sum_{i=1}^{n} Y_i\right)}{n}}{\sum_{i=1}^{n} X_i^2 - \frac{\left(\sum_{i=1}^{n} X_i\right)^2}{n}} = \frac{-3.7}{17.5} = -0.2114$$

and since

$$\overline{Y} = \frac{\sum_{i=1}^{n} Y_i}{n} = \frac{234.8}{6} = 39.13 \qquad \text{and} \qquad \overline{X} = \frac{\sum_{i=1}^{n} X_i}{n} = \frac{15}{6} = 2.5$$

then

$$b_0 = \overline{Y} - b_1\overline{X} = 39.13 - (-0.2114)(2.5) = 39.66$$

Thus the annually fitted trend line is given by

$$\hat{Y}_i = 39.66 - 0.2114X_i \tag{17.13}$$

where origin = 1971, and X units = 1 year.

To convert an anual trend model [Equation (17.13)] to a monthly basis we first divide the intercept by 12 and the slope by 144. This gives us

$$\hat{Y}_i = \frac{39.66}{12} - \frac{0.2114}{144}X_i = 3.3049 - 0.001468X_i \tag{17.14}$$

where origin = June 30–July 1, 1971 and X units = 1 month.

We may recall that when dealing with data on an annual basis, the data representing the entire year are recorded in the middle of the year. Hence when converting from an annual trend equation to a monthly trend equation, our resulting origin also falls in the middle of the year—between June 30 and July 1. Rather than stating that the monthly trend equation has an origin between the two months June and July, we merely shift the origin of the series to the middle of July by adding in one half the value of the slope to Equation (17.14), that is, to shift to July 15, 1971, we have

$$\begin{aligned}\hat{Y}_i &= 3.3049 - 0.001468(X_i + .5)\\ &= 3.3049 - 0.001468X_i - 0.000734\end{aligned}$$

so that

$$\hat{Y}_i = 3.3042 - 0.001468X_i \qquad (17.15)$$

where origin = July 15, 1971, and X units = 1 month.

For this series the new slope indicates that (on a monthly basis) the monetary values of residential construction contracts issued have been declining at a rate of 0.001468 billions of dollars (that is, 1.468 millions of dollars) per month. This is depicted in Figure 17.14 where the slope of the fitted monthly trend line [Equation (17.15)] exhibits a slight tendency to decline over time. Of course, Equation (17.15) may be used to project or forecast future monthly trend values in residential construction contracts. However, since such monthly time series are influenced by seasonal factors, we shall not make any future forecasts until we have developed a **seasonal index** which accounts for the month-to-month fluctuations. This will be accomplished in the following section.

17.5.2 Computing the Seasonal Index

It is important to isolate and study the seasonal movements in a monthly time series for two reasons. First, by knowing the value of the seasonal component for any particular month, the economist can easily adjust and improve upon trend projections for forecasting purposes. Second, by knowing the value of the seasonal component the economist can decompose the time series by eliminating its influences—along with those pertaining to trend and irregular fluctuations—and thereby concentrate on the cyclical movements of the series. If, as often it is assumed, the seasonal movements are fairly constant over time, the construction of a seasonal index may be illustrated from Tables 17.10 and 17.11.

To start, a series of 12-month moving totals is obtained. However, as depicted in column (3) of Table 17.10, when recording these moving totals the results are centered between the two middle months comprising each respective moving total. For example, the first moving total, which consists of the months of January 1971 through December 1971, is recorded between

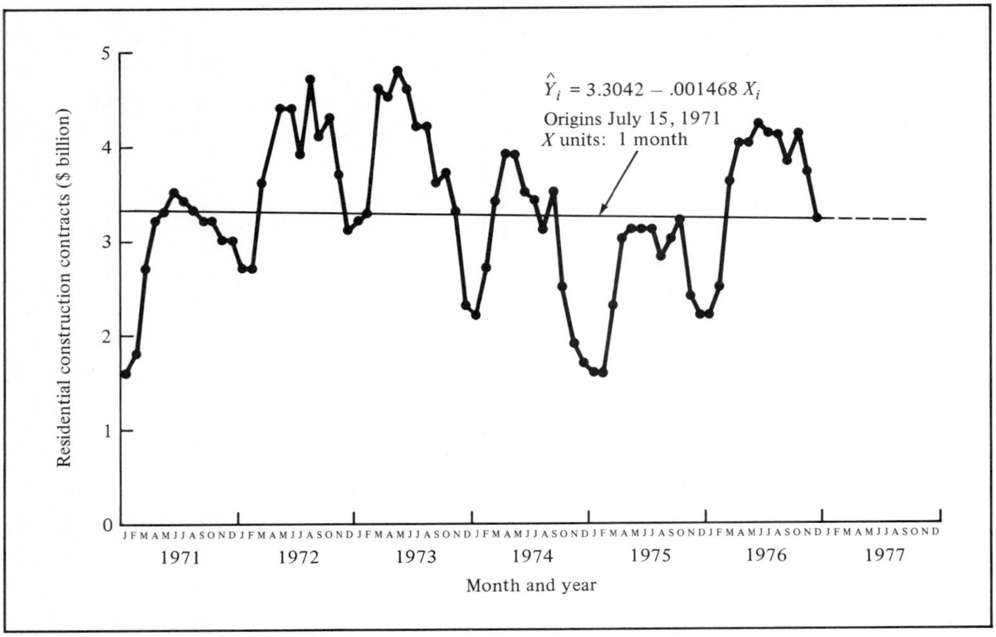

Figure 17.14 Fitting the least-squares trend line.

SOURCE: Data are taken from Tables 17.10 and 17.13.

June and July 1971; the second moving total, which consists of the months of February 1971 through January 1972, is recorded between July and August 1971; and so on. To center these results within a particular month, "2-month moving totals of the 12-month moving totals" are obtained as indicated in column (4) of Table 17.10. The first result, which consists of the total indicated between June and July plus that between July and August, is centered in July 1971.[7] By dividing these totals in column (4) by 24, the **centered moving averages** are obtained as shown in column (5). These centered moving averages are said to consist of the trend and cyclical components of the series. The original data [column (2)] are then divided by the respective centered moving averages [column (5)] yielding the **ratio to moving averages** depicted in column (6). Essentially, these ratios to moving averages represent the seasonal and irregular fluctuations in the series, since the division of the observed data [column (2)] by the centered moving averages [column (5)] effectively eliminates trend and cyclical influences as demonstrated in Equation (17.16):

[7] If we reflect on this computation, we will realize that these centered values presented in column (4) really represent a 13-month weighted moving total where the results are centered in the middle of the series (the seventh month). Moreover, the centered month receives a weight of 2, the 5 months preceding it receive a weight of 2, the 5 months following it receive a weight of 2, while the extremes—the first and last months in the moving total—receive a weight of 1.

Table 17.10
DEVELOPING THE SEASONAL INDEX

(1) Year and Month		(2) Residential Construction Contracts (billions of dollars)	(3) 12-Month Moving Totals	(4) 2-Month Moving Totals of 12-Month Moving Totals	(5) Centered 12-Month Moving Averages	(6) Ratios to Moving Average	(7) Seasonal Index	(8) Desea-sonalized Data
1971	Jan	1.6	—	—	—	—	0.701	2.3
	Feb	1.8	—	—	—	—	0.772	2.3
	Mar	2.7	—	—	—	—	1.045	2.6
	Apr	3.2	—	—	—	—	1.175	2.7
	May	3.3	—	—	—	—	1.215	2.7
	Jun	3.5	35.2	—	—	—	1.172	3.0
	Jul	3.4	36.3	71.5	2.98	1.141	1.140	3.0
	Aug	3.3	37.2	73.5	3.06	1.078	1.077	3.1
	Sep	3.2	38.1	75.3	3.14	1.019	1.042	3.1
	Oct	3.2	38.9	77.0	3.21	0.997	1.027	3.1
	Nov	3.0	40.0	78.9	3.29	0.912	0.911	3.3
	Dec	3.0	40.9	80.9	3.37	0.890	0.721	4.2
1972	Jan	2.7	41.4	82.3	3.43	0.787	0.701	3.9
	Feb	2.7	42.8	84.2	3.51	0.769	0.772	3.5
	Mar	3.6	43.7	86.5	3.60	1.000	1.045	3.4
	Apr	4.0	44.8	88.5	3.69	1.084	1.175	3.4
	May	4.4	45.5	90.3	3.76	1.170	1.215	3.6
	Jun	4.4	45.6	91.1	3.80	1.158	1.172	3.8
	Jul	3.9	46.1	91.7	3.82	1.021	1.140	3.4
	Aug	4.7	46.7	92.8	3.87	1.214	1.077	4.4
	Sep	4.1	47.7	94.4	3.93	1.043	1.042	3.9
	Oct	4.3	48.2	95.9	4.00	1.075	1.027	4.2
	Nov	3.7	48.6	96.8	4.03	0.918	0.911	4.1
	Dec	3.1	48.8	97.4	4.06	0.764	0.721	4.3
1973	Jan	3.2	49.1	97.9	4.08	0.784	0.701	4.6
	Feb	3.3	48.6	97.7	4.07	0.811	0.772	4.3
	Mar	4.6	48.1	96.7	4.03	1.141	1.045	4.4
	Apr	4.5	47.5	95.6	3.98	1.131	1.175	3.8
	May	4.8	47.1	94.6	3.94	1.218	1.215	4.0
	Jun	4.6	46.3	93.4	3.89	1.183	1.172	3.9
	Jul	4.2	45.3	91.6	3.82	1.099	1.140	3.7
	Aug	4.2	44.7	90.0	3.75	1.120	1.077	3.9
	Sep	3.6	43.5	88.2	3.68	0.978	1.042	3.5
	Oct	3.7	42.9	86.4	3.60	1.028	1.027	3.6
	Nov	3.3	42.0	84.9	3.54	0.932	0.911	3.6
	Dec	2.3	40.9	82.9	3.45	0.667	0.721	3.2
1974	Jan	2.2	40.1	81.0	3.38	0.651	0.701	3.1
	Feb	2.7	39.0	79.1	3.30	0.818	0.772	3.5
	Mar	3.4	38.9	77.9	3.25	1.046	1.045	3.3
	Apr	3.9	37.7	76.6	3.19	1.223	1.175	3.3
	May	3.9	36.3	74.0	3.08	1.266	1.215	3.2
	Jun	3.5	35.7	72.0	3.00	1.167	1.172	3.0
	Jul	3.4	35.1	70.8	2.95	1.153	1.140	3.0
	Aug	3.1	34.0	69.1	2.88	1.076	1.077	2.9
	Sep	3.5	32.9	66.9	2.79	1.254	1.042	3.4
	Oct	2.5	32.0	64.9	2.70	0.926	1.027	2.4
	Nov	1.9	31.2	63.2	2.63	0.722	0.911	2.1
	Dec	1.7	30.8	62.0	2.58	0.659	0.721	2.4
1975	Jan	1.6	30.5	61.3	2.55	0.627	0.701	2.3
	Feb	1.6	30.2	60.7	2.53	0.632	0.772	2.1
	Mar	2.3	29.7	59.9	2.50	0.920	1.045	2.2
	Apr	3.0	30.4	60.1	2.50	1.200	1.175	2.6
	May	3.1	30.9	61.3	2.55	1.216	1.215	2.6
	Jun	3.1	31.4	62.3	2.60	1.192	1.172	2.6
	Jul	3.1	32.0	63.4	2.64	1.174	1.140	2.7
	Aug	2.8	32.9	64.9	2.70	1.037	1.077	2.6
	Sep	3.0	34.2	67.1	2.80	1.071	1.042	2.9
	Oct	3.2	35.2	69.4	2.89	1.107	1.027	3.1
	Nov	2.4	36.1	71.3	2.97	0.808	0.911	2.6
	Dec	2.2	37.2	73.3	3.05	0.721	0.721	3.1
1976	Jan	2.2	38.2	75.4	3.14	0.701	0.701	3.1
	Feb	2.5	39.5	77.7	3.24	0.772	0.772	3.2
	Mar	3.6	40.3	79.8	3.32	1.084	1.045	3.4
	Apr	4.0	41.2	81.5	3.40	1.176	1.175	3.4
	May	4.0	42.5	83.7	3.49	1.146	1.215	3.3
	Jun	4.2	43.5	86.0	3.58	1.173	1.172	3.6
	Jul	4.1	—	—	—	—	1.140	3.6
	Aug	4.1	—	—	—	—	1.077	3.8
	Sep	3.8	—	—	—	—	1.042	3.6
	Oct	4.1	—	—	—	—	1.027	4.0
	Nov	3.7	—	—	—	—	0.911	4.1
	Dec	3.2	—	—	—	—	0.721	4.4

SOURCE: Data are taken from Series S-10, *Survey of Current Business* and *Business Statistics Supplement,* U.S. Department of Commerce.

Table 17.11
COMPUTING THE SEASONAL INDEX FROM THE MEDIAN OF MONTHLY RATIOS TO MOVING AVERAGES

Month	Year						Median	Seasonal Index
	1971	1972	1973	1974	1975	1976		
January	—	0.787	0.784	0.651	0.627	{0.701}	0.701	0.701
February	—	0.769	0.811	0.818	0.632	{0.772}	0.772	0.772
March	—	1.000	1.141	{1.046}	0.920	1.084	1.046	1.045
April	—	1.084	1.131	1.223	1.200	{1.176}	1.176	1.175
May	—	1.170	1.218	1.266	{1.216}	1.146	1.216	1.215
June	—	1.158	1.183	1.167	1.192	{1.173}	1.173	1.172
July	{1.141}	1.021	1.099	1.153	1.174	—	1.141	1.140
August	{1.078}	1.214	1.120	1.076	1.037	—	1.078	1.077
September	1.019	{1.043}	0.978	1.254	1.071	—	1.043	1.042
October	0.997	1.075	{1.028}	0.926	1.107	—	1.028	1.027
November	{0.912}	0.918	0.932	0.722	0.808	—	0.912	0.911
December	0.890	0.764	0.667	0.659	{0.721}	—	0.721	0.721
Total							12.007	11.998 ↓ 12.000

$$\text{Seasonal Index} = \frac{(12.000)(\text{Median})}{12.007}$$

SOURCE: Data are taken from Table 17.10.

$$\frac{Y_i}{\text{Centered Moving Average}} = \frac{\cancel{T_i} \cdot S_i \cdot \cancel{C_i} \cdot I_i}{\cancel{T_i} \cdot \cancel{C_i}} = S_i \cdot I_i \quad (17.16)$$

To form the seasonal index, the ratios to moving averages data from Table 17.10 are rearranged according to monthly values as depicted in Table 17.11.

From Table 17.11 it is seen that for each month the irregular variations can be eliminated if the median of the various obtained ratios to moving averages is used as an indicator of seasonal activity over time. As shown in Table 17.11, these median values are then adjusted so that the total value of the seasonal indexes over the year is 12.0 and the average value of each (monthly) seasonal index is 1.0. Thus we note that a seasonal index of .701 for the month of January indicates that the value of residential construction contracts issued in January is only 70.1% of the monthly average, while a seasonal index of 1.215 for the month of May indicates that the value of contracts issued in May is 21.5% better than average.

17.5.3 Using the Seasonal Index in Forecasting

To use the seasonal index to adjust a trend projection for forecasting purposes, we merely multiply the projected trend value for a particular month by the corresponding seasonal index for that month. For example, using Equation (17.15) the projected monthly trend values in residential construction contracts to be issued over the year 1979 are listed in column (1) of Table 17.12. The respective monthly seasonal indexes are displayed in column (2). Ad-

Table 17.12
ADJUSTING LEAST-SQUARES TREND PROJECTIONS BY SEASONAL INDEXES FOR FORECASTING PURPOSES

Month	(1) Monthly Trend Projection for Year 1979	(2) Seasonal Index	(3) Forecast
January	3.1721	.701	2.2236
February	3.1706	.772	2.4477
March	3.1691	1.045	3.3117
April	3.1677	1.175	3.7220
May	3.1662	1.215	3.8469
June	3.1647	1.172	3.7090
July	3.1633	1.140	3.6062
August	3.1618	1.077	3.4053
September	3.1603	1.042	3.2930
October	3.1588	1.027	3.2441
November	3.1574	.911	2.8764
December	3.1559	.721	2.2754

SOURCE: Data are taken from Table 17.10 and Equation (17.15).

justing for seasonal fluctuations, the product of the various projected monthly trend values with their respective seasonal indexes yields the set of monthly forecasts shown in column (3).

17.5.4 Deseasonalizing the Data

The seasonal index may also be used for isolating and removing the effects of seasonal influences on the data. When this is achieved in conjunction with the elimination of trend and irregular effects, the cyclical component may be examined. Hence from Table 17.10, to "deseasonalize" the data and thereby eliminate the seasonal effects, we merely divide each observed value in the monthly time series [column (2)] by the seasonal index for that month [column (7)]. The results are shown in column (8), and the deseasonalized series is plotted in Figure 17.15 along with the original series.

In terms of our classical multiplicative time-series model,

$$Y_i = T_i \cdot S_i \cdot C_i \cdot I_i$$

the deseasonalized series is given by

$$\frac{Y_i}{S_i} = \frac{T_i \cdot \cancel{S_i} \cdot C_i \cdot I_i}{\cancel{S_i}} = T_i \cdot C_i \cdot I_i \qquad (17.17)$$

Therefore if the trend component were also removed, we would be left with a series of **cyclical-irregular relatives**—$C_i \cdot I_i$. Hence from Table 17.13 the fitted trend values [column (3)], which were obtained from the linear trend model expressed by Equation (17.15), are divided into the deseasonalized series [column (2)] yielding the cyclical-irregular relatives in column (4), that is,

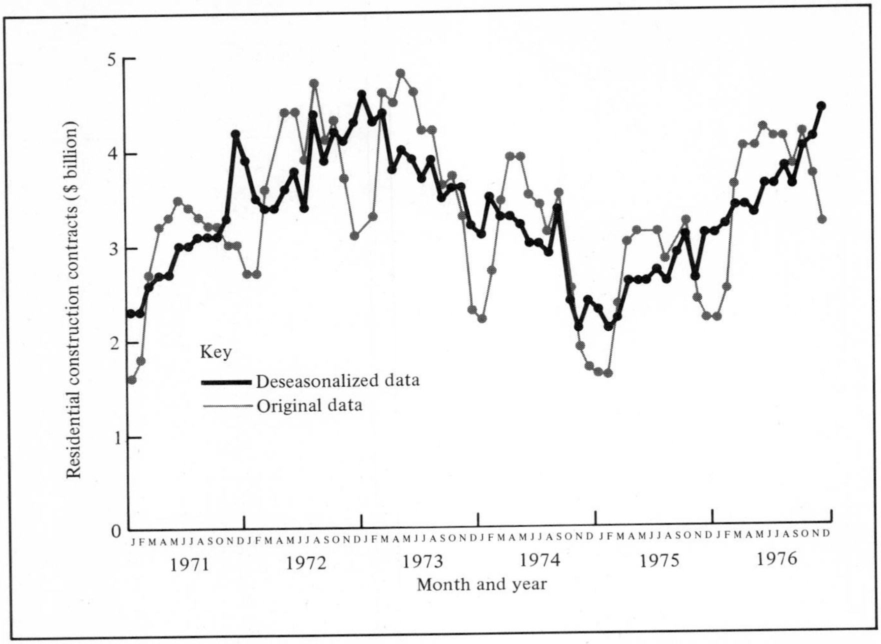

Figure 17.15 Value of residential construction contracts issued—original data and deseasonalized data.

SOURCE: Data are taken from Table 17.10.

$$\frac{T_i \cdot C_i \cdot I_i}{\hat{Y}_i} = \frac{\cancel{T_i} \cdot C_i \cdot I_i}{\cancel{T_i}} = C_i \cdot I_i \qquad (17.18)$$

17.5.5 Studying the Cyclical Component

Unlike annual data, the irregular fluctuations of a monthly time series are often removed by using 3-month weighted moving averages in which the middle value receives a weight of 2 while the two end values each receive a weight of 1. Thus from Table 17.13, for the series of cyclical-irregular relatives [column (4)], the set of 3-month weighted moving totals [column (5)] is obtained. These weighted moving totals are then divided by 4 to yield a series of isolated **cyclical relatives** [column (6)], and the time-series decomposition is completed, that is,

$$\frac{C_i \cdot \cancel{I_i}}{\cancel{I_i}} = C_i \qquad (17.19)$$

Table 17.13
ISOLATING THE CYCLICAL COMPONENT BY TIME-SERIES DECOMPOSITION

(1) Year and Month		(2) Deseasonalized Data	(3) Trend $\hat{Y}_i$	(4) Cyclical-Irregular Relatives	(5) 3-Month Weighted Moving Totals	(6) Cyclical Relatives
1971	Jan	2.3	3.3130	0.694	—	—
	Feb	2.3	3.3115	0.695	2.869	0.717
	Mar	2.6	3.3100	0.785	3.081	0.770
	Apr	2.7	3.3086	0.816	3.233	0.808
	May	2.7	3.3071	0.816	3.356	0.839
	Jun	3.0	3.3056	0.908	3.540	0.885
	Jul	3.0	3.3042	0.908	3.663	0.916
	Aug	3.1	3.3027	0.939	3.725	0.931
	Sep	3.1	3.3012	0.939	3.756	0.939
	Oct	3.1	3.2998	0.939	3.817	0.954
	Nov	3.3	3.2983	1.001	4.215	1.054
	Dec	4.2	3.2968	1.274	4.732	1.183
1972	Jan	3.9	3.2954	1.183	4.703	1.176
	Feb	3.5	3.2939	1.063	4.342	1.086
	Mar	3.4	3.2924	1.033	4.162	1.040
	Apr	3.4	3.2910	1.033	4.193	1.048
	May	3.6	3.2895	1.094	4.377	1.094
	Jun	3.8	3.2880	1.156	4.441	1.110
	Jul	3.4	3.2865	1.035	4.565	1.141
	Aug	4.4	3.2851	1.339	4.901	1.225
	Sep	3.9	3.2836	1.188	4.995	1.249
	Oct	4.2	3.2821	1.280	4.998	1.250
	Nov	4.1	3.2807	1.250	5.091	1.273
	Dec	4.3	3.2792	1.311	5.275	1.319
1973	Jan	4.6	3.2777	1.403	5.429	1.357
	Feb	4.3	3.2763	1.312	5.371	1.343
	Mar	4.4	3.2748	1.344	5.161	1.290
	Apr	3.8	3.2733	1.161	4.889	1.222
	May	4.0	3.2719	1.223	4.800	1.200
	Jun	3.9	3.2704	1.193	4.741	1.185
	Jul	3.7	3.2689	1.132	4.651	1.163
	Aug	3.9	3.2675	1.194	4.592	1.148
	Sep	3.5	3.2660	1.072	4.441	1.110
	Oct	3.6	3.2645	1.103	4.381	1.095
	Nov	3.6	3.2631	1.103	4.290	1.072
	Dec	3.2	3.2616	0.981	4.016	1.004
1974	Jan	3.1	3.2601	0.951	3.957	0.989
	Feb	3.5	3.2587	1.074	4.112	1.028
	Mar	3.3	3.2572	1.013	4.114	1.028
	Apr	3.3	3.2557	1.014	4.024	1.006
	May	3.2	3.2543	0.983	3.902	0.976
	Jun	3.0	3.2528	0.922	3.750	0.938
	Jul	3.0	3.2513	0.923	3.660	0.915
	Aug	2.9	3.2499	0.892	3.754	0.938
	Sep	3.4	3.2484	1.047	3.725	0.931
	Oct	2.4	3.2469	0.739	3.172	0.793
	Nov	2.1	3.2454	0.647	2.773	0.693
	Dec	2.4	3.2440	0.740	2.836	0.709
1975	Jan	2.3	3.2425	0.709	2.806	0.701
	Feb	2.1	3.2410	0.648	2.684	0.671
	Mar	2.2	3.2396	0.679	2.809	0.702
	Apr	2.6	3.2381	0.803	3.088	0.772
	May	2.6	3.2366	0.803	3.213	0.803
	Jun	2.6	3.2352	0.804	3.246	0.812
	Jul	2.7	3.2337	0.835	3.278	0.820
	Aug	2.6	3.2322	0.804	3.341	0.835
	Sep	2.9	3.2307	0.898	3.560	0.890
	Oct	3.1	3.2293	0.960	3.624	0.906
	Nov	2.6	3.2278	0.806	3.533	0.883
	Dec	3.1	3.2263	0.961	3.689	0.922
1976	Jan	3.1	3.2249	0.961	3.876	0.969
	Feb	3.2	3.2234	0.993	4.002	1.000
	Mar	3.4	3.2219	1.055	4.159	1.040
	Apr	3.4	3.2205	1.056	4.192	1.048
	May	3.3	3.2190	1.025	4.225	1.056
	Jun	3.6	3.2175	1.119	4.382	1.096
	Jul	3.6	3.2161	1.119	4.539	1.135
	Aug	3.8	3.2146	1.182	4.603	1.151
	Sep	3.6	3.2131	1.120	4.667	1.167
	Oct	4.0	3.2117	1.245	4.887	1.222
	Nov	4.1	3.2102	1.277	5.170	1.292
	Dec	4.4	3.2087	1.371	—	—

SOURCE: Data are taken from Table 17.10.

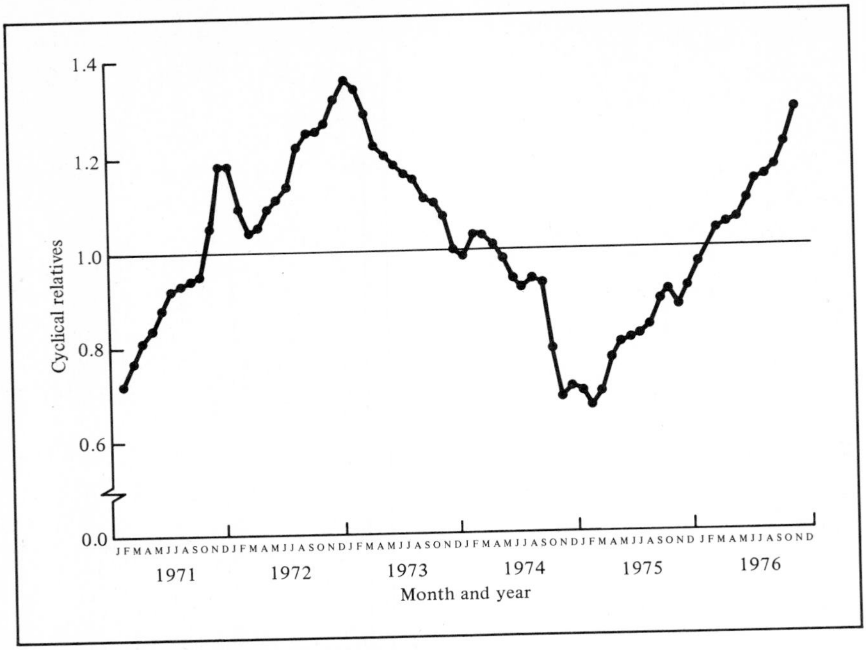

Figure 17.16 Plotting the cyclical relatives.
SOURCE: Data are taken from Table 17.13.

These cyclical relatives are plotted in Figure 17.16. Comparing the cyclical swings of the series to that of the overall economy, business forecasters have considered the series concerning monetary value of residential construction contracts to be a leading indicator of overall economic activity.

17.6 TIME-SERIES ANALYSIS: AN OVERVIEW

The value of such forecasting methodology as time series analysis, which utilizes past and present information as a guide to the future, was recognized and most eloquently expressed more than two centuries ago by the American statesman Patrick Henry who said:

> I have but one lamp by which my feet are guided, and that is the lamp of experience. I know no way of judging the future but by the past." (*Speech at Virginia Convention* (*Richmond*) *March 23, 1775.*)

If it were true (as time-series analysis assumes) that those factors which have affected particular patterns of economic activity in the past and present will continue to do so in a similar manner in the future, time-series analysis, by itself, would certainly be a most appropriate and effective forecasting tool as well as an aid in the managerial control of present activities.

On the other hand, critics of classical time-series methods have argued that these techniques are overly "naive" and "mechanical," that is, a mathematical model based on the past should not be utilized for mechanically extrapolating trends into the future without consideration as to personal judgments, business experiences, or changing technologies, habits, and needs. Thus in recent years econometricians have been concerned with including such factors in developing highly sophisticated computerized models of economic activity for forecasting purposes. Such forecasting methods, however, are beyond the scope of this text (References 5, 9, and 10).

Nevertheless, as we have seen from the previous sections of this chapter, time-series methods provide useful guides to business leaders as to projecting future trends (on a long-term or short-term basis) or as to measuring overall cyclical activity. If used properly—in conjunction with other forecasting methods as well as business judgment and experience—time-series methods will continue to be an excellent managerial tool for decision making.

PROBLEMS

17.1 The data given in Table 17.14 represent the annual incidence rates (per 100,000 persons) of reported acute poliomyelitis recorded over 5-year periods 1915–1955.

Table 17.14
INCIDENCE RATES OF REPORTED ACUTE POLIOMYELITIS

Year	1915	1920	1925	1930	1935	1940	1945	1950	1955
Rate	3.1	2.2	5.3	7.5	8.5	7.4	10.3	22.1	17.6

SOURCE: Data are taken from *The Statistical History of the United States from Colonial Times to the Present* (Series B303).

(a) Plot the data on a chart.

(b) Fit a least-squares trend line to the data and plot the line on your chart.

(c) What are your trend forecasts for the years 1960, 1965, and 1970?

(d) Go to your library and, using the above reference, look up the actually reported incidence rates of acute poliomyelitis for the years 1960, 1965, and 1970. Record your results.

(e) Why are the mechanical trend extrapolations from your least-squares model not useful? Discuss.

(f) Briefly discuss the advantages and disadvantages of time-series analysis as a forecasting tool.

17.2 As described in Section 17.1.1, if our dean and researcher wish to study whether there is an indication of persistent "grade inflation" at their business school over the past decade, the percentage of freshmen and sophomore students on the Dean's List may be examined on an annual basis (Table 17.15).

(a) Plot the data on a chart.

(b) Fit a least-squares trend line to the data and plot the line on your chart.

(c) What is your trend forecast for the year 1978?

(d) If you were the researcher, what would you tell the dean?

Table 17.15
PERCENTAGE OF STUDENTS ON DEAN'S LIST

Year	1968	1969	1970	1971	1972	1973	1974	1975	1976	1977
Percentage	3.9	3.9	5.5	7.2	9.4	9.2	13.8	11.0	11.3	11.5

17.3 The data given in Table 17.16 represent the annual passenger revenues (in billions of dollars) obtained by Eastern Air Lines, Inc., over the decade of 1967 through 1976.

Table 17.16
ANNUAL PASSENGER REVENUES

Year	1967	1968	1969	1970	1971	1972	1973	1974	1975	1976
Passenger Revenues	0.62	0.69	0.81	0.92	0.98	1.08	1.17	1.38	1.47	1.65

SOURCE: Data are taken from "Standard N.Y.S.E. Stock Reports," vol. 44, no. 179, Standard & Poor's Corp., September 1977.

(a) Plot the data on a chart.
(b) Fit a least-squares trend line to the data and plot the line on your chart.
(c) What are your trend forecasts for the years 1977, 1978, and 1979?
(d) Determine the cyclical-irregular relatives for the data and plot your results on a separate chart.
(e) What type of indicator of overall economic activity do you believe "passenger revenues" to be—leading, coinciding, or lagging? Discuss.

17.4 The data given in Table 17.17 represent the earnings per share of common stock (in dollars) held for the Minnesota Mining and Manufacturing Company over the 10-year period of 1967 through 1976.

Table 17.17
EARNINGS PER SHARE

Year	1967	1968	1969	1970	1971	1972	1973	1974	1975	1976
Earnings per Share	1.38	1.50	1.61	1.68	1.87	2.17	2.62	2.66	2.29	2.94

SOURCE: Data are taken from "Standard N.Y.S.E. Stock Reports," vol. 44, no. 173, Standard & Poor's Corp., September 1977.

(a) Plot the data on a chart.
(b) Fit a least-squares trend line to the data and plot the line on your chart.
(c) What are your trend forecasts for the years 1977, 1978, and 1979?
(d) What has been the annual growth rate in earnings per share over the decade?

* 17.5 The data given in Table 17.18 represent the annual net sales (in billions of dollars) obtained by General Motors Corporation from 1967 through 1976:

Table 17.18
ANNUAL NET SALES

Year	1967	1968	1969	1970	1971	1972	1973	1974	1975	1976
Net Sales	20.0	22.8	24.3	18.8	28.3	30.4	35.8	31.6	35.7	47.2

SOURCE: Data are taken from "Standard N.Y.S.E. Stock Reports," vol. 44, no. 157, Standard & Poor's Corp., August 1977.

(a) Plot the data on a chart.

(b) Fit a least-squares trend line to the data and plot the line on your chart.

(c) What are your trend forecasts for the years 1977, 1978, and 1979?

(d) Determine the cyclical-irregular relatives for the data and plot your results on a separate chart.

(e) What type of indicator of overall economic activity do you believe "net sales" to be—leading, coinciding, or lagging? Discuss.

17.6 The data given in Table 17.19 represent the annual gross revenues (in billions of dollars) obtained by the American Telephone and Telegraph Company over the decade of 1967 through 1976.

Table 17.19
ANNUAL GROSS REVENUES

Year	1967	1968	1969	1970	1971	1972	1973	1974	1975	1976
Gross Revenues	13.0	14.1	15.7	17.0	18.4	20.9	23.5	26.2	29.0	32.8

SOURCE: Data are taken from "Standard N.Y.S.E. Stock Reports," vol. 44, no. 149, Standard & Poor's Corp., August 1977.

(a) Plot the data on a chart.

(b) Fit a least-squares trend line to the data and plot the line on your chart.

(c) What are your trend forecasts for the years 1977, 1978, and 1979?

(d) What has been the annual growth rate in gross revenues over the decade?

17.7 The data given in Table 17.20 represent the annual total assets (in billions of dollars) of life insurance companies over the decade of 1967 through 1976.

Table 17.20
ANNUAL TOTAL ASSETS

Year	1967	1968	1969	1970	1971	1972	1973	1974	1975	1976
Assets	177.8	188.6	197.2	207.3	222.1	239.7	252.4	263.3	289.3	320.6

SOURCES: Data are taken from Series S-19, *Survey of Current Business* and *Business Statistics Supplement,* U.S. Department of Commerce.

(a) Plot the data on a chart.

(b) Fit a least-squares trend line to the data and plot the line on your chart.

(c) What are your trend forecasts for the years 1977, 1978, and 1979?

(d) Fit a 3-year moving average to the data and plot your results on your chart.

17.8 The data given in Table 17.21 represent the amount of consumer (short-term and intermediate-term) installment credit (in billions of dollars) held annually by finance companies from 1967 through 1976.

Table 17.21
CONSUMER INSTALLMENT CREDIT

Year	1967	1968	1969	1970	1971	1972	1973	1974	1975	1976
Credit Held	24.6	26.1	27.8	27.7	28.9	32.1	37.2	38.9	36.7	39.6

SOURCE: Data are taken from Series S-18, *Survey of Current Business* and *Business Statistics Supplement,* U.S. Department of Commerce.

(a) Plot the data on a chart.
(b) Fit a least-squares trend line to the data and plot the line on your chart.
(c) What are your trend forecasts for the years 1977, 1978, and 1979?
(d) Fit a 5-year moving average to the data and plot the results on your chart.
(e) Using a smoothing coefficient of .33, exponentially smooth the series and plot the results on your chart.
(f) What is your exponentially smoothed forecast for the trend in 1977?
(g) Compare and discuss your results in (b) and (c) with those in (f).

17.9 The data given in Table 17.22 represent the amount of consumer (short-term and intermediate-term) installment credit (in billions of dollars) held annually by commercial banks from 1966 through 1976.

Table 17.22
CONSUMER INSTALLMENT CREDIT

Year	1966	1967	1968	1969	1970	1971	1972	1973	1974	1975	1976
Credit Held	31.3	33.2	37.9	42.4	45.4	51.2	59.8	69.5	72.5	78.7	85.4

SOURCE: Data are taken from Series S-18, *Survey of Current Business* and *Business Statistics Supplement,* U.S. Department of Commerce.

(a) Plot the data on a chart.
(b) Fit a least-squares trend line to the data and plot the line on your chart.
(c) What are your trend forecasts for the years 1977, 1978 ,and 1979?
(d) Determine the cyclical-irregular relatives for the data and plot your results on a separate chart.
(e) What type of indicator of overall economic activity do you believe "consumer installment credit held" to be—leading, coinciding, or lagging? Discuss.

17.10 The data given in Table 17.23 represent the annual amount of corporate income taxes paid (in billions of dollars) to the federal government from 1966 through 1976.
(a) Plot the data on a chart.
(b) Fit a least-squares trend line to the data and plot the line on your chart.
(c) What is your least-squares trend forecast for the year 1977?

Table 17.23
CORPORATE INCOME TAXES PAID

Year	1966	1967	1968	1969	1970	1971	1972	1973	1974	1975	1976
Taxes Paid	30.1	34.0	28.7	36.7	32.8	26.8	32.2	36.2	38.6	42.6	55.7

SOURCE: Data are taken from Series S-19, *Survey of Current Business* and *Business Statistics Supplement,* U.S. Department of Commerce.

(d) Fit a 3-year moving average to the data and plot the results on your chart.

(e) Using a smoothing coefficient of .50, exponentially smooth the series and plot the results on your chart.

(f) What is your exponentially smoothed trend forecast for the year 1977?

(g) Compare and discuss your results in (b) and (c) with those in (f).

17.11 The data given in Table 17.24 represent the annual industrial and commercial failure rate (number per 10,000 concerns) recorded for the years of 1955 through 1976.

Table 17.24
ANNUAL INDUSTRIAL AND COMMERCIAL FAILURE RATES

Year	Failure Rate	Year	Failure Rate
1955	41.6	1966	51.6
1956	48.0	1967	49.0
1957	51.7	1968	38.6
1958	55.9	1969	37.3
1959	51.8	1970	43.8
1960	57.0	1971	41.7
1961	64.4	1972	38.3
1962	60.8	1973	36.4
1963	56.3	1974	38.4
1964	53.2	1975	42.6
1965	53.3	1976	34.8

SOURCE: Data are taken from Series S-7, *Survey of Current Business* and *Business Statistics Supplement,* U.S. Department of Commerce.

(a) Plot the data on a chart.

(b) Fit a 7-year moving average to the data and plot the results on your chart.

(c) Using a smoothing coefficient of .25, exponentially smooth the series and plot the results on your chart.

(d) What is your exponentially smoothed forecast for the trend in 1977?

17.12 The data given in Table 17.25 represent the annual revenues (in millions of dollars) from magazine advertising on beers, wines, and liquors from 1961 through 1976.

(a) Plot the data on a chart.

(b) Fit a 3-year moving average to the data and plot the results on your chart.

(c) Using a smoothing coefficient of .50, exponentially smooth the series and plot the results on your chart.

(d) What is your exponentially smoothed forecast for the trend in 1977?

Table 17.25
ANNUAL ADVERTISING REVENUES
(MILLIONS OF DOLLARS)

Year	Revenues	Year	Revenues
1961	51.0	1969	102.8
1962	54.1	1970	98.0
1963	56.4	1971	83.6
1964	58.1	1972	81.0
1965	69.5	1973	87.0
1966	79.2	1974	102.9
1967	89.2	1975	100.9
1968	93.0	1976	110.9

SOURCE: Data are taken from Series S-11, *Survey of Current Business* and *Business Statistics Supplement,* U.S. Department of Commerce.

* 17.13 The data given in Table 17.26 represent the annual number of employees (in millions) in transportation equipment for the years of 1961 through 1976.

(a) Plot the data on a chart.

(b) Fit a 3-year moving average to the data and plot the results on your chart.

(c) Using a smoothing coefficient of .50, exponentially smooth the series and plot the results on your chart.

(d) What is your exponentially smoothed forecast for the trend in 1977?

Table 17.26
EMPLOYEES

Year	Number	Year	Number
1961	1.45	1969	2.06
1962	1.55	1970	1.80
1963	1.61	1971	1.73
1964	1.60	1972	1.77
1965	1.74	1973	1.90
1966	1.92	1974	1.82
1967	1.95	1975	1.65
1968	2.04	1976	1.73

SOURCE: Data are taken from Series S-14, *Survey of Current Business* and *Business Statistics Supplement,* U.S. Department of Commerce.

17.14 The data given in Table 17.27 represent the average weekly gross hours per production worker in leather and leather products from 1961 through 1976.

(a) Plot the data on a chart.

(b) Using a smoothing coefficient of .10, exponentially smooth the series and plot the results on your chart.

(c) Using a smoothing coefficient of .30, exponentially smooth the series and plot the results on your chart.

(d) Using a smoothing coefficient of .50, exponentially smooth the series and plot the results on your chart.

(e) Based on your results in (b), (c), and (d), what are your exponentially smoothed forecasts for the trend in 1977?

(f) Comparing the three smoothed series based on past accuracy, which of your forecasts [part (e)] do you believe is most reliable? Go to your library and

Table 17.27
AVERAGE WEEKLY GROSS HOURS

Year	Hours	Year	Hours
1961	37.4	1969	37.2
1962	37.6	1970	37.2
1963	37.5	1971	37.7
1964	37.9	1972	38.3
1965	38.2	1973	37.9
1966	38.6	1974	37.2
1967	38.1	1975	37.4
1968	38.3	1976	37.3

SOURCE: Data are taken from Series S-15, *Survey of Current Business* and *Business Statistics Supplement*, U.S. Department of Commerce.

check your answer against the actual 1977 value presented in Series S-15 in the *Survey of Current Business.*

17.15 The following refers to the data presented in Problem 13.10.

(a) Plot the data on a chart.

(b) Fit a least-squares trend line and plot the line on your chart.

(c) What is your least-squares trend forecast for the year 1976?

(d) Using a smoothing coefficient of .20, exponentially smooth the series and plot the results on your chart.

(e) What is your exponentially smoothed trend forecast for the year 1976?

(f) Go to your library and compare your forecasts [parts (c) and (e)] against the actual 1976 value presented in Table 505 of the *Statistical Abstract of the United States.*

17.16 The following refers to the data presented in Problem 13.11.

(a) Plot the data on a chart.

(b) Fit a 5-year moving average to the data and plot the results on your chart.

(c) Fit a least-squares trend line and plot the line on your chart.

(d) What is your least-squares trend forecast for the year 1980?

17.17 The data given in Table 17.28 represent the number of persons (in millions) employed in agriculture on a monthly basis from January 1971 through December 1976.

(a) Plot the data on a chart.

(b) Compute the seasonal index.

(c) Fit a least-squares linear trend to the **average** annual employment.

(d) Convert the **annual** least-squares trend equation to a **monthly** trend equation. **Hint:** To convert an annual trend equation which is based on monthly averages to a monthly trend equation, keep the same intercept but divide the slope by 12. Adjust the results so as to center the origin within a particular month.

(e) Use the monthly trend equation and seasonal index to forecast the number of persons employed in agriculture for all twelve months of 1977 and 1978.

(f) Isolate and plot (on a separate chart) the cyclical relatives by detrending, deseasonalizing, and smoothing the irregular component with a 3-term weighted moving average.

Table 17.28
MONTHLY AGRICULTURAL EMPLOYMENT

Month	Year 1971	1972	1973	1974	1975	1976
January	2.877	2.869	2.955	3.197	2.888	2.853
February	2.846	2.909	2.956	3.283	2.890	2.802
March	3.042	3.094	3.131	3.334	2.988	2.897
April	3.505	3.287	3.295	3.437	3.171	3.273
May	3.598	3.531	3.467	3.604	3.622	3.415
June	3.920	3.976	4.053	3.895	3.869	3.780
July	3.971	4.061	4.165	4.024	4.090	3.931
August	3.764	4.031	3.826	3.851	3.886	3.842
September	3.444	3.658	3.436	3.563	3.626	3.396
October	3.470	3.721	3.525	3.536	3.524	3.447
November	3.262	3.363	3.419	3.224	3.156	3.081
December	2.948	3.163	3.202	2.959	2.856	2.850

SOURCE: Data are taken from Series S-13, *Survey of Current Business* and *Business Statistics Supplement,* U.S. Department of Commerce.

17.18 The data given in Table 17.29 represent the monthly average hourly gross earnings per production worker in apparel and other textile products from January 1971 through December 1976.

Table 17.29
MONTHLY AVERAGE HOURLY GROSS EARNINGS

Month	Year 1971	1972	1973	1974	1975	1976
January	2.45	2.55	2.73	2.85	3.14	3.33
February	2.47	2.58	2.72	2.86	3.13	3.33
March	2.47	2.57	2.74	2.87	3.16	3.37
April	2.46	2.58	2.75	2.89	3.16	3.37
May	2.46	2.57	2.74	2.96	3.15	3.38
June	2.47	2.60	2.76	2.98	3.16	3.40
July	2.47	2.58	2.75	3.01	3.16	3.39
August	2.49	2.62	2.79	3.05	3.16	3.42
September	2.52	2.65	2.84	3.09	3.22	3.49
October	2.51	2.67	2.85	3.10	3.24	3.49
November	2.51	2.69	2.86	3.10	3.25	3.50
December	2.54	2.70	2.84	3.11	3.27	3.52

SOURCE: Data are taken from Series S-16, *Survey of Current Business* and *Business Statistics Supplement,* U.S. Department of Commerce.

(a) Plot the data on a chart.
(b) Compute the seasonal index.
(c) Fit a least-squares trend line to the **mean** annual hourly gross earnings.
(d) Convert the **annual** least-squares trend equation to a **monthly** trend equation. **Hint:** To convert an annual trend equation which is based on monthly averages to a monthly trend equation, keep the same intercept but divide the slope by 12. Adjust the results so as to center the origin within a particular month.
(e) Use the monthly trend equation and seasonal index to forecast the average hourly gross earnings for all twelve months of 1977 and 1978.

(f) Isolate and plot (on a separate chart) the cyclical relatives by detrending, deseasonalizing, and smoothing the irregular component with a 3-term weighted moving average.

* 17.19 The data given in Table 17.30 represent the monthly outlays (in millions of dollars) to the National Aeronautics and Space Administration from January 1971 through December 1976.

Table 17.30
MONTHLY OUTLAYS TO NASA

Month	Year					
	1971	*1972*	*1973*	*1974*	*1975*	*1976*
January	262	259	271	251	298	260
February	295	276	241	231	283	291
March	333	310	301	252	315	307
April	252	238	265	293	287	293
May	274	270	255	278	301	279
June	245	292	301	447	185	287
July	377	289	278	216	368	344
August	291	289	262	247	310	359
September	273	273	246	267	313	250
October	266	271	249	281	312	368
November	286	272	246	297	325	359
December	285	284	221	288	326	345

SOURCE: Data are taken from Series S-19, *Survey of Current Business* and *Business Statistics Supplement,* U.S. Department of Commerce.

(a) Plot the data on a chart.
(b) Compute the seasonal index.
(c) Fit a least-squares trend line to the aggregate annual outlays.
(d) Use the monthly trend equation and seasonal index to forecast the monthly outlays for all twelve months of 1977 and 1978.
(e) Isolate and plot (on a separate chart) the cyclical relatives by detrending, deseasonalizing, and smoothing the irregular component with a 3-term weighted moving average.

17.20 The data given in Table 17.31 represent monthly recorded passenger miles (in billions) for scheduled domestic air operations from January 1971 through December 1976.
(a) Plot the data on a chart.
(b) Compute the seasonal index.
(c) Fit a least-squares trend line to the aggregate annual passenger miles.
(d) Use the monthly trend equation and seasonal index to forecast the monthly passenger miles for all twelve months of 1977 and 1978.
(e) Isolate and plot (on a separate chart) the cyclical relatives by detrending, deseasonalizing, and smoothing the irregular component with a 3-term weighted moving average.

17.21 The data given in Table 17.32 represent the monthly Dow-Jones averages for 30 industrial stocks from January 1971 through December 1976.

Table 17.31
MONTHLY PASSENGER MILES

Month	Year					
	1971	1972	1973	1974	1975	1976
January	8.44	9.31	9.80	10.26	12.64	13.94
February	7.20	8.19	8.80	9.45	11.01	12.75
March	8.17	9.60	10.26	11.16	13.30	14.19
April	9.02	9.59	10.44	11.08	12.19	14.67
May	8.40	9.15	10.11	10.67	12.91	14.66
June	9.45	10.68	11.55	12.00	14.90	16.31
July	10.31	11.28	12.00	12.07	16.10	17.72
August	10.76	11.93	13.00	13.18	17.30	18.15
September	8.33	9.22	9.86	9.86	12.90	14.19
October	8.62	9.50	10.13	10.19	13.36	14.32
November	8.04	9.25	9.77	9.05	12.26	12.99
December	9.67	10.42	10.58	10.76	13.93	15.19

SOURCE: Data are taken from Series S-24, *Survey of Current Business* and *Business Statistics Supplement,* U.S. Department of Commerce.

(a) Plot the data on a chart.

(b) Compute the seasonal index.

(c) Fit a least-squares trend line to the **mean** annual Dow-Jones (industrial) stock average.

(d) Convert the **annual** least-squares trend equation to a **monthly** trend equation. **Hint:** To convert an annual trend equation which is based on monthly averages to a monthly trend equation, keep the same intercept but divide the slope by 12. Adjust the results so as to center the origin within a particular month.

(e) Use the monthly trend equation and seasonal index to forecast the monthly Dow-Jones (industrial) stock average for all twelve months of 1977 and 1978.

(f) Isolate and plot (on a separate chart) the cyclical relatives by detrending, deseasonalizing, and smoothing the irregular component with a 3-term weighted moving average.

Table 17.32
MONTHLY DOW-JONES AVERAGES

Month	Year					
	1971	1972	1973	1974	1975	1976
January	849.04	904.65	1026.82	857.24	659.09	929.34
February	879.69	914.37	974.04	831.34	724.89	971.70
March	901.29	939.23	957.35	874.00	765.06	988.55
April	932.54	958.16	944.10	847.79	790.93	992.51
May	925.49	948.22	922.41	829.84	836.56	988.82
June	900.43	943.43	893.90	831.43	845.70	985.59
July	887.81	925.92	903.61	783.00	856.28	993.20
August	875.40	958.34	883.73	729.30	815.51	981.63
September	901.22	950.58	909.98	651.28	818.28	994.37
October	872.15	944.10	967.62	638.62	831.26	951.95
November	822.11	1001.19	878.98	642.10	845.51	944.58
December	869.90	1020.32	824.08	596.50	840.80	972.86

SOURCE: Data are taken from Series S-21, *Survey of Current Business* and *Business Statistics Supplement,* U.S. Department of Commerce.

17.22 The data given in Table 17.33 represent the wholesale prices of corn (in dollars per bushel) averaged over selected markets from January 1971 through December 1976.

Table 17.33
MONTHLY WHOLESALE CORN PRICES

	Year					
Month	*1971*	*1972*	*1973*	*1974*	*1975*	*1976*
January	1.51	1.22	1.57	2.80	3.12	2.63
February	1.50	1.21	1.57	3.02	2.90	2.63
March	1.52	1.21	1.56	2.95	2.88	2.70
April	1.48	1.23	1.65	2.64	2.95	2.66
May	1.54	1.23	2.02	2.61	2.90	2.80
June	1.52	1.20	2.30	2.80	2.86	2.87
July	1.43	1.22	2.33	3.27	2.93	2.94
August	1.29	1.21	2.70	3.53	3.15	2.79
September	1.13	1.28	2.40	3.46	2.95	2.71
October	1.11	1.28	2.35	3.69	2.73	3.46
November	1.09	1.30	2.39	3.46	2.58	2.40
December	1.20	1.54	2.58	3.42	2.57	2.48

SOURCE: Data are taken from Series S-27, *Survey of Current Business* and *Business Statistics Supplement*, U.S. Department of Commerce.

(a) Plot the data on a chart.
(b) Compute the seasonal index.
(c) Fit a least-squares trend line to the **mean** annual wholesale price of corn.
(d) Convert the **annual** least-squares trend equation to a **monthly** trend equation. **Hint:** To convert an annual trend equation which is based on monthly averages to a monthly trend equation, keep the same intercept but divide the slope by 12. Adjust the results so as to center the origin within a particular month.
(e) Use the monthly trend equation and seasonal index to forecast the wholesale price of corn for all twelve months of 1977 and 1978.
(f) Isolate and plot (on a separate chart) the cyclical relatives by detrending, deseasonalizing, and smoothing the irregular component with a 3-term weighted moving average.

REFERENCES

1. BROWN, R. G., *Smoothing, Forecasting, and Prediction* (Englewood Cliffs, N.J.: Prentice-Hall, 1963).
2. *Business Conditions Digest* (U.S. Department of Commerce, 1977).
3. *Business Statistics Supplement* (U.S. Department of Commerce, 1975).
4. CROXTON, F., D. COWDEN, AND S. KLEIN, *Applied General Statistics*, 3rd ed. (Englewood Cliffs, N.J.: Prentice-Hall, 1967).
5. CURTIS, G. A., *ESP: Economic Software Package* (Chicago, Ill.: Graduate School of Business, University of Chicago, January 1976).
6. GRIFFIN, J. I., *Statistics: Methods and Applications* (New York: Holt, Rinehart and Winston, 1962).
7. LEABO, D., *Basic Statistics*, 5th ed. (Homewood, Ill.: Richard D. Irwin, 1976).

8. MITCHELL, W. C., *Business Cycles: The Problem and its Setting* (New York: National Bureau of Economic Research, 1968).
9. PFAFFENBERGER, R., AND J. PATTERSON, *Statistical Methods for Business and Economics* (Homewood, Ill.: Richard D. Irwin, 1977).
10. RADUCHEL, W. J., *RAPFE:* "The Regression Analysis Program For Economists Reference Guide," Tech. Paper 10, Harvard Institute of Economic Research, Harvard University, Cambridge, Mass., April 1973.
11. *STATPACK: Statistical Package,* 2nd ed. (White Plains, N.Y.: IBM Corporation, February 1970).
12. *Survey of Current Business* (U.S. Department of Commerce, 1977).
13. WATTENBERG, B. J., ed., *The Statistical History of the United States from Colonial Times to the Present* (New York: Basic Books, 1976).

APPENDIX A

DATA BASE POPULATION

A.1 INTRODUCTION

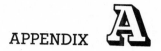

Throughout this text we have utilized a survey undertaken by a college administrator as a means of demonstrating the interrelationship of various statistical methodologies and concepts. The purpose of this appendix then is twofold. First, we must describe the scenario for this aforementioned survey which has served as our integrating theme. Second, we must present the complete set of 2,202 responses from the entire data base population so that students are able to draw their own random samples and work with their own particular set of data throughout their statistics course if they so desire.

A.2 SCENARIO DEVELOPMENT

Suppose the dean of a business school at a particular college is faced with the task of proposing and implementing policy which can be expected to have a broad effect on the quality of campus life. In order to obtain needed input information prior to making these decisions, the dean and a researcher develop a questionnaire (Figure 2.2) consisting of 26 quantitative and qualitative questions pertaining to socioeconomic and demographic characteristics, composition, physical attributes, attitudes, and interests of the student body.

They decide to conduct a mail survey at the beginning of the semester by drawing a random sample from the population of full-time, matriculated, day-session students currently enrolled in the business school (see Figure 2.6). According to the records contained in the files of the registrar's office, this population consists of 2,202 students who have declared business to be their overall major field of interest.[1]

A.3 POPULATION DATA BASE

The data presented in the following pages of this appendix contain the responses that would have been provided had each of the 2,202 students in the business school (population) been asked to fill out the questionnaire.

[1] Let us suppose that regulations at this particular college prohibit entering freshmen from making such a declaration and, furthermore, let us suppose that it is college policy to suggest that this decision be postponed until the student has achieved upper sophomore status.

File code number	Question number 1	2	3	4	5	6	7	8	9	10	11	12	13	14	15	16	17	18	19	20	21	22	23	24	25	26
0001	1	3	1	5	2.75	85	12000	200	3	2	2	2	1	1	26	66	170	2	1	2	4	1	2	4	6	1
0002	1	2	1	5	3.85	85	25000	2000	1	1	2	3	6	1	25	68	150	2	1	2	4	2	3	6	4	1
0003	1	3	3	2	3.00	75	12000	500	3	2	2	4	0	1	29	66	175	2	1	2	4	1	0	2	1	1
0004	1	3	3	2	2.22	70	12500	800	3	2	2	2	0	1	37	67	220	1	1	2	1	1	0	4	1	1
0005	2	3	1	3	2.00	80	14000	1000	1	2	2	2	3	1	28	63	110	2	1	2	1	1	0	4	4	2
0006	2	3	2	6	3.53	86	14000	100	1	2	5	4	4	1	27	60	100	2	1	1	2	2	0	4	4	2
0007	1	2	3	2	3.17	87	11000	700	2	2	1	2	9	1	19	73	175	2	1	1	3	1	0	2	4	1
0008	1	4	2	4	3.50	63	18000	1000	1	2	5	4	2	1	30	72	175	2	4	1	1	2	0	4	4	2
0009	1	1	3	4	2.47	82	20000	0	3	2	2	2	1	2	31	65	175	2	1	2	2	2	0	4	4	1
0010	1	4	2	5	2.05	84	10000	300	1	2	1	4	1	1	30	68	135	2	2	2	4	2	0	4	4	1
0011	1	3	3	2	2.42	78	10000	300	1	1	2	2	0	1	31	68	160	2	1	2	4	2	0	4	6	2
0012	2	3	3	1	2.25	85	12000	500	1	2	2	2	2	1	30	63	125	1	1	2	1	2	0	4	4	1
0013	1	4	3	4	3.06	76	12000	600	2	2	5	4	3	1	24	70	160	2	2	1	1	2	0	4	5	2
0014	1	4	3	6	3.20	81	9000	2000	1	2	2	2	1	1	29	69	135	2	1	2	3	1	0	4	5	1
0015	2	3	1	4	3.00	87	12000	0	1	2	2	2	1	2	25	63	105	2	1	2	1	1	0	1	4	2
0016	2	3	1	2	3.00	85	12000	0	1	2	2	2	2	1	22	62	105	2	1	1	1	2	0	1	4	2
0017	2	2	1	1	2.96	82	16000	700	1	2	1	3	8	2	24	68	120	2	1	1	4	1	0	4	4	2
0018	1	4	2	5	3.09	92	12500	900	1	2	2	4	4	1	24	71	135	2	3	1	4	2	0	4	4	2
0019	1	4	1	6	3.02	84	15000	1000	2	2	2	2	2	1	24	66	130	2	1	2	3	1	1	4	7	1
0020	2	4	1	5	3.60	83	20000	1000	2	2	3	3	5	1	21	67	135	2	1	2	4	1	3	3	7	1
0021	1	2	1	6	3.24	92	16000	1200	1	2	2	4	16	1	20	74	180	2	1	2	4	1	1	1	5	2
0022	2	1	2	4	3.05	83	12000	300	1	2	5	4	10	2	21	68	128	2	2	1	4	2	0	4	4	2
0023	2	4	2	5	2.30	87	15000	800	3	2	2	2	5	1	20	61	115	1	1	2	1	2	1	5	6	1
0024	1	1	1	5	3.20	78	11000	700	1	2	2	2	2	1	22	70	160	2	1	2	4	2	0	1	4	2
0025	1	4	3	3	3.21	89	12000	1000	2	2	2	2	3	1	22	68	165	2	3	1	3	2	1	4	5	1
0026	1	4	1	6	3.55	94	11000	2000	2	1	2	4	2	1	21	68	125	2	1	2	4	1	1	2	4	1
0027	1	3	1	6	2.80	85	15000	500	3	1	2	4	6	1	19	69	145	2	1	1	4	2	0	2	6	1
0028	2	4	1	5	3.63	83	15000	1000	2	2	3	3	4	1	21	67	135	2	1	2	4	1	3	2	6	1
0029	1	2	1	6	3.24	92	17500	1500	1	2	2	4	17	1	20	74	180	2	1	2	4	1	1	1	5	1
0030	2	3	1	6	2.90	89	15000	1000	2	2	2	4	10	1	21	66	120	2	3	1	4	1	0	4	4	1
0031	2	4	1	5	3.22	87	12000	60	2	1	2	2	9	1	21	65	106	2	2	1	4	2	1	4	3	1
0032	2	3	1	6	3.30	84	17000	2000	3	2	2	4	4	1	26	67	113	2	3	1	4	1	1	2	2	1
0033	2	3	1	7	2.94	85	15000	2000	3	2	2	4	5	2	21	61	105	2	1	2	3	1	1	4	5	2
0034	1	2	3	4	3.60	87	20000	600	2	2	1	2	5	1	19	72	220	2	1	2	4	2	0	2	4	1
0035	1	3	2	6	3.00	81	12000	1100	3	2	1	4	4	1	21	67	165	1	1	2	2	2	0	4	6	1
0036	1	4	1	2	3.50	88	11000	1000	2	2	1	4	4	1	21	71	130	2	1	2	4	1	2	2	6	1
0037	2	3	3	6	3.00	85	12000	100	3	2	2	2	8	1	21	63	100	2	1	2	4	2	1	3	5	2
0038	1	4	1	6	3.41	86	15000	250	2	2	1	3	5	1	21	70	165	2	1	2	3	1	0	4	3	1
0039	2	4	3	6	2.50	80	10000	500	3	2	2	4	3	1	28	62	130	2	2	2	3	2	1	6	5	1
0040	2	2	3	6	3.98	98	12000	900	2	2	2	4	0	1	19	62	98	2	1	2	3	2	2	2	5	1
0041	1	4	2	6	3.50	72	11000	500	3	2	2	4	4	1	22	65	140	2	1	2	2	2	0	4	6	1
0042	1	4	2	6	2.40	83	12000	100	2	2	2	4	5	1	21	68	150	1	1	1	3	2	0	4	4	1
0043	1	4	2	6	3.10	87	12000	1000	2	2	5	4	7	1	21	70	180	2	1	2	1	2	0	4	4	1
0044	2	4	2	6	3.30	91	10000	2000	2	2	2	4	3	2	21	60	111	2	1	2	3	1	0	4	5	2
0045	2	3	3	2	3.41	95	10000	1500	2	1	1	4	6	1	20	65	120	2	1	1	4	1	0	3	3	1
0046	1	3	3	2	4.00	87	12000	1200	2	1	2	4	5	1	21	75	185	2	1	2	4	1	2	5	5	1
0047	1	3	3	2	3.00	84	14000	100	3	2	2	2	2	1	34	68	146	2	1	2	4	1	1	1	4	1
0048	2	2	3	1	2.90	84	13500	1500	3	2	5	2	12	1	19	65	142	2	1	2	3	1	4	5	5	1
0049	1	2	3	2	3.62	85	20000	600	2	2	2	4	5	1	20	70	155	2	1	2	3	1	4	3	5	2
0050	1	3	2	2	2.50	76	12000	1000	3	2	1	3	4	1	24	73	170	2	3	2	1	1	2	1	5	1
0051	1	3	1	2	2.84	88	12500	1000	3	2	2	4	7	1	22	70	200	2	2	1	3	1	1	4	4	2
0052	1	3	3	2	3.00	87	12000	500	2	2	5	4	4	1	27	66	135	2	1	2	4	1	0	4	4	1
0053	2	2	3	2	2.80	88	16000	1000	2	2	2	4	0	1	19	62	93	2	1	2	4	1	1	5	7	2
0054	1	4	3	3	3.53	75	22000	1000	1	2	3	4	10	1	36	67	155	2	1	1	2	2	0	4	3	2
0055	1	1	3	2	2.55	91	10000	100	3	2	2	4	5	1	18	68	140	2	1	2	4	1	1	4	4	1
0056	1	3	2	2	2.50	80	12500	463	2	1	1	5	3	1	20	70	175	2	1	2	4	1	2	4	4	2
0057	2	3	3	4	2.51	82	12000	500	3	2	3	3	5	1	21	64	125	2	1	2	3	1	1	4	6	2
0058	1	4	3	4	2.06	70	9000	1000	2	2	2	2	1	1	22	70	160	2	4	1	3	2	0	1	4	1
0059	2	4	3	5	2.94	81	25000	900	2	2	2	4	10	1	21	64	142	2	1	1	3	2	0	4	5	2
0060	1	1	3	1	2.34	80	18000	800	3	2	2	4	20	1	18	68	137	2	2	1	3	2	0	4	4	2
0061	1	3	1	7	3.10	79	12500	463	2	2	3	4	5	1	20	71	165	2	1	1	3	1	4	2	2	2
0062	1	3	3	7	2.82	78	10000	1000	2	2	5	4	5	1	23	66	165	2	1	2	3	2	1	2	4	2
0063	1	3	3	2	3.57	85	15000	0	3	2	2	2	2	1	27	67	160	2	1	2	1	1	0	1	7	2
0064	2	1	2	4	2.91	82	14000	1000	2	2	1	3	10	1	18	63	110	2	1	2	4	1	0	3	2	2
0065	1	4	3	2	3.55	89	18000	100	1	2	2	2	15	1	26	68	170	2	1	2	1	1	0	4	4	2
0066	1	1	1	2	2.76	88	25000	0	2	1	5	4	3	1	18	71	178	2	1	2	4	2	0	4	4	2
0067	1	1	3	1	2.43	82	15000	500	3	2	2	4	4	1	18	71	155	2	1	2	3	2	0	4	4	1
0068	2	4	3	2	3.10	85	12000	950	2	1	2	4	4	1	21	60	105	2	1	2	3	1	0	4	4	2
0069	2	3	2	6	2.00	83	12000	400	2	2	2	4	7	1	23	63	100	2	1	2	1	2	0	4	5	1
0070	1	3	3	5	2.50	85	15000	300	3	2	2	2	0	2	30	60	148	2	1	2	1	2	0	4	4	1
0071	1	1	3	2	2.25	81	18000	800	3	2	3	4	6	2	18	72	155	2	1	2	4	1	0	4	4	2
0072	1	1	3	2	3.60	86	13000	1000	3	1	6	2	5	1	18	71	182	2	1	2	3	2	1	4	4	2
0073	1	3	3	6	2.50	82	10000	50	1	2	2	2	2	1	22	72	200	2	2	2	3	1	0	7	4	1
0074	1	4	1	2	2.86	90	12000	500	2	2	2	4	10	1	23	74	180	2	2	1	3	1	1	3	6	2
0075	1	3	3	1	3.10	86	7000	800	2	2	1	4	2	2	22	68	135	2	1	2	3	2	0	4	6	1
0076	2	3	1	4	2.00	75	13000	150	2	2	2	2	6	1	21	62	120	2	3	1	2	2	0	2	4	1
0077	1	3	2	6	2.50	85	12300	800	3	2	2	4	5	2	23	74	165	2	1	2	4	1	0	4	3	1
0078	1	3	2	6	3.06	80	8000	800	3	2	5	4	0	1	18	73	220	2	1	2	4	2	0	4	2	1
0079	2	3	1	3	3.00	80	25000	0	2	2	2	4	7	1	25	67	120	2	2	1	4	2	0	4	1	1
0080	1	3	2	6	2.50	80	11500	500	1	2	1	3	10	2	20	72	165	2	1	1	3	2	0	4	1	1
0081	1	3	1	5	2.75	80	15000	2000	2	2	3	4	15	1	21	63	135	2	2	1	1	2	3	1	1	1
0082	2	3	1	6	3.50	89	18000	1000	2	2	3	4	5	1	18	63	115	2	1	1	4	1	3	5	2	1
0083	2	3	1	2	3.00	88	13000	0	3	2	2	2	7	1	25	67	137	2	1	2	3	1	1	4	4	1
0084	1	4	3	6	3.08	75	12500	0	2	2	5	4	6	1	26	71	162	2	1	2	4	1	0	4	6	2
0085	1	3	3	2	3.04	85	12500	750	2	2	2	4	6	1	19	68	135	2	1	2	4	1	2	4	6	2
0086	1	3	1	2	2.32	75	10000	0	2	2	2	2	2	1	23	73	175	2	1	2	4	1	0	3	6	1
0087	2	3	2	2	3.53	90	10000	200	3	2	2	2	12	1	21	63	96	2	1	2	3	1	1	2	4	1
0088	2	2	3	2	2.28	80	10000	500	2	2	2	2	3	1	23	63	112	2	1	2	4	1	1	4	4	1
0089	1	3	2	2	3.51	90	13000	900	3	1	2	4	11	1	20	70	156	2	1	2	3	1	1	3	3	1
0090	1	3	3	4	2.91	88	12000	900	3	2	2	4	4	1	22	70	140	2	1	2	1	2	0	3	5	2
0091	1	3	1	2	3.30	91	13000	1000	2	2	2	2	5	1	19	70	195	2	1	2	4	1	6	1	3	1
0092	1	3	3	2	2.50	86	12000	500	2	2	2	4	7	1	21	71	169	2	1	2	2	1	0	4	4	1
0093	1	2	3	2	3.02	89	10000	1000	2	2	3	3	8	1	19	67	135	2	1	2	4	2	0	4	5	1
0094	1	3	3	2	2.80	85	12500	950	3	1	1	3	4	1	22	72	160	2	1	2	4	1	0	4	4	1
0095	2	4	2	2	3.61	94	13000	2000	2	2	4	4	4	1	21	67	125	2	2	2	4	2	1	6	4	1
0096	1	2	1	2	2.71	85	7000	1000	2	2	2	4	3	1	20	73	179	2	1	2	2	2	0	5	5	2
File code number	Sex	Class	Grad. school	Major	Grade-point index	H. S. average	Est. starting salary $	Tuition $	Employment	Sunday shop	Investment	Car pref.	No. jeans owned	Calculator	Age	Height	Weight	Blood pressure	Smoking status	Smoking belief	Entrance standards	Retention standards	No. clubs and groups	SPS rating	Library rating	Personal question

File code number	1	2	3	4	5	6	7	8	9	10	11	12	13	14	15	16	17	18	19	20	21	22	23	24	25	26
	Question number																									
0097	1	2	3	2	2.07	80	17000	1000	2	2	2	2	6	1	20	72	165	2	1	2	3	1	0	4	4	1
0098	1	3	1	2	2.04	74	8000	600	2	2	2	2	0	2	27	71	175	2	1	2	1	1	1	1	2	2
0099	1	3	2	7	2.25	77	12000	0	2	2	2	4	2	2	28	67	170	2	1	2	1	2	0	4	6	2
0100	1	2	3	1	2.55	78	10000	400	3	2	1	3	4	1	20	69	162	2	1	2	4	1	2	4	2	1
0101	1	4	1	2	2.30	84	14500	500	1	2	2	2	2	1	34	71	170	2	1	2	1	2	0	2	4	1
0102	1	2	3	2	2.00	80	10000	0	2	2	2	2	6	1	22	64	135	2	1	2	1	2	0	4	4	2
0103	1	2	1	2	3.86	87	15000	750	2	2	5	4	7	2	18	69	160	2	1	2	3	1	0	4	2	2
0104	1	4	2	2	3.53	90	13000	1000	2	2	3	4	4	1	22	72	150	2	1	2	4	1	0	5	6	1
0105	1	4	1	2	3.21	80	12500	1000	3	1	2	4	4	1	24	76	210	1	1	2	4	1	2	4	7	1
0106	1	3	1	2	3.38	86	10000	200	2	2	2	4	5	1	23	65	135	2	1	2	3	1	1	4	5	1
0107	2	3	2	2	3.00	80	10000	1000	3	2	2	2	3	1	21	62	125	2	1	1	1	2	0	4	4	1
0108	1	3	2	2	2.90	90	10000	800	1	1	2	2	6	1	20	67	125	2	1	2	4	2	0	1	1	2
0109	1	4	1	2	3.41	85	13000	2000	3	1	2	4	3	1	24	72	180	2	3	1	4	1	4	2	4	1
0110	1	4	3	2	3.05	80	12000	2000	2	2	3	3	5	1	21	70	155	2	1	2	4	1	2	4	5	1
0111	1	3	3	2	3.20	88	10000	900	1	2	2	2	5	1	21	70	165	2	1	2	3	2	0	5	6	2
0112	2	4	3	2	3.30	90	11000	1000	2	2	2	2	5	1	20	61	105	2	2	1	4	2	3	4	3	1
0113	2	3	3	2	3.40	90	12000	1000	3	2	2	2	8	1	22	60	105	2	1	2	1	2	1	4	4	2
0114	1	4	1	2	4.00	93	14000	1000	1	1	5	4	3	1	22	71	152	2	2	2	4	1	2	1	4	2
0115	1	3	3	2	3.90	91	11000	1000	2	2	1	4	3	1	21	69	145	2	1	2	4	1	2	4	4	1
0116	1	4	3	2	3.11	73	12000	1500	2	2	2	4	12	1	21	71	195	2	1	1	4	1	4	4	5	2
0117	1	2	3	2	2.80	89	10000	700	2	2	2	4	0	1	22	67	185	2	1	2	4	2	1	1	4	1
0118	1	4	3	2	3.35	87	11000	1000	3	2	2	4	1	1	21	66	155	2	1	2	3	2	1	4	6	2
0119	1	3	3	2	2.40	72	10000	1000	2	2	2	4	10	1	21	67	150	2	3	1	2	2	0	4	5	2
0120	1	3	3	2	3.10	84	12000	1000	2	2	1	3	4	1	22	71	188	2	1	2	4	2	1	4	4	2
0121	2	2	1	2	2.85	83	8500	1000	2	2	2	2	3	1	22	63	115	2	1	2	1	1	1	4	4	2
0122	1	4	3	2	3.00	75	13000	1500	1	1	2	4	5	2	31	70	180	1	1	2	4	2	0	4	4	1
0123	2	3	3	2	3.00	85	15000	1000	1	2	6	2	7	1	36	67	130	2	1	2	4	1	0	4	4	2
0124	1	4	2	2	3.22	79	15000	100	2	2	3	4	0	1	51	70	146	1	1	2	4	1	0	4	4	2
0125	2	2	1	1	2.85	87	18000	1200	1	2	2	2	4	1	30	66	112	2	1	2	4	2	2	5	6	1
0126	2	4	2	2	2.00	82	15000	1000	3	2	2	2	0	1	31	63	173	2	1	2	1	2	2	6	7	2
0127	1	2	1	5	3.00	75	12500	0	3	2	2	4	10	1	20	72	185	2	1	2	4	2	1	4	4	2
0128	2	3	2	2	3.00	90	15000	1000	2	2	2	4	12	1	30	63	120	2	2	2	4	1	0	4	7	1
0129	1	2	3	2	2.97	87	12000	350	2	1	2	2	3	1	20	72	164	2	1	2	3	2	0	4	6	1
0130	1	3	1	5	3.50	92	18000	200	3	2	2	2	0	1	22	70	158	2	1	2	2	1	2	4	4	2
0131	1	2	3	5	2.67	73	13000	600	2	2	2	2	4	1	27	71	163	2	2	2	3	2	1	7	4	2
0132	1	2	2	4	2.88	86	10000	900	3	2	2	3	10	2	21	70	130	2	1	2	4	2	0	4	4	1
0133	2	3	1	1	3.90	83	12000	1000	3	2	1	3	5	1	32	68	135	2	1	2	1	2	0	4	4	1
0134	1	2	3	2	2.80	80	10000	800	3	2	2	2	6	1	23	70	140	1	1	1	2	2	2	4	6	1
0135	1	2	2	2	3.20	85	12500	1200	3	1	1	4	5	1	19	71	135	2	1	2	4	1	1	1	3	2
0136	2	2	3	2	3.82	86	13450	900	2	2	1	4	0	1	21	60	98	2	1	2	1	1	0	4	6	1
0137	1	3	3	2	2.71	80	12750	850	3	2	2	4	4	2	21	68	135	2	1	2	1	2	1	4	4	2
0138	2	3	1	5	3.20	90	20000	100	3	2	2	2	2	1	20	67	155	2	1	2	4	2	1	2	3	2
0139	1	2	3	2	2.74	87	18000	350	3	2	2	2	5	1	20	68	150	2	1	1	3	1	1	4	5	1
0140	1	3	1	2	3.00	84	18000	200	2	2	2	4	8	1	20	71	145	1	3	1	1	1	0	4	1	1
0141	2	3	3	2	2.71	85	10000	500	3	2	2	4	8	1	22	66	113	1	1	2	3	1	4	5	5	1
0142	2	2	2	2	2.52	90	10000	1000	2	2	2	4	5	2	21	66	121	2	1	2	3	1	0	4	4	1
0143	2	2	2	2	3.31	90	22000	100	3	2	2	3	11	1	19	63	108	2	3	1	4	1	0	1	1	1
0144	1	3	2	1	3.00	83	9000	500	2	2	2	2	10	2	20	70	160	2	1	2	3	2	0	5	5	1
0145	2	2	1	4	3.06	85	12000	600	1	2	2	4	1	1	36	59	118	2	1	2	4	1	0	4	4	2
0146	1	3	3	3	3.22	80	14550	700	2	2	2	2	3	1	20	72	170	2	3	1	3	2	0	4	4	1
0147	1	3	3	2	2.57	73	10000	500	2	2	2	4	5	1	22	72	175	2	1	2	1	2	0	4	4	1
0148	2	3	2	2	2.63	70	10000	1000	3	2	2	4	2	1	22	64	110	2	1	2	2	1	1	5	6	1
0149	1	3	3	2	2.75	81	16200	300	2	2	3	3	5	1	21	70	145	2	1	2	3	2	1	1	4	1
0150	2	3	3	2	3.20	90	13000	500	2	2	1	3	3	1	21	64	130	2	2	2	4	2	1	4	4	1
0151	1	3	3	2	2.50	80	15000	1000	2	2	2	4	10	1	21	71	170	1	1	2	4	2	1	3	4	2
0152	1	2	1	2	3.08	95	10000	1000	1	2	2	2	0	2	36	70	140	1	1	2	4	1	0	4	4	1
0153	2	2	1	2	2.09	70	11500	0	3	2	2	4	10	1	38	65	138	1	2	1	4	1	0	4	7	1
0154	1	2	1	2	2.50	79	12500	500	1	2	1	4	2	1	23	69	158	2	1	2	1	2	0	3	3	2
0155	1	4	1	2	2.62	89	18000	1500	1	1	2	2	6	1	27	68	165	2	2	1	1	2	0	4	4	2
0156	1	4	1	3	1.93	68	23000	1000	1	2	2	4	4	1	31	75	240	2	1	1	3	2	0	4	2	1
0157	1	2	3	7	3.04	75	9000	100	3	2	3	2	4	1	23	73	190	2	3	2	3	2	1	4	7	2
0158	1	2	1	3	2.42	87	12000	600	1	2	5	4	5	1	21	64	240	2	3	1	1	2	0	7	7	1
0159	2	1	2	4	2.87	91	17500	200	1	2	2	4	0	1	44	62	135	2	3	1	1	1	0	4	5	2
0160	1	3	3	2	2.25	80	12000	200	2	2	2	2	1	1	32	66	160	2	1	2	1	2	1	5	6	1
0161	1	3	3	5	3.20	86	15000	800	1	2	2	4	2	1	26	75	227	1	3	2	3	1	0	4	6	2
0162	1	2	2	2	3.06	92	13000	300	2	2	2	2	4	1	19	70	170	2	2	1	2	2	0	4	4	1
0163	1	3	3	4	2.55	90	11000	600	1	2	2	2	5	1	25	72	170	2	1	2	1	1	0	4	7	2
0164	1	4	3	4	2.50	85	15000	500	1	2	2	2	2	1	37	67	135	2	2	1	1	2	0	1	4	2
0165	1	2	1	1	3.02	80	10000	600	1	2	6	2	2	1	19	66	125	2	3	1	3	1	0	3	6	1
0166	2	2	3	6	2.54	89	10000	900	3	2	2	1	3	2	23	67	130	2	1	2	3	2	0	4	3	1
0167	2	3	3	2	2.73	87	12000	1000	2	2	2	2	5	1	20	60	110	2	3	1	3	2	1	5	3	1
0168	1	2	2	2	2.70	83	12000	250	2	2	1	4	4	2	20	67	140	2	1	2	3	1	0	3	4	1
0169	1	3	2	2	2.04	85	13000	0	2	2	2	2	2	1	30	64	142	2	1	2	3	1	1	4	4	2
0170	2	4	1	6	3.23	85	18000	1000	3	1	3	3	4	1	24	62	117	2	1	2	3	2	2	4	4	2
0171	2	1	1	6	3.71	93	13500	1000	2	2	3	4	1	1	17	67	130	2	1	2	4	2	1	4	4	1
0172	1	4	1	4	2.80	88	11000	1000	2	2	2	2	3	2	22	69	145	2	2	1	4	1	0	4	4	1
0173	1	3	3	2	2.58	75	15000	300	3	2	2	2	2	2	24	70	180	1	1	2	1	2	1	4	2	1
0174	1	3	1	4	3.35	92	18000	500	2	2	2	2	4	1	20	70	160	2	1	2	2	1	0	2	6	2
0175	2	3	2	4	3.60	90	12000	1000	2	1	2	4	7	1	20	67	129	2	1	2	3	1	0	4	6	1
0176	1	2	1	5	3.05	95	19000	500	2	2	2	4	4	1	20	71	195	2	3	1	3	1	1	4	4	1
0177	1	3	2	2	3.42	80	10000	1500	3	2	2	4	8	1	21	70	145	2	1	2	1	2	0	4	4	2
0178	1	3	1	2	3.14	85	11000	750	2	2	1	4	6	1	20	69	160	2	1	2	4	1	1	3	6	1
0179	1	3	1	2	3.72	90	12000	800	2	2	5	4	5	1	25	68	150	2	1	2	4	1	2	4	6	2
0180	1	3	1	4	2.87	83	15000	1500	2	1	2	4	10	1	19	73	180	2	1	2	4	1	2	3	2	1
0181	2	2	2	2	3.21	90	12500	800	2	2	3	3	5	1	19	64	108	2	1	2	4	2	0	1	4	2
0182	1	3	2	6	2.81	84	13000	500	2	1	1	4	8	1	20	69	165	2	1	2	4	2	1	1	4	1
0183	1	3	3	2	3.03	85	11000	1000	2	2	2	4	4	1	20	72	190	2	1	2	3	2	0	4	5	2
0184	1	3	1	2	1.92	85	12000	150	2	2	2	2	0	1	29	66	128	2	1	2	1	2	0	4	4	1
0185	1	4	1	2	2.54	85	19000	600	3	2	2	2	0	1	32	67	156	2	4	1	4	2	1	2	1	2
0186	1	3	3	2	3.52	90	8000	500	2	2	2	2	2	1	19	71	168	2	1	2	4	1	0	2	6	1
0187	2	2	2	2	2.63	91	15000	750	1	2	2	3	0	1	24	60	105	2	1	2	4	2	0	4	4	2
0188	2	1	3	2	3.50	80	11500	800	1	2	2	4	2	1	33	62	117	2	1	2	3	2	0	4	4	2
0189	1	1	1	2	1.90	80	10000	0	1	2	2	4	7	2	27	62	130	2	1	2	1	1	0	4	4	2
0190	1	3	3	2	3.40	88	12000	600	2	2	2	4	4	1	20	64	142	2	1	2	4	1	0	4	4	1
0191	1	2	3	2	2.70	86	12000	1000	2	2	2	3	2	2	20	77	220	2	1	2	4	2	0	4	6	2
0192	1	3	2	2	2.85	84	15000	500	2	1	5	4	3	1	21	73	180	1	1	2	3	1	0	5	4	1
File code number	Sex	Class	Grad. school	Major	Grade-point index	H. S. average	Est. starting salary $	Tuition $	Employment	Sunday shop	Investment	Car pref.	No. jeans owned	Calculator	Age	Height	Weight	Blood pressure	Smoking status	Smoking belief	Entrance standards	Retention standards	No. clubs and groups	SPS rating	Library rating	Personal question

File code number	1 Sex	2 Class	3 Grad. school	4 Major	5 Grade-point index	6 H. S. average	7 Est. starting salary $	8 Tuition $	9 Employment	10 Sunday shop	11 Investment	12 Car pref.	13 No. jeans owned	14 Calculator	15 Age	16 Height	17 Weight	18 Blood pressure	19 Smoking status	20 Smoking belief	21 Entrance standards	22 Retention standards	23 No. clubs and groups	24 SPS rating	25 Library rating	26 Personal question
0193	1	3	3	2	3.02	85	15000	500	2	1	2	4	4	1	21	69	175	2	1	2	4	1	0	1	4	1
0194	1	3	1	2	3.06	90	12500	0	2	2	2	2	4	1	30	71	170	2	1	2	4	1	3	5	5	1
0195	2	3	3	2	3.00	87	11000	500	2	2	2	2	8	1	21	61	101	2	1	1	3	1	1	4	4	2
0196	2	2	2	2	3.33	92	15000	0	3	2	2	2	7	1	21	65	110	2	1	2	1	2	0	1	1	1
0197	2	3	3	4	3.04	82	10000	900	2	2	2	2	3	1	22	65	125	2	1	2	3	2	0	4	4	1
0198	1	1	1	6	3.21	90	12000	900	1	2	3	2	2	1	33	66	160	1	1	2	1	1	0	3	4	1
0199	1	2	1	4	3.30	88	10000	1000	1	2	2	4	4	1	25	71	180	2	2	1	4	1	0	4	4	1
0200	1	1	2	2	2.26	72	13500	1000	2	2	1	2	2	1	25	69	137	1	1	2	1	1	0	2	2	1
0201	1	1	3	1	2.50	88	10000	1000	2	2	2	4	10	1	19	73	179	2	1	2	2	2	0	4	2	1
0202	1	3	3	2	2.50	75	12000	1000	3	1	2	4	3	1	21	72	150	2	1	2	4	2	0	5	7	1
0203	1	3	3	2	3.20	78	12000	500	3	2	2	4	3	1	20	60	142	2	1	1	2	2	0	4	4	1
0204	1	3	3	5	2.34	65	8000	900	2	2	2	4	2	1	31	67	152	2	3	2	4	2	0	4	4	1
0205	1	3	1	1	2.91	76	12000	600	3	2	2	4	3	1	20	73	140	2	1	2	4	1	3	2	1	1
0206	1	3	1	5	2.31	80	15000	0	3	1	2	2	2	2	20	74	180	2	2	1	1	2	1	4	4	1
0207	1	4	2	5	2.06	75	12000	750	2	2	2	4	4	1	23	67	128	2	1	2	1	1	4	4	4	1
0208	1	2	3	2	2.94	80	10000	100	2	2	2	4	5	1	20	73	159	2	1	2	1	2	3	3	4	1
0209	2	2	2	5	3.48	92	9000	500	2	2	2	4	6	1	19	63	122	2	1	2	3	2	0	1	3	2
0210	1	4	3	4	2.82	78	10000	500	2	2	2	4	5	1	27	64	140	2	1	2	3	1	1	6	7	1
0211	2	4	3	5	3.37	91	9100	1200	3	2	3	4	4	2	20	66	120	2	1	2	4	1	2	4	6	1
0212	1	4	1	7	3.55	92	11000	2000	2	2	2	2	4	1	22	69	125	2	1	2	4	1	1	4	4	1
0213	1	4	2	4	3.01	84	12000	500	2	2	2	2	3	1	21	76	180	2	1	2	3	2	1	1	6	1
0214	1	3	3	2	3.02	85	13000	200	1	2	2	2	2	1	27	69	140	2	2	2	4	1	1	4	4	1
0215	1	3	2	2	3.13	82	16000	50	1	2	2	2	3	1	34	68	142	2	1	1	1	1	0	4	4	2
0216	2	4	3	5	3.00	92	10000	1000	2	2	2	2	1	1	22	61	110	2	1	2	1	2	1	4	4	1
0217	2	3	3	2	3.02	95	14000	100	3	2	2	2	1	1	22	66	125	2	1	2	4	1	0	4	4	1
0218	1	4	1	1	3.26	83	17500	3800	1	2	2	2	0	1	24	71	150	2	2	1	4	1	0	1	4	1
0219	1	3	3	4	2.50	75	10000	1000	2	2	2	4	4	2	21	74	195	2	1	2	4	1	1	4	4	1
0220	2	3	2	4	2.82	79	10000	500	3	2	3	3	6	1	21	62	115	2	1	2	3	2	1	1	2	2
0221	2	2	2	1	2.50	85	10000	1000	2	2	2	2	6	1	19	63	108	2	1	2	3	1	0	3	5	1
0222	2	1	1	6	3.07	86	11000	0	3	2	2	2	4	1	18	62	120	2	1	1	3	1	0	4	4	1
0223	1	3	1	7	2.32	73	10000	0	3	2	2	4	4	1	21	64	140	2	1	2	1	2	2	5	7	2
0224	2	2	3	2	2.46	85	12000	1000	3	2	2	2	1	1	20	66	125	1	2	1	2	2	0	1	4	1
0225	1	3	3	2	2.75	80	10000	400	2	2	5	2	6	1	22	74	185	2	1	2	3	2	1	3	4	1
0226	1	3	1	2	2.13	62	14000	2500	3	2	2	4	3	1	28	71	215	2	2	1	3	1	0	4	4	2
0227	1	2	1	2	2.82	88	11000	950	2	2	5	4	10	1	19	70	135	2	1	1	3	2	0	4	4	2
0228	1	3	2	2	3.12	87	12000	1500	3	2	1	2	11	1	20	67	145	2	1	2	3	1	1	4	5	1
0229	1	3	2	2	2.41	82	9000	1000	3	2	2	4	3	1	22	64	145	2	1	2	3	2	0	4	4	1
0230	1	3	2	2	3.40	85	18000	2000	3	2	3	4	0	1	20	70	140	2	1	2	4	1	0	1	4	2
0231	1	2	3	2	2.95	80	13000	1000	1	2	2	4	2	1	20	69	163	2	1	1	3	1	0	1	3	1
0232	1	3	3	2	2.64	73	12500	700	3	1	1	3	3	1	25	72	208	2	1	2	4	2	0	4	6	2
0233	1	2	1	2	2.82	80	13000	720	1	2	2	2	7	1	27	67	168	1	1	2	1	1	1	4	5	1
0234	1	3	3	2	3.13	88	8000	1200	1	2	2	2	0	1	19	68	142	2	1	2	1	2	0	4	6	2
0235	2	3	2	2	3.32	87	15000	0	3	2	2	2	0	1	25	63	115	2	1	2	1	2	1	2	4	2
0236	2	3	3	2	2.87	89	12000	1000	3	2	2	4	6	1	20	68	124	2	1	2	4	2	1	4	5	2
0237	1	2	1	2	2.75	85	15000	1000	3	2	2	2	11	1	25	72	175	1	1	1	4	2	0	4	5	2
0238	2	2	3	2	3.06	88	12000	0	3	1	2	4	5	1	20	64	112	2	1	2	4	2	1	3	4	1
0239	1	2	3	2	2.42	83	13000	1000	2	2	2	2	5	2	21	68	165	2	1	2	4	1	1	4	4	2
0240	1	3	3	2	2.51	77	15000	150	2	2	2	2	1	1	18	62	140	2	1	2	2	2	2	4	4	2
0241	1	3	3	5	2.60	83	10000	500	1	1	2	4	3	1	21	68	185	1	1	2	3	1	0	4	6	1
0242	1	4	3	7	3.20	81	10000	1000	2	2	5	4	3	1	22	72	176	2	2	1	3	1	0	2	3	1
0243	1	2	1	6	3.40	88	13000	500	2	2	4	4	4	1	19	69	145	2	1	2	3	1	2	3	4	2
0244	2	2	3	4	2.80	83	12000	1000	2	2	2	2	3	1	20	63	160	2	1	2	2	2	0	4	2	2
0245	1	2	3	2	2.64	87	14000	800	3	2	2	4	14	1	20	66	155	2	1	2	3	2	2	4	4	1
0246	1	3	1	4	3.00	85	12000	1500	3	2	1	4	5	1	19	70	165	2	3	1	3	1	2	7	5	1
0247	2	3	1	5	2.04	75	12000	800	2	2	2	4	2	1	22	64	140	2	2	1	2	1	1	4	3	1
0248	2	3	2	5	3.16	88	9000	1000	2	2	2	4	9	2	21	66	135	2	1	2	4	2	1	4	3	1
0249	2	4	1	7	3.32	83	8000	0	2	2	2	4	1	1	23	65	120	2	2	2	4	1	3	4	4	2
0250	1	4	3	5	3.25	85	12000	1000	2	1	2	4	4	1	21	72	140	2	2	1	3	2	0	4	4	2
0251	2	3	1	7	3.86	95	13000	0	3	2	1	3	10	1	21	60	105	1	1	2	4	1	0	1	2	1
0252	2	4	1	7	3.80	94	10000	450	2	2	5	3	4	2	39	65	120	2	1	2	3	2	0	4	1	2
0253	1	3	1	4	2.85	87	15000	1000	2	2	2	4	0	2	20	72	140	2	2	2	4	2	2	1	5	1
0254	1	4	1	2	3.32	81	10000	700	2	2	2	4	10	1	23	71	185	2	2	1	4	1	0	1	7	2
0255	1	3	2	5	2.91	80	12000	1000	2	2	2	4	2	1	20	69	155	2	1	2	4	1	0	2	4	1
0256	1	4	3	7	3.40	85	12000	0	2	2	4	4	7	1	20	66	145	2	1	2	4	2	0	3	4	2
0257	2	2	3	1	3.00	82	10000	0	2	2	2	2	6	2	20	65	105	2	1	1	1	2	0	3	3	2
0258	1	1	3	2	1.96	83	15000	500	3	2	2	4	3	1	18	72	186	1	1	2	4	2	0	4	5	2
0259	1	3	3	2	2.80	91	11000	150	2	2	6	2	5	2	21	68	169	2	1	2	3	2	2	4	4	1
0260	1	3	1	2	2.00	89	10000	500	2	2	2	4	5	1	22	70	150	2	1	1	4	1	0	5	4	1
0261	1	2	1	4	3.43	72	12000	2000	2	2	4	4	1	2	24	68	165	2	1	2	3	1	2	2	4	2
0262	1	3	3	1	2.51	78	10000	250	2	2	2	4	5	1	21	66	135	2	1	2	4	2	0	3	6	1
0263	1	2	2	2	2.60	89	10000	400	2	2	2	4	4	1	19	68	155	2	1	2	4	1	2	1	7	1
0264	1	4	3	5	2.00	75	13000	125	2	1	2	4	1	1	24	74	155	2	1	2	1	2	0	4	3	1
0265	2	2	3	2	3.50	95	12000	100	3	2	2	2	5	2	21	60	98	2	1	1	1	1	1	4	4	1
0266	1	3	1	2	2.31	95	12500	0	3	2	2	2	0	2	27	67	155	2	1	2	1	2	2	6	4	1
0267	1	3	2	5	2.84	83	10000	150	2	1	2	4	5	1	21	71	175	2	1	1	1	1	0	2	6	2
0268	1	2	3	2	3.82	94	10000	600	2	2	2	2	8	1	21	71	170	2	1	2	4	2	0	3	5	2
0269	2	1	3	6	2.50	85	15000	200	2	2	6	2	4	2	20	66	130	2	1	2	4	1	0	4	4	2
0270	2	3	3	5	2.75	80	11000	1000	3	2	2	4	7	1	20	66	127	2	1	2	1	1	0	4	4	1
0271	1	1	3	2	2.68	91	12000	500	2	2	2	2	2	1	20	68	150	2	1	2	3	2	0	4	5	1
0272	1	1	1	1	2.65	83	8000	0	1	2	2	4	7	2	22	64	160	2	3	1	1	1	0	2	6	1
0273	2	3	3	7	3.00	93	12000	1000	2	2	3	3	7	1	20	67	140	2	2	1	3	2	0	4	4	1
0274	1	3	3	2	2.97	88	12000	1000	2	2	3	1	5	1	20	72	170	2	1	2	3	1	0	3	6	1
0275	1	3	1	3	3.02	78	15000	100	3	2	6	2	4	1	30	71	184	2	1	2	4	2	0	4	4	2
0276	2	3	3	5	3.06	90	10000	0	2	2	2	4	5	1	21	62	100	2	1	1	3	2	1	2	2	1
0277	2	2	3	5	2.82	83	12000	100	1	2	2	4	10	1	20	62	103	2	2	1	4	2	2	4	5	1
0278	1	2	2	4	2.11	78	20000	250	2	2	3	4	4	1	21	73	175	2	1	2	3	2	0	2	1	1
0279	1	1	1	2	2.45	86	15000	0	3	2	2	4	6	1	18	73	185	2	1	1	3	1	0	4	6	1
0280	1	3	3	2	3.70	94	10000	900	2	1	5	4	2	1	21	68	140	2	2	1	3	1	0	4	4	1
0281	1	2	2	5	2.79	87	12000	1500	3	2	5	2	3	1	22	68	125	2	1	2	1	2	0	1	1	1
0282	1	4	1	7	2.82	80	10000	1000	3	1	1	4	6	1	21	69	165	2	2	2	3	2	3	4	4	1
0283	2	3	1	5	2.85	89	15000	800	3	2	2	2	4	2	19	60	105	2	1	2	3	1	1	4	4	1
0284	2	2	2	5	2.00	83	12000	0	3	2	2	2	6	1	21	63	115	2	1	1	4	2	0	5	5	1
0285	1	3	2	5	2.70	81	12000	1000	2	1	2	4	6	1	21	70	160	2	2	1	4	2	0	3	5	1
0286	1	3	3	2	3.62	88	15000	500	2	2	2	4	2	1	23	72	168	2	1	1	3	2	0	4	2	1
0287	1	4	3	5	2.66	76	10000	900	3	2	2	2	25	1	21	69	160	2	1	2	3	1	3	2	6	2
0288	2	3	3	2	3.00	90	12500	3000	2	2	2	4	2	1	18	63	125	2	3	1	4	2	0	4	5	2

File code number	Question number																									
	1	2	3	4	5	6	7	8	9	10	11	12	13	14	15	16	17	18	19	20	21	22	23	24	25	26
0289	1	3	3	5	2.00	75	12500	900	2	2	2	4	2	1	23	71	165	2	1	2	3	2	0	2	6	1
0290	1	4	3	5	2.67	84	10000	0	1	2	2	4	7	1	21	71	190	2	1	2	4	2	0	4	5	1
0291	1	3	3	2	2.08	80	13000	1000	2	2	2	2	0	1	29	72	185	2	1	2	4	2	1	1	4	2
0292	2	2	1	5	2.30	80	10000	1000	3	1	1	3	3	1	21	64	115	2	3	1	3	2	2	4	3	2
0293	2	2	1	2	2.80	87	18000	1000	2	2	2	2	2	1	22	62	86	2	1	2	2	2	2	4	5	1
0294	2	3	1	3	2.34	87	18000	0	3	2	1	2	4	1	21	67	120	1	1	2	1	2	1	4	5	1
0295	2	2	3	6	3.75	88	18000	1000	3	2	2	4	1	1	30	70	135	2	1	1	1	2	0	4	4	1
0296	1	3	1	2	3.42	83	10500	600	2	2	2	4	5	1	24	66	135	2	1	2	3	2	0	4	5	1
0297	1	2	2	2	2.50	84	12000	1000	2	2	5	4	0	1	20	67	165	2	1	2	3	2	1	4	5	1
0298	1	3	3	2	2.46	87	15000	750	1	2	2	4	2	1	23	70	163	2	1	2	3	1	0	4	2	1
0299	1	3	3	2	2.30	81	12000	400	3	2	2	5	10	1	22	70	150	2	1	2	4	1	0	5	4	2
0300	1	2	1	2	2.62	90	10000	950	2	2	2	4	7	1	20	72	155	2	3	2	4	1	0	1	4	1
0301	2	3	2	2	2.04	80	10000	200	1	2	2	2	15	1	22	62	110	2	1	2	1	1	0	3	3	1
0302	2	2	1	2	2.52	76	15000	0	2	1	2	4	0	1	20	68	150	2	1	2	1	1	1	3	4	1
0303	2	2	3	2	3.03	85	12000	0	2	2	2	4	3	1	21	70	113	2	1	2	1	2	1	1	4	1
0304	2	2	3	2	2.51	90	12000	400	2	2	2	2	11	1	21	60	110	2	1	2	1	1	2	4	4	1
0305	1	2	2	2	2.07	75	18000	0	3	2	2	4	6	1	33	64	160	2	1	2	2	2	0	4	3	1
0306	1	2	1	2	3.54	93	13500	500	3	2	2	4	15	1	19	72	185	2	1	2	4	1	4	4	5	1
0307	2	3	3	2	2.72	78	14000	0	3	2	2	4	5	1	21	64	128	2	1	2	1	2	2	4	3	1
0308	1	2	2	2	3.26	87	14000	1000	2	2	2	4	3	1	20	71	185	2	1	2	4	2	0	2	2	2
0309	1	3	1	2	3.20	87	12000	400	3	2	2	4	3	1	19	70	175	2	1	1	4	1	0	4	4	2
0310	1	4	3	2	2.31	80	11000	650	1	2	5	4	4	1	26	71	165	2	3	1	1	1	0	4	5	1
0311	1	1	1	2	2.42	85	8000	1000	3	2	2	5	0	2	19	71	205	2	1	1	1	2	0	5	6	1
0312	2	2	2	2	2.80	81	15000	0	3	1	2	4	0	2	23	69	138	2	1	1	3	2	1	5	6	2
0313	1	3	3	2	2.55	84	18000	200	3	2	2	4	4	1	22	66	129	2	1	2	1	1	1	4	4	1
0314	2	3	3	2	2.52	75	13000	0	3	2	2	4	5	2	21	66	135	2	1	2	1	2	1	3	3	1
0315	1	3	3	2	2.84	83	12000	1400	3	2	5	4	6	1	21	70	150	2	1	2	3	1	3	6	6	2
0316	1	3	3	2	2.58	81	13000	200	1	2	2	4	5	1	28	73	158	2	2	2	2	2	0	4	4	1
0317	2	3	1	2	3.02	85	13000	0	3	2	2	4	1	2	20	68	190	1	1	2	4	1	0	4	4	2
0318	1	2	1	2	3.01	83	15000	400	3	2	1	3	8	1	19	70	145	2	1	2	3	2	0	4	6	1
0319	1	3	3	2	2.93	86	12000	1000	1	2	2	4	2	1	22	71	170	2	2	2	4	1	0	6	6	1
0320	1	2	3	2	3.46	92	13500	1000	2	2	5	4	5	1	19	74	240	2	3	1	3	1	1	5	3	1
0321	1	2	1	2	3.10	75	12000	0	2	2	2	2	5	1	19	68	135	2	1	2	4	1	0	4	5	2
0322	1	2	3	2	3.00	85	12000	1000	2	2	2	2	3	1	20	70	140	2	1	2	3	1	0	4	4	1
0323	1	2	2	2	2.05	80	11500	850	2	2	1	3	12	1	21	72	185	2	3	1	4	1	0	4	4	2
0324	1	2	1	2	3.10	93	13000	1000	2	2	2	2	0	1	19	74	165	2	1	2	2	1	1	4	4	1
0325	1	2	1	2	2.77	83	10000	100	3	2	2	4	4	1	19	69	163	2	1	2	4	1	0	1	5	2
0326	2	3	2	2	2.67	81	15600	600	3	2	2	2	3	1	21	62	96	2	1	2	4	2	1	1	1	2
0327	1	3	1	5	2.96	86	15000	2000	2	1	2	4	2	1	20	78	150	2	1	2	4	1	0	2	2	1
0328	1	2	2	5	2.84	82	12000	500	3	2	1	3	5	1	20	74	140	2	1	1	3	2	0	4	6	1
0329	1	2	2	3	2.32	87	10000	1000	2	2	1	3	2	1	20	74	186	1	1	2	4	2	0	2	4	1
0330	2	3	3	1	3.06	85	12000	1000	2	2	5	3	5	1	21	66	130	1	2	1	3	2	0	4	4	2
0331	1	3	2	5	2.00	80	15000	300	2	2	5	2	4	1	20	71	190	2	3	1	1	2	0	2	5	1
0332	2	3	1	2	2.75	85	13000	0	3	2	2	2	0	1	32	64	125	2	1	2	4	2	0	2	2	2
0333	2	3	1	3	3.00	80	13000	200	2	2	1	3	1	1	19	63	130	2	1	2	4	2	2	2	1	1
0334	2	4	1	2	3.48	83	14500	926	1	2	5	5	6	1	26	69	112	2	2	1	4	1	2	2	4	2
0335	1	4	2	2	3.00	85	10000	450	3	2	5	4	1	1	24	68	130	2	1	2	2	2	0	4	4	1
0336	1	2	2	4	2.50	82	11000	1000	3	2	2	4	2	2	22	70	145	2	1	2	3	2	0	2	3	1
0337	2	2	1	2	3.75	85	16000	500	3	2	2	2	10	1	21	63	100	2	1	2	4	1	1	3	4	1
0338	1	1	1	6	2.97	85	15000	500	2	2	6	2	5	2	31	66	140	2	3	1	1	2	2	4	4	1
0339	1	2	3	2	2.56	76	7500	0	2	2	2	4	5	1	20	72	175	2	2	2	3	1	0	1	1	1
0340	1	3	2	2	2.96	89	10000	640	1	2	5	3	1	1	21	66	190	2	1	2	1	2	0	4	4	1
0341	2	3	3	2	3.02	85	15000	1000	1	2	2	3	3	1	26	63	125	2	1	2	4	1	0	4	4	1
0342	2	2	2	2	3.00	90	12000	2000	1	2	5	4	0	1	36	62	132	2	3	1	4	2	0	4	4	2
0343	2	3	2	2	3.23	91	15000	1500	1	2	2	4	5	1	31	65	117	2	2	2	1	2	0	4	5	2
0344	1	3	3	2	2.02	71	22000	800	3	2	5	4	6	1	23	73	245	1	1	2	3	2	0	4	2	2
0345	1	3	3	2	2.50	84	20000	200	2	2	2	2	4	1	23	60	167	2	1	1	4	2	0	4	4	1
0346	1	4	1	2	2.94	87	15000	500	1	1	2	4	4	1	24	72	164	2	1	2	1	2	0	4	5	1
0347	2	3	1	2	2.50	80	13000	400	2	2	2	2	0	1	25	64	125	2	1	2	1	2	4	4	4	2
0348	1	4	2	2	2.72	65	25000	0	3	2	1	3	5	1	30	69	230	2	1	2	4	2	0	1	7	2
0349	1	2	2	2	2.00	80	10000	300	1	2	2	2	3	1	26	64	123	2	1	2	4	2	0	1	2	1
0350	1	3	2	2	2.53	82	20000	200	1	2	2	4	5	1	26	65	150	2	1	2	3	2	0	4	5	1
0351	1	2	1	6	3.66	85	14000	620	1	2	2	2	1	1	31	68	153	2	1	2	4	2	0	4	4	2
0352	2	2	3	2	2.71	88	10000	0	2	2	2	4	5	2	21	65	110	2	1	2	1	2	0	4	6	2
0353	1	2	3	5	1.83	80	14500	1200	2	2	2	2	5	1	20	67	155	2	1	2	2	2	0	4	6	1
0354	1	1	1	3	2.67	81	20000	1000	3	2	2	4	2	1	17	73	175	2	1	1	3	1	0	4	6	1
0355	2	2	3	7	3.40	60	13000	200	2	1	2	4	7	1	23	63	105	2	1	1	4	1	0	4	2	2
0356	1	3	3	2	3.10	92	14000	1200	1	1	2	4	6	1	20	68	153	2	3	2	3	1	2	4	4	2
0357	2	3	3	5	3.25	90	9500	1000	2	2	2	4	7	1	22	65	117	2	3	1	4	1	2	2	4	2
0358	2	2	2	1	2.00	85	10000	200	3	2	2	2	16	1	19	64	100	2	1	2	4	1	1	3	4	1
0359	1	2	3	2	3.41	97	13000	400	1	2	2	2	20	1	21	68	142	2	1	2	1	1	2	3	3	1
0360	1	2	3	6	2.00	80	12000	1750	3	2	3	4	5	1	20	71	170	2	1	2	4	1	0	5	3	2
0361	1	2	2	5	3.13	85	8000	200	3	2	2	2	1	1	21	70	155	2	1	2	2	2	1	4	4	1
0362	2	2	1	6	3.56	92	15000	300	3	2	1	1	1	1	18	63	95	2	1	1	3	1	0	4	4	1
0363	1	3	3	5	2.82	85	13500	0	3	1	2	2	5	2	21	72	158	2	1	2	3	2	0	4	5	1
0364	2	4	3	7	3.80	83	8000	0	3	2	2	2	2	1	21	65	120	2	1	2	3	2	3	4	4	2
0365	1	2	3	2	2.40	78	13000	150	3	1	2	2	6	2	20	69	162	2	1	1	4	2	1	7	5	2
0366	2	3	1	5	2.62	80	12000	1000	2	2	1	3	2	1	29	69	130	2	3	2	4	1	0	4	4	2
0367	2	3	3	5	3.50	90	7000	0	3	1	2	4	4	1	20	63	126	2	1	1	3	2	1	1	4	2
0368	2	3	2	5	3.00	81	9000	0	2	1	2	4	10	1	20	63	135	2	1	2	4	1	0	1	4	2
0369	1	2	2	2	2.70	79	13000	200	3	1	2	4	5	1	20	70	180	2	2	1	3	2	1	2	4	2
0370	1	2	3	2	2.76	84	13500	900	2	2	1	3	2	2	21	70	160	2	2	2	3	1	0	4	4	1
0371	1	4	2	4	2.82	80	10000	500	3	2	1	4	4	1	26	70	170	2	1	2	1	2	1	3	4	1
0372	2	4	2	5	3.10	82	10000	400	2	2	2	4	10	1	22	62	140	2	3	1	1	2	2	2	6	2
0373	1	2	1	2	3.88	91	25000	1000	2	2	2	4	2	1	20	73	205	1	1	2	4	1	3	2	3	1
0374	1	3	2	5	2.75	85	11500	1200	2	2	2	4	8	1	20	70	150	2	1	1	4	2	0	5	6	2
0375	2	3	3	4	2.90	87	10000	975	2	2	1	4	3	2	21	66	135	2	1	2	3	1	0	4	5	2
0376	2	2	3	2	3.50	90	10000	0	3	2	2	4	8	1	22	65	150	2	1	2	4	1	0	4	6	1
0377	1	4	3	2	3.02	85	15000	1000	2	2	1	3	4	1	21	68	145	2	1	1	4	1	4	1	5	1
0378	1	3	1	2	2.09	85	15000	1000	3	1	2	4	6	2	40	71	185	2	1	2	4	2	0	2	4	2
0379	2	3	2	2	2.50	81	15000	200	1	2	2	3	3	2	23	66	137	2	1	1	1	2	0	4	4	1
0380	1	3	1	2	2.92	90	14000	500	1	2	2	2	4	1	25	72	150	1	1	2	1	2	0	4	4	1
0381	1	3	3	2	2.86	85	15000	200	1	2	2	4	4	1	25	72	185	2	1	2	4	2	0	4	4	1
0382	1	3	3	2	2.80	83	14000	300	1	1	2	4	4	1	28	69	170	2	1	2	4	1	0	7	4	2
0383	2	2	2	2	3.13	88	12000	600	1	2	3	3	0	1	29	66	150	2	1	2	4	2	0	2	4	2
0384	1	4	1	2	3.87	93	13000	1200	2	2	1	4	7	1	21	66	152	2	1	2	4	1	1	6	7	2
File code number	Sex	Class	Grad. school	Major	Grade-point index	H. S. average	Est. starting salary $	Tuition $	Employment	Sunday shop	Investment	Car pref.	No. jeans owned	Calculator	Age	Height	Weight	Blood pressure	Smoking status	Smoking belief	Entrance standards	Retention standards	No. clubs and groups	SPS rating	Library rating	Personal question

File code number	1	2	3	4	5	6	7	8	9	10	11	12	13	14	15	16	17	18	19	20	21	22	23	24	25	26
	Question number																									
0385	1	3	1	2	3.50	86	15000	4000	2	2	2	3	0	2	27	67	160	2	1	2	3	2	1	4	3	1
0386	2	4	1	2	3.00	80	20000	1000	1	2	3	4	4	1	35	65	120	2	1	2	4	2	0	5	5	1
0387	1	2	3	2	3.45	85	15000	1000	1	2	1	2	2	1	27	74	180	2	4	1	3	2	1	3	4	1
0388	1	3	1	2	3.72	86	18000	1000	1	2	4	4	0	1	30	73	265	1	1	2	4	1	0	4	4	1
0389	1	3	2	2	3.01	90	15000	500	1	2	3	4	0	1	47	72	205	2	1	2	4	2	0	1	1	2
0390	1	4	3	3	2.85	91	13000	700	1	1	2	4	5	1	25	70	165	2	1	2	4	1	0	4	4	2
0391	1	2	2	5	2.04	65	35000	1000	1	2	1	4	2	1	32	64	140	2	2	1	1	2	0	4	4	1
0392	2	2	1	6	3.53	89	16000	3500	2	2	2	4	3	2	19	63	117	2	1	2	4	1	0	3	3	1
0393	2	3	3	2	3.07	87	20000	1000	1	2	2	4	3	1	27	70	145	2	2	1	4	2	0	4	4	1
0394	1	3	3	6	3.05	90	12000	800	1	2	2	4	5	1	22	65	138	2	1	2	2	1	0	1	4	1
0395	1	3	1	2	2.04	90	15000	200	1	2	2	2	20	1	31	68	165	2	1	1	1	2	0	1	1	1
0396	1	3	3	2	3.00	83	15000	800	1	2	2	2	1	1	32	72	170	1	1	2	2	2	0	4	4	2
0397	2	3	2	4	3.32	86	15000	800	2	2	5	4	5	1	22	69	125	2	2	2	4	2	1	3	4	2
0398	2	3	2	1	1.84	80	8000	500	1	2	2	2	3	1	22	62	120	2	2	2	1	2	0	4	4	1
0399	1	3	3	2	3.20	90	12000	1500	1	1	2	2	5	1	31	68	150	2	1	2	1	2	0	4	4	2
0400	1	3	1	2	2.20	80	13000	1000	1	2	2	2	0	1	26	71	160	2	1	2	1	2	0	1	1	2
0401	1	4	3	2	3.04	82	18000	500	1	2	6	2	1	1	36	68	185	2	1	2	1	2	0	4	4	1
0402	1	2	3	2	3.07	88	10000	200	3	2	2	4	10	1	20	70	135	2	1	2	4	1	0	4	5	2
0403	1	2	1	4	2.81	83	13000	1500	2	2	2	4	4	1	20	72	168	1	1	2	1	2	0	1	4	2
0404	1	3	3	4	2.51	82	15000	160	2	2	2	4	5	1	22	71	170	2	1	2	4	1	0	4	4	2
0405	1	3	2	2	2.62	92	8000	400	2	2	2	4	8	1	21	65	140	2	1	2	3	2	0	4	5	1
0406	1	3	3	2	2.70	79	12000	900	1	2	2	2	6	1	20	74	165	2	2	1	3	2	2	5	5	1
0407	1	4	2	4	3.00	84	9000	80	2	1	3	4	12	1	22	71	140	2	1	2	1	2	2	2	4	2
0408	1	2	3	2	2.93	85	12500	1200	2	2	2	4	4	1	21	68	135	2	1	2	4	1	1	4	6	2
0409	1	1	1	2	3.10	90	10000	100	2	2	2	2	7	1	19	72	175	2	2	1	3	2	3	4	4	1
0410	1	3	3	2	2.64	81	12000	500	3	2	5	3	12	1	21	71	156	2	1	2	4	1	0	1	5	2
0411	1	3	1	2	2.02	76	25000	300	2	2	2	4	10	1	20	68	200	2	3	1	3	2	2	6	7	1
0412	1	3	2	5	2.13	79	15000	300	2	1	1	4	10	1	22	72	165	2	3	1	3	2	0	4	5	1
0413	1	2	2	2	2.21	75	12000	1000	2	2	2	4	5	1	22	67	145	2	1	2	2	2	0	4	5	2
0414	1	2	3	2	2.42	78	15000	0	2	2	5	4	5	1	19	69	180	2	1	1	3	2	0	4	5	1
0415	2	1	2	2	2.28	84	10000	0	1	2	2	2	1	1	20	61	93	2	1	2	2	1	0	4	4	1
0416	1	2	3	2	2.91	83	15000	1000	2	2	2	2	7	1	19	74	200	2	3	2	3	2	0	4	4	2
0417	2	3	1	3	3.74	89	18000	0	2	2	2	4	2	1	30	66	190	1	2	1	4	2	1	4	6	1
0418	2	1	3	2	2.89	89	12000	1000	2	2	6	2	4	1	20	63	110	2	1	2	3	2	0	4	4	1
0419	2	2	3	2	2.21	71	11000	300	2	2	2	4	3	2	22	65	135	2	1	2	3	2	0	4	4	1
0420	1	3	1	5	2.83	80	12000	80	2	2	5	3	0	2	22	72	195	2	1	2	4	2	1	4	2	2
0421	1	1	1	6	2.96	96	15000	600	3	2	2	2	3	1	18	70	132	2	1	2	4	1	0	4	4	1
0422	1	1	1	6	2.67	88	18000	500	3	2	2	4	2	1	17	69	178	2	1	2	3	1	0	5	5	1
0423	2	3	3	5	3.00	79	12500	600	3	2	2	4	6	1	24	66	135	2	1	1	4	2	0	4	6	2
0424	2	3	2	2	3.50	90	10500	850	3	2	2	4	8	1	20	61	102	2	1	2	4	1	0	4	5	2
0425	1	1	3	2	2.73	81	10000	0	3	2	2	2	5	1	20	70	125	2	1	1	3	1	2	4	5	2
0426	1	2	2	2	2.90	89	12000	1000	3	2	2	4	6	1	21	69	185	2	2	1	3	1	1	4	4	2
0427	1	3	1	3	3.46	85	11000	1500	3	2	1	1	2	1	26	69	180	2	1	2	4	1	0	4	6	1
0428	1	3	2	5	3.02	90	10000	1000	1	2	2	4	1	1	23	72	148	2	1	2	4	2	0	4	4	2
0429	2	3	2	6	3.00	88	12000	500	1	2	2	4	2	2	26	62	105	2	1	2	4	2	0	4	4	2
0430	1	3	3	5	3.53	92	10000	1000	1	2	2	4	8	1	21	64	130	2	2	1	2	2	0	3	4	1
0431	1	3	3	6	2.84	80	12000	1000	1	1	6	5	7	1	25	69	147	2	1	2	1	2	0	4	4	2
0432	2	3	2	5	2.36	76	10000	700	3	1	2	4	4	1	19	63	162	2	2	1	2	2	0	3	6	1
0433	1	3	1	2	3.20	75	13000	1500	1	2	3	3	4	1	29	68	145	2	3	2	1	1	0	2	6	2
0434	2	2	3	2	2.25	78	15000	0	1	2	2	3	2	1	26	68	138	2	3	1	4	2	0	4	4	1
0435	1	4	1	2	3.43	97	15000	1200	1	2	2	4	8	1	28	65	140	2	1	2	4	1	0	4	4	1
0436	2	3	2	2	3.09	71	14000	0	1	2	2	4	0	1	27	62	190	2	4	1	4	2	0	1	4	2
0437	1	4	3	2	2.51	75	11000	500	1	2	2	4	0	1	25	64	170	1	3	1	4	2	0	4	4	1
0438	2	4	1	2	2.86	92	15000	500	3	2	2	1	0	2	43	64	122	2	2	1	4	1	0	4	1	2
0439	1	4	1	4	2.50	76	10000	1800	1	2	2	2	0	1	29	68	150	2	1	2	1	1	3	4	7	2
0440	2	4	3	5	3.10	80	15000	750	3	2	4	4	10	1	21	67	130	2	1	2	4	1	2	4	6	2
0441	1	4	1	4	2.56	74	12000	0	1	2	5	2	4	1	23	68	140	2	1	2	1	2	0	1	4	2
0442	1	2	2	6	3.20	90	10000	500	1	2	2	2	3	1	20	73	182	2	1	2	3	2	0	4	4	2
0443	1	3	3	5	3.54	88	16000	500	3	2	3	4	6	1	21	74	190	2	1	2	4	1	2	4	3	1
0444	1	4	1	5	3.62	90	10000	0	1	1	2	4	0	1	23	70	150	2	1	1	2	2	1	4	6	1
0445	1	3	1	4	2.81	75	12000	0	1	2	2	4	1	1	23	72	165	2	1	2	1	1	0	2	3	1
0446	2	4	3	5	3.00	89	19000	0	1	2	2	4	2	1	28	63	125	2	1	2	4	1	0	1	1	2
0447	1	3	1	4	3.40	95	12000	0	1	2	2	2	0	1	28	72	200	1	3	1	1	1	0	1	1	1
0448	2	3	3	4	2.78	90	17000	0	1	2	2	2	1	1	37	66	150	2	1	2	4	1	0	1	4	1
0449	1	4	1	2	3.03	91	10000	0	1	2	2	2	1	1	23	69	145	2	1	2	4	2	0	1	4	1
0450	2	4	1	4	3.42	86	10000	2700	1	2	2	2	4	1	28	65	125	2	1	2	4	1	1	1	4	2
0451	2	4	1	7	2.65	88	20000	500	1	2	2	4	2	1	38	65	127	1	1	2	3	1	0	4	5	2
0452	2	4	1	4	3.00	83	15000	700	2	2	2	4	14	1	18	60	125	2	2	2	4	2	0	4	4	2
0453	1	4	3	2	3.02	83	11800	600	1	2	2	4	9	1	24	70	180	2	1	2	3	2	0	4	7	2
0454	2	3	3	2	2.46	80	10000	0	1	2	1	2	3	1	27	62	110	2	1	1	1	1	0	4	1	2
0455	1	4	2	2	2.23	85	13000	900	1	2	2	2	1	1	38	68	195	2	1	1	1	2	0	5	6	1
0456	1	4	3	2	3.33	93	13000	1200	1	2	3	3	2	1	26	67	180	2	3	1	3	1	0	1	5	2
0457	1	3	1	2	2.90	84	13500	0	2	2	6	2	2	2	26	74	205	2	1	2	3	2	0	4	4	1
0458	1	3	3	2	2.61	70	15000	480	1	2	2	4	0	2	48	72	150	1	1	2	1	2	1	4	7	2
0459	1	3	2	2	3.10	82	12000	1000	1	2	2	4	4	1	23	71	170	1	1	2	3	2	0	4	5	2
0460	1	2	1	1	2.97	86	10000	1500	1	2	5	3	0	1	28	71	190	2	3	1	4	1	0	4	4	2
0461	2	3	3	5	2.81	89	12000	2000	2	2	2	2	9	2	19	66	121	2	3	1	4	2	0	4	4	2
0462	1	3	1	2	3.80	93	12000	1300	2	2	2	2	3	1	20	70	175	2	1	2	3	1	1	4	2	1
0463	1	3	2	2	2.13	73	10000	0	3	2	4	4	0	1	21	71	167	2	1	2	4	2	0	4	4	2
0464	2	2	3	2	3.50	80	12000	0	2	1	2	4	5	1	32	64	117	2	1	1	4	2	1	4	4	2
0465	2	3	2	2	3.25	83	11500	950	1	2	2	4	2	1	35	66	150	1	1	2	3	1	1	4	4	1
0466	2	3	1	2	3.00	80	20000	0	1	2	1	3	1	1	25	62	105	2	3	1	1	2	0	5	4	1
0467	2	4	3	5	2.50	80	10000	800	3	2	1	4	5	2	21	63	125	2	1	2	4	1	2	1	4	2
0468	1	3	1	2	3.07	89	10000	0	2	2	2	2	3	1	21	69	150	2	1	2	2	1	1	4	6	1
0469	1	3	1	2	3.31	75	12000	1000	3	2	2	1	6	1	20	73	170	2	1	2	1	1	0	4	4	1
0470	1	3	3	2	2.80	92	12000	1500	2	2	2	4	4	1	20	74	240	2	1	2	3	1	1	4	5	2
0471	1	4	3	2	2.56	78	11000	0	3	2	2	4	2	1	25	70	160	2	2	2	4	2	0	4	4	2
0472	1	2	1	2	3.84	91	14000	2000	3	2	1	4	4	1	19	72	166	2	1	2	3	1	2	4	6	1
0473	1	3	1	2	3.70	92	15000	1500	2	2	2	4	4	1	20	68	145	1	1	2	1	1	2	7	7	2
0474	2	4	1	6	3.70	92	12000	1000	2	1	3	4	8	1	22	64	115	2	1	2	3	2	2	3	3	1
0475	1	3	1	6	3.00	75	20000	1000	1	2	2	4	2	1	30	70	142	2	1	2	4	1	1	4	3	1
0476	2	3	3	7	3.02	83	14000	1200	1	2	2	2	5	1	27	66	126	2	1	2	1	2	0	2	2	1
0477	2	4	1	6	2.06	82	17500	0	1	2	2	2	2	1	36	63	130	2	1	2	4	1	0	2	5	1
0478	1	4	3	6	3.54	82	9000	500	1	1	1	4	2	2	32	72	180	2	1	1	4	2	0	1	4	1
0479	1	4	3	6	3.03	90	13000	0	2	2	2	2	6	1	27	68	128	2	1	2	3	2	1	6	7	1
0480	1	4	3	6	2.89	84	20000	500	1	2	4	5	0	1	23	69	170	2	1	2	4	1	0	4	4	1
File code number	Sex	Class	Grad. school	Major	Grade-point index	H. S. average	Est. starting salary $	Tuition $	Employment	Sunday shop	Investment	Car pref.	No. jeans owned	Calculator	Age	Height	Weight	Blood pressure	Smoking status	Smoking belief	Entrance standards	Retention standards	No. clubs and groups	SPS rating	Library rating	Personal question

File code number	1 Sex	2 Class	3 Grad. school	4 Major	5 Grade-point index	6 H. S. average	7 Est. starting salary $	8 Tuition $	9 Employment	10 Sunday shop	11 Investment	12 Car pref.	13 No. jeans owned	14 Calculator	15 Age	16 Height	17 Weight	18 Blood pressure	19 Smoking status	20 Smoking belief	21 Entrance standards	22 Retention standards	23 No. clubs and groups	24 SPS rating	25 Library rating	26 Personal question
0481	1	3	1	2	3.70	85	10000	0	3	2	3	1	2	1	20	75	215	1	1	2	4	1	2	1	2	2
0482	1	4	2	2	3.31	86	15000	1500	1	2	2	4	0	1	23	71	175	2	1	2	3	1	0	4	5	1
0483	1	3	1	2	3.10	90	15000	900	1	2	6	2	1	1	44	66	128	2	2	2	1	2	1	4	6	2
0484	1	2	2	3	3.00	80	14000	1000	1	2	2	2	2	1	28	71	140	2	1	2	4	1	0	4	3	2
0485	2	4	1	2	3.50	95	13000	1000	2	2	2	2	4	1	22	70	145	2	1	2	1	2	0	4	2	1
0486	1	4	3	2	3.20	80	10000	1000	1	2	1	4	1	1	25	70	150	2	1	2	1	1	0	4	4	2
0487	1	4	1	2	3.21	84	12500	1200	1	2	2	5	0	1	26	66	150	2	2	2	1	2	0	4	4	2
0488	1	3	1	2	3.00	82	12500	2000	1	2	2	4	2	1	25	69	160	1	1	2	4	1	0	2	4	2
0489	2	2	1	4	2.69	70	20000	0	1	2	2	4	2	1	35	62	120	2	3	1	4	2	0	1	1	1
0490	1	4	1	2	3.61	91	10000	0	1	1	2	4	0	1	34	70	155	2	1	2	4	1	0	4	6	2
0491	1	4	3	4	3.20	81	12000	0	1	1	2	4	0	1	31	72	180	2	2	1	4	2	0	1	1	2
0492	1	3	2	6	3.24	70	14000	1000	1	2	5	2	3	1	27	73	195	2	4	1	3	2	0	6	6	1
0493	1	4	1	4	3.03	90	20000	0	1	1	2	2	0	1	39	71	210	2	1	2	4	2	2	4	4	2
0494	1	4	3	2	2.57	71	12000	500	3	2	2	2	3	1	26	70	185	2	1	2	1	2	1	5	5	2
0495	1	4	1	4	3.20	89	10500	1200	1	2	2	4	13	1	25	67	170	2	1	1	1	2	2	4	5	1
0496	2	3	1	3	3.86	93	10000	500	1	2	2	2	5	1	23	63	113	2	1	2	4	1	0	2	2	2
0497	2	3	1	2	3.41	90	12500	950	3	2	3	4	2	1	20	61	115	2	1	2	4	2	1	5	4	1
0498	1	4	3	3	2.98	79	10600	0	1	2	5	4	2	1	27	72	190	2	3	1	1	1	0	4	1	1
0499	1	4	1	3	3.20	80	10000	400	1	1	6	5	0	1	23	67	140	2	1	2	4	1	0	4	4	1
0500	2	3	1	7	2.00	76	15000	0	1	2	2	2	1	1	25	62	108	2	1	2	4	1	0	4	4	1
0501	1	2	1	3	3.50	78	12500	1500	2	2	2	4	2	1	36	66	165	2	1	2	3	1	0	4	4	1
0502	2	3	3	4	2.53	87	20000	100	1	2	2	4	3	1	27	65	120	2	1	2	4	1	0	4	4	1
0503	2	3	1	5	3.04	75	15000	2000	1	2	5	4	2	1	23	66	120	2	3	1	4	1	0	4	4	1
0504	1	4	1	5	3.25	85	15000	1000	1	2	2	5	3	1	33	70	167	2	3	2	4	2	0	4	4	2
0505	1	2	1	2	3.62	90	15000	1000	2	2	3	4	0	1	22	72	180	2	1	2	3	2	1	4	4	2
0506	1	2	2	2	3.03	90	12500	200	3	2	3	3	0	1	20	69	180	2	1	2	4	2	2	3	4	1
0507	1	3	3	2	3.02	79	10000	500	3	2	4	4	0	1	21	68	137	2	1	1	3	2	0	2	4	1
0508	1	3	1	2	2.94	81	14000	600	1	2	2	2	4	1	20	71	141	2	2	2	4	1	0	3	5	1
0509	1	2	3	2	3.05	80	21000	300	2	2	2	2	1	2	32	71	165	2	1	2	4	2	0	4	4	1
0510	2	2	1	5	3.04	85	30000	1000	1	2	2	4	10	1	27	66	109	2	2	1	2	2	0	1	4	1
0511	1	2	3	2	2.02	83	15500	1000	3	2	2	4	10	1	22	70	150	2	1	2	1	2	0	1	5	1
0512	1	2	1	2	3.88	96	13000	850	3	2	2	4	1	1	19	70	120	2	1	2	2	2	0	1	6	1
0513	1	3	3	7	2.67	75	10000	100	1	2	2	4	3	1	27	70	160	2	3	1	3	2	0	1	5	1
0514	1	2	3	2	3.20	80	10000	1500	3	2	2	2	7	1	19	76	197	2	1	2	2	1	0	4	6	2
0515	1	4	1	3	3.50	84	14000	0	1	2	5	2	2	1	25	66	135	2	1	2	4	1	0	4	4	1
0516	2	1	1	5	2.15	91	13000	0	3	2	6	2	3	1	24	60	99	2	1	2	1	2	0	5	4	2
0517	2	2	3	1	2.87	88	10000	800	2	2	2	2	0	2	18	66	126	2	1	2	4	1	0	4	5	1
0518	2	2	3	2	3.12	80	16000	600	1	2	2	2	6	1	29	66	120	2	1	2	2	1	0	4	4	1
0519	2	4	1	7	3.30	95	8000	250	2	2	2	4	7	2	39	64	105	2	1	1	4	2	2	1	4	2
0520	1	4	1	3	3.02	85	18000	2000	1	2	1	1	4	1	30	72	200	2	1	2	3	1	0	4	6	1
0521	1	3	1	2	3.51	85	12000	200	2	2	6	2	4	2	30	71	135	2	2	1	1	2	5	1	4	1
0522	2	3	3	4	3.00	87	15000	850	2	2	2	4	8	1	30	65	130	2	1	2	1	1	0	4	4	2
0523	2	3	3	4	2.56	70	15000	250	2	2	2	2	5	1	26	63	105	2	1	2	2	2	0	4	6	1
0524	2	3	1	2	3.02	85	10000	800	2	2	5	4	6	1	21	61	108	2	3	2	4	2	0	4	4	1
0525	2	2	3	2	3.13	93	12000	200	1	1	3	4	0	1	20	64	110	2	1	2	3	2	0	4	4	1
0526	1	2	2	2	3.26	77	15000	1000	1	2	2	2	2	1	38	68	154	2	1	2	4	1	1	4	4	2
0527	1	3	3	3	3.01	78	12000	200	1	2	2	4	15	1	39	71	180	2	1	2	2	2	0	4	4	1
0528	1	4	2	4	3.50	80	18000	1000	1	2	2	4	2	1	29	70	170	2	1	2	4	1	0	1	4	1
0529	2	3	1	5	2.81	80	12000	1000	2	2	2	4	2	1	22	60	102	2	1	2	3	2	0	3	5	1
0530	1	3	1	5	2.88	89	12000	1200	2	2	2	1	6	1	20	67	163	1	1	2	4	1	0	4	3	1
0531	2	3	3	5	2.04	84	15000	0	2	2	5	4	10	1	21	63	117	2	2	2	2	2	0	4	4	1
0532	2	3	2	7	2.52	89	10000	500	2	2	2	4	3	1	22	63	150	2	3	1	3	2	0	4	4	1
0533	1	3	3	5	2.92	79	11000	0	2	2	2	2	4	1	23	69	140	2	1	2	1	2	3	4	4	2
0534	1	3	1	5	2.63	76	10000	1000	1	2	1	3	5	1	20	68	148	2	1	2	3	1	1	4	1	1
0535	1	3	3	4	2.85	82	15000	300	2	2	2	4	10	1	22	71	150	2	1	2	1	2	0	4	5	2
0536	1	3	3	5	2.96	80	15000	700	3	2	1	4	5	2	21	70	145	2	1	2	2	1	2	3	4	1
0537	1	2	2	5	3.00	94	16000	750	1	1	5	4	10	1	19	72	180	2	1	2	3	1	0	2	2	1
0538	1	3	2	4	2.41	75	13000	0	2	2	5	4	4	1	30	70	160	2	1	2	1	2	0	3	4	1
0539	1	2	1	2	2.58	80	17500	500	2	2	2	2	5	1	19	69	158	1	1	2	3	2	1	1	1	2
0540	1	2	3	5	2.42	83	20000	2000	2	2	2	4	2	1	20	71	175	2	1	2	4	2	0	1	1	1
0541	1	2	3	5	2.50	78	12000	700	3	2	2	4	8	1	20	71	150	2	1	2	4	2	0	3	5	1
0542	1	3	3	5	2.80	80	12000	1500	2	1	2	2	4	1	21	70	180	2	1	2	3	2	1	3	4	1
0543	1	3	1	7	2.90	70	7000	1000	2	2	2	4	2	1	21	74	190	2	1	2	1	2	0	4	5	2
0544	1	2	3	4	2.91	79	12000	1200	2	2	3	4	5	1	20	70	170	2	2	2	4	1	0	4	6	1
0545	2	3	1	5	2.94	95	10000	1300	3	2	2	4	6	1	21	62	120	2	1	2	3	1	0	3	4	1
0546	1	3	2	2	3.00	85	12000	500	2	2	2	4	4	1	21	73	165	2	1	1	4	2	1	1	3	1
0547	1	3	1	2	2.75	89	10000	1000	2	2	4	4	2	1	23	70	170	2	1	2	3	1	2	4	6	1
0548	2	2	3	7	2.10	81	20000	0	2	2	2	2	10	2	20	63	110	2	1	2	1	1	3	1	1	2
0549	2	1	2	4	2.95	93	13000	1200	2	2	1	3	9	1	18	65	118	2	2	1	3	1	1	4	4	2
0550	2	3	3	4	3.45	92	10000	1200	3	2	2	2	5	1	22	62	106	2	1	2	1	1	0	4	4	1
0551	2	3	2	7	2.30	86	9500	600	3	2	5	4	5	2	20	65	125	2	2	2	4	2	0	4	4	2
0552	1	2	3	6	2.75	89	13000	0	2	2	2	2	6	1	21	68	148	2	2	1	4	2	2	5	6	1
0553	1	3	1	5	3.20	82	10000	400	2	2	2	2	3	1	20	71	170	2	1	2	3	2	2	2	5	1
0554	2	2	3	6	3.42	94	20000	900	2	2	3	4	2	2	19	68	123	2	1	2	4	1	0	4	4	1
0555	1	4	1	2	2.53	85	13000	0	1	2	2	2	1	1	26	72	190	1	3	1	1	2	0	1	4	2
0556	1	3	1	2	3.06	85	12000	1000	1	2	2	2	5	1	22	67	145	2	1	1	3	1	0	1	5	1
0557	2	2	1	2	2.84	80	10000	0	1	2	2	4	5	1	26	66	170	2	1	2	3	2	0	1	4	1
0558	2	4	3	2	3.02	81	13000	1000	1	2	1	3	1	1	25	65	116	2	2	2	4	1	0	4	2	2
0559	2	3	2	7	2.50	82	13000	600	3	2	2	4	5	1	20	60	115	2	2	2	3	2	3	2	4	1
0560	1	3	3	4	2.69	84	16000	400	3	2	5	4	2	2	20	69	150	2	2	1	2	2	4	4	6	2
0561	2	4	1	2	3.02	81	15000	0	1	2	2	4	7	1	28	64	145	2	1	2	1	2	0	4	4	2
0562	1	4	1	2	3.00	91	12000	0	3	2	2	2	3	1	35	73	176	2	1	2	1	2	2	1	1	1
0563	1	4	3	2	2.75	80	16000	0	1	2	2	2	1	1	39	65	145	2	1	2	1	2	0	4	4	1
0564	1	3	1	2	1.76	80	10000	700	1	1	3	4	0	1	27	74	200	2	1	1	4	2	0	4	6	2
0565	2	3	3	2	2.30	82	18000	0	1	2	2	2	6	1	32	66	180	2	1	2	3	2	0	1	1	1
0566	2	4	3	2	2.53	80	15000	0	1	2	2	4	2	1	36	62	168	2	1	2	1	2	0	4	4	2
0567	2	3	1	1	3.06	83	13000	1000	1	2	2	2	4	1	33	61	170	2	1	2	4	1	0	4	4	2
0568	1	4	1	2	3.10	70	12500	0	3	1	1	4	4	1	31	71	175	2	1	2	4	1	2	6	6	2
0569	1	3	1	2	3.02	80	10000	1000	2	2	2	2	0	1	24	65	145	2	1	2	3	1	1	4	4	1
0570	1	3	2	2	2.50	90	10000	400	1	2	2	4	1	1	23	74	210	2	1	2	4	2	0	4	2	1
0571	1	4	3	2	2.50	84	17000	0	3	1	2	4	0	2	49	65	150	1	1	2	1	2	0	1	1	1
0572	1	4	2	2	2.91	88	12000	1000	1	2	2	4	5	1	25	67	147	2	3	1	3	2	0	4	4	2
0573	1	4	1	2	3.54	92	18000	1000	3	2	2	2	0	1	23	74	180	2	1	2	4	1	0	4	5	2
0574	1	4	3	2	2.51	79	10000	750	2	2	2	2	3	1	25	70	199	2	1	2	3	2	1	3	6	1
0575	1	3	3	2	3.70	90	10500	60	1	1	2	4	5	1	22	66	145	2	1	2	1	2	0	3	2	1
0576	1	4	1	2	3.10	88	15000	2000	2	2	3	3	3	1	22	73	188	2	1	2	3	2	2	4	6	2

File code number	1 Sex	2 Class	3 Grad. school	4 Major	5 Grade-point index	6 H. S. average	7 Est. starting salary $	8 Tuition $	9 Employment	10 Sunday shop	11 Investment	12 Car pref	13 No. jeans owned	14 Calculator	15 Age	16 Height	17 Weight	18 Blood pressure	19 Smoking status	20 Smoking belief	21 Entrance standards	22 Retention standards	23 No. clubs and groups	24 SPS rating	25 Library rating	26 Personal question
0577	2	3	1	5	3.42	90	10000	500	3	2	1	4	4	1	27	63	135	2	2	1	4	1	0	4	6	1
0578	2	3	3	5	3.42	91	10500	0	2	1	2	2	9	1	20	60	130	2	2	1	3	2	2	4	2	2
0579	2	3	1	5	3.81	92	11000	1500	1	2	5	3	6	1	21	66	115	2	1	2	3	1	1	4	6	2
0580	2	1	2	6	2.84	89	11000	1000	1	2	2	2	0	1	20	63	140	2	4	1	4	2	0	4	4	2
0581	1	4	3	4	2.63	80	15000	1000	1	2	2	2	8	1	32	70	165	2	1	2	4	1	0	4	6	1
0582	1	4	1	7	2.04	75	12000	0	3	2	2	2	2	1	33	72	155	2	2	2	1	2	0	4	4	1
0583	1	3	2	5	2.41	87	9000	500	2	2	2	4	6	1	20	72	175	2	1	2	4	2	0	5	4	1
0584	1	3	2	3	3.02	85	13500	1200	3	2	1	4	6	2	20	71	165	2	1	2	3	1	2	1	3	2
0585	1	2	1	5	2.80	85	12333	1600	3	2	5	2	3	2	20	69	140	2	1	2	4	1	0	1	3	2
0586	1	4	2	5	2.70	80	12000	1000	3	2	6	2	0	1	30	67	123	2	1	2	4	1	1	1	2	2
0587	1	3	3	6	2.75	81	10000	1000	3	2	2	4	4	1	22	67	120	2	1	2	4	1	1	1	4	2
0588	1	1	3	1	3.20	86	10000	0	2	1	1	2	3	1	19	69	160	2	1	2	3	2	0	4	3	2
0589	1	3	1	5	3.51	85	12000	600	3	2	1	4	7	1	21	73	185	2	1	2	4	1	0	3	4	1
0590	1	4	1	5	2.77	85	11000	0	3	2	1	4	0	1	23	71	170	2	1	2	4	1	1	4	4	1
0591	1	4	3	5	3.60	70	14000	1000	3	2	2	5	6	1	31	66	165	2	1	1	1	2	0	1	4	1
0592	1	3	1	3	3.04	80	12000	250	2	2	2	2	2	2	25	72	180	2	3	1	2	1	1	1	4	2
0593	1	4	1	3	2.64	82	13000	1000	2	2	2	4	3	1	21	69	165	2	3	2	3	2	2	4	5	1
0594	1	4	3	5	2.74	74	5000	1000	1	2	2	4	4	1	22	72	145	2	1	1	3	2	1	7	4	2
0595	1	3	2	2	3.21	82	10000	463	2	2	5	2	4	1	19	66	140	2	1	2	3	1	1	4	6	2
0596	2	3	3	2	3.94	97	13000	0	2	2	2	4	3	1	20	67	128	2	1	2	4	1	1	4	4	1
0597	1	4	3	2	3.20	78	13000	2000	2	2	2	4	4	1	21	69	160	2	2	2	3	1	0	5	4	1
0598	1	4	2	2	3.05	86	12000	1200	3	2	2	4	8	1	21	67	135	2	1	2	3	2	1	4	4	2
0599	2	4	3	2	2.90	78	11000	1000	3	2	3	4	3	1	24	64	116	2	1	1	3	2	2	4	5	2
0600	2	3	3	5	2.90	85	10000	900	3	2	1	4	20	1	21	63	130	2	1	2	4	2	0	4	4	2
0601	1	3	3	2	3.30	91	12000	500	3	2	2	3	7	1	22	72	175	2	1	2	3	1	1	4	1	1
0602	1	4	1	2	2.74	88	11500	1000	3	2	2	1	10	1	24	63	150	2	1	2	3	2	0	5	6	2
0603	1	4	2	2	3.06	90	10000	700	3	2	5	1	4	1	22	67	135	2	1	1	4	2	1	2	6	2
0604	2	4	3	2	3.27	88	12000	500	2	2	2	4	2	2	24	67	145	1	2	2	3	2	0	2	6	1
0605	2	4	3	2	3.42	95	12000	0	3	2	2	2	2	2	29	68	175	2	1	2	1	2	1	4	1	1
0606	2	4	3	2	2.90	95	12000	0	2	2	2	2	2	2	27	65	120	2	1	1	4	1	0	4	5	2
0607	1	3	1	2	3.53	91	15000	800	3	2	2	4	10	1	27	62	120	2	1	2	2	2	0	4	4	1
0608	2	3	1	2	3.38	93	13000	1000	2	2	1	4	2	1	21	67	150	2	1	2	4	2	2	1	3	1
0609	2	3	1	2	2.81	93	10000	1500	2	2	2	4	4	1	22	60	110	2	1	2	3	2	0	4	2	1
0610	2	4	1	2	2.54	89	14000	484	2	2	5	4	7	1	21	61	110	2	1	2	3	2	0	4	6	1
0611	1	4	1	2	3.88	95	15000	2000	1	2	2	4	10	1	25	78	225	2	4	1	4	2	1	2	6	2
0612	2	3	2	2	3.30	86	10000	800	3	2	2	4	6	1	23	63	145	2	1	2	1	2	0	3	2	1
0613	1	3	3	2	2.64	84	14500	500	3	2	6	2	2	1	25	70	169	2	1	2	1	2	3	4	4	2
0614	1	4	1	2	2.62	89	10750	0	3	1	5	4	5	1	22	66	129	1	3	1	3	2	0	4	1	2
0615	2	3	1	2	2.81	86	18000	1000	1	2	1	3	2	1	38	63	140	2	2	2	2	2	0	4	5	2
0616	1	4	3	2	2.93	82	10000	900	1	2	2	4	0	1	39	66	180	1	3	1	4	1	0	4	4	2
0617	2	3	2	2	3.65	91	18000	1000	1	2	3	4	8	1	27	62	115	2	2	2	4	1	1	4	5	2
0618	1	4	1	2	3.23	80	16000	0	1	1	1	4	6	1	27	66	165	2	1	2	4	2	0	4	6	2
0619	1	3	3	2	2.67	81	10000	1000	1	2	5	3	6	2	26	74	185	2	3	2	4	2	0	4	2	1
0620	1	4	3	2	3.34	80	11000	3000	2	2	6	2	0	1	33	68	170	2	1	2	4	1	4	6	6	1
0621	1	3	1	2	3.00	84	12500	1200	1	1	2	4	8	1	20	69	185	2	1	2	3	1	0	4	4	2
0622	1	3	3	2	3.20	82	15000	0	1	2	2	2	3	1	31	70	140	2	3	2	1	2	2	1	1	2
0623	1	3	3	2	2.73	92	12000	1000	2	2	2	2	2	1	40	64	125	2	1	2	2	1	1	3	3	2
0624	1	4	1	2	3.02	78	15000	0	2	2	2	4	0	1	24	67	150	2	2	1	3	1	2	4	4	1
0625	1	3	1	2	2.50	75	20000	1000	1	2	5	4	5	1	32	66	180	2	1	1	4	2	0	4	4	1
0626	2	3	1	2	3.03	80	15000	1000	1	2	5	3	5	1	33	66	105	2	3	1	3	1	0	4	4	2
0627	1	4	2	2	3.08	80	13500	800	1	2	3	3	1	1	30	65	165	2	1	2	4	2	0	4	7	2
0628	1	3	1	2	3.00	87	17500	0	1	2	2	2	0	2	33	74	225	1	3	2	1	1	0	4	4	1
0629	1	4	1	2	3.24	85	13000	400	2	2	2	2	5	1	29	66	132	2	1	2	4	2	0	1	4	1
0630	1	3	1	2	3.06	88	15000	400	3	1	2	2	1	1	27	65	130	2	1	2	1	2	0	1	4	1
0631	1	4	1	2	2.92	79	9000	600	3	2	2	5	0	1	60	67	145	2	1	2	4	2	0	4	4	1
0632	1	4	1	2	3.20	90	13000	1000	1	2	2	4	2	1	24	73	160	2	1	2	4	1	1	6	3	2
0633	1	3	1	2	2.67	86	10000	1400	1	2	2	4	2	1	29	70	160	2	1	2	3	2	0	4	4	2
0634	2	3	1	2	2.95	88	25000	0	1	1	6	4	0	1	23	69	155	1	1	2	3	1	1	4	6	2
0635	1	4	3	2	2.88	81	10000	1000	1	1	1	3	1	1	32	69	172	1	1	1	3	2	0	4	4	1
0636	1	3	1	2	2.91	85	17500	0	3	2	2	3	0	1	31	68	165	2	4	2	1	2	0	1	2	1
0637	2	4	3	2	3.00	86	10000	0	1	2	6	2	15	1	25	61	96	2	1	1	1	2	0	4	4	2
0638	2	4	3	2	2.54	82	10000	0	1	2	6	2	3	2	26	63	120	2	1	2	1	2	0	4	4	2
0639	2	3	3	2	3.72	89	12500	500	1	2	2	2	0	1	25	69	128	2	1	2	1	2	0	4	4	1
0640	2	3	3	2	2.36	85	18000	1000	1	2	5	4	1	2	26	66	140	2	1	2	1	2	0	4	4	2
0641	1	4	2	7	3.57	88	20000	240	1	2	2	2	0	1	52	68	174	1	1	2	4	1	1	1	1	2
0642	1	3	2	6	2.70	73	12500	1200	3	2	5	2	5	1	21	72	173	2	1	1	4	1	0	6	7	1
0643	1	2	3	4	3.09	83	26000	1000	1	2	2	4	12	1	19	60	153	2	1	2	3	2	0	1	3	2
0644	1	2	3	6	2.43	80	14000	1000	2	1	2	4	15	1	20	67	140	2	1	2	3	1	1	3	4	1
0645	2	4	3	5	2.78	92	12000	1000	1	2	5	4	4	1	34	61	114	1	1	2	3	1	1	4	5	2
0646	1	4	1	5	2.66	78	14000	1000	2	2	2	2	3	1	22	69	152	2	2	2	2	2	0	3	6	1
0647	1	2	1	4	3.30	88	10000	700	2	2	1	3	2	1	19	69	153	2	1	2	4	1	0	4	2	2
0648	1	3	2	7	2.60	86	13500	463	2	2	2	2	0	1	23	68	293	1	1	2	1	2	1	1	4	1
0649	1	3	3	7	2.78	85	14375	463	2	2	1	3	7	1	22	72	155	2	1	2	1	2	1	1	3	2
0650	2	4	3	7	3.04	78	8000	0	2	2	1	4	3	2	25	61	130	2	1	2	4	1	2	1	4	1
0651	2	3	3	7	2.57	95	20000	1500	1	2	2	2	1	1	42	71	175	1	2	1	1	1	0	4	5	2
0652	2	4	1	7	3.03	86	10000	800	3	2	2	2	5	1	25	66	132	2	2	1	4	1	0	4	3	1
0653	2	3	1	7	3.52	93	10000	0	3	2	2	4	4	1	22	63	135	2	1	2	4	1	2	6	4	2
0654	1	3	1	5	3.61	82	25000	1500	2	2	1	4	6	1	21	68	149	2	2	2	4	1	0	3	5	2
0655	1	2	3	1	2.51	84	9000	1000	2	2	2	4	6	2	19	69	162	2	1	2	4	1	0	4	5	1
0656	2	3	3	5	3.06	90	12000	300	2	1	2	4	20	1	23	66	125	2	3	1	4	2	0	1	2	1
0657	1	2	3	2	3.83	96	10500	1000	3	2	2	2	0	1	19	63	110	2	1	2	4	1	0	7	7	2
0658	2	3	2	2	2.09	86	10000	600	3	2	2	4	0	1	22	62	154	2	1	1	2	1	0	4	5	1
0659	1	2	2	2	3.22	91	16000	1200	3	2	2	2	6	1	19	70	176	2	1	2	4	1	2	4	3	2
0660	1	4	3	5	2.84	81	9000	1250	2	2	2	4	1	1	22	72	164	2	1	1	4	1	0	3	5	2
0661	1	2	3	1	2.51	80	15000	900	2	1	2	4	4	1	20	70	145	2	1	2	3	2	1	5	6	2
0662	1	4	1	7	3.80	93	8000	0	2	1	2	4	4	1	22	75	200	2	1	2	4	2	1	4	4	2
0663	2	3	3	6	3.60	91	10000	1200	2	2	2	2	5	1	21	60	100	2	1	2	4	1	0	2	3	1
0664	2	4	1	7	3.93	90	8000	500	2	2	2	4	0	1	23	64	115	2	1	2	3	1	5	2	4	1
0665	1	4	1	6	3.02	89	12000	600	3	2	2	4	5	1	23	68	160	2	1	2	3	2	0	4	6	2
0666	1	4	1	5	2.81	65	10000	800	2	2	5	4	2	1	22	73	185	2	1	2	1	1	3	1	1	2
0667	1	2	1	2	3.20	92	12000	3000	3	2	1	4	10	1	20	69	155	2	1	2	3	1	0	4	2	2
0668	2	4	1	4	2.04	85	12000	600	2	2	2	4	13	1	24	62	110	2	2	1	4	2	0	4	4	1
0669	1	2	3	3	2.81	79	20000	0	1	2	2	2	0	1	27	70	141	2	1	2	1	2	0	4	6	2
0670	2	4	1	2	3.50	90	14000	1000	3	2	1	3	4	1	35	64	130	2	3	2	4	1	0	4	4	1
0671	2	2	3	3	3.20	78	12000	1000	2	2	2	2	10	1	25	64	114	2	1	2	4	2	0	4	3	2
0672	2	2	3	2	2.64	83	12000	800	3	2	2	2	6	2	20	66	123	2	1	2	3	2	0	3	4	1

File code number	Question number																									
	1	2	3	4	5	6	7	8	9	10	11	12	13	14	15	16	17	18	19	20	21	22	23	24	25	26
0673	1	3	3	2	2.13	82	15000	2000	2	2	2	2	1	1	22	66	130	2	1	1	1	2	0	4	4	2
0674	1	3	2	2	3.00	88	12500	900	3	2	1	4	3	1	22	71	180	2	1	2	3	1	1	4	5	2
0675	1	2	2	2	2.75	80	10000	600	2	2	2	4	10	1	25	67	135	1	2	2	4	2	0	4	4	1
0676	2	2	3	2	2.50	84	12000	800	3	2	2	3	10	1	20	64	122	2	3	1	4	2	0	4	4	2
0677	1	3	1	2	2.96	89	15000	1600	2	2	5	4	8	1	22	67	140	2	1	2	4	1	1	2	4	1
0678	2	3	2	5	2.70	87	10000	250	3	2	2	4	20	1	22	65	115	2	1	2	4	2	0	1	1	2
0679	2	2	3	1	2.60	84	11000	1200	2	2	1	3	5	1	20	65	103	2	2	1	3	1	0	6	3	2
0680	2	4	1	4	3.86	75	9500	600	2	2	1	3	5	1	23	65	130	2	1	2	3	1	3	2	6	1
0681	2	3	1	6	3.70	85	15000	1500	1	2	2	4	6	1	27	68	145	2	2	1	4	1	0	4	4	2
0682	1	4	2	7	2.58	75	10000	1050	1	2	5	4	4	2	26	70	180	2	1	1	4	2	0	5	6	2
0683	1	4	1	7	3.00	89	20000	400	3	2	2	2	2	1	25	73	168	2	1	2	4	1	1	6	5	2
0684	1	4	2	5	3.00	93	10000	0	1	2	2	4	1	1	26	72	175	2	1	2	1	2	0	4	4	1
0685	2	3	3	4	3.00	94	15000	0	1	2	2	2	5	2	25	63	105	2	1	2	4	1	0	4	3	1
0686	2	3	2	5	2.50	80	5000	0	2	2	2	4	4	1	21	67	112	2	1	2	4	2	0	1	6	2
0687	2	4	1	4	2.72	80	11000	0	1	2	1	4	4	1	30	69	135	2	1	1	4	2	0	4	4	2
0688	1	4	2	5	2.24	78	13000	0	2	1	1	4	5	1	25	70	135	2	1	2	1	2	0	4	5	1
0689	1	2	2	2	2.50	82	12800	0	2	2	2	4	5	2	21	68	140	2	1	2	3	2	0	2	4	1
0690	1	2	3	6	3.50	90	20000	875	1	2	1	4	18	1	23	71	164	2	1	2	4	1	0	1	4	1
0691	2	2	1	4	3.20	85	12000	800	2	2	5	4	3	1	21	64	138	2	1	2	4	2	0	4	4	2
0692	1	2	1	2	3.09	80	15000	1000	2	2	2	2	6	1	26	68	142	2	1	2	4	1	1	4	5	2
0693	1	3	1	2	3.40	90	14000	1000	1	2	2	4	6	1	21	72	170	2	2	1	4	1	0	4	3	2
0694	1	1	1	1	2.76	95	15000	1000	2	2	2	4	4	1	19	72	135	2	1	2	4	1	0	4	3	2
0695	1	3	1	3	3.60	85	12000	500	2	2	1	4	3	2	22	61	155	2	1	2	2	1	0	3	2	1
0696	1	2	2	2	2.70	85	13000	1500	2	2	2	4	5	1	21	70	160	2	1	1	3	2	1	4	6	1
0697	1	3	1	3	2.11	88	12000	3000	3	1	2	2	4	1	30	76	260	2	2	1	4	2	4	4	7	1
0698	2	4	2	2	2.90	78	11000	1000	3	2	1	4	1	1	24	61	105	2	2	1	3	1	1	5	4	2
0699	1	3	3	2	2.70	89	12000	326	2	2	1	4	5	1	23	73	195	2	1	1	4	1	0	4	4	2
0700	1	4	1	2	3.92	93	15000	500	1	2	2	4	14	2	26	78	225	2	4	1	4	1	0	3	4	2
0701	2	2	2	2	3.12	90	12000	1000	2	2	3	4	8	1	20	64	118	2	1	1	3	1	0	4	4	2
0702	2	2	2	2	3.35	90	13000	1000	2	2	5	4	3	1	20	68	153	2	2	2	4	2	0	4	4	2
0703	1	3	3	2	2.00	80	12000	1000	2	2	2	4	3	1	23	69	155	1	2	2	1	1	0	4	4	1
0704	1	3	2	2	2.68	86	12250	600	3	2	1	3	4	1	21	71	164	2	1	2	3	2	1	6	6	2
0705	2	2	2	1	3.00	83	12500	550	2	2	4	4	15	1	20	62	100	2	1	1	3	2	0	4	5	1
0706	2	4	2	2	3.40	93	12000	1000	3	2	2	4	5	1	29	64	110	2	1	1	4	1	0	4	4	2
0707	1	4	1	2	3.80	93	13000	800	2	2	2	2	3	1	20	70	140	2	1	2	3	1	2	4	5	1
0708	1	4	3	2	3.25	83	12500	1400	2	2	2	4	3	1	23	70	190	2	1	2	4	1	0	4	4	1
0709	1	3	1	2	2.75	82	12000	800	2	2	2	2	4	1	28	67	164	2	1	1	1	1	1	5	5	1
0710	1	3	3	2	3.50	90	13000	500	3	2	1	4	6	1	26	69	160	2	3	1	4	2	0	4	4	2
0711	2	4	1	2	3.70	86	10000	800	3	2	5	2	2	1	24	62	120	2	1	2	4	1	2	2	6	2
0712	2	2	2	2	3.20	89	15000	0	3	2	2	2	10	1	20	63	115	2	2	1	3	2	0	4	6	1
0713	2	4	3	2	2.50	80	10000	900	3	2	5	4	2	1	20	67	110	2	1	1	3	2	1	1	4	2
0714	2	2	3	2	2.79	87	10000	0	2	2	2	4	8	2	22	65	130	2	1	2	4	1	0	4	4	1
0715	1	3	1	2	3.15	87	11000	1500	2	2	1	3	7	1	21	66	137	2	1	2	4	1	2	4	5	1
0716	2	4	1	2	3.00	97	13000	0	3	2	2	4	0	1	25	67	150	2	1	2	1	2	0	3	3	2
0717	2	3	3	2	3.20	94	10000	600	3	2	2	4	7	1	23	65	104	2	1	2	3	2	1	4	1	1
0718	2	3	3	2	2.73	81	10000	300	2	2	6	2	2	1	24	59	80	2	1	2	4	2	0	3	6	1
0719	1	3	2	5	2.00	88	8000	600	2	1	5	4	5	1	19	69	145	2	2	1	4	2	0	1	2	1
0720	1	3	3	2	3.20	75	10000	1000	2	2	2	4	2	1	22	73	190	1	1	2	4	1	2	7	7	1
0721	2	3	1	7	3.70	80	10000	1000	2	2	3	4	8	1	20	66	125	2	1	2	2	2	2	4	4	2
0722	2	1	1	1	2.64	85	20000	2000	3	2	2	4	3	1	26	65	120	2	1	2	3	1	0	5	6	2
0723	2	1	1	3	1.85	82	19000	0	3	2	3	4	1	1	41	65	130	2	1	2	1	1	0	4	7	2
0724	1	3	1	5	3.00	83	12000	700	2	2	2	2	3	2	27	67	155	2	1	2	3	2	0	3	3	1
0725	1	3	3	2	2.66	71	11500	1200	2	2	2	4	2	1	23	71	174	2	1	2	3	2	0	5	5	2
0726	1	3	1	2	3.45	82	10000	0	2	2	1	3	6	1	21	69	180	2	1	2	4	1	1	1	6	1
0727	1	3	3	2	2.00	89	12000	1500	2	2	2	4	6	1	20	68	160	1	1	1	4	1	1	4	5	2
0728	1	3	3	4	2.20	83	11000	1500	2	2	1	1	5	2	20	69	180	2	1	2	4	1	0	4	5	1
0729	1	3	2	2	3.00	80	10000	1000	2	2	2	2	5	2	25	64	138	2	3	2	3	2	1	4	4	1
0730	1	2	2	2	2.70	82	15000	1000	3	2	2	4	10	1	19	67	160	2	1	2	2	1	0	4	5	1
0731	2	3	3	6	3.00	86	11000	1000	3	2	3	4	7	2	20	64	105	2	2	2	4	1	3	2	5	2
0732	1	3	2	2	3.75	90	10500	2000	2	1	2	4	4	1	21	72	155	2	1	2	3	2	0	4	4	1
0733	2	3	3	1	2.50	75	11000	1000	2	2	2	2	5	1	20	66	105	2	2	1	3	2	0	4	6	1
0734	2	3	3	2	2.31	83	12000	1000	3	2	2	2	2	1	27	68	110	2	1	2	1	1	1	4	4	1
0735	1	3	3	2	2.90	85	11550	1001	2	2	2	4	1	1	21	67	150	2	2	2	4	2	2	1	4	2
0736	1	3	3	2	3.00	88	12000	950	2	2	2	4	20	1	21	70	135	2	2	2	4	2	0	2	4	2
0737	2	3	2	4	3.20	93	11000	1000	3	2	2	2	15	1	20	59	91	2	3	1	4	2	0	4	3	1
0738	1	3	1	3	3.20	84	9600	2000	3	2	2	4	4	1	21	68	155	2	1	2	4	2	2	4	1	1
0739	1	2	2	2	2.90	88	8000	600	2	2	2	2	3	1	18	70	170	2	2	2	3	2	0	3	6	2
0740	1	3	3	4	3.00	78	14000	1000	3	2	2	2	10	1	24	71	155	2	2	1	4	1	0	4	5	1
0741	2	1	1	5	2.97	90	15700	2050	3	2	1	1	3	1	18	60	115	2	1	1	4	2	0	4	7	2
0742	2	3	3	2	3.40	94	16000	1500	3	2	5	4	6	1	20	66	125	2	1	2	3	1	0	4	4	1
0743	2	4	2	7	3.00	85	9500	1600	2	2	5	3	6	1	22	63	130	2	1	2	1	2	0	4	6	1
0744	1	3	3	1	3.00	85	14500	1000	2	1	5	3	3	1	21	69	155	2	1	2	1	2	1	4	5	1
0745	1	3	3	2	2.30	80	11600	982	2	2	1	4	6	1	20	74	160	2	1	2	1	1	1	4	4	1
0746	2	3	2	6	3.20	95	20000	0	3	2	2	4	8	1	23	64	100	2	1	2	1	1	0	5	6	1
0747	1	2	1	5	2.75	78	10000	300	2	2	5	4	5	1	23	74	165	2	3	1	4	2	1	4	3	2
0748	2	3	3	2	3.00	89	11000	10	1	2	5	4	0	1	21	64	120	2	1	2	3	1	2	6	2	1
0749	1	3	2	2	2.30	87	11500	1500	2	1	5	4	4	1	22	73	178	2	1	1	4	2	1	4	6	1
0750	1	3	3	2	2.65	74	10000	400	2	2	2	2	3	1	24	71	145	2	1	1	1	2	1	4	4	1
0751	2	3	3	2	3.10	95	12000	50	2	2	2	2	2	2	22	62	125	2	1	2	1	2	1	4	2	2
0752	1	2	2	2	2.30	80	14000	300	1	2	2	4	1	1	20	68	155	2	1	2	3	2	0	4	4	2
0753	2	4	1	2	2.90	95	13000	100	2	2	2	1	2	1	23	60	125	2	1	2	1	2	4	4	1	2
0754	1	3	2	2	3.30	90	12000	850	3	2	2	4	5	1	22	75	195	2	1	1	4	1	2	4	5	1
0755	2	3	3	2	2.20	72	11000	0	1	2	2	2	1	1	20	63	92	1	1	2	3	2	0	2	3	1
0756	2	3	2	2	2.70	82	10000	500	3	2	6	5	5	1	21	65	120	2	1	2	4	1	2	4	4	2
0757	1	3	3	2	3.20	87	11500	1000	2	2	1	5	3	1	21	65	135	2	1	1	4	2	1	4	6	1
0758	2	3	2	2	3.75	88	20000	400	3	2	2	2	10	1	21	61	115	2	1	2	4	1	0	4	4	1
0759	1	3	3	2	3.30	88	12000	500	2	2	3	2	8	1	21	74	180	2	1	2	3	1	3	4	6	1
0760	2	3	3	2	3.91	95	11500	2000	2	2	5	4	10	1	21	70	128	2	1	2	3	1	1	3	5	2
0761	1	3	2	1	2.30	75	10000	1200	3	2	2	4	2	1	23	68	155	2	1	2	4	2	1	1	1	2
0762	2	3	3	2	2.50	73	12000	100	3	2	5	2	0	1	38	66	165	2	1	2	1	1	0	4	4	1
0763	1	4	1	4	3.64	75	20000	0	1	2	1	4	3	1	29	72	220	1	1	2	4	1	0	4	5	1
0764	1	3	1	4	3.00	81	20000	100	1	2	2	5	3	1	28	69	162	2	3	1	4	2	0	4	3	2
0765	2	4	3	4	3.20	91	15000	3	1	2	2	2	1	1	40	67	150	1	1	2	4	2	1	4	4	1
0766	2	2	3	4	3.00	81	28000	300	3	1	6	2	2	1	35	67	160	2	1	2	1	2	1	6	6	2
0767	2	2	1	6	2.08	75	13000	0	2	2	2	2	3	1	21	62	115	2	1	1	3	2	0	4	2	2
0768	2	3	1	5	3.60	70	12000	800	2	2	2	4	2	1	27	68	112	2	2	1	4	1	0	4	4	1
File code number	Sex	Class	Grad. school	Major	Grade-point index	H. S. average	Est. starting salary $	Tuition $	Employment	Sunday shop	Investment	Car pref.	No. jeans owned	Calculator	Age	Height	Weight	Blood pressure	Smoking status	Smoking belief	Entrance standards	Retention standards	No. clubs and groups	SPS rating	Library rating	Personal question

File code number	Question number																									
	1	2	3	4	5	6	7	8	9	10	11	12	13	14	15	16	17	18	19	20	21	22	23	24	25	26
0769	1	3	1	4	2.50	77	9000	1000	3	2	1	3	3	1	28	67	145	2	1	2	3	1	2	3	5	1
0770	1	4	2	4	2.80	78	9000	1000	1	2	2	4	4	1	23	74	150	2	1	2	3	1	0	1	6	1
0771	2	3	1	5	3.91	90	8000	1000	2	2	6	4	10	1	20	65	118	2	3	2	4	1	1	4	4	1
0772	1	3	1	4	2.40	79	10000	480	3	2	2	4	10	1	21	70	158	2	2	1	4	2	0	4	6	2
0773	1	3	2	5	3.66	88	10000	500	2	2	1	4	0	2	25	68	180	2	1	2	4	1	0	1	4	2
0774	2	3	2	4	3.43	88	12000	1300	1	1	2	4	4	1	22	60	110	2	2	1	4	1	0	4	1	1
0775	1	3	3	5	2.75	84	11000	2100	2	2	2	4	4	1	22	75	213	2	1	2	3	2	1	4	4	2
0776	1	3	3	7	2.50	83	15000	1200	2	2	2	4	13	1	20	71	180	2	1	2	3	1	1	4	4	2
0777	1	4	1	7	3.00	79	5200	500	2	2	5	4	5	1	23	69	170	2	3	2	1	2	3	5	3	1
0778	2	3	3	5	3.20	88	15000	487	1	1	2	4	7	1	25	61	101	2	3	1	4	1	4	3	6	2
0779	1	2	2	2	2.90	84	14000	800	1	1	2	2	2	1	20	69	150	2	1	2	3	2	0	4	4	1
0780	1	3	2	2	2.75	82	12500	250	1	2	3	3	3	1	22	72	150	2	2	1	4	2	0	1	4	1
0781	2	2	1	2	3.00	90	10000	300	2	2	2	2	12	1	20	66	111	2	1	2	3	1	0	4	6	1
0782	2	3	3	3	2.50	85	15000	0	2	2	2	2	10	1	23	62	110	2	2	1	1	1	0	4	3	1
0783	2	2	2	2	3.46	83	12000	1500	3	2	2	4	4	1	31	66	133	2	2	2	4	1	0	4	4	2
0784	2	3	1	5	3.11	85	20000	250	2	2	2	4	4	2	24	67	130	2	1	2	3	2	0	3	5	1
0785	1	3	3	2	2.55	72	11000	400	3	2	1	4	3	2	19	69	175	1	1	?	2	2	2	2	6	2
0786	2	2	2	2	2.05	90	15000	900	2	2	2	2	4	1	21	60	120	2	1	1	1	2	0	3	5	1
0787	1	3	1	2	3.23	84	15000	250	1	2	2	2	0	1	53	65	160	2	3	2	4	1	0	4	4	2
0788	1	2	3	3	3.01	87	14000	800	3	2	2	4	4	1	20	67	180	1	2	2	4	2	0	4	5	1
0789	1	4	2	3	2.81	83	10000	500	1	2	1	2	0	1	30	66	150	2	1	2	2	2	1	4	4	2
0790	1	3	1	2	3.00	90	12000	500	1	2	6	2	5	2	22	66	125	2	2	2	2	2	0	5	6	1
0791	2	4	2	2	3.00	83	12000	600	1	2	1	4	5	1	26	62	115	2	1	2	3	2	0	3	5	2
0792	2	3	3	2	3.44	84	15000	1000	3	2	2	3	4	1	24	61	110	2	3	2	4	2	0	6	5	2
0793	2	3	3	4	3.02	85	12500	250	2	2	1	3	3	2	21	64	105	2	1	2	3	2	0	4	4	1
0794	1	3	3	5	2.73	82	13000	500	2	2	2	4	10	2	31	76	230	2	4	2	3	1	0	4	4	2
0795	2	3	3	4	3.04	75	18000	0	1	2	2	4	1	2	32	66	135	2	2	1	4	2	0	4	4	2
0796	1	4	2	5	2.56	76	18200	50	2	2	2	4	15	2	24	70	160	1	1	2	4	2	0	1	4	1
0797	2	2	1	5	3.02	85	30000	1000	1	2	2	4	7	1	27	66	109	2	2	1	4	1	0	3	2	1
0798	2	3	2	5	2.05	75	13000	1000	1	2	2	2	2	2	26	64	125	2	1	2	3	2	0	3	4	2
0799	2	3	2	1	3.08	89	20000	600	1	2	2	4	8	1	25	66	160	2	1	2	1	2	1	4	4	1
0800	1	3	2	2	3.33	81	12000	1000	1	2	1	4	3	1	28	69	210	1	1	2	4	1	0	4	4	2
0801	2	3	3	2	3.50	82	17000	1500	1	2	3	3	6	1	32	64	120	1	1	2	4	1	0	1	7	2
0802	1	3	3	5	2.50	72	11500	0	1	2	2	4	4	1	32	68	183	2	1	2	1	1	0	4	4	2
0803	2	3	2	4	3.65	90	18000	1500	1	2	2	2	3	1	28	63	135	2	3	2	4	2	0	4	4	2
0804	1	3	2	1	2.25	72	12000	300	1	2	2	2	2	2	30	66	190	1	1	2	1	2	0	4	4	1
0805	2	4	1	7	2.85	89	25000	600	1	2	1	3	2	1	24	64	110	2	3	2	4	2	0	3	4	1
0806	2	4	3	5	2.52	81	15000	1000	1	1	2	2	0	1	38	68	150	1	1	2	1	2	0	1	7	2
0807	1	3	1	5	2.60	78	15000	1800	1	2	5	4	3	1	31	65	150	2	3	1	3	2	0	2	4	1
0808	1	4	1	2	2.27	73	14000	0	1	2	3	4	5	1	28	70	145	2	1	2	4	1	0	4	7	2
0809	1	4	1	2	3.66	80	12480	390	1	1	3	4	8	1	25	68	145	2	1	1	4	2	0	1	4	2
0810	1	3	1	3	3.00	81	11000	800	1	2	2	4	5	1	27	70	160	2	1	2	3	2	0	4	5	2
0811	2	3	3	6	3.00	85	8500	700	1	2	2	2	5	1	23	63	165	2	1	2	3	1	2	6	6	2
0812	1	3	3	2	3.41	83	10100	800	1	2	2	1	0	1	38	68	195	2	1	2	3	1	0	4	4	2
0813	1	4	3	2	3.40	86	13000	0	1	2	2	2	3	1	30	69	172	2	1	2	1	1	2	4	4	2
0814	1	3	3	2	3.50	85	11000	500	1	2	1	4	3	1	25	70	150	2	1	2	4	1	0	4	4	2
0815	1	3	1	2	3.81	95	14000	1500	2	1	1	3	2	1	28	71	150	2	2	1	3	2	0	4	6	2
0816	1	4	1	2	3.00	90	12000	2000	3	2	2	4	0	1	22	70	150	2	1	2	1	1	0	4	4	2
0817	1	4	1	2	3.50	90	13000	3000	2	2	2	4	10	1	21	69	148	2	1	2	4	1	2	7	3	1
0818	1	4	1	2	3.04	93	12000	0	1	2	2	2	1	2	38	66	175	2	1	2	4	2	0	4	4	2
0819	1	4	3	2	3.42	80	17000	1500	1	2	2	2	0	1	24	68	180	2	1	2	4	1	0	4	4	2
0820	1	4	3	2	2.53	78	15000	500	1	2	2	2	0	1	42	66	150	2	1	2	4	2	4	4	4	2
0821	2	2	1	2	2.51	82	15000	800	1	2	2	2	3	1	30	64	118	2	1	2	1	2	0	3	6	2
0822	2	3	1	2	3.02	94	13000	500	1	1	2	2	5	1	26	62	102	2	1	2	4	1	0	1	1	1
0823	2	2	1	1	2.00	79	15000	500	2	2	1	4	6	1	27	63	133	2	1	2	3	2	2	1	4	1
0824	1	2	2	2	3.00	94	14500	350	3	2	2	2	4	1	19	65	116	2	1	2	3	1	1	4	5	2
0825	1	4	1	2	3.02	87	15000	0	3	2	2	2	1	1	20	66	135	2	1	2	4	2	2	2	4	1
0826	1	4	1	2	2.93	81	15000	300	1	1	5	4	4	1	22	69	150	2	3	1	4	1	0	3	4	1
0827	1	3	3	2	3.81	80	15000	100	1	2	2	2	4	1	23	67	165	2	1	2	3	2	1	4	5	1
0828	1	3	3	1	2.54	78	10250	1600	1	2	2	2	3	1	23	71	220	1	1	1	1	2	0	4	5	1
0829	1	4	1	2	3.17	92	15000	100	2	2	6	2	1	1	27	72	180	2	1	2	1	1	3	4	4	2
0830	1	4	1	2	2.70	83	12000	487	1	2	2	2	2	1	31	70	165	2	1	2	3	2	3	4	6	1
0831	1	4	1	2	2.22	79	20000	0	1	2	2	4	4	1	30	66	150	2	1	2	1	1	0	3	4	2
0832	1	4	2	2	3.12	78	10000	600	1	1	2	4	4	1	30	66	150	2	2	2	1	2	0	1	1	2
0833	1	4	2	2	2.24	84	18000	1000	1	2	1	4	6	1	27	69	180	2	2	1	3	2	0	4	4	1
0834	2	4	3	2	2.79	77	12500	0	1	2	2	4	8	1	29	65	132	2	1	2	2	2	0	2	3	1
0835	2	4	1	2	3.85	96	12000	0	2	2	2	2	3	1	24	63	135	2	2	1	4	1	0	4	4	2
0836	1	3	1	2	3.60	81	11600	0	3	2	2	2	2	1	22	72	160	2	2	1	3	1	1	4	5	2
0837	1	4	3	2	2.11	78	12500	1200	1	2	2	2	0	2	26	69	210	2	3	1	4	2	0	7	4	2
0838	2	2	3	4	3.20	85	14500	0	1	2	3	4	2	1	41	68	131	2	4	1	3	1	0	4	4	2
0839	1	2	1	3	2.78	76	15000	0	1	2	2	2	4	1	22	72	160	2	1	2	1	1	0	4	4	1
0840	2	2	1	2	3.54	90	12000	800	3	2	1	2	4	1	19	71	135	2	1	2	4	1	1	4	7	2
0841	1	2	3	2	2.20	80	13000	300	2	2	1	3	6	1	22	71	165	2	1	1	4	1	0	2	6	2
0842	1	4	3	4	3.90	90	14000	1500	1	2	2	4	2	1	48	66	165	2	1	2	4	2	1	4	2	2
0843	1	3	1	5	3.14	85	12000	2000	1	1	2	4	4	1	29	73	165	2	1	2	4	1	0	4	4	1
0844	2	2	3	2	2.05	78	10000	0	3	2	6	2	8	2	21	61	128	1	1	2	3	2	0	1	4	1
0845	2	1	3	2	3.22	79	16000	1000	1	2	2	2	2	1	28	65	125	2	1	2	4	1	0	4	4	2
0846	1	3	1	6	2.03	79	12000	0	2	2	2	2	6	1	21	69	140	2	1	2	4	1	2	4	4	1
0847	2	2	3	2	2.00	84	12000	1000	1	2	2	4	10	1	19	65	120	2	3	1	4	2	0	4	4	2
0848	2	3	1	2	3.06	89	15000	1000	1	2	1	4	2	1	20	65	150	1	1	2	3	1	0	4	5	1
0849	1	2	1	2	2.52	80	17000	1500	1	2	2	2	5	1	20	73	175	2	1	1	4	1	0	4	5	2
0850	1	3	3	2	3.24	87	11000	1000	3	1	3	4	0	1	19	69	165	2	1	1	4	2	2	4	6	1
0851	1	4	1	3	3.03	82	12000	1000	3	2	2	4	1	2	38	70	170	1	1	2	4	2	0	4	6	2
0852	1	2	3	4	3.66	73	12000	0	1	2	5	3	1	1	22	74	190	2	1	1	4	1	0	1	4	1
0853	1	3	3	2	2.50	74	14000	1200	1	2	5	4	12	2	21	71	165	2	3	1	4	1	0	4	4	2
0854	2	4	1	7	3.00	83	11000	0	3	2	2	4	6	1	22	68	105	2	2	1	4	1	0	2	4	2
0855	1	3	3	4	3.03	78	15000	1200	1	2	2	4	4	1	22	73	215	2	1	2	4	1	0	6	4	1
0856	1	3	1	4	3.75	94	15000	500	1	2	2	4	4	1	31	64	135	2	1	2	4	2	0	1	1	2
0857	2	3	1	6	2.51	75	17000	412	1	2	2	2	12	1	20	64	140	2	4	2	4	1	0	3	4	1
0858	1	3	3	2	3.17	96	12000	1200	1	2	2	2	1	1	35	70	150	2	3	2	3	1	0	4	6	1
0859	2	3	3	2	3.00	93	14000	1200	1	2	2	4	8	1	29	62	117	2	1	2	4	1	2	4	6	2
0860	1	4	2	5	2.31	72	14000	100	2	1	2	4	2	2	31	69	140	2	1	1	1	2	0	1	7	1
0861	1	3	3	4	3.82	75	18000	800	2	1	2	3	1	1	20	69	180	2	1	2	3	1	0	4	4	2
0862	1	2	3	3	2.03	80	18000	0	2	2	2	4	15	2	20	70	175	2	1	2	3	1	0	4	5	2
0863	1	2	1	3	3.01	85	12000	700	3	1	3	3	6	2	19	68	139	2	1	1	3	1	0	4	3	1
0864	1	3	3	6	2.62	86	13000	200	1	2	3	3	10	1	20	73	175	2	3	2	4	2	0	4	5	1
File code number	Sex	Class	Grad. school	Major	Grade-point index	H. S. average	Est. starting salary $	Tuition $	Employment	Sunday shop	Investment	Car pref.	No. jeans owned	Calculator	Age	Height	Weight	Blood pressure	Smoking status	Smoking belief	Entrance standards	Retention standards	No. clubs and groups	SPS rating	Library rating	Personal question

File code number	1 Sex	2 Class	3 Grad. school	4 Major	5 Grade-point index	6 H. S. average	7 Est. starting salary $	8 Tuition $	9 Employment	10 Sunday shop	11 Investment	12 Car pref.	13 No. jeans owned	14 Calculator	15 Age	16 Height	17 Weight	18 Blood pressure	19 Smoking status	20 Smoking belief	21 Entrance standards	22 Retention standards	23 No. clubs and groups	24 SPS rating	25 Library rating	26 Personal question
0865	2	3	2	4	2.24	90	11000	200	2	1	1	3	10	1	21	61	118	1	3	1	3	1	0	2	1	1
0866	2	3	1	5	3.02	85	14000	500	2	2	2	4	2	2	20	64	125	2	2	1	3	1	0	4	6	2
0867	2	3	3	5	3.07	92	10000	1000	2	2	2	4	2	1	19	62	115	2	1	2	4	2	0	4	6	2
0868	2	4	2	7	2.04	85	11000	800	2	2	2	4	10	1	21	64	110	2	2	1	4	1	0	3	2	1
0869	1	3	1	4	3.28	92	12000	2500	3	2	2	4	7	1	19	74	170	2	2	2	4	1	0	4	7	1
0870	1	3	1	2	2.51	84	13000	0	2	2	2	2	2	1	23	66	140	2	1	1	1	2	0	4	4	1
0871	2	2	3	3	3.07	92	15000	1000	3	2	5	4	6	1	19	65	123	2	2	2	4	2	0	2	3	1
0872	1	2	2	5	3.21	90	12000	500	3	2	1	5	6	1	19	69	170	2	2	1	4	1	1	3	5	1
0873	1	2	3	1	2.61	84	20000	250	2	2	1	4	8	1	19	74	170	2	1	2	4	2	0	1	4	2
0874	2	2	2	3	2.71	82	10000	800	3	2	1	3	8	2	20	64	120	2	3	1	1	2	1	4	3	2
0875	2	3	3	5	2.52	84	17000	300	2	1	2	4	3	2	19	66	110	2	2	1	3	1	1	4	4	1
0876	2	2	3	2	3.00	90	10000	0	3	2	2	2	7	1	20	60	120	2	1	2	1	1	2	4	4	1
0877	1	2	2	5	2.00	84	12000	600	3	2	2	2	3	2	22	76	195	2	2	1	4	1	0	4	4	2
0878	1	3	2	2	3.75	93	10000	800	3	2	3	5	5	1	19	70	160	2	1	2	2	2	0	3	3	2
0879	2	2	3	1	2.88	85	10000	500	2	2	2	2	6	2	20	61	125	2	1	2	1	2	0	4	5	1
0880	1	2	2	2	2.19	87	10500	1000	3	2	2	2	5	1	21	63	125	2	1	2	3	1	1	3	6	1
0881	1	2	3	2	3.00	89	13000	1000	2	1	5	4	5	1	20	71	160	1	1	2	1	1	0	1	5	1
0882	2	3	3	5	2.89	88	12000	975	3	2	2	4	2	2	23	67	132	2	1	1	1	1	1	4	4	2
0883	1	3	1	2	3.00	83	14000	400	3	2	2	2	0	1	20	69	149	2	1	2	4	1	2	4	6	2
0884	2	3	3	4	2.94	80	15000	0	2	1	2	2	4	1	24	62	121	2	1	1	1	1	1	5	5	1
0885	1	3	2	4	3.02	89	9000	600	3	2	2	2	2	2	27	72	185	2	1	2	3	2	0	4	4	1
0886	2	3	2	7	2.04	80	10000	1000	2	2	3	4	7	2	21	62	102	2	1	1	3	2	1	4	4	1
0887	2	3	3	7	3.92	97	10000	0	3	2	6	2	1	1	48	63	140	2	1	2	4	1	0	4	7	1
0888	2	4	3	2	3.71	90	12000	1000	3	2	2	4	10	1	23	68	125	2	1	2	4	1	0	1	4	2
0889	2	2	1	1	3.39	85	17000	1000	2	1	2	4	3	1	20	69	118	2	1	2	3	2	0	4	4	2
0890	2	4	1	5	3.04	85	9360	600	2	2	2	4	8	1	21	62	103	2	1	2	4	2	0	4	4	2
0891	2	2	2	2	3.02	86	20000	80	2	2	2	4	20	1	19	62	112	2	3	1	4	1	0	4	4	1
0892	2	2	1	4	2.84	84	12500	450	3	2	2	2	7	1	18	67	144	2	2	1	1	2	3	4	5	1
0893	1	4	3	5	2.58	75	12000	800	2	2	2	4	3	1	23	71	165	2	1	2	3	2	0	5	5	2
0894	1	1	2	6	2.72	78	6240	388	3	2	2	4	3	1	18	67	140	2	1	2	3	2	2	4	2	2
0895	2	4	1	4	2.28	87	12000	1000	3	2	2	4	3	2	26	65	120	2	1	2	3	2	0	4	5	2
0896	2	3	3	7	2.43	87	10000	800	2	2	2	5	10	2	22	65	130	2	1	2	4	1	1	4	4	2
0897	1	3	1	5	2.51	83	16000	200	3	2	2	1	5	1	21	75	175	2	1	2	1	2	5	2	6	1
0898	1	2	3	2	2.04	81	17000	1000	2	1	3	3	4	1	18	73	158	2	1	2	3	2	0	2	6	1
0899	1	2	2	2	2.10	77	12000	1000	3	2	2	4	6	1	18	64	145	2	1	1	3	2	0	4	1	1
0900	1	2	3	4	2.83	88	13000	500	2	2	2	2	5	2	21	64	140	2	1	1	3	1	2	6	4	1
0901	1	3	2	5	3.32	90	10000	1000	2	2	5	4	6	2	20	72	183	1	1	2	3	2	0	4	4	1
0902	1	4	3	4	3.06	81	14000	2000	1	2	2	2	2	1	28	68	150	2	2	2	1	1	1	4	5	1
0903	1	3	1	5	2.71	74	13500	0	2	2	3	4	5	1	29	67	190	2	3	1	4	1	0	4	6	2
0904	1	4	3	5	2.95	70	10000	0	2	1	1	1	10	1	23	66	155	2	1	2	3	1	0	4	4	2
0905	1	4	1	7	2.09	79	16000	0	3	2	2	2	4	1	37	69	130	1	1	2	3	1	2	4	4	2
0906	1	3	2	6	3.00	81	9000	500	1	2	1	4	6	1	22	65	125	2	1	2	1	2	0	2	2	2
0907	2	2	1	2	2.75	80	20000	387	1	2	2	4	0	1	21	63	110	2	1	2	4	1	0	2	4	2
0908	2	2	1	5	3.89	90	25000	3000	2	1	1	4	7	1	22	64	105	2	2	2	3	2	1	5	6	1
0909	1	3	3	3	2.50	81	14000	500	1	2	2	2	6	2	28	70	175	2	2	1	1	2	0	4	6	1
0910	2	3	2	5	2.92	83	10000	1000	1	2	2	4	4	1	19	64	117	2	1	2	3	2	0	2	6	1
0911	1	4	3	7	3.23	87	11000	0	1	2	2	4	3	2	29	69	180	2	1	2	1	2	0	4	4	2
0912	1	2	2	2	3.21	75	16000	0	2	2	2	2	10	1	22	66	120	2	1	2	3	1	0	3	5	1
0913	2	3	2	5	3.00	90	19000	500	1	2	2	4	10	2	27	67	125	2	3	1	4	1	0	4	4	1
0914	1	4	3	3	3.10	75	11750	1000	1	2	2	2	4	1	24	72	165	2	2	1	3	1	1	1	5	1
0915	2	4	1	4	3.90	98	15000	1000	1	1	2	1	5	2	22	66	120	2	3	2	3	2	0	2	4	1
0916	1	3	1	2	3.00	88	15000	200	1	2	5	4	7	1	19	71	165	2	1	2	4	1	4	2	5	2
0917	2	3	1	2	3.12	87	13000	0	3	2	2	4	1	1	22	68	116	2	1	2	3	2	0	4	4	2
0918	1	2	2	2	3.04	94	15750	975	3	2	2	2	4	1	19	65	117	2	1	2	4	1	2	4	5	2
0919	2	3	1	4	3.52	80	13000	900	1	2	2	5	2	1	30	62	108	2	1	2	3	2	0	4	4	1
0920	1	2	1	3	3.81	97	20000	0	3	2	6	2	0	2	19	71	178	2	1	2	4	1	1	1	6	1
0921	2	3	3	4	3.21	87	12000	0	3	2	2	4	2	1	33	64	125	2	1	2	1	2	0	4	4	2
0922	2	3	3	3	3.06	79	12500	800	3	2	5	1	2	2	25	62	114	2	1	2	4	2	0	5	3	2
0923	1	4	1	3	3.02	80	12000	1000	2	2	2	2	4	1	20	63	123	2	2	2	3	2	0	6	3	2
0924	1	4	1	4	3.03	83	18000	1000	1	2	2	4	6	1	29	73	200	2	1	2	4	1	0	4	3	1
0925	1	4	3	7	3.42	81	10000	1000	3	2	2	2	1	1	28	66	115	2	1	2	1	1	1	4	6	2
0926	2	2	3	5	3.56	87	15000	1500	1	2	1	3	2	1	22	65	130	1	3	2	4	1	0	3	1	1
0927	1	2	1	2	3.80	76	15600	0	1	2	2	2	0	2	22	68	125	2	2	2	4	1	0	4	4	1
0928	2	3	3	2	3.16	85	15000	1000	3	2	2	4	0	1	20	64	135	1	1	2	1	2	0	4	4	1
0929	2	3	3	2	2.92	95	10000	1200	3	2	2	2	5	1	21	63	170	2	1	2	2	2	1	3	5	1
0930	2	4	3	4	3.75	86	13500	1200	3	2	1	3	3	1	25	65	103	2	1	2	4	2	0	4	4	2
0931	1	2	1	2	2.70	86	10000	850	2	2	2	4	7	1	19	70	180	2	1	2	2	2	0	4	4	1
0932	2	2	3	2	2.90	86	12500	3000	2	2	2	2	4	1	19	60	103	2	1	2	3	2	3	4	6	1
0933	2	2	3	2	2.99	89	10000	0	2	2	2	2	6	1	19	66	118	2	1	2	3	1	1	4	5	2
0934	2	4	3	5	3.03	79	11000	1000	2	2	2	4	5	1	21	65	140	2	1	2	3	2	1	4	7	2
0935	2	3	3	5	2.91	83	15000	2400	2	2	5	4	0	1	21	63	121	2	1	2	4	2	1	5	6	1
0936	2	3	1	5	3.24	92	9000	2000	2	2	5	4	15	1	20	64	121	2	3	2	3	2	0	4	4	1
0937	2	3	2	5	2.51	85	15000	1500	3	2	2	4	6	1	21	63	110	2	1	2	3	2	1	5	5	1
0938	2	4	1	5	2.50	96	10000	1000	2	2	2	4	5	1	21	61	120	2	1	2	3	1	0	3	5	1
0939	2	4	3	5	2.24	73	12500	0	3	2	2	2	3	1	26	66	120	2	1	2	1	2	2	5	5	1
0940	1	3	1	2	2.90	85	15000	600	1	2	2	2	3	1	46	70	158	2	3	1	4	1	0	4	3	1
0941	1	3	3	2	3.00	81	15000	800	1	2	2	2	5	1	47	69	138	2	3	1	3	1	0	4	6	1
0942	1	4	1	2	2.95	86	13500	720	3	2	2	2	0	1	26	72	200	2	2	1	1	2	0	1	4	1
0943	2	3	1	2	3.70	85	12000	800	1	2	2	2	6	1	36	69	145	1	3	2	4	2	0	3	4	1
0944	1	3	1	2	3.30	85	16000	1500	1	2	2	4	2	1	27	71	140	2	2	2	3	2	1	4	5	1
0945	1	2	2	2	3.24	85	13000	412	3	2	2	4	12	1	20	68	145	2	1	2	4	1	0	4	5	2
0946	1	3	1	2	2.80	83	14000	975	2	1	2	4	3	1	20	68	155	2	1	2	4	1	0	3	5	1
0947	1	4	1	2	2.61	88	14000	2000	1	2	2	2	6	1	30	65	145	2	1	2	4	2	4	1	7	2
0948	1	2	3	2	3.10	87	15000	500	2	2	2	2	12	1	20	73	180	2	1	1	4	1	0	4	4	2
0949	1	3	3	2	2.63	79	12000	1500	3	2	1	3	8	1	21	70	160	2	1	2	4	2	1	5	4	2
0950	2	3	3	2	2.52	82	12000	1500	3	2	2	1	6	1	21	63	140	2	2	1	4	2	0	4	3	2
0951	2	3	3	5	3.04	90	11000	1000	2	1	2	4	4	1	21	64	130	2	3	1	4	1	0	5	3	1
0952	1	2	1	2	2.68	86	15000	1000	2	2	1	4	6	1	19	65	130	2	1	2	1	1	0	4	5	1
0953	1	2	1	2	3.20	80	13000	1500	1	2	2	2	0	1	19	74	160	2	1	2	4	2	0	4	4	2
0954	2	3	3	2	2.50	90	13000	1500	1	2	2	2	0	1	33	62	135	2	1	2	1	1	0	4	4	1
0955	1	4	1	3	2.80	78	12000	0	3	1	2	2	0	2	28	70	175	2	1	2	1	2	4	2	4	1
0956	1	3	3	2	3.00	87	20000	1500	2	2	5	4	2	1	20	64	135	2	1	2	3	1	0	4	6	1
0957	2	2	3	2	2.75	77	12000	500	3	2	2	2	2	1	22	62	100	2	1	2	4	2	1	4	2	2
0958	1	4	3	2	3.00	80	10000	1200	1	2	2	1	2	1	26	71	179	2	1	2	4	1	0	4	5	2
0959	2	2	3	1	2.51	80	15000	2000	2	2	2	4	3	1	22	70	145	2	1	2	3	2	0	4	4	1
0960	1	4	1	2	3.22	97	12500	600	1	2	1	3	9	1	20	67	145	2	1	1	4	2	3	4	5	2

File code number	Question number																									
	1	2	3	4	5	6	7	8	9	10	11	12	13	14	15	16	17	18	19	20	21	22	23	24	25	26
0961	1	3	3	2	2.07	91	10000	1500	2	2	2	4	5	1	21	68	150	1	3	1	4	1	1	2	3	1
0962	2	3	3	3	3.02	94	12000	1500	2	2	2	4	6	1	20	65	125	2	2	2	4	1	0	4	4	1
0963	1	4	3	3	3.01	73	10000	250	2	2	2	2	1	1	26	66	134	2	1	2	4	1	0	4	1	2
0964	1	3	1	5	3.62	87	15000	500	3	2	1	3	1	1	29	69	155	2	2	2	3	1	0	4	2	2
0965	1	3	3	2	3.13	82	15000	500	2	2	1	3	6	1	20	71	170	2	1	2	2	2	0	4	5	1
0966	2	2	3	7	3.06	85	10000	900	3	2	2	2	3	2	19	61	105	2	1	1	3	1	0	4	3	2
0967	1	1	1	5	2.53	84	8500	1000	2	2	2	4	5	1	18	70	160	2	1	2	3	1	1	6	6	2
0968	2	2	1	4	3.92	91	10000	1000	3	2	5	3	6	1	18	64	120	2	1	2	3	1	4	7	7	2
0969	1	3	2	2	2.62	83	12000	1000	2	1	2	2	2	2	23	64	175	2	1	2	3	2	1	2	2	2
0970	2	2	3	3	3.11	78	14500	1500	3	2	3	3	2	1	20	61	109	2	1	1	4	1	0	3	4	2
0971	2	4	1	5	2.79	90	10000	1000	3	2	2	4	4	2	25	60	113	2	1	2	3	1	1	4	4	1
0972	2	4	3	3	3.12	78	12000	800	2	2	2	2	10	1	22	61	95	2	1	2	3	2	2	4	4	2
0973	2	4	1	5	3.10	86	10000	1500	3	2	5	4	6	1	20	61	105	2	1	2	4	1	1	4	4	2
0974	2	3	2	7	3.92	84	11000	0	2	2	3	4	1	2	19	59	95	2	1	2	3	2	2	4	6	2
0975	2	3	3	4	3.40	96	10000	400	2	2	3	4	9	1	21	67	140	2	2	2	4	1	2	1	7	2
0976	1	4	1	5	2.75	83	9500	2000	3	2	2	4	8	1	22	75	220	2	1	2	3	2	0	4	1	2
0977	1	3	1	3	3.00	87	12000	100	3	2	2	2	3	1	24	66	153	2	1	2	4	2	0	4	6	1
0978	1	3	1	7	2.85	77	12000	0	3	2	2	4	4	2	22	69	160	2	1	2	4	1	5	2	4	1
0979	1	2	1	3	2.74	85	12000	1000	2	2	1	4	6	2	19	75	160	2	1	2	4	2	0	4	5	2
0980	1	2	1	4	3.57	87	12000	2000	2	1	2	4	5	1	18	63	140	1	1	2	4	1	0	4	4	2
0981	2	2	3	6	2.59	86	10000	600	2	2	2	2	5	2	20	60	87	2	1	2	3	1	0	4	4	1
0982	2	3	1	6	2.80	90	13000	1000	3	2	2	2	3	1	23	64	140	2	1	2	1	2	0	4	5	2
0983	2	3	2	2	3.62	92	11000	1500	2	2	2	3	10	1	19	60	115	2	1	2	3	2	1	4	6	1
0984	1	3	1	2	3.03	93	10000	1200	3	2	2	4	1	1	21	70	130	2	1	2	3	2	1	5	6	1
0985	1	3	3	2	2.80	80	12000	900	2	2	5	2	2	1	21	68	145	2	3	2	3	2	0	4	4	2
0986	2	3	3	2	3.06	89	10000	1000	3	2	3	4	9	2	20	64	105	2	1	2	3	2	0	4	5	1
0987	1	2	2	2	3.10	87	11000	1000	2	2	1	4	6	1	19	75	170	2	1	2	2	2	0	2	4	1
0988	2	4	2	2	2.50	67	10000	0	3	2	2	5	3	1	24	62	133	2	1	2	3	2	3	1	5	1
0989	1	2	3	3	2.85	78	10000	1000	1	2	2	2	5	1	28	68	165	2	1	2	3	1	0	4	4	2
0990	1	3	3	2	3.19	82	10000	0	3	2	1	4	3	2	28	65	135	2	3	2	3	2	0	4	6	1
0991	1	3	2	2	3.07	80	12000	500	2	2	2	3	8	1	20	74	160	2	1	2	4	2	0	4	4	1
0992	1	3	2	2	2.30	81	10000	1200	3	2	2	4	2	1	22	74	155	2	1	1	3	2	0	4	7	1
0993	1	2	3	2	3.06	88	13500	1500	2	2	2	4	3	2	19	71	130	2	1	2	4	2	0	4	2	1
0994	1	4	3	2	3.00	82	13000	0	3	2	2	4	5	1	20	71	150	2	1	2	4	1	1	4	3	2
0995	1	2	3	2	3.41	84	17500	2000	2	2	2	2	8	1	19	70	150	2	1	2	3	2	1	3	5	1
0996	1	3	1	2	2.50	75	12000	2000	3	1	2	4	3	1	21	72	160	2	1	2	4	2	1	4	4	1
0997	1	2	3	2	3.00	85	15000	500	2	2	2	2	2	1	19	68	150	2	1	2	3	2	1	4	4	1
0998	1	3	3	2	2.23	69	12000	2000	3	2	2	3	6	1	28	68	150	2	1	2	1	2	0	4	6	2
0999	2	2	1	2	2.52	91	15000	1000	3	2	2	2	4	1	19	63	110	2	1	2	4	1	1	4	4	2
1000	1	3	3	2	3.87	75	11000	1000	2	2	2	2	4	1	25	66	186	1	1	2	2	2	1	5	6	1
1001	1	3	1	6	2.70	85	15000	600	3	1	2	4	4	1	21	69	145	2	2	1	4	2	0	2	6	1
1002	2	4	3	6	3.00	93	10000	500	3	2	2	2	12	1	21	63	120	2	1	1	2	1	0	4	5	2
1003	2	4	1	6	3.00	78	11000	1000	2	2	2	4	1	1	26	68	120	2	1	2	3	2	0	4	4	1
1004	2	4	3	7	2.70	80	9000	1	2	2	2	2	3	2	22	63	101	2	1	2	1	2	0	2	6	1
1005	1	4	1	4	2.90	80	10000	80	3	2	6	2	0	1	26	62	135	2	1	2	4	1	0	4	6	1
1006	2	3	1	7	2.50	88	12000	1000	2	2	2	4	4	1	21	64	135	1	1	2	3	1	0	1	1	2
1007	2	3	2	6	2.80	83	10000	400	2	2	1	4	4	2	21	61	125	2	2	1	4	1	0	4	5	2
1008	1	3	3	6	3.18	90	12000	500	2	1	1	4	3	1	21	71	127	2	1	2	3	2	3	6	5	1
1009	2	4	1	5	3.00	87	11000	1200	3	2	1	4	10	1	22	64	140	2	1	2	4	1	3	3	5	1
1010	1	4	3	2	3.00	80	10000	1000	2	2	2	4	0	1	22	75	165	2	2	2	3	2	1	1	3	2
1011	1	4	1	2	2.80	91	10000	1000	2	2	1	3	20	1	21	72	165	2	1	1	4	1	1	4	3	1
1012	2	4	3	6	3.58	94	10000	900	2	2	2	4	0	2	22	63	120	2	1	2	4	2	2	4	6	1
1013	1	4	1	7	3.28	87	10500	1000	2	2	2	4	7	1	21	65	135	2	1	2	3	2	0	4	4	1
1014	1	2	1	6	3.36	80	15000	2000	2	1	5	2	2	1	20	72	158	2	1	2	4	1	0	1	6	2
1015	1	4	1	6	3.00	81	18000	2000	1	2	3	4	3	2	22	69	160	2	1	2	4	2	0	4	3	1
1016	1	2	1	6	3.25	79	13000	500	3	2	2	4	5	1	21	68	140	2	1	2	4	1	1	4	4	2
1017	2	2	1	6	3.50	92	11000	700	2	1	6	5	3	1	20	67	122	2	1	1	3	2	0	4	6	2
1018	1	1	1	1	2.38	89	14000	500	2	2	2	4	5	1	18	71	143	2	1	2	3	1	0	3	6	2
1019	2	2	1	7	2.50	75	10000	1000	3	2	2	2	2	1	20	63	125	2	1	2	4	1	0	4	6	1
1020	2	3	3	7	3.00	85	11000	1000	2	2	2	4	8	1	21	64	125	2	1	2	3	2	0	2	4	1
1021	2	3	1	7	3.50	75	12000	1000	3	1	2	4	14	1	22	67	129	2	1	2	4	2	0	2	5	1
1022	1	3	1	7	2.90	90	10000	600	1	2	2	2	6	1	28	77	175	2	2	1	4	2	0	1	6	1
1023	2	3	1	7	3.50	80	15000	1000	3	2	2	4	6	1	27	62	108	2	1	2	4	1	0	4	3	1
1024	2	3	1	7	3.50	85	12500	900	3	2	3	3	4	1	24	65	130	2	1	1	4	2	1	2	6	2
1025	1	4	1	7	3.00	85	12000	500	1	1	2	4	15	1	21	66	165	2	1	2	4	1	5	4	4	1
1026	2	4	3	7	2.80	85	8000	800	3	2	2	4	5	2	21	63	117	2	1	2	4	1	0	4	3	2
1027	1	4	1	7	3.30	76	15000	900	2	2	1	4	5	1	22	68	157	2	1	2	2	2	1	1	1	2
1028	1	4	1	2	3.00	80	11000	600	2	2	6	2	3	1	26	67	130	2	1	2	3	2	2	5	5	1
1029	2	2	2	2	2.80	82	10000	3000	3	2	2	2	0	1	22	59	100	2	1	2	4	2	2	4	3	1
1030	1	3	3	2	3.40	88	12000	750	3	2	2	4	4	1	22	74	175	2	1	2	4	1	1	4	3	2
1031	1	3	3	2	2.80	80	12000	900	2	2	1	4	5	1	21	72	175	2	2	2	4	1	0	3	5	2
1032	1	3	3	2	3.10	86	13800	487	2	2	2	3	12	1	21	71	160	2	1	2	4	2	3	1	4	1
1033	2	3	2	2	3.10	86	12000	1000	3	2	2	2	7	1	20	65	120	2	1	2	3	2	0	4	5	1
1034	1	4	1	2	3.69	85	13500	1000	2	2	2	4	6	1	22	73	155	2	3	1	1	1	1	4	4	1
1035	2	4	2	2	2.80	85	10000	1500	2	1	1	3	12	1	22	68	148	2	1	2	4	2	1	4	4	1
1036	1	4	3	2	3.71	86	12000	1000	2	2	2	4	2	1	23	68	140	2	1	2	4	2	2	4	4	1
1037	2	4	1	2	3.00	85	13000	2000	2	2	3	2	2	1	21	60	115	2	2	1	3	2	1	4	4	2
1038	2	3	3	2	3.10	92	12000	900	2	2	2	4	4	1	22	60	110	2	2	1	4	2	1	1	1	1
1039	1	4	3	2	3.30	85	10000	0	2	2	2	4	3	1	22	70	150	2	2	1	4	1	1	1	5	1
1040	1	3	1	2	3.70	86	14000	750	2	2	2	4	0	1	22	69	175	2	1	2	4	1	0	1	4	2
1041	1	4	2	2	3.00	75	12000	300	3	2	2	2	2	1	23	71	155	2	1	1	3	2	0	4	4	1
1042	1	1	1	5	3.25	72	16000	2000	2	2	5	4	1	2	18	69	150	2	3	1	1	1	1	4	5	1
1043	1	3	3	2	3.57	95	9000	500	2	1	2	4	4	1	22	72	160	2	1	2	4	1	2	4	4	2
1044	1	3	3	2	3.58	84	12000	800	2	2	1	3	2	1	21	68	128	2	1	2	4	1	0	6	3	2
1045	1	3	1	2	3.92	93	10000	2000	2	2	2	4	3	1	20	70	155	2	1	2	4	1	1	5	6	2
1046	1	3	3	2	2.80	75	11700	200	3	2	2	4	3	1	22	68	155	2	1	2	3	2	0	4	6	2
1047	1	4	2	2	2.90	87	11000	1500	2	2	2	4	12	1	22	70	190	2	1	2	3	2	0	4	5	1
1048	2	2	1	2	2.00	81	18000	1600	3	2	1	4	2	1	30	62	145	2	4	1	2	2	5	4	5	2
1049	1	4	1	5	3.00	80	15000	500	1	2	1	4	2	2	51	68	215	1	1	1	3	2	3	4	6	1
1050	1	3	1	4	3.10	82	10000	500	1	2	3	2	5	1	27	75	195	2	1	2	1	1	0	4	4	1
1051	1	2	2	1	2.50	75	8000	800	1	2	2	4	10	2	20	68	165	2	1	2	3	1	0	1	4	2
1052	2	1	1	3	2.48	80	15000	1200	2	1	1	3	3	2	21	66	135	2	1	2	4	1	0	4	4	1
1053	1	3	1	4	3.70	95	20000	500	1	2	2	2	2	1	28	73	193	2	1	2	4	1	0	4	6	2
1054	1	2	1	7	3.30	75	10000	1200	2	1	2	3	5	1	20	72	140	2	3	2	1	1	0	1	4	1
1055	1	1	1	6	3.00	85	10000	900	1	2	2	2	2	2	30	66	135	2	2	1	1	2	0	1	1	1
1056	2	4	1	7	3.42	77	7000	100	2	2	2	4	3	2	25	62	100	2	2	1	3	2	3	1	4	1
File code number	Sex	Class	Grad. school	Major	Grade-point index	H. S. average	Est. starting salary $	Tuition $	Employment	Sunday shop	Investment	Car pref.	No. jeans owned	Calculator	Age	Height	Weight	Blood pressure	Smoking status	Smoking belief	Entrance standards	Retention standards	No. clubs and groups	SPS rating	Library rating	Personal question

File code number	Question number																									
	1	2	3	4	5	6	7	8	9	10	11	12	13	14	15	16	17	18	19	20	21	22	23	24	25	26
1057	2	2	3	1	3.00	85	11000	1000	2	2	3	4	6	1	21	63	125	2	3	1	3	2	0	1	4	1
1058	2	2	1	4	3.75	75	15000	500	3	2	1	4	2	2	20	67	135	2	1	1	1	1	0	4	6	2
1059	2	2	1	1	3.40	88	10000	100	1	2	2	4	0	1	20	62	100	2	1	2	4	1	0	3	4	2
1060	1	4	1	4	2.70	75	12000	0	1	2	2	4	0	1	51	71	140	1	1	2	4	2	0	4	4	1
1061	2	3	1	5	3.40	85	18000	1000	1	2	1	4	10	1	25	63	100	2	1	2	1	2	0	3	4	1
1062	1	4	1	5	2.20	85	25000	1000	2	2	1	4	1	1	31	70	170	2	1	2	1	2	0	3	2	1
1063	2	4	2	5	2.81	75	12000	300	3	2	2	2	3	1	30	69	159	2	1	2	4	2	0	4	4	1
1064	2	3	2	4	3.40	85	12000	300	1	2	2	4	5	1	29	60	102	2	1	2	2	2	1	4	2	2
1065	2	4	3	4	2.00	80	18000	500	1	2	5	4	0	1	31	64	145	2	1	2	4	1	0	2	2	1
1066	2	1	2	6	2.80	92	15000	500	1	2	2	4	4	2	21	63	120	2	1	2	4	1	0	4	4	1
1067	2	4	3	6	3.00	93	12000	500	1	2	2	4	5	2	31	63	106	1	1	2	1	2	3	4	4	2
1068	2	1	2	2	2.62	85	15000	400	1	1	1	3	7	2	24	63	125	2	1	2	3	2	0	4	3	2
1069	1	3	2	4	2.60	83	30000	500	1	2	2	4	5	1	28	71	160	2	1	2	3	2	0	2	2	1
1070	2	3	3	5	3.00	85	20000	100	1	2	6	2	2	2	25	69	150	2	1	2	4	1	0	4	4	1
1071	1	2	2	2	2.80	87	13000	600	2	2	5	3	10	1	21	70	185	2	1	1	3	1	3	4	4	2
1072	2	2	2	7	3.50	88	10000	1500	1	2	5	4	4	1	23	62	118	2	1	1	4	1	0	4	4	2
1073	1	2	3	2	3.60	85	12000	900	3	2	1	4	10	1	19	71	160	2	1	2	3	1	1	4	4	1
1074	2	2	2	7	3.20	86	10000	200	3	2	5	4	6	1	21	61	123	2	2	1	3	1	1	3	2	1
1075	2	3	2	2	3.30	70	14000	1000	3	2	3	4	12	1	21	61	96	2	1	2	3	2	1	4	4	1
1076	2	3	3	2	3.00	92	15000	1000	2	2	5	4	10	1	21	65	104	2	1	2	4	1	0	4	6	1
1077	1	3	3	2	3.00	80	15000	200	2	2	2	2	6	1	36	67	150	2	2	1	1	1	1	4	4	2
1078	1	2	2	1	3.00	80	12000	1000	3	2	2	4	6	1	19	72	170	2	1	2	4	2	0	4	4	2
1079	1	2	2	2	3.57	92	10000	900	2	2	2	5	4	1	20	68	140	2	1	2	4	2	0	4	4	1
1080	2	3	3	2	3.50	80	12000	1500	3	2	2	4	1	1	25	64	110	2	1	2	3	2	1	1	4	2
1081	2	2	2	2	3.20	88	7800	650	2	2	2	4	3	1	20	61	115	2	1	2	4	1	3	5	6	1
1082	2	3	1	2	3.50	95	13000	1000	3	2	2	4	2	1	19	61	105	2	1	2	4	1	2	4	6	1
1083	1	2	3	4	2.30	80	10000	800	2	1	5	4	5	1	20	66	138	2	1	2	3	2	0	2	4	1
1084	1	3	3	2	2.20	85	13000	1000	2	2	2	4	25	1	21	69	155	2	1	2	4	2	0	4	5	2
1085	1	2	1	2	3.00	90	16000	737	3	2	2	2	4	1	28	70	145	2	1	2	1	1	2	2	3	2
1086	1	3	3	2	3.00	80	12000	500	2	2	2	2	3	1	27	68	140	2	1	2	3	1	2	3	4	1
1087	1	4	2	6	3.00	80	12000	1500	1	2	2	5	1	1	29	71	170	2	1	2	4	2	0	6	4	2
1088	1	3	2	2	3.00	90	10000	900	3	2	2	4	10	1	21	70	150	2	1	2	4	2	1	3	7	1
1089	1	4	1	2	3.85	95	13500	900	2	2	2	2	3	1	29	71	165	2	2	2	3	1	0	4	4	1
1090	2	3	2	2	3.80	90	13000	900	3	2	3	4	3	1	20	68	124	2	1	2	4	1	1	4	6	2
1091	1	3	3	2	3.30	89	11000	950	2	1	1	3	4	1	20	71	155	2	1	2	3	1	0	4	5	1
1092	2	4	3	2	2.80	90	10000	1500	2	2	2	2	6	1	21	60	95	2	1	2	3	2	1	3	4	1
1093	2	3	1	2	2.45	81	11000	950	3	2	2	4	0	1	21	62	145	2	2	1	2	2	0	4	4	1
1094	1	3	3	2	3.80	92	12500	1000	3	2	3	5	7	1	21	71	185	2	1	2	2	2	1	3	5	2
1095	1	4	3	2	3.00	80	10000	950	1	2	2	2	0	1	35	71	190	2	1	2	1	1	0	1	2	1
1096	2	3	2	6	3.00	80	12000	800	1	2	2	2	5	1	23	62	92	2	1	2	4	1	2	4	3	1
1097	1	3	3	2	2.00	80	30000	0	2	2	2	2	4	1	21	70	125	2	1	2	1	1	0	4	5	1
1098	1	3	1	2	3.20	80	11500	1500	2	2	1	4	2	1	21	70	150	2	1	2	3	2	0	5	6	1
1099	2	3	3	2	3.30	82	11000	300	3	1	2	2	2	1	20	63	125	2	1	2	4	1	1	4	4	2
1100	2	4	2	2	2.65	86	10000	400	3	2	2	4	8	1	21	66	130	2	1	2	3	2	0	3	5	2
1101	1	2	1	2	3.00	90	13000	100	2	2	2	2	10	1	20	70	160	2	2	2	3	1	1	3	5	1
1102	1	3	3	2	2.80	81	12000	1000	3	2	2	2	5	1	20	66	145	2	1	1	2	2	0	3	6	1
1103	2	2	2	1	2.90	90	10000	800	2	2	3	5	3	2	20	69	135	2	3	2	4	2	0	4	5	1
1104	1	2	3	6	2.28	78	10000	1000	3	2	2	2	4	1	23	66	125	2	1	2	4	1	2	4	3	2
1105	1	2	2	2	2.75	85	12000	1000	2	2	5	4	3	1	20	66	135	2	1	1	1	2	0	3	4	1
1106	2	2	2	2	3.80	97	13000	0	3	2	2	4	15	1	20	62	124	2	1	2	3	2	2	4	4	2
1107	1	2	3	2	3.55	85	13000	1000	2	2	2	4	2	1	19	70	150	2	1	2	4	2	1	4	4	1
1108	2	3	3	2	3.20	80	12500	500	2	2	2	4	10	1	19	63	130	2	1	2	1	2	0	4	4	1
1109	1	3	2	2	2.50	86	12000	600	1	2	1	3	6	1	25	59	130	2	3	1	3	2	0	4	4	1
1110	1	3	3	7	2.50	75	15000	1000	3	2	1	3	7	2	19	71	130	2	1	1	1	2	3	4	4	1
1111	1	2	2	2	2.70	80	14000	1500	2	2	2	2	6	1	22	72	180	1	1	2	1	1	1	6	6	2
1112	2	2	2	2	3.00	85	15600	180	3	2	2	4	10	1	21	66	125	2	1	2	3	1	3	1	4	1
1113	1	2	3	2	3.50	80	12000	1000	3	2	1	3	6	2	22	71	172	2	1	2	3	1	1	1	4	2
1114	1	2	3	1	2.00	77	12500	1400	3	2	2	4	4	1	20	74	172	2	2	1	1	2	0	4	4	2
1115	2	2	3	2	2.52	78	12000	1000	2	2	2	2	5	2	20	62	90	2	1	2	1	2	0	2	4	1
1116	1	2	3	2	2.00	70	7800	500	2	2	1	3	1	1	23	68	180	2	1	2	3	2	0	5	5	2
1117	1	2	3	2	2.90	85	15000	750	3	2	2	2	11	1	20	67	132	2	1	2	1	2	2	4	3	1
1118	1	2	1	2	3.00	80	12000	385	1	2	2	2	1	2	23	68	149	2	2	1	4	1	0	4	7	2
1119	1	2	3	4	2.10	82	12000	0	2	2	1	4	6	1	20	71	152	2	1	2	3	2	0	1	4	2
1120	1	2	1	2	3.00	88	10000	1000	3	2	2	2	4	1	20	66	120	2	1	2	4	1	0	4	5	2
1121	1	3	2	5	1.98	73	10000	975	2	2	2	4	10	1	21	71	155	2	1	2	1	2	0	2	6	1
1122	2	1	3	3	3.43	88	12000	1000	2	2	2	4	7	1	18	67	127	2	1	2	4	2	0	4	5	1
1123	1	1	2	4	3.50	86	12000	2000	2	2	2	4	7	1	19	72	165	2	1	2	3	2	0	4	4	2
1124	1	4	2	2	3.00	80	8000	500	1	2	2	2	0	1	39	67	153	1	3	1	4	2	0	4	4	2
1125	1	3	3	2	3.00	80	15000	600	1	2	2	2	1	1	29	68	167	2	1	2	1	2	0	4	5	1
1126	1	4	3	3	3.40	80	14000	1000	1	2	1	4	4	2	30	66	150	2	1	2	3	2	1	4	5	1
1127	2	4	3	3	3.00	80	15000	500	3	2	2	2	0	1	21	71	148	1	2	2	1	2	0	7	7	2
1128	1	4	3	2	3.50	90	15000	1000	1	2	2	2	1	1	35	75	205	1	3	2	3	1	0	2	4	2
1129	2	2	3	1	3.00	80	14000	500	1	2	1	3	0	1	18	62	120	2	1	2	1	1	1	4	4	1
1130	2	4	3	6	3.00	90	11000	1000	2	2	6	2	5	1	21	60	100	2	1	2	4	1	0	4	4	1
1131	2	3	2	2	2.91	85	32000	1000	2	2	2	4	4	1	20	61	145	2	1	1	4	2	0	4	4	1
1132	1	3	1	6	3.40	91	12000	500	2	2	2	4	1	1	20	69	210	2	1	2	4	1	4	5	1	1
1133	2	4	2	6	3.00	92	10000	500	3	2	2	2	13	1	21	64	120	2	1	1	2	2	0	4	4	1
1134	1	3	2	2	3.00	86	11000	400	1	2	5	4	2	1	31	72	183	2	1	2	1	2	0	4	1	2
1135	1	2	1	4	2.60	87	15000	500	1	2	1	4	2	1	23	70	150	2	1	2	3	2	0	1	1	1
1136	2	2	2	1	3.00	86	15000	0	1	2	2	3	4	1	25	63	98	2	1	2	3	1	0	2	2	2
1137	2	2	2	1	2.30	86	15000	1500	2	2	2	2	3	1	20	61	130	2	1	2	1	1	0	4	5	1
1138	2	2	1	1	3.23	85	10000	1500	3	1	2	4	5	1	19	63	130	2	1	2	2	1	3	1	4	1
1139	2	3	1	5	3.50	92	9000	1000	3	2	5	4	20	1	20	63	125	2	3	2	3	2	0	5	4	2
1140	1	3	3	5	2.70	82	13000	1500	2	2	2	4	6	1	21	73	205	2	2	2	3	2	1	4	1	1
1141	2	3	3	5	3.50	85	12000	1500	2	2	1	3	6	2	22	63	112	2	2	1	4	2	0	2	5	1
1142	1	2	3	2	2.80	92	12000	500	3	2	1	2	5	1	19	70	185	2	1	2	4	2	0	4	6	2
1143	1	3	3	5	3.20	88	14000	700	2	2	5	4	10	2	21	68	162	2	1	2	4	1	0	5	5	1
1144	2	4	1	5	2.50	75	12000	975	3	2	2	4	4	1	29	69	140	2	1	2	4	2	2	3	2	2
1145	2	3	3	5	3.30	82	11700	1000	2	2	3	4	8	2	21	68	123	2	1	2	3	2	0	6	5	1
1146	1	3	1	2	2.50	72	12000	500	2	2	2	4	8	1	22	67	152	2	3	1	3	2	0	4	6	2
1147	1	3	3	4	2.25	83	10000	100	2	2	2	4	2	1	27	75	175	1	1	2	4	2	0	1	5	2
1148	1	3	1	2	3.60	88	13000	550	3	2	1	4	5	1	19	66	145	2	1	2	1	2	0	1	3	1
1149	2	4	3	5	3.30	88	15000	300	2	2	1	4	5	2	22	61	98	2	1	2	4	1	0	4	4	2
1150	2	4	1	4	3.86	98	9000	600	2	2	1	4	6	1	21	65	127	2	1	2	3	2	2	4	6	2
1151	2	2	3	6	3.20	93	12000	400	3	2	2	2	6	1	20	63	115	2	1	2	2	2	2	2	2	1
1152	2	3	3	5	2.90	82	9360	1000	2	2	5	4	6	1	21	63	123	2	1	2	3	1	2	6	4	1
File code number	Sex	Class	Grad. school	Major	Grade-point index	H. S. average	Est. starting salary $	Tuition $	Employment	Sunday shop	Investment	Car pref.	No. jeans owned	Calculator	Age	Height	Weight	Blood pressure	Smoking status	Smoking belief	Entrance standards	Retention standards	No. clubs and groups	SPS rating	Library rating	Personal question

File code number	Question number																									
	1	2	3	4	5	6	7	8	9	10	11	12	13	14	15	16	17	18	19	20	21	22	23	24	25	26
1153	1	2	3	2	3.20	85	12000	1000	3	2	2	4	6	1	20	72	175	1	1	2	3	2	0	4	4	1
1154	2	3	3	5	3.20	94	12000	1000	2	1	2	4	6	1	20	67	112	2	1	2	4	2	0	3	1	1
1155	2	2	3	1	3.20	87	13500	413	2	1	1	4	5	1	23	68	135	2	1	2	4	1	0	1	5	2
1156	1	2	3	2	2.70	87	7800	1000	3	2	2	2	7	1	19	68	155	1	1	2	3	2	0	4	4	1
1157	1	3	1	2	2.70	85	12500	1000	2	1	2	4	0	1	30	71	210	2	1	2	4	1	0	4	6	2
1158	2	3	3	5	2.90	88	9600	500	3	2	2	2	15	2	22	68	140	2	2	1	1	2	0	4	4	2
1159	1	2	1	2	3.00	79	12000	1800	2	2	5	4	5	1	19	70	140	2	2	2	3	1	1	3	4	1
1160	1	3	1	2	2.75	85	12000	300	2	2	2	2	7	1	22	65	128	2	1	2	4	1	1	4	5	1
1161	2	3	1	4	2.50	75	10000	2000	2	2	2	4	6	2	21	68	160	1	2	1	4	1	0	2	6	1
1162	1	1	3	2	2.60	83	15000	975	2	1	3	4	8	1	19	70	140	2	1	1	2	2	0	4	4	1
1163	1	3	1	5	2.82	81	11000	600	3	2	2	4	5	2	22	64	155	2	1	2	3	1	0	4	4	1
1164	1	2	1	1	2.50	85	15000	0	2	1	2	4	3	2	21	65	130	2	1	2	1	2	3	7	5	1
1165	2	4	1	5	2.50	90	10000	0	3	2	2	2	0	1	28	60	111	1	2	2	1	1	0	4	4	1
1166	2	2	1	2	2.56	84	10000	2000	2	2	1	3	5	1	20	65	103	2	4	1	3	1	0	4	4	1
1167	1	3	3	5	2.80	85	13000	350	2	2	2	4	10	2	21	69	135	2	1	2	1	2	1	4	5	1
1168	1	4	2	2	2.45	87	12000	13	1	2	2	4	10	1	24	67	150	2	3	1	1	2	0	4	3	1
1169	2	4	2	6	2.90	92	12000	900	1	1	2	4	6	1	23	65	116	2	1	2	4	2	1	3	6	1
1170	2	2	2	2	2.28	75	15000	1000	3	2	2	4	6	1	17	68	132	1	2	2	3	2	1	4	4	1
1171	1	4	1	6	3.00	84	12000	900	2	1	2	4	4	1	22	66	130	1	1	1	4	1	0	1	4	2
1172	1	4	3	2	3.00	80	20000	750	1	2	2	2	4	1	37	71	165	2	1	2	3	1	0	4	3	1
1173	1	2	3	7	2.64	83	15000	200	2	2	2	4	4	1	20	69	162	2	1	2	3	2	1	4	4	2
1174	1	2	1	2	2.92	80	15000	660	1	2	2	4	3	1	23	68	170	2	1	2	3	2	0	1	5	2
1175	2	3	3	2	3.00	80	10000	840	2	2	2	2	3	2	23	63	120	2	1	2	1	2	0	1	1	1
1176	1	3	3	2	3.12	82	15000	975	1	2	2	2	0	1	21	62	150	2	1	2	4	1	0	4	5	1
1177	1	3	2	6	3.00	75	15000	1000	1	1	2	2	1	1	23	63	160	2	1	2	3	1	0	4	6	1
1178	1	3	2	6	3.26	82	12000	600	1	2	4	3	4	2	35	73	185	1	1	2	3	1	0	4	4	1
1179	1	2	1	2	3.70	86	6000	400	3	2	2	4	4	1	19	75	186	2	2	2	2	1	4	3	2	2
1180	1	2	3	2	3.20	87	17500	975	1	2	2	2	0	1	29	65	140	2	1	2	1	2	0	4	1	2
1181	1	2	1	4	3.70	92	11000	750	1	2	2	2	6	1	33	68	175	2	2	2	4	2	0	4	4	2
1182	1	2	3	2	3.00	85	12000	800	1	2	2	2	5	1	20	67	153	2	1	2	3	2	0	4	3	2
1183	2	2	3	5	3.00	92	15000	840	1	2	5	4	5	1	22	66	112	2	3	1	4	2	0	4	4	2
1184	2	2	3	6	3.00	90	15000	840	1	2	3	4	5	1	20	65	112	2	1	2	4	1	0	4	4	2
1185	1	3	3	2	3.40	75	17000	1200	1	1	2	4	2	1	34	70	150	2	1	2	3	2	0	4	4	1
1186	2	2	1	5	2.75	87	12000	1000	3	2	2	4	11	2	20	63	115	2	1	2	4	1	0	4	4	2
1187	2	3	3	5	3.20	88	12000	960	1	2	5	3	4	1	32	64	115	2	3	1	4	1	0	3	4	2
1188	1	3	2	1	3.00	80	10000	300	1	2	2	4	2	2	20	65	125	2	1	2	4	2	0	4	4	1
1189	2	4	3	6	3.00	88	11000	900	2	1	2	4	6	1	23	66	121	2	1	2	4	2	1	4	6	1
1190	1	4	3	6	2.75	80	10000	400	3	2	2	4	15	1	23	69	145	2	1	2	4	1	2	1	4	2
1191	1	4	2	6	2.50	75	15000	500	1	2	2	2	4	1	30	66	140	2	1	2	4	1	0	4	6	1
1192	2	4	3	2	2.50	80	16000	975	1	1	2	2	0	1	30	62	160	2	1	2	4	2	2	1	6	1
1193	1	1	1	2	2.88	92	15000	300	1	2	2	2	1	1	31	74	230	2	2	2	1	2	0	4	4	1
1194	2	3	1	2	2.97	90	12000	900	2	2	2	4	0	1	21	64	160	2	1	2	3	2	1	4	4	1
1195	1	4	1	6	2.90	85	12000	500	1	2	2	2	0	1	28	67	132	2	1	2	1	1	1	4	4	1
1196	1	4	3	6	2.80	90	12500	500	3	2	2	2	3	1	35	72	200	1	2	2	3	2	0	5	6	1
1197	1	4	3	4	2.00	80	15000	0	1	2	2	4	0	1	34	67	146	2	1	1	1	2	0	1	5	2
1198	1	3	1	6	3.10	90	16000	400	1	2	2	2	1	1	22	73	175	2	1	2	4	1	0	1	1	2
1199	1	3	1	6	3.10	85	16000	400	1	2	2	4	1	1	21	72	195	1	1	2	4	2	0	1	1	2
1200	1	3	1	2	2.60	80	12000	1500	2	2	2	2	5	2	22	67	155	2	1	2	4	1	3	1	7	1
1201	1	4	1	6	3.80	95	15000	975	2	2	2	4	3	2	20	71	150	2	1	2	4	2	1	4	6	1
1202	2	4	1	4	3.60	92	11000	1000	2	2	2	4	5	1	19	64	108	2	1	2	4	1	2	3	4	1
1203	1	3	2	5	2.20	87	18000	1200	2	2	2	4	12	1	21	72	145	2	1	1	4	2	1	2	2	1
1204	1	3	3	4	3.20	75	10000	500	2	2	2	4	2	1	22	72	190	2	1	2	4	1	2	7	7	1
1205	1	3	3	5	2.30	80	11000	0	3	1	2	2	4	1	22	69	140	2	1	2	1	2	4	1	7	1
1206	1	3	3	4	2.80	89	15000	800	2	1	2	2	10	1	21	67	155	2	1	1	4	1	0	5	6	1
1207	2	4	1	4	2.50	95	15000	0	2	2	6	2	6	1	24	65	130	2	1	2	4	2	0	2	2	2
1208	1	3	2	5	3.20	88	12000	800	2	2	5	4	4	1	20	74	185	2	1	2	4	2	0	4	5	1
1209	1	3	3	4	3.00	80	15000	0	2	2	2	2	5	1	20	70	130	2	2	1	4	1	2	4	6	1
1210	1	2	1	6	2.90	77	16000	0	3	1	5	4	3	1	24	66	130	2	1	2	1	2	3	1	4	2
1211	2	3	3	4	2.50	70	10000	0	3	2	2	2	3	2	22	68	116	2	1	1	1	2	1	4	5	1
1212	2	3	2	5	2.80	80	10000	400	3	2	2	4	5	2	21	63	125	2	1	2	4	1	0	4	5	1
1213	2	3	3	3	3.20	85	10400	1200	2	2	2	4	6	1	23	62	95	1	2	1	3	1	0	4	4	1
1214	2	4	1	5	2.90	80	10000	1000	2	2	2	4	5	1	21	65	123	2	1	2	4	1	1	4	4	2
1215	2	4	3	7	2.80	81	15000	0	3	2	2	4	6	2	23	63	110	2	1	2	3	1	0	4	6	1
1216	1	2	3	7	3.50	85	10700	1500	3	2	2	4	0	1	29	70	201	2	2	2	4	1	2	5	4	2
1217	1	3	1	3	2.80	80	11000	800	2	2	2	2	0	1	24	72	145	2	1	2	4	1	3	4	6	1
1218	1	4	1	5	3.00	70	15000	1000	1	2	1	4	0	1	33	74	210	2	4	1	3	1	0	4	3	2
1219	2	4	3	4	2.90	70	12000	80	2	1	2	4	15	1	24	67	150	2	1	2	4	2	1	4	6	2
1220	2	3	1	7	2.80	80	10000	200	3	1	2	2	0	1	20	62	118	2	1	2	2	2	0	3	5	1
1221	1	3	1	5	3.50	85	15000	0	1	2	2	4	1	2	25	67	160	2	1	2	3	1	5	1	4	1
1222	1	3	3	4	2.50	85	12000	500	3	2	1	4	20	1	22	74	200	2	3	1	3	2	0	3	6	2
1223	1	3	1	4	2.50	80	11000	100	3	2	2	2	2	2	26	67	150	2	3	1	3	1	4	5	4	1
1224	2	2	2	4	2.50	89	15000	0	2	2	2	2	5	1	20	63	112	2	1	2	1	2	0	1	4	1
1225	1	3	1	4	2.70	88	15000	1000	2	2	4	4	4	1	20	67	150	2	1	1	2	2	1	3	6	1
1226	1	4	2	5	2.58	76	9500	500	2	2	2	4	8	1	22	73	176	2	2	1	3	2	0	1	5	1
1227	2	2	2	2	2.90	85	10000	800	3	2	5	2	7	1	19	62	115	2	1	2	1	1	2	4	3	1
1228	1	3	3	5	1.90	83	12000	600	2	2	2	2	12	1	21	69	165	2	1	2	3	2	2	1	2	2
1229	1	3	3	5	2.75	80	12000	1000	2	1	2	2	4	2	22	69	165	2	2	2	3	1	1	5	5	1
1230	2	3	2	1	2.25	85	9500	800	2	2	2	2	4	2	21	65	140	2	1	2	4	1	0	2	3	2
1231	2	2	2	5	3.60	84	10000	1200	2	2	1	3	9	1	19	63	114	2	1	2	4	2	1	4	5	2
1232	2	3	2	4	3.40	85	12000	1500	2	2	1	4	4	2	20	68	135	2	1	2	3	2	0	2	6	1
1233	1	3	1	5	3.40	90	13000	4000	1	2	2	2	2	1	20	72	245	1	1	2	1	2	0	6	5	1
1234	2	2	1	4	3.00	90	17000	1500	3	2	6	2	10	2	19	66	125	2	1	2	3	1	0	4	4	1
1235	1	3	3	6	2.30	80	15000	2000	2	2	2	4	2	1	22	68	165	2	1	2	3	1	1	2	4	1
1236	1	3	3	1	3.00	82	10000	800	2	2	2	4	4	1	22	67	160	2	1	2	4	1	0	4	4	1
1237	2	4	3	5	2.00	75	10000	0	2	2	2	4	15	1	23	63	110	2	1	2	4	2	3	3	4	1
1238	1	3	3	2	3.50	93	10000	975	2	2	2	4	5	1	20	66	115	2	2	1	3	1	0	4	5	2
1239	1	3	3	2	2.92	88	12000	1000	2	2	5	4	3	1	24	71	175	2	1	1	3	1	0	4	5	1
1240	2	3	1	1	3.50	85	13000	1000	3	2	5	3	15	1	18	66	125	2	4	1	3	2	6	1	4	2
1241	2	3	3	2	3.30	81	12000	2000	2	2	1	2	2	1	20	62	120	2	1	2	4	2	1	7	6	2
1242	1	3	2	2	2.80	82	10000	2000	2	2	2	4	4	1	21	73	168	2	1	2	4	2	1	4	4	1
1243	1	3	3	5	2.00	80	12000	1500	1	2	2	4	2	1	27	71	180	2	3	1	4	2	0	4	4	1
1244	2	4	1	5	3.00	92	10500	980	2	2	3	4	0	1	21	66	124	2	1	2	4	2	0	4	2	2
1245	2	3	1	5	4.00	93	8000	2000	3	2	1	4	1	1	20	62	117	1	1	2	4	1	0	5	7	1
1246	1	2	3	2	2.70	70	12000	500	3	2	2	2	2	1	17	67	148	2	1	2	3	1	2	4	6	2
1247	1	2	3	4	2.70	79	10000	1000	3	2	2	2	6	1	19	70	149	2	1	2	4	2	0	4	4	1
1248	2	4	1	7	3.50	92	9000	500	2	2	2	2	6	1	23	62	100	2	1	2	2	1	2	1	4	1
File code number	Sex	Class	Grad. school	Major	Grade-point index	H. S. average	Est. starting salary $	Tuition $	Employment	Sunday shop	Investment	Car pref.	No. jeans owned	Calculator	Age	Height	Weight	Blood pressure	Smoking status	Smoking belief	Entrance standards	Retention standards	No. clubs and groups	SPS rating	Library rating	Personal question

File code number	Question number																									
	1	2	3	4	5	6	7	8	9	10	11	12	13	14	15	16	17	18	19	20	21	22	23	24	25	26
1249	2	3	1	5	3.00	95	20000	975	3	1	2	4	3	1	20	63	120	2	2	2	1	2	0	4	4	2
1250	1	3	3	5	2.80	90	10000	1500	2	2	2	4	6	1	20	71	180	2	1	2	4	1	2	2	4	2
1251	2	3	1	5	2.90	89	12000	487	2	2	2	4	7	2	20	65	104	2	1	2	4	1	0	2	5	1
1252	1	4	1	5	3.00	92	12000	975	3	2	2	2	0	1	25	68	170	2	1	1	1	1	2	5	6	1
1253	1	4	1	7	3.40	85	10000	1200	2	2	2	4	8	1	22	73	180	2	3	1	4	1	1	1	4	2
1254	2	3	2	5	3.40	85	10000	1000	2	2	1	4	6	1	21	65	115	2	1	2	3	2	0	1	1	1
1255	2	3	3	6	2.75	83	14000	200	2	2	2	2	2	2	20	64	125	2	2	1	2	1	0	4	6	2
1256	1	4	1	5	3.50	80	11000	840	2	2	3	4	5	2	23	68	160	2	1	2	3	1	0	3	6	2
1257	1	4	3	4	3.50	80	10500	500	3	2	2	4	10	1	21	61	115	1	1	2	4	1	1	4	5	1
1258	2	3	1	7	2.70	90	12000	1000	1	2	2	1	4	1	21	70	130	2	1	2	4	1	3	4	4	1
1259	1	3	3	7	2.70	80	18500	500	2	2	2	4	6	1	21	69	142	2	1	1	4	2	0	4	6	1
1260	2	3	2	7	2.00	80	11000	975	3	2	2	4	6	2	22	69	142	2	2	1	3	2	1	6	4	1
1261	2	4	2	5	3.40	91	11000	975	2	2	2	4	6	1	20	64	103	2	1	2	4	1	1	2	5	2
1262	2	2	3	2	3.18	90	10000	500	2	2	1	2	11	2	18	64	108	2	1	2	4	1	1	4	4	1
1263	1	4	1	5	3.40	94	12000	600	2	2	2	2	4	1	21	70	156	2	1	2	3	2	1	4	5	2
1264	2	4	2	4	3.10	98	13000	200	2	2	2	4	5	1	22	62	130	2	1	2	4	1	0	1	7	2
1265	2	3	1	2	3.20	80	10400	975	1	2	6	2	5	1	24	64	115	2	1	1	4	1	0	4	4	1
1266	1	4	3	5	2.50	73	5000	975	1	2	5	4	3	2	31	71	185	2	4	1	1	1	0	4	4	1
1267	1	4	1	3	2.80	84	17200	1000	1	2	1	4	5	2	30	72	210	2	1	2	2	1	0	4	5	2
1268	1	4	1	3	3.10	72	12000	2000	1	2	2	4	5	1	27	67	185	1	1	2	4	1	0	4	6	1
1269	1	3	1	3	3.25	85	10000	800	2	2	2	4	3	1	20	71	175	2	2	1	1	2	0	4	4	1
1270	1	3	2	4	3.00	87	10000	800	2	2	2	1	4	2	22	72	155	2	1	2	4	1	1	4	4	2
1271	1	4	1	3	3.30	80	12500	980	1	2	2	4	3	1	33	73	190	2	1	2	1	1	1	4	4	1
1272	1	4	3	7	3.20	89	16000	1500	1	2	1	4	4	1	29	72	190	2	2	2	3	1	0	4	6	2
1273	1	4	3	4	3.85	90	30000	1500	1	2	2	4	5	1	43	69	170	2	1	1	4	1	3	4	5	1
1274	2	3	1	2	3.40	93	12000	2000	2	2	2	2	0	1	21	63	129	2	1	2	1	2	0	4	3	1
1275	2	4	3	2	3.22	89	14790	988	2	2	3	2	0	1	24	64	135	2	1	2	1	2	0	4	4	1
1276	2	3	1	4	2.50	79	12000	600	2	2	2	4	8	1	21	61	115	2	1	2	4	1	2	7	5	2
1277	2	3	1	7	3.40	92	12500	1000	2	2	1	3	4	2	46	61	110	2	1	2	2	2	0	2	5	1
1278	2	3	1	5	3.00	96	20000	1000	2	2	5	4	10	1	23	63	115	2	1	2	4	2	0	5	5	2
1279	1	3	3	5	3.00	90	10000	1000	2	2	2	2	5	1	21	72	160	2	2	2	1	1	2	4	5	2
1280	1	3	2	5	3.50	86	10000	1000	1	2	2	4	2	2	25	68	170	2	1	2	4	1	0	4	4	2
1281	1	3	3	5	2.60	80	12000	1500	3	2	5	4	6	1	21	67	145	2	1	2	3	1	0	2	5	1
1282	2	3	3	3	3.30	80	20000	1000	3	2	4	2	8	1	25	63	95	2	1	2	3	2	0	4	4	2
1283	2	3	1	4	3.00	88	25000	1440	1	2	3	3	7	1	26	63	104	2	1	2	1	2	0	4	4	2
1284	2	4	3	4	2.80	80	10000	1200	1	2	2	2	3	1	24	61	125	2	1	2	3	2	0	4	4	2
1285	1	3	3	3	3.00	88	13500	1500	2	1	5	4	7	1	20	73	195	1	1	1	4	2	0	5	6	1
1286	1	3	1	4	3.50	88	14000	500	1	1	5	4	6	1	23	67	170	2	1	1	4	1	1	2	3	1
1287	1	4	3	5	2.20	88	10000	1000	3	2	2	2	5	1	33	65	130	2	1	2	4	2	2	4	4	1
1288	1	4	2	4	2.80	81	11000	985	2	2	3	3	0	1	22	72	175	1	1	2	1	1	0	4	6	1
1289	2	2	2	5	3.20	89	10000	825	2	2	2	2	8	1	21	66	126	2	1	2	1	2	0	4	5	2
1290	1	1	1	5	3.70	89	20000	1000	2	2	2	4	5	1	24	69	154	2	2	2	4	2	2	4	4	2
1291	1	4	2	5	3.00	80	10000	1200	2	2	2	2	3	1	33	68	120	2	1	2	1	2	2	4	4	1
1292	1	2	3	1	2.50	75	9000	600	2	2	2	2	7	1	23	68	150	2	1	2	2	2	0	2	1	2
1293	1	4	1	4	2.50	80	11500	400	3	2	2	2	4	1	36	70	150	2	1	2	4	1	1	4	7	1
1294	1	2	1	6	3.00	89	20000	100	3	2	2	2	4	1	25	66	115	2	3	1	3	2	0	4	5	1
1295	2	3	3	4	3.00	85	12000	1000	2	2	1	3	0	2	20	64	125	2	1	2	4	2	0	4	2	1
1296	1	3	1	5	3.15	85	11000	300	2	2	6	2	4	1	23	72	180	2	1	2	4	1	0	4	5	2
1297	2	2	2	5	3.15	88	15000	500	2	2	5	4	5	1	19	63	117	2	1	2	3	2	0	4	1	1
1298	2	3	1	1	2.40	85	12000	400	3	1	2	2	1	1	23	62	115	2	1	2	1	2	2	3	5	2
1299	2	2	2	5	2.80	83	10000	700	2	2	1	4	6	1	20	65	125	2	2	2	3	1	0	4	4	2
1300	2	4	1	5	2.80	80	11000	975	2	2	5	4	0	1	21	67	120	2	2	2	4	2	0	2	4	2
1301	1	4	3	5	3.10	80	10000	600	1	2	2	2	0	1	29	69	160	2	2	2	1	1	0	4	5	2
1302	2	4	1	4	3.80	90	8500	2000	3	2	2	4	4	1	24	65	127	2	3	2	3	1	1	4	6	1
1303	1	3	3	5	2.80	83	9500	500	2	2	2	2	4	1	23	67	147	2	1	2	3	1	0	4	4	1
1304	2	3	1	7	2.20	81	10000	400	3	2	2	4	0	1	21	62	150	2	2	1	2	2	0	4	4	2
1305	2	2	3	5	2.30	87	18000	0	3	2	2	2	9	1	29	63	98	2	3	2	4	2	1	4	5	1
1306	1	2	2	2	2.00	75	10000	1000	2	2	2	3	8	1	21	74	185	2	1	2	3	2	0	1	5	2
1307	2	3	2	5	3.20	92	12000	500	2	2	2	4	0	1	22	61	108	2	1	1	4	1	0	4	4	1
1308	1	2	3	5	2.76	85	10000	1000	3	2	2	2	5	1	19	67	140	2	1	2	4	1	3	4	6	2
1309	1	2	2	2	2.48	83	15000	1000	2	2	2	4	7	1	22	70	170	2	1	2	3	2	0	4	4	2
1310	1	2	1	4	3.40	85	25000	1400	2	2	2	4	7	1	26	73	175	2	2	2	4	1	0	4	4	2
1311	1	3	1	6	3.30	92	12000	975	2	2	2	2	5	2	25	72	165	2	2	2	4	1	0	4	4	1
1312	2	3	1	3	2.59	84	15000	400	2	2	2	1	1	1	23	67	160	2	1	2	4	1	2	4	6	1
1313	2	3	2	2	3.38	82	22000	250	2	2	2	5	8	1	21	64	115	2	1	2	2	2	0	4	4	1
1314	1	3	1	5	2.80	82	10400	500	2	2	5	4	8	1	20	69	160	2	1	2	3	2	0	4	6	2
1315	2	3	2	2	3.95	93	13000	500	2	2	4	4	4	1	21	65	150	2	3	1	4	2	0	4	4	1
1316	1	3	2	2	3.40	85	10000	1000	3	2	2	4	6	1	22	70	145	2	1	2	1	2	0	5	5	2
1317	2	3	3	4	3.20	86	12000	1000	2	2	2	4	5	1	22	60	125	2	2	1	3	2	1	4	4	1
1318	1	4	2	5	3.00	70	11000	1000	3	2	5	2	10	1	23	74	185	2	1	2	2	2	2	4	3	2
1319	1	3	1	2	2.30	74	12000	400	2	2	5	4	4	1	20	67	145	2	1	1	4	2	0	3	5	1
1320	2	3	1	7	3.00	80	13000	500	2	2	6	2	1	2	36	66	150	2	2	1	4	2	2	3	1	2
1321	2	3	3	4	3.50	85	9000	500	3	2	5	5	4	2	20	66	130	2	1	2	4	2	0	4	4	1
1322	2	3	3	5	3.10	86	8840	500	2	2	5	5	4	1	21	62	115	2	1	2	3	2	0	3	4	1
1323	1	3	3	2	2.66	79	15780	975	2	2	2	4	6	1	21	69	155	2	1	2	4	1	0	1	5	1
1324	2	4	3	7	3.00	80	15000	975	1	2	2	2	3	1	31	61	160	2	1	2	2	2	0	6	6	1
1325	1	3	3	2	2.50	75	15000	1000	2	2	3	4	6	2	21	69	165	2	1	2	1	2	0	2	4	2
1326	1	3	1	6	2.50	85	10500	1200	2	1	1	3	4	1	20	71	175	2	1	2	3	2	0	5	4	1
1327	2	3	3	3	3.15	95	11500	400	2	2	2	2	4	2	22	64	112	2	1	2	3	2	1	4	4	1
1328	2	4	2	7	2.40	93	10000	900	2	2	2	4	4	1	21	60	98	2	1	2	3	2	0	4	5	1
1329	1	2	2	6	3.00	80	18000	725	3	2	2	4	5	1	21	62	115	2	1	2	4	1	0	4	2	1
1330	2	2	1	2	3.00	80	12000	400	2	2	2	2	10	1	21	62	125	2	1	2	1	1	0	1	1	1
1331	1	3	3	5	3.20	65	10000	500	2	2	1	3	10	1	20	69	145	2	1	2	3	1	0	6	5	1
1332	1	3	3	4	2.80	78	10000	975	3	2	2	2	0	2	21	72	150	2	1	2	1	2	3	1	4	1
1333	1	3	1	1	3.20	70	10000	300	3	2	2	2	10	2	30	70	165	2	1	2	1	1	0	4	4	1
1334	1	3	3	5	2.90	85	14000	500	2	2	2	2	0	1	21	69	195	2	3	1	4	1	1	6	2	1
1335	1	1	1	3	3.00	80	12000	725	3	2	2	2	3	1	22	71	135	2	1	2	1	2	2	4	3	1
1336	1	4	1	2	3.00	83	12000	0	1	2	2	2	5	1	36	72	180	2	1	1	1	1	0	4	2	2
1337	2	1	1	2	2.83	80	8320	900	1	2	2	2	3	1	22	60	100	2	1	2	1	2	0	4	5	1
1338	1	1	1	1	2.90	98	12000	1000	2	2	2	4	0	1	18	74	240	1	1	2	4	2	1	4	4	1
1339	2	3	2	4	3.40	90	13000	300	1	2	2	4	1	1	35	64	131	2	1	2	3	2	0	4	4	1
1340	2	4	2	5	2.00	75	7800	500	1	2	2	2	10	2	24	62	105	2	1	2	2	1	0	4	5	2
1341	1	2	2	3	2.70	79	10000	1000	1	2	2	4	4	1	22	69	140	2	1	2	2	2	0	4	4	2
1342	1	4	2	4	2.80	75	15000	1000	1	2	2	2	4	1	43	68	190	2	1	1	3	2	0	5	7	2
1343	1	2	1	2	3.00	86	12000	485	3	2	1	3	6	1	22	71	175	2	1	2	1	1	0	4	4	1
1344	1	3	1	2	2.60	85	11000	1000	3	2	2	2	7	1	20	72	150	2	2	1	1	1	0	1	4	1
File code number	Sex	Class	Grad. school	Major	Grade-point index	H. S. average	Est. starting salary $	Tuition $	Employment	Sunday shop	Investment	Car pref.	No. jeans owned	Calculator	Age	Height	Weight	Blood pressure	Smoking status	Smoking belief	Entrance standards	Retention standards	No. clubs and groups	SPS rating	Library rating	Personal question

File code number	1 Sex	2 Class	3 Grad. school	4 Major	5 Grade-point index	6 H. S. average	7 Est. starting salary $	8 Tuition $	9 Employment	10 Sunday shop	11 Investment	12 Car pref.	13 No. jeans owned	14 Calculator	15 Age	16 Height	17 Weight	18 Blood pressure	19 Smoking status	20 Smoking belief	21 Entrance standards	22 Retention standards	23 No. clubs and groups	24 SPS rating	25 Library rating	26 Personal question
1345	1	3	1	2	2.50	88	13500	2560	1	2	2	4	1	1	27	64	160	2	1	2	4	1	0	4	4	1
1346	2	1	3	2	3.00	80	20000	1000	2	2	1	4	5	1	19	70	180	2	2	1	4	1	1	3	5	2
1347	1	1	3	2	3.00	80	13000	500	2	2	1	4	4	1	20	72	185	1	2	2	4	2	0	4	4	2
1348	1	1	3	2	3.80	85	11000	750	3	2	2	2	6	1	18	70	140	2	1	2	3	2	1	4	5	1
1349	1	2	1	2	3.85	90	20000	500	3	2	1	4	0	2	20	69	165	2	1	2	4	1	0	4	4	1
1350	1	3	1	4	3.00	86	15000	975	1	2	2	2	1	1	30	75	215	2	1	2	4	2	0	4	4	2
1351	2	3	1	2	3.70	85	16000	1000	2	2	2	4	2	2	28	70	165	2	1	2	4	1	0	4	4	1
1352	2	4	3	4	3.30	85	15000	975	1	2	6	2	0	1	51	63	120	2	1	2	1	2	0	1	1	2
1353	1	2	3	2	2.90	80	10000	725	3	2	2	2	2	1	24	68	160	2	2	2	1	1	0	4	4	1
1354	1	1	3	2	2.60	75	10000	675	2	2	2	4	0	1	21	68	140	2	2	1	1	2	0	4	7	1
1355	1	3	1	2	2.50	90	15000	1500	1	2	2	2	3	1	22	75	176	2	1	1	3	2	1	1	5	1
1356	2	3	3	2	3.00	80	15000	1000	1	2	2	2	3	1	28	62	130	2	1	2	1	2	0	4	6	2
1357	1	2	1	2	2.50	80	15000	500	2	2	2	2	2	2	22	72	165	2	1	2	3	1	0	4	4	1
1358	2	4	1	7	2.50	78	14000	2000	1	1	2	4	5	1	25	66	200	2	1	1	3	1	0	4	7	1
1359	1	3	1	2	3.20	85	14000	1000	1	2	2	4	2	1	28	71	140	2	2	2	3	2	1	4	5	2
1360	2	3	1	5	3.20	89	22000	400	1	2	2	2	3	1	21	63	103	2	1	2	4	1	2	4	6	1
1361	1	3	1	5	2.40	78	18000	500	1	2	1	4	3	2	26	70	165	2	1	2	1	2	0	4	5	2
1362	1	3	1	2	2.00	75	15000	500	1	2	2	4	4	2	36	65	125	2	2	1	1	2	0	4	4	2
1363	1	4	1	4	3.50	70	13000	600	1	2	2	4	2	1	32	64	165	2	1	1	1	2	0	4	7	1
1364	1	1	3	4	2.40	78	14000	300	3	2	2	4	2	1	35	61	160	2	3	1	4	1	0	2	1	1
1365	2	3	1	5	2.50	75	15000	1000	1	2	5	4	2	2	24	68	121	2	1	2	2	2	0	4	4	1
1366	1	3	3	3	3.00	87	12000	1000	1	2	1	4	10	1	28	72	190	1	2	1	4	2	0	4	4	1
1367	2	3	3	4	2.80	80	14000	2400	1	2	3	5	1	1	26	61	107	2	1	2	4	1	0	4	6	1
1368	1	3	3	2	3.60	85	11000	400	1	2	3	4	4	1	34	67	140	2	1	2	1	2	0	4	6	2
1369	1	3	3	2	2.00	80	24000	400	1	2	2	2	0	1	26	64	120	2	1	2	2	2	1	4	6	2
1370	1	4	1	2	2.80	84	20000	800	1	2	2	4	5	1	21	67	150	2	1	2	4	2	0	4	5	1
1371	1	2	3	2	3.80	90	11000	600	2	2	2	2	1	1	25	72	160	2	1	2	3	1	0	4	6	1
1372	1	4	1	4	2.90	85	13000	488	1	2	2	2	6	2	39	69	154	2	3	1	2	2	2	4	5	1
1373	1	4	1	5	3.50	90	18000	725	1	2	2	2	0	1	48	73	190	2	2	1	3	1	1	3	5	2
1374	2	3	1	4	3.00	94	12000	800	1	2	1	5	5	1	22	67	155	2	3	2	3	2	1	4	4	2
1375	1	3	2	5	2.75	76	12000	1600	1	2	2	2	3	2	23	70	170	1	1	2	3	1	0	4	5	2
1376	1	3	1	2	3.00	80	20000	975	1	2	2	2	0	1	45	69	165	2	1	2	1	2	0	5	6	2
1377	2	3	3	3	2.80	80	16000	975	1	2	2	4	6	1	23	61	105	2	1	2	4	2	0	1	4	2
1378	1	3	3	5	3.00	80	10000	100	1	2	2	2	6	1	21	70	175	2	2	2	2	2	0	4	5	1
1379	1	3	1	4	2.50	85	13500	400	1	2	6	2	4	1	30	72	165	2	3	1	1	2	1	5	2	1
1380	1	2	1	5	3.50	90	13500	675	2	1	2	4	10	1	20	68	145	2	2	1	3	1	2	3	2	1
1381	2	3	3	4	2.00	85	10000	0	1	2	2	2	1	2	25	67	130	2	1	2	1	2	0	4	6	2
1382	2	3	2	2	2.70	80	14500	936	3	2	5	4	2	1	28	60	110	2	1	2	4	1	0	4	4	2
1383	2	3	1	4	3.00	80	11000	450	1	2	2	2	2	1	38	65	125	2	1	2	1	1	0	4	4	1
1384	1	4	1	5	2.60	75	12000	800	1	2	2	2	6	1	28	68	135	2	1	2	4	1	0	4	6	1
1385	1	3	1	3	3.00	76	12000	1000	3	2	2	2	4	2	28	70	180	2	1	2	4	2	2	6	4	2
1386	1	4	1	2	2.50	75	11000	1000	2	2	2	4	3	1	22	70	169	2	2	2	3	2	0	4	4	1
1387	1	3	1	6	3.00	83	17500	1000	2	2	1	4	3	1	25	69	165	2	3	1	1	2	0	4	4	1
1388	1	2	3	2	3.00	80	10400	1000	1	2	2	2	2	1	26	64	120	2	1	2	1	2	0	4	4	1
1389	1	2	1	5	2.86	88	18000	600	3	2	2	4	11	1	18	72	152	2	1	2	2	2	3	4	6	1
1390	1	3	1	2	2.00	87	10000	1000	3	2	2	4	5	1	21	66	200	2	1	2	2	2	0	4	7	2
1391	1	2	3	2	2.54	79	10000	0	3	2	2	2	2	2	25	68	135	2	1	2	1	2	0	4	7	1
1392	1	4	3	7	2.60	78	18000	800	1	2	1	3	5	1	28	68	178	2	1	1	4	2	0	4	4	1
1393	1	2	1	2	2.00	82	16000	1000	2	1	2	4	4	1	21	70	159	2	2	1	2	2	0	4	3	1
1394	1	2	3	2	3.40	79	15000	1000	1	2	3	4	0	1	20	67	185	2	1	1	4	2	0	4	7	1
1395	2	3	1	2	3.65	85	13000	1200	3	2	2	4	5	2	35	66	120	2	1	2	4	1	1	4	4	2
1396	2	3	2	2	3.21	83	10000	100	2	2	2	2	4	1	21	62	108	2	1	1	4	2	0	3	3	1
1397	2	3	3	2	2.98	80	11000	550	3	2	2	4	8	1	23	60	105	2	1	2	3	2	0	2	5	2
1398	2	4	1	2	3.00	85	14000	1000	2	2	1	4	5	1	25	62	115	2	4	2	4	1	0	4	4	1
1399	1	3	1	2	3.30	85	12500	2500	2	1	1	5	5	1	21	72	190	2	1	2	2	2	1	6	6	2
1400	2	2	3	2	2.00	83	11000	412	3	2	2	4	1	1	20	61	116	2	1	2	2	2	0	4	3	2
1401	1	2	1	2	2.50	75	10000	975	3	2	3	4	2	2	19	74	150	2	1	2	1	2	1	4	4	1
1402	1	2	1	2	3.00	85	13000	0	2	2	2	2	10	1	20	66	120	2	1	2	4	1	2	4	4	1
1403	2	2	1	2	3.00	89	10000	800	2	2	2	2	26	1	19	60	93	2	2	2	3	1	1	5	6	1
1404	2	3	3	2	3.50	85	10000	500	2	2	2	4	1	1	24	64	113	1	2	2	4	1	0	3	3	1
1405	1	1	3	6	3.08	85	15000	500	3	2	1	4	7	1	18	72	160	2	1	2	4	1	2	4	4	2
1406	2	2	3	1	2.40	82	18000	700	3	1	2	2	9	1	18	67	131	2	1	1	3	2	2	2	4	2
1407	2	3	3	2	3.22	92	10500	2000	2	1	2	4	5	1	20	62	110	2	1	2	4	1	3	4	6	2
1408	1	4	1	2	3.20	93	13000	900	3	2	2	4	6	1	21	66	165	2	1	2	4	1	0	1	4	1
1409	2	3	1	2	3.50	80	12000	500	2	1	1	4	2	2	36	62	98	2	1	2	4	1	0	4	4	2
1410	2	2	2	2	3.00	82	18000	1000	3	2	1	4	1	1	23	67	125	2	2	2	3	2	0	4	5	2
1411	1	3	1	2	2.16	96	11000	500	3	2	6	2	4	1	20	76	240	1	2	2	3	2	10	1	4	1
1412	1	3	3	2	2.80	78	10000	500	3	1	2	4	8	1	24	68	178	2	2	1	1	2	2	3	2	1
1413	1	2	2	2	2.85	85	14000	1200	3	2	2	2	5	1	20	68	130	2	1	2	3	2	0	4	5	1
1414	1	2	1	4	3.00	80	13000	600	1	1	5	4	4	1	32	73	148	2	3	2	1	2	0	4	4	1
1415	2	3	1	3	2.00	80	15000	0	1	2	6	2	6	1	29	60	135	2	1	2	4	2	0	6	5	1
1416	1	2	1	6	3.50	86	15000	600	1	1	1	4	4	1	25	69	165	2	1	2	3	2	0	4	4	1
1417	1	4	1	5	2.00	72	17000	600	1	2	5	2	0	1	26	69	153	2	3	1	1	2	0	1	4	1
1418	1	1	1	1	3.60	87	8000	750	3	2	1	3	5	1	18	66	140	2	3	1	4	1	0	4	2	1
1419	1	4	2	4	2.00	70	25000	100	1	2	1	3	0	1	27	72	170	1	2	2	4	2	0	2	2	2
1420	1	3	1	6	2.00	75	12500	150	1	2	2	1	8	2	28	77	205	2	1	2	1	1	0	4	4	2
1421	1	3	3	4	1.90	87	12000	0	1	2	5	4	5	1	22	73	195	2	1	2	3	2	0	4	4	1
1422	1	2	2	2	2.00	89	11000	500	1	1	2	2	2	1	22	69	201	1	1	1	1	2	0	1	1	2
1423	1	1	1	5	2.47	80	30000	150	2	2	1	3	0	1	20	69	190	2	3	2	4	2	0	1	5	2
1424	2	3	3	2	2.00	80	15000	300	1	2	3	4	2	1	41	64	137	2	2	1	3	2	0	4	4	2
1425	1	2	1	2	2.00	78	15000	280	1	2	2	4	20	1	27	70	178	2	1	2	3	2	0	4	4	1
1426	2	2	1	5	3.20	88	15000	0	1	2	5	4	7	2	26	63	123	2	1	2	4	1	0	4	4	1
1427	1	4	1	5	2.90	75	15000	75	1	2	2	2	2	1	25	75	220	2	1	2	4	1	0	4	1	2
1428	1	3	3	2	3.00	80	15000	500	3	2	2	2	3	1	29	69	170	2	4	1	4	2	0	4	6	2
1429	2	1	1	4	3.00	97	13500	600	1	2	3	3	12	1	22	68	165	2	3	1	4	1	0	4	6	2
1430	2	1	2	7	3.00	80	13000	200	1	2	2	4	2	1	31	68	140	2	1	1	4	1	0	4	4	1
1431	1	3	1	3	3.40	90	18500	500	1	2	5	4	5	1	33	70	177	2	1	1	4	1	0	4	5	1
1432	1	3	3	1	2.00	79	15000	500	1	2	2	2	7	2	29	69	140	2	3	2	1	1	0	4	4	1
1433	1	3	3	2	3.00	78	18000	900	1	2	2	2	6	1	23	73	165	2	1	2	3	2	0	3	4	1
1434	2	4	1	7	3.00	90	13500	400	1	2	2	4	4	1	23	60	112	2	1	2	1	2	1	4	5	2
1435	2	1	1	4	3.00	80	15000	750	2	2	2	2	1	1	26	64	115	2	1	2	1	2	0	4	4	2
1436	2	4	3	2	2.00	70	15000	450	1	2	2	2	6	1	32	60	114	2	1	2	1	1	0	4	4	1
1437	2	2	3	3	2.75	85	15000	600	1	2	2	2	2	1	27	66	114	2	1	2	4	1	0	4	4	1
1438	2	4	3	5	2.75	74	10000	210	3	2	1	3	3	1	22	65	123	2	2	1	3	2	0	4	5	1
1439	1	2	1	2	3.50	89	10000	3000	2	2	3	3	1	1	20	72	135	2	1	2	3	1	0	4	6	1
1440	2	2	3	1	3.17	82	11700	1000	1	2	2	4	6	1	21	67	110	1	2	1	1	2	0	4	4	1

File code number	Question number																									
	1	2	3	4	5	6	7	8	9	10	11	12	13	14	15	16	17	18	19	20	21	22	23	24	25	26
1441	1	2	3	2	3.71	87	13000	800	2	2	1	4	7	1	20	73	177	2	1	2	4	2	0	1	2	1
1442	2	2	3	5	2.90	89	12500	250	2	2	2	2	5	1	20	61	100	2	1	2	4	1	0	4	4	1
1443	1	4	3	5	3.00	85	10000	1000	2	2	1	4	10	1	22	71	165	2	1	2	4	2	0	4	4	2
1444	1	2	1	2	3.20	88	17000	1000	2	2	2	4	5	1	20	72	180	2	1	2	4	1	1	4	3	1
1445	1	2	3	5	2.50	78	13000	200	2	2	2	4	10	1	20	68	180	2	1	1	3	1	1	4	4	1
1446	2	2	3	1	3.00	80	10400	413	3	2	3	4	6	1	19	65	110	2	1	1	2	2	0	5	3	2
1447	1	4	3	4	2.00	73	10000	300	3	2	2	4	7	1	23	74	160	2	1	2	3	2	2	6	6	1
1448	2	3	3	7	3.30	94	11700	1000	3	2	2	4	7	1	20	62	105	2	1	1	3	2	0	5	5	2
1449	1	2	1	2	1.75	75	13000	50	2	2	2	2	9	1	20	68	150	2	1	2	4	1	0	4	6	1
1450	1	3	3	5	2.80	84	10000	1000	2	2	5	5	5	2	21	73	170	2	1	2	3	1	0	4	6	2
1451	1	1	1	7	3.40	87	9000	2000	2	1	2	4	5	1	18	62	130	1	2	2	4	1	0	4	1	2
1452	1	2	3	2	3.30	86	15000	300	2	1	2	4	7	1	19	67	160	2	1	2	3	2	0	4	5	1
1453	2	2	1	5	3.50	89	12000	2000	2	2	2	2	2	1	19	63	120	2	1	2	2	1	0	4	5	1
1454	2	1	3	2	2.00	80	15000	600	1	2	2	2	6	2	20	59	90	2	1	2	1	1	0	4	4	1
1455	1	3	3	4	3.00	85	12000	1000	1	2	2	2	10	1	20	68	175	2	1	2	3	1	1	1	2	1
1456	1	2	2	2	2.80	91	12000	1000	2	2	1	4	6	1	19	70	220	2	2	1	1	1	1	4	4	1
1457	2	3	1	5	3.00	86	12000	1000	2	2	2	4	3	1	20	66	130	2	2	1	4	2	0	4	5	2
1458	1	2	2	4	2.50	78	10000	100	2	2	2	4	2	2	27	73	182	2	2	2	2	2	0	3	3	1
1459	1	3	2	3	2.50	77	11500	1000	3	2	1	4	1	1	21	68	146	2	1	2	2	2	0	4	3	2
1460	1	1	1	2	2.70	84	12000	800	2	2	1	4	8	1	18	70	210	2	1	2	4	1	0	4	6	2
1461	1	2	2	2	2.50	90	10000	462	2	2	1	3	2	2	30	67	135	1	3	2	3	2	2	4	4	1
1462	2	3	1	2	3.00	82	12000	500	2	2	2	4	0	2	51	66	140	2	1	2	4	2	0	4	4	1
1463	2	2	3	2	3.00	80	12000	400	3	2	2	4	5	1	31	65	105	2	1	2	4	1	0	4	5	1
1464	1	3	3	2	2.80	80	11000	1500	1	2	5	4	1	1	21	71	143	2	1	2	3	2	1	4	6	2
1465	2	3	2	1	3.00	80	10000	1000	1	2	2	4	0	2	20	67	148	2	1	2	1	2	0	4	4	1
1466	2	2	1	4	3.50	83	11000	975	1	2	2	3	5	2	28	66	122	1	4	1	3	2	0	4	4	2
1467	1	2	3	2	2.85	93	12000	833	1	2	2	2	2	2	20	67	122	2	1	2	4	2	0	1	4	2
1468	1	3	2	2	3.00	85	12000	1000	1	2	2	4	6	1	43	73	210	2	3	2	4	1	0	4	5	2
1469	2	3	3	2	3.50	87	15000	800	1	2	2	4	0	1	24	64	108	2	1	2	1	1	0	1	4	1
1470	1	2	1	2	3.80	75	10000	875	1	2	2	4	2	1	48	64	150	2	1	2	4	1	0	4	4	2
1471	2	3	1	4	3.00	80	14000	600	1	2	6	2	4	1	28	67	135	2	1	2	4	1	0	4	4	1
1472	1	2	1	5	2.50	75	15000	500	1	2	2	2	4	1	20	68	195	2	2	1	4	2	0	4	4	1
1473	1	2	1	2	3.00	80	10400	500	2	2	2	2	9	1	20	71	145	2	1	2	4	2	4	4	7	1
1474	2	2	3	2	3.14	83	10400	400	1	2	2	2	10	1	23	61	96	2	1	2	1	2	0	4	7	2
1475	1	3	1	2	2.76	88	12000	1800	2	2	2	4	12	1	19	67	140	2	1	2	4	1	0	4	4	2
1476	1	1	3	2	4.00	85	10000	450	1	1	2	2	4	1	27	79	300	1	1	2	3	2	0	1	3	2
1477	1	2	3	2	3.20	82	12000	1000	3	2	1	3	10	1	20	76	220	2	1	2	4	2	0	4	4	1
1478	2	1	2	1	3.00	80	10000	500	3	2	3	3	2	2	20	61	110	2	1	1	4	2	0	4	5	2
1479	1	2	2	5	3.20	90	15000	1400	3	2	4	4	8	1	20	69	137	2	1	2	4	1	0	4	4	1
1480	2	4	1	5	3.34	85	8000	400	2	2	2	2	4	2	22	61	110	2	1	2	4	2	2	4	4	1
1481	1	1	1	6	3.30	84	11500	1000	1	2	2	2	4	1	26	70	165	1	3	1	4	1	0	4	4	2
1482	2	1	3	1	3.20	89	10000	1800	1	2	5	4	5	2	20	64	135	2	1	2	4	1	0	4	5	2
1483	2	4	1	6	3.30	79	9000	400	3	2	3	4	1	2	23	64	115	2	1	2	4	2	0	4	4	2
1484	1	3	3	2	3.05	87	9000	1000	3	2	2	4	10	1	20	70	133	2	1	2	4	1	0	4	5	1
1485	2	3	2	2	2.97	85	10000	875	3	2	2	2	0	1	23	62	95	2	1	2	1	1	1	4	4	2
1486	1	3	2	2	2.70	85	13000	2000	2	2	1	3	2	1	20	75	210	2	3	1	4	1	2	2	5	2
1487	2	3	1	2	3.00	80	10000	800	3	2	2	4	3	1	34	60	102	2	1	2	1	1	0	4	4	2
1488	1	4	1	2	3.30	95	12000	500	2	2	2	2	6	1	32	73	195	2	1	2	4	1	2	1	2	2
1489	2	3	3	2	2.40	82	12000	2500	2	2	2	4	4	1	20	68	155	2	2	1	3	2	0	1	5	2
1490	1	3	2	2	3.20	85	12000	500	2	1	1	4	5	1	20	69	240	1	1	2	4	2	2	2	4	1
1491	1	3	2	2	2.68	83	12000	1000	3	2	3	4	6	1	20	71	160	2	2	1	3	2	0	1	4	2
1492	1	4	1	2	2.90	76	13000	2500	3	2	2	2	8	2	23	68	173	2	1	1	2	1	1	4	4	1
1493	1	4	1	2	2.70	95	13000	1000	2	2	2	4	5	1	21	70	200	2	1	2	1	2	0	2	5	1
1494	1	2	1	2	3.60	80	15000	1000	3	2	2	4	5	1	20	78	205	2	1	1	4	1	0	5	3	2
1495	2	3	2	4	3.51	88	15000	3000	2	2	1	4	15	2	20	62	98	2	1	2	4	1	2	4	4	1
1496	2	4	1	5	2.70	82	10000	975	2	2	2	4	2	1	35	62	130	2	1	2	4	2	0	1	4	1
1497	2	4	2	5	1.70	70	6240	100	3	2	2	4	2	1	22	64	145	2	1	2	3	2	0	4	3	1
1498	1	4	1	5	3.20	76	10000	500	2	1	2	2	4	1	24	65	140	2	1	2	4	2	0	4	4	2
1499	1	3	1	5	2.90	80	15000	450	3	2	1	3	0	1	22	67	205	2	1	2	4	1	0	4	6	2
1500	1	3	3	2	3.00	75	10000	1500	3	2	2	2	3	1	22	69	140	2	1	2	4	1	0	4	4	2
1501	1	2	3	2	3.80	97	12000	0	3	2	6	2	7	1	21	68	128	2	2	1	4	1	0	4	4	1
1502	2	2	3	2	3.00	90	11700	75	3	2	5	4	8	1	19	65	125	2	1	2	3	1	0	4	5	1
1503	2	4	2	5	2.50	70	9000	300	3	2	2	2	0	1	28	71	160	2	1	2	4	2	0	4	6	1
1504	2	3	2	2	3.44	94	7800	0	3	1	2	4	1	1	20	62	130	2	1	2	4	2	0	3	3	1
1505	2	1	3	6	2.71	94	10100	400	2	2	2	4	3	1	19	62	95	2	1	2	1	1	1	4	4	1
1506	1	3	3	4	2.50	75	12000	400	3	2	2	2	3	1	29	66	140	2	2	1	3	2	1	4	6	1
1507	1	4	1	7	3.20	85	15000	100	3	2	2	2	4	1	29	70	145	2	2	1	3	2	0	1	4	1
1508	1	3	3	5	2.80	85	12000	400	1	2	2	4	3	1	21	69	165	2	2	1	4	1	0	3	4	1
1509	2	3	2	2	2.90	85	11000	100	3	2	2	4	10	2	24	65	125	2	1	2	2	1	0	4	5	2
1510	1	3	2	2	2.50	80	12000	1200	2	2	2	4	4	1	20	68	155	2	1	2	4	2	2	4	5	1
1511	1	3	1	3	3.00	93	12000	1000	1	2	2	4	4	1	28	75	210	2	1	2	1	2	2	4	1	1
1512	1	4	3	4	2.80	70	10400	100	1	1	1	4	3	1	22	70	160	2	1	2	1	2	0	2	4	1
1513	1	3	1	5	3.00	80	12000	400	2	2	2	2	1	1	20	67	126	1	1	2	4	2	1	4	6	1
1514	1	1	1	2	3.20	93	11000	1200	2	2	1	4	5	2	18	68	155	2	1	2	3	1	0	4	6	1
1515	2	3	2	3	2.82	85	10000	500	2	2	1	3	3	1	23	69	140	2	1	2	3	1	1	4	4	1
1516	1	1	3	6	2.50	88	15000	1500	3	2	5	4	8	1	19	70	185	2	2	1	3	2	0	4	4	2
1517	1	3	2	4	2.20	81	10400	400	2	2	2	2	8	1	21	68	135	2	2	1	1	2	0	4	4	1
1518	2	4	3	5	3.30	80	12000	487	3	2	3	4	6	1	25	62	115	2	1	2	4	1	0	1	4	2
1519	1	2	3	1	3.50	87	12000	1000	2	2	1	2	3	1	19	72	200	2	1	2	4	1	0	4	6	1
1520	1	2	1	6	3.50	85	10000	0	2	2	2	4	4	2	23	74	165	2	1	2	1	2	0	1	4	1
1521	2	3	1	5	3.00	80	15000	200	2	2	2	4	10	2	20	75	150	2	2	1	3	2	0	4	4	1
1522	1	2	1	6	3.00	80	14000	300	1	2	6	2	5	1	24	65	125	2	3	2	1	2	0	1	1	1
1523	1	4	3	6	2.40	80	15000	800	2	2	2	2	1	1	27	71	195	2	1	2	2	1	0	5	6	1
1524	1	3	1	2	3.10	82	16000	1000	2	2	2	2	5	1	21	68	160	2	2	1	4	2	1	1	4	2
1525	1	3	1	2	3.60	92	10500	0	2	2	2	2	2	1	24	70	175	2	1	2	4	2	1	2	6	2
1526	1	3	3	2	2.50	75	17000	1000	1	2	2	2	2	1	31	72	175	2	1	2	4	2	0	5	5	2
1527	1	2	3	2	2.85	89	12000	500	2	2	5	4	8	1	19	71	155	2	1	1	3	2	0	4	4	1
1528	1	3	1	2	3.30	72	11000	600	3	2	2	4	5	1	21	73	170	2	1	2	3	2	1	4	4	1
1529	1	2	1	2	3.00	80	12000	1000	2	2	5	3	4	1	20	68	159	2	1	2	4	1	0	5	2	1
1530	1	3	2	2	3.00	78	13000	500	3	2	2	4	2	1	21	71	145	2	1	2	3	2	0	4	7	2
1531	2	2	2	2	3.30	90	16000	100	3	2	2	3	6	1	19	63	104	2	3	1	4	2	0	1	4	2
1532	1	2	3	2	2.90	85	13000	0	2	2	5	4	7	1	19	71	185	2	1	2	2	2	1	4	6	2
1533	2	4	1	2	3.00	90	13000	1000	2	2	2	4	4	1	21	63	110	2	1	2	3	1	1	4	5	1
1534	2	3	2	2	2.35	90	13000	500	2	2	2	4	5	1	20	65	145	2	3	2	1	2	2	1	2	2
1535	2	3	3	2	2.00	90	13000	0	3	2	2	2	0	2	23	64	120	2	1	2	1	2	0	4	4	2
1536	2	3	3	2	3.20	88	12000	1000	2	2	1	4	5	1	20	68	112	2	1	2	4	2	0	4	1	1
File code number	Sex	Class	Grad. school	Major	Grade-point index	H. S. average	Est. starting salary $	Tuition $	Employment	Sunday shop	Investment	Car pref.	No. jeans owned	Calculator	Age	Height	Weight	Blood pressure	Smoking status	Smoking belief	Entrance standards	Retention standards	No. clubs and groups	SPS rating	Library rating	Personal question

File code number	1	2	3	4	5	6	7	8	9	10	11	12	13	14	15	16	17	18	19	20	21	22	23	24	25	26
	Question number																									
1537	2	3	2	2	2.50	87	13000	500	2	2	5	3	10	1	21	71	140	2	1	1	4	1	0	4	4	1
1538	1	3	1	2	3.40	93	14000	850	2	1	2	4	3	1	21	74	175	2	1	2	3	1	0	7	7	1
1539	1	2	3	2	2.70	93	14000	350	1	2	3	3	8	1	20	71	168	2	1	2	3	1	0	2	4	2
1540	1	4	2	2	3.30	92	13000	600	3	2	2	4	6	1	21	66	165	2	1	2	4	1	0	1	5	1
1541	2	2	3	2	2.80	87	10000	900	2	2	5	4	5	1	20	63	115	2	1	2	4	2	0	4	4	2
1542	1	3	3	2	3.30	87	11000	1500	2	2	2	2	3	2	22	70	171	2	2	2	4	1	0	6	5	2
1543	1	4	3	2	3.00	70	10000	200	2	2	2	4	2	1	22	66	145	2	3	2	1	1	0	4	6	1
1544	1	4	3	4	2.67	77	15300	1500	1	2	2	2	1	1	47	69	185	2	1	2	4	1	0	4	4	2
1545	2	4	1	2	3.50	95	12575	100	2	1	5	5	0	1	19	60	105	2	1	1	1	2	6	4	4	2
1546	2	4	1	4	3.50	85	10000	1000	1	1	2	4	6	1	29	63	115	2	3	2	3	2	0	4	4	2
1547	1	3	1	7	2.90	78	12000	0	1	2	2	2	1	1	36	68	190	1	1	2	3	1	0	4	4	2
1548	1	2	3	1	3.27	83	10000	850	1	2	2	2	2	1	29	68	158	2	1	2	4	1	0	4	4	2
1549	1	2	2	2	3.41	90	12000	0	1	2	2	4	0	2	50	67	165	2	1	2	4	1	0	4	4	2
1550	1	4	3	5	3.60	96	13500	0	1	2	2	2	0	1	41	67	150	2	1	2	1	2	0	1	1	2
1551	2	4	3	4	3.40	93	18000	1000	1	2	2	4	2	1	32	70	185	2	1	2	4	2	1	4	4	2
1552	2	3	3	5	3.20	87	19000	500	1	2	5	4	2	1	29	65	120	2	1	2	4	1	0	3	3	2
1553	1	3	3	3	3.00	75	10000	1000	2	1	1	4	4	1	23	71	175	2	1	1	3	1	0	4	6	1
1554	1	3	1	5	3.06	90	10000	1000	2	2	1	4	3	1	21	74	150	2	1	2	3	1	1	3	4	1
1555	1	4	1	3	3.02	88	12000	1230	3	2	2	2	1	1	26	68	160	2	1	2	4	1	0	4	4	2
1556	1	2	1	5	3.01	85	15000	400	2	2	2	2	10	1	20	67	128	2	1	1	4	2	0	1	2	1
1557	1	3	3	4	2.56	92	12000	450	3	2	2	2	2	2	21	66	142	1	1	1	1	2	0	4	7	1
1558	1	1	1	1	3.75	93	12000	500	1	2	2	2	4	2	18	73	160	2	1	2	4	1	2	4	4	1
1559	2	2	3	2	3.08	91	10000	900	2	2	2	4	10	1	20	62	130	2	1	1	4	2	0	4	6	2
1560	1	2	1	2	3.32	90	14000	500	2	2	2	2	4	1	19	71	140	2	1	2	3	2	0	4	5	1
1561	1	3	3	2	3.04	83	11000	1800	3	2	2	4	15	1	20	71	150	2	1	2	3	2	1	4	4	1
1562	2	3	2	6	3.02	85	10000	200	3	1	2	4	10	1	23	65	97	2	1	2	4	2	0	4	3	1
1563	2	3	1	2	3.03	88	13500	1000	2	2	2	4	3	2	19	67	142	2	1	1	4	1	1	4	6	2
1564	2	3	1	4	2.73	76	10000	1000	3	2	2	4	4	1	23	71	121	2	1	2	3	1	0	4	4	2
1565	1	2	3	2	3.32	91	14000	250	2	1	2	2	6	2	19	72	170	2	1	2	3	1	0	2	6	1
1566	1	2	1	2	3.43	85	15000	2000	2	1	5	4	0	1	19	68	250	1	3	1	4	2	0	4	4	2
1567	1	3	2	2	3.66	89	15000	1000	2	2	1	3	4	1	18	70	160	2	1	2	4	1	0	4	4	1
1568	1	3	3	3	3.12	79	12000	600	2	2	2	4	7	1	21	65	155	2	1	1	3	2	0	2	4	2
1569	1	2	3	2	3.92	97	18000	2500	3	2	3	4	0	1	19	70	140	2	1	2	4	1	1	4	5	2
1570	2	4	3	4	2.50	73	13000	500	3	2	1	2	5	1	23	66	135	2	1	1	2	2	1	2	4	1
1571	2	2	2	1	2.08	90	15000	1000	3	2	2	2	5	1	19	61	115	2	1	2	1	1	1	5	4	2
1572	1	3	3	7	2.51	83	10400	300	2	2	2	4	5	1	20	70	150	2	1	1	1	2	0	2	6	1
1573	1	3	1	2	2.92	88	11500	1000	2	2	2	4	5	1	21	70	150	2	1	2	3	2	0	5	6	1
1574	2	3	1	4	2.95	78	13500	500	2	2	2	4	7	1	28	66	132	2	1	2	4	2	2	5	6	1
1575	2	2	2	1	2.04	75	10000	200	3	2	2	4	12	2	23	67	110	1	2	1	4	2	0	4	4	1
1576	1	2	3	2	3.00	87	12000	1000	1	2	5	4	6	1	28	71	155	2	1	2	3	1	0	4	4	1
1577	1	4	1	3	3.79	97	15000	200	1	2	6	2	0	1	29	68	149	2	1	2	4	1	0	3	3	2
1578	2	4	3	7	2.52	75	15000	1000	1	2	2	2	1	1	45	66	145	2	1	2	1	2	0	4	4	2
1579	2	4	1	7	3.04	78	15000	1000	1	2	2	2	4	1	26	66	200	2	1	1	4	1	0	4	7	2
1580	1	2	3	4	2.02	89	17000	1000	1	2	2	4	1	1	20	68	189	2	4	1	3	1	0	4	5	2
1581	1	4	3	4	2.61	77	11500	2000	1	2	2	5	3	1	23	67	140	2	1	1	3	2	0	4	6	2
1582	1	2	1	2	3.22	86	12000	1500	3	2	3	5	5	1	29	67	150	2	1	2	3	1	0	4	3	2
1583	1	4	1	3	3.50	79	11600	2300	3	2	2	2	4	1	26	71	145	2	2	2	1	1	1	4	5	2
1584	1	3	1	2	2.31	86	15000	975	1	2	2	2	3	1	26	74	175	1	2	2	1	2	0	1	4	1
1585	2	3	3	7	3.61	85	12000	1500	3	2	2	4	3	1	24	66	123	2	2	2	4	1	0	4	5	2
1586	1	2	2	2	3.88	90	12000	1500	2	2	2	4	3	1	19	75	160	2	1	2	3	1	0	3	5	2
1587	1	3	3	7	3.27	84	12000	900	3	1	2	4	2	1	22	72	200	1	2	1	4	2	0	1	4	2
1588	1	3	1	3	2.71	79	20000	700	1	1	2	2	1	2	28	68	145	2	1	2	1	2	0	4	4	2
1589	1	3	2	2	2.00	82	12000	1000	1	2	2	2	3	1	31	70	170	1	1	2	1	2	0	4	4	1
1590	1	2	1	6	3.00	89	10000	0	2	2	6	2	6	1	20	63	135	2	1	2	3	1	1	3	6	1
1591	1	3	3	2	3.52	85	14000	2000	1	2	1	4	4	1	23	68	175	2	1	2	4	2	0	1	5	2
1592	1	3	1	2	3.43	90	12000	500	1	2	2	2	0	1	23	64	150	2	1	2	1	1	1	4	4	1
1593	1	3	1	5	2.54	78	15000	500	1	2	2	2	2	1	37	68	162	2	1	2	4	1	0	4	4	2
1594	2	3	2	5	3.17	85	15000	100	2	2	2	4	2	1	20	65	120	2	2	2	1	2	2	2	5	1
1595	1	4	1	4	3.09	83	18200	1000	1	1	2	2	2	1	32	66	206	2	1	2	2	1	3	4	4	2
1596	2	4	3	2	3.50	90	12500	1200	2	2	5	2	4	1	22	62	98	2	3	2	4	1	1	4	4	1
1597	2	2	1	2	3.02	82	4250	0	3	2	2	4	3	1	33	64	150	2	3	2	4	1	0	4	4	2
1598	1	4	1	5	3.04	83	13000	975	1	2	2	4	4	1	38	73	185	2	1	2	1	2	2	5	2	1
1599	1	1	1	2	3.60	91	15000	0	2	2	2	4	2	2	24	69	155	2	1	2	1	1	0	6	7	1
1600	2	2	2	2	3.03	93	10000	1000	2	2	2	4	5	1	20	67	125	2	1	2	4	1	0	4	4	2
1601	1	4	1	5	2.82	84	12000	500	1	2	2	4	4	1	23	69	150	2	3	2	4	2	0	4	2	1
1602	2	2	1	2	2.91	91	10000	1000	3	1	2	4	1	2	21	64	119	2	1	2	3	2	0	4	5	1
1603	1	3	3	6	3.02	82	11000	900	3	2	2	4	5	1	29	69	150	2	4	1	4	1	0	4	4	1
1604	2	4	2	1	2.54	78	12000	200	2	1	2	4	3	2	22	63	135	2	1	2	2	2	0	4	4	1
1605	1	3	1	2	3.02	78	11000	600	2	2	2	4	12	1	24	71	165	2	1	2	4	1	0	4	5	2
1606	1	1	3	7	2.00	77	12760	1000	2	2	2	2	3	2	19	66	165	2	1	2	4	2	0	4	4	1
1607	1	4	3	2	2.89	85	25000	400	1	1	2	4	5	1	21	69	175	2	1	2	3	2	0	4	4	2
1608	1	3	2	5	2.22	78	12000	900	2	2	3	3	4	1	20	70	165	2	2	1	4	2	0	3	5	2
1609	1	3	1	7	2.73	83	9000	900	2	2	2	4	2	1	29	69	185	2	2	2	4	1	1	4	6	1
1610	2	3	3	7	2.00	77	12000	500	3	2	2	2	6	2	23	63	114	2	2	2	1	2	0	4	7	1
1611	2	4	1	7	2.95	93	10000	1000	2	2	2	4	4	2	20	67	113	2	2	2	4	1	0	4	4	1
1612	2	4	3	4	2.52	83	15000	0	3	2	2	2	2	1	30	66	138	2	1	2	1	1	0	1	4	2
1613	1	4	2	3	2.80	78	11000	1000	2	2	2	2	3	1	21	68	138	2	2	2	4	1	1	4	6	1
1614	1	4	1	5	2.54	75	12000	100	2	2	2	4	5	1	24	70	165	2	1	2	4	1	0	1	4	1
1615	2	3	2	1	3.00	87	9000	1000	2	2	2	2	15	1	20	62	115	2	1	1	3	1	2	5	4	1
1616	1	3	3	5	2.75	80	12000	1000	2	1	5	4	5	1	19	68	140	2	1	2	4	2	0	1	4	2
1617	1	4	1	7	3.20	95	8000	1000	2	2	1	3	1	2	26	70	140	2	2	2	4	1	2	4	6	1
1618	1	3	2	5	3.22	94	15000	1000	1	2	2	4	12	1	22	66	150	2	3	1	4	1	0	5	7	1
1619	2	2	3	2	3.50	89	15000	1000	2	2	2	4	6	1	19	60	121	2	1	2	4	1	0	3	6	2
1620	2	3	2	5	3.40	94	13000	0	2	2	5	3	6	1	18	66	135	2	1	1	4	2	1	4	1	2
1621	2	2	2	2	3.30	90	12500	600	3	2	1	3	0	1	18	67	140	2	1	2	4	1	0	4	4	1
1622	2	4	3	5	2.88	85	10000	0	2	2	2	4	10	1	23	66	140	2	2	2	4	2	0	4	1	2
1623	1	2	3	2	2.86	82	10000	1000	3	1	2	4	10	1	21	63	110	2	1	2	4	2	0	4	3	2
1624	1	4	3	7	3.24	94	12000	400	3	2	2	4	5	1	23	70	180	2	1	2	4	2	0	4	6	1
1625	1	4	1	2	3.02	87	11000	800	3	2	2	4	5	1	21	70	150	2	1	2	3	1	0	4	4	1
1626	1	3	1	2	3.20	91	12000	1000	1	2	5	3	2	1	20	69	190	1	1	2	4	1	4	3	5	2
1627	1	4	1	5	2.56	85	10400	500	1	2	1	4	10	1	21	72	168	2	2	1	3	1	0	1	7	1
1628	1	3	3	2	3.00	90	12000	500	2	1	2	3	6	2	29	69	160	2	1	2	4	2	0	4	4	2
1629	1	3	1	3	2.52	79	10400	400	2	2	2	2	5	1	20	69	146	2	1	2	3	1	0	3	6	1
1630	1	3	1	4	3.25	79	10000	1000	2	2	2	4	3	2	32	69	160	2	3	1	2	1	1	4	5	2
1631	1	4	1	7	3.23	90	10000	1200	2	2	2	2	4	1	24	65	170	2	2	1	4	1	3	6	6	1
1632	1	4	1	4	2.57	78	10000	0	3	2	1	3	4	1	29	68	145	2	1	2	2	1	2	4	5	1
File code number	Sex	Class	Grad. school	Major	Grade-point index	H. S. average	Est. starting salary $	Tuition $	Employment	Sunday shop	Investment	Car pref.	No. jeans owned	Calculator	Age	Height	Weight	Blood pressure	Smoking status	Smoking belief	Entrance standards	Retention standards	No. clubs and groups	SPS rating	Library rating	Personal question

File code number	Question number																									
	1	2	3	4	5	6	7	8	9	10	11	12	13	14	15	16	17	18	19	20	21	22	23	24	25	26
1633	1	3	3	5	3.08	83	14000	2000	1	2	2	2	4	2	27	70	172	2	2	2	4	1	0	1	4	1
1634	1	3	3	5	2.93	85	14000	900	2	1	2	4	5	1	21	71	170	2	1	2	1	2	0	2	6	2
1635	1	3	1	5	3.09	82	14000	2500	3	2	2	4	10	1	20	70	190	2	1	2	3	2	0	5	5	1
1636	2	3	2	5	3.00	76	14000	300	2	2	3	4	10	1	19	65	120	1	1	2	3	2	2	1	3	2
1637	1	4	2	5	2.80	79	11000	1500	1	2	2	2	0	1	21	69	140	2	1	1	4	1	0	1	2	2
1638	1	4	3	5	2.57	78	13500	1000	2	2	1	4	3	1	21	76	175	2	1	2	3	2	0	4	4	2
1639	1	3	3	5	3.51	90	12000	800	2	1	5	4	5	1	19	69	145	2	1	2	4	2	0	2	4	2
1640	1	4	3	5	3.05	79	14000	1000	3	2	5	2	4	1	30	66	145	1	1	2	3	1	0	4	6	1
1641	2	3	1	5	2.84	81	13500	0	2	2	2	4	2	2	22	64	115	2	1	1	1	2	0	3	2	1
1642	2	3	3	5	2.77	83	11000	1500	3	1	1	1	7	1	20	68	122	2	3	1	4	1	0	2	1	1
1643	2	3	2	5	2.51	85	12000	750	3	2	2	2	0	2	24	62	100	2	1	2	1	1	0	4	2	1
1644	1	4	2	5	2.50	81	15000	975	2	2	2	4	5	1	24	66	170	2	1	2	3	2	0	4	3	1
1645	2	4	1	5	3.70	95	12000	600	2	2	2	4	3	1	20	63	135	2	1	2	2	1	1	4	4	1
1646	1	3	3	2	3.62	80	13000	1200	2	1	3	2	3	1	23	60	150	2	2	2	1	2	3	4	5	1
1647	1	4	1	2	3.64	93	9000	1000	3	2	2	4	1	1	19	73	125	2	1	2	4	1	0	4	5	2
1648	1	2	2	2	2.52	75	14000	500	2	2	4	4	2	1	20	72	160	2	1	2	4	1	0	1	4	2
1649	1	2	1	2	3.53	89	20000	700	2	2	2	2	18	1	20	71	145	2	1	2	4	1	1	1	6	1
1650	1	3	2	2	2.98	80	10000	412	3	2	2	4	4	1	21	67	150	2	1	2	3	1	0	2	4	1
1651	2	2	3	2	3.10	91	14000	600	2	2	2	4	10	1	20	63	100	2	1	2	3	2	1	4	4	1
1652	2	4	1	2	3.40	91	13500	1000	1	2	5	4	10	1	20	60	110	2	1	1	3	1	4	4	5	2
1653	1	3	2	2	2.52	82	15000	1000	2	2	1	4	5	1	19	67	160	2	1	2	2	2	1	2	5	1
1654	1	3	2	2	3.42	87	14000	1000	3	2	2	4	7	1	20	72	175	2	1	2	4	2	0	3	3	2
1655	2	3	3	2	3.43	95	12000	750	3	2	6	2	0	1	20	66	145	2	1	2	1	2	0	4	5	1
1656	1	3	1	2	3.33	79	11600	975	3	2	2	4	2	1	21	71	150	2	1	1	3	2	0	4	4	2
1657	2	3	3	2	2.72	80	15000	1200	3	2	1	4	2	1	29	60	108	2	1	2	4	1	0	2	4	2
1658	2	4	2	5	2.90	80	10000	600	2	2	5	3	7	1	23	61	95	2	2	2	4	1	1	2	4	1
1659	2	2	3	2	3.21	85	11000	1000	2	2	2	2	5	1	20	62	105	2	1	2	2	2	2	4	5	1
1660	2	4	1	2	3.16	81	12000	1000	1	2	2	2	2	1	24	65	130	2	1	2	4	1	2	3	1	1
1661	1	3	1	3	3.84	80	11500	1000	3	2	1	3	4	1	22	70	175	2	1	2	4	1	1	1	3	2
1662	2	3	2	2	2.84	85	15000	0	2	2	2	2	6	1	21	60	115	2	1	2	1	2	0	3	4	1
1663	2	3	2	2	2.72	83	15600	0	3	2	2	2	5	1	22	60	95	2	1	2	3	2	0	4	4	2
1664	1	3	3	2	2.91	85	13500	3000	3	2	2	2	4	1	20	68	131	2	1	2	3	2	0	4	5	1
1665	2	3	3	2	2.53	84	12000	0	2	2	2	2	2	1	27	68	110	2	1	2	3	1	1	4	6	2
1666	1	2	1	4	3.06	80	13000	750	2	2	2	2	12	1	21	67	145	2	1	2	3	2	3	1	7	2
1667	1	2	3	6	3.92	93	14500	600	2	2	2	4	3	1	19	70	150	2	1	1	3	2	0	4	4	1
1668	2	4	2	7	2.28	80	11000	0	2	2	2	4	6	2	22	64	135	2	1	1	4	1	0	4	6	1
1669	1	3	3	6	2.97	83	11500	300	2	2	5	4	4	1	21	69	140	2	1	2	4	1	0	1	4	2
1670	1	2	3	6	2.04	80	14500	0	2	2	2	4	5	1	20	64	135	2	1	1	2	2	0	4	6	2
1671	1	2	2	5	2.35	84	16000	1000	2	2	2	4	2	1	16	67	135	2	1	2	4	2	0	4	4	1
1672	1	4	1	5	3.62	85	12000	400	2	2	1	4	3	1	22	73	155	2	1	2	3	1	0	4	1	2
1673	2	2	2	6	3.04	89	10000	200	1	2	2	2	8	1	20	63	108	2	1	2	3	1	0	2	3	1
1674	2	3	3	6	3.02	88	12000	1800	2	2	2	2	8	1	21	61	90	2	1	2	3	2	1	4	4	1
1675	1	3	1	5	2.63	81	13500	0	3	2	2	2	8	1	28	72	170	2	1	2	3	2	2	4	4	1
1676	1	3	3	3	3.54	87	12500	0	2	2	2	4	0	1	21	69	155	2	1	2	1	2	0	4	4	2
1677	1	4	1	5	3.61	85	12500	1000	3	2	2	4	1	1	29	72	160	2	4	1	4	1	0	4	6	2
1678	2	2	3	4	3.04	91	15000	500	3	2	3	4	2	1	20	64	160	2	1	2	3	2	0	4	7	1
1679	1	2	1	3	3.32	96	12500	0	2	2	2	4	4	1	19	72	178	2	1	2	4	1	3	4	4	1
1680	2	3	2	4	3.01	80	12000	500	3	2	1	3	10	1	19	71	155	2	1	1	4	1	0	6	7	2
1681	1	2	1	6	3.07	91	16500	0	2	2	6	2	6	1	20	63	120	2	1	2	3	1	1	1	5	1
1682	1	3	1	3	3.29	73	14500	900	2	2	3	2	2	1	34	68	161	2	1	2	4	2	0	4	4	2
1683	1	4	3	5	2.24	84	10000	300	2	2	2	2	3	1	23	64	160	2	3	1	3	2	1	4	3	1
1684	1	2	3	2	2.51	83	12000	1000	2	2	1	4	4	2	18	71	140	2	1	2	3	2	0	2	5	1
1685	1	2	1	4	2.52	85	15000	1000	2	2	2	4	5	1	22	72	175	1	2	1	4	1	3	4	5	2
1686	1	3	1	7	3.03	86	10000	200	2	2	2	4	6	2	22	75	270	1	4	1	3	1	0	4	6	1
1687	2	3	1	5	2.82	78	15000	0	3	2	2	2	4	1	27	66	105	2	2	1	4	2	3	3	5	1
1688	1	3	1	5	2.75	85	10000	750	3	2	2	4	4	1	20	70	150	2	2	2	1	1	0	5	6	2
1689	2	4	3	2	3.40	85	10500	0	3	2	2	2	8	1	22	66	117	2	1	2	3	2	4	6	6	1
1690	2	3	2	2	3.11	94	13000	2000	2	2	2	4	10	1	21	66	122	1	1	2	3	2	0	4	4	2
1691	2	3	3	2	3.10	90	10000	0	3	2	2	4	0	1	30	62	126	1	2	2	3	2	0	4	4	2
1692	1	3	2	2	2.82	88	12000	600	3	2	2	2	0	2	21	72	175	2	1	2	1	2	0	4	4	2
1693	1	3	3	2	3.37	82	11500	500	3	2	2	2	7	2	20	73	185	2	1	2	4	2	2	4	7	1
1694	1	3	2	2	2.55	85	11000	500	2	2	2	4	5	1	20	68	147	2	3	1	4	2	0	4	5	1
1695	1	3	1	5	2.75	81	15000	487	2	1	5	4	20	1	22	72	165	2	1	2	4	1	2	1	4	1
1696	2	3	1	5	3.50	75	12000	970	2	2	2	4	3	1	21	67	140	2	3	2	4	1	0	4	4	1
1697	2	2	3	2	2.00	80	15003	800	2	2	6	2	6	1	21	63	110	2	1	2	3	2	2	2	5	2
1698	1	3	3	4	2.30	85	13000	2000	1	2	2	4	5	1	21	71	165	2	1	2	4	2	0	4	5	1
1699	1	4	3	5	2.72	81	8000	1400	1	2	2	2	1	1	31	69	140	2	1	2	3	2	0	4	2	2
1700	2	3	3	5	2.81	86	15000	1000	3	2	1	4	5	1	22	64	115	2	3	2	3	1	0	1	5	2
1701	1	3	2	2	2.94	90	12000	1500	2	2	2	2	0	1	21	72	195	2	1	2	4	2	0	4	6	2
1702	1	3	3	5	2.87	84	10000	1000	2	2	2	2	4	1	20	69	155	2	1	2	4	2	3	4	4	1
1703	1	2	1	2	2.66	82	12000	1000	2	2	2	4	8	1	20	72	205	1	1	2	4	2	1	4	5	1
1704	2	4	2	7	3.03	72	8680	800	2	2	2	4	2	1	22	62	120	2	1	2	4	1	0	4	4	1
1705	1	1	2	4	3.44	88	12000	1000	3	2	2	4	8	1	19	72	165	2	1	2	3	2	0	4	3	2
1706	1	3	1	5	2.51	82	13000	1500	2	2	3	4	4	1	21	66	140	2	1	2	2	1	0	1	2	2
1707	1	3	3	6	2.80	85	15000	500	2	2	2	4	3	1	20	64	125	2	2	1	3	2	0	4	5	1
1708	1	3	1	5	2.89	92	16000	600	3	2	6	2	0	2	21	65	163	1	1	2	3	2	2	5	6	1
1709	1	3	3	7	3.14	76	10000	1800	2	2	2	4	3	1	22	76	210	2	3	1	4	2	0	4	6	2
1710	1	2	3	1	2.90	86	13500	850	3	2	5	4	10	2	23	73	200	2	3	2	4	2	0	1	3	2
1711	1	2	1	3	3.14	72	15000	0	1	2	2	4	6	1	20	72	180	2	1	2	3	1	1	1	4	2
1712	1	2	3	5	3.07	80	10000	412	2	2	2	4	5	2	21	67	140	2	2	1	1	2	1	4	4	1
1713	1	3	1	4	3.23	85	14000	100	2	2	2	2	4	1	26	72	178	2	3	1	1	2	1	5	6	1
1714	1	3	1	4	2.36	81	16000	1000	3	2	2	4	4	1	20	71	172	2	1	1	3	2	1	1	1	1
1715	2	3	3	5	3.47	88	10000	1000	2	2	2	4	6	2	21	62	108	2	1	1	4	2	1	4	4	1
1716	2	4	3	5	2.80	83	10000	500	2	2	2	4	5	2	22	64	130	2	1	2	4	1	1	2	4	1
1717	2	2	2	2	3.11	91	10000	400	2	2	2	4	15	1	20	63	98	2	1	1	3	2	0	4	4	1
1718	2	2	2	2	3.62	87	12000	2000	3	2	1	4	7	1	18	62	115	2	2	2	3	2	2	4	4	1
1719	1	2	3	2	3.20	89	12500	500	2	2	2	4	5	1	22	72	185	2	1	1	3	1	2	3	3	2
1720	2	3	1	2	2.50	85	12000	0	3	2	2	2	10	1	21	63	130	1	2	2	3	2	4	5	5	1
1721	2	3	2	2	3.04	87	10000	3000	3	2	2	2	4	2	22	63	107	2	1	2	2	2	1	3	4	2
1722	2	2	3	2	2.50	75	10000	1000	2	2	5	2	2	1	20	60	85	2	1	2	2	1	0	1	4	1
1723	1	4	3	4	2.02	70	10000	975	3	2	2	2	4	1	22	71	156	1	2	2	2	1	2	4	4	1
1724	1	2	3	2	3.52	90	10000	1000	3	2	2	4	3	1	18	70	153	2	1	2	2	2	0	4	5	1
1725	1	3	2	2	2.53	82	12000	450	2	2	3	4	1	2	20	68	130	2	1	2	3	2	0	4	4	2
1726	2	3	2	2	3.02	91	15000	1000	3	2	3	4	5	1	19	63	150	2	1	2	4	2	0	4	4	1
1727	1	2	1	2	3.33	78	9000	1200	2	2	3	4	8	1	19	66	145	2	1	2	4	1	0	2	1	1
1728	2	2	3	2	3.07	80	7000	675	2	2	2	2	7	1	21	65	123	2	1	2	1	1	0	4	4	1
File code number	Sex	Class	Grad. school	Major	Grade-point index	H. S. average	Est. starting salary $	Tuition $	Employment	Sunday shop	Investment	Car pref.	No. jeans owned	Calculator	Age	Height	Weight	Blood pressure	Smoking status	Smoking belief	Entrance standards	Retention standards	No. clubs and groups	SPS rating	Library rating	Personal question

File code number	Question number 1	2	3	4	5	6	7	8	9	10	11	12	13	14	15	16	17	18	19	20	21	22	23	24	25	26
1729	2	2	3	2	3.51	90	10000	200	3	2	2	2	7	1	21	62	130	2	1	1	1	2	0	4	5	2
1730	1	3	2	2	3.73	92	12000	1001	2	2	2	5	5	1	20	68	140	2	1	2	4	1	0	4	4	1
1731	1	3	1	2	3.02	85	16000	500	2	1	5	3	2	1	21	68	145	2	1	2	4	2	1	3	3	2
1732	1	2	1	2	3.38	89	12000	1200	3	2	2	4	2	2	20	72	166	2	2	2	2	1	0	5	5	1
1733	1	3	2	2	3.50	93	15000	250	3	1	1	4	10	1	20	70	155	2	1	2	1	2	1	1	1	1
1734	1	2	1	2	3.79	91	13000	500	3	2	1	4	0	1	20	69	160	2	1	2	4	1	0	1	4	2
1735	1	3	1	2	3.60	90	13000	1000	2	2	5	4	6	1	19	72	185	2	1	2	4	1	1	5	6	1
1736	2	3	3	2	2.54	88	10000	1000	2	2	2	2	12	1	19	64	115	2	1	1	3	2	0	4	4	1
1737	2	2	3	2	2.82	85	7000	250	2	2	6	2	6	1	21	65	113	2	1	2	2	2	1	3	4	1
1738	1	2	3	2	2.76	75	13000	250	2	1	1	3	10	1	19	67	120	2	3	1	4	2	0	4	6	2
1739	1	3	1	2	2.04	80	14000	0	3	2	2	2	3	2	27	72	210	2	1	2	1	1	0	4	7	1
1740	1	4	1	2	3.02	81	12500	1200	3	2	2	4	10	1	21	68	160	2	1	2	4	1	1	4	4	2
1741	1	4	1	2	2.50	85	12000	500	3	2	2	4	6	1	22	71	185	2	1	2	4	1	2	4	5	2
1742	1	2	1	2	2.51	80	13000	500	3	2	2	4	8	1	20	66	115	2	1	2	4	1	0	2	2	1
1743	1	3	2	2	3.20	88	14000	2000	3	1	2	2	4	1	21	70	140	2	1	2	1	2	0	4	4	1
1744	1	1	1	3	2.76	84	10000	1500	1	2	5	4	5	1	26	68	170	2	1	2	4	1	1	6	7	2
1745	1	3	2	5	2.54	85	10000	600	2	2	2	4	6	1	20	60	110	2	1	2	4	2	1	4	3	2
1746	1	3	2	4	2.42	78	13500	600	2	2	5	3	10	1	21	72	142	2	1	1	3	2	1	5	7	2
1747	2	3	3	1	2.67	85	10000	1000	2	2	2	4	2	1	20	66	145	2	1	2	4	1	1	3	3	2
1748	2	3	3	7	2.83	80	10000	1500	3	2	2	4	4	1	20	60	100	2	3	2	4	2	0	4	3	2
1749	1	3	1	6	3.00	94	16000	900	3	2	2	2	4	1	20	64	160	2	1	1	1	1	1	3	5	1
1750	1	2	3	2	2.54	80	12000	900	2	2	2	4	3	1	19	72	165	2	1	1	1	2	0	5	3	2
1751	2	1	3	1	3.90	94	11700	1200	2	2	1	4	10	1	19	61	110	2	1	2	3	2	0	5	5	1
1752	2	2	2	2	3.28	81	10500	2000	2	2	5	4	5	1	18	63	150	2	2	1	3	2	1	6	5	1
1753	1	2	3	5	3.02	72	25000	1200	2	2	3	3	6	1	19	72	175	2	1	2	4	1	0	1	6	1
1754	1	2	1	2	2.81	82	15000	400	1	2	2	2	4	1	29	64	168	2	1	2	2	1	2	6	6	2
1755	2	2	3	2	3.03	95	5000	100	1	2	6	2	3	1	26	66	115	1	1	2	4	1	0	4	5	1
1756	1	1	3	6	3.54	85	14000	700	2	2	2	2	2	1	21	69	150	2	1	1	4	1	0	4	5	1
1757	1	3	3	3	2.50	79	13000	200	1	2	2	2	2	2	40	65	130	2	1	2	4	2	0	4	6	1
1758	1	2	3	2	2.06	80	10400	0	3	2	5	4	1	1	20	73	174	2	2	1	3	1	0	5	6	1
1759	1	2	3	2	3.12	90	13500	800	3	2	2	4	5	2	20	72	135	2	1	2	3	1	0	2	6	2
1760	2	3	1	5	3.02	75	13500	1000	2	2	2	4	3	1	21	69	130	2	1	2	4	1	1	3	7	1
1761	2	3	2	5	2.30	85	15000	1500	3	2	2	3	5	1	21	63	108	2	1	2	4	2	1	4	5	1
1762	1	3	3	5	3.00	87	10000	1000	3	2	2	4	3	2	21	70	160	2	1	2	4	2	2	4	6	1
1763	1	1	1	2	3.14	75	13500	200	1	2	6	2	10	1	18	64	120	2	1	2	3	1	3	4	4	1
1764	2	3	3	4	2.80	81	12000	200	3	2	2	4	6	2	25	59	85	2	2	2	4	1	0	1	7	1
1765	1	3	3	2	3.41	79	9500	900	1	2	2	2	0	2	29	68	140	2	1	2	1	2	0	4	5	1
1766	1	3	3	5	2.92	78	11000	1000	2	2	2	2	3	1	26	68	155	2	1	2	1	2	0	4	4	1
1767	2	3	1	7	2.70	77	15000	500	3	2	2	4	10	1	30	66	160	1	2	1	1	2	2	4	7	1
1768	2	4	1	7	2.86	78	15000	1000	1	2	2	4	10	1	21	67	118	2	1	2	4	2	1	3	3	2
1769	2	3	2	5	3.12	91	11000	300	3	2	2	2	11	1	21	60	95	2	1	1	1	2	1	3	3	1
1770	2	3	3	5	3.21	94	10400	300	2	2	5	4	15	1	20	65	125	2	1	2	3	2	1	5	4	1
1771	1	3	2	2	2.74	83	10000	500	2	2	2	2	12	1	21	72	180	2	2	1	4	1	2	1	5	2
1772	1	2	1	2	2.93	88	10000	600	2	1	1	4	11	1	21	75	179	2	2	1	4	1	0	2	5	2
1773	1	4	1	2	2.37	72	18000	1000	1	2	2	2	12	1	31	67	130	2	3	2	4	2	0	4	1	1
1774	2	2	1	3	3.40	90	17000	1000	3	2	2	2	4	2	19	65	120	2	1	2	4	2	3	4	4	1
1775	1	3	3	4	2.40	80	10000	500	2	2	1	4	8	1	20	68	150	2	1	2	3	2	0	3	6	1
1776	1	2	1	1	2.88	83	20000	413	2	2	2	4	2	1	19	75	185	2	2	1	4	2	1	1	4	1
1777	1	3	2	5	2.39	82	10000	100	2	2	3	5	2	1	20	73	185	2	1	2	3	1	0	4	5	1
1778	2	2	3	5	2.60	75	8400	1600	3	2	2	4	5	2	21	70	131	2	1	2	3	2	0	4	4	1
1779	1	2	1	3	2.51	79	15000	1000	3	2	2	2	1	2	30	67	155	2	1	2	1	2	0	4	6	2
1780	1	4	2	4	2.00	71	15000	600	1	2	2	2	3	1	44	67	160	2	2	1	1	2	0	2	4	1
1781	2	3	3	2	2.81	78	18000	1000	1	2	1	3	0	1	49	61	145	2	1	2	4	1	0	4	5	2
1782	2	3	3	2	3.60	83	18000	500	1	2	2	3	6	1	24	67	130	2	1	1	4	2	2	4	4	1
1783	2	3	1	4	2.82	90	15000	0	1	2	2	2	0	2	32	66	150	2	1	2	4	2	0	4	4	2
1784	2	4	3	7	1.50	74	10000	500	3	2	2	4	0	2	25	60	125	2	1	2	1	2	0	3	4	1
1785	1	4	1	4	3.52	84	22000	1000	1	2	2	2	8	1	32	72	187	2	1	1	4	1	2	4	4	2
1786	2	2	1	5	3.00	86	12000	440	2	2	2	2	4	2	19	61	105	2	1	2	4	1	1	4	5	1
1787	1	3	1	4	3.03	79	12000	600	1	2	1	4	5	1	26	73	200	2	1	2	3	2	0	4	5	2
1788	1	3	3	4	2.45	89	11500	800	3	2	5	4	7	1	19	72	185	2	1	1	2	2	4	4	5	1
1789	1	2	1	5	2.45	80	15000	500	1	2	2	4	5	1	19	72	160	2	1	2	3	2	0	2	5	2
1790	2	3	3	5	3.64	87	16000	300	1	1	1	4	3	1	31	66	170	2	3	1	3	2	0	7	4	1
1791	2	4	1	2	2.52	81	20000	1000	1	2	3	3	4	1	41	70	170	2	1	2	4	1	0	1	4	1
1792	1	3	2	5	3.06	90	10000	1800	1	2	2	4	4	1	30	71	170	2	1	2	4	1	0	4	6	2
1793	2	3	1	2	2.80	85	12000	2000	3	2	2	4	4	1	25	60	115	2	2	1	3	1	0	4	6	2
1794	1	4	3	2	2.52	80	15000	600	1	2	2	4	0	1	29	67	150	2	1	2	4	1	0	4	4	2
1795	2	2	3	4	3.53	90	14000	500	1	2	6	2	1	1	31	64	125	2	1	2	3	2	0	2	5	1
1796	2	1	3	5	3.02	75	25000	400	1	2	2	2	2	2	31	63	125	1	2	1	1	2	0	4	2	2
1797	1	4	1	5	2.37	80	15000	1500	1	2	2	2	4	1	40	70	150	2	2	1	1	2	0	4	3	1
1798	2	3	3	5	2.81	90	17000	1500	1	2	2	3	3	1	28	64	115	2	1	2	3	2	0	2	4	1
1799	2	2	2	3	2.56	82	10000	550	2	2	2	2	9	1	19	64	112	2	2	1	3	2	0	4	4	1
1800	1	3	2	2	2.04	76	18500	750	1	2	2	2	0	1	33	73	180	2	3	2	1	2	0	1	7	2
1801	1	3	2	2	2.11	71	16000	100	1	2	6	2	3	1	50	65	190	2	1	2	1	2	0	4	4	1
1802	1	3	1	3	3.20	81	9000	500	1	2	2	2	8	1	21	69	140	2	1	2	3	2	2	4	3	1
1803	1	3	1	3	3.72	92	12000	1200	1	2	3	2	4	1	25	72	175	2	1	2	4	1	1	1	5	2
1804	2	3	1	4	2.50	80	20000	800	1	2	2	4	2	1	30	64	121	2	1	2	3	2	0	4	4	1
1805	2	3	1	6	3.03	79	15000	500	1	2	2	4	0	2	29	66	120	1	1	2	3	2	0	4	3	1
1806	1	3	1	2	2.75	82	12000	1200	2	1	2	4	7	1	24	68	155	2	1	2	4	2	0	2	4	2
1807	2	3	3	4	3.51	91	13000	500	1	2	2	4	5	1	29	63	115	2	1	2	4	1	0	5	5	1
1808	1	2	2	1	2.00	75	15000	778	3	2	2	4	5	1	19	73	205	2	1	1	1	1	1	3	4	2
1809	2	3	1	5	2.74	76	15000	0	2	2	2	2	2	1	22	66	102	2	1	2	1	2	1	2	6	1
1810	2	2	3	2	3.60	86	13000	1000	2	2	2	4	0	1	22	67	125	2	1	2	1	2	0	4	4	2
1811	2	3	3	4	3.03	79	15000	900	3	2	5	4	0	2	34	69	165	2	4	1	3	2	0	4	4	2
1812	2	3	1	2	3.75	81	15000	2000	3	2	1	4	3	1	20	62	120	2	1	2	4	2	3	4	7	2
1813	2	2	3	2	2.88	90	14600	975	2	2	5	4	12	1	19	72	157	2	2	2	4	2	0	4	3	1
1814	1	2	1	2	3.82	91	13500	1400	2	1	2	4	2	1	19	69	150	2	1	2	3	2	1	3	6	1
1815	2	2	3	5	3.70	92	13000	1500	2	2	3	4	5	2	18	63	118	2	1	2	3	2	1	4	4	2
1816	2	3	2	5	3.00	90	12000	1000	2	1	5	4	7	1	21	65	129	2	3	1	4	2	0	4	6	1
1817	2	2	2	4	3.24	87	20000	1000	2	1	2	4	4	1	20	63	120	2	1	1	3	1	1	1	3	2
1818	1	2	2	2	1.76	84	12500	0	2	2	2	4	6	1	19	69	160	2	1	2	1	1	0	2	4	1
1819	2	3	2	5	2.54	80	15000	975	2	2	2	4	4	1	21	67	110	2	1	2	4	2	0	1	6	2
1820	2	3	2	5	3.02	75	12000	1000	2	2	1	4	12	1	20	66	115	2	2	1	3	2	1	4	6	2
1821	1	4	1	5	2.80	88	10000	900	3	2	1	4	6	1	22	71	175	2	1	2	3	2	0	4	3	2
1822	1	1	1	1	3.21	79	10000	2000	2	2	2	2	0	1	19	73	175	1	1	2	1	1	2	4	4	2
1823	1	3	3	1	2.11	71	12000	1000	3	2	2	2	3	2	36	69	153	2	1	2	4	2	0	4	5	1
1824	1	2	1	7	2.67	75	20000	700	2	2	2	2	10	2	21	70	157	2	1	2	3	2	0	4	3	2
File code number	Sex	Class	Grad. school	Major	Grade-point index	H. S. average	Est. starting salary $	Tuition $	Employment	Sunday shop	Investment	Car pref.	No. jeans owned	Calculator	Age	Height	Weight	Blood pressure	Smoking status	Smoking belief	Entrance standards	Retention standards	No. clubs and groups	SPS rating	Library rating	Personal question

File code number	Question number 1	2	3	4	5	6	7	8	9	10	11	12	13	14	15	16	17	18	19	20	21	22	23	24	25	26
1825	1	2	2	2	3.60	93	15000	1000	2	2	2	4	4	1	19	69	155	2	1	2	2	1	0	1	2	1
1826	1	2	2	2	2.95	93	10000	900	3	2	1	3	5	1	19	70	170	2	1	2	2	2	0	4	6	1
1827	2	3	3	5	3.00	89	11000	1500	2	2	2	4	6	2	19	66	125	2	1	2	4	2	0	5	5	2
1828	1	4	1	5	2.80	78	12000	500	2	2	2	4	4	1	23	74	180	2	2	1	4	1	0	3	5	2
1829	1	1	1	3	2.66	85	13000	1000	2	2	1	4	6	2	19	75	160	2	1	2	3	1	0	4	4	2
1830	1	3	2	2	2.83	84	10000	712	2	2	5	4	6	1	20	70	155	2	1	2	4	2	0	4	5	1
1831	1	2	1	2	2.05	87	12000	200	3	2	2	2	20	1	22	64	110	2	2	2	4	2	0	1	5	2
1832	2	3	3	2	3.60	92	15000	500	2	2	2	2	2	1	20	63	130	1	1	2	4	2	1	1	4	1
1833	1	4	2	2	2.73	75	11000	1500	3	2	2	4	1	1	27	70	169	2	1	1	3	2	1	4	4	1
1834	1	3	2	2	2.71	80	10400	0	3	2	2	4	0	2	21	69	160	2	1	1	1	2	0	1	4	2
1835	1	3	2	2	2.80	81	12000	700	2	2	2	4	3	1	21	72	155	2	1	2	1	2	0	2	4	1
1836	1	3	3	6	3.06	80	11000	700	1	2	2	2	5	2	32	65	150	2	1	2	1	2	1	4	4	2
1837	2	2	3	2	2.77	75	12000	80	3	2	2	2	3	2	20	68	140	2	1	2	1	2	1	4	1	2
1838	1	3	3	3	3.08	89	8500	450	3	1	1	4	3	1	21	73	230	2	1	2	4	1	0	4	3	1
1839	2	3	1	4	3.20	89	15000	1000	2	2	2	2	8	2	20	62	100	2	2	1	2	2	0	4	4	1
1840	2	2	1	6	3.26	90	15000	875	3	1	2	4	4	1	28	67	125	2	2	2	4	1	0	1	4	2
1841	2	2	1	6	3.12	93	12500	1000	3	2	4	4	4	1	20	70	140	2	1	2	3	2	2	5	3	2
1842	1	3	2	6	2.24	85	10000	100	3	2	2	4	5	1	21	68	178	2	1	2	4	2	0	4	4	1
1843	2	2	2	6	3.60	88	12000	1000	2	2	3	3	2	1	33	65	120	2	2	2	3	1	0	6	4	1
1844	1	3	1	2	3.98	97	13500	1000	2	2	2	4	10	1	20	71	135	2	1	2	4	1	2	5	5	2
1845	1	2	2	4	3.20	90	10000	1000	2	2	2	4	0	1	20	70	163	2	1	2	4	1	0	4	5	1
1846	1	3	3	6	2.84	84	16000	800	3	1	1	4	10	1	20	69	160	2	1	2	4	1	2	1	1	2
1847	2	4	1	2	3.22	82	12000	975	3	2	2	4	0	2	37	61	130	2	1	2	4	1	0	4	4	1
1848	2	3	3	7	3.06	81	20000	1000	1	2	2	2	5	2	35	64	120	2	1	2	4	1	1	1	4	2
1849	1	3	3	2	2.05	87	15000	0	1	1	2	4	0	1	30	69	180	2	1	2	2	2	0	4	4	1
1850	1	3	3	1	3.03	85	13500	0	2	1	2	2	3	1	28	75	175	2	2	1	4	1	0	4	4	1
1851	2	3	2	4	3.06	79	16500	0	1	2	3	4	4	2	28	62	140	2	2	2	4	1	0	5	4	2
1852	1	2	1	2	3.20	91	18000	1200	1	2	2	4	5	1	30	70	167	1	1	2	4	2	0	4	4	1
1853	1	2	2	2	2.75	75	13000	1000	3	2	3	4	4	1	20	71	195	2	1	2	4	2	0	2	4	2
1854	2	3	3	2	3.07	82	12000	1000	1	2	5	2	2	1	30	65	140	2	1	2	4	1	0	3	5	1
1855	2	3	3	2	3.82	93	12000	900	1	1	2	2	4	1	26	62	103	1	1	2	4	1	0	1	1	1
1856	1	3	1	5	3.06	85	12000	600	2	2	2	2	2	1	37	68	165	2	1	2	3	2	1	2	5	2
1857	1	3	1	2	3.04	92	14000	2500	1	2	2	2	5	1	30	60	180	2	1	1	1	1	0	4	7	1
1858	1	3	1	3	3.02	90	20000	1000	1	2	2	2	1	1	35	69	175	2	1	2	4	2	0	1	5	2
1859	2	2	1	1	2.12	73	17000	0	1	2	2	4	0	1	22	63	135	2	3	1	1	2	1	5	4	1
1860	1	3	1	2	3.91	87	15000	0	2	1	1	4	4	1	25	71	153	2	1	1	3	1	0	4	4	2
1861	1	4	1	4	3.35	87	14000	600	1	2	6	2	0	1	21	73	175	2	2	1	4	1	3	4	2	2
1862	1	3	1	2	3.37	90	16500	1000	1	2	2	4	5	1	29	72	185	2	1	2	1	1	0	4	4	1
1863	1	4	1	3	2.73	85	12000	0	1	2	6	2	0	1	22	68	145	2	1	2	1	2	0	4	4	2
1864	2	3	1	2	2.64	73	15000	0	1	2	2	2	0	1	29	70	170	2	1	2	4	2	0	4	4	1
1865	2	2	3	2	3.82	94	12000	0	2	2	6	4	6	1	21	61	108	2	1	2	4	1	0	4	4	1
1866	1	3	1	2	2.88	78	20000	0	2	2	2	2	0	1	28	69	140	2	2	1	1	2	0	4	4	2
1867	2	3	1	2	2.07	81	20000	300	3	2	2	4	5	1	28	66	122	2	1	2	1	2	0	4	4	2
1868	2	2	3	6	3.20	83	15000	1050	1	2	2	2	0	1	32	65	150	2	1	2	1	1	0	4	4	2
1869	1	3	1	5	2.81	75	12000	0	1	2	2	4	3	1	29	68	155	2	3	2	4	1	1	1	4	2
1870	1	2	3	2	3.02	80	15000	100	1	2	2	4	2	1	23	71	204	2	3	1	3	1	0	4	4	2
1871	2	3	1	5	2.84	81	12000	400	3	2	2	2	0	1	26	67	180	2	1	2	3	1	0	4	6	1
1872	1	3	1	4	2.09	91	16000	0	1	2	2	2	0	2	46	71	190	1	1	2	1	1	2	5	6	2
1873	1	3	1	2	3.60	87	12000	1000	1	2	2	4	2	1	20	73	205	2	1	2	4	1	2	2	6	1
1874	2	3	1	5	3.24	85	15000	1000	2	2	2	4	10	1	19	67	119	2	2	2	4	2	0	1	4	1
1875	2	3	3	7	3.10	82	25000	450	1	2	2	4	3	1	16	66	177	2	1	2	4	2	0	3	5	2
1876	2	2	1	6	3.18	89	9000	1000	2	2	3	4	4	1	20	66	128	2	1	2	1	1	4	5	6	1
1877	1	3	1	2	3.00	79	12000	2800	2	2	5	2	3	2	21	68	170	2	1	2	1	2	0	4	4	1
1878	1	2	3	6	3.45	86	10000	1000	2	1	3	4	5	1	21	70	175	2	1	2	4	2	0	1	6	1
1879	1	3	3	6	2.80	88	13000	2000	1	1	2	4	10	1	21	60	200	2	1	2	3	2	0	6	6	1
1880	2	3	2	6	3.60	90	12000	1500	2	2	1	4	5	1	20	64	110	2	1	2	3	2	0	5	6	1
1881	1	3	2	6	2.82	75	10500	1000	3	2	2	1	4	1	21	68	160	2	1	2	3	2	0	4	2	2
1882	2	3	1	7	3.00	85	12000	900	2	1	1	4	5	1	19	66	150	2	2	1	4	2	1	6	5	2
1883	2	3	1	7	2.33	85	12000	1000	3	2	2	4	1	1	19	67	128	1	3	1	4	2	1	3	2	1
1884	1	3	1	7	3.28	86	15000	412	2	2	2	3	7	1	19	73	155	2	2	2	4	1	0	2	3	1
1885	2	2	1	7	2.60	80	10500	900	3	1	2	4	2	1	34	61	145	1	1	2	4	2	0	4	4	1
1886	2	2	3	7	3.60	90	14000	1200	1	2	5	2	2	2	20	63	118	2	1	2	3	1	2	3	5	1
1887	2	4	3	7	2.75	86	18000	1000	1	2	2	2	6	1	43	61	123	2	1	2	3	1	0	4	3	2
1888	2	1	3	7	3.00	87	14500	1050	1	2	2	4	7	1	23	59	120	2	3	1	4	1	0	4	3	2
1889	2	2	2	7	2.82	85	15000	1000	1	2	3	4	5	1	24	61	132	2	1	2	4	1	0	4	4	1
1890	2	4	1	7	3.43	82	15000	1200	1	2	4	4	4	1	33	64	120	2	1	2	4	1	1	6	6	1
1891	2	3	1	7	3.40	80	13000	900	1	1	2	2	0	1	32	64	129	2	2	1	1	2	0	1	1	2
1892	2	3	1	7	3.45	85	14000	1000	1	2	2	4	5	1	35	69	164	2	1	2	3	1	0	4	4	2
1893	1	3	2	4	2.80	91	15000	2000	2	2	2	2	4	1	20	67	200	1	1	2	4	1	0	4	6	1
1894	1	3	3	5	3.00	72	25000	1000	2	2	5	4	1	1	21	68	145	2	2	1	3	2	0	4	4	1
1895	1	3	3	5	3.00	80	14000	1000	3	1	5	4	10	1	22	72	162	2	1	2	4	1	0	4	5	1
1896	1	4	2	5	2.20	76	9000	500	2	2	2	2	9	1	22	73	185	2	1	2	3	1	0	4	7	1
1897	1	4	2	7	3.10	80	10000	2000	3	2	2	4	5	1	26	67	150	2	1	2	4	1	2	4	4	1
1898	2	4	3	5	3.20	81	12000	2000	2	2	5	3	5	1	23	65	125	2	1	1	4	2	1	4	5	2
1899	1	4	2	5	2.30	78	10500	1200	2	2	2	2	3	1	22	72	156	2	1	2	4	2	2	2	7	1
1900	1	4	1	4	3.20	72	11500	400	2	2	2	4	5	1	23	72	175	2	1	2	3	1	2	4	6	1
1901	1	3	1	5	2.80	82	14000	1000	2	2	5	4	4	1	21	66	143	2	1	2	3	1	0	1	1	2
1902	1	4	1	4	2.50	77	11000	100	3	2	2	2	4	1	33	70	165	2	2	2	1	1	2	6	7	1
1903	1	4	1	5	3.80	90	10000	1000	1	2	5	4	9	1	36	77	200	2	4	1	1	2	0	7	3	2
1904	1	3	1	7	2.50	74	9000	600	2	2	2	4	2	1	27	68	185	2	2	2	4	1	2	5	6	2
1905	1	3	3	5	3.00	76	13000	0	3	2	2	2	3	1	26	69	152	2	1	1	1	2	0	4	4	1
1906	1	3	2	5	3.50	95	12000	700	2	1	3	4	1	1	27	68	135	2	4	2	2	2	4	1	4	1
1907	1	3	1	5	2.50	81	14500	1000	3	2	3	4	2	2	20	71	230	2	1	1	3	1	2	1	2	1
1908	1	3	2	1	2.30	76	13000	250	3	2	6	2	1	2	20	64	130	2	1	2	1	2	0	2	4	1
1909	1	4	1	5	2.50	84	10000	500	2	2	2	2	1	1	24	69	144	2	1	2	1	1	2	4	5	1
1910	2	3	2	5	2.00	77	9100	0	2	2	2	2	2	1	24	62	90	2	1	2	1	2	0	5	7	1
1911	2	4	2	7	3.00	78	10000	1000	3	2	2	4	5	1	20	60	96	2	3	1	4	1	0	4	4	2
1912	1	3	3	4	3.00	82	10000	1000	3	2	1	4	1	1	22	71	150	2	1	2	4	2	0	4	5	2
1913	1	4	1	4	3.10	85	12000	2500	2	2	1	5	5	1	20	71	175	2	1	2	4	1	1	5	5	1
1914	2	4	1	7	3.80	78	12500	500	3	2	5	4	3	1	27	64	115	2	1	2	4	1	0	2	5	1
1915	1	4	2	5	2.50	80	8500	500	2	2	5	4	5	1	22	68	157	2	1	1	1	2	0	4	4	2
1916	1	1	3	5	2.36	81	11500	440	1	2	3	2	4	1	27	71	150	2	1	2	1	2	0	4	4	1
1917	1	4	3	5	2.00	79	12000	800	1	2	2	2	1	2	48	67	160	2	1	2	4	2	2	3	5	2
1918	1	4	1	3	2.00	73	14000	1000	1	2	2	2	10	2	33	68	190	2	1	2	4	1	0	4	6	2
1919	2	1	3	2	1.91	84	20000	1000	1	2	2	2	6	1	29	66	160	2	1	2	1	1	0	4	4	1
1920	1	4	1	4	2.74	76	15000	2500	1	2	2	2	0	1	33	62	130	2	1	1	4	1	0	5	5	2
File code number	Sex	Class	Grad. school	Major	Grade-point index	H. S. average	Est. starting salary $	Tuition $	Employment	Sunday shop	Investment	Car pref.	No. jeans owned	Calculator	Age	Height	Weight	Blood pressure	Smoking status	Smoking belief	Entrance standards	Retention standards	No. clubs and groups	SPS rating	Library rating	Personal question

File code number	Question number: 1 Sex	2 Class	3 Grad. school	4 Major	5 Grade-point index	6 H. S. average	7 Est. starting salary $	8 Tuition $	9 Employment	10 Sunday shop	11 Investment	12 Car pref.	13 No. jeans owned	14 Calculator	15 Age	16 Height	17 Weight	18 Blood pressure	19 Smoking status	20 Smoking belief	21 Entrance standards	22 Retention standards	23 No. clubs and groups	24 SPS rating	25 Library rating	26 Personal question
1921	1	1	3	6	2.59	78	12000	700	1	2	2	2	0	1	26	66	147	2	1	2	1	2	0	4	4	2
1922	1	4	3	3	2.00	73	16000	800	2	2	2	2	2	1	27	76	185	2	1	2	4	1	0	4	4	1
1923	2	3	1	7	3.00	82	16000	1500	1	1	3	4	3	1	31	63	116	2	1	2	3	1	0	4	4	2
1924	1	4	1	7	2.00	75	13000	400	2	2	2	2	0	1	29	67	160	2	1	2	3	2	0	5	6	1
1925	2	3	1	7	3.00	85	12000	900	1	2	2	4	5	1	27	66	149	2	1	2	4	2	0	1	6	1
1926	1	4	1	5	3.00	81	20000	1750	1	1	6	2	1	1	39	71	200	2	1	2	4	1	0	5	5	2
1927	1	4	2	2	2.50	85	14500	975	1	2	1	4	1	1	24	69	135	1	1	2	4	1	0	4	4	2
1928	2	4	2	7	2.00	77	12000	0	1	2	6	2	0	1	24	62	145	2	1	2	4	2	0	4	4	2
1929	2	3	3	4	3.20	85	20000	700	1	2	2	4	2	1	40	60	155	2	1	2	4	1	0	5	6	2
1930	1	1	3	7	2.96	90	15000	675	1	1	2	4	2	1	29	68	143	1	1	2	4	1	0	4	4	2
1931	1	3	1	5	2.80	77	16000	500	1	2	2	2	2	1	24	67	145	2	1	2	2	1	4	4	2	2
1932	1	1	3	6	3.00	83	15000	0	1	2	2	2	2	1	26	66	135	2	1	2	4	2	0	4	4	1
1933	1	4	3	7	2.90	81	12000	0	1	2	2	2	4	1	38	67	132	2	1	2	4	2	0	4	4	2
1934	1	4	3	3	3.60	75	11000	1200	1	1	1	4	6	1	29	70	178	2	1	2	3	2	0	4	4	1
1935	1	4	1	5	2.70	75	15000	500	1	2	2	4	9	1	28	69	155	1	1	1	3	2	0	1	3	1
1936	1	4	1	3	2.80	80	11000	1800	1	2	3	2	6	1	22	73	195	2	2	1	4	1	0	4	4	2
1937	1	4	2	4	3.60	90	18000	1000	1	2	5	4	4	1	31	69	155	1	1	2	1	2	0	4	7	2
1938	1	2	1	5	3.50	92	14000	300	2	1	2	4	0	1	21	68	163	2	3	1	4	1	0	4	5	2
1939	2	4	2	7	3.90	96	5000	500	2	2	1	3	4	2	25	68	130	2	1	2	4	1	0	4	4	1
1940	1	4	3	4	3.10	84	12000	1000	1	2	2	4	4	1	21	68	162	2	1	2	3	2	1	4	3	1
1941	1	3	3	4	3.99	76	12500	3000	1	2	5	4	3	1	27	67	165	2	1	2	1	1	0	4	2	1
1942	1	2	1	2	2.75	81	12600	450	3	2	2	2	3	1	37	66	125	2	2	2	1	2	0	1	1	2
1943	2	4	3	7	2.21	80	15000	1000	1	2	2	4	3	1	28	64	135	2	2	2	1	2	0	1	4	1
1944	2	4	3	5	3.49	86	12000	1500	1	2	5	4	4	1	23	65	125	2	1	2	4	2	0	3	4	1
1945	1	4	3	5	2.73	85	20000	1000	1	2	3	4	8	1	26	68	165	2	3	1	4	1	0	4	4	2
1946	1	2	3	1	2.50	78	20000	675	1	2	5	4	8	1	31	69	182	2	3	1	4	1	2	3	1	1
1947	1	2	1	1	2.21	72	9000	0	1	2	4	2	2	1	36	61	150	2	2	2	4	2	2	4	4	1
1948	1	3	3	5	2.20	85	12000	1500	1	2	2	2	6	1	22	68	145	2	1	2	4	2	0	1	2	2
1949	1	4	3	3	2.98	85	15000	1000	1	2	1	4	12	1	29	72	187	1	3	1	4	1	0	4	4	1
1950	1	4	1	3	3.40	88	16000	0	1	2	5	2	4	1	26	66	140	2	1	2	4	1	0	4	4	2
1951	2	3	3	7	2.80	85	10000	1200	2	2	2	2	10	1	20	62	110	2	1	2	3	1	0	4	4	1
1952	1	4	1	4	3.52	82	20000	3000	1	2	2	2	2	1	35	69	165	2	3	1	1	2	2	5	6	1
1953	2	3	2	4	3.50	84	16000	985	1	2	2	2	1	1	21	66	124	2	1	2	1	2	0	5	4	1
1954	1	4	3	3	2.90	86	10000	2000	2	2	1	4	7	1	20	60	215	2	1	2	1	1	0	4	3	1
1955	2	1	3	1	2.65	82	20000	0	1	2	2	2	2	1	24	65	125	2	1	2	2	2	0	4	4	2
1956	1	2	1	1	2.96	79	14000	600	1	2	1	3	6	1	22	69	145	2	1	2	1	2	1	3	3	2
1957	1	3	3	3	3.04	85	10000	1500	1	2	2	4	5	1	24	70	180	2	1	2	3	2	0	5	5	1
1958	2	1	3	5	2.82	80	14100	988	1	2	2	2	4	2	29	65	123	2	1	2	1	1	0	5	4	2
1959	1	4	1	5	3.63	87	15000	1800	1	2	5	4	3	1	31	65	150	2	3	1	1	2	0	2	4	1
1960	1	4	3	5	2.85	79	13500	1350	1	2	1	4	4	1	25	70	220	2	1	2	1	2	0	3	3	2
1961	1	4	1	5	2.75	85	20000	600	2	2	2	4	3	1	21	66	130	2	1	2	4	2	0	2	4	1
1962	2	2	2	2	3.00	85	11020	25	3	1	2	2	8	1	20	65	112	2	1	2	1	2	2	3	3	1
1963	2	2	2	2	3.46	95	12000	100	3	2	2	2	18	1	19	63	95	2	1	2	3	1	2	4	3	1
1964	1	4	3	5	3.00	81	10000	100	2	1	2	4	6	1	21	71	150	2	2	1	4	2	0	1	4	1
1965	2	1	3	2	3.83	81	8000	200	2	2	2	2	0	1	19	63	97	2	1	2	2	2	0	5	4	1
1966	1	2	3	2	3.00	79	13500	1400	2	2	5	4	1	1	19	70	155	2	1	2	3	2	2	1	6	1
1967	1	2	2	2	3.25	87	14000	1000	2	1	2	2	6	1	20	73	185	2	1	2	3	2	2	3	5	1
1968	1	1	3	2	3.80	87	17000	4000	2	2	1	4	1	1	18	69	142	2	1	2	3	2	3	6	6	1
1969	1	2	3	2	3.00	85	12000	1000	2	2	2	4	6	1	19	70	165	2	1	2	3	1	0	4	4	1
1970	2	2	1	2	3.67	93	12000	1500	2	2	2	4	8	1	19	66	130	2	1	2	4	1	0	4	6	1
1971	1	2	1	2	3.48	82	11000	1500	2	2	3	4	8	1	19	76	190	2	1	2	4	1	0	4	6	1
1972	1	4	2	7	2.89	92	20000	350	2	2	2	4	7	1	21	74	200	2	4	1	3	1	2	1	5	1
1973	2	2	1	2	3.50	91	11500	1000	2	2	2	1	7	1	19	67	125	2	3	2	3	1	1	4	4	1
1974	1	2	3	2	4.00	92	12500	1000	2	2	2	4	6	2	19	67	135	2	1	1	3	2	0	4	5	1
1975	2	2	3	1	3.11	85	9500	800	2	2	5	2	9	1	19	62	105	2	1	1	3	2	2	4	5	2
1976	1	3	2	2	2.00	88	13000	700	3	2	2	2	4	1	22	68	160	2	1	2	4	2	1	4	4	1
1977	1	2	2	4	2.40	88	20000	1000	2	2	2	4	7	1	19	68	151	2	1	2	3	2	1	1	4	1
1978	1	4	1	3	3.20	85	11000	1000	1	2	2	2	3	1	23	66	130	2	1	2	2	2	0	2	4	1
1979	1	2	2	2	3.13	88	13000	1000	2	1	2	2	2	1	20	72	173	2	1	1	4	2	1	2	4	1
1980	2	2	3	1	3.45	90	13000	1000	2	2	2	3	4	1	19	60	108	2	4	2	3	2	0	4	4	2
1981	2	2	3	1	3.76	92	12000	1200	3	1	1	4	9	1	19	61	107	2	1	2	4	1	1	4	5	2
1982	2	2	3	2	2.66	88	12000	1500	2	2	2	4	9	1	19	66	105	2	1	2	3	1	1	4	3	2
1983	2	2	3	1	3.27	94	11000	1500	2	1	1	4	9	1	19	69	130	2	1	2	3	1	1	4	4	2
1984	2	2	2	6	1.90	82	10000	300	3	2	2	2	2	1	19	70	133	2	1	2	1	1	1	4	4	2
1985	1	2	3	1	3.00	88	15000	2000	3	2	2	4	3	1	19	70	160	2	1	2	4	2	0	4	5	2
1986	2	4	1	4	3.60	75	14000	500	3	2	2	4	4	1	30	63	110	2	1	2	1	2	0	2	3	1
1987	1	1	3	2	2.75	74	13000	800	2	2	2	2	6	1	19	73	183	2	1	2	3	1	0	3	6	1
1988	2	3	3	2	3.42	94	12000	300	3	2	1	3	1	1	25	65	110	2	1	2	4	1	2	3	2	2
1989	2	2	2	2	2.20	87	15000	1200	3	2	1	3	9	1	20	70	140	2	1	1	4	1	0	4	5	2
1990	1	2	1	6	3.49	91	11000	2000	3	2	5	4	6	1	19	70	177	1	1	2	3	2	1	1	2	2
1991	1	3	3	2	2.70	85	8500	1500	2	2	2	4	3	1	20	72	175	2	1	2	2	2	1	2	4	2
1992	2	3	3	4	3.00	90	13000	1000	2	2	2	1	10	1	19	62	100	2	2	2	4	2	0	4	4	1
1993	1	2	2	6	2.75	78	10000	900	3	2	2	2	4	1	22	65	122	2	2	2	1	1	4	4	4	1
1994	1	3	1	5	3.00	80	12000	1000	2	2	2	4	4	2	21	71	155	2	1	2	4	1	0	2	5	1
1995	2	4	3	5	3.40	84	11500	200	2	2	4	4	10	1	20	66	124	2	1	2	4	2	1	4	2	1
1996	1	2	1	6	3.53	89	13000	1000	3	2	1	2	6	1	19	72	160	1	1	2	3	2	0	4	4	1
1997	1	3	1	3	3.33	85	14000	900	1	2	2	2	6	1	29	70	148	2	1	1	2	2	2	2	5	1
1998	2	3	1	5	3.75	90	10000	1000	3	2	1	4	2	1	19	67	110	2	1	2	4	1	2	2	2	1
1999	2	3	2	6	3.00	89	12000	1000	3	2	2	2	5	2	23	62	93	2	1	2	4	1	0	4	4	2
2000	1	3	1	5	3.17	82	10000	1500	2	2	5	4	3	1	20	68	165	2	1	1	3	1	0	4	4	1
2001	2	3	1	6	3.12	88	12000	1000	2	2	2	2	5	1	20	62	120	2	1	1	3	2	0	4	6	2
2002	1	3	1	6	2.54	81	9000	1400	1	2	2	2	7	1	22	66	110	2	1	2	2	2	1	4	4	1
2003	2	3	3	5	3.20	85	10000	1000	3	2	2	4	5	1	20	63	105	2	1	2	3	1	1	4	5	1
2004	1	2	3	4	2.60	80	15000	2000	2	1	2	2	3	1	26	70	165	2	1	2	1	2	2	4	4	1
2005	1	1	1	2	2.40	89	12000	1200	2	1	2	2	8	1	18	71	190	2	1	2	3	1	0	2	4	1
2006	1	1	3	1	2.00	93	18000	1500	2	2	2	4	8	1	18	67	134	2	1	2	3	2	3	4	4	1
2007	2	4	3	4	2.83	80	12000	1000	3	2	2	3	4	1	22	68	129	2	1	1	3	2	1	4	3	2
2008	2	2	3	2	2.70	80	10000	800	3	2	1	3	2	2	19	64	115	2	1	2	1	2	1	4	2	2
2009	1	2	3	2	2.30	89	12000	1000	2	2	2	4	7	1	18	74	180	2	3	1	4	1	1	1	4	2
2010	2	3	3	1	2.60	80	12000	500	2	2	1	4	5	2	23	68	121	2	1	2	1	2	1	4	4	2
2011	1	3	3	6	3.50	92	12000	900	2	1	1	3	10	1	19	74	150	2	2	1	1	2	1	4	4	2
2012	1	2	3	5	2.60	86	10000	1000	2	2	5	3	7	2	20	74	180	2	3	1	3	2	1	2	3	1
2013	2	2	2	2	3.60	85	13000	800	2	2	2	4	3	1	20	68	160	2	1	2	3	2	0	4	6	2
2014	2	3	1	6	3.20	90	11000	1500	2	2	5	4	3	1	20	60	110	2	1	2	4	1	1	4	5	1
2015	2	3	3	6	3.00	75	12000	0	2	1	2	4	5	1	23	63	110	2	1	2	3	2	1	4	5	1
2016	1	2	3	5	2.70	83	10000	1000	2	2	2	4	6	1	18	70	150	2	1	2	4	2	2	4	5	1

File code number	1 Sex	2 Class	3 Grad. school	4 Major	5 Grade-point index	6 H. S. average	7 Est. starting salary $	8 Tuition $	9 Employment	10 Sunday shop	11 Investment	12 Car pref.	13 No. jeans owned	14 Calculator	15 Age	16 Height	17 Weight	18 Blood pressure	19 Smoking status	20 Smoking belief	21 Entrance standards	22 Retention standards	23 No. clubs and groups	24 SPS rating	25 Library rating	26 Personal question
2017	1	3	3	5	2.80	79	11500	650	1	2	5	4	18	2	27	69	168	2	1	2	4	2	0	4	4	1
2018	1	1	3	2	2.00	90	11000	1500	3	2	5	4	7	2	18	68	140	2	1	2	4	2	0	4	4	2
2019	2	1	2	7	2.00	90	10000	1000	3	2	2	4	6	1	18	63	111	2	1	2	1	2	0	4	4	2
2020	1	2	3	2	3.18	86	18000	1200	3	2	1	4	3	1	19	71	160	1	1	2	3	1	2	4	5	1
2021	1	3	2	5	2.75	83	11000	900	2	1	5	4	6	1	24	72	195	1	3	1	3	1	1	2	3	1
2022	1	3	1	2	3.50	86	12500	1500	3	2	1	4	9	1	19	69	160	2	1	2	4	1	2	1	3	2
2023	2	1	1	4	2.00	90	16000	2000	2	2	2	4	4	1	17	62	115	2	1	2	4	1	0	5	6	1
2024	1	1	3	2	2.00	90	12000	1000	2	2	2	2	8	2	17	67	142	2	1	1	1	2	0	4	4	2
2025	1	1	2	2	2.00	83	15000	1000	3	2	2	4	5	1	18	68	160	2	1	2	3	1	2	4	4	1
2026	1	4	1	7	3.35	95	20000	400	2	2	2	4	10	1	20	67	150	2	1	1	3	2	3	2	3	1
2027	1	1	2	1	3.48	89	15000	1000	2	2	1	4	3	2	19	71	134	2	1	2	1	2	0	3	6	2
2028	2	1	1	2	2.00	85	10000	200	3	2	2	2	7	2	18	66	130	2	1	2	4	1	0	4	4	1
2029	1	3	3	4	2.57	78	11000	487	2	2	2	4	7	2	24	70	157	2	2	1	4	2	0	3	6	2
2030	2	2	1	4	3.02	77	22000	1500	1	2	1	3	3	1	28	64	122	2	1	2	2	2	0	4	6	2
2031	1	1	1	2	2.00	90	14000	1500	2	2	2	4	8	1	18	74	178	2	1	2	4	2	0	4	4	1
2032	1	1	1	2	2.00	85	15000	1000	2	2	2	2	12	2	18	67	150	2	1	2	1	1	3	6	4	2
2033	2	1	2	6	2.00	90	10000	900	3	2	1	3	2	2	18	72	150	2	1	2	1	2	0	5	6	1
2034	2	4	2	6	2.75	85	10000	925	3	2	2	2	2	1	21	65	110	2	1	2	3	2	1	4	5	2
2035	1	3	1	3	2.50	71	9500	500	3	1	1	4	1	2	28	68	160	2	1	1	3	2	3	2	6	1
2036	1	1	3	1	2.00	86	14000	1000	3	2	2	2	4	2	18	71	168	2	1	2	4	2	0	4	4	2
2037	1	1	3	6	2.00	83	9000	2000	2	2	2	2	5	1	18	68	139	1	1	1	3	1	0	4	6	1
2038	2	1	1	1	2.00	90	30000	950	2	1	5	4	4	2	17	65	130	2	1	1	1	2	1	4	4	2
2039	1	3	2	3	2.80	84	11000	1200	2	2	1	4	8	1	19	74	180	2	1	2	1	2	0	5	3	2
2040	2	3	1	3	3.40	84	10000	2000	3	2	2	2	3	1	22	65	130	2	1	2	4	2	0	4	6	2
2041	2	2	2	5	3.15	94	12000	400	1	2	5	4	5	1	19	61	102	2	2	1	3	2	0	2	4	2
2042	1	2	2	4	2.80	90	13000	1200	2	2	2	2	4	1	18	71	175	2	1	1	2	1	0	4	7	1
2043	1	1	1	2	3.15	88	15000	1400	2	2	5	4	12	1	19	64	125	2	1	1	3	1	0	4	4	1
2044	2	2	3	6	3.64	95	10000	1000	3	1	3	3	7	1	18	64	118	2	1	1	3	2	0	2	3	1
2045	2	2	1	7	3.15	85	11000	1500	3	2	2	4	7	2	19	65	125	2	1	1	4	1	1	4	2	2
2046	2	3	3	4	3.79	85	10000	1000	2	2	2	2	2	2	31	63	110	2	1	2	4	1	2	3	6	1
2047	1	3	1	4	3.87	90	12000	1500	2	2	2	3	2	1	39	67	150	2	1	2	4	1	3	2	5	1
2048	1	2	3	2	2.77	85	14000	1500	2	2	3	4	4	1	19	68	138	2	1	2	3	2	0	4	5	2
2049	1	3	3	4	3.20	90	12000	500	3	2	2	4	5	2	21	74	240	1	1	2	4	1	0	3	6	2
2050	1	3	3	5	2.70	81	18000	2000	1	2	2	4	8	2	21	60	167	2	3	1	3	1	0	4	4	2
2051	1	3	2	3	2.00	89	10000	1000	2	2	2	2	3	2	28	72	152	2	1	2	4	2	0	4	4	2
2052	1	1	1	2	2.20	74	11364	1000	2	2	2	4	5	1	18	72	173	1	1	2	3	2	0	4	4	1
2053	2	4	3	2	3.00	85	12000	1000	2	1	2	4	2	1	21	61	105	2	1	2	3	1	1	4	4	2
2054	2	3	3	5	3.00	87	12500	2000	3	2	2	4	5	1	18	63	120	2	1	1	2	2	4	3	3	2
2055	1	1	1	1	2.27	93	15000	1600	2	2	2	2	7	1	26	72	166	2	1	2	3	1	0	1	4	1
2056	1	2	1	2	3.24	95	13000	2000	1	2	2	2	13	1	27	66	150	2	2	2	1	1	3	4	4	2
2057	1	4	1	7	3.00	85	9000	1000	2	2	2	4	3	1	21	66	120	2	3	1	1	2	0	4	4	1
2058	2	3	2	2	2.90	80	9000	80	1	2	2	2	5	1	21	68	170	1	2	1	1	1	0	4	4	1
2059	1	3	1	3	2.80	84	17000	2000	1	2	5	2	2	1	26	72	180	1	1	2	4	1	0	4	4	2
2060	2	3	2	4	3.00	87	16000	1000	1	2	6	2	5	1	24	65	132	2	1	2	3	2	0	4	6	1
2061	1	3	1	2	2.00	85	12000	1000	3	2	2	4	3	1	22	72	185	2	1	1	3	2	0	4	4	2
2062	1	3	1	3	2.80	85	12730	750	1	2	5	4	1	1	21	65	140	1	1	2	4	1	0	4	4	1
2063	1	3	1	1	3.60	75	15000	750	1	2	6	2	0	1	32	70	175	2	1	2	1	1	0	4	4	1
2064	1	1	3	2	2.70	80	15000	1000	3	2	2	2	3	2	17	71	152	2	1	2	2	1	0	4	5	1
2065	1	2	1	5	3.00	89	16000	800	1	2	2	2	3	1	30	66	145	2	1	2	1	1	0	2	6	1
2066	1	3	1	5	3.04	88	20000	1200	1	2	2	2	5	1	40	65	148	2	3	1	1	1	0	4	7	2
2067	1	4	2	5	2.50	75	15000	1000	1	2	1	4	2	1	33	68	187	2	2	1	3	2	0	4	4	2
2068	1	3	2	2	2.72	85	18000	1000	3	2	1	4	0	1	21	73	148	2	1	2	3	2	0	4	6	2
2069	1	4	1	2	3.00	82	30000	1000	1	2	1	4	3	1	28	70	200	2	1	1	4	1	0	4	6	2
2070	1	2	1	2	3.90	80	12000	1000	3	1	1	4	3	1	21	69	135	2	1	2	3	1	3	3	6	1
2071	1	4	3	5	3.20	79	16000	1000	1	2	2	2	2	1	27	74	220	2	2	1	3	2	0	4	1	1
2072	1	4	1	2	3.20	80	15000	900	1	2	2	2	6	1	31	67	150	2	2	1	2	1	0	4	5	1
2073	2	3	3	5	3.10	88	13000	500	3	2	2	4	7	2	27	64	105	2	1	1	4	2	0	4	4	2
2074	1	4	2	5	2.80	84	12000	900	1	2	1	2	3	1	27	65	190	2	1	2	3	2	0	4	5	2
2075	2	4	3	4	3.10	87	15000	800	1	2	2	4	10	2	25	61	103	2	2	2	3	1	0	4	4	1
2076	2	1	3	2	2.00	88	15000	1000	1	2	2	2	4	2	20	64	132	1	2	1	4	1	0	2	4	2
2077	2	4	1	6	2.06	95	12000	1000	3	2	2	2	2	1	22	65	260	1	2	1	4	2	1	3	4	1
2078	2	2	2	6	3.00	92	12000	1000	3	1	2	4	10	1	18	65	120	2	1	2	3	2	0	2	4	1
2079	1	2	1	7	3.00	88	13520	1100	1	2	1	4	7	1	28	64	125	2	1	2	3	1	1	4	4	2
2080	1	3	3	5	2.50	80	10000	2000	3	2	2	4	3	1	22	67	155	2	1	1	1	1	0	6	6	1
2081	2	3	2	6	2.80	78	15000	1000	2	2	5	4	4	1	20	66	140	2	3	1	4	1	0	5	6	2
2082	2	3	2	7	3.52	85	12000	1000	2	2	2	2	5	1	20	64	110	2	1	2	4	2	0	4	3	2
2083	2	3	1	7	3.00	85	11865	900	3	2	2	1	10	2	21	63	138	2	1	1	1	2	1	4	4	1
2084	1	2	2	6	3.50	91	14000	900	2	2	2	2	6	1	19	67	155	2	1	2	3	1	1	4	5	1
2085	1	3	2	6	2.10	83	14000	800	2	2	5	4	10	1	20	68	120	2	1	2	3	2	0	4	5	1
2086	1	2	1	6	3.00	87	10400	0	2	2	2	4	7	1	20	69	197	2	3	1	4	2	0	3	4	1
2087	1	3	3	6	2.60	89	12000	150	2	2	2	2	8	1	20	69	148	2	2	1	4	1	1	4	4	1
2088	1	3	1	5	3.20	83	10000	1000	3	2	2	2	9	1	20	64	134	1	1	2	3	2	1	1	2	1
2089	1	2	3	6	3.04	87	10000	200	3	1	1	3	10	1	18	66	135	2	2	1	4	1	1	4	3	1
2090	2	3	3	5	3.20	81	11000	500	3	2	5	4	6	1	21	66	130	2	3	2	3	2	0	4	4	2
2091	2	3	1	6	2.75	81	12000	1000	3	2	5	3	2	1	21	61	100	2	1	2	3	1	0	4	4	1
2092	2	4	1	1	3.10	95	10000	1500	2	2	6	2	2	2	61	60	90	2	1	2	1	1	0	3	4	1
2093	2	2	1	6	3.76	94	15000	500	2	2	2	4	5	1	19	64	107	2	1	2	3	2	3	1	5	2
2094	1	4	1	6	2.75	82	15000	1700	1	2	2	4	4	1	32	73	170	2	1	2	4	1	1	4	6	2
2095	1	2	2	6	3.20	75	10500	1500	1	2	5	4	1	1	31	73	245	2	3	1	3	1	1	5	6	1
2096	1	4	1	6	3.00	93	10000	2000	1	2	2	4	5	2	27	68	160	2	1	2	4	1	0	4	4	2
2097	1	4	1	6	3.57	82	8000	2000	2	2	6	2	5	2	24	64	115	2	1	2	3	1	0	4	6	1
2098	1	4	1	4	3.00	76	8000	1000	1	1	2	2	4	1	34	63	110	2	2	2	4	1	1	6	6	2
2099	2	3	1	7	3.78	94	10000	1500	2	2	2	4	4	2	23	60	102	2	1	2	4	1	0	2	4	2
2100	1	4	1	5	3.80	84	10000	1000	1	1	1	4	3	1	20	70	145	2	1	2	4	1	0	4	3	2
2101	1	4	2	6	3.00	90	10000	1000	3	2	6	2	1	1	31	69	145	2	3	1	3	2	0	4	6	2
2102	1	4	1	5	3.50	90	12000	2000	2	2	2	2	2	1	27	71	142	2	1	2	4	1	0	4	4	1
2103	1	4	1	7	3.00	83	10000	1000	3	1	2	4	5	2	47	70	150	2	1	1	3	2	0	4	4	2
2104	1	3	1	3	3.00	93	7800	1000	1	2	2	4	3	1	47	68	180	2	1	2	4	1	0	1	4	2
2105	1	4	1	7	3.30	88	14000	1500	3	2	2	4	5	1	28	72	146	1	1	1	4	1	0	4	6	1
2106	2	4	1	5	3.00	88	9000	3000	2	2	4	4	2	1	32	61	105	2	1	2	4	1	0	4	3	2
2107	1	3	1	6	3.47	88	13000	1000	1	2	2	4	8	1	24	66	160	1	3	1	4	1	3	5	4	2
2108	2	2	1	6	3.33	86	13000	950	3	2	1	4	1	1	18	67	130	2	1	2	3	1	1	4	5	2
2109	2	4	1	7	3.00	72	10000	2000	2	1	6	2	1	1	35	62	162	2	3	2	1	1	0	4	4	2
2110	1	3	3	5	3.16	84	13000	975	3	2	2	4	5	1	20	74	172	2	1	1	3	2	0	4	5	1
2111	2	3	1	5	3.04	78	18000	500	1	2	2	4	0	1	32	65	150	2	3	1	1	2	0	6	3	1
2112	1	3	1	5	2.08	73	11500	0	1	2	2	2	4	1	30	68	150	2	1	2	1	2	0	3	1	2

File code number	Question number 1	2	3	4	5	6	7	8	9	10	11	12	13	14	15	16	17	18	19	20	21	22	23	24	25	26
2113	2	2	3	2	3.20	90	15000	0	2	2	2	4	2	1	19	58	90	2	1	2	4	2	0	1	4	1
2114	2	3	1	3	3.02	90	15000	2500	1	1	3	4	5	2	26	62	110	2	1	2	3	2	0	4	4	2
2115	1	3	2	2	3.25	87	11000	800	1	2	3	4	0	1	38	72	150	2	1	2	2	2	0	5	5	2
2116	1	4	3	2	3.80	90	13740	120	2	2	2	4	7	1	22	67	150	2	1	2	4	1	2	4	6	1
2117	1	4	2	2	2.00	84	20000	550	1	2	6	2	1	1	35	66	130	2	1	2	1	2	0	4	4	2
2118	2	3	1	5	3.02	94	13000	1500	3	1	5	4	10	2	19	67	150	2	1	2	4	2	2	1	6	1
2119	1	3	3	5	3.65	81	35000	2500	1	1	2	2	5	1	30	68	140	2	3	1	4	1	0	4	5	2
2120	2	3	2	2	2.96	70	14000	500	3	2	2	2	0	2	21	66	120	2	1	2	3	1	1	4	5	1
2121	2	2	3	1	3.42	92	13000	1000	2	2	1	1	8	2	20	64	110	2	1	2	3	2	1	4	4	1
2122	2	3	3	2	3.06	80	12000	0	1	2	5	4	0	2	27	68	128	2	3	2	1	2	0	4	4	1
2123	2	3	2	1	4.00	96	15000	20	2	1	1	1	3	1	27	74	185	2	4	1	4	1	2	4	6	1
2124	1	4	2	5	3.26	82	22500	500	1	2	2	4	3	2	30	66	115	2	1	2	2	2	0	3	4	2
2125	1	1	1	4	2.35	80	10000	1000	3	2	5	4	5	1	20	66	145	1	1	2	4	1	1	4	6	1
2126	1	4	2	2	2.40	86	11000	1000	2	1	4	3	4	1	21	70	160	2	1	2	3	1	2	3	3	1
2127	1	3	1	2	2.27	78	14000	0	3	2	2	2	2	1	29	70	160	2	1	2	1	2	0	4	4	1
2128	1	4	1	5	2.43	75	10000	500	1	2	2	4	4	1	21	68	146	2	1	2	3	2	0	4	3	2
2129	2	2	1	3	2.50	76	12000	1000	3	2	2	1	2	2	23	62	135	2	1	2	1	1	1	4	5	1
2130	1	3	1	5	3.41	94	10000	200	1	1	1	4	17	1	24	70	170	2	3	1	4	2	1	1	4	1
2131	1	3	2	2	2.63	80	8000	0	2	2	2	2	4	1	27	69	170	2	1	2	1	2	0	4	4	1
2132	1	3	3	2	3.20	77	13000	2000	3	2	2	2	5	1	20	67	135	2	1	1	4	2	2	5	6	2
2133	1	3	3	4	3.02	83	12000	1000	2	2	2	2	4	1	21	67	165	2	1	2	1	2	0	1	6	2
2134	1	3	3	2	2.87	84	15000	500	3	1	2	2	7	1	21	70	143	2	1	2	2	2	1	1	6	1
2135	2	3	3	7	2.00	78	12500	250	1	2	2	2	4	2	26	62	160	2	1	2	1	1	0	4	6	2
2136	2	3	1	2	3.66	82	10000	0	1	2	1	4	2	1	33	64	112	2	1	2	1	1	0	4	5	1
2137	1	4	3	5	2.52	81	12500	0	2	2	2	4	10	1	24	75	165	2	3	1	4	1	0	4	2	2
2138	1	4	1	6	1.79	82	9500	100	2	2	2	4	0	1	29	68	150	2	4	2	1	2	0	4	6	2
2139	1	2	1	4	2.72	81	12000	1000	2	2	2	2	6	1	20	71	165	2	1	2	2	1	0	5	6	2
2140	1	2	3	1	2.83	84	11000	400	2	2	1	4	2	1	21	76	184	2	1	1	3	2	0	4	6	1
2141	1	4	2	5	2.68	75	10000	700	2	2	5	4	10	2	23	72	160	2	2	1	1	2	0	3	5	1
2142	2	4	1	7	2.55	87	11000	400	1	2	2	2	3	1	31	61	113	2	1	2	1	2	0	4	6	1
2143	2	4	1	4	3.00	85	15000	500	1	2	1	3	3	1	22	66	118	2	3	1	1	2	0	4	4	1
2144	1	3	3	5	3.04	80	15000	800	1	1	2	2	10	1	23	65	175	2	3	1	4	2	1	4	5	2
2145	1	3	1	5	3.10	75	9000	1000	3	2	5	4	7	1	22	69	200	2	3	1	4	1	0	1	2	1
2146	1	4	3	6	2.00	75	11000	500	2	2	1	2	5	1	22	72	183	2	1	1	3	2	0	4	4	1
2147	1	2	1	4	2.56	75	9500	0	2	2	2	4	4	1	25	68	163	2	1	2	4	1	2	1	1	1
2148	1	4	1	4	2.97	70	12000	400	1	2	1	4	2	1	48	70	160	2	1	2	4	2	0	4	6	1
2149	1	3	1	7	3.30	82	12000	488	1	1	1	4	10	1	28	69	137	2	2	2	4	1	0	4	5	2
2150	1	3	1	4	2.64	74	30000	1500	1	2	2	4	3	1	29	70	190	2	1	1	1	1	0	1	4	2
2151	1	2	3	1	3.52	89	12500	2000	2	1	2	4	6	1	19	72	185	2	1	2	4	1	0	4	4	2
2152	1	1	3	1	3.00	86	15000	150	1	2	2	2	5	1	21	68	145	2	1	2	4	2	1	4	5	1
2153	1	3	1	2	3.73	82	13000	1800	2	2	2	4	9	1	20	69	165	2	3	1	3	1	0	4	4	1
2154	2	3	1	4	3.50	99	10500	1000	3	2	2	4	4	2	21	62	120	2	2	1	4	2	0	4	4	1
2155	1	2	2	2	2.00	80	13500	1000	2	2	2	4	7	1	21	73	173	2	2	1	3	2	0	4	6	2
2156	1	3	1	2	3.00	85	10400	240	2	1	1	2	3	2	20	67	140	2	2	2	1	2	0	4	4	2
2157	2	3	2	2	3.00	91	12000	0	3	2	2	3	5	1	20	71	110	2	1	2	2	2	0	4	2	2
2158	2	3	3	4	2.91	82	11500	100	2	2	2	2	2	1	21	70	127	2	1	2	1	2	2	4	5	1
2159	1	2	1	2	3.04	84	12000	500	3	2	2	2	4	1	24	66	150	2	1	1	4	1	0	4	4	1
2160	1	2	3	4	2.72	85	18000	1000	1	2	2	2	0	1	25	68	140	2	1	2	4	1	0	4	4	1
2161	1	3	1	4	3.11	92	20000	400	1	2	2	2	4	2	32	67	147	2	2	2	4	2	0	4	5	2
2162	1	4	3	4	2.71	81	10000	200	2	1	1	4	6	1	22	70	200	2	1	2	4	2	1	1	6	1
2163	1	3	3	5	2.50	77	7800	4000	3	2	2	2	4	2	21	72	145	2	1	2	4	2	0	4	4	2
2164	1	3	3	1	2.84	75	11000	1000	2	1	2	4	3	2	21	73	175	2	1	2	4	1	0	3	3	1
2165	1	3	1	2	2.92	75	13500	1500	2	2	2	4	5	2	20	72	165	2	3	1	4	2	1	2	3	2
2166	1	2	1	2	3.40	82	25000	1000	1	2	2	4	3	1	23	67	183	2	1	2	3	1	0	4	4	2
2167	2	4	1	7	2.54	80	10000	0	3	2	2	4	10	2	22	66	135	1	1	2	4	2	1	6	5	1
2168	1	3	1	4	2.86	77	8000	0	3	2	2	2	2	2	21	71	145	2	1	2	1	2	0	4	4	1
2169	1	3	2	7	2.56	83	10500	80	2	2	2	4	3	1	22	71	195	1	1	2	3	1	0	4	4	2
2170	1	3	1	2	3.40	91	13000	700	2	2	2	4	9	1	22	68	140	2	1	2	4	1	1	4	3	1
2171	1	3	1	2	3.00	84	18000	200	2	2	2	4	8	1	20	71	145	1	3	1	1	1	0	4	1	1
2172	2	4	3	2	2.00	71	12500	300	2	2	2	2	3	1	40	64	135	2	2	2	4	2	0	1	7	1
2173	1	2	1	2	3.10	89	14000	100	2	2	2	2	10	1	19	70	160	2	1	2	3	2	1	3	5	1
2174	1	3	3	5	2.88	76	16700	0	2	2	2	2	5	1	24	60	170	2	2	2	3	1	3	3	5	2
2175	2	2	1	6	3.30	92	15000	875	3	2	1	4	2	2	18	63	120	2	1	2	4	1	2	3	2	1
2176	1	3	3	2	3.30	80	10000	2000	2	2	2	4	6	1	21	76	200	2	1	2	4	1	0	4	4	1
2177	1	3	1	6	3.40	90	10000	975	3	2	2	2	2	2	25	72	145	2	3	1	3	1	0	4	6	1
2178	2	2	3	2	3.00	80	9875	975	3	2	2	2	0	1	22	63	105	2	1	2	1	2	2	4	4	1
2179	1	3	1	2	2.50	80	11500	1000	2	2	2	4	2	1	24	73	175	2	1	2	4	1	2	5	5	1
2180	1	3	2	7	3.00	70	10000	1200	2	2	2	2	7	1	20	68	160	2	1	2	1	2	0	4	4	2
2181	1	2	3	4	2.40	78	16600	1000	2	1	2	4	3	1	20	70	155	2	1	2	3	2	0	4	4	2
2182	2	1	3	1	3.70	90	15000	500	3	2	5	4	7	1	18	63	125	2	1	1	2	1	2	6	6	2
2183	2	4	3	2	2.50	82	15000	600	1	2	2	4	1	1	40	64	125	2	1	2	3	2	0	4	4	2
2184	1	4	3	6	3.20	87	12000	900	2	2	3	4	10	1	22	68	190	2	1	2	3	1	2	4	5	2
2185	2	2	3	2	3.50	90	8000	2000	3	2	2	4	0	1	18	62	110	2	1	2	4	2	0	4	4	1
2186	1	2	3	2	3.25	87	11000	0	1	2	2	2	0	1	26	66	140	2	3	2	1	2	0	4	5	2
2187	1	3	1	6	2.90	75	20000	1000	2	1	1	2	10	1	21	72	210	2	1	1	3	1	0	5	5	1
2188	2	3	2	5	2.80	85	12000	1000	3	2	1	4	5	2	20	65	125	2	1	1	4	2	0	4	6	2
2189	2	3	3	5	3.11	85	15000	1000	1	2	2	2	3	1	23	69	143	2	2	1	1	2	0	4	4	2
2190	2	2	1	2	2.80	85	9000	0	3	2	2	2	2	2	23	65	118	2	1	2	4	2	0	4	4	1
2191	2	1	3	1	3.00	89	12000	2000	2	2	2	2	10	1	20	63	115	2	1	1	4	1	0	1	6	1
2192	2	2	3	2	3.60	80	12000	100	3	2	2	2	8	1	23	63	98	2	1	2	4	1	1	5	6	1
2193	1	3	3	3	2.70	85	17000	160	1	2	2	4	1	1	21	70	150	2	3	1	3	1	0	2	3	1
2194	2	4	3	5	2.80	83	15000	1000	1	2	2	2	2	2	48	68	160	2	1	2	3	1	0	3	5	2
2195	1	1	1	2	3.20	80	9600	0	3	2	2	4	2	1	20	67	140	2	1	1	4	1	2	4	7	1
2196	2	1	1	2	3.50	83	12000	1000	2	2	2	2	6	1	23	62	120	2	1	2	4	1	0	4	7	1
2197	2	2	3	2	2.00	81	13600	675	3	2	2	2	3	2	20	62	115	2	2	2	4	2	0	2	4	2
2198	2	3	2	7	2.70	80	9000	350	2	2	5	2	5	2	21	64	115	2	1	1	4	1	0	4	6	1
2199	1	3	1	7	3.00	82	10000	1200	1	2	5	4	3	1	24	71	150	2	1	2	3	2	1	4	4	2
2200	1	3	1	2	3.23	90	13000	900	3	2	2	2	3	1	28	67	150	2	1	2	3	1	2	4	4	2
2201	2	3	2	6	3.25	86	10000	400	3	1	2	4	10	1	23	65	98	2	1	2	4	1	0	4	4	1
2202	1	4	2	5	3.00	80	8000	900	3	2	2	4	3	1	23	72	210	2	1	2	3	1	0	6	3	2
File code number	Sex	Class	Grad. school	Major	Grade-point index	H. S. average	Est. starting salary $	Tuition $	Employment	Sunday shop	Investment	Car pref.	No. jeans owned	Calculator	Age	Height	Weight	Blood pressure	Smoking status	Smoking belief	Entrance standards	Retention standards	No. clubs and groups	SPS rating	Library rating	Personal question

APPENDIX **B**

REVIEW OF ARITHMETIC, ALGEBRA, AND SUMMATION NOTATION

B.1 RULES FOR ARITHMETIC OPERATIONS

The following is a summary of various rules for arithmetic operations with each rule illustrated by a numerical example.

RULE	EXAMPLE
1. $a + b = c$ and $b + a = c$	$2 + 1 = 3$ and $1 + 2 = 3$
2. $a + (b + c) = (a + b) + c$	$5 + (7 + 4) = (5 + 7) + 4 = 16$
3. $a - b = c$ but $b - a \neq c$	$9 - 7 = 2$ but $7 - 9 = -2$
4. $a \times b = b \times a$	$7 \times 6 = 6 \times 7 = 42$
5. $a \times (b + c) = (a \times b) + (a \times c)$	$2 \times (3 + 5) = (2 \times 3) + (2 \times 5) = 16$
6. $a \div b \neq b \div a$	$12 \div 3 \neq 3 \div 12$
7. $\frac{a + b}{c} = \frac{a}{c} + \frac{b}{c}$	$\frac{7 + 3}{2} = \frac{7}{2} + \frac{3}{2} = 5$
8. $\frac{a}{b + c} \neq \frac{a}{b} + \frac{a}{c}$	$\frac{3}{4 + 5} \neq \frac{3}{4} + \frac{3}{5}$

9. $\frac{1}{a} + \frac{1}{b} = \frac{b + a}{ab}$ $\qquad$ $\frac{1}{3} + \frac{1}{5} = \frac{5 + 3}{(3)(5)} = \frac{8}{15}$

10. $\frac{a}{b} \times \frac{c}{d} = \frac{a \times c}{b \times d}$ $\qquad$ $\frac{2}{3} \times \frac{6}{7} = \frac{2 \times 6}{3 \times 7} = \frac{12}{21}$

11. $\frac{a}{b} \div \frac{c}{d} = \frac{a \times d}{b \times c}$ $\qquad$ $\frac{5}{8} \div \frac{3}{7} = \frac{5 \times 7}{8 \times 3} = \frac{35}{24}$

B.2 RULES FOR ALGEBRA: EXPONENTS AND SQUARE ROOTS

The following is a summary of various rules for algebraic operations with each rule illustrated by a numerical example.

RULE	EXAMPLE
1. $X^a \cdot X^b = X^{a+b}$	$4^2 \cdot 4^3 = 4^5$
2. $(X^a)^b = X^{ab}$	$(2^2)^3 = 2^6$
3. $(X^a/X^b) = X^{a-b}$	$\frac{3^5}{3^3} = 3^2$
4. $\frac{X^a}{X^a} = X^0 = 1$	$\frac{3^4}{3^4} = 3^0 = 1$
5. $\sqrt{XY} = \sqrt{X}\sqrt{Y}$	$\sqrt{(25)(4)} = \sqrt{25}\sqrt{4} = 10$
6. $\sqrt{\frac{X}{Y}} = \frac{\sqrt{X}}{\sqrt{Y}}$	$\sqrt{\frac{16}{100}} = \frac{\sqrt{16}}{\sqrt{100}} = .40$

B.3 SUMMATION NOTATION

Since the operation of addition occurs so frequently in statistics, the special symbol $\sum$ (sigma) is used to denote "taking the sum of." Suppose, for example, we have a set of n values for some variable X. The expression $\sum_{i=1}^{n} X_i$ means that these n values are to be added together. Thus,

$$\sum_{i=1}^{n} X_i = X_1 + X_2 + X_3 + \cdots + X_n$$

The use of the summation notation can be illustrated in the following problem. Suppose we have five observations of a variable X: $X_1 = 2$, $X_2 = 0$, $X_3 = -1$, $X_4 = 5$, and $X_5 = 7$. Thus,

$$\sum_{i=1}^{5} X_i = X_1 + X_2 + X_3 + X_4 + X_5$$

$$= 2 + 0 + (-1) + 5 + 7 = 13$$

In statistics we are also frequently involved with summing the squared values of a variable. Thus,

$$\sum_{i=1}^{n} X_i^2 = X_1^2 + X_2^2 + X_3^2 + \cdots + X_n^2$$

and, in our example, we have

$$\begin{aligned}\sum_{i=1}^{5} X_i^2 &= X_1^2 + X_2^2 + X_3^2 + X_4^2 + X_5^2 \\ &= 2^2 + 0^2 + (-1)^2 + 5^2 + 7^2 \\ &= 4 + 0 + 1 + 25 + 49 \\ &= 79\end{aligned}$$

We should realize here that $\sum_{i=1}^{n} X_i^2$, the summation of the squares, is not the same as $\left(\sum_{i=1}^{n} X_i\right)^2$, the square of the sum, that is,

$$\sum_{i=1}^{n} X_i^2 \neq \left(\sum_{i=1}^{n} X_i\right)^2$$

In our example the summation of squares is equal to 79. This is not equal to the square of the sum which is $13^2 = 169$.

Another frequently used operation involves the summation of the product. That is, suppose we have two variables, X and Y, each having n observations. Then,

$$\sum_{i=1}^{n} X_i Y_i = X_1 Y_1 + X_2 Y_2 + X_3 Y_3 + \cdots + X_n Y_n$$

Continuing with our previous example, suppose that there is also a second variable Y whose 5 values are $Y_1 = 1$, $Y_2 = 3$, $Y_3 = -2$, $Y_4 = 4$, and $Y_5 = 3$. Then,

$$\begin{aligned}\sum_{i=1}^{5} X_i Y_i &= X_1 Y_1 + X_2 Y_2 + X_3 Y_3 + X_4 Y_4 + X_5 Y_5 \\ &= 2(1) + 0(3) + (-1)(-2) + 5(4) + 7(3) \\ &= 2 + 0 + 2 + 20 + 21 \\ &= 45\end{aligned}$$

In computing $\sum_{i=1}^{n} X_i Y_i$ we must realize that the first value of X is multiplied by the first value of Y, the second value of X is multiplied by the second value of Y, etc. These cross products are then summed in order to obtain the desired result. However, we should note here that the summation of cross products is not equal to the product of the individual sums, that is,

$$\sum_{i=1}^{n} X_i Y_i \neq \left(\sum_{i=1}^{n} X_i\right)\left(\sum_{i=1}^{n} Y_i\right)$$

In our example, $\sum_{i=1}^{5} X_i = 13$ and $\sum_{i=1}^{5} Y_i = 1 + 3 + (-2) + 4 + 3 = 9$, so that $\left(\sum_{i=1}^{5} X_i\right)\left(\sum_{i=1}^{5} Y_i\right) = 13(9) = 117$. This is not the same as $\sum_{i=1}^{5} X_i Y_i$ which equals 45.

Before studying the four basic rules of performing operations with summation notation, it would be helpful to present the values for each of the five observations of X and Y in a tabular format.

Observation	X_i	Y_i
1	2	1
2	0	3
3	−1	−2
4	5	4
5	7	3
	$\sum_{i=1}^{5} X_i = 13$	$\sum_{i=1}^{5} Y_i = 9$

Rule 1: The summation of the values of two variables is equal to the sum of the values of each summed variable.

$$\sum_{i=1}^{n} (X_i + Y_i) = \sum_{i=1}^{n} X_i + \sum_{i=1}^{n} Y_i$$

Thus, in our example,

$$\begin{aligned}\sum_{i=1}^{5} (X_i + Y_i) &= (2 + 1) + (0 + 3) + (-1 + (-2)) + (5 + 4) + (7 + 3)\\ &= 3 + 3 + (-3) + 9 + 10\\ &= 22 = \sum_{i=1}^{5} X_i + \sum_{i=1}^{5} Y_i = 13 + 9 = 22\end{aligned}$$

Rule 2: The summation of a difference between the values of two variables is equal to the difference between the summed values of the variables.

$$\sum_{i=1}^{n} (X_i - Y_i) = \sum_{i=1}^{n} X_i - \sum_{i=1}^{n} Y_i$$

Thus, in our example,

$$\begin{aligned}\sum_{i=1}^{5} (X_i - Y_i) &= (2-1) + (0-3) + (-1-(-2)) + (5-4) + (7-3) \\ &= 1 + (-3) + 1 + 1 + 4 \\ &= 4 = \sum_{i=1}^{5} X_i - \sum_{i=1}^{5} Y_i = 13 - 9 = 4\end{aligned}$$

Rule 3: The summation of a constant times a variable is equal to that constant times the summation of the values of the variable.

$$\sum_{i=1}^{n} CX_i = C \sum_{i=1}^{n} X_i$$

where C is a constant.

Thus, in our example, if $C = 2$,

$$\begin{aligned}\sum_{i=1}^{5} CX_i = \sum_{i=1}^{5} 2X_i &= 2(2) + 2(0) + 2(-1) + 2(5) + 2(7) \\ &= 4 + 0 + (-2) + 10 + 14 \\ &= 26 = 2 \sum_{i=1}^{5} X_i = 2(13) = 26\end{aligned}$$

Rule 4: A constant summed n times will be equal to n times the value of the constant.

$$\sum_{i=1}^{n} C = nC$$

where C is a constant.

Thus, if the constant $C = 2$ is summed 5 times, we would have

$$\begin{aligned}\sum_{i=1}^{5} C &= 2 + 2 + 2 + 2 + 2 \\ &= 10 = 5(2) = 10\end{aligned}$$

To illustrate how these summation rules are used, we may demonstrate one of the mathematical properties pertaining to the average or arithmetic mean (see Section 4.3.1), that is,

$$\sum_{i=1}^{n} (X_i - \bar{X}) = 0$$

This property states that the summation of the differences between each observation and the arithmetic mean is zero. This can be proven mathematically in the following manner:

1. From Equation (4.1),

$$\bar{X} = \frac{\sum_{i=1}^{n} X_i}{n}$$

Thus, using summation rule 2, we have

$$\sum_{i=1}^{n} (X_i - \bar{X}) = \sum_{i=1}^{n} X_i - \sum_{i=1}^{n} \bar{X}$$

2. Since, for any fixed set of data, $\bar{X}$ can be considered a constant, from summation rule 4 we have

$$\sum_{i=1}^{n} \bar{X} = n\bar{X}$$

Therefore,

$$\sum_{i=1}^{n} (X_i - \bar{X}) = \sum_{i=1}^{n} X_i - n\bar{X}$$

3. However, from Equation (4.1), since

$$\bar{X} = \frac{\sum_{i=1}^{n} X_i}{n} \quad \text{then} \quad n\bar{X} = \sum_{i=1}^{n} X_i$$

Therefore,

$$\sum_{i=1}^{n} (X_i - \bar{X}) = \sum_{i=1}^{n} X_i - \sum_{i=1}^{n} X_i$$

and, thus, we have shown that

$$\sum_{i=1}^{n} (X_i - \bar{X}) = 0$$

PROBLEM

Suppose there are six observations for the variables X and Y such that $X_1 = 2$, $X_2 = 1$, $X_3 = 5$, $X_4 = -3$, $X_5 = 1$, and $X_6 = -2$, and $Y_1 = 4$, $Y_2 = 0$, $Y_3 = -1$, $Y_4 = 2$, $Y_5 = 7$, and $Y_6 = -3$. Compute each of the following:

(a) $\sum_{i=1}^{6} X_i$

(b) $\sum_{i=1}^{6} Y_i$

(c) $\sum_{i=1}^{6} X_i^2$

(d) $\sum_{i=1}^{6} Y_i^2$

(e) $\sum_{i=1}^{6} X_i Y_i$

(f) $\sum_{i=1}^{6} (X_i + Y_i)$

(g) $\sum_{i=1}^{6} (X_i - Y_i)$

(h) $\sum_{i=1}^{6} (X_i - 3Y_i + 2X_i^2)$

(i) $\sum_{i=1}^{6} (CX_i)$ where $C = -1$

(j) $\sum_{i=1}^{6} (X_i - 3Y_i + C)$ where $C = +3$

REFERENCE

BASHAW, W. L., *Mathematics for Statistics* (New York: Wiley, 1969).

APPENDIX **C**

STATISTICAL SYMBOLS AND GREEK ALPHABET

C.1 STATISTICAL SYMBOLS

$+$	add	$\times$	multiply
$-$	subtract	$\div$	divide
$=$	equals	$\neq$	not equal
$\cong$	approximately equal to		
$>$	greater than	$<$	less than
$\geq$ or $\geqq$	greater than or equal to		
$\leq$ or $\leqq$	less than or equal to		

C.2 GREEK ALPHABET

Greek letter	Greek name	English equivalent	Greek letter	Greek name	English equivalent
Α α	Alpha	a	Ν ν	Nu	n
Β β	Beta	b	Ξ ξ	Xi	x
Γ γ	Gamma	g	Ο o	Omicron	ŏ
Δ δ	Delta	d	Π π	Pi	p
Ε ε	Epsilon	ĕ	Ρ ρ	Rho	r
Ζ ζ	Zeta	z	Σ σ s	Sigma	s
Η η	Eta	ē	Τ τ	Tau	t
Θ θ ϑ	Theta	th	Υ υ	Upsilon	u
Ι ι	Iota	i	Φ ϕ φ	Phi	ph
Κ κ	Kappa	k	Χ χ	Chi	ch
Λ λ	Lambda	l	Ψ ψ	Psi	ps
Μ μ	Mu	m	Ω ω	Omega	ō

APPENDIX D

THE METRIC SYSTEM

The current system of measurement used in the United States is called the English system. The purpose of this appendix is to become familiarized with the much more widely used Metric system of measurement.

There are several reasons for adopting the metric system of measurement. One major advantage of converting to the metric system is found in the area of international trade and commerce. The manufacturing process will become much more efficient if, in the production of goods for export, different machinery and/or different production setups were not needed. A second advantage to the metric system is found in the area of travel and communications. Whether they be reasons of business or pleasure, as more and more Americans spend time traveling abroad, and, at the same time, we accept visitors from other countries to our shores, this "exchange process" would greatly be facilitated if we used the same system of measurement as is used throughout most of the world. A third major advantage to using the metric system is its simplicity. The metric system of measurement is a decimal system, and, therefore, is much simpler to use and work with than our current English system.

METRIC CONVERSIONS

Type of Measurement	From Metric to Metric	From English to Metric	From Metric to English
Length (linear measure)	1 micrometer = $\frac{1}{1,000,000}$ meter 1 millimeter = $\frac{1}{1,000}$ meter 1 centimeter = $\frac{1}{100}$ meter 1 hectometer = 100 meters 1 kilometer = 1,000 meters 1 megameter = 1,000,000 meters	1 inch = 25.4 millimeters 1 inch = 2.54 centimeters 1 foot = 0.3 meter 1 yard = 0.9 meter 1 mile = 1.6 kilometers	1 millimeter = 0.04 inch 1 centimeter = 0.4 inch 1 meter = 3.3 feet 1 meter = 1.1 yards 1 kilometer = 0.621 mile
Area (square measure)		1 square inch = 6.5 square centimeters 1 square foot = 0.09 square meter 1 square yard = 0.8 square meter 1 acre = 0.4 square hectometer	1 square centimeter = 0.16 square inch 1 square meter = 11 square feet 1 square meter = 1.2 square yards 1 square hectometer = 2.5 acres
Weight (mass)	1 microgram = $\frac{1}{1,000,000}$ gram 1 milligram = $\frac{1}{1,000}$ gram 1 centigram = $\frac{1}{100}$ gram 1 hectogram = 100 grams 1 kilogram = 1,000 grams 1 megagram = 1,000,000 grams	1 ounce = 28.3 grams 1 pound = 0.45 kilogram	1 gram = 0.035 ounce 1 kilogram = 2.2 pounds
Volume (capacity)	1 microliter = $\frac{1}{1,000,000}$ liter 1 milliliter = $\frac{1}{1,000}$ liter 1 centiliter = $\frac{1}{100}$ liter 1 hectoliter = 100 liters 1 kiloliter = 1,000 liters 1 megaliter = 1,000,000 liters	1 quart = 0.9463 liter 1 gallon = 3.7853 liters	1 liter = 1.06 quarts 1 liter = 0.26 gallon

APPENDIX **E**

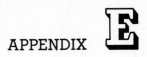

TABLES

Table E.1
TABLE OF RANDOM NUMBERS

	Column							
Row	00000 12345	00001 67890	11111 12345	11112 67890	22222 12345	22223 67890	33333 12345	33334 67890
01	66194	28926	99547	16625	45515	67953	12108	57846
02	78240	43195	24837	32511	70880	22070	52622	61881
03	00833	88000	67299	68215	11274	55624	32991	17436
04	12111	86683	61270	58036	64192	90611	15145	01748
05	47189	99951	05755	03834	43782	90599	40282	51417
06	76396	72486	62423	27618	84184	78922	73561	52818
07	46409	17469	32483	09083	76175	19985	26309	91536
08	74626	22111	87286	46772	42243	68046	44250	42439
09	34450	81974	93723	49023	58432	67083	36876	93391
10	36327	72135	33005	28701	34710	49359	50693	89311
11	74185	77536	84825	09934	99103	09325	67389	45869
12	12296	41623	62873	37943	25584	09609	63360	47270
13	90822	60280	88925	99610	42772	60561	76873	04117
14	72121	79152	96591	90305	10189	79778	68016	13747
15	95268	41377	25684	08151	61816	58555	54305	86189
16	92603	09091	75884	93424	72586	88903	30061	14457
17	18813	90291	05275	01223	79607	95426	34900	09778
18	38840	26903	28624	67157	51986	42865	14508	49315
19	05959	33836	53758	16562	41081	38012	41230	20528
20	85141	21155	99212	32685	51403	31926	69813	58781
21	75047	59643	31074	38172	03718	32119	69506	67143
22	30752	95260	68032	62871	58781	34143	68790	69766
23	22986	82575	42187	62295	84295	30634	66562	31442
24	99439	86692	90348	66036	48399	73451	26698	39437
25	20389	93029	11881	71685	65452	89047	63669	02656
26	39249	05173	68256	36359	20250	68686	05947	09335
27	96777	33605	29481	20063	09398	01843	35139	61344
28	04860	32918	10798	50492	52655	33359	94713	28393
29	41613	42375	00403	03656	77580	87772	86877	57085
30	17930	00794	53836	53692	67135	98102	61912	11246
31	24649	31845	25736	75231	83808	98917	93829	99430
32	79899	34061	54308	59358	56462	58166	97302	86828
33	76801	49594	81002	30397	52728	15101	72070	33706
34	36239	63636	38140	65731	39788	06872	38971	53363
35	07392	64449	17886	63632	53995	17574	22247	62607
36	67133	04181	33874	98835	67453	59734	76381	63455
37	77759	31504	32832	70861	15152	29733	75371	39174
38	85992	72268	42920	20810	29361	51423	90306	73574
39	79553	75952	54116	65553	47139	60579	09165	85490
40	41101	17336	48951	53674	17880	45260	08575	49321
41	36191	17095	32123	91576	84221	78902	82010	30847
42	62329	63898	23268	74283	26091	68409	69704	82267
43	14751	13151	93115	01437	56945	89661	67680	79790
44	48462	59278	44185	29616	76537	19589	83139	28454
45	29435	88105	59651	44391	74588	55114	80834	85686
46	28340	29285	12965	14821	80425	16602	44653	70467
47	02167	58940	27149	80242	10587	79786	34959	75339
48	17864	00991	39557	54981	23588	81914	37609	13128
49	79675	80605	60059	35862	00254	36546	21545	78179
50	72335	82037	92003	34100	29879	46613	89720	13274

Table E.1 (*Continued*)

	Column							
Row	00000 12345	00001 67890	11111 12345	11112 67890	22222 12345	22223 67890	33333 12345	33334 67890
51	49280	88924	35779	00283	81163	07275	89863	02348
52	61870	41657	07468	08612	98083	97349	20775	45091
53	43898	65923	25078	86129	78496	97653	91550	08078
54	62993	93912	30454	84598	56095	20664	12872	64647
55	33850	58555	51438	85507	71865	79488	76783	31708
56	55336	71264	88472	04334	63919	36394	11095	92470
57	70543	29776	10087	10072	55980	64688	68239	20461
58	89382	93809	00796	95945	34101	81277	66090	88872
59	37818	72142	67140	50785	22380	16703	53362	44940
60	60430	22834	14130	96593	23298	56203	92671	15925
61	82975	66158	84731	19436	55790	69229	28661	13675
62	39087	71938	40355	54324	08401	26299	49420	59208
63	55700	24586	93247	32596	11865	63397	44251	43189
64	14756	23997	78643	75912	83832	32768	18928	57070
65	32166	53251	70654	92827	63491	04233	33825	69662
66	23236	73751	31888	81718	06546	83246	47651	04877
67	45794	26926	15130	82455	78305	55058	52551	47182
68	09893	20505	14225	68514	46427	56788	96297	78822
69	54382	74598	91499	14523	68479	27686	46162	83554
70	94750	89923	37089	20048	80336	94598	26940	36858
71	70297	34135	53140	33340	42050	82341	44104	82949
72	85157	47954	32979	26575	57600	40881	12250	73742
73	11100	02340	12860	74697	96644	89439	28707	25815
74	36871	50775	30592	57143	17381	68856	25853	35041
75	23913	48357	63308	16090	51690	54607	72407	55538
76	79348	36085	27973	65157	07456	22255	25626	57054
77	92074	54641	53673	54421	18130	60103	69593	49464
78	06873	21440	75593	41373	49502	17972	82578	16364
79	12478	37622	99659	31065	83613	69889	58869	29571
80	57175	55564	65411	42547	70457	03426	72937	83792
81	91616	11075	80103	07831	59309	13276	26710	73000
82	78025	73539	14621	39044	47450	03197	12787	47709
83	27587	67228	80145	10175	12822	86687	65530	49325
84	16690	20427	04251	64477	73709	73945	92396	68263
85	70183	58065	65489	31833	82093	16747	10386	59293
86	90730	35385	15679	99742	50866	78028	75573	67257
87	10934	93242	13431	24590	02770	48582	00906	58595
88	82462	30166	79613	47416	13389	80268	05085	96666
89	27463	10433	07606	16285	93699	60912	94532	95632
90	02979	52997	09079	92709	90110	47506	53693	49892
91	46888	69929	75233	52507	32097	37594	10067	67327
92	53638	83161	08289	12639	08141	12640	28437	09268
93	82433	61427	17239	89160	19666	08814	37841	12847
94	35766	31672	50082	22795	66948	65581	84393	15890
95	10853	42581	08792	13257	61973	24450	52351	16602
96	20341	27398	72906	63955	17276	10646	74692	48438
97	54458	90542	77563	51839	52901	53355	83281	19177
98	26337	66530	16687	35179	46560	00123	44546	79896
99	34314	23729	85264	05575	96855	23820	11091	79821
00	28603	10708	68933	34189	92166	15181	66628	58599

SOURCE: Partially extracted from The Rand Corporation, *A Million Random Digits with 100,000 Normal Deviates* (Glencoe, Ill.: The Free Press, 1955).

Table E.2
THE STANDARDIZED NORMAL DISTRIBUTION

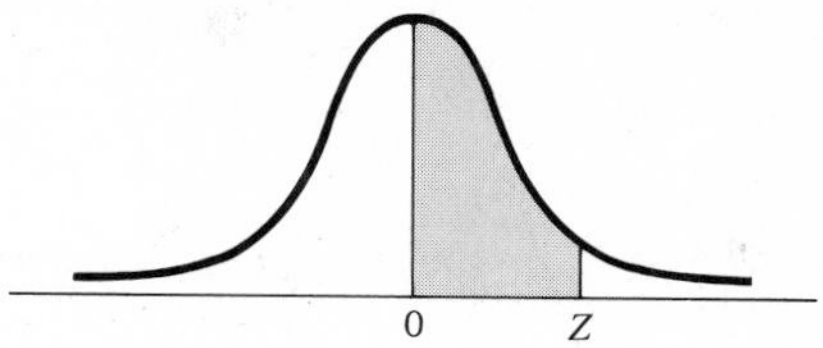

Entry represents area under the standardized normal distribution from the mean to Z

Z	.00	.01	.02	.03	.04	.05	.06	.07	.08	.09
0.0	.0000	.0040	.0080	.0120	.0160	.0199	.0239	.0279	.0319	.0359
0.1	.0398	.0438	.0478	.0517	.0557	.0596	.0636	.0675	.0714	.0753
0.2	.0793	.0832	.0871	.0910	.0948	.0987	.1026	.1064	.1103	.1141
0.3	.1179	.1217	.1255	.1293	.1331	.1368	.1406	.1443	.1480	.1517
0.4	.1554	.1591	.1628	.1664	.1700	.1736	.1772	.1808	.1844	.1879
0.5	.1915	.1950	.1985	.2019	.2054	.2088	.2123	.2157	.2190	.2224
0.6	.2257	.2291	.2324	.2357	.2389	.2422	.2454	.2486	.2518	.2549
0.7	.2580	.2612	.2642	.2673	.2704	.2734	.2764	.2794	.2823	.2852
0.8	.2881	.2910	.2939	.2967	.2995	.3023	.3051	.3078	.3106	.3133
0.9	.3159	.3186	.3212	.3238	.3264	.3289	.3315	.3340	.3365	.3389
1.0	.3413	.3438	.3461	.3485	.3508	.3531	.3554	.3577	.3599	.3621
1.1	.3643	.3665	.3686	.3708	.3729	.3749	.3770	.3790	.3810	.3830
1.2	.3849	.3869	.3888	.3907	.3925	.3944	.3962	.3980	.3997	.4015
1.3	.4032	.4049	.4066	.4082	.4099	.4115	.4131	.4147	.4162	.4177
1.4	.4192	.4207	.4222	.4236	.4251	.4265	.4279	.4292	.4306	.4319
1.5	.4332	.4345	.4357	.4370	.4382	.4394	.4406	.4418	.4429	.4441
1.6	.4452	.4463	.4474	.4484	.4495	.4505	.4515	.4525	.4535	.4545
1.7	.4554	.4564	.4573	.4582	.4591	.4599	.4608	.4616	.4625	.4633
1.8	.4641	.4649	.4656	.4664	.4671	.4678	.4686	.4693	.4699	.4706
1.9	.4713	.4719	.4726	.4732	.4738	.4744	.4750	.4756	.4761	.4767
2.0	.4772	.4778	.4783	.4788	.4793	.4798	.4803	.4808	.4812	.4817
2.1	.4821	.4826	.4830	.4834	.4838	.4842	.4846	.4850	.4854	.4857
2.2	.4861	.4864	.4868	.4871	.4875	.4878	.4881	.4884	.4887	.4890
2.3	.4893	.4896	.4898	.4901	.4904	.4906	.4909	.4911	.4913	.4916
2.4	.4918	.4920	.4922	.4925	.4927	.4929	.4931	.4932	.4934	.4936
2.5	.4938	.4940	.4941	.4943	.4945	.4946	.4948	.4949	.4951	.4952
2.6	.4953	.4955	.4956	.4957	.4959	.4960	.4961	.4962	.4963	.4964
2.7	.4965	.4966	.4967	.4968	.4969	.4970	.4971	.4972	.4973	.4974
2.8	.4974	.4975	.4976	.4977	.4977	.4978	.4979	.4979	.4980	.4981
2.9	.4981	.4982	.4982	.4983	.4984	.4984	.4985	.4985	.4986	.4986
3.0	.49865	.49869	.49874	.49878	.49882	.49886	.49889	.49893	.49897	.49900
3.1	.49903	.49906	.49910	.49913	.49916	.49918	.49921	.49924	.49926	.49929
3.2	.49931	.49934	.49936	.49938	.49940	.49942	.49944	.49946	.49948	.49950
3.3	.49952	.49953	.49955	.49957	.49958	.49960	.49961	.49962	.49964	.49965
3.4	.49966	.49968	.49969	.49970	.49971	.49972	.49973	.49974	.49975	.49976
3.5	.49977	.49978	.49978	.49979	.49980	.49981	.49981	.49982	.49983	.49983
3.6	.49984	.49985	.49985	.49986	.49986	.49987	.49987	.49988	.49988	.49989
3.7	.49989	.49990	.49990	.49990	.49991	.49991	.49992	.49992	.49992	.49992
3.8	.49993	.49993	.49993	.49994	.49994	.49994	.49994	.49995	.49995	.49995
3.9	.49995	.49995	.49996	.49996	.49996	.49996	.49996	.49996	.49997	.49997

Table E.3
CRITICAL VALUES OF t

For a particular number of degrees of freedom, entry represents the critical value of t corresponding to a specified upper tail area α

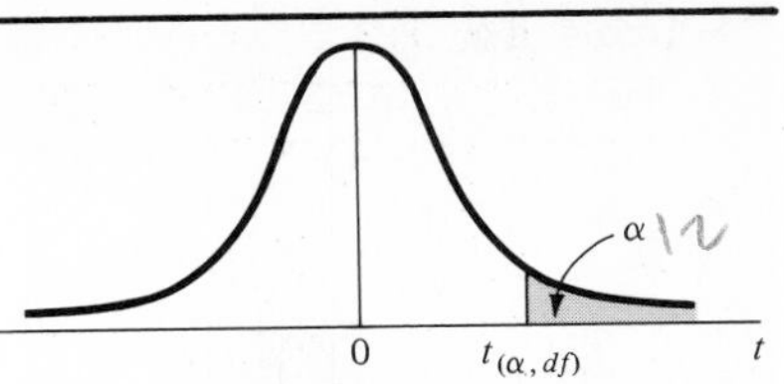

Degrees of Freedom	Upper Tail Areas .25	.10	.05	.025	.01	.005
1	1.0000	3.0777	6.3138	12.7062	31.8207	63.6574
2	0.8165	1.8856	2.9200	4.3027	6.9646	9.9248
3	0.7649	1.6377	2.3534	3.1824	4.5407	5.8409
4	0.7407	1.5332	2.1318	2.7764	3.7469	4.6041
5	0.7267	1.4759	2.0150	2.5706	3.3649	4.0322
6	0.7176	1.4398	1.9432	2.4469	3.1427	3.7074
7	0.7111	1.4149	1.8946	2.3646	2.9980	3.4995
8	0.7064	1.3968	1.8595	2.3060	2.8965	3.3554
9	0.7027	1.3830	1.8331	2.2622	2.8214	3.2498
10	0.6998	1.3722	1.8125	2.2281	2.7638	3.1693
11	0.6974	1.3634	1.7959	2.2010	2.7181	3.1058
12	0.6955	1.3562	1.7823	2.1788	2.6810	3.0545
13	0.6938	1.3502	1.7709	2.1604	2.6503	3.0123
14	0.6924	1.3450	1.7613	2.1448	2.6245	2.9768
15	0.6912	1.3406	1.7531	2.1315	2.6025	2.9467
16	0.6901	1.3368	1.7459	2.1199	2.5835	2.9208
17	0.6892	1.3334	1.7396	2.1098	2.5669	2.8982
18	0.6884	1.3304	1.7341	2.1009	2.5524	2.8784
19	0.6876	1.3277	1.7291	2.0930	2.5395	2.8609
20	0.6870	1.3253	1.7247	2.0860	2.5280	2.8453
21	0.6864	1.3232	1.7207	2.0796	2.5177	2.8314
22	0.6858	1.3212	1.7171	2.0739	2.5083	2.8188
23	0.6853	1.3195	1.7139	2.0687	2.4999	2.8073
24	0.6848	1.3178	1.7109	2.0639	2.4922	2.7969
25	0.6844	1.3163	1.7081	2.0595	2.4851	2.7874
26	0.6840	1.3150	1.7056	2.0555	2.4786	2.7787
27	0.6837	1.3137	1.7033	2.0518	2.4727	2.7707
28	0.6834	1.3125	1.7011	2.0484	2.4671	2.7633
29	0.6830	1.3114	1.6991	2.0452	2.4620	2.7564
30	0.6828	1.3104	1.6973	2.0423	2.4573	2.7500
31	0.6825	1.3095	1.6955	2.0395	2.4528	2.7440
32	0.6822	1.3086	1.6939	2.0369	2.4487	2.7385
33	0.6820	1.3077	1.6924	2.0345	2.4448	2.7333
34	0.6818	1.3070	1.6909	2.0322	2.4411	2.7284
35	0.6816	1.3062	1.6896	2.0301	2.4377	2.7238
36	0.6814	1.3055	1.6883	2.0281	2.4345	2.7195
37	0.6812	1.3049	1.6871	2.0262	2.4314	2.7154
38	0.6810	1.3042	1.6860	2.0244	2.4286	2.7116
39	0.6808	1.3036	1.6849	2.0227	2.4258	2.7079
40	0.6807	1.3031	1.6839	2.0211	2.4233	2.7045
41	0.6805	1.3025	1.6829	2.0195	2.4208	2.7012
42	0.6804	1.3020	1.6820	2.0181	2.4185	2.6981
43	0.6802	1.3016	1.6811	2.0167	2.4163	2.6951
44	0.6801	1.3011	1.6802	2.0154	2.4141	2.6923
45	0.6800	1.3006	1.6794	2.0141	2.4121	2.6896

Table E.3 (*Continued*)

Degrees of Freedom	Upper Tail Areas .25	.10	.05	.025	.01	.005
46	0.6799	1.3002	1.6787	2.0129	2.4102	2.6870
47	0.6797	1.2998	1.6779	2.0117	2.4083	2.6846
48	0.6796	1.2994	1.6772	2.0106	2.4066	2.6822
49	0.6795	1.2991	1.6766	2.0096	2.4049	2.6800
50	0.6794	1.2987	1.6759	2.0086	2.4033	2.6778
51	0.6793	1.2984	1.6753	2.0076	2.4017	2.6757
52	0.6792	1.2980	1.6747	2.0066	2.4002	2.6737
53	0.6791	1.2977	1.6741	2.0057	2.3988	2.6718
54	0.6791	1.2974	1.6736	2.0049	2.3974	2.6700
55	0.6790	1.2971	1.6730	2.0040	2.3961	2.6682
56	0.6789	1.2969	1.6725	2.0032	2.3948	2.6665
57	0.6788	1.2966	1.6720	2.0025	2.3936	2.6649
58	0.6787	1.2963	1.6716	2.0017	2.3924	2.6633
59	0.6787	1.2961	1.6711	2.0010	2.3912	2.6618
60	0.6786	1.2958	1.6706	2.0003	2.3901	2.6603
61	0.6785	1.2956	1.6702	1.9996	2.3890	2.6589
62	0.6785	1.2954	1.6698	1.9990	2.3880	2.6575
63	0.6784	1.2951	1.6694	1.9983	2.3870	2.6561
64	0.6783	1.2949	1.6690	1.9977	2.3860	2.6549
65	0.6783	1.2947	1.6686	1.9971	2.3851	2.6536
66	0.6782	1.2945	1.6683	1.9966	2.3842	2.6524
67	0.6782	1.2943	1.6679	1.9960	2.3833	2.6512
68	0.6781	1.2941	1.6676	1.9955	2.3824	2.6501
69	0.6781	1.2939	1.6672	1.9949	2.3816	2.6490
70	0.6780	1.2938	1.6669	1.9944	2.3808	2.6479
71	0.6780	1.2936	1.6666	1.9939	2.3800	2.6469
72	0.6779	1.2934	1.6663	1.9935	2.3793	2.6459
73	0.6779	1.2933	1.6660	1.9930	2.3785	2.6449
74	0.6778	1.2931	1.6657	1.9925	2.3778	2.6439
75	0.6778	1.2929	1.6654	1.9921	2.3771	2.6430
76	0.6777	1.2928	1.6652	1.9917	2.3764	2.6421
77	0.6777	1.2926	1.6649	1.9913	2.3758	2.6412
78	0.6776	1.2925	1.6646	1.9908	2.3751	2.6403
79	0.6776	1.2924	1.6644	1.9905	2.3745	2.6395
80	0.6776	1.2922	1.6641	1.9901	2.3739	2.6387
81	0.6775	1.2921	1.6639	1.9897	2.3733	2.6379
82	0.6775	1.2920	1.6636	1.9893	2.3727	2.6371
83	0.6775	1.2918	1.6634	1.9890	2.3721	2.6364
84	0.6774	1.2917	1.6632	1.9886	2.3716	2.6356
85	0.6774	1.2916	1.6630	1.9883	2.3710	2.6349
86	0.6774	1.2915	1.6628	1.9879	2.3705	2.6342
87	0.6773	1.2914	1.6626	1.9876	2.3700	2.6335
88	0.6773	1.2912	1.6624	1.9873	2.3695	2.6329
89	0.6773	1.2911	1.6622	1.9870	2.3690	2.6322
90	0.6772	1.2910	1.6620	1.9867	2.3685	2.6316

Table E.3 (*Continued*)

Degrees of Freedom	Upper Tail Areas					
	.25	*.10*	*.05*	*.025*	*.01*	*.005*
91	0.6772	1.2909	1.6618	1.9864	2.3680	2.6309
92	0.6772	1.2908	1.6616	1.9861	2.3676	2.6303
93	0.6771	1.2907	1.6614	1.9858	2.3671	2.6297
94	0.6771	1.2906	1.6612	1.9855	2.3667	2.6291
95	0.6771	1.2905	1.6611	1.9853	2.3662	2.6286
96	0.6771	1.2904	1.6609	1.9850	2.3658	2.6280
97	0.6770	1.2903	1.6607	1.9847	2.3654	2.6275
98	0.6770	1.2902	1.6606	1.9845	2.3650	2.6269
99	0.6770	1.2902	1.6604	1.9842	2.3646	2.6264
100	0.6770	1.2901	1.6602	1.9840	2.3642	2.6259
102	0.6769	1.2899	1.6599	1.9835	2.3635	2.6249
104	0.6769	1.2897	1.6596	1.9830	2.3627	2.6239
106	0.6768	1.2896	1.6594	1.9826	2.3620	2.6230
108	0.6768	1.2894	1.6591	1.9822	2.3614	2.6221
110	0.6767	1.2893	1.6588	1.9818	2.3607	2.6213
112	0.6767	1.2892	1.6586	1.9814	2.3601	2.6204
114	0.6766	1.2890	1.6583	1.9810	2.3595	2.6196
116	0.6766	1.2889	1.6581	1.9806	2.3589	2.6189
118	0.6766	1.2888	1.6579	1.9803	2.3584	2.6181
120	0.6765	1.2886	1.6577	1.9799	2.3578	2.6174
122	0.6765	1.2885	1.6574	1.9796	2.3573	2.6167
124	0.6765	1.2884	1.6572	1.9793	2.3568	2.6161
126	0.6764	1.2883	1.6570	1.9790	2.3563	2.6154
128	0.6764	1.2882	1.6568	1.9787	2.3558	2.6148
130	0.6764	1.2881	1.6567	1.9784	2.3554	2.6142
132	0.6764	1.2880	1.6565	1.9781	2.3549	2.6136
134	0.6763	1.2879	1.6563	1.9778	2.3545	2.6130
136	0.6763	1.2878	1.6561	1.9776	2.3541	2.6125
138	0.6763	1.2877	1.6560	1.9773	2.3537	2.6119
140	0.6762	1.2876	1.6558	1.9771	2.3533	2.6114
142	0.6762	1.2875	1.6557	1.9768	2.3529	2.6109
144	0.6762	1.2875	1.6555	1.9766	2.3525	2.6104
146	0.6762	1.2874	1.6554	1.9763	2.3522	2.6099
148	0.6762	1.2873	1.6552	1.9761	2.3518	2.6095
150	0.6761	1.2872	1.6551	1.9759	2.3515	2.6090
∞	0.6745	1.2816	1.6449	1.9600	2.3263	2.5758

SOURCE: Donald B. Owen, *Handbook of Statistical Tables*, © 1962, Addison-Wesley, Reading, Massachusetts. Extracted from Table 2.1. Reprinted by permission of the U.S. Department of Energy.

Table E.4

CRITICAL VALUES OF χ^2

For a particular number of degrees of freedom, entry represents the critical value of χ^2 corresponding to a specified percentile or lower tail area $(1 - \alpha)$

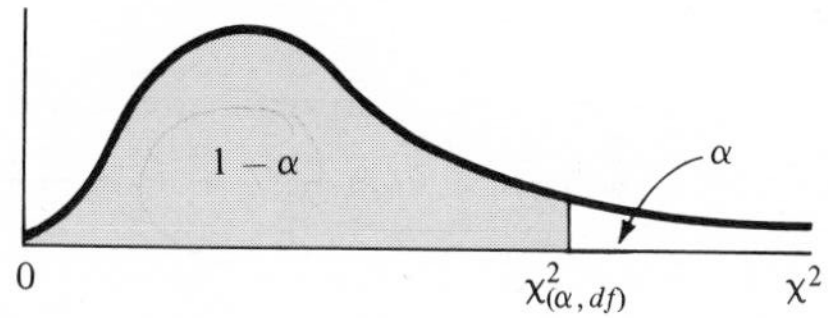

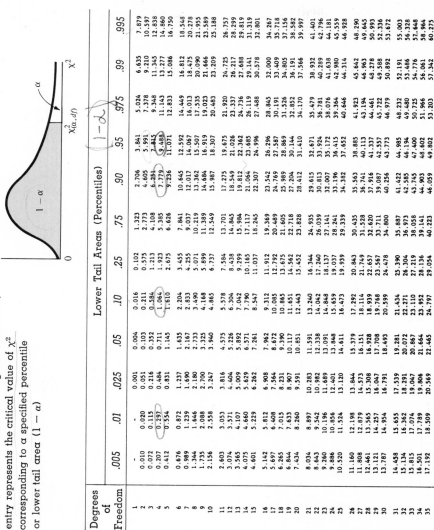

Degrees of Freedom	Lower Tail Areas (Percentiles)											
	.005	.01	.025	.05	.10	.25	.75	.90	.95	.975	.99	.995
1	-	-	0.001	0.004	0.016	0.102	1.323	2.706	3.841	5.024	6.635	7.879
2	0.010	0.020	0.051	0.103	0.211	0.575	2.773	4.605	5.991	7.378	9.210	10.597
3	0.072	0.115	0.216	0.352	0.584	1.213	4.108	6.251	7.815	9.348	11.345	12.838
4	0.207	0.297	0.484	0.711	1.064	1.923	5.385	7.779	9.488	11.143	13.277	14.860
5	0.412	0.554	0.831	1.145	1.610	2.675	6.626	9.236	11.071	12.833	15.086	16.750
6	0.676	0.872	1.237	1.635	2.204	3.455	7.841	10.645	12.592	14.449	16.812	18.548
7	0.989	1.239	1.690	2.167	2.833	4.255	9.037	12.017	14.067	16.013	18.475	20.278
8	1.344	1.646	2.180	2.733	3.490	5.071	10.219	13.362	15.507	17.535	20.090	21.955
9	1.735	2.088	2.700	3.325	4.168	5.899	11.389	14.684	16.919	19.023	21.666	23.589
10	2.156	2.558	3.247	3.940	4.865	6.737	12.549	15.987	18.307	20.483	23.209	25.188
11	2.603	3.053	3.816	4.575	5.578	7.584	13.701	17.275	19.675	21.920	24.725	26.757
12	3.074	3.571	4.404	5.226	6.304	8.438	14.845	18.549	21.026	23.337	26.217	28.299
13	3.565	4.107	5.009	5.892	7.042	9.299	15.984	19.812	22.362	24.736	27.688	29.819
14	4.075	4.660	5.629	6.571	7.790	10.165	17.117	21.064	23.685	26.119	29.141	31.319
15	4.601	5.229	6.262	7.261	8.547	11.037	18.245	22.307	24.996	27.488	30.578	32.801
16	5.142	5.812	6.908	7.962	9.312	11.912	19.369	23.542	26.296	28.845	32.000	34.267
17	5.697	6.408	7.564	8.672	10.085	12.792	20.489	24.769	27.587	30.191	33.409	35.718
18	6.265	7.015	8.231	9.390	10.865	13.675	21.605	25.989	28.869	31.526	34.805	37.156
19	6.844	7.633	8.907	10.117	11.651	14.562	22.718	27.204	30.144	32.852	36.191	38.582
20	7.434	8.260	9.591	10.851	12.443	15.452	23.828	28.412	31.410	34.170	37.566	39.997
21	8.034	8.897	10.283	11.591	13.240	16.344	24.935	29.615	32.671	35.479	38.932	41.401
22	8.643	9.542	10.982	12.338	14.042	17.240	26.039	30.813	33.924	36.781	40.289	42.796
23	9.260	10.196	11.689	13.091	14.848	18.137	27.141	32.007	35.172	38.076	41.638	44.181
24	9.886	10.856	12.401	13.848	15.659	19.037	28.241	33.196	36.415	39.364	42.980	45.559
25	10.520	11.524	13.120	14.611	16.473	19.939	29.339	34.382	37.652	40.646	44.314	46.928
26	11.160	12.198	13.844	15.379	17.292	20.843	30.435	35.563	38.885	41.923	45.642	48.290
27	11.808	12.879	14.573	16.151	18.114	21.749	31.528	36.741	40.113	43.194	46.963	49.645
28	12.461	13.565	15.308	16.928	18.939	22.657	32.620	37.916	41.337	44.461	48.278	50.993
29	13.121	14.257	16.047	17.708	19.768	23.567	33.711	39.087	42.557	45.722	49.588	52.336
30	13.787	14.954	16.791	18.493	20.599	24.478	34.800	40.256	43.773	46.979	50.892	53.672
31	14.458	15.655	17.539	19.281	21.434	25.390	35.887	41.422	44.985	48.232	52.191	55.003
32	15.134	16.362	18.291	20.072	22.271	26.304	36.973	42.585	46.194	49.480	53.486	56.328
33	15.815	17.074	19.047	20.867	23.110	27.219	38.058	43.745	47.400	50.725	54.776	57.648
34	16.501	17.789	19.806	21.664	23.952	28.136	39.141	44.903	48.602	51.966	56.061	58.964
35	17.192	18.509	20.569	22.465	24.797	29.054	40.223	46.059	49.802	53.203	57.342	60.275

Table E.4 (*Continued*)

Degrees of Freedom	Lower Tail Areas (Percentiles)											
	.005	.01	.025	.05	.10	.25	.75	.90	.95	.975	.99	.995
36	17.887	19.233	21.336	23.269	25.643	29.973	41.304	47.212	50.998	54.437	58.619	61.581
37	18.586	19.960	22.106	24.075	26.492	30.893	42.383	48.363	52.192	55.668	59.892	62.883
38	19.289	20.691	22.878	24.884	27.343	31.815	43.462	49.513	53.384	56.896	61.162	64.181
39	19.996	21.426	23.654	25.695	28.196	32.737	44.539	50.660	54.572	58.120	62.428	65.476
40	20.707	22.164	24.433	26.509	29.051	33.660	45.616	51.805	55.758	59.342	63.691	66.766
41	21.421	22.906	25.215	27.326	29.907	34.585	46.692	52.949	56.942	60.561	64.950	68.053
42	22.138	23.650	25.999	28.144	30.765	35.510	47.766	54.090	58.124	61.777	66.206	69.336
43	22.859	24.398	26.785	28.965	31.625	36.436	48.840	55.230	59.304	62.990	67.459	70.616
44	23.584	25.148	27.575	29.787	32.487	37.363	49.913	56.369	60.481	64.201	68.710	71.893
45	24.311	25.901	28.366	30.612	33.350	38.291	50.985	57.505	61.656	65.410	69.957	73.166
46	25.041	26.657	29.160	31.439	34.215	39.220	52.056	58.641	62.830	66.617	71.201	74.437
47	25.775	27.416	29.956	32.268	35.081	40.149	53.127	59.774	64.001	67.821	72.443	75.704
48	26.511	28.177	30.755	33.098	35.949	41.079	54.196	60.907	65.171	69.023	73.683	76.969
49	27.249	28.941	31.555	33.930	36.818	42.010	55.265	62.038	66.339	70.222	74.919	78.231
50	27.991	29.707	32.357	34.764	37.689	42.942	56.334	63.167	67.505	71.420	76.154	79.490
51	28.735	30.475	33.162	35.600	38.560	43.874	57.401	64.295	68.669	72.616	77.386	80.747
52	29.481	31.246	33.968	36.437	39.433	44.808	58.468	65.422	69.832	73.810	78.616	82.001
53	30.230	32.018	34.776	37.276	40.308	45.741	59.534	66.548	70.993	75.002	79.843	83.253
54	30.981	32.793	35.586	38.116	41.183	46.676	60.600	67.673	72.153	76.192	81.069	84.502
55	31.735	33.570	36.398	38.958	42.060	47.610	61.665	68.796	73.311	77.380	82.292	85.749
56	32.490	34.350	37.212	39.801	42.937	48.546	62.729	69.919	74.468	78.567	83.513	36.994
57	33.248	35.131	38.027	40.646	43.816	49.482	63.793	71.040	75.624	79.752	84.733	88.236
58	34.008	35.913	38.844	41.492	44.696	50.419	64.857	72.160	76.778	80.936	85.950	89.477
59	34.770	36.698	39.662	42.339	45.577	51.356	65.919	73.279	77.931	82.117	87.166	90.715
60	35.534	37.485	40.482	43.188	46.459	52.294	66.981	74.397	79.082	83.298	88.379	91.952
61	36.300	38.273	41.303	44.038	47.342	53.232	68.043	75.514	80.232	84.476	89.591	93.186
62	37.068	39.063	42.126	44.889	48.226	54.171	69.104	76.630	81.381	85.654	90.802	94.419
63	37.838	39.855	42.950	45.741	49.111	55.110	70.165	77.745	82.529	86.830	92.010	95.649
64	38.610	40.649	43.776	46.595	49.996	56.050	71.225	78.860	83.675	88.004	93.217	96.878
65	39.383	41.444	44.603	47.450	50.883	56.990	72.285	79.973	84.821	89.177	94.422	98.105
66	40.158	42.240	45.431	48.305	51.770	57.931	73.344	81.085	85.965	90.349	95.626	99.330
67	40.935	43.038	46.261	49.162	52.659	58.872	74.403	82.197	87.108	91.519	96.828	100.554
68	41.713	43.838	47.092	50.020	53.548	59.814	75.461	83.308	88.250	92.689	98.028	101.776
69	42.494	44.639	47.924	50.879	54.438	60.756	76.519	84.418	89.391	93.856	99.228	102.996
70	43.275	45.442	48.758	51.739	55.329	61.698	77.577	85.527	90.531	95.023	100.425	104.215
71	44.058	46.246	49.592	52.600	56.221	62.641	78.634	86.635	91.670	96.189	101.621	105.432
72	44.843	47.051	50.428	53.462	57.113	63.585	79.690	87.743	92.808	97.353	102.816	106.648
73	45.629	47.858	51.265	54.325	58.006	64.528	80.747	88.850	93.945	98.516	104.010	107.862
74	46.417	48.666	52.103	55.189	58.900	65.472	81.803	89.956	95.081	99.678	105.202	109.074
75	47.206	49.475	52.942	56.054	59.795	66.417	82.858	91.061	96.217	100.839	106.393	110.286
76	47.997	50.286	53.782	56.920	60.690	67.362	83.913	92.166	97.351	101.999	107.583	111.495
77	48.788	51.097	54.623	57.786	61.586	68.307	84.968	93.270	98.484	103.158	108.771	112.704
78	49.582	51.910	55.466	58.654	62.483	69.252	86.022	94.374	99.617	104.316	109.958	113.911
79	50.376	52.725	56.309	59.522	63.380	70.198	87.077	95.476	100.749	105.473	111.144	115.117
80	51.172	53.540	57.153	60.391	64.278	71.145	88.130	96.578	101.879	106.629	112.329	116.321
81	51.969	54.357	57.998	61.261	65.176	72.091	89.184	97.680	103.010	107.783	113.512	117.524
82	52.767	55.174	58.845	62.132	66.076	73.038	90.237	98.780	104.139	108.937	114.695	118.726
83	53.567	55.993	59.692	63.004	66.976	73.985	91.289	99.880	105.267	110.090	115.876	119.927
84	54.368	56.813	60.540	63.876	67.876	74.933	92.342	100.980	106.395	111.242	117.057	121.126
85	55.170	57.634	61.389	64.749	68.777	75.881	93.394	102.079	107.522	112.393	118.236	122.325

Table E.4 (*Continued*)

Degrees of Freedom	Lower Tail Areas (Percentiles)											
	.005	.01	.025	.05	.10	.25	.75	.90	.95	.975	.99	.995
86	55.973	58.456	62.239	65.623	69.679	76.829	94.446	103.177	108.648	113.544	119.414	123.522
87	56.777	59.279	63.089	66.498	70.581	77.777	95.497	104.275	109.773	114.693	120.591	124.718
88	57.582	60.103	63.941	67.373	71.484	78.726	96.548	105.372	110.898	115.841	121.767	125.913
89	58.389	60.928	64.793	68.249	72.387	79.675	97.599	106.469	112.022	116.989	122.942	127.106
90	59.196	61.754	65.647	69.126	73.291	80.625	98.650	107.565	113.145	118.136	124.116	128.299
91	60.005	62.581	66.501	70.003	74.196	81.574	99.700	108.661	114.268	119.282	125.289	129.491
92	60.815	63.409	67.356	70.882	75.100	82.524	100.750	109.756	115.390	120.427	126.462	130.681
93	61.625	64.238	68.211	71.760	76.006	83.474	101.800	110.850	116.511	121.571	127.633	131.871
94	62.437	65.068	69.068	72.640	76.912	84.425	102.850	111.944	117.632	122.715	128.803	133.059
95	63.250	65.898	69.925	73.520	77.818	85.376	103.899	113.038	118.752	123.858	129.973	134.247
96	64.063	66.730	70.783	74.401	78.725	86.327	104.948	114.131	119.871	125.000	131.141	135.433
97	64.878	67.562	71.642	75.282	79.633	87.278	105.997	115.223	120.990	126.141	132.309	136.619
98	65.694	68.396	72.501	76.164	80.541	88.229	107.045	116.315	122.108	127.282	133.476	137.803
99	66.510	69.230	73.361	77.046	81.449	89.181	108.093	117.407	123.225	128.422	134.642	138.987
100	67.328	70.065	74.222	77.929	82.358	90.133	109.141	118.498	124.342	129.561	135.807	140.169
102	68.965	71.737	75.946	79.697	84.177	92.038	111.236	120.679	126.574	131.838	138.134	142.532
104	70.606	73.413	77.672	81.468	85.998	93.944	113.331	122.858	128.804	134.111	140.459	144.891
106	72.251	75.092	79.401	83.240	87.821	95.850	115.424	125.035	131.031	136.382	142.780	147.247
108	73.899	76.774	81.133	85.015	89.645	97.758	117.517	127.211	133.257	138.651	145.099	149.599
110	75.550	78.458	82.867	86.792	91.471	99.666	119.608	129.385	135.480	140.917	147.414	151.948
112	77.204	80.146	84.604	88.570	93.299	101.575	121.699	131.558	137.701	143.180	149.727	154.294
114	78.862	81.836	86.342	90.351	95.128	103.485	123.789	133.729	139.921	145.441	152.037	156.637
116	80.522	83.529	88.084	92.134	96.958	105.396	125.878	135.898	142.138	147.700	154.344	158.977
118	82.185	85.225	89.827	93.918	98.790	107.307	127.967	138.066	144.354	149.957	156.648	161.314
120	83.852	86.923	91.573	95.705	100.624	109.220	130.055	140.233	146.567	152.211	158.950	163.648
122	85.520	88.624	93.320	97.493	102.458	111.133	132.142	142.398	148.779	154.464	161.250	165.980
124	87.192	90.327	95.070	99.283	104.295	113.046	134.228	144.562	150.989	156.714	163.546	168.308
126	88.866	92.033	96.822	101.074	106.132	114.961	136.313	146.724	153.198	158.962	165.841	170.634
128	90.543	93.741	98.576	102.867	107.971	116.876	138.398	148.885	155.405	161.209	168.133	172.957
130	92.222	95.451	100.331	104.662	109.811	118.792	140.482	151.045	157.610	163.453	170.423	175.278
132	93.904	97.163	102.089	106.459	111.652	120.708	142.566	153.204	159.814	165.696	172.711	177.597
134	95.588	98.878	103.848	108.257	113.495	122.625	144.649	155.361	162.016	167.936	174.996	179.913
136	97.275	100.595	105.609	110.056	115.338	124.543	146.731	157.518	164.216	170.175	177.280	182.226
138	98.964	102.314	107.372	111.857	117.183	126.461	148.813	159.673	166.415	172.412	179.561	184.538
140	100.655	104.034	109.137	113.659	119.029	128.380	150.894	161.827	168.613	174.648	181.840	186.847
142	102.348	105.757	110.903	115.463	120.876	130.299	152.975	163.980	170.809	176.882	184.118	189.154
144	104.044	107.482	112.671	117.268	122.724	132.219	155.055	166.132	173.004	179.114	186.393	191.458
146	105.741	109.209	114.441	119.075	124.574	134.140	157.134	168.283	175.198	181.344	188.666	193.761
148	107.441	110.937	116.212	120.883	126.424	136.061	159.213	170.432	177.390	183.573	190.938	196.062
150	109.142	112.668	117.985	122.692	128.275	137.983	161.291	172.581	179.581	185.800	193.208	198.360
200	152.241	156.432	162.728	168.279	174.835	186.172	213.102	226.021	233.994	241.058	249.445	255.264
250	196.161	200.939	208.098	214.392	221.806	234.577	264.697	279.050	287.882	295.689	304.940	311.346
300	240.663	245.972	253.912	260.878	269.068	283.135	316.138	331.789	341.395	349.874	359.906	366.844
400	330.903	337.155	346.482	354.641	364.207	380.577	418.697	436.649	447.632	457.305	468.724	476.606
500	422.303	429.388	439.936	449.147	459.926	478.323	520.950	540.930	553.127	563.852	576.493	585.207
600	514.529	522.365	534.019	544.180	556.056	576.286	622.988	644.800	658.094	669.769	683.516	692.982
700	607.380	615.907	628.577	639.613	652.497	674.413	724.861	748.359	762.661	775.211	789.974	800.131
800	700.725	709.897	723.513	735.362	749.185	772.669	826.604	851.671	866.911	880.275	895.984	906.786
900	794.475	804.252	818.756	831.370	846.075	871.032	928.241	954.782	970.904	985.032	1001.630	1013.036
1000	888.564	898.912	914.257	927.594	943.133	969.484	1029.790	1057.724	1074.679	1089.531	1106.969	1118.948

SOURCE: Donald B. Owen, *Handbook of Statistical Tables*, © 1962, Addison-Wesley, Reading, Massachusetts. Extracted from Table 3.1. Reprinted by permission of the U.S. Department of Energy.

Table E.5
CRITICAL VALUES OF F

For a particular combination of numerator and denominator degrees of freedom, entry represents the critical values of F corresponding to a specified upper tail area (α).

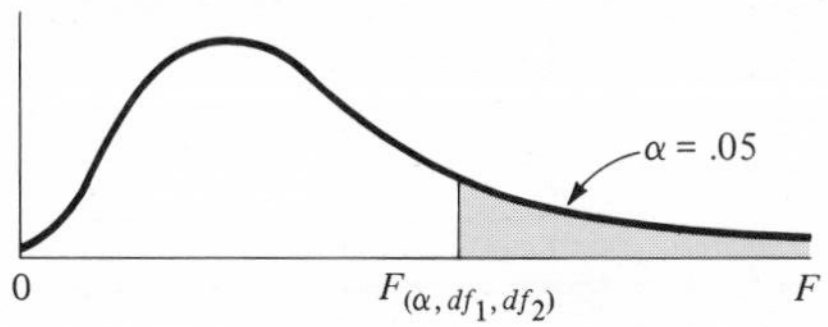

Numerator df_1

Denominator df_2	1	2	3	4	5	6	7	8	9	10	12	15	20	24	30	40	60	120	∞
1	161.4	199.5	215.7	224.6	230.2	234.0	236.8	238.9	240.5	241.9	243.9	245.9	248.0	249.1	250.1	251.1	252.2	253.3	254.3
2	18.51	19.00	19.16	19.25	19.30	19.33	19.35	19.37	19.38	19.40	19.41	19.43	19.45	19.45	19.46	19.47	19.48	19.49	19.50
3	10.13	9.55	9.28	9.12	9.01	8.94	8.89	8.85	8.81	8.79	8.74	8.70	8.66	8.64	8.62	8.59	8.57	8.55	8.53
4	7.71	6.94	6.59	6.39	6.26	6.16	6.09	6.04	6.00	5.96	5.91	5.86	5.80	5.77	5.75	5.72	5.69	5.66	5.63
5	6.61	5.79	5.41	5.19	5.05	4.95	4.88	4.82	4.77	4.74	4.68	4.62	4.56	4.53	4.50	4.46	4.43	4.40	4.36
6	5.99	5.14	4.76	4.53	4.39	4.28	4.21	4.15	4.10	4.06	4.00	3.94	3.87	3.84	3.81	3.77	3.74	3.70	3.67
7	5.59	4.74	4.35	4.12	3.97	3.87	3.79	3.73	3.68	3.64	3.57	3.51	3.44	3.41	3.38	3.34	3.30	3.27	3.23
8	5.32	4.46	4.07	3.84	3.69	3.58	3.50	3.44	3.39	3.35	3.28	3.22	3.15	3.12	3.08	3.04	3.01	2.97	2.93
9	5.12	4.26	3.86	3.63	3.48	3.37	3.29	3.23	3.18	3.14	3.07	3.01	2.94	2.90	2.86	2.83	2.79	2.75	2.71
10	4.96	4.10	3.71	3.48	3.33	3.22	3.14	3.07	3.02	2.98	2.91	2.85	2.77	2.74	2.70	2.66	2.62	2.58	2.54
11	4.84	3.98	3.59	3.36	3.20	3.09	3.01	2.95	2.90	2.85	2.79	2.72	2.65	2.61	2.57	2.53	2.49	2.45	2.40
12	4.75	3.89	3.49	3.26	3.11	3.00	2.91	2.85	2.80	2.75	2.69	2.62	2.54	2.51	2.47	2.43	2.38	2.34	2.30
13	4.67	3.81	3.41	3.18	3.03	2.92	2.83	2.77	2.71	2.67	2.60	2.53	2.46	2.42	2.38	2.34	2.30	2.25	2.21
14	4.60	3.74	3.34	3.11	2.96	2.85	2.76	2.70	2.65	2.60	2.53	2.46	2.39	2.35	2.31	2.27	2.22	2.18	2.13
15	4.54	3.68	3.29	3.06	2.90	2.79	2.71	2.64	2.59	2.54	2.48	2.40	2.33	2.29	2.25	2.20	2.16	2.11	2.07
16	4.49	3.63	3.24	3.01	2.85	2.74	2.66	2.59	2.54	2.49	2.42	2.35	2.28	2.24	2.19	2.15	2.11	2.06	2.01
17	4.45	3.59	3.20	2.96	2.81	2.70	2.61	2.55	2.49	2.45	2.38	2.31	2.23	2.19	2.15	2.10	2.06	2.01	1.96
18	4.41	3.55	3.16	2.93	2.77	2.66	2.58	2.51	2.46	2.41	2.34	2.27	2.19	2.15	2.11	2.06	2.02	1.97	1.92
19	4.38	3.52	3.13	2.90	2.74	2.63	2.54	2.48	2.42	2.38	2.31	2.23	2.16	2.11	2.07	2.03	1.98	1.93	1.88
20	4.35	3.49	3.10	2.87	2.71	2.60	2.51	2.45	2.39	2.35	2.28	2.20	2.12	2.08	2.04	1.99	1.95	1.90	1.84
21	4.32	3.47	3.07	2.84	2.68	2.57	2.49	2.42	2.37	2.32	2.25	2.18	2.10	2.05	2.01	1.96	1.92	1.87	1.81
22	4.30	3.44	3.05	2.82	2.66	2.55	2.46	2.40	2.34	2.30	2.23	2.15	2.07	2.03	1.98	1.94	1.89	1.84	1.78
23	4.28	3.42	3.03	2.80	2.64	2.53	2.44	2.37	2.32	2.27	2.20	2.13	2.05	2.01	1.96	1.91	1.86	1.81	1.76
24	4.26	3.40	3.01	2.78	2.62	2.51	2.42	2.36	2.30	2.25	2.18	2.11	2.03	1.98	1.94	1.89	1.84	1.79	1.73
25	4.24	3.39	2.99	2.76	2.60	2.49	2.40	2.34	2.28	2.24	2.16	2.09	2.01	1.96	1.92	1.87	1.82	1.77	1.71
26	4.23	3.37	2.98	2.74	2.59	2.47	2.39	2.32	2.27	2.22	2.15	2.07	1.99	1.95	1.90	1.85	1.80	1.75	1.69
27	4.21	3.35	2.96	2.73	2.57	2.46	2.37	2.31	2.25	2.20	2.13	2.06	1.97	1.93	1.88	1.84	1.79	1.73	1.67
28	4.20	3.34	2.95	2.71	2.56	2.45	2.36	2.29	2.24	2.19	2.12	2.04	1.96	1.91	1.87	1.82	1.77	1.71	1.65
29	4.18	3.33	2.93	2.70	2.55	2.43	2.35	2.28	2.22	2.18	2.10	2.03	1.94	1.90	1.85	1.81	1.75	1.70	1.64
30	4.17	3.32	2.92	2.69	2.53	2.42	2.33	2.27	2.21	2.16	2.09	2.01	1.93	1.89	1.84	1.79	1.74	1.68	1.62
40	4.08	3.23	2.84	2.61	2.45	2.34	2.25	2.18	2.12	2.08	2.00	1.92	1.84	1.79	1.74	1.69	1.64	1.58	1.51
60	4.00	3.15	2.76	2.53	2.37	2.25	2.17	2.10	2.04	1.99	1.92	1.84	1.75	1.70	1.65	1.59	1.53	1.47	1.39
120	3.92	3.07	2.68	2.45	2.29	2.17	2.09	2.02	1.96	1.91	1.83	1.75	1.66	1.61	1.55	1.50	1.43	1.35	1.25
∞	3.84	3.00	2.60	2.37	2.21	2.10	2.01	1.94	1.88	1.83	1.75	1.67	1.57	1.52	1.46	1.39	1.32	1.22	1.00

Table E.5 (Continued)

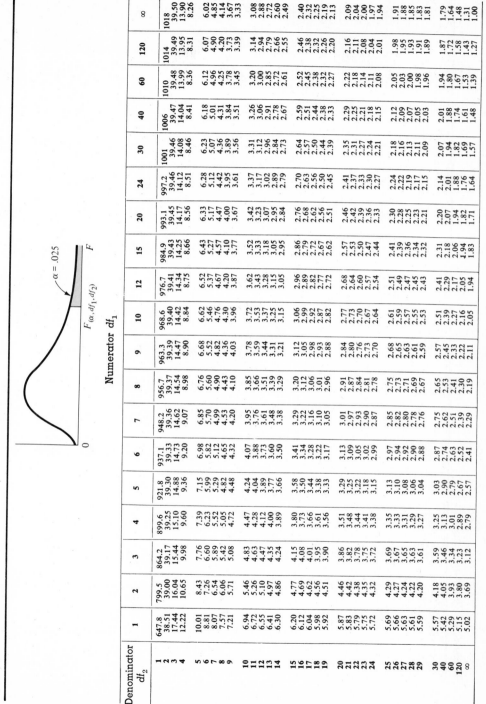

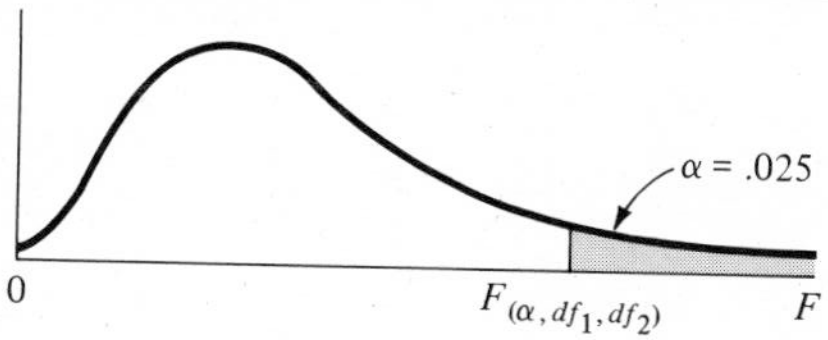

Numerator df_1

Denominator df_2	1	2	3	4	5	6	7	8	9	10	12	15	20	24	30	40	60	120	∞
1	647.8	799.5	864.2	899.6	921.8	937.1	948.2	956.7	963.3	968.6	976.7	984.9	993.1	997.2	1001	1006	1010	1014	1018
2	38.51	39.00	39.17	39.25	39.30	39.33	39.36	39.37	39.39	39.40	39.41	39.43	39.45	39.46	39.46	39.47	39.48	39.49	39.50
3	17.44	16.04	15.44	15.10	14.88	14.73	14.62	14.54	14.47	14.42	14.34	14.25	14.17	14.12	14.08	14.04	13.99	13.95	13.90
4	12.22	10.65	9.98	9.60	9.36	9.20	9.07	8.98	8.90	8.84	8.75	8.66	8.56	8.51	8.46	8.41	8.36	8.31	8.26
5	10.01	8.43	7.76	7.39	7.15	6.98	6.85	6.76	6.68	6.62	6.52	6.43	6.33	6.28	6.23	6.18	6.12	6.07	6.02
6	8.81	7.26	6.60	6.23	5.99	5.82	5.70	5.60	5.52	5.46	5.37	5.27	5.17	5.12	5.07	5.01	4.96	4.90	4.85
7	8.07	6.54	5.89	5.52	5.29	5.12	4.99	4.90	4.82	4.76	4.67	4.57	4.47	4.42	4.36	4.31	4.25	4.20	4.14
8	7.57	6.06	5.42	5.05	4.82	4.65	4.53	4.43	4.36	4.30	4.20	4.10	4.00	3.95	3.89	3.84	3.78	3.73	3.67
9	7.21	5.71	5.08	4.72	4.48	4.32	4.20	4.10	4.03	3.96	3.87	3.77	3.67	3.61	3.56	3.51	3.45	3.39	3.33
10	6.94	5.46	4.83	4.47	4.24	4.07	3.95	3.85	3.78	3.72	3.62	3.52	3.42	3.37	3.31	3.26	3.20	3.14	3.08
11	6.72	5.26	4.63	4.28	4.04	3.88	3.76	3.66	3.59	3.53	3.43	3.33	3.23	3.17	3.12	3.06	3.00	2.94	2.88
12	6.55	5.10	4.47	4.12	3.89	3.73	3.61	3.51	3.44	3.37	3.28	3.18	3.07	3.02	2.96	2.91	2.85	2.79	2.72
13	6.41	4.97	4.35	4.00	3.77	3.60	3.48	3.39	3.31	3.25	3.15	3.05	2.95	2.89	2.84	2.78	2.72	2.66	2.60
14	6.30	4.86	4.24	3.89	3.66	3.50	3.38	3.29	3.21	3.15	3.05	2.95	2.84	2.79	2.73	2.67	2.61	2.55	2.49
15	6.20	4.77	4.15	3.80	3.58	3.41	3.29	3.20	3.12	3.06	2.96	2.86	2.76	2.70	2.64	2.59	2.52	2.46	2.40
16	6.12	4.69	4.08	3.73	3.50	3.34	3.22	3.12	3.05	2.99	2.89	2.79	2.68	2.63	2.57	2.51	2.45	2.38	2.32
17	6.04	4.62	4.01	3.66	3.44	3.28	3.16	3.06	2.98	2.92	2.82	2.72	2.62	2.56	2.50	2.44	2.38	2.32	2.25
18	5.98	4.56	3.95	3.61	3.38	3.22	3.10	3.01	2.93	2.87	2.77	2.67	2.56	2.50	2.44	2.38	2.32	2.26	2.19
19	5.92	4.51	3.90	3.56	3.33	3.17	3.05	2.96	2.88	2.82	2.72	2.62	2.51	2.45	2.39	2.33	2.27	2.20	2.13
20	5.87	4.46	3.86	3.51	3.29	3.13	3.01	2.91	2.84	2.77	2.68	2.57	2.46	2.41	2.35	2.29	2.22	2.16	2.09
21	5.83	4.42	3.82	3.48	3.25	3.09	2.97	2.87	2.80	2.73	2.64	2.53	2.42	2.37	2.31	2.25	2.18	2.11	2.04
22	5.79	4.38	3.78	3.44	3.22	3.05	2.93	2.84	2.76	2.70	2.60	2.50	2.39	2.33	2.27	2.21	2.14	2.08	2.00
23	5.75	4.35	3.75	3.41	3.18	3.02	2.90	2.81	2.73	2.67	2.57	2.47	2.36	2.30	2.24	2.18	2.11	2.04	1.97
24	5.72	4.32	3.72	3.38	3.15	2.99	2.87	2.78	2.70	2.64	2.54	2.44	2.33	2.27	2.21	2.15	2.08	2.01	1.94
25	5.69	4.29	3.69	3.35	3.13	2.97	2.85	2.75	2.68	2.61	2.51	2.41	2.30	2.24	2.18	2.12	2.05	1.98	1.91
26	5.66	4.27	3.67	3.33	3.10	2.94	2.82	2.73	2.65	2.59	2.49	2.39	2.28	2.22	2.16	2.09	2.03	1.95	1.88
27	5.63	4.24	3.65	3.31	3.08	2.92	2.80	2.71	2.63	2.57	2.47	2.36	2.25	2.19	2.13	2.07	2.00	1.93	1.85
28	5.61	4.22	3.63	3.29	3.06	2.90	2.78	2.69	2.61	2.55	2.45	2.34	2.23	2.17	2.11	2.05	1.98	1.91	1.83
29	5.59	4.20	3.61	3.27	3.04	2.88	2.76	2.67	2.59	2.53	2.43	2.32	2.21	2.15	2.09	2.03	1.96	1.89	1.81
30	5.57	4.18	3.59	3.25	3.03	2.87	2.75	2.65	2.57	2.51	2.41	2.31	2.20	2.14	2.07	2.01	1.94	1.87	1.79
40	5.42	4.05	3.46	3.13	2.90	2.74	2.62	2.53	2.45	2.39	2.29	2.18	2.07	2.01	1.94	1.88	1.80	1.72	1.64
60	5.29	3.93	3.34	3.01	2.79	2.63	2.51	2.41	2.33	2.27	2.17	2.06	1.94	1.88	1.82	1.74	1.67	1.58	1.48
120	5.15	3.80	3.23	2.89	2.67	2.52	2.39	2.30	2.22	2.16	2.05	1.94	1.82	1.76	1.69	1.61	1.53	1.43	1.31
∞	5.02	3.69	3.12	2.79	2.57	2.41	2.29	2.19	2.11	2.05	1.94	1.83	1.71	1.64	1.57	1.48	1.39	1.27	1.00

Table E.5 (*Continued*)

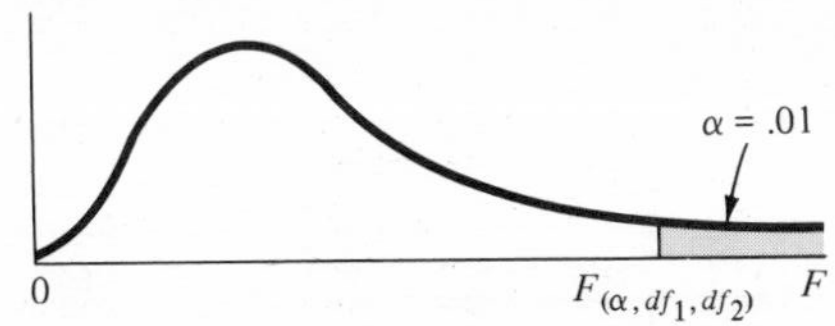

Numerator df_1

Denominator df_2	1	2	3	4	5	6	7	8	9	10	12	15	20	24	30	40	60	120	∞
1	4052	4999.5	5403	5625	5764	5859	5928	5982	6022	6056	6106	6157	6209	6235	6261	6287	6313	6339	6366
2	98.50	99.00	99.17	99.25	99.30	99.33	99.36	99.37	99.39	99.40	99.42	99.43	99.45	99.46	99.47	99.47	99.48	99.49	99.50
3	34.12	30.82	29.46	28.71	28.24	27.91	27.67	27.49	27.35	27.23	27.05	26.87	26.69	26.60	26.50	26.41	26.32	26.22	26.13
4	21.20	18.00	16.69	15.98	15.52	15.21	14.98	14.80	14.66	14.55	14.37	14.20	14.02	13.93	13.84	13.75	13.65	13.56	13.46
5	16.26	13.27	12.06	11.39	10.97	10.67	10.46	10.29	10.16	10.05	9.89	9.72	9.55	9.47	9.38	9.29	9.20	9.11	9.02
6	13.75	10.92	9.78	9.15	8.75	8.47	8.26	8.10	7.98	7.87	7.72	7.56	7.40	7.31	7.23	7.14	7.06	6.97	6.88
7	12.25	9.55	8.45	7.85	7.46	7.19	6.99	6.84	6.72	6.62	6.47	6.31	6.16	6.07	5.99	5.91	5.82	5.74	5.65
8	11.26	8.65	7.59	7.01	6.63	6.37	6.18	6.03	5.91	5.81	5.67	5.52	5.36	5.28	5.20	5.12	5.03	4.95	4.86
9	10.56	8.02	6.99	6.42	6.06	5.80	5.61	5.47	5.35	5.26	5.11	4.96	4.81	4.73	4.65	4.57	4.48	4.40	4.31
10	10.04	7.56	6.55	5.99	5.64	5.39	5.20	5.06	4.94	4.85	4.71	4.56	4.41	4.33	4.25	4.17	4.08	4.00	3.91
11	9.65	7.21	6.22	5.67	5.32	5.07	4.89	4.74	4.63	4.54	4.40	4.25	4.10	4.02	3.94	3.86	3.78	3.69	3.60
12	9.33	6.93	5.95	5.41	5.06	4.82	4.64	4.50	4.39	4.30	4.16	4.01	3.86	3.78	3.70	3.62	3.54	3.45	3.36
13	9.07	6.70	5.74	5.21	4.86	4.62	4.44	4.30	4.19	4.10	3.96	3.82	3.66	3.59	3.51	3.43	3.34	3.25	3.17
14	8.86	6.51	5.56	5.04	4.69	4.46	4.28	4.14	4.03	3.94	3.80	3.66	3.51	3.43	3.35	3.27	3.18	3.09	3.00
15	8.68	6.36	5.42	4.89	4.56	4.32	4.14	4.00	3.89	3.80	3.67	3.52	3.37	3.29	3.21	3.13	3.05	2.96	2.87
16	8.53	6.23	5.29	4.77	4.44	4.20	4.03	3.89	3.78	3.69	3.55	3.41	3.26	3.18	3.10	3.02	2.93	2.84	2.75
17	8.40	6.11	5.18	4.67	4.34	4.10	3.93	3.79	3.68	3.59	3.46	3.31	3.16	3.08	3.00	2.92	2.83	2.75	2.65
18	8.29	6.01	5.09	4.58	4.25	4.01	3.84	3.71	3.60	3.51	3.37	3.23	3.08	3.00	2.92	2.84	2.75	2.66	2.57
19	8.18	5.93	5.01	4.50	4.17	3.94	3.77	3.63	3.52	3.43	3.30	3.15	3.00	2.92	2.84	2.76	2.67	2.58	2.49
20	8.10	5.85	4.94	4.43	4.10	3.87	3.70	3.56	3.46	3.37	3.23	3.09	2.94	2.86	2.78	2.69	2.61	2.52	2.42
21	8.02	5.78	4.87	4.37	4.04	3.81	3.64	3.51	3.40	3.31	3.17	3.03	2.88	2.80	2.72	2.64	2.55	2.46	2.36
22	7.95	5.72	4.82	4.31	3.99	3.76	3.59	3.45	3.35	3.26	3.12	2.98	2.83	2.75	2.67	2.58	2.50	2.40	2.31
23	7.88	5.66	4.76	4.26	3.94	3.71	3.54	3.41	3.30	3.21	3.07	2.93	2.78	2.70	2.62	2.54	2.45	2.35	2.26
24	7.82	5.61	4.72	4.22	3.90	3.67	3.50	3.36	3.26	3.17	3.03	2.89	2.74	2.66	2.58	2.49	2.40	2.31	2.21
25	7.77	5.57	4.68	4.18	3.85	3.63	3.46	3.32	3.22	3.13	2.99	2.85	2.70	2.62	2.54	2.45	2.36	2.27	2.17
26	7.72	5.53	4.64	4.14	3.82	3.59	3.42	3.29	3.18	3.09	2.96	2.81	2.66	2.58	2.50	2.42	2.33	2.23	2.13
27	7.68	5.49	4.60	4.11	3.78	3.56	3.39	3.26	3.15	3.06	2.93	2.78	2.63	2.55	2.47	2.38	2.29	2.20	2.10
28	7.64	5.45	4.57	4.07	3.75	3.53	3.36	3.23	3.12	3.03	2.90	2.75	2.60	2.52	2.44	2.35	2.26	2.17	2.06
29	7.60	5.42	4.54	4.04	3.73	3.50	3.33	3.20	3.09	3.00	2.87	2.73	2.57	2.49	2.41	2.33	2.23	2.14	2.03
30	7.56	5.39	4.51	4.02	3.70	3.47	3.30	3.17	3.07	2.98	2.84	2.70	2.55	2.47	2.39	2.30	2.21	2.11	2.01
40	7.31	5.18	4.31	3.83	3.51	3.29	3.12	2.99	2.89	2.80	2.66	2.52	2.37	2.29	2.20	2.11	2.02	1.92	1.80
60	7.08	4.98	4.13	3.65	3.34	3.12	2.95	2.82	2.72	2.63	2.50	2.35	2.20	2.12	2.03	1.94	1.84	1.73	1.60
120	6.85	4.79	3.95	3.48	3.17	2.96	2.79	2.66	2.56	2.47	2.34	2.19	2.03	1.95	1.86	1.76	1.66	1.53	1.38
∞	6.63	4.61	3.78	3.32	3.02	2.80	2.64	2.51	2.41	2.32	2.18	2.04	1.88	1.79	1.70	1.59	1.47	1.32	1.00

Table E.5 (Continued)

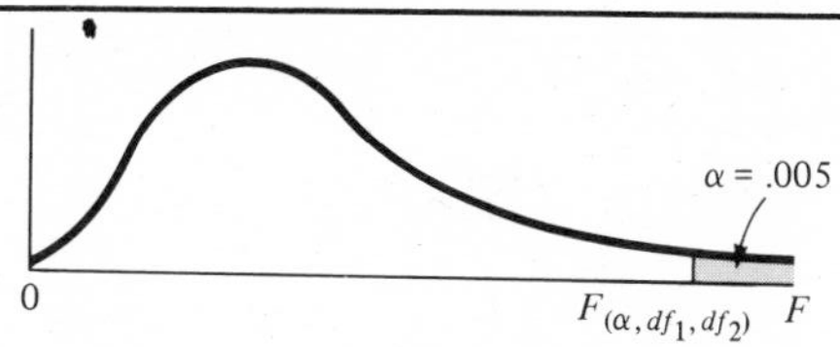

Numerator df_1

Denominator df_2	1	2	3	4	5	6	7	8	9	10	12	15	20	24	30	40	60	120	∞
1	16211	20000	21615	22500	23056	23437	23715	23925	24091	24224	24426	24630	24836	24940	25044	25148	25253	25359	25465
2	198.5	199.0	199.2	199.2	199.3	199.3	199.4	199.4	199.4	199.4	199.4	199.4	199.4	199.5	199.5	199.5	199.5	199.5	199.5
3	55.55	49.80	47.47	46.19	45.39	44.84	44.43	44.13	43.88	43.69	43.39	43.08	42.78	42.62	42.47	42.31	42.15	41.99	41.83
4	31.33	26.28	24.26	23.15	22.46	21.97	21.62	21.35	21.14	20.97	20.70	20.44	20.17	20.03	19.89	19.75	19.61	19.47	19.32
5	22.78	18.31	16.53	15.56	14.94	14.51	14.20	13.96	13.77	13.62	13.38	13.15	12.90	12.78	12.66	12.53	12.40	12.27	12.14
6	18.63	14.54	12.92	12.03	11.46	11.07	10.79	10.57	10.39	10.25	10.03	9.81	9.59	9.47	9.36	9.24	9.12	9.00	8.88
7	16.24	12.40	10.88	10.05	9.52	9.16	8.89	8.68	8.51	8.38	8.18	7.97	7.75	7.65	7.53	7.42	7.31	7.19	7.08
8	14.69	11.04	9.60	8.81	8.30	7.95	7.69	7.50	7.34	7.21	7.01	6.81	6.61	6.50	6.40	6.29	6.18	6.06	5.95
9	13.61	10.11	8.72	7.96	7.47	7.13	6.88	6.69	6.54	6.42	6.23	6.03	5.83	5.73	5.62	5.52	5.41	5.30	5.19
10	12.83	9.43	8.08	7.34	6.87	6.54	6.30	6.12	5.97	5.85	5.66	5.47	5.27	5.17	5.07	4.97	4.86	4.75	4.64
11	12.23	8.91	7.60	6.88	6.42	6.10	5.86	5.68	5.54	5.42	5.24	5.05	4.86	4.76	4.65	4.55	4.44	4.34	4.23
12	11.75	8.51	7.23	6.52	6.07	5.76	5.52	5.35	5.20	5.09	4.91	4.72	4.53	4.43	4.33	4.23	4.12	4.01	3.90
13	11.37	8.19	6.93	6.23	5.79	5.48	5.25	5.08	4.94	4.82	4.64	4.46	4.27	4.17	4.07	3.97	3.87	3.76	3.65
14	11.06	7.92	6.68	6.00	5.56	5.26	5.03	4.86	4.72	4.60	4.43	4.25	4.06	3.96	3.86	3.76	3.66	3.55	3.44
15	10.80	7.70	6.48	5.80	5.37	5.07	4.85	4.67	4.54	4.42	4.25	4.07	3.88	3.79	3.69	3.58	3.48	3.37	3.26
16	10.58	7.51	6.30	5.64	5.21	4.91	4.69	4.52	4.38	4.27	4.10	3.92	3.73	3.64	3.54	3.44	3.33	3.22	3.11
17	10.38	7.35	6.16	5.50	5.07	4.78	4.56	4.39	4.25	4.14	3.97	3.79	3.61	3.51	3.41	3.31	3.21	3.10	2.98
18	10.22	7.21	6.03	5.37	4.96	4.66	4.44	4.28	4.14	4.03	3.86	3.68	3.50	3.40	3.30	3.20	3.10	2.99	2.87
19	10.07	7.09	5.92	5.27	4.85	4.56	4.34	4.18	4.04	3.93	3.76	3.59	3.40	3.31	3.21	3.11	3.00	2.89	2.78
20	9.94	6.99	5.82	5.17	4.76	4.47	4.26	4.09	3.96	3.85	3.68	3.50	3.32	3.22	3.12	3.02	2.92	2.81	2.69
21	9.83	6.89	5.73	5.09	4.68	4.39	4.18	4.01	3.88	3.77	3.60	3.43	3.24	3.15	3.05	2.95	2.84	2.73	2.61
22	9.73	6.81	5.65	5.02	4.61	4.32	4.11	3.94	3.81	3.70	3.54	3.36	3.18	3.08	2.98	2.88	2.77	2.66	2.55
23	9.63	6.73	5.58	4.95	4.54	4.26	4.05	3.88	3.75	3.64	3.47	3.30	3.12	3.02	2.92	2.82	2.71	2.60	2.48
24	9.55	6.66	5.52	4.89	4.49	4.20	3.99	3.83	3.69	3.59	3.42	3.25	3.06	2.97	2.87	2.77	2.66	2.55	2.43
25	9.48	6.60	5.46	4.84	4.43	4.15	3.94	3.78	3.64	3.54	3.37	3.20	3.01	2.92	2.82	2.72	2.61	2.50	2.38
26	9.41	6.54	5.41	4.79	4.38	4.10	3.89	3.73	3.60	3.49	3.33	3.15	2.97	2.87	2.77	2.67	2.56	2.45	2.33
27	9.34	6.49	5.36	4.74	4.34	4.06	3.85	3.69	3.56	3.45	3.28	3.11	2.93	2.83	2.73	2.63	2.52	2.41	2.29
28	9.28	6.44	5.32	4.70	4.30	4.02	3.81	3.65	3.52	3.41	3.25	3.07	2.89	2.79	2.69	2.59	2.48	2.37	2.25
29	9.23	6.40	5.28	4.66	4.26	3.98	3.77	3.61	3.48	3.38	3.21	3.04	2.86	2.76	2.66	2.56	2.45	2.33	2.21
30	9.18	6.35	5.24	4.62	4.23	3.95	3.74	3.58	3.45	3.34	3.18	3.01	2.82	2.73	2.63	2.52	2.42	2.30	2.18
40	8.83	6.07	4.98	4.37	3.99	3.71	3.51	3.35	3.22	3.12	2.95	2.78	2.60	2.50	2.40	2.30	2.18	2.06	1.93
60	8.49	5.79	4.73	4.14	3.76	3.49	3.29	3.13	3.01	2.90	2.74	2.57	2.39	2.29	2.19	2.08	1.96	1.83	1.69
120	8.18	5.54	4.50	3.92	3.55	3.28	3.09	2.93	2.81	2.71	2.54	2.37	2.19	2.09	1.98	1.87	1.75	1.61	1.43
∞	7.88	5.30	4.28	3.72	3.35	3.09	2.90	2.74	2.62	2.52	2.36	2.19	2.00	1.90	1.79	1.67	1.53	1.36	1.00

SOURCE: Reprinted from E. S. Pearson and H. O. Hartley (eds.), *Biometrika Tables for Statisticians* (3rd ed., 1966), by permission of the *Biometrika* Trustees.

Table E.6
TABLE OF POISSON PROBABILITIES

For a given value of μ, entry indicates the probability of obtaining a specified value of X

X	μ = 0.1	0.2	0.3	0.4	0.5	0.6	0.7	0.8	0.9	1.0
0	.9048	.8187	.7408	.6703	.6065	.5488	.4966	.4493	.4066	.3679
1	.0905	.1637	.2222	.2681	.3033	.3293	.3476	.3595	.3659	.3679
2	.0045	.0164	.0333	.0536	.0758	.0988	.1217	.1438	.1647	.1839
3	.0002	.0011	.0033	.0072	.0126	.0198	.0284	.0383	.0494	.0613
4	.0000	.0001	.0003	.0007	.0016	.0030	.0050	.0077	.0111	.0153
5	.0000	.0000	.0000	.0001	.0002	.0004	.0007	.0012	.0020	.0031
6	.0000	.0000	.0000	.0000	.0000	.0000	.0001	.0002	.0003	.0005
7	.0000	.0000	.0000	.0000	.0000	.0000	.0000	.0000	.0000	.0001

X	μ = 1.1	1.2	1.3	1.4	1.5	1.6	1.7	1.8	1.9	2.0
0	.3329	.3012	.2725	.2466	.2231	.2019	.1827	.1653	.1496	.1353
1	.3662	.3614	.3543	.3452	.3347	.3230	.3106	.2975	.2842	.2707
2	.2014	.2169	.2303	.2417	.2510	.2584	.2640	.2678	.2700	.2707
3	.0738	.0867	.0998	.1128	.1255	.1378	.1496	.1607	.1710	.1804
4	.0203	.0260	.0324	.0395	.0471	.0551	.0636	.0723	.0812	.0902
5	.0045	.0062	.0084	.0111	.0141	.0176	.0216	.0260	.0309	.0361
6	.0008	.0012	.0018	.0026	.0035	.0047	.0061	.0078	.0098	.0120
7	.0001	.0002	.0003	.0005	.0008	.0011	.0015	.0020	.0027	.0034
8	.0000	.0000	.0001	.0001	.0001	.0002	.0003	.0005	.0006	.0009
9	.0000	.0000	.0000	.0000	.0000	.0000	.0001	.0001	.0001	.0002

X	μ = 2.1	2.2	2.3	2.4	2.5	2.6	2.7	2.8	2.9	3.0
0	.1225	.1108	.1003	.0907	.0821	.0743	.0672	.0608	.0550	.0498
1	.2572	.2438	.2306	.2177	.2052	.1931	.1815	.1703	.1596	.1494
2	.2700	.2681	.2652	.2613	.2565	.2510	.2450	.2384	.2314	.2240
3	.1890	.1966	.2033	.2090	.2138	.2176	.2205	.2225	.2237	.2240
4	.0992	.1082	.1169	.1254	.1336	.1414	.1488	.1557	.1622	.1680
5	.0417	.0476	.0538	.0602	.0668	.0735	.0804	.0872	.0940	.1008
6	.0146	.0174	.0206	.0241	.0278	.0319	.0362	.0407	.0455	.0504
7	.0044	.0055	.0068	.0083	.0099	.0118	.0139	.0163	.0188	.0216
8	.0011	.0015	.0019	.0025	.0031	.0038	.0047	.0057	.0068	.0081
9	.0003	.0004	.0005	.0007	.0009	.0011	.0014	.0018	.0022	.0027
10	.0001	.0001	.0001	.0002	.0002	.0003	.0004	.0005	.0006	.0008
11	.0000	.0000	.0000	.0000	.0000	.0001	.0001	.0001	.0002	.0002
12	.0000	.0000	.0000	.0000	.0000	.0000	.0000	.0000	.0000	.0001

X	μ = 3.1	3.2	3.3	3.4	3.5	3.6	3.7	3.8	3.9	4.0
0	.0450	.0408	.0369	.0334	.0302	.0273	.0247	.0224	.0202	.0183
1	.1397	.1304	.1217	.1135	.1057	.0984	.0915	.0850	.0789	.0733
2	.2165	.2087	.2008	.1929	.1850	.1771	.1692	.1615	.1539	.1465
3	.2237	.2226	.2209	.2186	.2158	.2125	.2087	.2046	.2001	.1954
4	.1734	.1781	.1823	.1858	.1888	.1912	.1931	.1944	.1951	.1954
5	.1075	.1140	.1203	.1264	.1322	.1377	.1429	.1477	.1522	.1563
6	.0555	.0608	.0662	.0716	.0771	.0826	.0881	.0936	.0989	.1042
7	.0246	.0278	.0312	.0348	.0385	.0425	.0466	.0508	.0551	.0595
8	.0095	.0111	.0129	.0148	.0169	.0191	.0215	.0241	.0269	.0298
9	.0033	.0040	.0047	.0056	.0066	.0076	.0089	.0102	.0116	.0132

Table E.6 (*Continued*)

X	3.1	3.2	3.3	3.4	3.5	3.6	3.7	3.8	3.9	4.0
					μ					
10	.0010	.0013	.0016	.0019	.0023	.0028	.0033	.0039	.0045	.0053
11	.0003	.0004	.0005	.0006	.0007	.0009	.0011	.0013	.0016	.0019
12	.0001	.0001	.0001	.0002	.0002	.0003	.0003	.0004	.0005	.0006
13	.0000	.0000	.0000	.0000	.0001	.0001	.0001	.0001	.0002	.0002
14	.0000	.0000	.0000	.0000	.0000	.0000	.0000	.0000	.0000	.0001

X	4.1	4.2	4.3	4.4	4.5	4.6	4.7	4.8	4.9	5.0
					μ					
0	.0166	.0150	.0136	.0123	.0111	.0101	.0091	.0082	.0074	.0067
1	.0679	.0630	.0583	.0540	.0500	.0462	.0427	.0395	.0365	.0337
2	.1393	.1323	.1254	.1188	.1125	.1063	.1005	.0948	.0894	.0842
3	.1904	.1852	.1798	.1743	.1687	.1631	.1574	.1517	.1460	.1404
4	.1951	.1944	.1933	.1917	.1898	.1875	.1849	.1820	.1789	.1755
5	.1600	.1633	.1662	.1687	.1708	.1725	.1738	.1747	.1753	.1755
6	.1093	.1143	.1191	.1237	.1281	.1323	.1362	.1398	.1432	.1462
7	.0640	.0686	.0732	.0778	.0824	.0869	.0914	.0959	.1002	.1044
8	.0328	.0360	.0393	.0428	.0463	.0500	.0537	.0575	.0614	.0653
9	.0150	.0168	.0188	.0209	.0232	.0255	.0280	.0307	.0334	.0363
10	.0061	.0071	.0081	.0092	.0104	.0118	.0132	.0147	.0164	.0181
11	.0023	.0027	.0032	.0037	.0043	.0049	.0056	.0064	.0073	.0082
12	.0008	.0009	.0011	.0014	.0016	.0019	.0022	.0026	.0030	.0034
13	.0002	.0003	.0004	.0005	.0006	.0007	.0008	.0009	.0011	.0013
14	.0001	.0001	.0001	.0001	.0002	.0002	.0003	.0003	.0004	.0005
15	.0000	.0000	.0000	.0000	.0001	.0001	.0001	.0001	.0001	.0002

X	5.1	5.2	5.3	5.4	5.5	5.6	5.7	5.8	5.9	6.0
					μ					
0	.0061	.0055	.0050	.0045	.0041	.0037	.0033	.0030	.0027	.0025
1	.0311	.0287	.0265	.0244	.0225	.0207	.0191	.0176	.0162	.0149
2	.0793	.0746	.0701	.0659	.0618	.0580	.0544	.0509	.0477	.0446
3	.1348	.1293	.1239	.1185	.1133	.1082	.1033	.0985	.0938	.0892
4	.1719	.1681	.1641	.1600	.1558	.1515	.1472	.1428	.1383	.1339
5	.1753	.1748	.1740	.1728	.1714	.1697	.1678	.1656	.1632	.1606
6	.1490	.1515	.1537	.1555	.1571	.1584	.1594	.1601	.1605	.1606
7	.1086	.1125	.1163	.1200	.1234	.1267	.1298	.1326	.1353	.1377
8	.0692	.0731	.0771	.0810	.0849	.0887	.0925	.0962	.0998	.1033
9	.0392	.0423	.0454	.0486	.0519	.0552	.0586	.0620	.0654	.0688
10	.0200	.0220	.0241	.0262	.0285	.0309	.0334	.0359	.0386	.0413
11	.0093	.0104	.0116	.0129	.0143	.0157	.0173	.0190	.0207	.0225
12	.0039	.0045	.0051	.0058	.0065	.0073	.0082	.0092	.0102	.0113
13	.0015	.0018	.0021	.0024	.0028	.0032	.0036	.0041	.0046	.0052
14	.0006	.0007	.0008	.0009	.0011	.0013	.0015	.0017	.0019	.0022
15	.0002	.0002	.0003	.0003	.0004	.0005	.0006	.0007	.0008	.0009
16	.0001	.0001	.0001	.0001	.0001	.0002	.0002	.0002	.0003	.0003
17	.0000	.0000	.0000	.0000	.0000	.0000	.0001	.0001	.0001	.0001

Table E.6 (*Continued*)

X	μ = 6.1	6.2	6.3	6.4	6.5	6.6	6.7	6.8	6.9	7.0
0	.0022	.0020	.0018	.0017	.0015	.0014	.0012	.0011	.0010	.0009
1	.0137	.0126	.0116	.0106	.0098	.0090	.0082	.0076	.0070	.0064
2	.0417	.0390	.0364	.0340	.0318	.0296	.0276	.0258	.0240	.0223
3	.0848	.0806	.0765	.0726	.0688	.0652	.0617	.0584	.0552	.0521
4	.1294	.1249	.1205	.1162	.1118	.1076	.1034	.0992	.0952	.0912
5	.1579	.1549	.1519	.1487	.1454	.1420	.1385	.1349	.1314	.1277
6	.1605	.1601	.1595	.1586	.1575	.1562	.1546	.1529	.1511	.1490
7	.1399	.1418	.1435	.1450	.1462	.1472	.1480	.1486	.1489	.1490
8	.1066	.1099	.1130	.1160	.1188	.1215	.1240	.1263	.1284	.1304
9	.0723	.0757	.0791	.0825	.0858	.0891	.0923	.0954	.0985	.1014
10	.0441	.0469	.0498	.0528	.0558	.0588	.0618	.0649	.0679	.0710
11	.0245	.0265	.0285	.0307	.0330	.0353	.0377	.0401	.0426	.0452
12	.0124	.0137	.0150	.0164	.0179	.0194	.0210	.0227	.0245	.0264
13	.0058	.0065	.0073	.0081	.0089	.0098	.0108	.0119	.0130	.0142
14	.0025	.0029	.0033	.0037	.0041	.0046	.0052	.0058	.0064	.0071
15	.0010	.0012	.0014	.0016	.0018	.0020	.0023	.0026	.0029	.0033
16	.0004	.0005	.0005	.0006	.0007	.0008	.0010	.0011	.0013	.0014
17	.0001	.0002	.0002	.0002	.0003	.0003	.0004	.0004	.0005	.0006
18	.0000	.0001	.0001	.0001	.0001	.0001	.0001	.0002	.0002	.0002
19	.0000	.0000	.0000	.0000	.0000	.0000	.0000	.0001	.0001	.0001

X	μ = 7.1	7.2	7.3	7.4	7.5	7.6	7.7	7.8	7.9	8.0
0	.0008	.0007	.0007	.0006	.0006	.0005	.0005	.0004	.0004	.0003
1	.0059	.0054	.0049	.0045	.0041	.0038	.0035	.0032	.0029	.0027
2	.0208	.0194	.0180	.0167	.0156	.0145	.0134	.0125	.0116	.0107
3	.0492	.0464	.0438	.0413	.0389	.0366	.0345	.0324	.0305	.0286
4	.0874	.0836	.0799	.0764	.0729	.0696	.0663	.0632	.0602	.0573
5	.1241	.1204	.1167	.1130	.1094	.1057	.1021	.0986	.0951	.0916
6	.1468	.1445	.1420	.1394	.1367	.1339	.1311	.1282	.1252	.1221
7	.1489	.1486	.1481	.1474	.1465	.1454	.1442	.1428	.1413	.1396
8	.1321	.1337	.1351	.1363	.1373	.1382	.1388	.1392	.1395	.1396
9	.1042	.1070	.1096	.1121	.1144	.1167	.1187	.1207	.1224	.1241
10	.0740	.0770	.0800	.0829	.0858	.0887	.0914	.0941	.0967	.0993
11	.0478	.0504	.0531	.0558	.0585	.0613	.0640	.0667	.0695	.0722
12	.0283	.0303	.0323	.0344	.0366	.0388	.0411	.0434	.0457	.0481
13	.0154	.0168	.0181	.0196	.0211	.0227	.0243	.0260	.0278	.0296
14	.0078	.0086	.0095	.0104	.0113	.0123	.0134	.0145	.0157	.0169
15	.0037	.0041	.0046	.0051	.0057	.0062	.0069	.0075	.0083	.0090
16	.0016	.0019	.0021	.0024	.0026	.0030	.0033	.0037	.0041	.0045
17	.0007	.0008	.0009	.0010	.0012	.0013	.0015	.0017	.0019	.0021
18	.0003	.0003	.0004	.0004	.0005	.0006	.0006	.0007	.0008	.0009
19	.0001	.0001	.0001	.0002	.0002	.0002	.0003	.0003	.0003	.0004
20	.0000	.0000	.0001	.0001	.0001	.0001	.0001	.0001	.0001	.0002
21	.0000	.0000	.0000	.0000	.0000	.0000	.0000	.0000	.0001	.0001

Table E.6 (*Continued*)

	μ									
X	8.1	8.2	8.3	8.4	8.5	8.6	8.7	8.8	8.9	9.0
0	.0003	.0003	.0002	.0002	.0002	.0002	.0002	.0002	.0001	.0001
1	.0025	.0023	.0021	.0019	.0017	.0016	.0014	.0013	.0012	.0011
2	.0100	.0092	.0086	.0079	.0074	.0068	.0063	.0058	.0054	.0050
3	.0269	.0252	.0237	.0222	.0208	.0195	.0183	.0171	.0160	.0150
4	.0544	.0517	.0491	.0466	.0443	.0420	.0398	.0377	.0357	.0337
5	.0882	.0849	.0816	.0784	.0752	.0722	.0692	.0663	.0635	.0607
6	.1191	.1160	.1128	.1097	.1066	.1034	.1003	.0972	.0941	.0911
7	.1378	.1358	.1338	.1317	.1294	.1271	.1247	.1222	.1197	.1171
8	.1395	.1392	.1388	.1382	.1375	.1366	.1356	.1344	.1332	.1318
9	.1256	.1269	.1280	.1290	.1299	.1306	.1311	.1315	.1317	.1318
10	.1017	.1040	.1063	.1084	.1104	.1123	.1140	.1157	.1172	.1186
11	.0749	.0776	.0802	.0828	.0853	.0878	.0902	.0925	.0948	.0970
12	.0505	.0530	.0555	.0579	.0604	.0629	.0654	.0679	.0703	.0728
13	.0315	.0334	.0354	.0374	.0395	.0416	.0438	.0459	.0481	.0504
14	.0182	.0196	.0210	.0225	.0240	.0256	.0272	.0289	.0306	.0324
15	.0098	.0107	.0116	.0126	.0136	.0147	.0158	.0169	.0182	.0194
16	.0050	.0055	.0060	.0066	.0072	.0079	.0086	.0093	.0101	.0109
17	.0024	.0026	.0029	.0033	.0036	.0040	.0044	.0048	.0053	.0058
18	.0011	.0012	.0014	.0015	.0017	.0019	.0021	.0024	.0026	.0029
19	.0005	.0005	.0006	.0007	.0008	.0009	.0010	.0011	.0012	.0014
20	.0002	.0002	.0002	.0003	.0003	.0004	.0004	.0005	.0005	.0006
21	.0001	.0001	.0001	.0001	.0001	.0002	.0002	.0002	.0002	.0003
22	.0000	.0000	.0000	.0000	.0001	.0001	.0001	.0001	.0001	.0001

	μ									
X	9.1	9.2	9.3	9.4	9.5	9.6	9.7	9.8	9.9	10
0	.0001	.0001	.0001	.0001	.0001	.0001	.0001	.0001	.0001	.0000
1	.0010	.0009	.0009	.0008	.0007	.0007	.0006	.0005	.0005	.0005
2	.0046	.0043	.0040	.0037	.0034	.0031	.0029	.0027	.0025	.0023
3	.0140	.0131	.0123	.0115	.0107	.0100	.0093	.0087	.0081	.0076
4	.0319	.0302	.0285	.0269	.0254	.0240	.0226	.0213	.0201	.0189
5	.0581	.0555	.0530	.0506	.0483	.0460	.0439	.0418	.0398	.0378
6	.0881	.0851	.0822	.0793	.0764	.0736	.0709	.0682	.0656	.0631
7	.1145	.1118	.1091	.1064	.1037	.1010	.0982	.0955	.0928	.0901
8	.1302	.1286	.1269	.1251	.1232	.1212	.1191	.1170	.1148	.1126
9	.1317	.1315	.1311	.1306	.1300	.1293	.1284	.1274	.1263	.1251
10	.1198	.1210	.1219	.1228	.1235	.1241	.1245	.1249	.1250	.1251
11	.0991	.1012	.1031	.1049	.1067	.1083	.1098	.1112	.1125	.1137
12	.0752	.0776	.0799	.0822	.0844	.0866	.0888	.0908	.0928	.0948
13	.0526	.0549	.0572	.0594	.0617	.0640	.0662	.0685	.0707	.0729
14	.0342	.0361	.0380	.0399	.0419	.0439	.0459	.0479	.0500	.0521
15	.0208	.0221	.0235	.0250	.0265	.0281	.0297	.0313	.0330	.0347
16	.0118	.0127	.0137	.0147	.0157	.0168	.0180	.0192	.0204	.0217
17	.0063	.0069	.0075	.0081	.0088	.0095	.0103	.0111	.0119	.0128
18	.0032	.0035	.0039	.0042	.0046	.0051	.0055	.0060	.0065	.0071
19	.0015	.0017	.0019	.0021	.0023	.0026	.0028	.0031	.0034	.0037

Table E.6 (*Continued*)

X	μ = 9.1	9.2	9.3	9.4	9.5	9.6	9.7	9.8	9.9	10
20	.0007	.0008	.0009	.0010	.0011	.0012	.0014	.0015	.0017	.0019
21	.0003	.0003	.0004	.0004	.0005	.0006	.0006	.0007	.0008	.0009
22	.0001	.0001	.0002	.0002	.0002	.0002	.0003	.0003	.0004	.0004
23	.0000	.0001	.0001	.0001	.0001	.0001	.0001	.0001	.0002	.0002
24	.0000	.0000	.0000	.0000	.0000	.0000	.0000	.0001	.0001	.0001

X	μ = 11	12	13	14	15	16	17	18	19	20
0	.0000	.0000	.0000	.0000	.0000	.0000	.0000	.0000	.0000	.0000
1	.0002	.0001	.0000	.0000	.0000	.0000	.0000	.0000	.0000	.0000
2	.0010	.0004	.0002	.0001	.0000	.0000	.0000	.0000	.0000	.0000
3	.0037	.0018	.0008	.0004	.0002	.0001	.0000	.0000	.0000	.0000
4	.0102	.0053	.0027	.0013	.0006	.0003	.0001	.0001	.0000	.0000
5	.0224	.0127	.0070	.0037	.0019	.0010	.0005	.0002	.0001	.0001
6	.0411	.0255	.0152	.0087	.0048	.0026	.0014	.0007	.0004	.0002
7	.0646	.0437	.0281	.0174	.0104	.0060	.0034	.0018	.0010	.0005
8	.0888	.0655	.0457	.0304	.0194	.0120	.0072	.0042	.0024	.0013
9	.1085	.0874	.0661	.0473	.0324	.0213	.0135	.0083	.0050	.0029
10	.1194	.1048	.0859	.0663	.0486	.0341	.0230	.0150	.0095	.0058
11	.1194	.1144	.1015	.0844	.0663	.0496	.0355	.0245	.0164	.0106
12	.1094	.1144	.1099	.0984	.0829	.0661	.0504	.0368	.0259	.0176
13	.0926	.1056	.1099	.1060	.0956	.0814	.0658	.0509	.0378	.0271
14	.0728	.0905	.1021	.1060	.1024	.0930	.0800	.0655	.0514	.0387
15	.0534	.0724	.0885	.0989	.1024	.0992	.0906	.0786	.0650	.0516
16	.0367	.0543	.0719	.0866	.0960	.0992	.0963	.0884	.0772	.0646
17	.0237	.0383	.0550	.0713	.0847	.0934	.0963	.0936	.0863	.0760
18	.0145	.0256	.0397	.0554	.0706	.0830	.0909	.0936	.0911	.0844
19	.0084	.0161	.0272	.0409	.0557	.0699	.0814	.0887	.0911	.0888
20	.0046	.0097	.0177	.0286	.0418	.0559	.0692	.0798	.0866	.0888
21	.0024	.0055	.0109	.0191	.0299	.0426	.0560	.0684	.0783	.0846
22	.0012	.0030	.0065	.0121	.0204	.0310	.0433	.0560	.0676	.0769
23	.0006	.0016	.0037	.0074	.0133	.0216	.0320	.0438	.0559	.0669
24	.0003	.0008	.0020	.0043	.0083	.0144	.0226	.0328	.0442	.0557
25	.0001	.0004	.0010	.0024	.0050	.0092	.0154	.0237	.0336	.0446
26	.0000	.0002	.0005	.0013	.0029	.0057	.0101	.0164	.0246	.0343
27	.0000	.0001	.0002	.0007	.0016	.0034	.0063	.0109	.0173	.0254
28	.0000	.0000	.0001	.0003	.0009	.0019	.0038	.0070	.0117	.0181
29	.0000	.0000	.0001	.0002	.0004	.0011	.0023	.0044	.0077	.0125
30	.0000	.0000	.0000	.0001	.0002	.0006	.0013	.0026	.0049	.0083
31	.0000	.0000	.0000	.0000	.0001	.0003	.0007	.0015	.0030	.0054
32	.0000	.0000	.0000	.0000	.0001	.0001	.0004	.0009	.0018	.0034
33	.0000	.0000	.0000	.0000	.0000	.0001	.0002	.0005	.0010	.0020
34	.0000	.0000	.0000	.0000	.0000	.0000	.0001	.0002	.0006	.0012
35	.0000	.0000	.0000	.0000	.0000	.0000	.0000	.0001	.0003	.0007
36	.0000	.0000	.0000	.0000	.0000	.0000	.0000	.0001	.0002	.0004
37	.0000	.0000	.0000	.0000	.0000	.0000	.0000	.0000	.0001	.0002
38	.0000	.0000	.0000	.0000	.0000	.0000	.0000	.0000	.0000	.0001
39	.0000	.0000	.0000	.0000	.0000	.0000	.0000	.0000	.0000	.0001

SOURCE: Extracted from William H. Beyer (ed.), *CRC Basic Statistical Tables* (Cleveland: The Chemical Rubber Co., 1971).

Table E.7
TABLE OF BINOMIAL PROBABILITIES

For a given combination of n and p, entry indicates the probability of obtaining a specified value of X. To locate entry: **when $p \leq .50$,** read p across the top heading and both n and X down the left margin; **when $p \geq .50$,** read p across the bottom heading and both n and X up the right margin

		P																										
n	X	0.01	0.02	0.03	0.04	0.05	0.06	0.07	0.08	0.09	0.10	0.11	0.12	0.13	0.14	0.15	0.16	0.17	0.18	0.19	0.20	0.21	0.22	0.23	0.24	0.25	X	n
2	0	0.9801	0.9604	0.9409	0.9216	0.9025	0.8836	0.8649	0.8464	0.8281	0.8100	0.7921	0.7744	0.7569	0.7396	0.7225	0.7056	0.6889	0.6724	0.6561	0.6400	0.6241	0.6084	0.5929	0.5776	0.5625	2	
	1	0.0198	0.0392	0.0582	0.0768	0.0950	0.1128	0.1302	0.1472	0.1638	0.1800	0.1958	0.2112	0.2262	0.2408	0.2550	0.2688	0.2822	0.2952	0.3078	0.3200	0.3318	0.3432	0.3542	0.3648	0.3750	1	
	2	0.0001	0.0004	0.0009	0.0016	0.0025	0.0036	0.0049	0.0064	0.0081	0.0100	0.0121	0.0144	0.0169	0.0196	0.0225	0.0256	0.0289	0.0324	0.0361	0.0400	0.0441	0.0484	0.0529	0.0576	0.0625	0	2
3	0	0.9703	0.9412	0.9127	0.8847	0.8574	0.8306	0.8044	0.7787	0.7536	0.7290	0.7050	0.6815	0.6585	0.6361	0.6141	0.5927	0.5718	0.5514	0.5314	0.5120	0.4930	0.4746	0.4565	0.4390	0.4219	3	
	1	0.0294	0.0576	0.0847	0.1106	0.1354	0.1590	0.1816	0.2031	0.2236	0.2430	0.2614	0.2788	0.2952	0.3106	0.3251	0.3387	0.3513	0.3631	0.3740	0.3840	0.3932	0.4015	0.4091	0.4159	0.4219	2	
	2	0.0003	0.0012	0.0026	0.0046	0.0071	0.0102	0.0137	0.0177	0.0221	0.0270	0.0323	0.0380	0.0441	0.0506	0.0574	0.0645	0.0720	0.0797	0.0877	0.0960	0.1045	0.1133	0.1222	0.1313	0.1406	1	
	3	0.0000	0.0000	0.0000	0.0001	0.0001	0.0002	0.0003	0.0005	0.0007	0.0010	0.0013	0.0017	0.0022	0.0027	0.0034	0.0041	0.0049	0.0058	0.0069	0.0080	0.0093	0.0106	0.0122	0.0138	0.0156	0	3
4	0	0.9606	0.9224	0.8853	0.8493	0.8145	0.7807	0.7481	0.7164	0.6857	0.6561	0.6274	0.5997	0.5729	0.5470	0.5220	0.4979	0.4746	0.4521	0.4305	0.4096	0.3895	0.3702	0.3515	0.3336	0.3164	4	
	1	0.0388	0.0753	0.1095	0.1416	0.1715	0.1993	0.2252	0.2492	0.2713	0.2916	0.3102	0.3271	0.3424	0.3562	0.3685	0.3793	0.3888	0.3970	0.4039	0.4096	0.4142	0.4176	0.4200	0.4214	0.4219	3	
	2	0.0006	0.0023	0.0051	0.0088	0.0135	0.0191	0.0254	0.0325	0.0402	0.0486	0.0575	0.0669	0.0767	0.0870	0.0975	0.1084	0.1195	0.1307	0.1421	0.1536	0.1651	0.1767	0.1882	0.1996	0.2109	2	
	3	0.0000	0.0000	0.0001	0.0002	0.0005	0.0008	0.0013	0.0019	0.0027	0.0036	0.0047	0.0061	0.0076	0.0094	0.0115	0.0138	0.0163	0.0191	0.0222	0.0256	0.0293	0.0332	0.0375	0.0420	0.0469	1	
	4	0.0000	0.0000	0.0000	0.0000	0.0000	0.0000	0.0000	0.0000	0.0001	0.0001	0.0001	0.0002	0.0003	0.0004	0.0005	0.0007	0.0008	0.0010	0.0013	0.0016	0.0019	0.0023	0.0028	0.0033	0.0039	0	4
5	0	0.9510	0.9039	0.8587	0.8154	0.7738	0.7339	0.6957	0.6591	0.6240	0.5905	0.5584	0.5277	0.4984	0.4704	0.4437	0.4182	0.3939	0.3707	0.3487	0.3277	0.3077	0.2887	0.2707	0.2536	0.2373	5	
	1	0.0480	0.0922	0.1328	0.1699	0.2036	0.2342	0.2618	0.2866	0.3086	0.3280	0.3451	0.3598	0.3724	0.3829	0.3915	0.3983	0.4034	0.4069	0.4089	0.4096	0.4090	0.4072	0.4043	0.4003	0.3955	4	
	2	0.0010	0.0038	0.0082	0.0142	0.0214	0.0299	0.0394	0.0498	0.0610	0.0729	0.0853	0.0981	0.1113	0.1247	0.1382	0.1517	0.1652	0.1786	0.1919	0.2048	0.2174	0.2297	0.2415	0.2529	0.2637	3	
	3	0.0000	0.0001	0.0003	0.0006	0.0011	0.0019	0.0030	0.0043	0.0060	0.0081	0.0105	0.0134	0.0166	0.0203	0.0244	0.0289	0.0338	0.0392	0.0450	0.0512	0.0578	0.0648	0.0721	0.0798	0.0879	2	
	4	0.0000	0.0000	0.0000	0.0000	0.0000	0.0001	0.0001	0.0002	0.0003	0.0004	0.0007	0.0009	0.0012	0.0017	0.0022	0.0028	0.0035	0.0043	0.0053	0.0064	0.0077	0.0091	0.0108	0.0126	0.0146	1	
	5	0.0000	0.0000	0.0000	0.0000	0.0000	0.0000	0.0000	0.0000	0.0000	0.0000	0.0000	0.0000	0.0000	0.0001	0.0001	0.0001	0.0001	0.0002	0.0002	0.0003	0.0004	0.0005	0.0006	0.0008	0.0010	0	5
6	0	0.9415	0.8858	0.8330	0.7828	0.7351	0.6899	0.6470	0.6064	0.5679	0.5314	0.4970	0.4644	0.4336	0.4046	0.3771	0.3513	0.3269	0.3040	0.2824	0.2621	0.2431	0.2252	0.2084	0.1927	0.1780	6	
	1	0.0571	0.1085	0.1546	0.1957	0.2321	0.2642	0.2922	0.3164	0.3370	0.3543	0.3685	0.3800	0.3888	0.3952	0.3993	0.4015	0.4018	0.4004	0.3975	0.3932	0.3877	0.3811	0.3735	0.3651	0.3560	5	
	2	0.0014	0.0055	0.0120	0.0204	0.0305	0.0422	0.0550	0.0688	0.0833	0.0984	0.1139	0.1295	0.1452	0.1608	0.1762	0.1912	0.2057	0.2197	0.2331	0.2458	0.2577	0.2687	0.2789	0.2882	0.2966	4	
	3	0.0000	0.0002	0.0005	0.0011	0.0021	0.0036	0.0055	0.0080	0.0110	0.0146	0.0188	0.0236	0.0289	0.0349	0.0415	0.0486	0.0562	0.0643	0.0729	0.0819	0.0913	0.1011	0.1111	0.1214	0.1318	3	
	4	0.0000	0.0000	0.0000	0.0000	0.0001	0.0002	0.0003	0.0005	0.0008	0.0012	0.0017	0.0024	0.0032	0.0043	0.0055	0.0069	0.0086	0.0106	0.0128	0.0154	0.0182	0.0214	0.0249	0.0287	0.0330	2	
	5	0.0000	0.0000	0.0000	0.0000	0.0000	0.0000	0.0000	0.0000	0.0000	0.0001	0.0001	0.0001	0.0002	0.0003	0.0004	0.0005	0.0007	0.0009	0.0012	0.0015	0.0019	0.0024	0.0030	0.0036	0.0044	1	
	6	0.0000	0.0000	0.0000	0.0000	0.0000	0.0000	0.0000	0.0000	0.0000	0.0000	0.0000	0.0000	0.0000	0.0000	0.0000	0.0000	0.0000	0.0000	0.0000	0.0001	0.0001	0.0001	0.0001	0.0002	0.0002	0	6
7	0	0.9321	0.8681	0.8080	0.7514	0.6983	0.6485	0.6017	0.5578	0.5168	0.4783	0.4423	0.4087	0.3773	0.3479	0.3206	0.2951	0.2714	0.2493	0.2288	0.2097	0.1920	0.1757	0.1605	0.1465	0.1335	7	
	1	0.0659	0.1240	0.1749	0.2192	0.2573	0.2897	0.3170	0.3396	0.3578	0.3720	0.3827	0.3901	0.3946	0.3965	0.3960	0.3935	0.3891	0.3830	0.3756	0.3670	0.3573	0.3468	0.3356	0.3237	0.3115	6	
	2	0.0020	0.0076	0.0162	0.0274	0.0406	0.0555	0.0716	0.0886	0.1061	0.1240	0.1419	0.1596	0.1769	0.1936	0.2097	0.2248	0.2391	0.2523	0.2643	0.2753	0.2850	0.2935	0.3007	0.3067	0.3115	5	
	3	0.0000	0.0003	0.0008	0.0019	0.0036	0.0059	0.0090	0.0128	0.0175	0.0230	0.0292	0.0363	0.0441	0.0525	0.0617	0.0714	0.0816	0.0923	0.1033	0.1147	0.1263	0.1379	0.1497	0.1614	0.1730	4	
	4	0.0000	0.0000	0.0000	0.0001	0.0002	0.0004	0.0007	0.0011	0.0017	0.0026	0.0036	0.0049	0.0066	0.0086	0.0109	0.0136	0.0167	0.0203	0.0242	0.0287	0.0336	0.0389	0.0447	0.0510	0.0577	3	
	5	0.0000	0.0000	0.0000	0.0000	0.0000	0.0000	0.0000	0.0001	0.0001	0.0002	0.0003	0.0004	0.0006	0.0008	0.0012	0.0016	0.0021	0.0027	0.0034	0.0043	0.0054	0.0066	0.0080	0.0097	0.0115	2	
	6	0.0000	0.0000	0.0000	0.0000	0.0000	0.0000	0.0000	0.0000	0.0000	0.0000	0.0000	0.0000	0.0000	0.0000	0.0001	0.0001	0.0001	0.0002	0.0003	0.0004	0.0005	0.0006	0.0008	0.0010	0.0013	1	
	7	0.0000	0.0000	0.0000	0.0000	0.0000	0.0000	0.0000	0.0000	0.0000	0.0000	0.0000	0.0000	0.0000	0.0000	0.0000	0.0000	0.0000	0.0000	0.0000	0.0000	0.0000	0.0000	0.0000	0.0000	0.0001	0	7
8	0	0.9227	0.8508	0.7837	0.7214	0.6634	0.6096	0.5596	0.5132	0.4703	0.4305	0.3937	0.3596	0.3282	0.2992	0.2725	0.2479	0.2252	0.2044	0.1853	0.1678	0.1517	0.1370	0.1236	0.1113	0.1001	8	
	1	0.0746	0.1389	0.1939	0.2405	0.2793	0.3113	0.3370	0.3570	0.3721	0.3826	0.3892	0.3923	0.3923	0.3897	0.3847	0.3777	0.3691	0.3590	0.3477	0.3355	0.3226	0.3092	0.2953	0.2812	0.2670	7	
	2	0.0026	0.0099	0.0210	0.0351	0.0515	0.0695	0.0888	0.1087	0.1288	0.1488	0.1684	0.1872	0.2052	0.2220	0.2376	0.2518	0.2646	0.2758	0.2855	0.2936	0.3002	0.3052	0.3087	0.3108	0.3115	6	
	3	0.0001	0.0004	0.0013	0.0029	0.0054	0.0089	0.0134	0.0189	0.0255	0.0331	0.0416	0.0511	0.0613	0.0723	0.0839	0.0959	0.1084	0.1211	0.1339	0.1468	0.1596	0.1722	0.1844	0.1963	0.2076	5	
	4	0.0000	0.0000	0.0001	0.0002	0.0004	0.0007	0.0013	0.0021	0.0031	0.0046	0.0064	0.0087	0.0115	0.0147	0.0185	0.0228	0.0277	0.0332	0.0393	0.0459	0.0530	0.0607	0.0689	0.0775	0.0865	4	
	5	0.0000	0.0000	0.0000	0.0000	0.0000	0.0000	0.0001	0.0001	0.0002	0.0004	0.0006	0.0009	0.0014	0.0019	0.0026	0.0035	0.0045	0.0058	0.0074	0.0092	0.0113	0.0137	0.0165	0.0196	0.0231	3	
	6	0.0000	0.0000	0.0000	0.0000	0.0000	0.0000	0.0000	0.0000	0.0000	0.0000	0.0000	0.0001	0.0001	0.0002	0.0002	0.0003	0.0005	0.0006	0.0009	0.0011	0.0015	0.0019	0.0025	0.0031	0.0038	2	
	7	0.0000	0.0000	0.0000	0.0000	0.0000	0.0000	0.0000	0.0000	0.0000	0.0000	0.0000	0.0000	0.0000	0.0000	0.0000	0.0000	0.0000	0.0000	0.0001	0.0001	0.0001	0.0002	0.0002	0.0003	0.0004	1	
	8	0.0000	0.0000	0.0000	0.0000	0.0000	0.0000	0.0000	0.0000	0.0000	0.0000	0.0000	0.0000	0.0000	0.0000	0.0000	0.0000	0.0000	0.0000	0.0000	0.0000	0.0000	0.0000	0.0000	0.0000	0.0000	0	8
9	0	0.9135	0.8337	0.7602	0.6925	0.6302	0.5730	0.5204	0.4722	0.4279	0.3874	0.3504	0.3165	0.2855	0.2573	0.2316	0.2082	0.1869	0.1676	0.1501	0.1342	0.1199	0.1069	0.0952	0.0846	0.0751	9	
	1	0.0830	0.1531	0.2116	0.2597	0.2985	0.3292	0.3525	0.3695	0.3809	0.3874	0.3897	0.3884	0.3840	0.3770	0.3679	0.3569	0.3446	0.3312	0.3169	0.3020	0.2867	0.2713	0.2558	0.2404	0.2253	8	
	2	0.0034	0.0125	0.0262	0.0433	0.0629	0.0840	0.1061	0.1285	0.1507	0.1722	0.1927	0.2119	0.2295	0.2455	0.2597	0.2720	0.2823	0.2908	0.2973	0.3020	0.3049	0.3061	0.3056	0.3037	0.3003	7	
	3	0.0001	0.0006	0.0019	0.0042	0.0077	0.0125	0.0186	0.0261	0.0348	0.0446	0.0556	0.0674	0.0800	0.0933	0.1069	0.1209	0.1349	0.1489	0.1627	0.1762	0.1891	0.2014	0.2130	0.2238	0.2336	6	
	4	0.0000	0.0000	0.0001	0.0003	0.0006	0.0012	0.0021	0.0034	0.0052	0.0074	0.0103	0.0138	0.0179	0.0228	0.0283	0.0345	0.0415	0.0490	0.0573	0.0661	0.0754	0.0852	0.0954	0.1060	0.1168	5	
	5	0.0000	0.0000	0.0000	0.0000	0.0000	0.0001	0.0002	0.0003	0.0005	0.0008	0.0013	0.0019	0.0027	0.0037	0.0050	0.0066	0.0085	0.0108	0.0134	0.0165	0.0200	0.0240	0.0285	0.0335	0.0389	4	
	6	0.0000	0.0000	0.0000	0.0000	0.0000	0.0000	0.0000	0.0000	0.0000	0.0001	0.0001	0.0002	0.0003	0.0004	0.0006	0.0008	0.0012	0.0016	0.0021	0.0028	0.0036	0.0045	0.0057	0.0070	0.0087	3	
	7	0.0000	0.0000	0.0000	0.0000	0.0000	0.0000	0.0000	0.0000	0.0000	0.0000	0.0000	0.0000	0.0000	0.0000	0.0000	0.0000	0.0001	0.0001	0.0002	0.0003	0.0004	0.0005	0.0007	0.0010	0.0012	2	
	8	0.0000	0.0000	0.0000	0.0000	0.0000	0.0000	0.0000	0.0000	0.0000	0.0000	0.0000	0.0000	0.0000	0.0000	0.0000	0.0000	0.0000	0.0000	0.0000	0.0000	0.0000	0.0000	0.0001	0.0001	0.0001	1	
	9	0.0000	0.0000	0.0000	0.0000	0.0000	0.0000	0.0000	0.0000	0.0000	0.0000	0.0000	0.0000	0.0000	0.0000	0.0000	0.0000	0.0000	0.0000	0.0000	0.0000	0.0000	0.0000	0.0000	0.0000	0.0000	0	9
n	X	0.99	0.98	0.97	0.96	0.95	0.94	0.93	0.92	0.91	0.90	0.89	0.88	0.87	0.86	0.85	0.84	0.83	0.82	0.81	0.80	0.79	0.78	0.77	0.76	0.75	X	n
		P																										

Table E.7 (Continued)

n	X	0.26	0.27	0.28	0.29	0.30	0.31	0.32	0.33	0.34	0.35	0.36	0.37	0.38	0.39	0.40	0.41	0.42	0.43	0.44	0.45	0.46	0.47	0.48	0.49	0.50	X	n
2	0	0.5476	0.5329	0.5184	0.5041	0.4900	0.4761	0.4624	0.4489	0.4356	0.4225	0.4096	0.3969	0.3844	0.3721	0.3600	0.3481	0.3364	0.3249	0.3136	0.3025	0.2916	0.2809	0.2704	0.2601	0.2500	2	
	1	0.3848	0.3942	0.4032	0.4118	0.4200	0.4278	0.4352	0.4422	0.4488	0.4550	0.4608	0.4662	0.4712	0.4758	0.4800	0.4838	0.4872	0.4902	0.4928	0.4950	0.4968	0.4982	0.4992	0.4998	0.5000	1	
	2	0.0676	0.0729	0.0784	0.0841	0.0900	0.0961	0.1024	0.1089	0.1156	0.1225	0.1296	0.1369	0.1444	0.1521	0.1600	0.1681	0.1764	0.1849	0.1936	0.2025	0.2116	0.2209	0.2304	0.2401	0.2500	0	2
3	0	0.4052	0.3890	0.3732	0.3579	0.3430	0.3285	0.3144	0.3008	0.2875	0.2746	0.2621	0.2500	0.2383	0.2270	0.2160	0.2054	0.1951	0.1852	0.1756	0.1664	0.1575	0.1489	0.1406	0.1327	0.1250	3	
	1	0.4271	0.4316	0.4355	0.4386	0.4410	0.4428	0.4439	0.4444	0.4443	0.4436	0.4424	0.4406	0.4382	0.4354	0.4320	0.4282	0.4239	0.4191	0.4140	0.4084	0.4024	0.3961	0.3894	0.3823	0.3750	2	
	2	0.1501	0.1597	0.1693	0.1791	0.1890	0.1989	0.2089	0.2189	0.2289	0.2389	0.2488	0.2587	0.2686	0.2783	0.2880	0.2975	0.3069	0.3162	0.3252	0.3341	0.3428	0.3512	0.3594	0.3674	0.3750	1	
	3	0.0176	0.0197	0.0220	0.0244	0.0270	0.0298	0.0328	0.0359	0.0393	0.0429	0.0467	0.0507	0.0549	0.0593	0.0640	0.0689	0.0741	0.0795	0.0852	0.0911	0.0973	0.1038	0.1106	0.1176	0.1250	0	3
4	0	0.2999	0.2840	0.2687	0.2541	0.2401	0.2267	0.2138	0.2015	0.1897	0.1785	0.1678	0.1575	0.1478	0.1385	0.1296	0.1212	0.1132	0.1056	0.0983	0.0915	0.0850	0.0789	0.0731	0.0677	0.0625	4	
	1	0.4214	0.4201	0.4180	0.4152	0.4116	0.4074	0.4025	0.3970	0.3910	0.3845	0.3775	0.3701	0.3623	0.3541	0.3456	0.3368	0.3278	0.3185	0.3091	0.2995	0.2897	0.2799	0.2700	0.2600	0.2500	3	
	2	0.2221	0.2331	0.2439	0.2544	0.2646	0.2745	0.2841	0.2933	0.3021	0.3105	0.3185	0.3260	0.3330	0.3396	0.3456	0.3511	0.3560	0.3604	0.3643	0.3675	0.3702	0.3723	0.3738	0.3747	0.3750	2	
	3	0.0520	0.0575	0.0632	0.0693	0.0756	0.0822	0.0891	0.0963	0.1038	0.1115	0.1194	0.1276	0.1361	0.1447	0.1536	0.1627	0.1719	0.1813	0.1908	0.2005	0.2102	0.2201	0.2300	0.2400	0.2500	1	
	4	0.0046	0.0053	0.0061	0.0071	0.0081	0.0092	0.0105	0.0119	0.0134	0.0150	0.0168	0.0187	0.0209	0.0231	0.0256	0.0283	0.0311	0.0342	0.0375	0.0410	0.0448	0.0488	0.0531	0.0576	0.0625	0	4
5	0	0.2219	0.2073	0.1935	0.1804	0.1681	0.1564	0.1454	0.1350	0.1252	0.1160	0.1074	0.0992	0.0916	0.0845	0.0778	0.0715	0.0656	0.0602	0.0551	0.0503	0.0459	0.0418	0.0380	0.0345	0.0312	5	
	1	0.3898	0.3834	0.3762	0.3685	0.3601	0.3513	0.3421	0.3325	0.3226	0.3124	0.3020	0.2914	0.2808	0.2700	0.2592	0.2484	0.2376	0.2270	0.2164	0.2059	0.1956	0.1854	0.1755	0.1657	0.1562	4	
	2	0.2739	0.2836	0.2926	0.3010	0.3087	0.3157	0.3220	0.3275	0.3323	0.3364	0.3397	0.3423	0.3441	0.3452	0.3456	0.3452	0.3442	0.3424	0.3400	0.3369	0.3332	0.3289	0.3240	0.3185	0.3125	3	
	3	0.0962	0.1049	0.1138	0.1229	0.1323	0.1418	0.1515	0.1613	0.1712	0.1811	0.1911	0.2010	0.2109	0.2207	0.2304	0.2399	0.2492	0.2583	0.2671	0.2757	0.2838	0.2916	0.2990	0.3060	0.3125	2	
	4	0.0169	0.0194	0.0221	0.0251	0.0283	0.0319	0.0357	0.0397	0.0441	0.0488	0.0537	0.0590	0.0646	0.0706	0.0768	0.0834	0.0902	0.0974	0.1049	0.1128	0.1209	0.1293	0.1380	0.1470	0.1562	1	
	5	0.0012	0.0014	0.0017	0.0021	0.0024	0.0029	0.0034	0.0039	0.0045	0.0053	0.0060	0.0069	0.0079	0.0090	0.0102	0.0116	0.0131	0.0147	0.0165	0.0185	0.0206	0.0229	0.0255	0.0282	0.0312	0	5
6	0	0.1642	0.1513	0.1393	0.1281	0.1176	0.1079	0.0989	0.0905	0.0827	0.0754	0.0687	0.0625	0.0568	0.0515	0.0467	0.0422	0.0381	0.0343	0.0308	0.0277	0.0248	0.0222	0.0198	0.0176	0.0156	6	
	1	0.3462	0.3358	0.3251	0.3139	0.3025	0.2909	0.2792	0.2673	0.2555	0.2437	0.2319	0.2203	0.2089	0.1976	0.1866	0.1759	0.1654	0.1552	0.1454	0.1359	0.1267	0.1179	0.1095	0.1014	0.0937	5	
	2	0.3041	0.3105	0.3160	0.3206	0.3241	0.3267	0.3284	0.3292	0.3290	0.3280	0.3261	0.3235	0.3201	0.3159	0.3110	0.3055	0.2994	0.2928	0.2856	0.2780	0.2699	0.2615	0.2527	0.2436	0.2344	4	
	3	0.1424	0.1531	0.1639	0.1746	0.1852	0.1957	0.2061	0.2162	0.2260	0.2355	0.2446	0.2533	0.2616	0.2693	0.2765	0.2831	0.2891	0.2945	0.2992	0.3032	0.3065	0.3091	0.3110	0.3121	0.3125	3	
	4	0.0375	0.0425	0.0478	0.0535	0.0595	0.0660	0.0727	0.0799	0.0873	0.0951	0.1032	0.1116	0.1202	0.1291	0.1382	0.1475	0.1570	0.1666	0.1763	0.1861	0.1958	0.2056	0.2153	0.2249	0.2344	2	
	5	0.0053	0.0063	0.0074	0.0087	0.0102	0.0119	0.0137	0.0157	0.0180	0.0205	0.0232	0.0262	0.0295	0.0330	0.0369	0.0410	0.0455	0.0503	0.0554	0.0609	0.0667	0.0729	0.0795	0.0864	0.0937	1	
	6	0.0003	0.0004	0.0005	0.0006	0.0007	0.0009	0.0011	0.0013	0.0015	0.0018	0.0022	0.0026	0.0030	0.0035	0.0041	0.0048	0.0055	0.0063	0.0073	0.0083	0.0095	0.0108	0.0122	0.0138	0.0156	0	6
7	0	0.1215	0.1105	0.1003	0.0910	0.0824	0.0745	0.0672	0.0606	0.0546	0.0490	0.0440	0.0394	0.0352	0.0314	0.0280	0.0249	0.0221	0.0195	0.0173	0.0152	0.0134	0.0117	0.0103	0.0090	0.0078	7	
	1	0.2989	0.2860	0.2731	0.2600	0.2471	0.2342	0.2215	0.2090	0.1967	0.1848	0.1732	0.1619	0.1511	0.1407	0.1306	0.1211	0.1119	0.1032	0.0950	0.0872	0.0798	0.0729	0.0664	0.0604	0.0547	6	
	2	0.3150	0.3174	0.3186	0.3186	0.3177	0.3156	0.3127	0.3088	0.3040	0.2985	0.2922	0.2853	0.2778	0.2698	0.2613	0.2524	0.2431	0.2336	0.2239	0.2140	0.2040	0.1940	0.1840	0.1740	0.1641	5	
	3	0.1845	0.1956	0.2065	0.2169	0.2269	0.2363	0.2452	0.2535	0.2610	0.2679	0.2740	0.2793	0.2838	0.2875	0.2903	0.2923	0.2934	0.2937	0.2932	0.2918	0.2897	0.2867	0.2830	0.2786	0.2734	4	
	4	0.0648	0.0724	0.0803	0.0886	0.0972	0.1062	0.1154	0.1248	0.1345	0.1442	0.1541	0.1640	0.1739	0.1838	0.1935	0.2031	0.2125	0.2216	0.2304	0.2388	0.2468	0.2543	0.2612	0.2676	0.2734	3	
	5	0.0137	0.0161	0.0187	0.0217	0.0250	0.0286	0.0326	0.0369	0.0416	0.0466	0.0520	0.0578	0.0640	0.0705	0.0774	0.0847	0.0923	0.1003	0.1086	0.1172	0.1261	0.1353	0.1447	0.1543	0.1641	2	
	6	0.0016	0.0020	0.0024	0.0030	0.0036	0.0043	0.0051	0.0061	0.0071	0.0084	0.0098	0.0113	0.0131	0.0150	0.0172	0.0196	0.0223	0.0252	0.0284	0.0320	0.0358	0.0400	0.0445	0.0494	0.0547	1	
	7	0.0001	0.0001	0.0001	0.0002	0.0002	0.0003	0.0003	0.0004	0.0005	0.0006	0.0008	0.0009	0.0011	0.0014	0.0016	0.0019	0.0023	0.0027	0.0032	0.0037	0.0044	0.0051	0.0059	0.0068	0.0078	0	7
8	0	0.0899	0.0806	0.0722	0.0646	0.0576	0.0514	0.0457	0.0406	0.0360	0.0319	0.0281	0.0248	0.0218	0.0192	0.0168	0.0147	0.0128	0.0111	0.0097	0.0084	0.0072	0.0062	0.0053	0.0046	0.0039	8	
	1	0.2527	0.2386	0.2247	0.2110	0.1977	0.1847	0.1721	0.1600	0.1484	0.1373	0.1267	0.1166	0.1071	0.0981	0.0896	0.0816	0.0742	0.0672	0.0608	0.0548	0.0493	0.0442	0.0395	0.0352	0.0312	7	
	2	0.3108	0.3089	0.3058	0.3017	0.2965	0.2904	0.2835	0.2758	0.2675	0.2587	0.2494	0.2397	0.2297	0.2194	0.2090	0.1985	0.1880	0.1776	0.1672	0.1569	0.1469	0.1371	0.1275	0.1183	0.1094	6	
	3	0.2184	0.2285	0.2379	0.2464	0.2541	0.2609	0.2668	0.2717	0.2756	0.2786	0.2805	0.2815	0.2815	0.2806	0.2787	0.2759	0.2723	0.2679	0.2627	0.2568	0.2503	0.2431	0.2355	0.2273	0.2187	5	
	4	0.0959	0.1056	0.1156	0.1258	0.1361	0.1465	0.1569	0.1673	0.1775	0.1875	0.1973	0.2067	0.2157	0.2242	0.2322	0.2397	0.2465	0.2526	0.2580	0.2627	0.2665	0.2695	0.2717	0.2730	0.2734	4	
	5	0.0270	0.0313	0.0360	0.0411	0.0467	0.0527	0.0591	0.0659	0.0732	0.0808	0.0888	0.0971	0.1058	0.1147	0.1239	0.1332	0.1428	0.1525	0.1622	0.1719	0.1816	0.1912	0.2006	0.2098	0.2187	3	
	6	0.0047	0.0058	0.0070	0.0084	0.0100	0.0118	0.0139	0.0162	0.0188	0.0217	0.0250	0.0285	0.0324	0.0367	0.0413	0.0463	0.0517	0.0575	0.0637	0.0703	0.0774	0.0848	0.0926	0.1008	0.1094	2	
	7	0.0005	0.0006	0.0008	0.0010	0.0012	0.0015	0.0019	0.0023	0.0028	0.0033	0.0040	0.0048	0.0057	0.0067	0.0079	0.0092	0.0107	0.0124	0.0143	0.0164	0.0188	0.0215	0.0244	0.0277	0.0312	1	
	8	0.0000	0.0000	0.0000	0.0001	0.0001	0.0001	0.0001	0.0001	0.0002	0.0002	0.0003	0.0004	0.0004	0.0005	0.0007	0.0008	0.0010	0.0012	0.0014	0.0017	0.0020	0.0024	0.0028	0.0033	0.0039	0	8
9	0	0.0665	0.0589	0.0520	0.0458	0.0404	0.0355	0.0311	0.0272	0.0238	0.0207	0.0180	0.0156	0.0135	0.0117	0.0101	0.0087	0.0074	0.0064	0.0054	0.0046	0.0039	0.0033	0.0028	0.0023	0.0020	9	
	1	0.2104	0.1960	0.1820	0.1685	0.1556	0.1433	0.1317	0.1206	0.1102	0.1004	0.0912	0.0826	0.0747	0.0673	0.0605	0.0542	0.0484	0.0431	0.0383	0.0339	0.0299	0.0263	0.0231	0.0202	0.0176	8	
	2	0.2957	0.2899	0.2831	0.2754	0.2668	0.2576	0.2478	0.2376	0.2270	0.2162	0.2052	0.1941	0.1831	0.1721	0.1612	0.1506	0.1402	0.1301	0.1204	0.1110	0.1020	0.0934	0.0853	0.0776	0.0703	7	
	3	0.2424	0.2502	0.2569	0.2624	0.2668	0.2701	0.2721	0.2731	0.2729	0.2716	0.2693	0.2660	0.2618	0.2567	0.2508	0.2442	0.2369	0.2291	0.2207	0.2119	0.2027	0.1933	0.1837	0.1739	0.1641	6	
	4	0.1278	0.1388	0.1499	0.1608	0.1715	0.1820	0.1921	0.2017	0.2109	0.2194	0.2272	0.2344	0.2407	0.2462	0.2508	0.2545	0.2573	0.2592	0.2601	0.2600	0.2590	0.2571	0.2543	0.2506	0.2461	5	
	5	0.0449	0.0513	0.0583	0.0657	0.0735	0.0818	0.0904	0.0994	0.1086	0.1181	0.1278	0.1376	0.1475	0.1574	0.1672	0.1769	0.1863	0.1955	0.2044	0.2128	0.2207	0.2280	0.2347	0.2408	0.2461	4	
	6	0.0105	0.0127	0.0151	0.0179	0.0210	0.0245	0.0284	0.0326	0.0373	0.0424	0.0479	0.0539	0.0603	0.0671	0.0743	0.0819	0.0900	0.0983	0.1070	0.1160	0.1253	0.1348	0.1445	0.1542	0.1641	3	
	7	0.0016	0.0020	0.0025	0.0031	0.0039	0.0047	0.0057	0.0069	0.0082	0.0098	0.0116	0.0136	0.0158	0.0184	0.0212	0.0244	0.0279	0.0318	0.0360	0.0407	0.0458	0.0512	0.0571	0.0635	0.0703	2	
	8	0.0001	0.0002	0.0002	0.0003	0.0004	0.0005	0.0007	0.0008	0.0011	0.0013	0.0016	0.0020	0.0024	0.0029	0.0035	0.0042	0.0051	0.0060	0.0071	0.0083	0.0097	0.0114	0.0132	0.0153	0.0176	1	
	9	0.0000	0.0000	0.0000	0.0000	0.0000	0.0000	0.0000	0.0000	0.0001	0.0001	0.0001	0.0001	0.0002	0.0002	0.0003	0.0003	0.0004	0.0005	0.0006	0.0008	0.0009	0.0011	0.0014	0.0016	0.0020	0	9
n	X	0.74	0.73	0.72	0.71	0.70	0.69	0.68	0.67	0.66	0.65	0.64	0.63	0.62	0.61	0.60	0.59	0.58	0.57	0.56	0.55	0.54	0.53	0.52	0.51	0.50	X	n

(Column headings are values of P: top row for P ≤ 0.50 with X read at left; bottom row for P ≥ 0.50 with X read at right.)

Table E.7 (Continued)

		P																										
n	X	0.01	0.02	0.03	0.04	0.05	0.06	0.07	0.08	0.09	0.10	0.11	0.12	0.13	0.14	0.15	0.16	0.17	0.18	0.19	0.20	0.21	0.22	0.23	0.24	0.25	X	n
10	0	0.9044	0.8171	0.7374	0.6648	0.5987	0.5386	0.4840	0.4344	0.3894	0.3487	0.3118	0.2785	0.2484	0.2213	0.1969	0.1749	0.1552	0.1374	0.1216	0.1074	0.0947	0.0834	0.0733	0.0643	0.0563	10	
	1	0.0914	0.1667	0.2281	0.2770	0.3151	0.3438	0.3643	0.3777	0.3851	0.3874	0.3854	0.3798	0.3712	0.3603	0.3474	0.3331	0.3178	0.3017	0.2852	0.2684	0.2517	0.2351	0.2188	0.2030	0.1877	9	
	2	0.0042	0.0153	0.0317	0.0519	0.0746	0.0988	0.1234	0.1478	0.1714	0.1937	0.2143	0.2330	0.2496	0.2639	0.2759	0.2856	0.2929	0.2980	0.3010	0.3020	0.3011	0.2984	0.2942	0.2885	0.2816	8	
	3	0.0001	0.0008	0.0026	0.0058	0.0105	0.0168	0.0248	0.0343	0.0452	0.0574	0.0706	0.0847	0.0995	0.1146	0.1298	0.1450	0.1600	0.1745	0.1883	0.2013	0.2134	0.2244	0.2343	0.2429	0.2503	7	
	4	0.0000	0.0000	0.0001	0.0004	0.0010	0.0019	0.0033	0.0052	0.0078	0.0112	0.0153	0.0202	0.0260	0.0326	0.0401	0.0483	0.0573	0.0670	0.0773	0.0881	0.0993	0.1108	0.1225	0.1343	0.1460	6	
	5	0.0000	0.0000	0.0000	0.0000	0.0001	0.0001	0.0003	0.0005	0.0009	0.0015	0.0023	0.0033	0.0047	0.0064	0.0085	0.0111	0.0141	0.0177	0.0218	0.0264	0.0317	0.0375	0.0439	0.0509	0.0584	5	
	6	0.0000	0.0000	0.0000	0.0000	0.0000	0.0000	0.0000	0.0000	0.0001	0.0001	0.0002	0.0004	0.0006	0.0009	0.0012	0.0018	0.0024	0.0032	0.0043	0.0055	0.0070	0.0088	0.0109	0.0134	0.0162	4	
	7	0.0000	0.0000	0.0000	0.0000	0.0000	0.0000	0.0000	0.0000	0.0000	0.0000	0.0000	0.0000	0.0000	0.0001	0.0001	0.0002	0.0003	0.0004	0.0006	0.0008	0.0011	0.0014	0.0019	0.0024	0.0031	3	
	8	0.0000	0.0000	0.0000	0.0000	0.0000	0.0000	0.0000	0.0000	0.0000	0.0000	0.0000	0.0000	0.0000	0.0000	0.0000	0.0000	0.0000	0.0000	0.0001	0.0001	0.0001	0.0002	0.0002	0.0003	0.0004	2	
	9	0.0000	0.0000	0.0000	0.0000	0.0000	0.0000	0.0000	0.0000	0.0000	0.0000	0.0000	0.0000	0.0000	0.0000	0.0000	0.0000	0.0000	0.0000	0.0000	0.0000	0.0000	0.0000	0.0000	0.0000	0.0000	1	
	10	0.0000	0.0000	0.0000	0.0000	0.0000	0.0000	0.0000	0.0000	0.0000	0.0000	0.0000	0.0000	0.0000	0.0000	0.0000	0.0000	0.0000	0.0000	0.0000	0.0000	0.0000	0.0000	0.0000	0.0000	0.0000	0	10
11	0	0.8953	0.8007	0.7153	0.6382	0.5688	0.5063	0.4501	0.3996	0.3544	0.3138	0.2775	0.2451	0.2161	0.1903	0.1673	0.1469	0.1288	0.1127	0.0985	0.0859	0.0748	0.0650	0.0564	0.0489	0.0422	11	
	1	0.0995	0.1798	0.2433	0.2925	0.3293	0.3555	0.3727	0.3823	0.3855	0.3835	0.3773	0.3676	0.3552	0.3408	0.3248	0.3078	0.2901	0.2721	0.2541	0.2362	0.2187	0.2017	0.1854	0.1697	0.1549	10	
	2	0.0050	0.0183	0.0376	0.0609	0.0867	0.1135	0.1403	0.1662	0.1906	0.2131	0.2332	0.2507	0.2654	0.2774	0.2866	0.2932	0.2971	0.2987	0.2980	0.2953	0.2907	0.2845	0.2768	0.2680	0.2581	9	
	3	0.0002	0.0011	0.0035	0.0076	0.0137	0.0217	0.0317	0.0434	0.0566	0.0710	0.0865	0.1025	0.1190	0.1355	0.1517	0.1675	0.1826	0.1967	0.2097	0.2215	0.2318	0.2407	0.2481	0.2539	0.2581	8	
	4	0.0000	0.0000	0.0002	0.0006	0.0014	0.0028	0.0048	0.0075	0.0112	0.0158	0.0214	0.0280	0.0356	0.0441	0.0536	0.0638	0.0748	0.0864	0.0984	0.1107	0.1232	0.1358	0.1482	0.1603	0.1721	7	
	5	0.0000	0.0000	0.0000	0.0000	0.0001	0.0002	0.0005	0.0009	0.0015	0.0025	0.0037	0.0053	0.0074	0.0101	0.0132	0.0170	0.0214	0.0265	0.0323	0.0388	0.0459	0.0536	0.0620	0.0709	0.0803	6	
	6	0.0000	0.0000	0.0000	0.0000	0.0000	0.0000	0.0000	0.0001	0.0002	0.0003	0.0005	0.0007	0.0011	0.0016	0.0023	0.0032	0.0044	0.0058	0.0076	0.0097	0.0122	0.0151	0.0185	0.0224	0.0268	5	
	7	0.0000	0.0000	0.0000	0.0000	0.0000	0.0000	0.0000	0.0000	0.0000	0.0000	0.0000	0.0001	0.0001	0.0002	0.0003	0.0004	0.0006	0.0009	0.0013	0.0017	0.0023	0.0030	0.0039	0.0050	0.0064	4	
	8	0.0000	0.0000	0.0000	0.0000	0.0000	0.0000	0.0000	0.0000	0.0000	0.0000	0.0000	0.0000	0.0000	0.0000	0.0000	0.0000	0.0001	0.0001	0.0001	0.0002	0.0003	0.0004	0.0006	0.0008	0.0011	3	
	9	0.0000	0.0000	0.0000	0.0000	0.0000	0.0000	0.0000	0.0000	0.0000	0.0000	0.0000	0.0000	0.0000	0.0000	0.0000	0.0000	0.0000	0.0000	0.0000	0.0000	0.0000	0.0000	0.0001	0.0001	0.0001	2	
	10	0.0000	0.0000	0.0000	0.0000	0.0000	0.0000	0.0000	0.0000	0.0000	0.0000	0.0000	0.0000	0.0000	0.0000	0.0000	0.0000	0.0000	0.0000	0.0000	0.0000	0.0000	0.0000	0.0000	0.0000	0.0000	1	
	11	0.0000	0.0000	0.0000	0.0000	0.0000	0.0000	0.0000	0.0000	0.0000	0.0000	0.0000	0.0000	0.0000	0.0000	0.0000	0.0000	0.0000	0.0000	0.0000	0.0000	0.0000	0.0000	0.0000	0.0000	0.0000	0	11
12	0	0.8864	0.7847	0.6938	0.6127	0.5404	0.4759	0.4186	0.3677	0.3225	0.2824	0.2470	0.2157	0.1880	0.1637	0.1422	0.1234	0.1069	0.0924	0.0798	0.0687	0.0591	0.0507	0.0434	0.0371	0.0317	12	
	1	0.1074	0.1922	0.2575	0.3064	0.3413	0.3645	0.3781	0.3837	0.3827	0.3766	0.3663	0.3529	0.3372	0.3197	0.3012	0.2821	0.2627	0.2434	0.2245	0.2062	0.1885	0.1717	0.1557	0.1407	0.1267	11	
	2	0.0060	0.0216	0.0438	0.0702	0.0988	0.1280	0.1565	0.1835	0.2082	0.2301	0.2490	0.2647	0.2771	0.2863	0.2924	0.2955	0.2960	0.2939	0.2897	0.2835	0.2756	0.2663	0.2558	0.2444	0.2323	10	
	3	0.0002	0.0015	0.0045	0.0098	0.0173	0.0272	0.0393	0.0532	0.0686	0.0852	0.1026	0.1203	0.1380	0.1553	0.1720	0.1876	0.2021	0.2151	0.2265	0.2362	0.2442	0.2503	0.2547	0.2573	0.2581	9	
	4	0.0000	0.0001	0.0003	0.0009	0.0021	0.0039	0.0067	0.0104	0.0153	0.0213	0.0285	0.0369	0.0464	0.0569	0.0683	0.0804	0.0931	0.1062	0.1195	0.1329	0.1460	0.1589	0.1712	0.1828	0.1936	8	
	5	0.0000	0.0000	0.0000	0.0001	0.0002	0.0004	0.0008	0.0014	0.0024	0.0038	0.0056	0.0081	0.0111	0.0148	0.0193	0.0245	0.0305	0.0373	0.0449	0.0532	0.0621	0.0717	0.0818	0.0924	0.1032	7	
	6	0.0000	0.0000	0.0000	0.0000	0.0000	0.0000	0.0001	0.0001	0.0003	0.0005	0.0008	0.0013	0.0019	0.0028	0.0040	0.0054	0.0073	0.0096	0.0123	0.0155	0.0193	0.0236	0.0285	0.0340	0.0401	6	
	7	0.0000	0.0000	0.0000	0.0000	0.0000	0.0000	0.0000	0.0000	0.0000	0.0000	0.0001	0.0001	0.0002	0.0004	0.0006	0.0009	0.0013	0.0018	0.0025	0.0033	0.0044	0.0057	0.0073	0.0092	0.0115	5	
	8	0.0000	0.0000	0.0000	0.0000	0.0000	0.0000	0.0000	0.0000	0.0000	0.0000	0.0000	0.0000	0.0000	0.0000	0.0001	0.0001	0.0002	0.0002	0.0004	0.0005	0.0007	0.0010	0.0014	0.0018	0.0024	4	
	9	0.0000	0.0000	0.0000	0.0000	0.0000	0.0000	0.0000	0.0000	0.0000	0.0000	0.0000	0.0000	0.0000	0.0000	0.0000	0.0000	0.0000	0.0000	0.0000	0.0001	0.0001	0.0001	0.0002	0.0003	0.0004	3	
	10	0.0000	0.0000	0.0000	0.0000	0.0000	0.0000	0.0000	0.0000	0.0000	0.0000	0.0000	0.0000	0.0000	0.0000	0.0000	0.0000	0.0000	0.0000	0.0000	0.0000	0.0000	0.0000	0.0000	0.0000	0.0000	2	
	11	0.0000	0.0000	0.0000	0.0000	0.0000	0.0000	0.0000	0.0000	0.0000	0.0000	0.0000	0.0000	0.0000	0.0000	0.0000	0.0000	0.0000	0.0000	0.0000	0.0000	0.0000	0.0000	0.0000	0.0000	0.0000	1	
	12	0.0000	0.0000	0.0000	0.0000	0.0000	0.0000	0.0000	0.0000	0.0000	0.0000	0.0000	0.0000	0.0000	0.0000	0.0000	0.0000	0.0000	0.0000	0.0000	0.0000	0.0000	0.0000	0.0000	0.0000	0.0000	0	12
13	0	0.8775	0.7690	0.6730	0.5882	0.5133	0.4474	0.3893	0.3383	0.2935	0.2542	0.2198	0.1898	0.1636	0.1408	0.1209	0.1037	0.0887	0.0758	0.0646	0.0550	0.0467	0.0396	0.0334	0.0282	0.0238	13	
	1	0.1152	0.2040	0.2706	0.3186	0.3512	0.3712	0.3809	0.3824	0.3773	0.3672	0.3532	0.3364	0.3178	0.2979	0.2774	0.2567	0.2362	0.2163	0.1970	0.1787	0.1613	0.1450	0.1299	0.1159	0.1029	12	
	2	0.0070	0.0250	0.0502	0.0797	0.1109	0.1422	0.1720	0.1995	0.2239	0.2448	0.2619	0.2753	0.2849	0.2910	0.2937	0.2934	0.2903	0.2848	0.2773	0.2680	0.2573	0.2455	0.2328	0.2195	0.2059	11	
	3	0.0003	0.0019	0.0057	0.0122	0.0214	0.0333	0.0475	0.0636	0.0812	0.0997	0.1187	0.1376	0.1561	0.1737	0.1900	0.2049	0.2180	0.2293	0.2385	0.2457	0.2508	0.2539	0.2550	0.2542	0.2517	10	
	4	0.0000	0.0001	0.0004	0.0013	0.0028	0.0053	0.0089	0.0138	0.0201	0.0277	0.0367	0.0469	0.0583	0.0707	0.0838	0.0976	0.1116	0.1258	0.1399	0.1535	0.1667	0.1790	0.1904	0.2007	0.2097	9	
	5	0.0000	0.0000	0.0000	0.0001	0.0003	0.0006	0.0012	0.0022	0.0036	0.0055	0.0082	0.0115	0.0157	0.0207	0.0266	0.0335	0.0412	0.0497	0.0591	0.0691	0.0797	0.0909	0.1024	0.1141	0.1258	8	
	6	0.0000	0.0000	0.0000	0.0000	0.0000	0.0001	0.0001	0.0003	0.0005	0.0008	0.0013	0.0021	0.0031	0.0045	0.0063	0.0085	0.0112	0.0145	0.0185	0.0230	0.0283	0.0342	0.0408	0.0480	0.0559	7	
	7	0.0000	0.0000	0.0000	0.0000	0.0000	0.0000	0.0000	0.0000	0.0000	0.0001	0.0002	0.0003	0.0005	0.0007	0.0011	0.0016	0.0023	0.0032	0.0043	0.0058	0.0075	0.0096	0.0122	0.0152	0.0186	6	
	8	0.0000	0.0000	0.0000	0.0000	0.0000	0.0000	0.0000	0.0000	0.0000	0.0000	0.0000	0.0000	0.0001	0.0001	0.0001	0.0002	0.0004	0.0005	0.0008	0.0011	0.0015	0.0020	0.0027	0.0036	0.0047	5	
	9	0.0000	0.0000	0.0000	0.0000	0.0000	0.0000	0.0000	0.0000	0.0000	0.0000	0.0000	0.0000	0.0000	0.0000	0.0000	0.0000	0.0000	0.0001	0.0001	0.0001	0.0002	0.0003	0.0005	0.0006	0.0009	4	
	10	0.0000	0.0000	0.0000	0.0000	0.0000	0.0000	0.0000	0.0000	0.0000	0.0000	0.0000	0.0000	0.0000	0.0000	0.0000	0.0000	0.0000	0.0000	0.0000	0.0000	0.0000	0.0000	0.0001	0.0001	0.0001	3	
	11	0.0000	0.0000	0.0000	0.0000	0.0000	0.0000	0.0000	0.0000	0.0000	0.0000	0.0000	0.0000	0.0000	0.0000	0.0000	0.0000	0.0000	0.0000	0.0000	0.0000	0.0000	0.0000	0.0000	0.0000	0.0000	2	
	12	0.0000	0.0000	0.0000	0.0000	0.0000	0.0000	0.0000	0.0000	0.0000	0.0000	0.0000	0.0000	0.0000	0.0000	0.0000	0.0000	0.0000	0.0000	0.0000	0.0000	0.0000	0.0000	0.0000	0.0000	0.0000	1	
	13	0.0000	0.0000	0.0000	0.0000	0.0000	0.0000	0.0000	0.0000	0.0000	0.0000	0.0000	0.0000	0.0000	0.0000	0.0000	0.0000	0.0000	0.0000	0.0000	0.0000	0.0000	0.0000	0.0000	0.0000	0.0000	0	13
14	0	0.8687	0.7536	0.6528	0.5647	0.4877	0.4205	0.3620	0.3112	0.2670	0.2288	0.1956	0.1670	0.1423	0.1211	0.1028	0.0871	0.0736	0.0621	0.0523	0.0440	0.0369	0.0309	0.0258	0.0214	0.0178	14	
	1	0.1229	0.2153	0.2827	0.3294	0.3593	0.3758	0.3815	0.3788	0.3698	0.3559	0.3385	0.3188	0.2977	0.2759	0.2539	0.2322	0.2112	0.1910	0.1719	0.1539	0.1372	0.1218	0.1077	0.0948	0.0832	13	
	2	0.0081	0.0286	0.0568	0.0892	0.1229	0.1559	0.1867	0.2141	0.2377	0.2570	0.2720	0.2826	0.2892	0.2919	0.2912	0.2875	0.2811	0.2725	0.2620	0.2501	0.2371	0.2234	0.2091	0.1946	0.1802	12	
	3	0.0003	0.0023	0.0070	0.0149	0.0259	0.0398	0.0562	0.0745	0.0940	0.1142	0.1345	0.1542	0.1728	0.1901	0.2056	0.2190	0.2303	0.2393	0.2459	0.2501	0.2521	0.2520	0.2499	0.2459	0.2402	11	
	4	0.0000	0.0001	0.0006	0.0017	0.0037	0.0070	0.0116	0.0178	0.0256	0.0349	0.0457	0.0578	0.0710	0.0851	0.0998	0.1147	0.1297	0.1444	0.1586	0.1720	0.1843	0.1955	0.2052	0.2135	0.2202	10	
	5	0.0000	0.0000	0.0000	0.0001	0.0004	0.0009	0.0018	0.0031	0.0051	0.0078	0.0113	0.0158	0.0212	0.0277	0.0352	0.0437	0.0531	0.0634	0.0744	0.0860	0.0980	0.1103	0.1226	0.1348	0.1468	9	
	6	0.0000	0.0000	0.0000	0.0000	0.0000	0.0001	0.0002	0.0004	0.0008	0.0013	0.0021	0.0032	0.0048	0.0068	0.0093	0.0125	0.0163	0.0209	0.0262	0.0322	0.0391	0.0466	0.0549	0.0639	0.0734	8	
	7	0.0000	0.0000	0.0000	0.0000	0.0000	0.0000	0.0000	0.0000	0.0001	0.0002	0.0003	0.0005	0.0008	0.0013	0.0019	0.0027	0.0038	0.0052	0.0070	0.0092	0.0119	0.0150	0.0188	0.0231	0.0280	7	
	8	0.0000	0.0000	0.0000	0.0000	0.0000	0.0000	0.0000	0.0000	0.0000	0.0000	0.0000	0.0001	0.0001	0.0002	0.0003	0.0005	0.0007	0.0010	0.0014	0.0020	0.0028	0.0037	0.0049	0.0064	0.0082	6	
	9	0.0000	0.0000	0.0000	0.0000	0.0000	0.0000	0.0000	0.0000	0.0000	0.0000	0.0000	0.0000	0.0000	0.0000	0.0000	0.0001	0.0001	0.0001	0.0002	0.0003	0.0005	0.0007	0.0010	0.0013	0.0018	5	
	10	0.0000	0.0000	0.0000	0.0000	0.0000	0.0000	0.0000	0.0000	0.0000	0.0000	0.0000	0.0000	0.0000	0.0000	0.0000	0.0000	0.0000	0.0000	0.0000	0.0000	0.0001	0.0001	0.0001	0.0002	0.0003	4	
	11	0.0000	0.0000	0.0000	0.0000	0.0000	0.0000	0.0000	0.0000	0.0000	0.0000	0.0000	0.0000	0.0000	0.0000	0.0000	0.0000	0.0000	0.0000	0.0000	0.0000	0.0000	0.0000	0.0000	0.0000	0.0000	3	
	12	0.0000	0.0000	0.0000	0.0000	0.0000	0.0000	0.0000	0.0000	0.0000	0.0000	0.0000	0.0000	0.0000	0.0000	0.0000	0.0000	0.0000	0.0000	0.0000	0.0000	0.0000	0.0000	0.0000	0.0000	0.0000	2	
	13	0.0000	0.0000	0.0000	0.0000	0.0000	0.0000	0.0000	0.0000	0.0000	0.0000	0.0000	0.0000	0.0000	0.0000	0.0000	0.0000	0.0000	0.0000	0.0000	0.0000	0.0000	0.0000	0.0000	0.0000	0.0000	1	
	14	0.0000	0.0000	0.0000	0.0000	0.0000	0.0000	0.0000	0.0000	0.0000	0.0000	0.0000	0.0000	0.0000	0.0000	0.0000	0.0000	0.0000	0.0000	0.0000	0.0000	0.0000	0.0000	0.0000	0.0000	0.0000	0	14
n	X	0.99	0.98	0.97	0.96	0.95	0.94	0.93	0.92	0.91	0.90	0.89	0.88	0.87	0.86	0.85	0.84	0.83	0.82	0.81	0.80	0.79	0.78	0.77	0.76	0.75	X	n
		P																										

Table E.7 (Continued)

n	X	0.26	0.27	0.28	0.29	0.30	0.31	0.32	0.33	0.34	0.35	0.36	0.37	0.38	0.39	0.40	0.41	0.42	0.43	0.44	0.45	0.46	0.47	0.48	0.49	0.50	X	n
10	0	0.0492	0.0430	0.0374	0.0326	0.0282	0.0245	0.0211	0.0182	0.0157	0.0135	0.0115	0.0098	0.0084	0.0071	0.0060	0.0051	0.0043	0.0036	0.0030	0.0025	0.0021	0.0017	0.0014	0.0012	0.0010	10	
	1	0.1730	0.1590	0.1456	0.1330	0.1211	0.1099	0.0995	0.0898	0.0808	0.0725	0.0649	0.0578	0.0514	0.0456	0.0403	0.0355	0.0312	0.0273	0.0238	0.0207	0.0180	0.0155	0.0133	0.0114	0.0098	9	
	2	0.2735	0.2646	0.2548	0.2444	0.2335	0.2222	0.2107	0.1990	0.1873	0.1757	0.1642	0.1529	0.1419	0.1312	0.1209	0.1111	0.1017	0.0927	0.0843	0.0763	0.0688	0.0619	0.0554	0.0494	0.0439	8	
	3	0.2563	0.2609	0.2642	0.2662	0.2668	0.2662	0.2644	0.2614	0.2573	0.2522	0.2462	0.2394	0.2319	0.2237	0.2150	0.2058	0.1963	0.1865	0.1765	0.1665	0.1564	0.1464	0.1364	0.1267	0.1172	7	
	4	0.1576	0.1689	0.1798	0.1903	0.2001	0.2093	0.2177	0.2253	0.2320	0.2377	0.2424	0.2461	0.2487	0.2503	0.2508	0.2503	0.2488	0.2462	0.2427	0.2384	0.2331	0.2271	0.2204	0.2130	0.2051	6	
	5	0.0664	0.0750	0.0839	0.0933	0.1029	0.1128	0.1229	0.1332	0.1434	0.1536	0.1636	0.1734	0.1829	0.1920	0.2007	0.2087	0.2162	0.2229	0.2289	0.2340	0.2383	0.2417	0.2441	0.2456	0.2461	5	
	6	0.0195	0.0231	0.0272	0.0317	0.0368	0.0422	0.0482	0.0547	0.0616	0.0689	0.0767	0.0849	0.0934	0.1023	0.1115	0.1209	0.1304	0.1401	0.1499	0.1596	0.1692	0.1786	0.1878	0.1966	0.2051	4	
	7	0.0039	0.0049	0.0060	0.0074	0.0090	0.0108	0.0130	0.0154	0.0181	0.0212	0.0247	0.0285	0.0327	0.0374	0.0425	0.0480	0.0540	0.0604	0.0673	0.0746	0.0824	0.0905	0.0991	0.1080	0.1172	3	
	8	0.0005	0.0007	0.0009	0.0011	0.0014	0.0018	0.0023	0.0028	0.0035	0.0043	0.0052	0.0063	0.0075	0.0090	0.0106	0.0125	0.0147	0.0171	0.0198	0.0229	0.0263	0.0301	0.0343	0.0389	0.0439	2	
	9	0.0000	0.0001	0.0001	0.0001	0.0001	0.0002	0.0002	0.0003	0.0004	0.0005	0.0006	0.0008	0.0010	0.0013	0.0016	0.0019	0.0024	0.0029	0.0035	0.0042	0.0050	0.0059	0.0070	0.0083	0.0098	1	
	10	0.0000	0.0000	0.0000	0.0000	0.0000	0.0000	0.0000	0.0000	0.0000	0.0000	0.0000	0.0000	0.0001	0.0001	0.0001	0.0001	0.0002	0.0002	0.0003	0.0003	0.0004	0.0005	0.0006	0.0008	0.0010	0	10
11	0	0.0364	0.0314	0.0270	0.0231	0.0198	0.0169	0.0144	0.0122	0.0104	0.0088	0.0074	0.0062	0.0052	0.0044	0.0036	0.0030	0.0025	0.0021	0.0017	0.0014	0.0011	0.0009	0.0008	0.0006	0.0005	11	
	1	0.1408	0.1276	0.1153	0.1038	0.0932	0.0834	0.0744	0.0662	0.0587	0.0518	0.0457	0.0401	0.0351	0.0306	0.0266	0.0231	0.0199	0.0171	0.0147	0.0125	0.0107	0.0090	0.0076	0.0064	0.0054	10	
	2	0.2474	0.2360	0.2242	0.2121	0.1998	0.1874	0.1751	0.1630	0.1511	0.1395	0.1284	0.1177	0.1075	0.0978	0.0887	0.0801	0.0721	0.0646	0.0577	0.0513	0.0454	0.0401	0.0352	0.0308	0.0269	9	
	3	0.2608	0.2619	0.2616	0.2599	0.2568	0.2526	0.2472	0.2408	0.2335	0.2254	0.2167	0.2074	0.1977	0.1876	0.1774	0.1670	0.1566	0.1462	0.1359	0.1259	0.1161	0.1067	0.0976	0.0888	0.0806	8	
	4	0.1832	0.1937	0.2035	0.2123	0.2201	0.2269	0.2326	0.2372	0.2406	0.2428	0.2436	0.2436	0.2423	0.2399	0.2365	0.2321	0.2267	0.2206	0.2136	0.2060	0.1978	0.1892	0.1801	0.1707	0.1611	7	
	5	0.0901	0.1003	0.1108	0.1214	0.1321	0.1427	0.1533	0.1636	0.1735	0.1830	0.1920	0.2003	0.2079	0.2148	0.2207	0.2258	0.2299	0.2329	0.2350	0.2360	0.2359	0.2348	0.2327	0.2296	0.2256	6	
	6	0.0317	0.0371	0.0431	0.0496	0.0566	0.0641	0.0721	0.0806	0.0894	0.0985	0.1080	0.1176	0.1274	0.1373	0.1471	0.1569	0.1664	0.1757	0.1846	0.1931	0.2010	0.2083	0.2148	0.2206	0.2256	5	
	7	0.0079	0.0098	0.0120	0.0145	0.0173	0.0206	0.0242	0.0283	0.0329	0.0379	0.0434	0.0494	0.0558	0.0627	0.0701	0.0779	0.0861	0.0947	0.1036	0.1128	0.1223	0.1319	0.1416	0.1514	0.1611	4	
	8	0.0014	0.0018	0.0023	0.0030	0.0037	0.0046	0.0057	0.0070	0.0085	0.0102	0.0122	0.0145	0.0171	0.0200	0.0234	0.0271	0.0312	0.0357	0.0407	0.0462	0.0521	0.0585	0.0654	0.0727	0.0806	3	
	9	0.0002	0.0002	0.0003	0.0004	0.0005	0.0007	0.0009	0.0011	0.0015	0.0018	0.0023	0.0028	0.0035	0.0043	0.0052	0.0063	0.0075	0.0090	0.0107	0.0126	0.0148	0.0173	0.0201	0.0233	0.0269	2	
	10	0.0000	0.0000	0.0000	0.0000	0.0000	0.0001	0.0001	0.0001	0.0001	0.0002	0.0003	0.0003	0.0004	0.0005	0.0007	0.0009	0.0011	0.0014	0.0017	0.0021	0.0025	0.0031	0.0037	0.0045	0.0054	1	
	11	0.0000	0.0000	0.0000	0.0000	0.0000	0.0000	0.0000	0.0000	0.0000	0.0000	0.0000	0.0000	0.0000	0.0000	0.0000	0.0001	0.0001	0.0001	0.0001	0.0002	0.0002	0.0002	0.0003	0.0004	0.0005	0	11
12	0	0.0270	0.0229	0.0194	0.0164	0.0138	0.0116	0.0098	0.0082	0.0068	0.0057	0.0047	0.0039	0.0032	0.0027	0.0022	0.0018	0.0014	0.0012	0.0010	0.0008	0.0006	0.0005	0.0004	0.0003	0.0002	12	
	1	0.1137	0.1016	0.0906	0.0804	0.0712	0.0628	0.0552	0.0484	0.0422	0.0368	0.0319	0.0276	0.0237	0.0204	0.0174	0.0148	0.0126	0.0106	0.0090	0.0075	0.0063	0.0052	0.0043	0.0036	0.0029	11	
	2	0.2197	0.2068	0.1937	0.1807	0.1678	0.1552	0.1429	0.1310	0.1197	0.1088	0.0986	0.0890	0.0800	0.0716	0.0639	0.0567	0.0502	0.0442	0.0388	0.0339	0.0294	0.0255	0.0220	0.0189	0.0161	10	
	3	0.2573	0.2549	0.2511	0.2460	0.2397	0.2324	0.2241	0.2151	0.2055	0.1954	0.1849	0.1742	0.1634	0.1526	0.1419	0.1314	0.1211	0.1111	0.1015	0.0923	0.0836	0.0754	0.0676	0.0604	0.0537	9	
	4	0.2034	0.2122	0.2197	0.2261	0.2311	0.2349	0.2373	0.2384	0.2382	0.2367	0.2340	0.2302	0.2254	0.2195	0.2128	0.2054	0.1973	0.1886	0.1794	0.1700	0.1602	0.1504	0.1405	0.1306	0.1208	8	
	5	0.1143	0.1255	0.1367	0.1477	0.1585	0.1688	0.1787	0.1879	0.1963	0.2039	0.2106	0.2163	0.2210	0.2246	0.2270	0.2284	0.2285	0.2276	0.2256	0.2225	0.2184	0.2134	0.2075	0.2008	0.1934	7	
	6	0.0469	0.0542	0.0620	0.0704	0.0792	0.0885	0.0981	0.1079	0.1180	0.1281	0.1382	0.1482	0.1580	0.1675	0.1766	0.1851	0.1931	0.2003	0.2068	0.2124	0.2171	0.2208	0.2234	0.2250	0.2256	6	
	7	0.0141	0.0172	0.0207	0.0246	0.0291	0.0341	0.0396	0.0456	0.0521	0.0591	0.0666	0.0746	0.0830	0.0918	0.1009	0.1103	0.1198	0.1295	0.1393	0.1489	0.1585	0.1678	0.1768	0.1853	0.1934	5	
	8	0.0031	0.0040	0.0050	0.0063	0.0078	0.0096	0.0116	0.0140	0.0168	0.0199	0.0234	0.0274	0.0318	0.0367	0.0420	0.0479	0.0542	0.0611	0.0684	0.0762	0.0844	0.0930	0.1020	0.1113	0.1208	4	
	9	0.0005	0.0007	0.0009	0.0011	0.0015	0.0019	0.0024	0.0031	0.0038	0.0048	0.0059	0.0071	0.0087	0.0104	0.0125	0.0148	0.0175	0.0205	0.0239	0.0277	0.0319	0.0367	0.0418	0.0475	0.0537	3	
	10	0.0001	0.0001	0.0001	0.0001	0.0002	0.0003	0.0003	0.0005	0.0006	0.0008	0.0010	0.0013	0.0016	0.0020	0.0025	0.0031	0.0038	0.0046	0.0056	0.0068	0.0082	0.0098	0.0116	0.0137	0.0161	2	
	11	0.0000	0.0000	0.0000	0.0000	0.0000	0.0000	0.0000	0.0000	0.0001	0.0001	0.0001	0.0001	0.0002	0.0002	0.0003	0.0004	0.0005	0.0006	0.0008	0.0010	0.0013	0.0016	0.0019	0.0024	0.0029	1	
	12	0.0000	0.0000	0.0000	0.0000	0.0000	0.0000	0.0000	0.0000	0.0000	0.0000	0.0000	0.0000	0.0000	0.0000	0.0000	0.0000	0.0000	0.0000	0.0001	0.0001	0.0001	0.0001	0.0001	0.0002	0.0002	0	12
13	0	0.0200	0.0167	0.0140	0.0117	0.0097	0.0080	0.0066	0.0055	0.0045	0.0037	0.0030	0.0025	0.0020	0.0016	0.0013	0.0010	0.0008	0.0007	0.0005	0.0004	0.0003	0.0003	0.0002	0.0002	0.0001	13	
	1	0.0911	0.0804	0.0706	0.0619	0.0540	0.0469	0.0407	0.0351	0.0302	0.0259	0.0221	0.0188	0.0159	0.0135	0.0113	0.0095	0.0079	0.0066	0.0054	0.0045	0.0037	0.0030	0.0024	0.0020	0.0016	12	
	2	0.1921	0.1784	0.1648	0.1516	0.1388	0.1265	0.1148	0.1037	0.0933	0.0836	0.0746	0.0663	0.0586	0.0516	0.0453	0.0395	0.0344	0.0298	0.0256	0.0220	0.0188	0.0160	0.0135	0.0114	0.0095	11	
	3	0.2475	0.2419	0.2351	0.2271	0.2181	0.2084	0.1981	0.1874	0.1763	0.1651	0.1538	0.1427	0.1317	0.1210	0.1107	0.1007	0.0913	0.0823	0.0739	0.0660	0.0587	0.0519	0.0457	0.0401	0.0349	10	
	4	0.2174	0.2237	0.2285	0.2319	0.2337	0.2341	0.2331	0.2307	0.2270	0.2222	0.2163	0.2095	0.2018	0.1934	0.1845	0.1750	0.1653	0.1553	0.1451	0.1350	0.1250	0.1151	0.1055	0.0962	0.0873	9	
	5	0.1375	0.1489	0.1600	0.1705	0.1803	0.1893	0.1974	0.2045	0.2105	0.2154	0.2190	0.2215	0.2227	0.2226	0.2214	0.2189	0.2154	0.2108	0.2053	0.1989	0.1917	0.1838	0.1753	0.1664	0.1571	8	
	6	0.0644	0.0734	0.0829	0.0928	0.1030	0.1134	0.1239	0.1343	0.1446	0.1546	0.1643	0.1734	0.1820	0.1898	0.1968	0.2029	0.2080	0.2121	0.2151	0.2169	0.2177	0.2173	0.2158	0.2131	0.2095	7	
	7	0.0226	0.0272	0.0323	0.0379	0.0442	0.0509	0.0583	0.0662	0.0745	0.0833	0.0924	0.1019	0.1115	0.1213	0.1312	0.1410	0.1506	0.1600	0.1690	0.1775	0.1854	0.1927	0.1992	0.2048	0.2095	6	
	8	0.0060	0.0075	0.0094	0.0116	0.0142	0.0172	0.0206	0.0244	0.0288	0.0336	0.0390	0.0449	0.0513	0.0582	0.0656	0.0735	0.0818	0.0905	0.0996	0.1089	0.1185	0.1282	0.1379	0.1476	0.1571	5	
	9	0.0012	0.0015	0.0020	0.0026	0.0034	0.0043	0.0054	0.0067	0.0082	0.0101	0.0122	0.0146	0.0175	0.0207	0.0243	0.0284	0.0329	0.0379	0.0435	0.0495	0.0561	0.0631	0.0707	0.0788	0.0873	4	
	10	0.0002	0.0002	0.0003	0.0004	0.0006	0.0008	0.0010	0.0013	0.0017	0.0022	0.0027	0.0034	0.0043	0.0053	0.0065	0.0079	0.0095	0.0114	0.0137	0.0162	0.0191	0.0224	0.0261	0.0303	0.0349	3	
	11	0.0000	0.0000	0.0000	0.0000	0.0001	0.0001	0.0001	0.0002	0.0002	0.0003	0.0004	0.0006	0.0007	0.0009	0.0012	0.0015	0.0019	0.0024	0.0029	0.0036	0.0044	0.0054	0.0066	0.0079	0.0095	2	
	12	0.0000	0.0000	0.0000	0.0000	0.0000	0.0000	0.0000	0.0000	0.0000	0.0000	0.0000	0.0001	0.0001	0.0001	0.0001	0.0002	0.0002	0.0003	0.0004	0.0005	0.0006	0.0008	0.0010	0.0013	0.0016	1	
	13	0.0000	0.0000	0.0000	0.0000	0.0000	0.0000	0.0000	0.0000	0.0000	0.0000	0.0000	0.0000	0.0000	0.0000	0.0000	0.0000	0.0000	0.0000	0.0000	0.0000	0.0000	0.0001	0.0001	0.0001	0.0001	0	13
14	0	0.0148	0.0122	0.0101	0.0083	0.0068	0.0055	0.0045	0.0037	0.0030	0.0024	0.0019	0.0016	0.0012	0.0010	0.0008	0.0006	0.0005	0.0004	0.0003	0.0002	0.0002	0.0001	0.0001	0.0001	0.0001	14	
	1	0.0726	0.0632	0.0548	0.0473	0.0407	0.0349	0.0298	0.0253	0.0215	0.0181	0.0152	0.0128	0.0106	0.0088	0.0073	0.0060	0.0049	0.0040	0.0033	0.0027	0.0021	0.0017	0.0014	0.0011	0.0009	13	
	2	0.1659	0.1519	0.1385	0.1256	0.1134	0.1018	0.0911	0.0811	0.0719	0.0634	0.0557	0.0487	0.0424	0.0367	0.0317	0.0272	0.0233	0.0198	0.0168	0.0141	0.0118	0.0099	0.0082	0.0068	0.0056	12	
	3	0.2331	0.2248	0.2154	0.2052	0.1943	0.1830	0.1715	0.1598	0.1481	0.1366	0.1253	0.1144	0.1039	0.0940	0.0845	0.0757	0.0674	0.0597	0.0527	0.0462	0.0403	0.0350	0.0303	0.0260	0.0222	11	
	4	0.2252	0.2286	0.2304	0.2305	0.2290	0.2261	0.2219	0.2164	0.2098	0.2022	0.1938	0.1848	0.1752	0.1652	0.1549	0.1446	0.1342	0.1239	0.1138	0.1040	0.0945	0.0854	0.0768	0.0687	0.0611	10	
	5	0.1583	0.1691	0.1792	0.1883	0.1963	0.2032	0.2088	0.2132	0.2161	0.2178	0.2181	0.2170	0.2147	0.2112	0.2066	0.2009	0.1943	0.1869	0.1788	0.1701	0.1610	0.1515	0.1418	0.1320	0.1222	9	
	6	0.0834	0.0938	0.1045	0.1153	0.1262	0.1369	0.1474	0.1575	0.1670	0.1759	0.1840	0.1912	0.1974	0.2026	0.2066	0.2094	0.2111	0.2115	0.2108	0.2088	0.2057	0.2015	0.1963	0.1902	0.1833	8	
	7	0.0335	0.0397	0.0464	0.0538	0.0618	0.0703	0.0793	0.0886	0.0983	0.1082	0.1183	0.1283	0.1383	0.1480	0.1574	0.1663	0.1747	0.1824	0.1892	0.1952	0.2003	0.2043	0.2071	0.2089	0.2095	7	
	8	0.0103	0.0128	0.0158	0.0192	0.0232	0.0276	0.0326	0.0382	0.0443	0.0510	0.0582	0.0659	0.0742	0.0828	0.0918	0.1011	0.1107	0.1204	0.1301	0.1398	0.1493	0.1585	0.1673	0.1756	0.1833	6	
	9	0.0024	0.0032	0.0041	0.0052	0.0066	0.0083	0.0102	0.0125	0.0152	0.0183	0.0218	0.0258	0.0303	0.0353	0.0408	0.0469	0.0534	0.0605	0.0682	0.0762	0.0848	0.0937	0.1030	0.1125	0.1222	5	
	10	0.0004	0.0006	0.0008	0.0011	0.0014	0.0019	0.0024	0.0031	0.0039	0.0049	0.0061	0.0076	0.0093	0.0113	0.0136	0.0163	0.0193	0.0228	0.0268	0.0312	0.0361	0.0415	0.0475	0.0540	0.0611	4	
	11	0.0001	0.0001	0.0001	0.0002	0.0002	0.0003	0.0004	0.0006	0.0007	0.0010	0.0013	0.0016	0.0021	0.0026	0.0033	0.0041	0.0051	0.0063	0.0076	0.0093	0.0112	0.0134	0.0160	0.0189	0.0222	3	
	12	0.0000	0.0000	0.0000	0.0000	0.0000	0.0000	0.0000	0.0001	0.0001	0.0001	0.0002	0.0002	0.0003	0.0004	0.0005	0.0007	0.0009	0.0012	0.0015	0.0019	0.0024	0.0030	0.0037	0.0045	0.0056	2	
	13	0.0000	0.0000	0.0000	0.0000	0.0000	0.0000	0.0000	0.0000	0.0000	0.0000	0.0000	0.0000	0.0000	0.0000	0.0001	0.0001	0.0001	0.0001	0.0002	0.0002	0.0003	0.0004	0.0005	0.0007	0.0009	1	
	14	0.0000	0.0000	0.0000	0.0000	0.0000	0.0000	0.0000	0.0000	0.0000	0.0000	0.0000	0.0000	0.0000	0.0000	0.0000	0.0000	0.0000	0.0000	0.0000	0.0000	0.0000	0.0000	0.0000	0.0000	0.0001	0	14
n	X	0.74	0.73	0.72	0.71	0.70	0.69	0.68	0.67	0.66	0.65	0.64	0.63	0.62	0.61	0.60	0.59	0.58	0.57	0.56	0.55	0.54	0.53	0.52	0.51	0.50	X	n
		P																										

Table E.7 (Continued)

n	X	P 0.01	0.02	0.03	0.04	0.05	0.06	0.07	0.08	0.09	0.10	0.11	0.12	0.13	0.14	0.15	0.16	0.17	0.18	0.19	0.20	0.21	0.22	0.23	0.24	0.25	X	n
18	0	0.8345	0.6951	0.5780	0.4796	0.3972	0.3283	0.2708	0.2229	0.1831	0.1501	0.1227	0.1002	0.0815	0.0662	0.0536	0.0434	0.0349	0.0281	0.0225	0.0180	0.0144	0.0114	0.0091	0.0072	0.0056	18	
	1	0.1517	0.2554	0.3217	0.3597	0.3763	0.3772	0.3669	0.3489	0.3260	0.3002	0.2731	0.2458	0.2193	0.1940	0.1704	0.1486	0.1288	0.1110	0.0951	0.0811	0.0687	0.0580	0.0487	0.0407	0.0338	17	
	2	0.0130	0.0443	0.0846	0.1274	0.1683	0.2047	0.2348	0.2579	0.2741	0.2835	0.2869	0.2850	0.2785	0.2685	0.2556	0.2407	0.2243	0.2071	0.1897	0.1723	0.1553	0.1390	0.1236	0.1092	0.0958	16	
	3	0.0007	0.0048	0.0140	0.0283	0.0473	0.0697	0.0942	0.1196	0.1446	0.1680	0.1891	0.2072	0.2220	0.2331	0.2406	0.2445	0.2450	0.2425	0.2373	0.2297	0.2202	0.2091	0.1969	0.1839	0.1704	15	
	4	0.0000	0.0004	0.0016	0.0044	0.0093	0.0167	0.0266	0.0390	0.0536	0.0700	0.0877	0.1060	0.1244	0.1423	0.1592	0.1746	0.1882	0.1996	0.2087	0.2153	0.2195	0.2212	0.2205	0.2177	0.2130	14	
	5	0.0000	0.0000	0.0001	0.0005	0.0014	0.0030	0.0056	0.0095	0.0148	0.0218	0.0303	0.0405	0.0520	0.0649	0.0787	0.0931	0.1079	0.1227	0.1371	0.1507	0.1634	0.1747	0.1845	0.1925	0.1988	13	
	6	0.0000	0.0000	0.0000	0.0000	0.0002	0.0004	0.0009	0.0018	0.0032	0.0052	0.0081	0.0120	0.0168	0.0229	0.0301	0.0384	0.0479	0.0584	0.0697	0.0816	0.0941	0.1067	0.1194	0.1317	0.1436	12	
	7	0.0000	0.0000	0.0000	0.0000	0.0000	0.0000	0.0001	0.0003	0.0005	0.0010	0.0017	0.0028	0.0043	0.0064	0.0091	0.0126	0.0168	0.0220	0.0280	0.0350	0.0429	0.0516	0.0611	0.0713	0.0820	11	
	8	0.0000	0.0000	0.0000	0.0000	0.0000	0.0000	0.0000	0.0000	0.0001	0.0002	0.0003	0.0005	0.0009	0.0014	0.0022	0.0033	0.0047	0.0066	0.0090	0.0120	0.0157	0.0200	0.0251	0.0310	0.0376	10	
	9	0.0000	0.0000	0.0000	0.0000	0.0000	0.0000	0.0000	0.0000	0.0000	0.0000	0.0000	0.0001	0.0001	0.0003	0.0004	0.0007	0.0011	0.0016	0.0024	0.0033	0.0046	0.0063	0.0083	0.0109	0.0139	9	
	10	0.0000	0.0000	0.0000	0.0000	0.0000	0.0000	0.0000	0.0000	0.0000	0.0000	0.0000	0.0000	0.0000	0.0000	0.0001	0.0001	0.0002	0.0003	0.0005	0.0008	0.0011	0.0016	0.0022	0.0031	0.0042	8	
	11	0.0000	0.0000	0.0000	0.0000	0.0000	0.0000	0.0000	0.0000	0.0000	0.0000	0.0000	0.0000	0.0000	0.0000	0.0000	0.0000	0.0000	0.0001	0.0001	0.0001	0.0002	0.0003	0.0005	0.0007	0.0010	7	
	12	0.0000	0.0000	0.0000	0.0000	0.0000	0.0000	0.0000	0.0000	0.0000	0.0000	0.0000	0.0000	0.0000	0.0000	0.0000	0.0000	0.0000	0.0000	0.0000	0.0000	0.0000	0.0001	0.0001	0.0001	0.0002	6	
	13	0.0000	0.0000	0.0000	0.0000	0.0000	0.0000	0.0000	0.0000	0.0000	0.0000	0.0000	0.0000	0.0000	0.0000	0.0000	0.0000	0.0000	0.0000	0.0000	0.0000	0.0000	0.0000	0.0000	0.0000	0.0000	5	
	14	0.0000	0.0000	0.0000	0.0000	0.0000	0.0000	0.0000	0.0000	0.0000	0.0000	0.0000	0.0000	0.0000	0.0000	0.0000	0.0000	0.0000	0.0000	0.0000	0.0000	0.0000	0.0000	0.0000	0.0000	0.0000	4	
	15	0.0000	0.0000	0.0000	0.0000	0.0000	0.0000	0.0000	0.0000	0.0000	0.0000	0.0000	0.0000	0.0000	0.0000	0.0000	0.0000	0.0000	0.0000	0.0000	0.0000	0.0000	0.0000	0.0000	0.0000	0.0000	3	
	16	0.0000	0.0000	0.0000	0.0000	0.0000	0.0000	0.0000	0.0000	0.0000	0.0000	0.0000	0.0000	0.0000	0.0000	0.0000	0.0000	0.0000	0.0000	0.0000	0.0000	0.0000	0.0000	0.0000	0.0000	0.0000	2	
	17	0.0000	0.0000	0.0000	0.0000	0.0000	0.0000	0.0000	0.0000	0.0000	0.0000	0.0000	0.0000	0.0000	0.0000	0.0000	0.0000	0.0000	0.0000	0.0000	0.0000	0.0000	0.0000	0.0000	0.0000	0.0000	1	
	18	0.0000	0.0000	0.0000	0.0000	0.0000	0.0000	0.0000	0.0000	0.0000	0.0000	0.0000	0.0000	0.0000	0.0000	0.0000	0.0000	0.0000	0.0000	0.0000	0.0000	0.0000	0.0000	0.0000	0.0000	0.0000	0	18
19	0	0.8262	0.6812	0.5606	0.4604	0.3774	0.3086	0.2519	0.2051	0.1666	0.1351	0.1092	0.0881	0.0709	0.0569	0.0456	0.0364	0.0290	0.0230	0.0182	0.0144	0.0113	0.0089	0.0070	0.0054	0.0042	19	
	1	0.1586	0.2642	0.3294	0.3645	0.3774	0.3743	0.3602	0.3389	0.3131	0.2852	0.2565	0.2284	0.2014	0.1761	0.1529	0.1318	0.1129	0.0961	0.0813	0.0685	0.0573	0.0477	0.0396	0.0326	0.0268	18	
	2	0.0144	0.0485	0.0917	0.1367	0.1787	0.2150	0.2440	0.2652	0.2787	0.2852	0.2854	0.2803	0.2708	0.2581	0.2428	0.2259	0.2081	0.1898	0.1717	0.1540	0.1371	0.1212	0.1064	0.0927	0.0803	17	
	3	0.0008	0.0056	0.0161	0.0323	0.0533	0.0778	0.1041	0.1307	0.1562	0.1796	0.1999	0.2166	0.2293	0.2381	0.2428	0.2439	0.2415	0.2361	0.2282	0.2182	0.2065	0.1937	0.1800	0.1659	0.1517	16	
	4	0.0000	0.0005	0.0020	0.0054	0.0112	0.0199	0.0313	0.0455	0.0618	0.0798	0.0988	0.1181	0.1371	0.1550	0.1714	0.1858	0.1979	0.2073	0.2141	0.2182	0.2196	0.2185	0.2151	0.2096	0.2023	15	
	5	0.0000	0.0000	0.0002	0.0007	0.0018	0.0038	0.0071	0.0119	0.0183	0.0266	0.0366	0.0483	0.0614	0.0757	0.0907	0.1062	0.1216	0.1365	0.1507	0.1636	0.1751	0.1849	0.1928	0.1986	0.2023	14	
	6	0.0000	0.0000	0.0000	0.0001	0.0002	0.0006	0.0012	0.0024	0.0042	0.0069	0.0106	0.0154	0.0214	0.0288	0.0374	0.0472	0.0581	0.0699	0.0825	0.0955	0.1086	0.1217	0.1343	0.1463	0.1574	13	
	7	0.0000	0.0000	0.0000	0.0000	0.0000	0.0001	0.0002	0.0004	0.0008	0.0014	0.0024	0.0039	0.0059	0.0087	0.0122	0.0167	0.0221	0.0285	0.0359	0.0443	0.0536	0.0637	0.0745	0.0858	0.0974	12	
	8	0.0000	0.0000	0.0000	0.0000	0.0000	0.0000	0.0000	0.0001	0.0001	0.0002	0.0004	0.0008	0.0013	0.0021	0.0032	0.0048	0.0068	0.0094	0.0126	0.0166	0.0214	0.0270	0.0334	0.0406	0.0487	11	
	9	0.0000	0.0000	0.0000	0.0000	0.0000	0.0000	0.0000	0.0000	0.0000	0.0000	0.0001	0.0001	0.0002	0.0004	0.0007	0.0011	0.0017	0.0025	0.0036	0.0051	0.0069	0.0093	0.0122	0.0157	0.0198	10	
	10	0.0000	0.0000	0.0000	0.0000	0.0000	0.0000	0.0000	0.0000	0.0000	0.0000	0.0000	0.0000	0.0000	0.0001	0.0001	0.0002	0.0003	0.0006	0.0009	0.0013	0.0018	0.0026	0.0036	0.0050	0.0066	9	
	11	0.0000	0.0000	0.0000	0.0000	0.0000	0.0000	0.0000	0.0000	0.0000	0.0000	0.0000	0.0000	0.0000	0.0000	0.0000	0.0000	0.0001	0.0001	0.0002	0.0003	0.0004	0.0006	0.0009	0.0013	0.0018	8	
	12	0.0000	0.0000	0.0000	0.0000	0.0000	0.0000	0.0000	0.0000	0.0000	0.0000	0.0000	0.0000	0.0000	0.0000	0.0000	0.0000	0.0000	0.0000	0.0000	0.0000	0.0001	0.0001	0.0002	0.0003	0.0004	7	
	13	0.0000	0.0000	0.0000	0.0000	0.0000	0.0000	0.0000	0.0000	0.0000	0.0000	0.0000	0.0000	0.0000	0.0000	0.0000	0.0000	0.0000	0.0000	0.0000	0.0000	0.0000	0.0000	0.0000	0.0000	0.0001	6	
	14	0.0000	0.0000	0.0000	0.0000	0.0000	0.0000	0.0000	0.0000	0.0000	0.0000	0.0000	0.0000	0.0000	0.0000	0.0000	0.0000	0.0000	0.0000	0.0000	0.0000	0.0000	0.0000	0.0000	0.0000	0.0000	5	
	15	0.0000	0.0000	0.0000	0.0000	0.0000	0.0000	0.0000	0.0000	0.0000	0.0000	0.0000	0.0000	0.0000	0.0000	0.0000	0.0000	0.0000	0.0000	0.0000	0.0000	0.0000	0.0000	0.0000	0.0000	0.0000	4	
	16	0.0000	0.0000	0.0000	0.0000	0.0000	0.0000	0.0000	0.0000	0.0000	0.0000	0.0000	0.0000	0.0000	0.0000	0.0000	0.0000	0.0000	0.0000	0.0000	0.0000	0.0000	0.0000	0.0000	0.0000	0.0000	3	
	17	0.0000	0.0000	0.0000	0.0000	0.0000	0.0000	0.0000	0.0000	0.0000	0.0000	0.0000	0.0000	0.0000	0.0000	0.0000	0.0000	0.0000	0.0000	0.0000	0.0000	0.0000	0.0000	0.0000	0.0000	0.0000	2	
	18	0.0000	0.0000	0.0000	0.0000	0.0000	0.0000	0.0000	0.0000	0.0000	0.0000	0.0000	0.0000	0.0000	0.0000	0.0000	0.0000	0.0000	0.0000	0.0000	0.0000	0.0000	0.0000	0.0000	0.0000	0.0000	1	
	19	0.0000	0.0000	0.0000	0.0000	0.0000	0.0000	0.0000	0.0000	0.0000	0.0000	0.0000	0.0000	0.0000	0.0000	0.0000	0.0000	0.0000	0.0000	0.0000	0.0000	0.0000	0.0000	0.0000	0.0000	0.0000	0	19
20	0	0.8179	0.6676	0.5438	0.4420	0.3585	0.2901	0.2342	0.1887	0.1516	0.1216	0.0972	0.0776	0.0617	0.0490	0.0388	0.0306	0.0241	0.0189	0.0148	0.0115	0.0090	0.0069	0.0054	0.0041	0.0032	20	
	1	0.1652	0.2725	0.3364	0.3683	0.3774	0.3703	0.3526	0.3282	0.3000	0.2702	0.2403	0.2115	0.1844	0.1595	0.1368	0.1165	0.0986	0.0829	0.0693	0.0576	0.0477	0.0392	0.0321	0.0261	0.0211	19	
	2	0.0159	0.0528	0.0988	0.1458	0.1887	0.2246	0.2521	0.2711	0.2818	0.2852	0.2822	0.2740	0.2618	0.2466	0.2293	0.2109	0.1919	0.1730	0.1545	0.1369	0.1204	0.1050	0.0910	0.0783	0.0669	18	
	3	0.0010	0.0065	0.0183	0.0364	0.0596	0.0860	0.1139	0.1414	0.1672	0.1901	0.2093	0.2242	0.2347	0.2409	0.2428	0.2410	0.2358	0.2278	0.2175	0.2054	0.1920	0.1777	0.1631	0.1484	0.1339	17	
	4	0.0000	0.0006	0.0024	0.0065	0.0133	0.0233	0.0364	0.0523	0.0703	0.0898	0.1099	0.1299	0.1491	0.1666	0.1821	0.1951	0.2053	0.2125	0.2168	0.2182	0.2169	0.2131	0.2070	0.1991	0.1897	16	
	5	0.0000	0.0000	0.0002	0.0009	0.0022	0.0048	0.0088	0.0145	0.0222	0.0319	0.0435	0.0567	0.0713	0.0868	0.1028	0.1189	0.1345	0.1493	0.1627	0.1746	0.1845	0.1923	0.1979	0.2012	0.2023	15	
	6	0.0000	0.0000	0.0000	0.0001	0.0003	0.0008	0.0017	0.0032	0.0055	0.0089	0.0134	0.0193	0.0266	0.0353	0.0454	0.0566	0.0689	0.0819	0.0954	0.1091	0.1226	0.1356	0.1478	0.1589	0.1686	14	
	7	0.0000	0.0000	0.0000	0.0000	0.0000	0.0001	0.0002	0.0005	0.0011	0.0020	0.0033	0.0053	0.0080	0.0115	0.0160	0.0216	0.0282	0.0360	0.0448	0.0545	0.0652	0.0765	0.0883	0.1003	0.1124	13	
	8	0.0000	0.0000	0.0000	0.0000	0.0000	0.0000	0.0000	0.0001	0.0002	0.0004	0.0007	0.0012	0.0019	0.0030	0.0046	0.0067	0.0094	0.0128	0.0171	0.0222	0.0282	0.0351	0.0429	0.0515	0.0609	12	
	9	0.0000	0.0000	0.0000	0.0000	0.0000	0.0000	0.0000	0.0000	0.0000	0.0001	0.0001	0.0002	0.0004	0.0007	0.0011	0.0017	0.0026	0.0038	0.0053	0.0074	0.0100	0.0132	0.0171	0.0217	0.0271	11	
	10	0.0000	0.0000	0.0000	0.0000	0.0000	0.0000	0.0000	0.0000	0.0000	0.0000	0.0000	0.0000	0.0001	0.0001	0.0002	0.0004	0.0006	0.0009	0.0014	0.0020	0.0029	0.0041	0.0056	0.0075	0.0099	10	
	11	0.0000	0.0000	0.0000	0.0000	0.0000	0.0000	0.0000	0.0000	0.0000	0.0000	0.0000	0.0000	0.0000	0.0000	0.0000	0.0001	0.0001	0.0002	0.0003	0.0005	0.0007	0.0010	0.0015	0.0022	0.0030	9	
	12	0.0000	0.0000	0.0000	0.0000	0.0000	0.0000	0.0000	0.0000	0.0000	0.0000	0.0000	0.0000	0.0000	0.0000	0.0000	0.0000	0.0000	0.0000	0.0001	0.0001	0.0001	0.0002	0.0003	0.0005	0.0008	8	
	13	0.0000	0.0000	0.0000	0.0000	0.0000	0.0000	0.0000	0.0000	0.0000	0.0000	0.0000	0.0000	0.0000	0.0000	0.0000	0.0000	0.0000	0.0000	0.0000	0.0000	0.0000	0.0000	0.0001	0.0001	0.0002	7	
	14	0.0000	0.0000	0.0000	0.0000	0.0000	0.0000	0.0000	0.0000	0.0000	0.0000	0.0000	0.0000	0.0000	0.0000	0.0000	0.0000	0.0000	0.0000	0.0000	0.0000	0.0000	0.0000	0.0000	0.0000	0.0000	6	
	15	0.0000	0.0000	0.0000	0.0000	0.0000	0.0000	0.0000	0.0000	0.0000	0.0000	0.0000	0.0000	0.0000	0.0000	0.0000	0.0000	0.0000	0.0000	0.0000	0.0000	0.0000	0.0000	0.0000	0.0000	0.0000	5	
	16	0.0000	0.0000	0.0000	0.0000	0.0000	0.0000	0.0000	0.0000	0.0000	0.0000	0.0000	0.0000	0.0000	0.0000	0.0000	0.0000	0.0000	0.0000	0.0000	0.0000	0.0000	0.0000	0.0000	0.0000	0.0000	4	
	17	0.0000	0.0000	0.0000	0.0000	0.0000	0.0000	0.0000	0.0000	0.0000	0.0000	0.0000	0.0000	0.0000	0.0000	0.0000	0.0000	0.0000	0.0000	0.0000	0.0000	0.0000	0.0000	0.0000	0.0000	0.0000	3	
	18	0.0000	0.0000	0.0000	0.0000	0.0000	0.0000	0.0000	0.0000	0.0000	0.0000	0.0000	0.0000	0.0000	0.0000	0.0000	0.0000	0.0000	0.0000	0.0000	0.0000	0.0000	0.0000	0.0000	0.0000	0.0000	2	
	19	0.0000	0.0000	0.0000	0.0000	0.0000	0.0000	0.0000	0.0000	0.0000	0.0000	0.0000	0.0000	0.0000	0.0000	0.0000	0.0000	0.0000	0.0000	0.0000	0.0000	0.0000	0.0000	0.0000	0.0000	0.0000	1	
	20	0.0000	0.0000	0.0000	0.0000	0.0000	0.0000	0.0000	0.0000	0.0000	0.0000	0.0000	0.0000	0.0000	0.0000	0.0000	0.0000	0.0000	0.0000	0.0000	0.0000	0.0000	0.0000	0.0000	0.0000	0.0000	0	20
n	X	0.99	0.98	0.97	0.96	0.95	0.94	0.93	0.92	0.91	0.90	0.89	0.88	0.87	0.86	0.85	0.84	0.83	0.82	0.81	0.80	0.79	0.78	0.77	0.76	0.75 P	X	n

Table E.7 (Continued)

														P													
n	X	0.26	0.27	0.28	0.29	0.30	0.31	0.32	0.33	0.34	0.35	0.36	0.37	0.38	0.39	0.40	0.41	0.42	0.43	0.44	0.45	0.46	0.47	0.48	0.49	0.50	X	n
18	0	0.0044	0.0035	0.0027	0.0021	0.0016	0.0013	0.0010	0.0007	0.0006	0.0004	0.0003	0.0002	0.0002	0.0001	0.0001	0.0001	0.0001	0.0000	0.0000	0.0000	0.0000	0.0000	0.0000	0.0000	0.0000	18	
	1	0.0280	0.0231	0.0189	0.0155	0.0126	0.0102	0.0082	0.0066	0.0052	0.0042	0.0033	0.0026	0.0020	0.0016	0.0012	0.0009	0.0007	0.0005	0.0004	0.0003	0.0002	0.0002	0.0001	0.0001	0.0001	17	
	2	0.0836	0.0725	0.0626	0.0537	0.0458	0.0388	0.0327	0.0275	0.0229	0.0190	0.0157	0.0129	0.0105	0.0086	0.0069	0.0055	0.0044	0.0035	0.0028	0.0022	0.0017	0.0013	0.0010	0.0008	0.0006	16	
	3	0.1567	0.1431	0.1298	0.1169	0.1046	0.0930	0.0822	0.0722	0.0630	0.0547	0.0471	0.0404	0.0344	0.0292	0.0246	0.0206	0.0171	0.0141	0.0116	0.0095	0.0077	0.0062	0.0050	0.0039	0.0031	15	
	4	0.2065	0.1985	0.1892	0.1790	0.1681	0.1567	0.1450	0.1333	0.1217	0.1104	0.0994	0.0890	0.0791	0.0699	0.0614	0.0536	0.0464	0.0400	0.0342	0.0291	0.0246	0.0206	0.0172	0.0142	0.0117	14	
	5	0.2031	0.2055	0.2061	0.2048	0.2017	0.1971	0.1911	0.1838	0.1755	0.1664	0.1566	0.1463	0.1358	0.1252	0.1146	0.1042	0.0941	0.0844	0.0753	0.0666	0.0586	0.0512	0.0444	0.0382	0.0327	13	
	6	0.1546	0.1647	0.1736	0.1812	0.1873	0.1919	0.1948	0.1962	0.1959	0.1941	0.1908	0.1862	0.1803	0.1734	0.1655	0.1569	0.1477	0.1380	0.1281	0.1181	0.1081	0.0983	0.0887	0.0796	0.0708	12	
	7	0.0931	0.1044	0.1157	0.1269	0.1376	0.1478	0.1572	0.1656	0.1730	0.1792	0.1840	0.1875	0.1895	0.1900	0.1892	0.1869	0.1833	0.1785	0.1726	0.1657	0.1579	0.1494	0.1404	0.1310	0.1214	11	
	8	0.0450	0.0531	0.0619	0.0713	0.0811	0.0913	0.1017	0.1122	0.1226	0.1327	0.1423	0.1514	0.1597	0.1671	0.1734	0.1786	0.1825	0.1852	0.1864	0.1864	0.1850	0.1822	0.1782	0.1731	0.1669	10	
	9	0.0176	0.0218	0.0267	0.0323	0.0386	0.0456	0.0532	0.0614	0.0701	0.0794	0.0890	0.0988	0.1087	0.1187	0.1284	0.1379	0.1469	0.1552	0.1628	0.1694	0.1751	0.1795	0.1828	0.1848	0.1855	9	
	10	0.0056	0.0073	0.0094	0.0119	0.0149	0.0184	0.0225	0.0272	0.0325	0.0385	0.0450	0.0522	0.0600	0.0683	0.0771	0.0862	0.0957	0.1054	0.1151	0.1248	0.1342	0.1433	0.1519	0.1598	0.1669	8	
	11	0.0014	0.0020	0.0026	0.0035	0.0046	0.0060	0.0077	0.0097	0.0122	0.0151	0.0184	0.0223	0.0267	0.0318	0.0374	0.0436	0.0504	0.0578	0.0658	0.0742	0.0831	0.0924	0.1020	0.1117	0.1214	7	
	12	0.0003	0.0004	0.0006	0.0008	0.0012	0.0016	0.0021	0.0028	0.0037	0.0047	0.0060	0.0076	0.0096	0.0118	0.0145	0.0177	0.0213	0.0254	0.0301	0.0354	0.0413	0.0478	0.0549	0.0626	0.0708	6	
	13	0.0000	0.0001	0.0001	0.0002	0.0002	0.0003	0.0005	0.0006	0.0009	0.0012	0.0016	0.0021	0.0027	0.0035	0.0045	0.0057	0.0071	0.0089	0.0109	0.0134	0.0162	0.0196	0.0234	0.0278	0.0327	5	
	14	0.0000	0.0000	0.0000	0.0000	0.0000	0.0001	0.0001	0.0001	0.0002	0.0002	0.0003	0.0004	0.0006	0.0008	0.0011	0.0014	0.0018	0.0024	0.0031	0.0039	0.0049	0.0062	0.0077	0.0095	0.0117	4	
	15	0.0000	0.0000	0.0000	0.0000	0.0000	0.0000	0.0000	0.0000	0.0000	0.0000	0.0000	0.0001	0.0001	0.0001	0.0002	0.0003	0.0004	0.0005	0.0006	0.0009	0.0011	0.0015	0.0019	0.0024	0.0031	3	
	16	0.0000	0.0000	0.0000	0.0000	0.0000	0.0000	0.0000	0.0000	0.0000	0.0000	0.0000	0.0000	0.0000	0.0000	0.0000	0.0000	0.0000	0.0001	0.0001	0.0001	0.0002	0.0002	0.0003	0.0004	0.0006	2	
	17	0.0000	0.0000	0.0000	0.0000	0.0000	0.0000	0.0000	0.0000	0.0000	0.0000	0.0000	0.0000	0.0000	0.0000	0.0000	0.0000	0.0000	0.0000	0.0000	0.0000	0.0000	0.0000	0.0000	0.0000	0.0001	1	
	18	0.0000	0.0000	0.0000	0.0000	0.0000	0.0000	0.0000	0.0000	0.0000	0.0000	0.0000	0.0000	0.0000	0.0000	0.0000	0.0000	0.0000	0.0000	0.0000	0.0000	0.0000	0.0000	0.0000	0.0000	0.0000	0	18
19	0	0.0033	0.0025	0.0019	0.0015	0.0011	0.0009	0.0007	0.0005	0.0004	0.0003	0.0002	0.0002	0.0001	0.0001	0.0001	0.0000	0.0000	0.0000	0.0000	0.0000	0.0000	0.0000	0.0000	0.0000	0.0000	19	
	1	0.0219	0.0178	0.0144	0.0116	0.0093	0.0074	0.0059	0.0046	0.0036	0.0029	0.0022	0.0017	0.0013	0.0010	0.0008	0.0006	0.0004	0.0003	0.0002	0.0002	0.0001	0.0001	0.0001	0.0001	0.0000	18	
	2	0.0692	0.0592	0.0503	0.0426	0.0358	0.0299	0.0249	0.0206	0.0169	0.0138	0.0112	0.0091	0.0073	0.0058	0.0046	0.0037	0.0029	0.0022	0.0017	0.0013	0.0010	0.0008	0.0006	0.0004	0.0003	17	
	3	0.1377	0.1240	0.1109	0.0985	0.0869	0.0762	0.0664	0.0574	0.0494	0.0422	0.0358	0.0302	0.0253	0.0211	0.0175	0.0144	0.0118	0.0096	0.0077	0.0062	0.0049	0.0039	0.0031	0.0024	0.0018	16	
	4	0.1935	0.1835	0.1726	0.1610	0.1491	0.1370	0.1249	0.1131	0.1017	0.0909	0.0806	0.0710	0.0621	0.0540	0.0467	0.0400	0.0341	0.0289	0.0243	0.0203	0.0168	0.0138	0.0113	0.0092	0.0074	15	
	5	0.2040	0.2036	0.2013	0.1973	0.1916	0.1846	0.1764	0.1672	0.1572	0.1468	0.1360	0.1251	0.1143	0.1036	0.0933	0.0834	0.0741	0.0653	0.0572	0.0497	0.0429	0.0368	0.0313	0.0265	0.0222	14	
	6	0.1672	0.1757	0.1827	0.1880	0.1916	0.1935	0.1936	0.1921	0.1890	0.1844	0.1785	0.1714	0.1634	0.1546	0.1451	0.1353	0.1252	0.1150	0.1049	0.0949	0.0853	0.0761	0.0674	0.0593	0.0518	13	
	7	0.1091	0.1207	0.1320	0.1426	0.1525	0.1615	0.1692	0.1757	0.1808	0.1844	0.1865	0.1870	0.1860	0.1835	0.1797	0.1746	0.1683	0.1611	0.1530	0.1443	0.1350	0.1254	0.1156	0.1058	0.0961	12	
	8	0.0575	0.0670	0.0770	0.0874	0.0981	0.1088	0.1195	0.1298	0.1397	0.1489	0.1573	0.1647	0.1710	0.1760	0.1797	0.1820	0.1829	0.1823	0.1803	0.1771	0.1725	0.1668	0.1601	0.1525	0.1442	11	
	9	0.0247	0.0303	0.0366	0.0436	0.0514	0.0597	0.0687	0.0782	0.0880	0.0980	0.1082	0.1182	0.1281	0.1375	0.1464	0.1546	0.1618	0.1681	0.1732	0.1771	0.1796	0.1808	0.1806	0.1791	0.1762	10	
	10	0.0087	0.0112	0.0142	0.0178	0.0220	0.0268	0.0323	0.0385	0.0453	0.0528	0.0608	0.0694	0.0785	0.0879	0.0976	0.1074	0.1172	0.1268	0.1361	0.1449	0.1530	0.1603	0.1667	0.1721	0.1762	9	
	11	0.0025	0.0034	0.0045	0.0060	0.0077	0.0099	0.0124	0.0155	0.0191	0.0233	0.0280	0.0334	0.0394	0.0460	0.0532	0.0611	0.0694	0.0783	0.0875	0.0970	0.1066	0.1163	0.1259	0.1352	0.1442	8	
	12	0.0006	0.0008	0.0012	0.0016	0.0022	0.0030	0.0039	0.0051	0.0066	0.0083	0.0105	0.0131	0.0161	0.0196	0.0237	0.0283	0.0335	0.0394	0.0458	0.0529	0.0606	0.0688	0.0775	0.0866	0.0961	7	
	13	0.0001	0.0002	0.0002	0.0004	0.0005	0.0007	0.0010	0.0014	0.0018	0.0024	0.0032	0.0041	0.0053	0.0067	0.0085	0.0106	0.0131	0.0160	0.0194	0.0233	0.0278	0.0328	0.0385	0.0448	0.0518	6	
	14	0.0000	0.0000	0.0000	0.0001	0.0001	0.0001	0.0002	0.0003	0.0004	0.0006	0.0008	0.0010	0.0014	0.0018	0.0024	0.0032	0.0041	0.0052	0.0065	0.0082	0.0101	0.0125	0.0152	0.0185	0.0222	5	
	15	0.0000	0.0000	0.0000	0.0000	0.0000	0.0000	0.0000	0.0000	0.0001	0.0001	0.0001	0.0002	0.0003	0.0004	0.0005	0.0007	0.0010	0.0013	0.0017	0.0022	0.0029	0.0037	0.0047	0.0059	0.0074	4	
	16	0.0000	0.0000	0.0000	0.0000	0.0000	0.0000	0.0000	0.0000	0.0000	0.0000	0.0000	0.0000	0.0000	0.0001	0.0001	0.0001	0.0002	0.0002	0.0003	0.0005	0.0006	0.0008	0.0011	0.0014	0.0018	3	
	17	0.0000	0.0000	0.0000	0.0000	0.0000	0.0000	0.0000	0.0000	0.0000	0.0000	0.0000	0.0000	0.0000	0.0000	0.0000	0.0000	0.0000	0.0000	0.0000	0.0001	0.0001	0.0001	0.0002	0.0002	0.0003	2	
	18	0.0000	0.0000	0.0000	0.0000	0.0000	0.0000	0.0000	0.0000	0.0000	0.0000	0.0000	0.0000	0.0000	0.0000	0.0000	0.0000	0.0000	0.0000	0.0000	0.0000	0.0000	0.0000	0.0000	0.0000	0.0000	1	
	19	0.0000	0.0000	0.0000	0.0000	0.0000	0.0000	0.0000	0.0000	0.0000	0.0000	0.0000	0.0000	0.0000	0.0000	0.0000	0.0000	0.0000	0.0000	0.0000	0.0000	0.0000	0.0000	0.0000	0.0000	0.0000	0	19
20	0	0.0024	0.0018	0.0014	0.0011	0.0008	0.0006	0.0004	0.0003	0.0002	0.0002	0.0001	0.0001	0.0001	0.0001	0.0000	0.0000	0.0000	0.0000	0.0000	0.0000	0.0000	0.0000	0.0000	0.0000	0.0000	20	
	1	0.0170	0.0137	0.0109	0.0087	0.0068	0.0054	0.0042	0.0033	0.0025	0.0020	0.0015	0.0011	0.0009	0.0007	0.0005	0.0004	0.0003	0.0002	0.0001	0.0001	0.0001	0.0001	0.0000	0.0000	0.0000	19	
	2	0.0569	0.0480	0.0403	0.0336	0.0278	0.0229	0.0188	0.0153	0.0124	0.0100	0.0080	0.0064	0.0050	0.0040	0.0031	0.0024	0.0018	0.0014	0.0011	0.0008	0.0006	0.0005	0.0003	0.0002	0.0002	18	
	3	0.1199	0.1065	0.0940	0.0823	0.0716	0.0619	0.0531	0.0453	0.0383	0.0323	0.0270	0.0224	0.0185	0.0152	0.0123	0.0100	0.0080	0.0064	0.0051	0.0040	0.0031	0.0024	0.0019	0.0014	0.0011	17	
	4	0.1790	0.1675	0.1553	0.1429	0.1304	0.1181	0.1062	0.0947	0.0839	0.0738	0.0645	0.0559	0.0482	0.0412	0.0350	0.0295	0.0247	0.0206	0.0170	0.0139	0.0113	0.0092	0.0074	0.0059	0.0046	16	
	5	0.2013	0.1982	0.1933	0.1868	0.1789	0.1698	0.1599	0.1493	0.1384	0.1272	0.1161	0.1051	0.0945	0.0843	0.0746	0.0656	0.0573	0.0496	0.0427	0.0365	0.0309	0.0260	0.0217	0.0180	0.0148	15	
	6	0.1768	0.1833	0.1879	0.1907	0.1916	0.1907	0.1881	0.1839	0.1782	0.1712	0.1632	0.1543	0.1447	0.1347	0.1244	0.1140	0.1037	0.0936	0.0839	0.0746	0.0658	0.0577	0.0501	0.0432	0.0370	14	
	7	0.1242	0.1356	0.1462	0.1558	0.1643	0.1714	0.1770	0.1811	0.1836	0.1844	0.1836	0.1812	0.1774	0.1722	0.1659	0.1585	0.1502	0.1413	0.1318	0.1221	0.1122	0.1023	0.0925	0.0830	0.0739	13	
	8	0.0709	0.0815	0.0924	0.1034	0.1144	0.1251	0.1354	0.1450	0.1537	0.1614	0.1678	0.1730	0.1767	0.1790	0.1797	0.1790	0.1768	0.1732	0.1683	0.1623	0.1553	0.1474	0.1388	0.1296	0.1201	12	
	9	0.0332	0.0402	0.0479	0.0563	0.0654	0.0750	0.0849	0.0952	0.1056	0.1158	0.1259	0.1354	0.1444	0.1526	0.1597	0.1658	0.1707	0.1742	0.1763	0.1771	0.1763	0.1742	0.1708	0.1661	0.1602	11	
	10	0.0128	0.0163	0.0205	0.0253	0.0308	0.0370	0.0440	0.0516	0.0598	0.0686	0.0779	0.0875	0.0974	0.1073	0.1171	0.1268	0.1359	0.1446	0.1524	0.1593	0.1652	0.1700	0.1734	0.1755	0.1762	10	
	11	0.0041	0.0055	0.0072	0.0094	0.0120	0.0151	0.0188	0.0231	0.0280	0.0336	0.0398	0.0467	0.0542	0.0624	0.0710	0.0801	0.0895	0.0991	0.1089	0.1185	0.1280	0.1370	0.1455	0.1533	0.1602	9	
	12	0.0011	0.0015	0.0021	0.0029	0.0039	0.0051	0.0066	0.0085	0.0108	0.0136	0.0168	0.0206	0.0249	0.0299	0.0355	0.0417	0.0486	0.0561	0.0642	0.0727	0.0818	0.0911	0.1007	0.1105	0.1201	8	
	13	0.0002	0.0003	0.0005	0.0007	0.0010	0.0014	0.0019	0.0026	0.0034	0.0045	0.0058	0.0074	0.0094	0.0118	0.0146	0.0178	0.0217	0.0260	0.0310	0.0366	0.0429	0.0497	0.0572	0.0653	0.0739	7	
	14	0.0000	0.0001	0.0001	0.0001	0.0002	0.0003	0.0005	0.0006	0.0009	0.0012	0.0016	0.0022	0.0029	0.0038	0.0049	0.0062	0.0078	0.0098	0.0122	0.0150	0.0183	0.0221	0.0264	0.0314	0.0370	6	
	15	0.0000	0.0000	0.0000	0.0000	0.0000	0.0001	0.0001	0.0001	0.0002	0.0003	0.0004	0.0005	0.0007	0.0010	0.0013	0.0017	0.0023	0.0030	0.0038	0.0049	0.0062	0.0078	0.0098	0.0121	0.0148	5	
	16	0.0000	0.0000	0.0000	0.0000	0.0000	0.0000	0.0000	0.0000	0.0000	0.0000	0.0001	0.0001	0.0001	0.0002	0.0003	0.0004	0.0005	0.0007	0.0009	0.0013	0.0017	0.0022	0.0028	0.0036	0.0046	4	
	17	0.0000	0.0000	0.0000	0.0000	0.0000	0.0000	0.0000	0.0000	0.0000	0.0000	0.0000	0.0000	0.0000	0.0000	0.0000	0.0001	0.0001	0.0001	0.0002	0.0002	0.0003	0.0005	0.0006	0.0008	0.0011	3	
	18	0.0000	0.0000	0.0000	0.0000	0.0000	0.0000	0.0000	0.0000	0.0000	0.0000	0.0000	0.0000	0.0000	0.0000	0.0000	0.0000	0.0000	0.0000	0.0000	0.0000	0.0000	0.0001	0.0001	0.0001	0.0002	2	
	19	0.0000	0.0000	0.0000	0.0000	0.0000	0.0000	0.0000	0.0000	0.0000	0.0000	0.0000	0.0000	0.0000	0.0000	0.0000	0.0000	0.0000	0.0000	0.0000	0.0000	0.0000	0.0000	0.0000	0.0000	0.0000	1	
	20	0.0000	0.0000	0.0000	0.0000	0.0000	0.0000	0.0000	0.0000	0.0000	0.0000	0.0000	0.0000	0.0000	0.0000	0.0000	0.0000	0.0000	0.0000	0.0000	0.0000	0.0000	0.0000	0.0000	0.0000	0.0000	0	20
n	X	0.74	0.73	0.72	0.71	0.70	0.69	0.68	0.67	0.66	0.65	0.64	0.63	0.62	0.61	0.60	0.59	0.58	0.57	0.56	0.55	0.54	0.53	0.52	0.51	0.50	X	n
														P														

Table E.8

LOWER AND UPPER CRITICAL VALUES *U* FOR THE RUNS TEST FOR RANDOMNESS

Part 1. Lower Tail
($\alpha = .025$)

n_1 \ n_2	2	3	4	5	6	7	8	9	10	11	12	13	14	15	16	17	18	19	20
2											2	2	2	2	2	2	2	2	2
3					2	2	2	2	2	2	2	2	2	3	3	3	3	3	3
4				2	2	2	3	3	3	3	3	3	3	3	4	4	4	4	4
5			2	2	3	3	3	3	3	4	4	4	4	4	4	4	5	5	5
6		2	2	3	3	3	3	4	4	4	4	5	5	5	5	5	5	6	6
7		2	2	3	3	3	4	4	5	5	5	5	5	6	6	6	6	6	6
8		2	3	3	3	4	4	5	5	5	6	6	6	6	6	7	7	7	7
9		2	3	3	4	4	5	5	5	6	6	6	7	7	7	7	8	8	8
10		2	3	3	4	5	5	5	6	6	7	7	7	7	8	8	8	8	9
11		2	3	4	4	5	5	6	6	7	7	7	8	8	8	9	9	9	9
12	2	2	3	4	4	5	6	6	7	7	7	8	8	8	9	9	9	10	10
13	2	2	3	4	5	5	6	6	7	7	8	8	9	9	9	10	10	10	10
14	2	2	3	4	5	5	6	7	7	8	8	9	9	9	10	10	10	11	11
15	2	3	3	4	5	6	6	7	7	8	8	9	9	10	10	11	11	11	12
16	2	3	4	4	5	6	6	7	8	8	9	9	10	10	11	11	11	12	12
17	2	3	4	4	5	6	7	7	8	9	9	10	10	11	11	11	12	12	13
18	2	3	4	5	5	6	7	8	8	9	9	10	10	11	11	12	12	13	13
19	2	3	4	5	6	6	7	8	8	9	10	10	11	11	12	12	13	13	13
20	2	3	4	5	6	6	7	8	9	9	10	10	11	12	12	13	13	13	14

Part 2. Upper Tail
($\alpha = .025$)

n_1 \ n_2	2	3	4	5	6	7	8	9	10	11	12	13	14	15	16	17	18	19	20
2																			
3																			
4				9	9														
5			9	10	10	11	11												
6			9	10	11	12	12	13	13	13	13								
7				11	12	13	13	14	14	14	14	15	15	15					
8				11	12	13	14	14	15	15	16	16	16	16	17	17	17	17	17
9					13	14	14	15	16	16	16	17	17	18	18	18	18	18	18
10					13	14	15	16	16	17	17	18	18	18	19	19	19	20	20
11					13	14	15	16	17	17	18	19	19	19	20	20	20	21	21
12					13	14	16	16	17	18	19	19	20	20	21	21	21	22	22
13						15	16	17	18	19	19	20	20	21	21	22	22	23	23
14						15	16	17	18	19	20	20	21	22	22	23	23	23	24
15						15	16	18	18	19	20	21	22	22	23	23	24	24	25
16							17	18	19	20	21	21	22	23	23	24	25	25	25
17							17	18	19	20	21	22	23	23	24	25	25	26	26
18							17	18	19	20	21	22	23	24	25	25	26	26	27
19							17	18	20	21	22	23	23	24	25	26	26	27	27
20							17	18	20	21	22	23	24	25	25	26	27	27	28

SOURCE: Adapted from F. S. Swed and C. Eisenhart, *Ann. Math. Statist.*, vol. 14 (1943) pp. 83–86.

Table E.9
LOWER AND UPPER CRITICAL VALUES V OF COX–STUART UNWEIGHTED SIGN TEST FOR TREND

Number of Untied Pairs n	One-Tailed: $\alpha = .05$ Two-Tailed: $\alpha = .10$	$\alpha = .025$ $\alpha = .05$	$\alpha = .01$ $\alpha = .02$	$\alpha = .005$ $\alpha = .01$
	(Lower, Upper)			
5	0,5	—,—	—,—	—,—
6	0,6	0,6	—,—	—,—
7	0,7	0,7	0,7	—,—
8	1,7	0,8	0,8	0,8
9	1,8	1,8	0,9	0,9
10	1,9	1,9	0,10	0,10
11	2,9	1,10	1,10	0,11
12	2,10	2,10	1,11	1,11
13	3,10	2,11	1,12	1,12
14	3,11	2,12	2,12	1,13
15	3,12	3,12	2,13	2,13
16	4,12	3,13	2,14	2,14
17	4,13	4,13	3,14	2,15
18	5,13	4,14	3,15	3,15
19	5,14	4,15	4,15	3,16
20	5,15	5,15	4,16	3,17

Table E.10

LOWER AND UPPER CRITICAL VALUES *W* OF WILCOXON ONE-SAMPLE SIGNED RANKS TEST

n	One-Tailed: $\alpha = .05$ / Two-Tailed: $\alpha = .10$	One-Tailed: $\alpha = .025$ / Two-Tailed: $\alpha = .05$	One-Tailed: $\alpha = .01$ / Two-Tailed: $\alpha = .02$	One-Tailed: $\alpha = .005$ / Two-Tailed: $\alpha = .01$
	(Lower, Upper)			
5	0,15	—,—	—,—	—,—
6	2,19	0,21	—,—	—,—
7	3,25	2,26	0,28	—,—
8	5,31	3,33	1,35	0,36
9	8,37	5,40	3,42	1,44
10	10,45	8,47	5,50	3,52
11	13,53	10,56	7,59	5,61
12	17,61	13,65	10,68	7,71
13	21,70	17,74	12,79	10,81
14	25,80	21,84	16,89	13,92
15	30,90	25,95	19,101	16,104
16	35,101	29,107	23,113	19,117
17	41,112	34,119	27,126	23,130
18	47,124	40,131	32,139	27,144
19	53,137	46,144	37,153	32,158
20	60,150	52,158	43,167	37,173

SOURCE: Adapted from Table 2 of F. Wilcoxon and R. A. Wilcox, *Some Rapid Approximate Statistical Procedures* (Pearl River, N.Y.: Lederle Laboratories, 1964) with permission of the American Cyanamid Company.

Table E.11
LOWER AND UPPER CRITICAL VALUES T_{n_1}
OF WILCOXON RANK SUM TEST

n_2	α One-Tailed	α Two-Tailed	n_1 = 4	5	6	7	8	9	10
4	.05	.10	11,25						
	.025	.05	10,26						
	.01	.02	—,—						
	.005	.01	—,—						
5	.05	.10	12,28	19,36					
	.025	.05	11,29	17,38					
	.01	.02	10,30	16,39					
	.005	.01	—,—	15,40					
6	.05	.10	13,31	20,40	28,50				
	.025	.05	12,32	18,42	26,52				
	.01	.02	11,33	17,43	24,54				
	.005	.01	10,34	16,44	23,55				
7	.05	.10	14,34	21,44	29,55	39,66			
	.025	.05	13,35	20,45	27,57	36,69			
	.01	.02	11,37	18,47	25,59	34,71			
	.005	.01	10,38	16,49	24,60	32,73			
8	.05	.10	15,37	23,47	31,59	41,71	51,85		
	.025	.05	14,38	21,49	29,61	38,74	49,87		
	.01	.02	12,40	19,51	27,63	35,77	45,91		
	.005	.01	11,41	17,53	25,65	34,78	43,93		
9	.05	.10	16,40	24,51	33,63	43,76	54,90	66,105	
	.025	.05	14,42	22,53	31,65	40,79	51,93	62,109	
	.01	.02	13,43	20,55	28,68	37,82	47,97	59,112	
	.005	.01	11,45	18,57	26,70	35,84	45,99	56,115	
10	.05	.10	17,43	26,54	35,67	45,81	56,96	69,111	82,128
	.025	.05	15,45	23,57	32,70	42,84	53,99	65,115	78,132
	.01	.02	13,47	21,59	29,73	39,87	49,103	61,119	74,136
	.005	.01	12,48	19,61	27,75	37,89	47,105	58,122	71,139

SOURCE: Adapted from Table 1 of F. Wilcoxon and R. A. Wilcox, *Some Rapid Approximate Statistical Procedures* (Pearl River, N.Y.: Lederle Laboratories, 1964) with permission of the American Cyanamid Company.

Table E.12

TABLE OF SQUARES AND SQUARE ROOTS

N	$\sqrt{N}$	N^2	$\sqrt{10N}$	N	$\sqrt{N}$	N^2	$\sqrt{10N}$	N	$\sqrt{N}$	N^2	$\sqrt{10N}$	N	$\sqrt{N}$	N^2	$\sqrt{10N}$
1	1.00000	1	3.16228	51	7.14143	2601	22.58318	101	10.04988	10201	31.78050	151	12.28821	22801	38.85872
2	1.41421	4	4.47214	52	7.21110	2704	22.80351	102	10.09950	10404	31.93744	152	12.32883	23104	38.98718
3	1.73205	9	5.47723	53	7.28011	2809	23.02173	103	10.14889	10609	32.09361	153	12.36932	23409	39.11521
4	2.00000	16	6.32456	54	7.34847	2916	23.23790	104	10.19804	10816	32.24903	154	12.40967	23716	39.24283
5	2.23607	25	7.07107	55	7.41620	3025	23.45208	105	10.24695	11025	32.40370	155	12.44990	24025	39.37004
6	2.44949	36	7.74597	56	7.48331	3136	23.66432	106	10.29563	11236	32.55764	156	12.49000	24336	39.49684
7	2.64575	49	8.36660	57	7.54983	3249	23.87467	107	10.34408	11449	32.71085	157	12.52996	24649	39.62323
8	2.82843	64	8.94427	58	7.61577	3364	24.08319	108	10.39230	11664	32.86335	158	12.56981	24964	39.74921
9	3.00000	81	9.48683	59	7.68115	3481	24.28992	109	10.44031	11881	33.01515	159	12.60952	25281	39.87480
10	3.16228	100	10.00000	60	7.74597	3600	24.49490	110	10.48809	12100	33.16625	160	12.64911	25600	40.00000
11	3.31662	121	10.48809	61	7.81025	3721	24.69818	111	10.53565	12321	33.31666	161	12.68858	25921	40.12481
12	3.46410	144	10.95445	62	7.87401	3844	24.89980	112	10.58301	12544	33.46640	162	12.72792	26244	40.24922
13	3.60555	169	11.40175	63	7.93725	3969	25.09980	113	10.63015	12769	33.61547	163	12.76715	26569	40.37326
14	3.74166	196	11.83216	64	8.00000	4096	25.29822	114	10.67708	12996	33.76389	164	12.80625	26896	40.49691
15	3.87298	225	12.24745	65	8.06226	4225	25.49510	115	10.72381	13225	33.91165	165	12.84523	27225	40.62019
16	4.00000	256	12.64911	66	8.12404	4356	25.69047	116	10.77033	13456	34.05877	166	12.88410	27556	40.74310
17	4.12311	289	13.03840	67	8.18535	4489	25.88436	117	10.81665	13689	34.20526	167	12.92285	27889	40.86563
18	4.24264	324	13.41641	68	8.24621	4624	26.07681	118	10.86278	13924	34.35113	168	12.96148	28224	40.98780
19	4.35890	361	13.78405	69	8.30662	4761	26.26785	119	10.90871	14161	34.49638	169	13.00000	28561	41.10961
20	4.47214	400	14.14214	70	8.36660	4900	26.45751	120	10.95445	14400	34.64102	170	13.03840	28900	41.23106
21	4.58258	441	14.49138	71	8.42615	5041	26.64583	121	11.00000	14641	34.78505	171	13.07670	29241	41.35215
22	4.69042	484	14.83240	72	8.48528	5184	26.83282	122	11.04536	14884	34.92850	172	13.11488	29584	41.47288
23	4.79583	529	15.16575	73	8.54400	5329	27.01851	123	11.09054	15129	35.07136	173	13.15295	29929	41.59327
24	4.89898	576	15.49193	74	8.60233	5476	27.20294	124	11.13553	15376	35.21363	174	13.19091	30276	41.71331
25	5.00000	625	15.81139	75	8.66025	5625	27.38613	125	11.18034	15625	35.35534	175	13.22876	30625	41.83300
26	5.09902	676	16.12452	76	8.71780	5776	27.56810	126	11.22497	15876	35.49648	176	13.26650	30976	41.95235
27	5.19615	729	16.43168	77	8.77496	5929	27.74887	127	11.26943	16129	35.63706	177	13.30413	31329	42.07137
28	5.29150	784	16.73320	78	8.83176	6084	27.92848	128	11.31371	16384	35.77709	178	13.34166	31684	42.19005
29	5.38516	841	17.02939	79	8.88819	6241	28.10694	129	11.35782	16641	35.91657	179	13.37909	32041	42.30839
30	5.47723	900	17.32051	80	8.94427	6400	28.28427	130	11.40175	16900	36.05551	180	13.41641	32400	42.42641
31	5.56776	961	17.60682	81	9.00000	6561	28.46050	131	11.44552	17161	36.19392	181	13.45362	32761	42.54409
32	5.65685	1024	17.88854	82	9.05539	6724	28.63564	132	11.48913	17424	36.33180	182	13.49074	33124	42.66146
33	5.74456	1089	18.16590	83	9.11043	6889	28.80972	133	11.53256	17689	36.46917	183	13.52775	33489	42.77850
34	5.83095	1156	18.43909	84	9.16515	7056	28.98275	134	11.57584	17956	36.60601	184	13.56466	33856	42.89522
35	5.91608	1225	18.70829	85	9.21954	7225	29.15476	135	11.61895	18225	36.74235	185	13.60147	34225	43.01163
36	6.00000	1296	18.97367	86	9.27362	7396	29.32576	136	11.66190	18496	36.87818	186	13.63818	34596	43.12772
37	6.08276	1369	19.23538	87	9.32738	7569	29.49576	137	11.70470	18769	37.01351	187	13.67479	34969	43.24350
38	6.16441	1444	19.49359	88	9.38083	7744	29.66479	138	11.74734	19044	37.14835	188	13.71131	35344	43.35897
39	6.24500	1521	19.74842	89	9.43398	7921	29.83287	139	11.78983	19321	37.28270	189	13.74773	35721	43.47413
40	6.32456	1600	20.00000	90	9.48683	8100	30.00000	140	11.83216	19600	37.41657	190	13.78405	36100	43.58899
41	6.40312	1681	20.24846	91	9.53939	8281	30.16621	141	11.87434	19881	37.54997	191	13.82027	36481	43.70355
42	6.48074	1764	20.49390	92	9.59166	8464	30.33150	142	11.91638	20164	37.68289	192	13.85641	36864	43.81780
43	6.55744	1849	20.73644	93	9.64365	8649	30.49590	143	11.95826	20449	37.81534	193	13.89244	37249	43.93177
44	6.63325	1936	20.97618	94	9.69536	8836	30.65942	144	12.00000	20736	37.94733	194	13.92839	37636	44.04543
45	6.70820	2025	21.21320	95	9.74679	9025	30.82207	145	12.04159	21025	38.07887	195	13.96424	38025	44.15880
46	6.78233	2116	21.44761	96	9.79796	9216	30.98387	146	12.08305	21316	38.20995	196	14.00000	38416	44.27189
47	6.85565	2209	21.67948	97	9.84886	9409	31.14482	147	12.12436	21609	38.34058	197	14.03567	38809	44.38468
48	6.92820	2304	21.90890	98	9.89949	9604	31.30495	148	12.16553	21904	38.47077	198	14.07125	39204	44.49719
49	7.00000	2401	22.13594	99	9.94987	9801	31.46427	149	12.20656	22201	38.60052	199	14.10674	39601	44.60942
50	7.07107	2500	22.36068	100	10.00000	10000	31.62278	150	12.24745	22500	38.72983	200	14.14214	40000	44.72136

Table E.12 (*Continued*)

N	$\sqrt{N}$	N^2	$\sqrt{10N}$	N	$\sqrt{N}$	N^2	$\sqrt{10N}$	N	$\sqrt{N}$	N^2	$\sqrt{10N}$	N	$\sqrt{N}$	N^2	$\sqrt{10N}$
201	14.17745	40401	44.83302	251	15.84298	63001	50.09990	301	17.34935	90601	54.86347	351	18.73499	123201	59.24525
202	14.21267	40804	44.94441	252	15.87451	63504	50.19960	302	17.37815	91204	54.95453	352	18.76166	123904	59.32959
203	14.24781	41209	45.05552	253	15.90597	64009	50.29911	303	17.40690	91809	55.04544	353	18.78829	124609	59.41380
204	14.28286	41616	45.16636	254	15.93738	64516	50.39841	304	17.43560	92416	55.13620	354	18.81489	125316	59.49790
205	14.31782	42025	45.27693	255	15.96872	65025	50.49752	305	17.46425	93025	55.22681	355	18.84144	126025	59.58188
206	14.35270	42436	45.38722	256	16.00000	65536	50.59644	306	17.49286	93636	55.31727	356	18.86796	126736	59.66574
207	14.38749	42849	45.49725	257	16.03122	66049	50.69517	307	17.52142	94249	55.40758	357	18.89444	127449	59.74948
208	14.42221	43264	45.60702	258	16.06238	66564	50.79370	308	17.54993	94864	55.49775	358	18.92089	128164	59.83310
209	14.45683	43681	45.71652	259	16.09348	67081	50.89204	309	17.57840	95481	55.58777	359	18.94730	128881	59.91661
210	14.49138	44100	45.82576	260	16.12452	67600	50.99020	310	17.60682	96100	55.67764	360	18.97367	129600	60.00000
211	14.52584	44521	45.93474	261	16.15549	68121	51.08816	311	17.63519	96721	55.76737	361	19.00000	130321	60.08328
212	14.56022	44944	46.04346	262	16.18641	68644	51.18594	312	17.66352	97344	55.85696	362	19.02630	131044	60.16644
213	14.59452	45369	46.15192	263	16.21727	69169	51.28353	313	17.69181	97969	55.94640	363	19.05256	131769	60.24948
214	14.62874	45796	46.26013	264	16.24808	69696	51.38093	314	17.72005	98596	56.03570	364	19.07878	132496	60.33241
215	14.66288	46225	46.36809	265	16.27882	70225	51.47815	315	17.74824	99225	56.12486	365	19.10497	133225	60.41523
216	14.69694	46656	46.47580	266	16.30951	70756	51.57519	316	17.77639	99856	56.21388	366	19.13113	133956	60.49793
217	14.73092	47089	46.58326	267	16.34013	71289	51.67204	317	17.80449	100489	56.30275	367	19.15724	134689	60.58052
218	14.76482	47524	46.69047	268	16.37071	71824	51.76872	318	17.83255	101124	56.39149	368	19.18333	135424	60.66300
219	14.79865	47961	46.79744	269	16.40122	72361	51.86521	319	17.86057	101761	56.48008	369	19.20937	136161	60.74537
220	14.83240	48400	46.90416	270	16.43168	72900	51.96152	320	17.88854	102400	56.56854	370	19.23538	136900	60.82763
221	14.86607	48841	47.01064	271	16.46208	73441	52.05766	321	17.91647	103041	56.65686	371	19.26136	137641	60.90977
222	14.89966	49284	47.11688	272	16.49242	73984	52.15362	322	17.94436	103684	56.74504	372	19.28730	138384	60.99180
223	14.93318	49729	47.22288	273	16.52271	74529	52.24940	323	17.97220	104329	56.83309	373	19.31321	139129	61.07373
224	14.96663	50176	47.32864	274	16.55295	75076	52.34501	324	18.00000	104976	56.92100	374	19.33908	139876	61.15554
225	15.00000	50625	47.43416	275	16.58312	75625	52.44044	325	18.02776	105625	57.00877	375	19.36492	140625	61.23724
226	15.03330	51076	47.53946	276	16.61325	76176	52.53570	326	18.05547	106276	57.09641	376	19.39072	141376	61.31884
227	15.06652	51529	47.64452	277	16.64332	76729	52.63079	327	18.08314	106929	57.18391	377	19.41649	142129	61.40033
228	15.09967	51984	47.74935	278	16.67333	77284	52.72571	328	18.11077	107584	57.27128	378	19.44222	142884	61.48170
229	15.13275	52441	47.85394	279	16.70329	77841	52.82045	329	18.13836	108241	57.35852	379	19.46792	143641	61.56298
230	15.16575	52900	47.95832	280	16.73320	78400	52.91503	330	18.16590	108900	57.44563	380	19.49359	144400	61.64414
231	15.19868	53361	48.06246	281	16.76305	78961	53.00943	331	18.19341	109561	57.53260	381	19.51922	145161	61.72520
232	15.23155	53824	48.16638	282	16.79286	79524	53.10367	332	18.22087	110224	57.61944	382	19.54482	145924	61.80615
233	15.26434	54289	48.27007	283	16.82260	80089	53.19774	333	18.24829	110889	57.70615	383	19.57039	146689	61.88699
234	15.29706	54756	48.37355	284	16.85230	80656	53.29165	334	18.27567	111556	57.79273	384	19.59592	147456	61.96773
235	15.32971	55225	48.47680	285	16.88194	81225	53.38539	335	18.30301	112225	57.87918	385	19.62142	148225	62.04837
236	15.36229	55696	48.57983	286	16.91153	81796	53.47897	336	18.33030	112896	57.96551	386	19.64688	148996	62.12890
237	15.39480	56169	48.68265	287	16.94107	82369	53.57238	337	18.35756	113569	58.05170	387	19.67232	149769	62.20932
238	15.42725	56644	48.78524	288	16.97056	82944	53.66563	338	18.38478	114244	58.13777	388	19.69772	150544	62.28965
239	15.45962	57121	48.88763	289	17.00000	83521	53.75872	339	18.41195	114921	58.22371	389	19.72308	151321	62.36986
240	15.49193	57600	48.98979	290	17.02939	84100	53.85165	340	18.43909	115600	58.30952	390	19.74842	152100	62.44998
241	15.52417	58081	49.09175	291	17.05872	84681	53.94442	341	18.46619	116281	58.39521	391	19.77372	152881	62.52999
242	15.55635	58564	49.19350	292	17.08801	85264	54.03702	342	18.49324	116964	58.48077	392	19.79899	153664	62.60990
243	15.58846	59049	49.29503	293	17.11724	85849	54.12947	343	18.52026	117649	58.56620	393	19.82423	154449	62.68971
244	15.62050	59536	49.39636	294	17.14643	86436	54.22177	344	18.54724	118336	58.65151	394	19.84943	155236	62.76942
245	15.65248	60025	49.49747	295	17.17556	87025	54.31390	345	18.57418	119025	58.73670	395	19.87461	156025	62.84903
246	15.68439	60516	49.59839	296	17.20465	87616	54.40588	346	18.60108	119716	58.82176	396	19.89975	156816	62.92853
247	15.71623	61009	49.69909	297	17.23369	88209	54.49771	347	18.62794	120409	58.90671	397	19.92486	157609	63.00794
248	15.74802	61504	49.79960	298	17.26268	88804	54.58938	348	18.65476	121104	58.99152	398	19.94994	158404	63.08724
249	15.77973	62001	49.89990	299	17.29162	89401	54.68089	349	18.68154	121801	59.07622	399	19.97498	159201	63.16645
250	15.81139	62500	50.00000	300	17.32051	90000	54.77226	350	18.70829	122500	59.16080	400	20.00000	160000	63.24555

Table E.12 (*Continued*)

N	$\sqrt{N}$	N^2	$\sqrt{10N}$
401	20.02498	160801	63.32456
402	20.04994	161604	63.40347
403	20.07486	162409	63.48228
404	20.09975	163216	63.56099
405	20.12461	164025	63.63961
406	20.14944	164836	63.71813
407	20.17424	165649	63.79655
408	20.19901	166464	63.87488
409	20.22375	167281	63.95311
410	20.24846	168100	64.03124
411	20.27313	168921	64.10928
412	20.29778	169744	64.18723
413	20.32240	170569	64.26508
414	20.34699	171396	64.34283
415	20.37155	172225	64.42049
416	20.39608	173056	64.49806
417	20.42058	173889	64.57554
418	20.44505	174724	64.65292
419	20.46949	175561	64.73021
420	20.49390	176400	64.80741
421	20.51828	177241	64.88451
422	20.54264	178084	64.96153
423	20.56696	178929	65.03845
424	20.59126	179776	65.11528
425	20.61553	180625	65.19202
426	20.63977	181476	65.26868
427	20.66398	182329	65.34524
428	20.68816	183184	65.42171
429	20.71232	184041	65.49809
430	20.73644	184900	65.57439
431	20.76054	185761	65.65059
432	20.78461	186624	65.72671
433	20.80865	187489	65.80274
434	20.83267	188356	65.87868
435	20.85665	189225	65.95453
436	20.88061	190096	66.03030
437	20.90454	190969	66.10598
438	20.92845	191844	66.18157
439	20.95233	192721	66.25708
440	20.97618	193600	66.33250
441	21.00000	194481	66.40783
442	21.02380	195364	66.48308
443	21.04757	196249	66.55825
444	21.07131	197136	66.63332
445	21.09502	198025	66.70832
446	21.11871	198916	66.78323
447	21.14237	199809	66.85806
448	21.16601	200704	66.93280
449	21.18962	201601	67.00746
450	21.21320	202500	67.08204

N	$\sqrt{N}$	N^2	$\sqrt{10N}$
451	21.23676	203401	67.15653
452	21.26029	204304	67.23095
453	21.28380	205209	67.30527
454	21.30728	206116	67.37952
455	21.33073	207025	67.45369
456	21.35416	207936	67.52777
457	21.37756	208849	67.60178
458	21.40093	209764	67.67570
459	21.42429	210681	67.74954
460	21.44761	211600	67.82330
461	21.47091	212521	67.89698
462	21.49419	213444	67.97058
463	21.51743	214369	68.04410
464	21.54066	215296	68.11755
465	21.56386	216225	68.19091
466	21.58703	217156	68.26419
467	21.61018	218089	68.33740
468	21.63331	219024	68.41053
469	21.65641	219961	68.48357
470	21.67948	220900	68.55655
471	21.70253	221841	68.62944
472	21.72556	222784	68.70226
473	21.74856	223729	68.77500
474	21.77154	224676	68.84766
475	21.79449	225625	68.92024
476	21.81742	226576	68.99275
477	21.84033	227529	69.06519
478	21.86321	228484	69.13754
479	21.88607	229441	69.20983
480	21.90890	230400	69.28203
481	21.93171	231361	69.35416
482	21.95450	232324	69.42622
483	21.97726	233289	69.49820
484	22.00000	234256	69.57011
485	22.02272	235225	69.64194
486	22.04541	236196	69.71370
487	22.06808	237169	69.78539
488	22.09072	238144	69.85700
489	22.11334	239121	69.92853
490	22.13594	240100	70.00000
491	22.15852	241081	70.07139
492	22.18107	242064	70.14271
493	22.20360	243049	70.21396
494	22.22611	244036	70.28513
495	22.24860	245025	70.35624
496	22.27106	246016	70.42727
497	22.29350	247009	70.49823
498	22.31591	248004	70.56912
499	22.33831	249001	70.63993
500	22.36068	250000	70.71068

N	$\sqrt{N}$	N^2	$\sqrt{10N}$
501	22.38303	251001	70.78135
502	22.40536	252004	70.85196
503	22.42766	253009	70.92249
504	22.44994	254016	70.99296
505	22.47221	255025	71.06335
506	22.49444	256036	71.13368
507	22.51666	257049	71.20393
508	22.53886	258064	71.27412
509	22.56103	259081	71.34424
510	22.58318	260100	71.41428
511	22.60531	261121	71.48426
512	22.62742	262144	71.55418
513	22.64950	263169	71.62402
514	22.67157	264196	71.69379
515	22.69361	265225	71.76350
516	22.71563	266256	71.83314
517	22.73763	267289	71.90271
518	22.75961	268324	71.97222
519	22.78157	269361	72.04165
520	22.80351	270400	72.11103
521	22.82542	271441	72.18033
522	22.84732	272484	72.24957
523	22.86919	273529	72.31874
524	22.89105	274576	72.38784
525	22.91288	275625	72.45688
526	22.93469	276676	72.52586
527	22.95648	277729	72.59477
528	22.97825	278784	72.66361
529	23.00000	279841	72.73239
530	23.02173	280900	72.80110
531	23.04344	281961	72.86975
532	23.06513	283024	72.93833
533	23.08679	284089	73.00685
534	23.10844	285156	73.07530
535	23.13007	286225	73.14369
536	23.15167	287296	73.21202
537	23.17326	288369	73.28028
538	23.19483	289444	73.34848
539	23.21637	290521	73.41662
540	23.23790	291600	73.48469
541	23.25941	292681	73.55270
542	23.28089	293764	73.62065
543	23.30236	294849	73.68853
544	23.32381	295936	73.75636
545	23.34524	297025	73.82412
546	23.36664	298116	73.89181
547	23.38803	299209	73.95945
548	23.40940	300304	74.02702
549	23.43075	301401	74.09453
550	23.45208	302500	74.16198

N	$\sqrt{N}$	N^2	$\sqrt{10N}$
551	23.47339	303601	74.22937
552	23.49468	304704	74.29670
553	23.51595	305809	74.36397
554	23.53720	306916	74.43118
555	23.55844	308025	74.49832
556	23.57965	309136	74.56541
557	23.60085	310249	74.63243
558	23.62202	311364	74.69940
559	23.64318	312481	74.76630
560	23.66432	313600	74.83315
561	23.68544	314721	74.89993
562	23.70654	315844	74.96666
563	23.72762	316969	75.03333
564	23.74868	318096	75.09993
565	23.76973	319225	75.16648
566	23.79075	320356	75.23297
567	23.81176	321489	75.29940
568	23.83275	322624	75.36577
569	23.85372	323761	75.43209
570	23.87467	324900	75.49834
571	23.89561	326041	75.56454
572	23.91652	327184	75.63068
573	23.93742	328329	75.69676
574	23.95830	329476	75.76279
575	23.97916	330625	75.82875
576	24.00000	331776	75.89466
577	24.02082	332929	75.96052
578	24.04163	334084	76.02631
579	24.06242	335241	76.09205
580	24.08319	336400	76.15773
581	24.10394	337561	76.22336
582	24.12468	338724	76.28892
583	24.14539	339889	76.35444
584	24.16609	341056	76.41989
585	24.18677	342225	76.48529
586	24.20744	343396	76.55064
587	24.22808	344569	76.61593
588	24.24871	345744	76.68116
589	24.26932	346921	76.74634
590	24.28992	348100	76.81146
591	24.31049	349281	76.87652
592	24.33105	350464	76.94154
593	24.35159	351649	77.00649
594	24.37212	352836	77.07140
595	24.39262	354025	77.13624
596	24.41311	355216	77.20104
597	24.43358	356409	77.26578
598	24.45404	357604	77.33046
599	24.47448	358801	77.39509
600	24.49490	360000	77.45967

Table E.12 (Continued)

N	$\sqrt{N}$	N^2	$\sqrt{10N}$	N	$\sqrt{N}$	N^2	$\sqrt{10N}$	N	$\sqrt{N}$	N^2	$\sqrt{10N}$	N	$\sqrt{N}$	N^2	$\sqrt{10N}$
601	24.51530	361201	77.52419	651	25.51470	423801	80.68457	701	26.47640	491401	83.72574	751	27.40438	564001	86.66026
602	24.53569	362404	77.58866	652	25.53429	425104	80.74652	702	26.49528	492804	83.78544	752	27.42262	565504	86.71793
603	24.55606	363609	77.65307	653	25.55386	426409	80.80842	703	26.51415	494209	83.84510	753	27.44085	567009	86.77557
604	24.57641	364816	77.71744	654	25.57342	427716	80.87027	704	26.53300	495616	83.90471	754	27.45906	568516	86.83317
605	24.59675	366025	77.78175	655	25.59297	429025	80.93207	705	26.55184	497025	83.96428	755	27.47726	570025	86.89074
606	24.61707	367236	77.84600	656	25.61250	430336	80.99383	706	26.57066	498436	84.02381	756	27.49545	571536	86.94826
607	24.63737	368449	77.91020	657	25.63201	431649	81.05554	707	26.58947	499849	84.08329	757	27.51363	573049	87.00575
608	24.65766	369664	77.97435	658	25.65151	432964	81.11720	708	26.60827	501264	84.14274	758	27.53180	574564	87.06320
609	24.67793	370881	78.03845	659	25.67100	434281	81.17881	709	26.62705	502681	84.20214	759	27.54995	576081	87.12061
610	24.69818	372100	78.10250	660	25.69047	435600	81.24038	710	26.64583	504100	84.26150	760	27.56810	577600	87.17798
611	24.71841	373321	78.16649	661	25.70992	436921	81.30191	711	26.66458	505521	84.32082	761	27.58623	579121	87.23531
612	24.73863	374544	78.23043	662	25.72936	438244	81.36338	712	26.68333	506944	84.38009	762	27.60435	580644	87.29261
613	24.75884	375769	78.29432	663	25.74879	439569	81.42481	713	26.70206	508369	84.43933	763	27.62245	582169	87.34987
614	24.77902	376996	78.35815	664	25.76820	440896	81.48620	714	26.72078	509796	84.49852	764	27.64055	583696	87.40709
615	24.79919	378225	78.42194	665	25.78759	442225	81.54753	715	26.73948	511225	84.55767	765	27.65863	585225	87.46428
616	24.81935	379456	78.48567	666	25.80698	443556	81.60882	716	26.75818	512656	84.61678	766	27.67671	586756	87.52143
617	24.83948	380689	78.54935	667	25.82634	444889	81.67007	717	26.77686	514089	84.67585	767	27.69476	588289	87.57854
618	24.85961	381924	78.61298	668	25.84570	446224	81.73127	718	26.79552	515524	84.73488	768	27.71281	589824	87.63561
619	24.87971	383161	78.67655	669	25.86503	447561	81.79242	719	26.81418	516961	84.79387	769	27.73085	591361	87.69265
620	24.89980	384400	78.74008	670	25.88436	448900	81.85353	720	26.83282	518400	84.85281	770	27.74887	592900	87.74964
621	24.91987	385641	78.80355	671	25.90367	450241	81.91459	721	26.85144	519841	84.91172	771	27.76689	594441	87.80661
622	24.93993	386884	78.86698	672	25.92296	451584	81.97561	722	26.87006	521284	84.97058	772	27.78489	595984	87.86353
623	24.95997	388129	78.93035	673	25.94224	452929	82.03658	723	26.88866	522729	85.02941	773	27.80288	597529	87.92042
624	24.97999	389376	78.99367	674	25.96151	454276	82.09750	724	26.90725	524176	85.08819	774	27.82086	599076	87.97727
625	25.00000	390625	79.05694	675	25.98076	455625	82.15838	725	26.92582	525625	85.14693	775	27.83882	600625	88.03408
626	25.01999	391876	79.12016	676	26.00000	456976	82.21922	726	26.94439	527076	85.20563	776	27.85678	602176	88.09086
627	25.03997	393129	79.18333	677	26.01922	458329	82.28001	727	26.96294	528529	85.26429	777	27.87472	603729	88.14760
628	25.05993	394384	79.24645	678	26.03843	459684	82.34076	728	26.98148	529984	85.32292	778	27.89265	605284	88.20431
629	25.07987	395641	79.30952	679	26.05763	461041	82.40146	729	27.00000	531441	85.38150	779	27.91057	606841	88.26098
630	25.09980	396900	79.37254	680	26.07681	462400	82.46211	730	27.01851	532900	85.44004	780	27.92848	608400	88.31761
631	25.11971	398161	79.43551	681	26.09598	463761	82.52272	731	27.03701	534361	85.49854	781	27.94638	609961	88.37420
632	25.13961	399424	79.49843	682	26.11513	465124	82.58329	732	27.05550	535824	85.55700	782	27.96426	611524	88.43076
633	25.15949	400689	79.56130	683	26.13427	466489	82.64381	733	27.07397	537289	85.61542	783	27.98214	613089	88.48729
634	25.17936	401956	79.62412	684	26.15339	467856	82.70429	734	27.09243	538756	85.67380	784	28.00000	614656	88.54377
635	25.19921	403225	79.68689	685	26.17250	469225	82.76473	735	27.11088	540225	85.73214	785	28.01785	616225	88.60023
636	25.21904	404496	79.74961	686	26.19160	470596	82.82512	736	27.12932	541696	85.79044	786	28.03569	617796	88.65664
637	25.23886	405769	79.81228	687	26.21068	471969	82.88546	737	27.14774	543169	85.84870	787	28.05352	619369	88.71302
638	25.25866	407044	79.87490	688	26.22975	473344	82.94577	738	27.16616	544644	85.90693	788	28.07134	620944	88.76936
639	25.27845	408321	79.93748	689	26.24881	474721	83.00602	739	27.18455	546121	85.96511	789	28.08914	622521	88.82567
640	25.29822	409600	80.00000	690	26.26785	476100	83.06624	740	27.20294	547600	86.02325	790	28.10694	624100	88.88194
641	25.31798	410881	80.06248	691	26.28688	477481	83.12641	741	27.22132	549081	86.08136	791	28.12472	625681	88.93818
642	25.33772	412164	80.12490	692	26.30589	478864	83.18654	742	27.23968	550564	86.13942	792	28.14249	627264	88.99438
643	25.35744	413449	80.18728	693	26.32489	480249	83.24662	743	27.25803	552049	86.19745	793	28.16026	628849	89.05055
644	25.37716	414736	80.24961	694	26.34388	481636	83.30666	744	27.27636	553536	86.25543	794	28.17801	630436	89.10668
645	25.39685	416025	80.31189	695	26.36285	483025	83.36666	745	27.29469	555025	86.31338	795	28.19574	632025	89.16277
646	25.41653	417316	80.37413	696	26.38181	484416	83.42661	746	27.31300	556516	86.37129	796	28.21347	633616	89.21883
647	25.43619	418609	80.43631	697	26.40076	485809	83.48653	747	27.33130	558009	86.42916	797	28.23119	635209	89.27486
648	25.45584	419904	80.49845	698	26.41969	487204	83.54639	748	27.34959	559504	86.48699	798	28.24889	636804	89.33085
649	25.47548	421201	80.56054	699	26.43861	488601	83.60622	749	27.36786	561001	86.54479	799	28.26659	638401	89.38680
650	25.49510	422500	80.62258	700	26.45751	490000	83.66600	750	27.38613	562500	86.60254	800	28.28427	640000	89.44272

Table E.12 (*Continued*)

N	$\sqrt{N}$	N^2	$\sqrt{10N}$	N	$\sqrt{N}$	N^2	$\sqrt{10N}$	N	$\sqrt{N}$	N^2	$\sqrt{10N}$	N	$\sqrt{N}$	N^2	$\sqrt{10N}$
801	28.30194	641601	89.49860	851	29.17190	724201	92.24966	901	30.01666	811801	94.92102	951	30.83829	904401	97.51923
802	28.31960	643204	89.55445	852	29.18904	725904	92.30385	902	30.03331	813604	94.97368	952	30.85450	906304	97.57049
803	28.33725	644809	89.61027	853	29.20616	727609	92.35800	903	30.04996	815409	95.02631	953	30.87070	908209	97.62172
804	28.35489	646416	89.66605	854	29.22328	729316	92.41212	904	30.06659	817216	95.07891	954	30.88689	910116	97.67292
805	28.37252	648025	89.72179	855	29.24038	731025	92.46621	905	30.08322	819025	95.13149	955	30.90307	912025	97.72410
806	28.39014	649636	89.77750	856	29.25748	732736	92.52027	906	30.09983	820836	95.18403	956	30.91925	913936	97.77525
807	28.40775	651249	89.83318	857	29.27456	734449	92.57429	907	30.11644	822649	95.23655	957	30.93542	915849	97.82638
808	28.42534	652864	89.88882	858	29.29164	736164	92.62829	908	30.13304	824464	95.28903	958	30.95158	917764	97.87747
809	28.44293	654481	89.94443	859	29.30870	737881	92.68225	909	30.14963	826281	95.34149	959	30.96773	919681	97.92855
810	28.46050	656100	90.00000	860	29.32576	739600	92.73618	910	30.16621	828100	95.39392	960	30.98387	921600	97.97959
811	28.47806	657721	90.05554	861	29.34280	741321	92.79009	911	30.18278	829921	95.44632	961	31.00000	923521	98.03061
812	28.49561	659344	90.11104	862	29.35984	743044	92.84396	912	30.19934	831744	95.49869	962	31.01612	925444	98.08160
813	28.51315	660969	90.16651	863	29.37686	744769	92.89779	913	30.21589	833569	95.55103	963	31.03224	927369	98.13256
814	28.53069	662596	90.22195	864	29.39388	746496	92.95160	914	30.23243	835396	95.60335	964	31.04835	929296	98.18350
815	28.54820	664225	90.27735	865	29.41088	748225	93.00538	915	30.24897	837225	95.65563	965	31.06445	931225	98.23441
816	28.56571	665856	90.33272	866	29.42788	749956	93.05912	916	30.26549	839056	95.70789	966	31.08054	933156	98.28530
817	28.58321	667489	90.38805	867	29.44486	751689	93.11283	917	30.28201	840889	95.76012	967	31.09662	935089	98.33616
818	28.60070	669124	90.44335	868	29.46184	753424	93.16652	918	30.29851	842724	95.81232	968	31.11270	937024	98.38699
819	28.61818	670761	90.49862	869	29.47881	755161	93.22017	919	30.31501	844561	95.86449	969	31.12876	938961	98.43780
820	28.63564	672400	90.55385	870	29.49576	756900	93.27379	920	30.33150	846400	95.91663	970	31.14482	940900	98.48858
821	28.65310	674041	90.60905	871	29.51271	758641	93.32738	921	30.34798	848241	95.96874	971	31.16087	942841	98.53933
822	28.67054	675684	90.66422	872	29.52965	760384	93.38094	922	30.36445	850084	96.02083	972	31.17691	944784	98.59006
823	28.68798	677329	90.71935	873	29.54657	762129	93.43447	923	30.38092	851929	96.07289	973	31.19295	946729	98.64076
824	28.70540	678976	90.77445	874	29.56349	763876	93.48797	924	30.39737	853776	96.12492	974	31.20897	948676	98.69144
825	28.72281	680625	90.82951	875	29.58040	765625	93.54143	925	30.41381	855625	96.17692	975	31.22499	950625	98.74209
826	28.74022	682276	90.88454	876	29.59730	767376	93.59487	926	30.43025	857476	96.22889	976	31.24100	952576	98.79271
827	28.75761	683929	90.93954	877	29.61419	769129	93.64828	927	30.44667	859329	96.28084	977	31.25700	954529	98.84331
828	28.77499	685584	90.99451	878	29.63106	770884	93.70165	928	30.46309	861184	96.33276	978	31.27299	956484	98.89388
829	28.79236	687241	91.04944	879	29.64793	772641	93.75500	929	30.47950	863041	96.38465	979	31.28898	958441	98.94443
830	28.80972	688900	91.10434	880	29.66479	774400	93.80832	930	30.49590	864900	96.43651	980	31.30495	960400	98.99495
831	28.82707	690561	91.15920	881	29.68164	776161	93.86160	931	30.51229	866761	96.48834	981	31.32092	962361	99.04544
832	28.84441	692224	91.21403	882	29.69848	777924	93.91486	932	30.52868	868624	96.54015	982	31.33688	964324	99.09591
833	28.86174	693889	91.26883	883	29.71532	779689	93.96808	933	30.54505	870489	96.59193	983	31.35283	966289	99.14636
834	28.87906	695556	91.32360	884	29.73214	781456	94.02127	934	30.56141	872356	96.64368	984	31.36877	968256	99.19677
835	28.89637	697225	91.37833	885	29.74895	783225	94.07444	935	30.57777	874225	96.69540	985	31.38471	970225	99.24717
836	28.91366	698896	91.43304	886	29.76575	784996	94.12757	936	30.59412	876096	96.74709	986	31.40064	972196	99.29753
837	28.93095	700569	91.48770	887	29.78255	786769	94.18068	937	30.61046	877969	96.79876	987	31.41656	974169	99.34787
838	28.94823	702244	91.54234	888	29.79933	788544	94.23375	938	30.62679	879844	96.85040	988	31.43247	976144	99.39819
839	28.96550	703921	91.59694	889	29.81610	790321	94.28680	939	30.64311	881721	96.90201	989	31.44837	978121	99.44848
840	28.98275	705600	91.65151	890	29.83287	792100	94.33981	940	30.65942	883600	96.95360	990	31.46427	980100	99.49874
841	29.00000	707281	91.70605	891	29.84962	793881	94.39280	941	30.67572	885481	97.00515	991	31.48015	982081	99.54898
842	29.01724	708964	91.76056	892	29.86637	795664	94.44575	942	30.69202	887364	97.05668	992	31.49603	984064	99.59920
843	29.03446	710649	91.81503	893	29.88311	797449	94.49868	943	30.70831	889249	97.10819	993	31.51190	986049	99.64939
844	29.05168	712336	91.86947	894	29.89983	799236	94.55157	944	30.72458	891136	97.15966	994	31.52777	988036	99.69955
845	29.06888	714025	91.92388	895	29.91655	801025	94.60444	945	30.74085	893025	97.21111	995	31.54362	990025	99.74969
846	29.08608	715716	91.97826	896	29.93326	802816	94.65728	946	30.75711	894916	97.26253	996	31.55947	992016	99.79980
847	29.10326	717409	92.03260	897	29.94996	804609	94.71008	947	30.77337	896809	97.31393	997	31.57531	994009	99.84989
848	29.12044	719104	92.08692	898	29.96665	806404	94.76286	948	30.78961	898704	97.36529	998	31.59114	996004	99.89995
849	29.13760	720801	92.14120	899	29.98333	808201	94.81561	949	30.80584	900601	97.41663	999	31.60696	998001	99.94999
850	29.15476	722500	92.19544	900	30.00000	810000	94.86833	950	30.82207	902500	97.46794	1000	31.62278	1000000	100.00000

ANSWERS TO SELECTED ODD-NUMBERED PROBLEMS (*)

CHAPTER 2

2.3 (a) Quantitative (continuous); (b) quantitative (continuous); (c) qualitative; (d) quantitative (discrete); (e) quantitative (continuous) or qualitative; (f) quantitative (discrete).

2.11 12 11 18 66 61 27 05 36 64 19 29 06 11.

2.15 Line 1—column 7 has a 4 "punch" (no code); line 2—columns 13, 14 have a 42 "punch" for high school average; line 3—column 6 has a 6 "punch" (no code); line 4—columns 33, 34 have a 98 "punch" for height; line 5—column 42 has a 3 "punch" (no code).

CHAPTER 3

3.7

Response	No.	%
Yes	293	69.93
No	80	19.09
Do not know or refused to answer	46	10.98
	419	100.00

(a) (1)

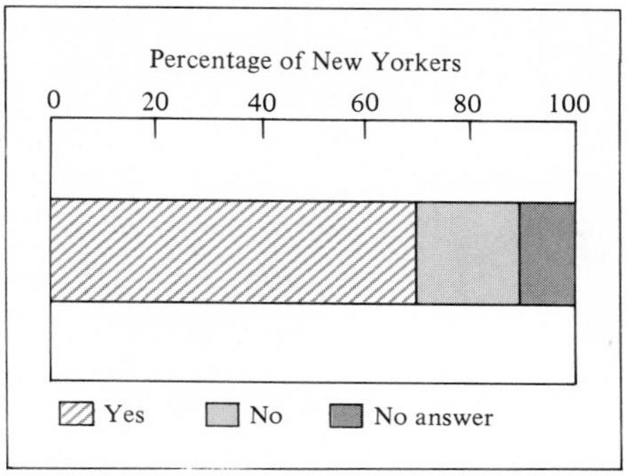

Percentage component bar chart of attitudes toward adequacy of fire and police protection.

(a) (2)

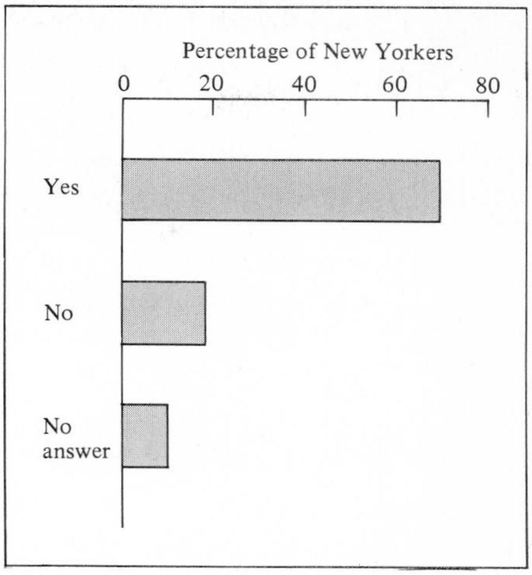

Percentage bar chart of attitudes toward adequacy of fire and police protection.

(a) (3)

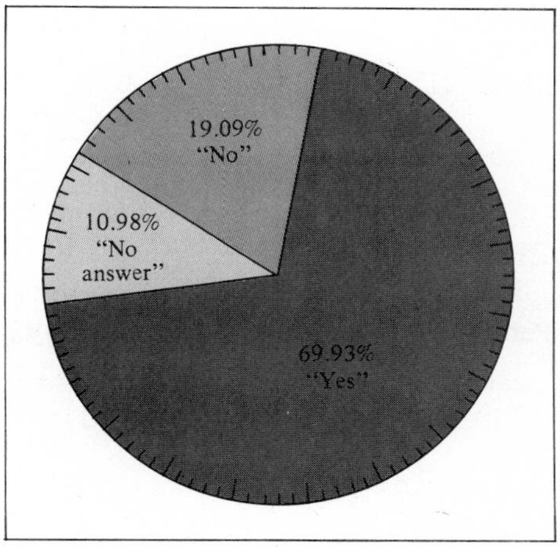

Percentage pie diagram of attitudes toward adequacy of fire and police protection.

3.21 (b) Stem-and-leaf display of high school averages of 94 students:

7^L	2 0 2 2
7^U	6 5 5 8 6 5 8 8 7 9 5
8^L	4 0 2 1 4 0 3 3 0 0 4 3 3 2 0 3 0 1 4 2 3 0 1 0 3 0 2 0 4 4 1
8^U	5 7 8 7 6 8 8 5 9 8 7 7 8 5 5 5 6 9 5 8 5 9 7 5 5 6 9 8 5 7 7 9 7 7 7 5 9 5
9^L	3 1 2 2 2 0 0 0 0
9^U	8

Note: L = unit digits 0, 1, 2, 3, 4 and U = unit digits 5, 6, 7, 8, 9.

(c) Frequency and percentage distributions of high school averages:

High School Averages	No. of Students	% of Students
70 but less than 75	4	4.3
75 but less than 80	11	11.7
80 but less than 85	31	33.0
85 but less than 90	38	40.4
90 but less than 95	9	9.6
95 but less than 100	1	1.1
Totals	94	100.1*

SOURCE: Figure 2.6, data for question 6.
* Error due to rounding.

(d) Frequency histogram of high school averages of 94 students:

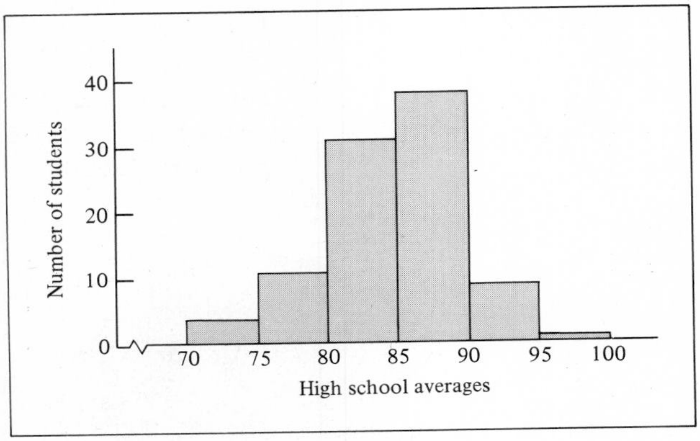

(d)

(e) Percentage polygon of high school averages:

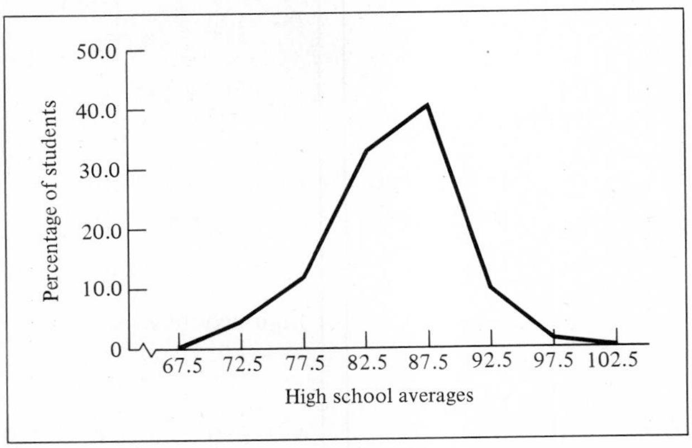

(e)

(f) Cumulative percentage distribution of high school averages of 94 students:

High School Average	Percentage of Students	
	< Value Indicated	≥ Value Indicated
70	0.0	100.0
75	4.3	95.7
80	16.0	84.0
85	49.0	51.0
90	89.4	10.6
95	99.0	1.0
100	100.0	0.0

SOURCE: Figure 2.6, data for question 6.

(g) Cumulative percentage polygon of high school averages of 94 students:

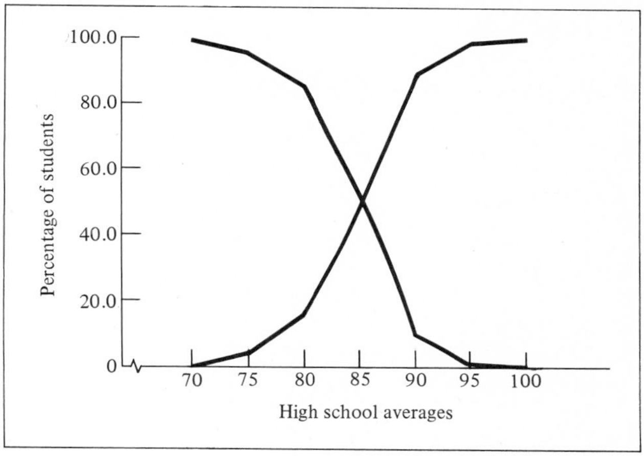

(g)

(h) Ungrouped data, 85; grouped data, 85 to under 90.
(i) Ungrouped data, 28; grouped data, 30.
(j) Ungrouped data, 16.0%; grouped data, 16.0%.
(k) Ungrouped data, approx. 89; grouped data, approx. 89.5.

CHAPTER 4

4.9 (a) $\overline{X}$ = \$15,333.33, median = \$16,000, S = \$3,674.23.
(b) Total ≃ \$4.6 million, S_{total} ≃ \$1.1 million. 367,423
(c) Total ≃ \$18.4 million, S_{total} ≃ \$4.4 million. 1,469,692

4.11 (a) $\overline{X}$ = 9.8, median = 9, S = 3.7.
(b) $\overline{X}$ = 10.3, median = 9.9, S = 3.7
using class limits 4 but less than 8, 8 but less than 12, etc.
(c) Right or positive skewed.
(d) Ungrouped: 2.4 to 17.2.
(e) 96.0%.

CHAPTER 5

5.1 (a) 120/200; (b) 70/200; (c) 160/200; (d) 70/120;
(e) 70/200 ≠ (120/200)(110/200), not statistically independent.

5.5 (a) 60/200; (b) 48/200; (c) 83/200; (d) 48/60; (e) $(48/200) \neq (165/200)(60/200)$, not statistically independent.

5.9 (a) .366; (b) .25; (c) .08; (d) .432; (e) 36/150.

5.21 (a) $(2/7)(1/6) = 2/42$; (b) $(2/7)(5/6) = 10/42$.

5.23 .1168.

5.25 (a) .75; (b) .64.

5.27 (a) .4615; (b) .325.

5.31 (a) $2^7 = 128$; (b) $6^7 = 279{,}936$.

5.33 $10 \times 13 = 130$.

5.35 $7! = 5{,}040$.

5.37 $12!/9! = 1{,}320$.

5.39 $7!/4!3! = 35$.

CHAPTER 6

6.1 (a) $\mu = 7$; (b) $\sigma = 2.42$;
(c)

X	$P(X)$
$-1	20/36
+1	14/36
+2	2/36
	1

(d) $\mu = -.056$; (e) player loses 5.6 cents per bet; (f) house wins 5.6 cents per bet.

6.9 (a) .1488; (b) .8131; (c) right-skewed.

6.15 (a) .4691; (b) .5854; (c) probability of audit increases.

6.17 (a) .6626; (b) .0338; (c) .6964; (d) .3036.

6.21 (a)(1) .00001, (2) .59049, (3) .9914.
(b)(1) .0002, (2) .6065, (3) .9856.
(c)(1) .1494, (2) .0498, (3) .4232.

6.25 (a) .7605; (b) .6599; (c) .1814; (d) 6.31 minutes (using $Z = 1.645$).

6.27 (a) 95.35%; (b) .8164; (c) 258 calls; (d) 21.10%; (e) 59.6 seconds; (f) .4347.

6.41 (a) .9573 without continuity correction; (b) .6730 without continuity correction; (c) $\mu = 5$.

6.43 (a)(1) .0821, (2) .5438.
(b)(1) .1587 without continuity correction, (2) .6826 without continuity correction.

CHAPTER 7

7.1 (a) .1915; (b) .0254; (c) 5 and .25; (d) normal; (e) .4772; (f) .00013.

7.3 6.726.

7.5 (a) \$12.33; (b) .9544; (c) central limit theorem holds; (d) \$11.17; (e) central limit theorem holds.

7.7 (a) .9938; (b) .0062; (c) central limit theorem holds; (d) .99865 and .00135.

7.13 (a) .2486; (b) .0918; (c) .1293 and .2514.

7.17 .9957 and .0043.

CHAPTER 8

8.1 $\{510.03 \leq \mu \leq 569.97\}$.

8.3 $\{\$17{,}922.32 \leq \mu \leq \$18{,}497.68\}$.

8.5 (a) $\{12.547 \leq \mu \leq 18.119\}$; (b) $\{3{,}764.1 \leq \tau \leq 5{,}435.9\}$.

8.9 $\{.227 \leq p \leq .493\}$.

8.11 $\{.1096 \leq p \leq .1904\}$.

8.13 $n = 246$.

8.15 $n = 27$.

8.17 $n = 107$.

8.19 $n = 323$.

8.21 $n = 2{,}305$.

8.23 (a) $\{510.51 \leq \mu \leq 569.49\}$; (b) $n = 92$.

8.25 $\{.2284 \leq p \leq .3716\}$; (b) $n = 214$.

CHAPTER 9

9.3 $t_{99} = 4.0 > 2.6264$, reject H_0; credit balance is not \$30.

9.5 $t_{24} = -9.25 < -1.7109$, reject H_0; process is not working properly.

9.11 (a) $t_{11} = -1.84 > -2.7181$; do not reject H_0; no evidence that the travel time is less than 60 minutes.
(b) Population is normally distributed.

9.13 $Z = -1.33 > -2.33$; do not reject H_0; no evidence that the claim is invalid.

9.15 $Z = -6.124 < -1.645$; reject H_0; the proportion is different from .60.

9.17 (a) Power = .6387, β = .3613; (b) power = .9908, β = .0092; (c) n = 64.

9.19 (a) power = .8037, β = .1963; (b) power = .9996, β = .0004.

9.27 (a) .8051; (b) .9909; (c) n = 14.

9.29 (a) .6387; (b) .9400.

CHAPTER 10

10.1 $t_{198} = +1.91 > 1.645$; reject H_0; there is a difference between shifts.

10.3 $t_{28} = 1.714 < 2.4671$; do not reject H_0; no evidence that the expense vouchers are higher in Dept. I.

10.7 (a) $t_6 = 3.03 < 3.143$; do not reject H_0; no evidence that the new system uses less processing time.
(b) The population is normally distributed.

10.9 (a) $t_{14} = -.616 > -2.9768$; do not reject H_0; no evidence of a difference in earnings per share in the 2 years.
(b) The population is normally distributed.

10.13 (a) $Z = +4.629 > 2.58$ or $\chi_1^2 = 21.4286 > 6.635$; reject H_0.
(b) $Z^2 = (4.629)^2 \simeq \chi_1^2 = 21.4286$.
(c) The Z test as a one-tailed test.

10.17 (a) {\$104.64 $\leq \mu \leq$ \$115.36}
(b) {.00 $\leq p \leq$ .4064}
(c) $t_{24} = -10.0 < -2.7969$; reject H_0; the average monthly balance of plan A is not \$105.
(d) $Z = 1.443 < 2.58$; do not reject H_0; no evidence the proportion is different from .4.

(e) $t_{73} = 9.903 > 2.6449$; reject H_0; there is a difference in average monthly balance between Plan A and Plan B.

(f) $Z = 2.5 < 2.58$ or $\chi_1^2 = 6.25 < 6.635$; do not reject H_0; no evidence of a difference between Plan A and Plan B.

10.21 $\chi_2^2 = 1.125 < 9.21$; do not reject H_0; no evidence of a difference in attitude.

10.23 $\chi_2^2 = 19.358 > 5.99$; reject H_0; there is a difference in the proportion of women shoppers.

10.25 $\chi_4^2 = 84.751 > 13.277$; reject H_0; there is an association between interest in statistics and ability in mathematics.

10.29 $\chi_4^2 = 5.038 < 7.779$; do not reject H_0; no evidence of a difference between areas.

10.35 $\chi_5^2 = 1.249 < 15.086$; do not reject H_0; no evidence that the waiting line of customers follows other than a Poisson distribution. **Note:** One $f_t < 5$.

10.41 (a) $.831 < \chi_5^2 = 4.208 < 12.833$; do not reject H_0; no evidence that σ is different from \$2.

(b) $-2.179 < t_{12} = 1.242 < 2.179$; do not reject H_0; no evidence of a difference in price between the two cities.

(c) Populations are normally distributed and $\sigma_1^2 = \sigma_2^2$.

(d) $.146 < F_{5,7} = .591 < 5.29$; do not reject H_0.

(e) Populations are normally distributed.

10.43 $17.192 < \chi_{35}^2 = 42.35 < 60.275$; do not reject H_0; no evidence that σ is different from \$3,000.

10.45 $F_{14,14} = 1.5625 < 2.48$; do not reject H_0; no evidence that the variance is greater in department 2.

CHAPTER 11

11.1 (a) Sell ice cream; (c) EOL (Soda) = \$15, EOL (Ice Cream) = \$8; (d) EVPI = \$8; (e) she would be willing to pay up to \$8 for perfect information; (f) P(Cool) = .64, P(Warm) = .36; (g) Sell soda.

11.3 (a) Small factory; (c) EOL (Small) = \$90,000, EOL (Large) = \$140,000; (d) willing to pay up to \$90,000 for perfect information.

11.5 (a) Purchase 1,000 pounds; (c) EOL (500) = \$400, EOL (1,000) = \$275, EOL (20,000) = \$525; (d) willing to pay up to \$275 for perfect information.

11.11 (a) Yes, institute this service.

CHAPTER 12

12.1 $F_{3,12} = 4.088 > 3.49$; reject H_0; there is a difference between brands.

12.7 $F_{2,12} = 1.597 < 6.93$; do not reject H_0; no evidence of a difference between languages.

12.9 $F_{3,16} = 9.236 > 5.29$; reject H_0; there is a difference in profitability between strategies.

CHAPTER 13

13.7 $U_L = 10 < U = 15 < U_U = 22$; do not reject H_0; no evidence that the sequence is not random.

13.11 $V = 13 > V_U = 10$; reject H_0; there is evidence of a "+" trend.

13.13 $W = 16 < W_L = 25$; reject H_0; there is evidence that $M < 30$.

13.23 $T_{n_1} = 62 < T_{n_{1L}} = 69$; reject H_0; there is evidence that $M_W < M_M$.

13.37 $K = 6.64$, $Z = 2.45$; reject H_0; there is evidence that $M_D > 0.0$.

13.45 $H = 22.735 > \chi^2 = 9.488$; reject H_0; there is a difference in product perception.

CHAPTER 14

14.1 (b) $b_0 = +1.45$; $b_1 = +0.074$.
(c) for each increase of 1 foot of shelf space, sales will increase by \$7.40 per week.
(d) $\hat{Y}_i = 2.042$.
(e) $S_{YX} = .308$.
(f) $r^2 = .684$; 68.4% of the variation in sales can be explained by variation in shelf space.
(g) $r = +.827$.
(h) $\{1.835 \leq \mu_{YX} \leq 2.249\}$.
(i) $t_{10} = 4.653 > 1.8125$; reject H_0; there is a linear relationship.

14.3 (b) $b_0 = 21.9256$; $b_1 = +2.0687$.
(c) If the cars have no options, delivery time averages approximately 22 days; for each option ordered delivery time increases by 2.0687 days.
(d) 55.0248 days.
(e) $S_{YX} = 3.0448$.
(f) $r^2 = .9575$; 95.75% of the variation in delivery time can be explained by variation in the number of options ordered.
(g) $r = +.9785$.
(h) $\{53.1115 \leq \mu_{YX} \leq 56.9381\}$.
(i) $t_{14} = 17.769 > 2.1448$; reject H_0; there is a linear relationship.
(j) $\{+1.8187 \leq \beta_1 \leq +2.3187\}$.

14.7 (b) $b_0 = 6.9$; $b_1 = .64$.

(c) If no fertilizer is applied, the average yield is predicted to be 6.9 lbs.; for each additional pound of fertilizer applied, yield increases by .64 lbs.

(d) 16.5 lbs.

(e) $S_{YX} = 2.089$.

(f) $r^2 = .959$; 95.9% of the variation in yield can be explained by variation in the amount of fertilizer.

(g) $r = +.979$.

(h) $\{15.197 \leq \mu_{YX} \leq 17.803\}$.

(i) $t_8 = 13.703 > 1.8595$; reject H_0; there is a significant linear relationship.

(j) $\{.5532 \leq \beta_1 \leq .7268\}$.

14.11 (a) $r = +.523$; (b) $t_6 = +1.503 < 2.4469$; do not reject H_0; there is no evidence of a linear relationship between test score and final rating.

CHAPTER 15

15.1 (a) $b_0 = 25.64085$, $b_1 = -.1616572$, $b_2 = -.6252808$ (X_1 = age and X_2 = annual salary).

(b) For each increase of one year in age, for a given annual salary, the average days absent decreases by .16 days; for each increase of $1,000 in annual salary, days absent decreases by .625 days for a given age.

(c) 11.046 days.

(d) $F_{2,7} = 140.1696 > 4.74$; there is a significant relationship.

(e) $r_{Y.12}{}^2 = .97564$; 97.564% of the variation in days absent can be explained by variation in age and annual salary.

(f) $F_{1,7} = 39.613 > 5.59$ and $F_{1,7} = 30.458 > 5.59$; each independent variable makes a significant contribution and should be included in the model.

(g) $S_{YX} = 0.74645$.

(h) $P\{-.2224 \leq \beta_1 \leq -.1009\} = .95$.

(i) $P\{10.477 \leq \mu_{YX} \leq 11.615\} = .95$.

(j) $r_{Y1.2}{}^2 = 0.85$; $r_{Y2.1}{}^2 = 0.813$; for a given annual salary, 85% of the variation in days absent can be explained by variation in age. For a given age, 81.3% of the variation in days absent can be explained by variation in annual salary.

15.3 (a) $b_0 = 440.9463$, $b_1 = -9.0487$, $b_2 = 37.18082$ (X_1 = age and X_2 = number of rooms).

(b) For each increase of one year in age, with a fixed number of rooms, the average county taxes decrease by $9.05; for each additional room in a house of a given age, county taxes increase by $37.18.

(c) $647.91.

(d) $F_{2,16} = 47.97585 > 3.63$; there is a significant relationship.

(e) $r_{Y.12}{}^2 = 0.85959$; 85.959% of the variation in county taxes can be explained by variation in the age and number of rooms in the house.

(f) $F_{1,16} = 59.029 > 4.49$ and $F_{1,16} = 6.423 > 4.49$; each independent variable makes a significant contribution and should be included in the model.

(g) S_{YX} = $71.44868.

(h) $P\{-\$11.55 \le \beta_1 \le -\$6.55\} = .95$.

(i) $P\{\$483.65 \le \mu_{YX} \le \$558.79\}$.

(j) $r_{Y1.2}{}^2 = 0.7867$, $r_{Y2.1}{}^2 = .2852$; for a given number of rooms, 78.67% of the variation in county taxes can be explained by variation in the age of houses. For a given age of houses, 28.52% of the variation in county taxes can be explained by variation in the number of rooms.

15.9 (b) $b_0 = 185.8241$, $b_1 = 3.314196$, $b_{11} = -.01629$.

(c) \$245.59.

(d) $F_{2,10} = 13.47 > 4.10$; there is a significant curvilinear relationship.

(e) $R^2 = .7293$; 72.93% of the variation in weekly salary can be explained by the curvilinear relationship with length of employment.

(f) $F_{1,10} = 7.082 > 4.96$; reject H_0; the curvilinear model is a significantly better fit than the linear model.

CHAPTER 16

16.7 $\sum_{i=1}^{n} P_i^{(t)}$: 1950 = 167.71, 1955 = 215.35, 1960 = 260.52, 1965 = 312.52, 1970 = 411.65, 1975 = 563.08.

(a) $I_{SA}^{(t)}$: 1950 = 64.38, 1955 = 82.66, 1960 = 100.00, 1965 = 119.96, 1970 = 158.01, 1975 = 216.14.

(b) $I_{SM}^{(t)}$: 1950 = 65.13, 1955 = 83.07, 1960 = 100.00, 1965 = 119.20, 1970 = 154.93, 1975 = 211.10.

(c) 1975: $I_{LA} = 208.00$, $I_{PA} = 206.53$, $I_{M\text{-}EA} = 207.19$, $I_{FIA} = 207.26$, $I_{FWA} = 207.45$.

(d) 1975: $I_{LA} = 137.92$, $I_{PA} = 137.59$, $I_{M\text{-}EA} = 137.75$, $I_{FIA} = 137.75$, $I_{FWA} = 137.92$.

(e) (1) 280.81%, (2) 1950: \$96.64, 1975: \$164.61, (3) 70.33%, (4) purchasing power is increased by 70.33%.

CHAPTER 17

17.5 (b) $\hat{Y}_i = 17.91 + 2.57X_i$, where origin = 1967 and X units = 1 year.

(c) 1977: 43.65, 1978: 46.22, 1979: 48.79.

(d)

X_i	Y_i	$\hat{Y}_i$	$Y_i/\hat{Y}_i$
0.0	20.00	17.91	1.117
1.0	22.80	20.48	1.113
2.0	24.30	23.06	1.054
3.0	18.80	25.63	0.734
4.0	28.30	28.20	1.003
5.0	30.40	30.78	0.988
6.0	35.80	33.35	1.073
7.0	31.60	35.92	0.880
8.0	35.70	38.50	0.927
9.0	47.20	41.07	1.149

(e) According to Series 54 and 55 of *Business Conditions Digest,* this series may be considered as either a leading or coinciding indicator.

17.13 (b) and (c)

Pd.	Year	Y_i	3-Year Moving Total	3-Year Moving Avg.	(W = .50) ε_i
1	1961	1.45	*.*	*.*	1.45
2	62	1.55	4.61	1.54	1.50
3	63	1.61	4.76	1.59	1.55
4	64	1.60	4.95	1.65	1.58
5	65	1.74	5.26	1.75	1.66
6	66	1.92	5.61	1.87	1.79
7	67	1.95	5.91	1.97	1.87
8	68	2.04	6.05	2.02	1.95
9	69	2.06	5.90	1.97	2.01
10	70	1.80	5.59	1.86	1.90
11	71	1.73	5.30	1.77	1.82
12	72	1.77	5.40	1.80	1.79
13	73	1.90	5.49	1.83	1.85
14	74	1.82	5.37	1.79	1.83
15	75	1.65	5.20	1.73	1.74
16	76	1.73	*.*	*.*	1.74

(d) $\hat{Y}_{1977} = \varepsilon_{1976} = 1.74.$

17.19 (c) $\hat{Y}_i = 266.54 + 0.556X_i$, where origin = January 15, 1971 and X units = 1 month.

(d)

Month and Year	S_i	$\hat{Y}_i$	Forecast
1977			
Jan.	0.933	306.54	286.071
Feb.	0.950	307.10	291.674
Mar.	1.078	307.65	331.514
Apr.	0.971	308.21	299.302
May	0.971	308.76	299.907
Jun.	1.053	309.32	325.661
Jul.	1.066	309.88	330.228
Aug.	1.018	310.43	315.921
Sep.	0.961	310.99	298.741
Oct.	0.972	311.54	302.697
Nov.	1.012	312.10	315.959
Dec.	1.016	312.66	317.730
1978			
Jan.	0.933	313.21	292.293
Feb.	0.950	313.76	298.006
Mar.	1.078	314.32	338.698
Apr.	0.971	314.88	305.777
May	0.971	315.43	306.383
Jun.	1.053	315.99	332.680
Jul.	1.066	316.54	337.333
Aug.	1.018	317.10	322.706
Sep.	0.961	317.65	305.145
Oct.	0.972	318.21	309.174
Nov.	1.012	318.77	322.708
Dec.	1.016	319.32	324.506

(e) Obtaining the cyclical relatives (C_i) for 1975 and 1976.

Month and Year	Y_i	S_i	$T_iC_iI_i$	$\hat{Y}_i$	C_iI_i	Weighted Moving Total	C_i
1975							
Jan.	298.00	0.933	319.32	293.21	1.089	4.161	1.040
Feb.	283.00	0.950	297.97	293.76	1.014	4.111	1.028
Mar.	315.00	1.078	292.33	294.32	0.993	4.003	1.001
Apr.	287.00	0.971	295.54	294.88	1.002	4.047	1.012
May	301.00	0.971	309.89	295.43	1.049	3.694	0.923
Jun.	185.00	1.053	175.72	295.99	0.594	3.401	0.850
Jul.	368.00	1.066	345.32	296.54	1.164	3.948	0.987
Aug.	310.00	1.018	304.61	297.10	1.025	4.310	1.077
Sep.	313.00	0.961	325.83	297.65	1.095	4.291	1.073
Oct.	312.00	0.972	321.12	298.21	1.077	4.323	1.081
Nov.	325.00	1.012	321.03	298.76	1.075	4.298	1.074
Dec.	326.00	1.016	320.79	299.32	1.072	4.147	1.037
1976							
Jan.	260.00	0.933	278.60	299.88	0.929	3.950	0.987
Feb.	291.00	0.950	306.39	300.43	1.020	3.915	0.979
Mar.	307.00	1.078	284.90	300.99	0.947	3.914	0.978
Apr.	293.00	0.971	301.72	301.54	1.001	3.899	0.975
May	279.00	0.971	287.24	302.10	0.951	3.803	0.951
Jun.	287.00	1.053	272.60	302.65	0.901	3.817	0.954
Jul.	344.00	1.066	322.80	303.21	1.065	4.191	1.048
Aug.	359.00	1.018	352.76	303.76	1.161	4.242	1.061
Sep.	250.00	0.961	260.25	304.32	0.855	4.114	1.028
Oct.	368.00	0.972	378.75	304.88	1.242	4.501	1.125
Nov.	359.00	1.012	354.61	305.43	1.161	4.674	1.168
Dec.	345.00	1.016	339.49	305.99	1.109	*.*	*.*

INDEX